T0212655

Lecture Notes in Artificial Intelligence   **9345**

Subseries of Lecture Notes in Computer Science

More information about this series at http://www.springer.com/series/1244

Francesco Calimeri · Giovambattista Ianni
Miroslaw Truszczynski (Eds.)

Logic Programming and Nonmonotonic Reasoning

13th International Conference, LPNMR 2015
Lexington, KY, USA, September 27–30, 2015
Proceedings

 Springer

Editors
Francesco Calimeri
University of Calabria
Rende
Italy

Giovambattista Ianni
University of Calabria
Rende
Italy

Miroslaw Truszczynski
Department of Computer Science
University of Kentucky
Lexington, KY
USA

ISSN 0302-9743 ISSN 1611-3349 (electronic)
Lecture Notes in Artificial Intelligence
ISBN 978-3-319-23263-8 ISBN 978-3-319-23264-5 (eBook)
DOI 10.1007/978-3-319-23264-5

Library of Congress Control Number: 2015946999

LNCS Sublibrary: SL7 – Artificial Intelligence

Springer Cham Heidelberg New York Dordrecht London

Printed on acid-free paper

Springer International Publishing AG Switzerland is part of Springer Science+Business Media
(www.springer.com)

Preface

This volume contains the papers presented at the 13th International Conference on Logic Programming and Nonmonotonic Reasoning (LPNMR 2015) held September 27–30, 2015, in Lexington, Kentucky, USA.

LPNMR is a forum for exchanging ideas on declarative logic programming, nonmonotonic reasoning, and knowledge representation. The aim of the LPNMR conferences is to facilitate interactions between researchers interested in the design and implementation of logic-based programming languages and database systems, and researchers who work in the areas of knowledge representation and nonmonotonic reasoning. LPNMR strives to encompass theoretical and experimental studies that have led or will lead to the construction of practical systems for declarative programming and knowledge representation. LPNMR 2015 was the 13th event in the series. Past editions were held in Washington, D.C., USA (1991), Lisbon, Portugal (1993), Lexington, Kentucky, USA (1995), Dagstuhl, Germany (1997), El Paso, Texas, USA (1999), Vienna, Austria (2001), Fort Lauderdale, Florida, USA (2004), Diamante, Italy (2005), Tempe, Arizona, USA (2007), Potsdam, Germany (2009), Vancouver, Canada (2011), and Corunna, Spain (2013).

LPNMR 2015 received 60 submissions in three categories (technical papers, applications, and system descriptions) and two different formats (long and short papers). Each submission was reviewed by at least three Program Committee members. The final list of 40 accepted papers consists of 29 long and 11 short contributions, of which 21 were technical papers (16 long and 5 short), 13 were application papers (9 long, 4 short), and 6 were system descriptions (4 long, 2 short).

The LPNMR 2015 program was anchored by the invited talks by Jérôme Lang, Nada Lavrač, and Pedro Cabalar. It also included presentations of the technical papers mentioned above, a session dedicated to the 6th Answer Set Programming Competition (a report on the results and the award ceremony), a doctoral consortium held jointly with the 4th International Conference on Algorithmic Decision Theory, ADT 2015, and four workshops.

The conference proceedings include papers for the three invited talks, the 40 technical papers, the paper reporting on the Answer Set Programming competition, and four papers presented by LPNMR student attendees at the doctoral consortium.

Many people and organizations contributed to the success of LPNMR 2015. Victor Marek, the conference general chair, oversaw all organization efforts and led interactions with the organizers of ADT 2015. The members of the Program Committee and the additional reviewers worked diligently to produce fair and thorough evaluations of the submitted papers. Martin Gebser, Marco Maratea, and Francesco Ricca organized and ran the programming competition, which has grown to be one of the major driving forces of our field. Esra Erdem and Nick Mattei put together an excellent doctoral consortium program, focused on the development of young logic programming, nonmonotonic reasoning and algorithmic decision theory researchers. Yuliya Lierler, the workshop chair, spearheaded the effort to expand the scope of the conference with an exciting selection of workshops, and all workshop organizers worked hard to make their workshops interesting and relevant. Most importantly, the invited speakers, and the authors of the accepted papers provided the conference with the state-of-the-art technical substance. We thank all of them!

Furthermore, we gratefully acknowledge our sponsors for their generous support: the *Artificial Intelligence* journal, the Association of Logic Programming (ALP), the Association for the Advancement of Artificial Intelligence (AAAI), the European Coordinating Committee for Artificial Intelligence (ECCAI), the Knowledge Representation and Reasoning, Incorporated Foundation (KR, Inc.), the National Science Foundation (NSF USA), the University of Kentucky, and the University of Calabria.

Last, but not least, we thank the people of EasyChair for providing resources and a marvelous conference management system.

July 2015   Francesco Calimeri
                                                                                   Giovambattista Ianni
                                                                                   Miroslaw Truszczynski

Organization

General Chair

Victor Marek   University of Kentucky, KY, USA

Program Committee Chairs

Giovambattista Ianni   Università della Calabria, Italy
Mirek Truszczynski   University of Kentucky, KY, USA

Workshops Chair

Yuliya Lierler   University of Nebraska at Omaha, NE, USA

Publicity Chair

Francesco Calimeri   Università della Calabria, Italy

Joint LPNMR-ADT Doctoral Consortium Chair

Esra Erdem (LPNMR)   Sabanci University, Turkey
Nick Mattei (ADT)   NICTA, Australia

Program Committee

Jose Julio Alferes   Universidade NOVA de Lisboa, Portugal
Saadat Anwar   Arizona State University, AZ, USA
Marcello Balduccini   Drexel University, PA, USA
Chitta Baral   Arizona State University, AZ, USA
Bart Bogaerts   KU Leuven, Belgium
Gerhard Brewka   Leipzig University, Germany
Pedro Cabalar   University of Corunna, Spain
Francesco Calimeri   Università della Calabria, Italy
Stefania Costantini   Università di L'Aquila, Italy
Marina De Vos   University of Bath, UK
James Delgrande   Simon Fraser University, Canada
Marc Denecker   K.U. Leuven, Belgium
Agostino Dovier   Università di Udine, Italy
Thomas Eiter   Vienna University of Technology, Austria
Esra Erdem   Sabanci University, Istanbul, Turkey
Wolfgang Faber   University of Huddersfield, UK

Michael Fink	Vienna University of Technology, Austria
Paul Fodor	Stony Brook University, NY, USA
Andrea Formisano	Università di Perugia, Italy
Martin Gebser	Aalto University, Finland
Michael Gelfond	Texas Tech University, TX, USA
Giovanni Grasso	Oxford University, UK
Daniela Inclezan	Miami University, OH, USA
Tomi Janhunen	Aalto University, Finland
Matthias Knorr	NOVA-LINCS, Universidade Nova de Lisboa, Portugal
Joohyung Lee	Arizona State University, AZ, USA
Yuliya Lierler	University of Nebraska at Omaha, NE, USA
Vladimir Lifschitz	University of Texas at Austin, TX, USA
Fangzhen Lin	Hong Kong University of Science and Technology, Hong Kong, SAR China
Alessandra Mileo	INSIGHT Centre for Data Analytics, National University of Ireland, Galway, Ireland
Emilia Oikarinen	Aalto University, Finland
Mauricio Osorio	Fundacion de la Universidad de las Americas, Puebla, Mexico
David Pearce	Universidad Politécnica de Madrid, Spain
Axel Polleres	Vienna University of Economics and Business, Austria
Enrico Pontelli	New Mexico State University, NM, USA
Christoph Redl	Vienna University of Technology, Austria
Orkunt Sabuncu	University of Potsdam, Germany
Chiaki Sakama	Wakayama University, Japan
Torsten Schaub	University of Potsdam, Germany
Tran Cao Son	New Mexico State University, NM, USA
Hannes Strass	Leipzig University, Germany
Terrance Swift	NOVA-LINCS, Universidade Nova de Lisboa, Portugal
Eugenia Ternovska	Simon Fraser University, Canada
Hans Tompits	Vienna University of Technology, Austria
Agustín Valverde	Universidad de Màlaga, Spain
Kewen Wang	Griffith University, Australia
Yisong Wang	Guizhou University, China
Stefan Woltran	Vienna University of Technology, Austria
Fangkai Yang	Schlumberger Ltd., TX, USA
Jia You	University of Alberta, Canada
Yi Zhou	University of Western Sydney, Australia

Sponsors

Artificial Intelligence, Elsevier
Association of Logic Programming (ALP)
Association for the Advancement of Artificial Intelligence (AAAI)
European Coordinating Committee for Artificial Intelligence (ECCAI)

Knowledge Representation and Reasoning, Incorporated Foundation (KR, Inc.)
National Science Foundation (NSF USA)
University of Kentucky, USA
University of Calabria, Italy

Additional Reviewers

Acosta-Guadarrama, Juan C.
Alviano, Mario
Andres, Benjamin
Balduccini, Marcello
Beck, Harald
Berger, Gerald
Bi, Yi
Binnewies, Sebastian
Charwat, Guenther
Chowdhury, Md. Solimul
Dasseville, Ingmar
Devriendt, Jo
Erdoğan, Selim
Fandinno, Jorge
Fichte, Johannes Klaus
Fuscà, Davide
Gavanelli, Marco
Gebser, Martin
Havur, Giray
Huang, Yi
Jansen, Joachim
Kaufmann, Benjamin

Kiesl, Benjamin
Le, Tiep
Leblanc, Emily
Lindauer, Marius
Liu, Guohua
Nieves, Juan Carlos
Pfandler, Andreas
Pührer, Jörg
Romero, Javier
Schulz-Hanke, Christian
Stepanova, Daria
Steyskal, Simon
Susman, Benjamin
Thorstensen, Evgenij
Uridia, Levan
Van Hertum, Pieter
Veltri, Pierfrancesco
Viegas Damásio, Carlos
Wallner, Johannes P.
Weinzierl, Antonius
Zangari, Jessica
Zepeda Cortes, Claudia

Contents

XII   Contents

Stable Models for Temporal Theories
— Invited Talk —

Pedro Cabalar[✉]

Department of Computer Science, University of Corunna, A Coruña, Spain
cabalar@udc.es

Abstract. This work makes an overview on an hybrid formalism that combines the syntax of Linear-time Temporal Logic (LTL) with a non-monotonic selection of models based on Equilibrium Logic. The resulting approach, called Temporal Equilibrium Logic, extends the concept of a stable model for any arbitrary modal temporal theory, constituting a suitable formal framework for the specification and verification of dynamic scenarios in Answer Set Programming (ASP). We will recall the basic definitions of this logic and explain their effects on some simple examples. After that, we will proceed to summarize the advances made so far, both in the fundamental realm and in the construction of reasoning tools. Finally, we will explain some open topics, many of them currently under study, and foresee potential challenges for future research.

1 Introduction

The birth of Non-Monotonic Reasoning (NMR) in the 1980s was intimately related to temporal reasoning in action domains. The solution to the *frame problem* [1] (the unfeasibility of explicitly specifying all the non-effects of an action) played a central role in research on NMR formalisms capable of representing defaults. In particular, the area of reasoning about actions and change was initially focused on properly capturing the *inertia law*, a dynamic default which can be phrased as "fluent values remain unchanged along time, unless there is evidence on the contrary." NMR was also essential to deal with other typical representational problems in action theories, such as the *ramification* and the *qualification* problems.

The combination of temporal reasoning and NMR in action theories was typically done inside the realm of first order logic. Classical action languages such as *Situation Calculus* [1] or *Event Calculus* [2] have combined some NMR technique, usually predicate circumscription [3], with a first-order formalisation of time using temporal predicates and objects (situations or events, respectively). In this way, we get very rich and expressive formalisms without limitations on the quantification of temporal terms or the construction of arbitrary expressions involving them, although we inherit the undecidability of first order logic in the general case.

This research was partially supported by Spanish MEC project TIN2013-42149-P and Xunta de Galicia grant GPC 2013/070.

F. Calimeri et al. (Eds.): LPNMR 2015, LNAI 9345, pp. 1–13, 2015.
DOI: 10.1007/978-3-319-23264-5_1

Another way of dealing with temporal reasoning in NMR approaches has been the use of modal temporal logic, a combination perhaps less popular[1], but not unfrequent in the literature [5–7]. But probably, the simplest treatment of time we find in action theories is the use of an integer index to denote situations, as done for instance in [8] for reasoning about actions using Logic Programming (LP), and in the family of action languages [9] inspired on that methodology.

With the consolidation of Answer Set Programming (ASP) [10,11] as a successful paradigm for practical NMR, many examples and benchmarks formalising dynamic scenarios became available. ASP inherited the treatment of time as an integer index from LP-based action languages but, in practice, it further restricted all reasoning tasks to finite narratives, something required for grounding time-related variables. To illustrate this orientation, consider an extremely simple ASP program where a fluent p represents that a switch is on and q represents that it is off. Moreover, suppose we have freedom to arbitrarily fix p true at any moment and that either p or q holds initially. A typical ASP representation of this problem could look like this:

$$p(0) \lor q(0) \tag{1}$$
$$p(I{+}1) \leftarrow p(I), not\, q(I{+}1), sit(I) \tag{2}$$
$$q(I{+}1) \leftarrow q(I), not\, p(I{+}1), sit(I) \tag{3}$$
$$p(I) \lor not\, p(I) \leftarrow sit(I) \tag{4}$$

where (1) describes the initial state, (2) and (3) are the inertia rules for p and q, and (4) acts as a choice rule[2] allowing the introduction of p at any situation. Predicate sit would have some finite domain $0 \ldots n$ for some constant $n \geq 0$. A planning problem can be solved incrementally [12], using an iterative deepening strategy similar to SAT-based planning [13]. If we want to reach a state satisfying $p \land \neg q$, we would include two constraints for the last situation:

$$\bot \leftarrow not\, p(n) \qquad\qquad \bot \leftarrow q(n)$$

and go increasing n until a solution is found. However, this strategy falls short for many temporal reasoning problems that involve dealing with infinite time such as proving the non-existence of a plan or checking the satisfaction of temporal properties of a given dynamic system. For instance, questions such as "is there a reachable state in which both p and q are false?" or "can we show that whenever p is true it will remain so forever?" can be answered by an analytical inspection of our simple program, but cannot be solved in an automated way.

---

[1] John McCarthy, the founder of logical knowledge representation and commonsense reasoning, showed in several occasions an explicit disapproval of modal logics. See for instance his position paper with the self-explanatory title "Modality, si! Modal logic, no!" [4].

[2] Generally speaking, a disjunction of the form $\varphi \lor not\, \varphi$ in ASP is not a tautology. When included in a rule head it is usually written as $\{\ \varphi\ \}$ and acts as a non-deterministic choice possibly allowing the derivation of φ.

In principle, one may think that this kind of problems dealing with infinite time are typically best suited for modal temporal logics, whose expressive power, computation methods (usually decidable) and associated complexity have been extensively well-studied. Unfortunately, as happens with SAT in the non-temporal case, temporal logics are not designed for Knowledge Representation (KR). For instance, the best known temporal logics are monotonic, so that the frame and ramification problems constantly manifest in their applications, even for very simple scenarios.

In this work, we make a general overview on *Temporal Equilibrium Logic* [14], to the best of our knowledge, the first non-monotonic approach that fully covers the syntax of some standard modal temporal logic, providing a logic programming semantics that properly extends *stable models* [15], the foundational basis of ASP. TEL shares the syntax of *Linear-time Temporal Logic* (LTL) [16,17] which is perhaps the simplest, most used and best known temporal logic in Theoretical Computer Science. The main difference of TEL with respect to LTL lies in its non-monotonic entailment relation (obtained by a models selection criterion) and in its semantic interpretation of implication and negation, closer to intuitionistic logic. These two properties are actually inherited from the fact that TEL is a temporal extension of *Equilibrium Logic* [18], a non-monotonic formalism that generalises stable models to the case of arbitrary propositional formulas. This semantic choice is a valuable feature because, on the one hand, it provides a powerful connection to a successful practical KR paradigm like ASP, and on the other hand, unlike the original definition of stable models, the semantics of Equilibrium Logic does not depend on syntactic transformations but, on the contrary, is just a simple minimisation criterion for an intermediate logic (the logic of *Here-and-There* [19]). This purely logical definition provides an easier and more homogeneous way to extend the formalism, using standard techniques from other hybrid logical approaches.

As an example, the ASP program (1)–(4) would be represented in TEL as:

$$p \vee q \tag{5}$$
$$\Box(p \wedge \neg \bigcirc q \to \bigcirc p) \tag{6}$$
$$\Box(q \wedge \neg \bigcirc p \to \bigcirc q) \tag{7}$$
$$\Box(p \vee \neg p) \tag{8}$$

where, as usual in LTL, '$\Box$' stands for "always" and '$\bigcirc$' stands for "next." Checking whether p and q can be eventually false would correspond to look for a plan satisfying the constraint $\neg\Diamond(\neg p \wedge \neg q) \to \bot$ with '$\Diamond$' meaning "eventually." Similarly, to test whether p remains true after becoming true we would add the constraint $\Box(p \to \Box p) \to \bot$ and check that, indeed, no temporal stable model exists.

The rest of the paper is organised as follows. In Sect. 2 we recall the basic definitions of TEL and explain their effects on some simple examples. In Sect. 3 we summarize some fundamental properties whereas in Sect. 4 we explain some aspects related to computation. Finally, Sect. 5 concludes the paper and explains some open topics. For a more detailed survey, see [20].

2   Syntax and Semantics

The syntax is defined as in propositional LTL. A temporal *formula* φ can be expressed following the grammar shown below:

$$\varphi :: = \bot \mid p \mid \alpha \wedge \beta \mid \alpha \vee \beta \mid \alpha \rightarrow \beta \mid \bigcirc \alpha \mid \alpha \, \mathcal{U} \, \beta \mid \alpha \, \mathcal{R} \, \beta$$

where p is an atom of some finite signature At, and α and β are temporal formulas in their turn. The formula $\alpha \, \mathcal{U} \, \beta$ stands for "α *until* β" whereas $\alpha \, \mathcal{R} \, \beta$ is read as "α *release* β" and is the dual of "until." Derived operators such as $\Box$ ("always") and $\Diamond$ ("at some future time") are defined as $\Box\varphi \stackrel{\text{def}}{=} \bot \, \mathcal{R} \, \varphi$ and $\Diamond\varphi \stackrel{\text{def}}{=} \top \, \mathcal{U} \, \varphi$. Other usual propositional operators are defined as follows: $\neg\varphi \stackrel{\text{def}}{=} \varphi \rightarrow \bot$, $\top \stackrel{\text{def}}{=} \neg\bot$ and $\varphi \leftrightarrow \psi \stackrel{\text{def}}{=} (\varphi \rightarrow \psi) \wedge (\psi \rightarrow \varphi)$.

Given a finite propositional signature At, an LTL-*interpretation* $\mathbf{T}$ is an infinite sequence of sets of atoms, $T_0, T_1, \ldots$ with $T_i \subseteq At$ for all $i \geq 0$. Given two LTL-interpretations $\mathbf{H}, \mathbf{T}$ we define $\mathbf{H} \leq \mathbf{T}$ as: $H_i \subseteq T_i$ for all $i \geq 0$.

The next step is defining a semantics for the temporal extension of the intermediate logic of Here-and-There, we will call *Temporal Here-and-There*[3] (THT). A THT-*interpretation* $\mathbf{M}$ for At is a pair of LTL-interpretations $\langle \mathbf{H}, \mathbf{T} \rangle$ satisfying $\mathbf{H} \leq \mathbf{T}$. A THT-interpretation is said to be *total* when $\mathbf{H} = \mathbf{T}$.

Definition 1 (THT satisfaction). *Given an interpretation* $\mathbf{M} = \langle \mathbf{H}, \mathbf{T} \rangle$, *we recursively define when* $\mathbf{M}$ *satisfies a temporal formula* φ *at some state* $i \in \mathbb{N}$ *as:*

- $\mathbf{M}, i \models p$ *iff* $p \in H_i$ *with* p *an atom*
- $\wedge, \vee, \bot$ *as usual*
- $\mathbf{M}, i \models \varphi \rightarrow \psi$ *iff for all* $w \in \{\mathbf{H}, \mathbf{T}\}$, $\langle w, \mathbf{T} \rangle, i \not\models \varphi$ *or* $\langle w, \mathbf{T} \rangle, i \models \psi$
- $\mathbf{M}, i \models \bigcirc \varphi$ *iff* $\mathbf{M}, i{+}1 \models \varphi$
- $\mathbf{M}, i \models \varphi \, \mathcal{U} \, \psi$ *iff* $\exists k \geq i$ *such that* $\mathbf{M}, k \models \psi$ *and* $\forall j \in \{i, \ldots, k\text{-}1\}, \mathbf{M}, j \models \varphi$
- $\mathbf{M}, i \models \varphi \mathcal{R} \psi$ *iff* $\forall k \geq i$ *such that* $\mathbf{M}, k \not\models \psi$ *then* $\exists j \in \{i, \ldots, k\text{-}1\}, \mathbf{M}, j \models \varphi$. $\quad\Box$

We say that $\langle \mathbf{H}, \mathbf{T} \rangle$ is a *model* of a theory Γ, written $\langle \mathbf{H}, \mathbf{T} \rangle \models \Gamma$, iff $\langle \mathbf{H}, \mathbf{T} \rangle, 0 \models \alpha$ for all formulas $\alpha \in \Gamma$.

Proposition 1 (from [20]). *The following properties are satisfied:*

(i) $\langle \mathbf{T}, \mathbf{T} \rangle, i \models \varphi$ *in THT iff* $\mathbf{T}, i \models \varphi$ *in LTL.*
(ii) $\langle \mathbf{H}, \mathbf{T} \rangle, i \models \varphi$ *implies* $\langle \mathbf{T}, \mathbf{T} \rangle, i \models \varphi$ *(that is,* $\mathbf{T}, i \models \varphi$*).*

In other words, (i) means that, when restricting to total interpretations, THT collapses to LTL, whereas (ii) means that the $\mathbf{T}$ component of a THT model is also an LTL-model.

Definition 2 (Temporal Equilibrium/Stable Model). *An interpretation* $\mathbf{M}$ *is a* temporal equilibrium model *of a theory* Γ *if it is a total model of* Γ, *that is,* $\mathbf{M} = \langle \mathbf{T}, \mathbf{T} \rangle \models \Gamma$, *and there is no* $\mathbf{H} < \mathbf{T}$ *such that* $\langle \mathbf{H}, \mathbf{T} \rangle \models \Gamma$. *An LTL-interpretation* $\mathbf{T}$ *is a* temporal stable model *(TS-model) of a theory* Γ *iff* $\langle \mathbf{T}, \mathbf{T} \rangle$ *is a temporal equilibrium model of* Γ. $\quad\Box$

---

[3] The axiomatisation of THT is currently under study [21].

By Proposition 1 (i) it is easy to see that any TS-model of a temporal theory Γ is also an LTL-model of Γ. As happens in LTL, the set of TS-models of a theory Γ can be captured by a Büchi automaton [22], a kind of finite automaton that accepts words of infinite length. In this case, the alphabet of the automaton would be the set of states (classical propositional interpretations) and the acceptance condition is that a word (a sequence of states) is accepted iff it corresponds to a run of the automaton that visits some acceptance state an infinite number of times. As an example, Fig. 1 shows the TS-models for the theory (5)–(8) which coincide with sequences of states of the forms $\{q\}^*\{p\}^\omega$ or $\{q\}^\omega$. Notice how p and q are never true simultaneously, whereas once p becomes true, it remains true forever.

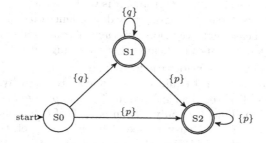

Fig. 1. Temporal stable models of theory (5)–(8).

Let us discuss next some simpler examples of the behaviour of this semantics. As a first example, consider the formula

$$\Box(\neg p \to \bigcirc p) \tag{9}$$

Its intuitive meaning corresponds to the logic program consisting of rules of the form: $p(s(X)) \leftarrow not\ p(X)$ where time has been reified as an extra parameter $X = 0, s(0), s(s(0)), \ldots$. Notice that the interpretation of $\neg$ is that of default negation not in logic programming. In this way, (9) is saying that, at any situation, if there is no evidence on p, then p will become true in the next state. In the initial state, we have no evidence on p, so this will imply $\bigcirc p$. As a result $\bigcirc \bigcirc p$ will have no applicable rule and thus will be false by default, and so on. It is easy to see that the unique temporal stable model of (9) is captured by the formula $\neg p \wedge \Box(\neg p \leftrightarrow \bigcirc p)$ and is shown in the automaton of Fig. 2(a).

As a second example, take the formula $\Diamond p$. This formula informally corresponds to an infinite disjunction $p \vee \bigcirc p \vee \bigcirc \bigcirc p \vee \ldots$. Again, as happens in disjunctive logic programming, in TEL we have a truth minimality condition that will make true the formula with as little information as possible. As a result, it is easy to see that the temporal stable models of $\Diamond p$ are captured by the formula $\neg p \, \mathcal{U} (p \wedge \bigcirc \Box \neg p)$ whose models are those where p holds true at exactly one position – see automaton in Fig. 2(b).

(a) TS-models of (9)   (b) TS-models of $\Diamond p$

Fig. 2. A pair of Büchi automata showing TS-models.

It is worth noting that an LTL satisfiable formula may have no temporal stable model. As a simple example (well-known from non-temporal ASP) the logic program rule $\neg p \rightarrow p$, whose only (classical) model is $\{p\}$, has no stable models. When dealing with logic programs, it is well-known that non-existence of stable models is always due to a kind of cyclic dependence on default negation like this. In the temporal case, however, non-existence of temporal stable models may also be due to a lack of a finite justification for satisfying the criterion of minimal knowledge. As an example, consider the formula $\alpha \stackrel{\text{def}}{=} \Box(\neg\bigcirc p \rightarrow p) \wedge \Box(\bigcirc p \rightarrow p)$. This formula has no temporal equilibrium models. To see why, note that α is LTL-equivalent (and THT-equivalent) to $\Box(\neg\bigcirc p \vee \bigcirc p \rightarrow p)$ that, in its turn, is LTL-equivalent to $\Box p$. Thus, the only LTL-model $\mathbf{T}$ of α has the form $T_i = \{p\}$ for any $i \geq 0$. However, it is easy to see that the interpretation $\langle \mathbf{H}, \mathbf{T} \rangle$ with $H_i = \emptyset$ for all $i \geq 0$ is also a THT model, whereas $\mathbf{H} < \mathbf{T}$.

Another example of TEL-unsatisfiable formula is $\Box\Diamond p$, typically used in LTL to assert that property p occurs infinitely often. This formula has no temporal stable models: all models must contain infinite occurrences of p and there is no way to establish a minimal $\mathbf{H}$ among them. Thus, formula $\Box\Diamond p$ is LTL satisfiable but it has no temporal stable model. This example does not mean a lack of expressiveness[4] of TEL: we can still check or force atoms to occur infinitely often by including formulas like $\Box\Diamond p$ in the antecedent of implications or in the scope of negation. As an example, take the formula:

$$\neg\Box\Diamond q \rightarrow \Diamond(q \,\mathcal{U}\, p) \tag{10}$$

An informal reading of (10) is: if we cannot prove that q occurs infinitely often ($\neg\Box\Diamond q$) then make q until p ($q \,\mathcal{U}\, p$) at some arbitrary future point. As we minimize truth, we may then assume q false at all states, and then $\Diamond(q \,\mathcal{U}\, p)$ collapses to $\Diamond(\bot \,\mathcal{U}\, p) = \Diamond(\Diamond p) = \Diamond p$. As a result, its TS-models also correspond to the Büchi automaton depicted in Fig. 2(b) we obtained for $\Diamond p$.

3 Fundamental Properties

We first begin providing some translation results relating TEL and LTL.

---

[4] In fact, Theorem 1 in the next section shows that LTL can be encoded into TEL by adding a simple axiom schema.

Proposition 2 (from [23]). *The LTL models of a formula φ for signature At coincide with (the THT) and the TEL models of the theory φ plus an axiom $\Box(p \vee \neg p)$ for each atom p in the signature At.* □

The translation from THT to LTL is not so straightforward. It requires adding an auxiliary atom p' by each atom p in the signature, so that the former captures the truth at component **H** in a THT model $\langle \mathbf{H}, \mathbf{T} \rangle$ while the latter represents truth at **T**. Given a propositional signature At, let us denote $At^* = At \cup \{p' \mid p \in At\}$. For any temporal formula φ we define its translation φ^* as follows:

1. $\bot^* \overset{\text{def}}{=} \bot$
2. $p^* \overset{\text{def}}{=} p'$ for any $p \in \Sigma$
3. $(\otimes \varphi)^* \overset{\text{def}}{=} \otimes \varphi^*$, for any unary operator $\otimes \in \{\Box, \Diamond, \bigcirc\}$
4. $(\varphi \oplus \psi)^* \overset{\text{def}}{=} \varphi^* \oplus \psi^*$ for any binary operator $\oplus \in \{\wedge, \vee, \mathcal{U}, \mathcal{R}\}$
5. $(\varphi \to \psi)^* \overset{\text{def}}{=} (\varphi \to \psi) \wedge (\varphi^* \to \psi^*)$

We associate to any THT interpretation $\mathbf{M} = \langle \mathbf{H}, \mathbf{T} \rangle$ the LTL interpretation $\mathbf{M}^t = \overline{I}$ in LTL defined as the sequence of sets of atoms $I_i = \{p' \mid p \in H_i\} \cup T_i$, for any $i \geq 0$.

Theorem 1 (from [23]). *Let φ' be the formula $\varphi^* \wedge \bigwedge_{p \in At} \Box(p' \to p)$. Then the set of LTL models for the formula φ' corresponds to the set of THT models for the temporal formula φ.* □

Theories like (5)–(8) have a strong resemblance to logic programs. For instance, a rule preceded by $\Box$ like (9) can be seen as an infinite set of rules of the form $\neg \bigcirc^i p \to \bigcirc^{i+1} p$ where we could understand expressions like '$\bigcirc^i p$' as an infinite propositional signature. In [24] it was recently proved that, in fact, we can use this understanding of modal operators as formulas in Infinitary Equilibrium Logic (see [25] for further detail) in the general case.

Definition 3. *The translation of φ into infinitary HT (HT^∞) up to level $k \geq 0$, written $\langle \varphi \rangle_k$, is recursively defined as follows:*

$\langle \bot \rangle_k \overset{\text{def}}{=} \emptyset^\vee$

$\langle p \rangle_k \overset{\text{def}}{=} \bigcirc^k p$, with $p \in At$. $\langle \varphi \to \psi \rangle_k \overset{\text{def}}{=} \langle \varphi \rangle_k \to \langle \psi \rangle_k$

$\langle \bigcirc \varphi \rangle_k \overset{\text{def}}{=} \langle \varphi \rangle_{k+1}$ $\langle \varphi \, \mathcal{U} \, \psi \rangle_k \overset{\text{def}}{=} \{\{\langle \psi \rangle_i, \langle \varphi \rangle_j \mid k \leq j < i\}^\wedge \mid k \leq i\}^\vee$

$\langle \varphi \wedge \psi \rangle_k \overset{\text{def}}{=} \{\langle \varphi \rangle_k, \langle \psi \rangle_k\}^\wedge$ $\langle \varphi \, \mathcal{R} \, \psi \rangle_k \overset{\text{def}}{=} \{\{\langle \psi \rangle_i, \langle \varphi \rangle_j \mid k \leq j < i\}^\vee \mid k \leq i\}^\wedge$

$\langle \varphi \vee \psi \rangle_k \overset{\text{def}}{=} \{\langle \varphi \rangle_k, \langle \psi \rangle_k\}^\vee$

It is easy to see that the derived operators $\Box$ and $\Diamond$ are then translated as follows: $\langle \Diamond \varphi \rangle_k = \{\langle \varphi \rangle_i \mid k \leq i\}^\vee$ and $\langle \Box \varphi \rangle_k = \{\langle \varphi \rangle_i \mid k \leq i\}^\wedge$. For instance, the translations for our examples $\langle \Diamond p \rangle_0$ and $\langle (10) \rangle_0$ respectively correspond to:

$$\{\neg \bigcirc^i p \to \bigcirc^{i+1} p \mid i \geq 0\}^\wedge$$
$$\{\{\{\bigcirc^k q \mid j \leq k\}^\vee \mid i \leq j\}^\wedge \to \{\{\bigcirc^k p, \bigcirc^h q \mid j \leq h < k\}^\wedge \mid i \leq j \leq k\}^\vee \mid i \geq 0\}^\wedge$$

Theorem 2 (from [24]). *Let φ be a temporal formula, $\mathbf{M} = \langle \mathbf{H}, \mathbf{T} \rangle$ a THT interpretation and $M^\infty = \langle H^\infty, T^\infty \rangle$ its corresponding HT interpretation where $\bigcirc^i p$ are considered as propositional atoms. For all $i \in \mathbb{N}$, it holds that:*

(i) $\mathbf{M}, i \models \varphi$ if and only if $M^\infty \models \langle \varphi \rangle_i$.
(ii) $\mathbf{M}$ is a temporal equilibrium model of φ if and only if M^∞ is an (infinitary) equilibrium model of $\langle \varphi \rangle_0$. $\qquad\square$

In [24] it was also proved that Kamp's translation from LTL to First Order Logic is sound for translating TEL into Quantified Equilibrium Logic [26] too. This means that there always exists a way of resorting to first-order ASP and reifying time as an argument, as we did before with $p(i)$ or $p(i+1)$, so that modal operators are replaced by standard quantifiers.

Definition 4 (Kamp's translation). *Kamp's translation for a temporal formula φ and a timepoint $t \in \mathbb{N}$, denoted by $[\varphi]_t$, is recursively defined as follows:*

$$[\bot]_t \overset{\text{def}}{=} \bot$$
$$[p]_t \overset{\text{def}}{=} p(t), \text{ with } p \in At.$$
$$[\neg\alpha]_t \overset{\text{def}}{=} \neg[\alpha]_t$$
$$[\alpha \wedge \beta]_t \overset{\text{def}}{=} [\alpha]_t \wedge [\beta]_t$$
$$[\alpha \vee \beta]_t \overset{\text{def}}{=} [\alpha]_t \vee [\beta]_t$$

$$[\alpha \to \beta]_t \overset{\text{def}}{=} [\alpha]_t \to [\beta]_t$$
$$[\bigcirc\alpha]_t \overset{\text{def}}{=} [\alpha]_{t+1}$$
$$[\alpha \,\mathcal{U}\, \beta]_t \overset{\text{def}}{=} \exists x \geq t.\, ([\beta]_x \wedge \forall y \in [t, x).\, [\alpha]_y)$$
$$[\alpha \,\mathcal{R}\, \beta]_t \overset{\text{def}}{=} \forall x \geq t.\, ([\beta]_x \vee \exists y \in [t, x).\, [\alpha]_y)$$

where $[\alpha]_{t+1}$ is an abbreviation of $\exists y \geq t.\, \neg\exists z \in [t, y).\, (t < z \wedge [\alpha]_y)$. $\qquad\square$

Note how, per each atom $p \in At$ in the temporal formula φ, we get a monadic predicate $p(x)$ in the translation. The effect of this translation on the derived operators $\Diamond$ and $\square$ yields the quite natural expressions $[\square\alpha]_t \equiv \forall x \geq t.\, [\alpha]_t$ and $[\Diamond\alpha]_t \equiv \exists x \geq t.\, [\alpha]_t$. For instance, the translations of our running examples (9) and (10) for $t = 0$ respectively correspond to:

$$\forall x \geq 0.\, (\neg p(x) \to p(x+1)) \tag{11}$$

$$\forall x \geq 0.\, \Big(\forall y \geq x.\, \exists z \geq y.\, q(z) \to \exists y \geq x.\, \exists z \geq y.\, \big(p(z) \wedge \forall t \geq y.\, zq(t)\big) \Big) \tag{12}$$

Theorem 3 (from [24]). *Let φ be a THT formula built on a set of atoms At, $\mathbf{M} = \langle \mathbf{H}, \mathbf{T} \rangle$ a THT-interpretation on At and $\mathcal{M} = \langle \mathcal{H}, \mathcal{T} \rangle$ its corresponding Quantified HT-interpretation. It holds that $\mathbf{M}, i \models \varphi$ in THT iff $\mathcal{M} \models [\varphi]_i$ in Quantified Here-and-There. Moreover, $\mathbf{T}$ is a TS-model of φ iff $\mathcal{T}$ is a stable model of $[\varphi]_0$ in Quantified Equilibrium Logic.* $\qquad\square$

Another group of properties is related to comparison among temporal theories and subclasses of theories. For instance, in NMR, the regular equivalence, understood as a mere coincidence of selected models, is too weak to consider that one theory Γ_1 can be safely replaced by a second one Γ_2 since the addition of a context Γ may make them behave in a different way due to non-monotonicity.

Formally, we say that Γ_1 and Γ_2 are *strongly equivalent* when, for any arbitrary theory Γ, both $\Gamma_1 \cup \Gamma$ and $\Gamma_2 \cup \Gamma$ have the same selected models (in this case, stable models). [27] proved that checking equivalence in the logic of Here-and-There is a necessary and sufficient condition for strong equivalence in Equilibrium Logic, that is, Γ_1 and Γ_2 are strongly equivalent iff $\Gamma_1 \equiv_{HT} \Gamma_2$. It must be noticed that one direction of this result, the sufficient condition, is actually trivial. As HT is monotonic, $\Gamma_1 \equiv_{HT} \Gamma_2$ implies $\Gamma_1 \cup \Gamma \equiv_{HT} \Gamma_2 \cup \Gamma$ and so, their selected models will also coincide. The real significant result is the opposite direction, namely, that HT-equivalence is also a necessary condition for strong equivalence, as it shows that HT is strong enough as a monotonic basis for Equilibrium Logic. In [28] it was shown that something similar happens in the temporal case, namely:

Theorem 4 (from [28]). *Two temporal formulas α and β are strongly equivalent in TEL iff they are THT-equivalent.* □

Another interesting result related to equivalence is the existence of normal forms for THT and TEL. In the case of Equilibrium Logic, it has been already proved [29] that any arbitrary propositional theory is strongly equivalent to a logic program (allowing disjunction and negation in the head). Similarly, in the case of (monotonic) LTL, an implicational clause-like normal form introduced in [30] was used for designing a temporal resolution method.

Following [31], TEL can be similarly reduced (under strong equivalence) to a normal form, called *temporal logic programs* (TLP), consisting of a set of implications (embraced by a necessity operator) quite close to logic program rules. The obtained normal form considerably reduces the possible uses of modal operators and, as we will see later, has became useful for a practical computation of TEL models. The definitions are as follows. Given a signature At, we define a *temporal literal* as any expression in the set $\{p, \bigcirc p, \neg p, \neg \bigcirc p \mid p \in At\}$.

Definition 5 (Temporal rule). *A temporal rule is either:*

1. *an* initial rule *of the form* $B_1 \wedge \cdots \wedge B_n \rightarrow C_1 \vee \cdots \vee C_m$ *where all the B_i and C_j are temporal literals, $n \geq 0$ and $m \geq 0$.*
2. *a* dynamic rule *of the form* $\Box r$, *where r is an initial rule.*
3. *a* fulfillment rule *like* $\Box(\Box p \rightarrow q)$ *or like* $\Box(p \rightarrow \Diamond q)$ *with p, q atoms.* □

In the three cases, we respectively call rule *body* and rule *head* to the antecedent and consequent of the (unique) rule implication. In initial (resp. dynamic) rules, we may have an empty head $m = 0$ corresponding to $\bot$ – if so, we talk about an *initial* (resp. *dynamic*) *constraint*. A *temporal logic program*[5] (TLP for short) is a finite set of temporal rules. The reduction into TLP normal form introduces an auxiliary atom per each subformula in the original theory and applies the inductive definitions of temporal operators used for LTL in [30]. We will not enter into further details (see [31]) but the obtained reduction into TLP is modular, polynomial and strongly faithful (that is, it preserves strong equivalence, if auxiliary atoms are ignored).

---

[5] In fact, as shown in [31], this normal form can be even more restrictive: initial rules can be replaced by atoms, and we can avoid the use of literals of the form $\neg \bigcirc p$.

4   Computation

Computation of TS-models is a complex task. THT-satisfiability has been classified [23] as PSPACE-complete, that is, the same complexity as LTL-satisfiability, whereas TEL-satisfiability rises to EXPSPACE-completeness, as recently proved in [32]. In this way, we face a similar situation as in the non-temporal case where HT-satisfiability is NP-complete like SAT, whereas existence of equilibrium model (for arbitrary theories) is Σ_2^P-complete (like disjunctive ASP).

There exists a pair of tools, STeLP [33] and ABSTEM [28], that allow computing temporal stable models (represented as Büchi automata). These tools can be used to check verification properties that are usual in LTL, like the typical safety, liveness and fairness conditions, but in the context of temporal ASP. Moreover, they can also be applied for planning problems that involve an indeterminate or even infinite number of steps, such as the non-existence of a plan.

The first tool, STeLP, accepts a strict subset of the TLP normal form called *splittable* temporal formulas (STF) which will be of one of the following types:

$$B \wedge N \to H \tag{13}$$

$$B \wedge \bigcirc B' \wedge N \wedge \bigcirc N' \to \bigcirc H' \tag{14}$$

$$\Box(B \wedge \bigcirc B' \wedge N \wedge \bigcirc N' \to \bigcirc H') \tag{15}$$

where B and B' are conjunctions of atomic formulas, N and N' are conjunctions of $\neg p$, being p an atomic formula and H and H' are disjunctions of atomic formulas.

The name *splittable* refers to the fact that these programs can be splitted using [34] thanks to the property that rule heads never refer to a time point previous to those referred in the body. As we can see above, the main property of a splittable temporal rule is that, informally speaking, *past never depends on the future*, that is, we never get references to $\bigcirc$ in the rule bodies unless all atoms in the head are also in the scope of $\bigcirc$. As shown in [35], when the input temporal program is splittable, it is possible to extend the technique of *loop formulas* [36] to temporal theories so that it is always possible to capture the TS-models of a theory Γ as the LTL-models of another theory Γ' obtained from Γ together with its loop formulas. Although splittable theories do not cover the full expressiveness of TEL, most action domains represented in ASP are indeed splittable. To cover an ASP-like syntax, STeLP further allows the use of variables: a preliminary grounding method was presented in [37], proving its correctness.

The tool ABSTEM, on the contrary, accepts any arbitrary temporal theory as an input, although it does not accept variables. It relies on an automata-based transformation described in [23] and it not only allows computing the TS-models of a temporal theory, but also accepts pairs of theories to decide different types of equivalence: LTL-equivalence, TEL-equivalence (i.e. coincidence in the set of TS-models) and strong equivalence (i.e., THT-equivalence). Moreover, when strong equivalence fails, ABSTEM obtains a context, that is, an additional formula that added to the compared theories makes them behave differently.

5   Conclusions and Future Work

In this survey we have summarised the basic results on Temporal Equilibrium Logic obtained so far, showing that it can be used as a powerful tool for combining temporal reasoning tasks with Answer Set Programming. Still, there are many open topics that deserve to be studied. For instance, in the theoretical setting, we still miss a complete axiomatisation of THT. Another open question is that, although we know that Kamp's translation from LTL into First Order Logic also works for translating TEL into Quantified Equilibrium Logic, we ignore whether the other direction of Kamp's theorem also holds in this case. Namely, we ignore whether any theory in Monadic Quantified Equilibrium Logic for a linear order relation $<$ can be represented in TEL. A possibly related question is whether the set of TS-models of a temporal theory can be captured as the set of LTL-models of another theory. This holds in the case of splittable temporal logic programs, but is open in the general case[6].

An interesting research line is the extension of TEL with past operators, since they seem more natural for rule bodies that describe the transitions of a dynamic system. Besides, following similar steps as those done in TEL, other hybrid approaches can be explored. For instance, [39] has considered the combination of Dynamic LTL with Equilibrium Logic. Similarly, other temporal approaches can be treated in an analogous way, such as CTL, CTL*, Dynamic Logic or μ-calculus. Other open topics are related to potential applications including translation of different action languages, policy languages with preferences [40] or planning with (temporal) control rules.

Acknowledgements. This research is part of a long term project developed during the last eight years in the KR group from the University of Corunna and, especially, in close cooperation with Felicidad Aguado, Martín Diéguez, Gilberto Pérez and Concepción Vidal together with the regular collaborators David Pearce and Luis Fariñas. I am also especially thankful to Stèphane Demri, Philippe Balbiani, Andreas Herzig, Laura Bozzelli, Manuel Ojeda, Agustín Valverde, Stefania Costantini, Michael Fisher, Mirosław Truszczyński, Vladimir Lifschitz and Torsten Schaub for their useful discussions and collaboration at different moments on specific topics of this work.

References

1. McCarthy, J., Hayes, P.J.: Some philosophical problems from the standpoint of artificial intelligence. In: Meltzer, B., Michie, D. (eds.) Machine Intelligence, pp. 463–502. Edinburgh University Press, Edinburgh (1969)
2. Kowalski, R., Sergot, M.: A logic-based calculus of events. New Gener. Comput. **4**, 67–95 (1986)
3. McCarthy, J.: Circumscription: a form of non-monotonic reasoning. Artif. Intell. **13**, 27–39 (1980)
4. Mccarthy, J.: Modality, si! modal logic, no!. Stud. Logica. **59**(1), 29–32 (1997)

---

[6] An incorrect proof was published in [38], but we still conjecture that the result might be positive.

5. Castilho, M.A., Gasquet, O., Herzig, A.: Formalizing action and change in modal logic I: the frame problem. J. Logic Comput. **9**(5), 701–735 (1999)
6. Giordano, L., Martelli, A., Schwind, C.: Ramification and causality in a modal action logic. J. Logic Comput. **10**(5), 625–662 (2000)
7. Baral, C., Zhao, J.: Nonmonotonic temporal logics for goal specification. In: Proceedings of the 20th International Joint Conference on Artificial Intelligence (IJCAI 2007), pp. 236–242 (2007)
8. Gelfond, M., Lifschitz, V.: Representing action and change by logic programs. J. Logic Program. **17**, 301–321 (1993)
9. Gelfond, M., Lifschitz, V.: Action languages. Linköping Electron. Art. Comput. Inf. Sci. **3**(16) (1998). http://www.ep.liu.se/ea/cis/1998/016
10. Niemelä, I.: Logic programs with stable model semantics as a constraint programming paradigm. Ann. Math. Artif. Intell. **25**(3–4), 241–273 (1999)
11. Marek, V., Truszczyński, M.: Stable models and an alternative logic programming paradigm. In: Apt, K.R., et al. (eds.) The Logic Programming Paradigm, pp. 169–181. Springer, Heidelberg (1999)
12. Gebser, M., Sabuncu, O., Schaub, T.: An incremental answer set programming based system for finite model computation. AI Commun. **24**(2), 195–212 (2011)
13. Kautz, H.A., Selman, B.: Planning as satisfiability. In: Proceedings of the European Conference on Artificial Intelligence (ECAI 1992), pp. 359–363 (1992)
14. Cabalar, P., Pérez Vega, G.: Temporal equilibrium logic: a first approach. In: Moreno Díaz, R., Pichler, F., Quesada Arencibia, A. (eds.) EUROCAST 2007. LNCS, vol. 4739, pp. 241–248. Springer, Heidelberg (2007)
15. Gelfond, M., Lifschitz, V.: The stable model semantics for logic programming. In: Proceedings of the 5th International Conference on Logic Programming (ICLP 1988), Seattle, Washington, pp. 1070–1080 (1988)
16. Prior, A.: Past, Present and Future. Oxford University Press, Oxford (1967)
17. Kamp, H.: Tense logic and the theory of linear order. Ph.D. thesis, UCLA (1968)
18. Pearce, D.: A new logical characterisation of stable models and answer sets. In: Proceedings of Non-Monotonic Extensions of Logic Programming (NMELP 1996), Bad Honnef, Germany, pp. 57–70 (1996)
19. Heyting, A.: Die formalen Regeln der intuitionistischen Logik. Sitzungsberichte der Preussischen Akademie der Wissenschaften, Physikalisch-mathematische Klasse (1930)
20. Aguado, F., Cabalar, P., Diéguez, M., Pérez, G., Vidal, C.: Temporal equilibrium logic: a survey. J. Appl. Non-classical Logics **23**(1–2), 2–24 (2013)
21. Balbiani, P., Diéguez, M.: An axiomatisation of the logic of temporal here-and-there (2015) (unpublished draft)
22. Büchi, J.R.: On a decision method in restricted second-order arithmetic. In: International Congress on Logic, Methodology, and Philosophy of Science, pp. 1–11 (1962)
23. Cabalar, P., Demri, S.: Automata-based computation of temporal equilibrium models. In: Vidal, G. (ed.) LOPSTR 2011. LNCS, vol. 7225, pp. 57–72. Springer, Heidelberg (2012)
24. Cabalar, P., Diéguez, M., Vidal, C.: An infinitary encoding of temporal equilibrium logic. In: Proceedings of the 31st International Conference on Logic Programming (ICLP 2015) (2015)
25. Harrison, A., Lifschitz, V., Pearce, D., Valverde, A.: Infinitary equilibrium logic. In: Working Notes of Workshop on Answer Set Programming and Other Computing Paradigms (ASPOCP 2014) (2014)

26. Pearce, D.J., Valverde, A.: Quantified equilibrium logic and foundations for answer set programs. In: Garcia de la Banda, M., Pontelli, E. (eds.) ICLP 2008. LNCS, vol. 5366, pp. 546–560. Springer, Heidelberg (2008)
27. Lifschitz, V., Pearce, D., Valverde, A.: Strongly equivalent logic programs. Comput. Logic **2**(4), 526–541 (2001)
28. Cabalar, P., Diéguez, M.: Strong equivalence of non-monotonic temporal theories. In: Proceedings of the 14th International Conference on Principles of Knowledge Representation and Reasoning (KR 2014), Vienna, Austria (2014)
29. Cabalar, P., Ferraris, P.: Propositional theories are strongly equivalent to logic programs. Theor. Pract. Logic Program. **7**(6), 745–759 (2007)
30. Fisher, M.: A resolution method for temporal logic. In: Proceedings of the 12th International Joint Conference on Artificial Intelligence (IJCAI 1991), pp. 99–104. Morgan Kaufmann Publishers Inc. (1991)
31. Cabalar, P.: A normal form for linear temporal equilibrium logic. In: Janhunen, T., Niemelä, I. (eds.) JELIA 2010. LNCS, vol. 6341, pp. 64–76. Springer, Heidelberg (2010)
32. Bozzelli, L., Pearce, D.: On the complexity of temporal equilibrium logic. In: Proceedings of the 30th Annual ACM/IEEE Symposium of Logic in Computer Science (LICS 2015), Kyoto, Japan (2015, to appear)
33. Cabalar, P., Diéguez, M.: STeLP – a tool for temporal answer set programming. In: Delgrande, J.P., Faber, W. (eds.) LPNMR 2011. LNCS, vol. 6645, pp. 370–375. Springer, Heidelberg (2011)
34. Lifschitz, V., Turner, H.: Splitting a logic program. In: Proceedings of the 11th International Conference on Logic programming (ICLP 1994), pp. 23–37 (1994)
35. Aguado, F., Cabalar, P., Pérez, G., Vidal, C.: Loop formulas for splitable temporal logic programs. In: Delgrande, J.P., Faber, W. (eds.) LPNMR 2011. LNCS, vol. 6645, pp. 80–92. Springer, Heidelberg (2011)
36. Ferraris, P., Lee, J., Lifschitz, V.: A generalization of the Lin-Zhao theorem. Ann. Math. Artif. Intell. **47**, 79–101 (2006)
37. Aguado, F., Cabalar, P., Diéguez, M., Pérez, G., Vidal, C.: Paving the way for temporal grounding. In: Proc. of the 28th International Conference on Logic Programming (ICLP 2012) (2012)
38. Cabalar, P., Diéguez, M.: Temporal stable models are LTL-representable. In: Proceedings of the 7th International Workshop on Answer Set Programming and Other Computing Paradigms (ASPOCP 2014) (2014)
39. Aguado, F., Pérez, G., Vidal, C.: Integrating temporal extensions of answer set programming. In: Cabalar, P., Son, T.C. (eds.) LPNMR 2013. LNCS, vol. 8148, pp. 23–35. Springer, Heidelberg (2013)
40. Bertino, E., Mileo, A., Provetti, A.: PDL with preferences. In: 6th IEEE International Workshop on Policies for Distributed Systems and Networks (POLICY 2005), pp. 213–222 (2005)

Algorithmic Decision Theory Meets Logic
— Invited Talk —

Jérôme Lang$^{(\boxtimes)}$

LAMSADE, CNRS and Université Paris Dauphine, Paris, France
`lang@lamsade.dauphine.fr`

1 Introduction

Algorithmic decision theory can be roughly defined as the design and study of languages and methods for expressing and solving various classes of decision problems, including: decision under uncertainty, sequential decision making, multicriteria decision making, collective decision making, and strategic interactions in distributed decision making. A decision problem is specified by two main components: the *preferences* of the agent(s); and the *beliefs* the agent(s) has (have) about the initial state of the world and its evolution, and possibly about the beliefs and preferences of other agents. Computational tasks involve, among others: the construction and refinement of the problem, through learning and elicitation tasks; the search for a centralized decision (for an agent or a group of agents); the impact of selfish behaviour in decentralized, multi-agent decision contexts.

Logic in algorithmic decision theory can be useful as a *declarative representation language* for the various components of the problems, and as a *generic problem solving tool*. The combination of both allow for representing *and* solving complex decision making problems. Below I point to some research issues at the meeting point of logic and algorithmic decision theory. The list is certainly not exhaustive, and it is biased towards my own work.

2 Representing and Reasoning with Preferences

2.1 Compact Representation

Domains of solutions in algorithmic decision theory often have a *combinatorial* structure of the form $A = D_1 \times \ldots \times D_p$, where each D_i is a finite set of values associated with a variable X_i. A can for instance be the set of all alternatives to choose from in many voting contexts[1] such as multiple referenda or committee elections,

Expressing preferences on such domains by listing or ranking explicitly all alternatives or solutions is practically infeasible as soon as the number of variables is more than a few units, because it puts too much communication burden

---

[1] See [1] for an survey of voting in combinatorial domains.

© Springer International Publishing Switzerland 2015
F. Calimeri et al. (Eds.): LPNMR 2015, LNAI 9345, pp. 14–19, 2015.
DOI: 10.1007/978-3-319-23264-5_2

on the agents. The AI community has produced a considerable amount of work on *compact representation languages* for preferences, aiming at expressing and processing preferences over large combinatorial domains using as few computational resources (space and time) as possible. Many of these languages are based on logic (see [2] for a survey). The most elementary language consists in specifying dichotomous preferences via propositional formulae; extensions to nondichotomous preferences consist in associating priorities or weights with formulas, using distances between interpretations, or expressing preferences between propositional formulas using a *ceteris paribus* completion principle. Logic programming languages, especially answer set programming, are also very useful: see [3–5] for surveys and [6] for a very recent development.

2.2   Preference Logics

While compact preference representation languages primarily aim at expressing succinctly preferences over combinatorial domains of alternatives, *preference logics* aim at *reasoning about* preferences. A preference logic consists of a semantics and/or a formal system for interpreting relative preferences between logical formulas, or monadic, absolute preferences over formulas. The starting point of preference logics is that individuals often express relative or absolute preferences that refer not to isolated alternatives, but to logical formulas representing sets of alternatives. The central component of a preference logic is the *lifting* operator inducing preferences between formulas from preferences over single alternatives. At least two families of preference logics have been developed:

Logics of *ceteris paribus* Preferences. When an agent expresses a preference statement such as "I prefer to spend my summer holiday in Kentucky than in California", they surely do not mean that they prefer *any* summer holiday in Kentucky to *any* summer holiday in California; the preference statement does not preclude that they would prefer a sunny holiday in California to a rainy one in Kentucky. The principle at work when interpreting such preference statements is that the alternatives should be compared *all other things being equal (ceteris paribus)*, or more generally, all irrelevant properties being equal. A few milestones in *ceteris paribus* preference logics are [7–10]. Note that [10] also compares and attempts to reunify compact representation languages and preference logics.

Defeasible Preferences and Conditional Preference Logics. Consider the following statements: (1) I'd like to spend my weekend in Lexington; (2) if there is a storm warning on Lexington next weekend, then I prefer not to go. Statement (1) corresponds to a *defeasible, default* preference: it applies not only if we know that there is no risk of storm but more generally if there is no specific information about the weather forecast. This corresponds to assuming that the state of the world is *normal* (no storm warning); upon receiving the storm forecast, (1) is overridden by the more specific statement (2). Defeasible preferences fit the intuition as well as the natural language expression of preferences, and allow for their succinct and modular representation: succinct because they avoid to specify explicitly all the exceptional conditions in which a preference statement does not apply; modular because a set of such preference statements can be completed at any time, without generating an inconsistency — coming back to the latter example, if to (1) and (2) we later add (3) *if there is an exciting joint conference on algorithmic decision theory and logic programming in Lexington then I want to be there (independently of the weather)*, it will have higher priority than (2) in the 'doubly exceptional' circumstance "stormy weather and ADT-LPNMR". A neat way of formalizing these defeasible, conditional preferences consists in using *conditional logics*; some key papers on conditional preferences are [11–13].

3   Representing and Reasoning with Beliefs

Logic (and in particular, doxastic and epistemic logics) allow for distinguishing between objective facts and subjective beliefs. As there are numerous classes of problems in algorithmic decision theory where the decision maker(s) do not have a complete knowledge of the situation, logic has definitely a role to play here.

3.1   The External Perspective: Incomplete Knowledge of Agents' Preferences

Let us consider the point of view of an external agent that has to make decisions or to make predictions based on an incomplete, partial view of the agents' preferences: for instance, a recommender system in decision aid, a central authority in group decision making, or the modeller in game theory.

For the sake of brevity, let us focus on social choice. Often, the central authority in charge of computing the outcome (the winner of an election, an allocation of resources, etc.) has an incomplete knowledge of the agents' preferences, perhaps because the elicitation process was not conducted until its end, or because the voters could not report complete preferences. The central authority sees a set of possible worlds, each corresponding to a complete preference profile; an alternative is a possible (resp. necessary) winner if it is a winner in some (resp. all) possible worlds(s) [14].[2] Similar notions have been studied in fair division. It is clear that these notions originate in epistemic logic.

3.2   The Internal Perspective: Beliefs and Strategic Behaviour

A crucial issue in distributed multiagent systems is the impact of strategic behaviour on the 'social quality' of the reached state. Focusing on social choice, a tremendously high number of papers examine the conditions under which a mechanism that takes as input the agents' declared preferences can be manipulated by them, the computational complexity of finding a manipulation, and the impact of manipulation on social welfare. The assumption typically made is that agents have complete knowledge of the others' preferences. What if they have complex mutual beliefs, weaker than common knowledge, but stronger than zero

---

[2] See [15] for a review of existing work along this line.

knowledge? First steps towards handling such mutual beliefs in social choice have been made [16–19], but they remain preliminary. On the other hand, on reasoning about mutual beliefs in game theory there is an abundant literature, and even a series of workshops (*Logical Foundations of Decision and Game Theory*).

4   Logic for Problem Solving

4.1   Logical Encoding and Resolution of Decision Problems

Probably the widest use of logic in decision making contexts takes place in sequential decision making settings, or more generally in contexts where one has to search in a combinatorial space of solutions. The paradigmatic example is *planning as satisfiability* [20]: the planning problem (initial state, action effects, goal, horizon) is translated into a set of propositional clauses, which is fed to a SAT solver; the model found by the solver (if any) is translated back to a plan. The framework has been extended to planning with nondeterministic actions.[3] Answer set programming is also a natural and efficient tool for expressing and solving planning problems [22,23] and for multicriteria optimisation [24].

4.2   Automated Theorem Proving and Discovery

Open research questions can be addressed using computer-aided theorem proving techniques. The role of computer science here is not to help solving a decision making problem, but to (re)prove theorems and/or discover new ones. Automated theorem proving and discovery is especially helpful in branches of decision making where combinatorial structures prevail, such as decision theory over discrete domains, social choice theory, cooperative or noncooperative game theory. Some examples are an automated proof of Arrow's theorem [25,26], impossibility theorems about pure Nash equilibria in two-person games [27], about ranking sets of alternatives [28], or about strategyproofness and participation in voting [29,30]. Also, the modelling of social choice mechanisms in modal logic [31,32] is related to this research line.

References

1. Lang, J., Xia, L.: Voting in combinatorial domains. In: Brandt, F., Conitzer, V., Endriss, U., Lang, J., Procaccia, A.D. (eds.) Handbook of Computational Social Choice. Cambridge University Press (2015, in Press)
2. Lang, J.: Logical representation of preferences. In: Bouyssou, D., Dubois, D., Prade, H., Pirlot, M. (eds.) Decision Making Process: Concepts and Methods, pp. 321–364. Wiley-ISTE, London (2009)
3. Delgrande, J.P., Schaub, T., Tompits, H., Wang, K.: A classification and survey of preference handling approaches in nonmonotonic reasoning. Comput. Intell. **20**, 308–334 (2004)

---

[3] See [21] for a survey.

4. Brewka, G.: Preferences in answer set programming. In: Marín, R., Onaindía, E., Bugarín, A., Santos, J. (eds.) CAEPIA 2005. LNCS (LNAI), vol. 4177, pp. 1–10. Springer, Heidelberg (2006)
5. Brewka, G., Niemelä, I., Truszczynski, M.: Preferences and nonmonotonic reasoning. AI Mag. **29**, 69–78 (2008)
6. Brewka, G., Delgrande, J.P., Romero, J., Schaub, T.: asprin: customizing answer set preferences without a headache. In: Proceedings of the Twenty-Ninth AAAI Conference on Artificial Intelligence, Austin, Texas, USA, 25–30 January 2015, pp. 1467–1474 (2015)
7. von Wright, G.: The Logic of Preference. Edinburgh University Press, Edinburgh (1963)
8. Hansson, S.O.: The Structure of Values and Norms. Cambridge University Press, Cambridge (2001)
9. van Benthem, J., Girard, P., Roy, O.: Everything else being equal: a modal logic for Ceteris Paribus preferences. J. Philos. Logic **38**, 83–125 (2009)
10. Bienvenu, M., Lang, J., Wilson, N.: From preference logics to preference languages, and back. In: Principles of Knowledge Representation and Reasoning: Proceedings of the Twelfth International Conference, KR 2010, Toronto, Ontario, Canada, 9–13 May 2010
11. Boutilier, C.: Toward a logic for qualitative decision theory. In: Proceedings of the 4th International Conference on Principles of Knowledge Representation and Reasoning (KR 1994), Bonn, Germany, 24–27 May 1994, pp. 75–86 (1994)
12. Lang, J.: Conditional desires and utilities: an alternative logical approach to qualitative decision theory. In: Proceedings of the 12th European Conference on Artificial Intelligence, Budapest, Hungary, 11–16 August 1996, pp. 318–322 (1996)
13. Lang, J., van der Torre, L.W.N., Weydert, E.: Hidden uncertainty in the logical representation of desires. In: IJCAI 2003, Proceedings of the Eighteenth International Joint Conference on Artificial Intelligence, Acapulco, Mexico, 9–15 August 2003, pp. 685–690 (2003)
14. Konczak, K., Lang, J.: Voting procedures with incomplete preferences. In: Multidisciplinary Workshop on Advances in Preference Handling, pp. 124–129 (2005)
15. Boutilier, C., Rosenschein, J.: Incomplete information and communication in voting. In: Brandt, F., Conitzer, V., Endriss, U., Lang, J., Procaccia, A.D. (eds.) Handbook of Computational Social Choice. Cambridge University Press (2015, in Press)
16. Chopra, S., Pacuit, E., Parikh, R.: Knowledge-theoretic properties of strategic voting. In: Alferes, J.J., Leite, J. (eds.) JELIA 2004. LNCS (LNAI), vol. 3229, pp. 18–30. Springer, Heidelberg (2004)
17. van Ditmarsch, H., Lang, J., Saffidine, A.: Strategic voting and the logic of knowledge. In: Proceedings of 14th TARK, ACM (2013)
18. Meir, R., Lev, O., Rosenschein, J.: A local-dominance theory of voting equilibria. In: ACM Conference on Economics and Computation, EC 2014, Stanford, CA, USA, 8–12 June 2014, pp. 313–330 (2014)
19. Conitzer, V., Walsh, T., Xia, L.: Dominating manipulations in voting with partial information. In: Proceedings of the Twenty-Fifth AAAI Conference on Artificial Intelligence, AAAI 2011, San Francisco, California, USA, 7–11 August 2011
20. Kautz, H., Selman, B.: Planning as satisfiability. In: ECAI, pp. 359–363 (1992)
21. Rintanen, J.: Planning and sat. In: Biere, A., Heule, M., van Maaren, H., Walsh, T. (eds.) Handbook of Satisfiability, pp. 483–504. IOS Press, Amsterdam (2009)
22. Lifschitz, V.: Answer set programming and plan generation. Artif. Intell. **138**, 39–54 (2002)

23. Eiter, T., Faber, W., Leone, N., Pfeifer, G., Polleres, A.: A logic programming approach to knowledge-state planning, II: the DLV^k system. Artif. Intell. **144**, 157–211 (2003)
24. Gebser, M., Kaminski, R., Kaufmann, B., Schaub, T.: Multi-criteria optimization in answer set programming. In: Technical Communications of the 27th International Conference on Logic Programming, p. 1 (2011)
25. Nipkow, T.: Social choice theory in HOL: Arrow and Gibbard-Satterthwaite. J. Autom. Reasoning **43**, 289–304 (2009)
26. Tang, P., Lin, F.: Computer-aided proofs of Arrow's and other impossibility theorems. Artif. Intell. **173**, 1041–1053 (2009)
27. Tang, P., Lin, F.: Discovering theorems in game theory: two-person games with unique pure nash equilibrium payoffs. Artif. Intell. **175**, 2010–2020 (2011)
28. Geist, C., Endriss, U.: Automated search for impossibility theorems in social choice theory: ranking sets of objects. J. Artif. Intell. Res. **40**, 143–174 (2011)
29. Brandt, F., Geist, C.: Finding strategyproof social choice functions via SAT solving. In: AAMAS 2014, pp. 1193–1200. IFAAMAS (2014)
30. Brandl, F., Brandt, F., Geist, C., Hofbauer, J.: Strategic abstention based on preference extensions: positive results and computer-generated impossibilities. In: IJCAI 2015 (2015)
31. Ågotnes, T., van der Hoek, W., Wooldridge, M.: On the logic of preference and judgment aggregation. Auton. Agents Multi-Agent Syst. **22**, 4–30 (2011)
32. Troquard, N., van der Hoek, W., Wooldridge, M.: Reasoning about social choice functions. J. Philos. Logic **40**, 473–498 (2011)

Relational and Semantic Data Mining
— Invited Talk —

Nada Lavrač [1,2,3](✉) and Anže Vavpetič[1,2]

[1] Jožef Stefan Institute, Jamova 39,
1000 Ljubljana, Slovenia
[2] Jožef Stefan International Postgraduate School, Jamova 39,
1000 Ljubljana, Slovenia
{nada.lavrac,anze.vavpetic}@ijs.si
[3] University of Nova Gorica, Nova Gorica, Slovenia

Abstract. Inductive Logic Programming (ILP) and Relational Data Mining (RDM) address the task of inducing models or patterns from multi-relational data. One of the established approaches to RDM is propositionalization, characterized by transforming a relational database into a single-table representation. After introducing ILP and RDM, the paper provides an overview of propositionalization algorithms, which have been made publicly available through the web-based Clowd-Flows data mining platform. The paper concludes by presenting recent advances in Semantic Data Mining, characterized by exploiting relational background knowledge in the form of domain ontologies in the process of model and pattern construction.

Keywords: Inductive Logic Programming · Relational Data Mining · Semantic Data Mining · Propositionalization

1 Introduction

Standard machine learning and data mining algorithms induce hypotheses in the form of models or propositional patterns learned from a given data table, where one example corresponds to a single row in the table. Most types of propositional models and patterns have corresponding relational counterparts, such as relational classification rules, relational regression trees, relational association rules. Inductive Logic Programming (ILP) [23] and Relational Data Mining (RDM) [4,6] algorithms can be used to induce such relational models and patterns from multi-relational data, e.g., data stored in a relational database.

For relational databases in which data instances are clearly identifiable (the so-called individual-centered representation [7]), various techniques can be used for transforming a relational database into a propositional single-table representation [14]. After performing such a transformation [18], usually named *propositionalization* [12], standard propositional learners can be used, including decision tree and classification rule learners.

© Springer International Publishing Switzerland 2015
F. Calimeri et al. (Eds.): LPNMR 2015, LNAI 9345, pp. 20–31, 2015.
DOI: 10.1007/978-3-319-23264-5_3

The first part of the paper presents a survey of the state-of-the-art propositionalization techniques. Following an introduction to the propositionalization problem and a description of a number of propositionalization methods, we translate and unify the terminology, using a language that should be familiar to an analyst working with relational databases. Furthermore, we provide an empirical comparison of freely available propositionalization algorithms. Finally, we present our approach to making the use of propositionalization algorithms easier for non-experts, as well as making the experiments shareable and repeatable. The freely available state-of-the-art methods discussed in this paper were wrapped as reusable components in the web-based data mining platform ClowdFlows [13], together with the utilities for working with a relational database management system (RDBMS).

The second part of the paper addresses a more recent ILP setting, named *semantic data mining* (SDM), characterized by exploiting relational background knowledge in the form of domain ontologies in the process of model and pattern construction. The development of SDM techniques is motivated by the availability of large amounts of semantically annotated data in all domains of science, and biology in particular, posing requirements for new data mining approaches which need to deal with increased data complexity, the relational character of semantic representations, as well as the reasoning capacities of the underlying ontologies. The paper briefly introduces the task of semantic data mining, followed by a short overview of the state-of-the-art approaches. Finally, the paper presents the Hedwig semantic subgroup discovery algorithm [1,33] developed by the authors of this paper.

The paper is structured as follows. Section 2 gives an introduction to the propositionalization task, describes the state-of-the-art methods, and presents a number of reusable propositionalization workflows implemented in the Clowd-Flows data mining platform. In Sect. 3 we introduce the SDM task, a quick state-of-the-art overview, and a recent semantic subgroup discovery approach Hedwig. Section 4 concludes the paper with a brief summary.

2 Propositionalization

Propositional representations (a single table format) impose the constraint that each training example is represented as a single fixed-length tuple. Due to the nature of some relational problems, there exists no elegant propositional encoding; for example, a citation network in general cannot be represented in a propositional format without loss of information, since each author can have any number of co-authors and papers. The problem is naturally represented using multiple relations, e.g., including the *author* and the *paper* relations.

Problems characterized by multiple relations can be tackled in two different ways: (1) by using a relational learner such as Progol [22] or Aleph [30], which can build a model or induce a set of patterns directly, or (2) by constructing complex relational features used to transform the relational representation into a propositional format and then applying a propositional learner on the transformed single-table representation. In this paper we focus on the latter approach, called *propositionalization*. Propositionalization is a form of *constructive induction*, since it involves changing the representation for learning. As we noted before, propositionalization cannot always be done without loss of information, but it can be a powerful method when a suitable relational learner is not available, when a non-conventional ILP task needs to be performed on data from a given relational database (e.g., clustering), and when the problem at hand is *individual-centered* [7]. Such problems have a clear notion of an individual and the learning occurs only at the level of (sets of) individual instances rather than the (network of) relationships between the instances. As an example consider the problem of classifying authors into research fields given a citation network; in this case the author is an individual and learning occurs at the author level, i.e. assigning class labels to authors, rather than classifying the authors in terms of their citations in the citation network of other authors.

To illustrate the propositionalization scenario, consider a simplified multi-relational problem, where the data to be mined is a database of authors and their papers, with the task of assigning a research field to unseen authors. In essence, a complete propositional representation of the problem (shown in Table 1) would be a set of queries $q \in Q$ (complex relational features) that return value *true* or *false* for a

Table 1. A sample propositional representation of authors table.

Author	q_1	q_2	$\ldots$	q_m	Class
A1	1	1	$\ldots$	1	C_1
A2	0	1	$\ldots$	0	C_1
A3	1	0	$\ldots$	0	C_2
$\ldots$	$\ldots$	$\ldots$	$\ldots$	$\ldots$	$\ldots$
An	0	1	0	0	C_1

given author. Each query describes a property of an author. The property can involve a rather complex query, involving multiple relations as long as that query returns either true or false. For example, a query could be "does author X have a paper published at the ECML/PKDD conference?".

While this transformation could be done by hand by a data miner, we are only interested in automated propositionalization methods. Furthermore, the transformation into a propositional representation can be done with essentially any ML or DM task in mind: classification, association discovery, clustering, etc.

2.1   Relational Data Mining Task Formulation

A relational data mining task can be formally defined as follows.

Given:

- evidence E (examples, given extensionally),
- an initial theory B (background knowledge, given extensionally or as sets of clauses over the set of background relations).

Find:

- a theory H (hypothesis, in the form of a set of logical clauses) that together with B explains the target properties of E.

where the target property can be a selected class label (the target class) or some other property of interest.

This is a typical ILP definition of the problem, given that numerous existing approaches to relational data mining and propositionalization were developed within the field of ILP. However, since real-world data is in most cases stored in some Relational Database Management System (RDBMS), we try to unify the terminology used across various approaches to be as familiar as possible also to researchers working with databases—which are likely the ones most interested in propositionalization techniques. The definition of a relational data mining task using a more conventional database terminology is given below.

Given:

- target table t, where each row is one example,
- related tables T, connected to t via foreign keys.

Find:

- a query Q (a set of sub-queries) that together with T describes the target properties of t.

In the rest of this paper we will focus on the classification (and subgroup discovery) tasks with a clear notion of the target property of interest (a selected class label), since we can effectively compare different approaches via the performance of the resulting classifier. Using propositionalization to tackle classification tasks must involve two independent steps: (1) preparing a single-table representation of the input database, and (2) applying a propositional learner on that table. In contrast, learners that directly use the multi-relational representations intertwine feature construction and model construction. In propositionalization, these two steps are separated. The workload of finding good features (which have large coverage of instances, or which best separate between different classes) is done by the propositionalization algorithm, while the work of combining these features to produce a good classification model is offloaded to the propositional learner.

The actual art of propositionalization is to generate a number of good, potentially complex features (binary queries), to be evaluated as *true* or *false* for each individual, which the learner will use to construct a classifier. In the model construction phase, the learner exploits these queries about each individual as features used to construct the model. For example, if a decision tree model is constructed, each node in the tree will contain a single query, with the two values (*true* and *false*) on the outgoing branches of this node. Note that propositionalization is not limited only to binary features—many approaches (e.g., [15] and [11]) also use aggregation functions to calculate feature values.

To classify unseen individuals, the classifier must then evaluate the queries that are found in the decision tree nodes on the unseen example and follow the branches according to their answers to arrive at a classification in the leave of the decision tree.

2.2 Overview of Propositionalization Algorithms

The best known propositionalization algorithms are first briefly described, followed by the experimental evaluation of the ones which are publicly available.

LINUS [18] is one of the first propositionalization approaches. It generates features that do not allow recursion and newly introduced variables. The second limitation is more serious and means that the queries cannot contain joins. An improvement of LINUS is SINUS [19] which incorporates more advanced feature construction techniques inspired by feature construction implemented in 1BC [7].

Aleph [30] is an ILP toolkit with many modes of functionality: learning theories, feature construction, incremental learning, etc. In this paper we are interested in its feature construction facility which can be used as a tool for propositionalization. Aleph uses mode declarations to define the syntactic bias. Input relations are defined as Prolog clauses: either extensionally or intensionally.

RSD [36] is a relational subgroup discovery algorithm composed of two main steps: the propositionalization step and the subgroup discovery step. The output of the propositionalization step can be used also as input to other propositional learners. RSD effectively produces an exhaustive list of first-order features that comply with the user-defined mode constraints, similar to those of Progol [22] and Aleph [30]. Furthermore, RSD features satisfy the connectivity requirement, which imposes that no feature can be decomposed into a conjunction of two or more features. Mode declarations define the algorithm's syntactic bias, i.e. the space of possible features.

HiFi [17] is a propositionalization approach that constructs first-order features with hierarchical structure. Due to this feature property, the algorithm performs the transformation in polynomial time of the maximum feature length. Furthermore, the resulting features are the smallest in their semantic equivalence class. The algorithm is shown to perform several orders of magnitude faster than RSD for higher feature lengths.

RelF [16] constructs a set of tree-like relational features by combining smaller conjunctive blocks. The novelty is that RelF preserves the monotonicity of feature reducibility and redundancy (instead of the typical monotonicity of frequency), which allows the algorithm to scale far better than other state-of-the-art propositionalization algorithms.

RELAGGS [15], which stands for *rel*ational *agg*regation, is a propositionalization approach that uses the input relational database schema as a basis for a declarative bias and it aims to use optimization techniques usually used in relational databases (e.g., indexes). Furthermore, the approach employs aggregation functions in order to summarize non-target relations with respect to the individuals in the target table.

Stochastic propositionalization [12] employs a search strategy similar to random mutation hill-climbing: the algorithm iterates over generations of individuals, which are added and removed with a probability proportional to the fitness of individuals, where the fitness function used is based on the Minimum Description Length (MDL) principle.

Safarii [11] is a commercial multi-relation data mining tool.[1] It offers a unique pattern language that merges ILP-style structural descriptions as well as

---

[1] http://www.kiminkii.com/safarii.html.

aggregations. Safarii comes with a tool called ProSafarii, which offers several pre-processing utilities—including propositionalization via aggregation.

Wordification [26, 27] is a propositionalization method inspired by text mining, which can be seen as a transformation of a relational database into a corpus of text documents. Wordification aims at constructing simple, easy to understand features, acting as words in the transformed Bag-Of-Words representation.

Extensive description of the experimental evaluation of the available propositionalization algorithm is presented in [28]. Fully reproducing the experimental results is outside the scope of this paper. The evaluation of different propositionalization approaches was performed on binary classification tasks using seven datasets from five different relational domains. The Friedman test [8] using significance level $\alpha = 0.05$ and the corresponding Nemenyi post-hoc test [24] were applied. This evaluation approach was used as an alternative to the t-test, which is proven to be inappropriate for testing multiple algorithms on multiple datasets [5]. A birds's eye view of the results is shown in Fig. 1.

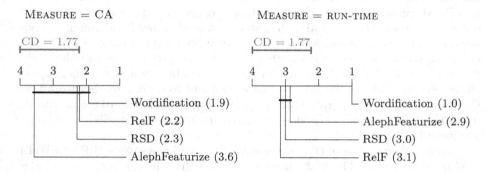

Fig. 1. Critical distance diagram for the reported classification accuracy (left; not enough evidence to prove that any algorithm performs better) and run-time (right; significant differences for $\alpha = 0.05$) results. The numbers in parentheses are the average ranks.

The statistical test was first performed using the J48 decision tree learner for classification accuracy and run-time. For classification accuracy, there is not enough evidence to prove that any propositionalization algorithm on average performs better than the others (Fig. 1 left, for significance level $\alpha = 0.05$), even though wordification achieves the best results on five out of seven benchmarks. We repeated the same statistical analysis for the LibSVM results, where the conclusion ended up the same. For run-time, however, the results are statistically significant in favor of wordification; see the critical distance diagram in the right part of Fig. 1. The diagram tells us that the wordification approach performs statistically significantly faster than other approaches, under the significance level $\alpha = 0.05$. Other approaches fall within the same critical distance and no statistically significant difference was detected.

2.3   ILP in the ClowdFlows Platform

The ClowdFlows platform [13] is an open-source, web-based data mining platform that supports the construction and execution of scientific workflows. This web application can be accessed and controlled from anywhere while the processing is performed in a cloud of computing nodes. A public installation of Clowd-Flows is accessible at http://clowdflows.org. For a developer, the graphical user interface supports simple operations that enable workflow construction: adding workflow components (widgets) on a canvas and creating connections between the components to form an executable workflow, which can be shared by other users or developers. Upon registration, the user can access, execute, modify, and store the modified workflows, enabling their sharing and reuse. On the other hand, by using anonymous login, the user can execute a predefined workflow, while any workflow modifications would be lost upon logout.

We have extended ClowdFlows with the implementation of an ILP toolkit, including the popular ILP system Aleph [30] together with its feature construction component, as well as RSD [36], RelF [16] and Wordification [26] propositionalization engines. Construction of RDM workflows is supported by other specialized RDM components (e.g., the MySQL package providing access to a relational database by connecting to a MySQL database server), other data mining components (e.g., the Weka [34] classifiers) and other supporting components (including cross-validation), accessible from other ClowdFlows modules. Each public workflow is assigned a unique URL that can be accessed by any user to either repeat the experiment, or use the workflow as a template to design another workflow. Consequently, the incorporated RDM algorithms become handy to use in real-life data analytics, which may therefore contribute to improved accessibility and popularity of ILP and RDM.

Figure 2 shows some of the implemented ILP workflows using ILP and Weka module components. The first workflow assumes that the user uploads the files required by RSD as Prolog programs. Workflows constructed for the other three propositionalization approaches Aleph, RelF and Wordification, which are also made publicly available, assume that the training data is read from a MySQL database.

In terms of workflows reusability, accessible by a single click on a web page where a workflow is exposed, the implemented propositionalization toolkit is a significant step towards making the ILP legacy accessible to the research community in a systematic and user-friendly way. To the best of our knowledge, this is the only workflow-based implementation of ILP and RDM algorithms in a platform accessible through a web browser, enabling simple workflow adaptation to the user's needs. Moreover, the ILP toolkit widgets actually use a Python library called `python-rdm` which is available on GitHub[2]. The authors welcome extensions and improvements from the community.

---

[2] https://github.com/anzev/rdm/.

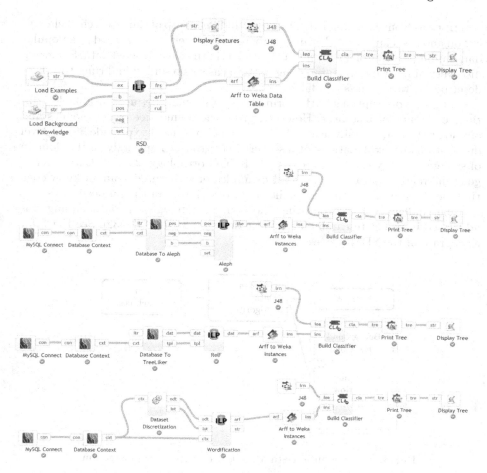

Fig. 2. First: RSD propositionalization workflow using ILP and Weka components is available online at http://clowdflows.org/workflow/471/ (the same RSD workflow, extended by accessing the training data using a MySQL database, is available at http://clowdflows.org/workflow/611/). Second: Aleph workflow available at http://clowdflows.org/workflow/2224/. Third: RelF workflow available at http://clowdflows.org/workflow/2227/. Fourth: Wordification workflow available at http://clowdflows.org/workflow/2222/.

3   Semantic Data Mining

Rule learning, which was initially focused on building predictive models formed of sets of classification rules, has recently shifted its focus to descriptive pattern mining. Well-known pattern mining techniques in the literature are based on association rule learning [2,29]. While the initial studies in association rule mining have focused on finding interesting patterns from large datasets in an unsupervised setting, association rules have been used also in a supervised setting, to learn pattern

descriptions from class-labeled data [20]. Building on top of the research in classification and association rule learning, subgroup discovery has emerged as a popular
data mining methodology for finding patterns in the class-labeled data. Subgroup
discovery aims at finding interesting patterns as sets of individual rules that best
describe the target class [10, 35].

Subgroup descriptions in the form of propositional rules are suitable descriptions of groups of instances. However, given the abundance of taxonomies and
ontologies that are readily available, these can also be used to provide higher-level
descriptors and explanations of discovered subgroups. Especially in the domain
of systems biology the GO ontology [3], KEGG orthology [25] and Entrez gene–
gene interaction data [21] are good examples of structured domain knowledge
that can be used as additional higher-level descriptors in the induced rules.

The challenge of incorporating the domain ontologies in data mining was
addressed in recent research on semantic data mining (SDM) [32]. See Fig. 3 for
a diagram of the SDM process.

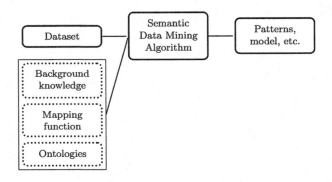

Fig. 3. The Semantic Data Mining (SDM) process illustration.

In [32] we described and evaluated the SDM toolkit that includes two semantic data mining systems: SDM-SEGS and SDM-Aleph. SDM-SEGS is an extension of the earlier domain-specific algorithm SEGS [31] which allows for semantic
subgroup discovery in gene expression data. SEGS constructs gene sets as combinations of GO ontology [3] terms, KEGG orthology [25] terms, and terms
describing gene–gene interactions obtained from the Entrez database [21]. SDM-
SEGS extends and generalizes this approach by allowing the user to input any set
of ontologies in the OWL ontology specification language and an empirical data
collection which is annotated by domain ontology terms. SDM-SEGS employs
ontologies to constrain and guide the top-down search of a hierarchically structured space of induced hypotheses. SDM-Aleph, which is built using the popular
inductive logic programming system Aleph [30] does not have the limitations of
SDM-SEGS, imposed by the domain-specific algorithm SEGS, and can accept
any number of OWL ontologies as background knowledge which is then used in
the learning process.

Based on the lessons learned in [32], we introduced a new system Hedwig in [33]. The system takes the best from both SDM-SEGS and SDM-Aleph. It uses a search mechanism tailored to exploit the hierarchical nature of ontologies. Furthermore, Hedwig can take into account background knowledge in the form of RDF triplets. Compared to [33], the current version of the system uses better redundancy pruning and significance tests based on [9]. Furthermore, the new version also support negations of unary predicates. Apart from the financial domain in [33], the approach was also applied on a multi-resolution dataset of chromosome abberrations in [1].

Hedwig is open-source software available on GitHub[3] and the authors welcome improvements from the community.

4   Conclusions

This paper addresses two lines of research of the authors, the propositionalization approach and the semantic data mining approach to RDM.

First, ILP and RDM are introduced, together with an overview of popular propositionalization algorithms. Next, the paper briefly presents the results of an experimental comparison of several such algorithms on several relational databases. These approaches have been made available through the web-based ClowdFlows data mining platform, together with repeatable and reusable workflows. The paper concludes by presenting recent advances in Semantic Data Mining, characterized by exploiting relational background knowledge in the form of domain ontologies in the process of model and pattern construction.

In further work, we will combine ILP and RDM approaches with the approaches developed in the network mining community, to address open challenges in linked data and heterogeneous information network analysis.

Acknowledgments. This work was supported by the Slovenian Ministry of Higher Education, Science and Technology [grant number P2-0103], the Slovenian Research Agency [grant number PR-04431], and the SemDM project (Development and application of new semantic data mining methods in life sciences) [grant number J2-5478].

References

1. Adhikari, P.R., Vavpetič, A., Kralj, J., Lavrač, N., Hollmén, J.: Explaining mixture models through semantic pattern mining and banded matrix visualization. In: Džeroski, S., Panov, P., Kocev, D., Todorovski, L. (eds.) DS 2014. LNCS, vol. 8777, pp. 1–12. Springer, Heidelberg (2014)
2. Agrawal, R., Srikant, R.: Fast algorithms for mining association rules in large databases. In: Bocca, J.B., Jarke, M., Zaniolo, C. (eds.) Proceedings of the 20th International Conference on Very Large Data Bases, pp. 487–499. Morgan Kaufmann Publishers Inc., San Francisco (1994)

---

[3] https://github.com/anzev/hedwig.

3. Gene Ontology Consortium: the Gene Ontology project in 2008. Nucleic Acids Res. **36**(Database-Issue), 440–444 (2008)
4. De Raedt, L.: Logical and relational learning. In: Zaverucha, G., da Costa, A.L. (eds.) SBIA 2008. LNCS (LNAI), vol. 5249, pp. 1–1. Springer, Heidelberg (2008)
5. Demšar, J.: Statistical comparison of classifiers over multiple data sets. J. Mach. Learn. Res. **7**, 1–30 (2006)
6. Džeroski, S., Lavrač, N. (eds.): Relational Data Mining. Springer, Heidelberg (2001)
7. Flach, P.A., Lachiche, N.: 1BC: a first-order Bayesian classifier. In: Džeroski, S., Flach, P.A. (eds.) ILP 1999. LNCS (LNAI), vol. 1634, pp. 92–103. Springer, Heidelberg (1999)
8. Friedman, M.: The use of ranks to avoid the assumption of normality implicit in the analysis of variance. J. Am. Stat. Assoc. **32**, 675–701 (1937)
9. Hämäläinen, W.: Efficient search for statistically significant dependency rules in binary data. Ph.D. thesis, Department of Computer Science, University of Helsinki, Finland (2010)
10. Klösgen, W.: Explora: a multipattern and multistrategy discovery assistant. In: Fayyad, U.M., Piatetsky-Shapiro, G., Smyth, P., Uthurusamy, R. (eds.) Advances in Knowledge Discovery and Data Mining, pp. 249–271. American Association for Artificial Intelligence, Menlo Park (1996)
11. Knobbe, A.J. (ed.): Multi-Relational Data Mining. Frontiers in Artificial Intelligence and Applications, vol. 145. IOS Press, Amestardam (2005)
12. Kramer, S., Pfahringer, B., Helma, C.: Stochastic propositionalization of non-determinate background knowledge. In: Page, D.L. (ed.) ILP 1998. LNCS, vol. 1446. Springer, Heidelberg (1998)
13. Kranjc, J., Podpečan, V., Lavrač, N.: ClowdFlows: a cloud based scientific workflow platform. In: Flach, P.A., De Bie, T., Cristianini, N. (eds.) ECML PKDD 2012, Part II. LNCS, vol. 7524, pp. 816–819. Springer, Heidelberg (2012)
14. Krogel, M.-A., Rawles, S., Železný, F., Flach, P.A., Lavrač, N., Wrobel, S.: Comparative evaluation of approaches to propositionalization. In: Horváth, T., Yamamoto, A. (eds.) ILP 2003. LNCS (LNAI), vol. 2835, pp. 197–214. Springer, Heidelberg (2003)
15. Krogel, M.-A., Wrobel, S.: Transformation-based learning using multirelational aggregation. In: Rouveirol, C., Sebag, M. (eds.) ILP 2001. LNCS (LNAI), vol. 2157, pp. 142–155. Springer, Heidelberg (2001)
16. Kuželka, O., Železný, F.: Block-wise construction of tree-like relational features with monotone reducibility and redundancy. Mach. Learn. **83**(2), 163–192 (2011)
17. Kuželka, O., Železný, F.: Hifi: tractable propositionalization through hierarchical feature construction. In: Železný, F., Lavrač, N. (eds.) Late Breaking Papers, the 18th International Conference on Inductive Logic Programming (2008)
18. Lavrač, N., Džeroski, S., Grobelnik, M.: Learning nonrecursive definitions of relations with LINUS. In: Kodratoff, Y. (ed.) EWSL 1991. LNCS, vol. 482, pp. 265–281. Springer, Heidelberg (1991)
19. Lavrač, N., Flach, P.A.: An extended transformation approach to Inductive Logic Programming. ACM Trans. Comput. Logic **2**(4), 458–494 (2001)
20. Liu, B., Hsu, W., Ma, Y.: Integrating classification and association rule mining. In: Proceedings of the 4th International Conference on Knowledge Discovery and Data mining (KDD 1998), pp. 80–86. AAAI Press, August 1998
21. Maglott, D., Ostell, J., Pruitt, K.D., Tatusova, T.: Entrez Gene: gene-centered information at NCBI. Nucleic Acids Res. **33**(Database issue), D54–D58 (2005)

22. Muggleton, S.: Inverse entailment and Progol. New Gener. Comput. **13**(3–4), 245–286 (1995). Special issue on Inductive Logic Programming
23. Muggleton, S. (ed.): Inductive Logic Programming. Academic Press, London (1992)
24. Nemenyi, P.B.: Distribution-free multiple comparisons. Ph.D. thesis (1963)
25. Ogata, H., Goto, S., Sato, K., Fujibuchi, W., Bono, H., Kanehisa, M.: KEGG: kyoto encyclopedia of genes and genomes. Nucleic Acids Res. **27**(1), 29–34 (1999)
26. Perovšek, M., Vavpetič, A., Cestnik, B., Lavrač, N.: A wordification approach to relational data mining. In: Fürnkranz, J., Hüllermeier, E., Higuchi, T. (eds.) DS 2013. LNCS, vol. 8140, pp. 141–154. Springer, Heidelberg (2013)
27. Perovšek, M., Vavpetič, A., Lavrač, N.: A wordification approach to relational data mining: early results. In: Riguzzi, F., Železný, F. (eds.) ILP 2012 Proceedings of Late Breaking Papers of the 22nd International Conference on Inductive Logic Programming, Dubrovnik, Croatia, 17–19 September 2012. CEUR Workshop Proceedings, vol. 975, pp. 56–61. CEUR-WS.org (2012)
28. Perovšek, M., Vavpetič, A., Kranjc, J., Cestnik, B., Lavrač, N.: Wordification: propositionalization by unfolding relational data into bags of words. Expert Syst. Appl. **42**(17–18), 6442–6456 (2015)
29. Piatetsky-Shapiro, G.: Discovery, analysis, and presentation of strong rules. In: Piatetsky-Shapiro, G., Frawley, W.J. (eds.) Knowledge Discovery in Databases. AAAI/MIT Press, Menlo Park (1991)
30. Srinivasan, A.: Aleph manual, March 2007. http://www.cs.ox.ac.uk/activities/machinelearning/Aleph/
31. Trajkovski, I., Lavrač, N., Tolar, J.: SEGS: search for enriched gene sets in microarray data. J. Biomed. Inform. **41**(4), 588–601 (2008)
32. Vavpetič, A., Lavrač, N.: Semantic subgroup discovery systems and workflows in the SDM-toolkit. Comput. J. **56**(3), 304–320 (2013)
33. Vavpetič, A., Novak, P.K., Grčar, M., Mozetič, I., Lavrač, N.: Semantic data mining of financial news articles. In: Fürnkranz, J., Hüllermeier, E., Higuchi, T. (eds.) DS 2013. LNCS, vol. 8140, pp. 294–307. Springer, Heidelberg (2013)
34. Witten, I.H., Frank, E., Hall, M.A.: Data Mining: Practical Machine Learning Tools and Techniques, 3rd edn. Morgan Kaufmann, Amsterdam (2011)
35. Wrobel, S.: An algorithm for multi-relational discovery of subgroups. In: Komorowski, J., Żytkow, J.M. (eds.) PKDD 1997. LNCS, vol. 1263, pp. 78–87. Springer, Heidelberg (1997)
36. Železný, F., Lavrač, N.: Propositionalization-based relational subgroup discovery with RSD. Mach. Learn. **62**(1–2), 33–63 (2006)

Shift Design with Answer Set Programming

Michael Abseher[1], Martin Gebser[2,3], Nysret Musliu[1], Torsten Schaub[3,4][✉],
and Stefan Woltran[1]

[1] TU Wien, Vienna, Austria
{abseher,musliu,woltran}@dbai.tuwien.ac.at
[2] Aalto University, HIIT, Espoo, Finland
[3] University of Potsdam, Potsdam, Germany
{gebser,torsten}@cs.uni-potsdam.de
[4] INRIA Rennes, Rennes, France

Abstract. Answer Set Programming (ASP) is a powerful declarative programming paradigm that has been successfully applied to many different domains. Recently, ASP has also proved successful for hard optimization problems like course timetabling. In this paper, we approach another important task, namely, the shift design problem, aiming at an alignment of a minimum number of shifts in order to meet required numbers of employees (which typically vary for different time periods) in such a way that over- and understaffing is minimized. We provide an ASP encoding of the shift design problem, which, to the best of our knowledge, has not been addressed by ASP yet.

1 Introduction

Answer Set Programming (ASP) [4] is a declarative formalism for solving hard computational problems. Thanks to the power of modern ASP technology [8], ASP was successfully used in various application areas, including product configuration [13], decision support for space shuttle flight controllers [11], team building and scheduling [12], and bio-informatics [9]. Recently, ASP also proved successful for optimization problems that had not been amenable to complete methods before, for instance in the domain of timetabling [2].

In this paper, we investigate the application of ASP to another important domain, namely, workforce scheduling [3]. Finding appropriate staff schedules is of great relevance because work schedules influence health, social life, and motivation of employees at work. Furthermore, organizations in the commercial and public sector must meet their workforce requirements and ensure the quality of their services and operations. Such problems appear especially in situations where the required number of employees fluctuates throughout time periods, while operations dealing with critical tasks are performed around the clock. Examples include air traffic control, personnel working in emergency services, call centers, etc. In fact, the general employee scheduling problem includes several

Torsten Schaub—Affiliated with Simon Fraser University, Canada, and IIIS Griffith University, Australia.

F. Calimeri et al. (Eds.): LPNMR 2015, LNAI 9345, pp. 32–39, 2015.
DOI: 10.1007/978-3-319-23264-5_4

subtasks. Usually, in the first stage, the temporal requirements are determined based on tasks that need to be performed. Further, the total number of employees is determined and the shifts are designed. In the last phase, the shifts and/or days off are assigned to the employees. For shift design [10], employee requirements for a period of time, constraints about the possible start and length of shifts, and limits for the average number of duties per week are considered. The aim is to generate solutions consisting of shifts (and the number of employees per shift) that fulfill all hard constraints, while minimizing the number of distinct shifts as well as over- and understaffing. This problem has been addressed by local search techniques, including a min-cost max-flow approach [10] and a hybrid method combining network flow with local search [6]. These techniques have been used to successfully solve randomly generated examples and problems arising in real-world applications. A detailed overview of previous work on shift design is given in [7].

Although the aforementioned state-of-the-art approaches for the shift design problem are able to provide optimal solutions in many cases, obtaining optimal solutions for large problems is still a challenging task. Indeed, for several instances the best solutions are still unknown. Therefore, the application of exact techniques like ASP is an important research target. More generally, it is interesting to see how far an elaboration-tolerant, general-purpose approach such as ASP can compete with dedicated methods when tackling industrial problems. Our ASP solution is based on the first author's master thesis [1] and relies on sophisticated modeling and solving techniques, whose application provides best practice examples for addressing similarly demanding use-cases. In particular, we demonstrate how order encoding techniques [5] can be used in ASP for modeling complex interval constraints. Experimental results and further details are provided in the full version[1] of this paper.

2   The Shift Design Problem

To begin with, let us introduce the shift design problem. Our problem formulation follows the one in [10]. As input, we are given the following:

- consecutive *time slots* sharing the same length. Each time slot is associated with a number of employees that should be present during the slot.
- *shift types* with associated parameters *min-start* and *max-start*, representing the earliest and latest start, and *min-length* and *max-length*, representing the minimum and maximum length of a shift.

The aim is to generate a collection of k shifts $s_1, \ldots, s_k$. Each shift s_i is completely determined by its *start* and *length*, which must belong to some shift type. Additionally, each shift s_i is associated with parameters indicating the number of employees assigned to s_i during each day of the planning period. Note that we consider cyclic planning periods where the successor of the last time slot is equal to the first time slot. An example of employee requirements and a corresponding (optimal) schedule are shown in Fig. 1.

---

[1] www.dbai.tuwien.ac.at/proj/Rota/ShiftDesignASP.pdf.

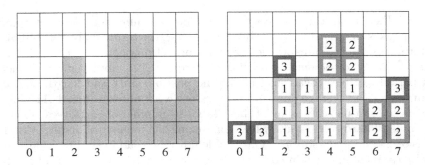

Fig. 1. Work demands over a day (left) and the unique optimal schedule (right) with shifts starting at slot 2, 4, or 7, respectively, indicated by different boxes, while other kinds of shifts are unused

In analogy to [6], we investigate the optimization of the following criteria: sum of *shortages* of workers in each time slot during the planning period, sum of *excesses* of workers in each time slot during the planning period, and the *number of shifts*.[2] Traditionally, the objective function is a weighted sum of the three components (although this kind of aggregation is not mandatory with ASP).

3   Shift Design in ASP

An instance like the one shown in Fig. 1 is specified by facts as in Fig. 2. Facts of the form $time(S, T)$ associate each slot S with a day time T. Our instance includes one day, divided into eight slots denoted by the times $0, \ldots, 7$. Instances of $next(S', S)$ provide predecessor or successor slots, respectively, where S is usually $S'+1$, except for the last slot whose successor is 0. (When another day is added, the slots $8, \ldots, 15$ would also be mapped to day times $0, \ldots, 7$, $next(7, 0)$ would be replaced by $next(7, 8)$, and $next(15, 0)$ would connect the new last slot to 0 instead.) For each slot S, a fact $work(S, N)$ gives the number N of desired employees, and $exceed(E)$ as well as $shorten(F)$ may limit the amount of employees at duty to at most $E+N$ or at least $N-F$, respectively. For instance, we obtain the upper bound 4 and the lower bound 2 for employees engaged in slot 7. Facts of the form $range(S, L, 1), \ldots, range(S, L, M)$ provide potential amounts of shifts of length L that can start from slot S, where M is the maximum number of desired employees over all slots within the horizon of the shift. For shifts starting from slot 7, those of length 2 or 3 stretch to slot 0 or 1, respectively, and the corresponding maximum number of desired employees is 3 in slot 7 itself; unlike that, shifts of length 4 also include slot 2 in which 4 employees shall be at duty.

Moreover, facts $opt(shortage, P, W)$, $opt(excess, P, W)$, and $opt(select, P, W)$ specify optimization criteria in terms of a priority P and a penalty weight W incurred in case of violations. The priorities in Fig. 2 tell us that the desired

---

[2] In [10], additionally, the average number of duties per week is considered.

$$\left\{ \begin{array}{l} time(0,0), time(1,1), \ldots, time(7,7), next(0,1), next(1,2), \ldots, next(7,0), \\ work(0,1), work(1,1), work(2,4), work(3,3), work(4,5), work(5,5), \\ work(6,2), work(7,3), exceed(1), shorten(1), opt(shortage,3,1), \\ opt(excess,2,1), opt(select,1,1), range(2,2,1), \ldots, range(2,2,4), \\ range(2,3,1), \ldots, range(2,3,5), range(2,4,1), \ldots, range(2,4,5), \\ range(4,2,1), \ldots, range(4,2,5), range(4,3,1), \ldots, range(4,3,5), \\ range(4,4,1), \ldots, range(4,4,5), range(5,2,1), \ldots, range(5,2,5), \\ range(5,3,1), \ldots, range(5,3,5), range(5,4,1), \ldots, range(5,4,5), \\ range(6,2,1), \ldots, range(6,2,3), range(6,3,1), \ldots, range(6,3,3), \\ range(6,4,1), \ldots, range(6,4,3), range(7,2,1), \ldots, range(7,2,3), \\ range(7,3,1), \ldots, range(7,3,3), range(7,4,1), \ldots, range(7,4,4) \end{array} \right\}$$

Fig. 2. ASP facts specifying an instance of the shift design problem

number of employees shall be present in the first place, then the amount of additional employees ought to be minimal, and third the number of utilized shifts in terms of day time and length should be as small as feasible. Given that the criteria are already distinguished by priority, the penalty weight of a violation of either kind is 1, thus counting particular violations to assess schedules.

$$\{run(S,L,I)\} \leftarrow range(S,L,I) \tag{1}$$

$$run(S,L,I) \leftarrow run(S',L{+}1,I), next(S',S), 0 < L \tag{2}$$

$$run(S,L,I) \leftarrow run(S,L{+}1,I), 0 < L \tag{3}$$

$$run(S,L,I{+}J) \leftarrow run(S,L{+}1,I), shift(S,L,J) \tag{4}$$

$$\leftarrow run(S,L,I{+}1), 0 < I, {\sim}run(S,L,I) \tag{5}$$

$$\leftarrow work(S,N), exceed(E), run(S,1,N{+}E{+}1) \tag{6}$$

$$\leftarrow work(S,N), shorten(F), F < N, {\sim}run(S,1,N{-}F) \tag{7}$$

$$length(S,L,I,1) \leftarrow range(S,L,I), run(S,L,I), {\sim}run(S,L{+}1,I) \tag{8}$$

$$length(S,L,I,J) \leftarrow length(S,L,I{+}1,J{-}1), 0 < I, {\sim}run(S,L{+}1,I) \tag{9}$$

$$shift(S,L,J) \leftarrow length(S,L,I,J) \tag{10}$$

$$shift(S,L,J) \leftarrow shift(S',L{+}1,J), next(S',S), 0 < L \tag{11}$$

$$start(S,L,J) \leftarrow range(S,L,J), next(S',S), shift(S,L,J), {\sim}shift(S',L{+}1,J) \tag{12}$$

$$W@P,S,I,shortage \leftsquigarrow opt(shortage,P,W), work(S,N), I \in [1,N), {\sim}run(S,1,I) \tag{13}$$

$$W@P,S,I,excess \leftsquigarrow opt(excess,P,W), work(S,N), run(S,1,I), N < I \tag{14}$$

$$W@P,T,L,select \leftsquigarrow opt(select,P,W), start(S,L,J), time(S,T) \tag{15}$$

Fig. 3. ASP encoding of the shift design problem

Our ASP encoding of the shift design problem is shown in Fig. 3. For a slot S, the intuitive reading of the predicate $run(S,L,I)$ is that at least I shifts including S and $L{-}1$ or more successor slots are scheduled. This is further refined by $length(S,L,I,J)$, telling that $1,\ldots,J$ of the scheduled shifts of exact length L

may start from S, where $I-1$ shifts that include at least $L-1$ successor slots are scheduled in addition. The predicate $shift(S, L, J)$ expresses that at least J of the scheduled shifts stretch to S and exactly $L-1$ successor slots, and $start(S, L, J)$ indicates that the J-th instance of such a shift indeed starts from S. A schedule is thus characterized by the number of (true) atoms of the form $start(S, L, J)$, yielding the amount of shifts of length L starting from slot S. For example, the schedule displayed in Fig. 1 is described by a stable model containing $start(2, 4, 1)$, $start(2, 4, 2)$, $start(2, 4, 3)$, $start(4, 4, 1)$, $start(4, 4, 2)$, and $start(7, 4, 1)$. When further scheduling a shift of length 2 to start from slot 6, it would be indicated by $start(6, 2, 3)$, as it adds to the two shifts from slot 4 stretching to slot 6 and 7 as well. However, the displayed schedule is the unique optimal solution, given that it matches the desired employees and uses a minimum number of shifts, viz. shifts of length 4 starting from slot 2, 4, or 7, respectively.

In more detail, the potential start of an instance I of a shift of length L from slot S is reflected by the choice rule (1) in Fig. 3. Rule (2) propagates the start of a shift to its $L-1$ successor slots, where the residual length is decreased down to 1 in the last slot of the shift. For shifts with longer residual length L, rule (3) closes the interval between 1 and L, thus overturning any choice rules for potential starts of shifts of shorter length. Moreover, this allows for pushing the J-th instance of a shift stretching to slot S to the position $I+J$ when I instances of shifts longer than the residual length L are scheduled, as expressed by rule (4). The integrity constraint (5) asserts that the positions associated with scheduled shifts must be ordered by residual length. This condition eliminates guesses on instances I of starting shifts, and it also provides a shortcut making interconnections between positions of scheduled shifts explicit, which turned out as effective to improve search performance. The additional integrity constraints (6) and (7) are applicable whenever the deviation from numbers of desired employees is bounded above or below, respectively. Note that it is sufficient to inspect atoms of the form $run(S, 1, I)$ for appropriate positions I, given that residual lengths are propagated via rule (3).

In order to derive the amount of scheduled shifts of exact residual length L, rule (8) marks positions I, where instances may start, with 1 when the length L matches. Instances associated with smaller positions then count on by means of rule (9) unless their positions are occupied by shifts with longer residual length. By projecting the positions out, rule (10) yields that $1, \ldots, J$ shifts of length L may start from slot S. In addition, longer shifts whose residual length decreases to L in S are propagated via rule (11). Finally, rule (12) compares instances that may start to propagated shifts and indicates the ones that indeed start from S. As a consequence, a stable model represents a schedule in terms of sequences of the form $start(S, L, m), \ldots, start(S, L, n)$, expressing that $n+1-m$ instances of a shift of length L start from slot S. It remains to assess the quality of a schedule, which is accomplished by means of the weak constraints (13), (14), and (15) for the three optimization criteria at hand. The penalty for deviating from a number of desired employees is characterized in terms of the priority P and weight W given in facts, a position I pointing to under- or overstaffing in a slot S, and the corresponding keyword *shortage* or *excess*, respectively, for avoiding clashes with penalties due to the utilization of shifts. The latter include the keyword *select* and map the slot S of a starting shift of length L to its day time T, so that the penalty $W@P$ is incurred at most once for a shift with particular parameters, no matter how many instances are actually utilized.

A prevalent feature of our ASP encoding in Fig. 3 is the use of closed intervals (starting from 1) to represent quantitative values such as residual lengths or instances of shifts. The basic idea is similar to the so-called *order encoding* [5], which has been successfully applied to solve constraint satisfaction problems by means of SAT [14]. In our ASP encoding, rule (4), (8), and (9) take particular advantage of the order encoding approach by referring to one value, viz. $L+1$, for testing whether any shift with longer residual length than L is scheduled. Likewise, the integrity constraints (6) and (7) as well as the weak constraints (13) and (14) focus on value 1, standing for any residual length, to determine the amount of employees at duty. That is, the order encoding approach enables a compact formulation of existence tests and general conditions, which then propagate to all target values greater or smaller than a certain threshold.

4 Discussion

In this work, we presented a novel approach to tackle the shift design problem by using ASP. Finding good solutions for shift design problems is of great importance in different organizations. However, such problems are very challenging due to the huge search space and conflicting constraints. Our work contributes to better understanding the strengths of ASP technology in this domain and extends the state of the art for the shift design problem by providing new optimal solutions for benchmark instances first presented in [6]. Below we summarize the main observations of our experiments, detailed in the full paper (see Footnote 1), regarding the application of ASP to the shift design problem:

– ASP shows very good results for shift design problems that have solutions without over- and understaffing. Our proposed ASP approach could provide optimal solutions for almost all such benchmark instances (DataSet1 and DataSet2 in [6]).
– The first results for problems that do not have solutions without over- or understaffing are promising. Although our current approach could not reproduce best known solutions for several problems, we were able to provide global optima for four hard instances (from DataSet3 in [6]), not previously solved to the optimum.
– Our experimental evaluation indicates that our approach could also be used in combination with other search techniques. For example, solutions computed by metaheuristic methods or min-cost max-flow techniques could be further improved by ASP.
– In general, the computational results show that ASP has the potential to provide good solutions in this domain. Therefore, our results open up the area of workforce scheduling, which is indeed challenging for state-of-the-art

ASP solvers. This is most probably caused by the nature of the shift design problem, as there are few hard constraints involved that could help to restrict the search space.

Concerning related work, we mention an ASP implementation of a problem from the domain of workforce management [12], where the focus is on the allocation of employees of different qualifications to tasks requiring different skills. The resulting system is tailored to the specific needs of the seaport of Gioia Tauro. From the conceptual point of view, the main difference to our work is that the encoded problem of [12] is a classical allocation problem with optimization towards work balance, while the problem we tackle here aims at an optimal alignment of shifts.

As future work, we plan to tackle the problem of optimization in shift design by combining ASP with domain-specific heuristics in order to better guide the search, but also exploiting off-the-shelf heuristics is a promising target for further investigation. We are confident that ASP combined with heuristics is a powerful tool for tackling problems in the area of workforce scheduling. This fact is already underlined by significantly improved results obtained for branch-and-bound based optimization when activating particular off-the-shelf heuristics. By using customized heuristics, tailored to the specific problem at hand, the chance for further improvements is thus high.

Acknowledgments. This work was funded by AoF (251170), DFG (550/9), and FWF (P25607-N23, P24814-N23, Y698-N23).

References

1. Abseher, M.: Solving shift design problems with answer set programming. Master's thesis, Technische Universität Wien (2013)
2. Banbara, M., Soh, T., Tamura, N., Inoue, K., Schaub, T.: Answer set programming as a modeling language for course timetabling. Theor. Pract. Logic Program. **13**(4–5), 783–798 (2013)
3. den Bergh, J., Beliën, J., Bruecker, P., Demeulemeester, E., Boeck, L.: Personnel scheduling: a literature review. Eur. J. Oper. Res. **226**(3), 367–385 (2013)
4. Brewka, G., Eiter, T., Truszczyński, M.: Answer set programming at a glance. Commun. ACM **54**(12), 92–103 (2011)
5. Crawford, J., Baker, A.: Experimental results on the application of satisfiability algorithms to scheduling problems. In: Proceedings of AAAI 1994, pp. 1092–1097. AAAI Press (1994)
6. Di Gaspero, L., Gärtner, J., Kortsarz, G., Musliu, N., Schaerf, A., Slany, W.: The minimum shift design problem. Ann. Oper. Res. **155**, 79–105 (2007)
7. Di Gaspero, L., Gärtner, J., Musliu, N., Schaerf, A., Schafhauser, W., Slany, W.: Automated shift design and break scheduling. In: Uyar, A.S., Ozcan, E., Urquhart, N. (eds.) Automated Scheduling and Planning. SCI, vol. 505, pp. 109–127. Springer, Heidelberg (2013)
8. Gebser, M., Kaminski, R., Kaufmann, B., Schaub, T.: Answer Set Solving in Practice. Morgan & Claypool Publishers, San Rafael (2012)

9. Guziolowski, C., Videla, S., Eduati, F., Thiele, S., Cokelaer, T., Siegel, A., Saez-Rodriguez, J.: Exhaustively characterizing feasible logic models of a signaling network using answer set programming. Bioinformatics **29**(18), 2320–2326 (2014)
10. Musliu, N., Schaerf, A., Slany, W.: Local search for shift design. Eur. J. Oper. Res. **153**(1), 51–64 (2004)
11. Nogueira, M., Balduccini, M., Gelfond, M., Watson, R., Barry, M.: An A-prolog decision support system for the space shuttle. In: Ramakrishnan, I.V. (ed.) PADL 2001. LNCS, vol. 1990, pp. 169–183. Springer, Heidelberg (2001)
12. Ricca, F., Grasso, G., Alviano, M., Manna, M., Lio, V., Iiritano, S., Leone, N.: Team-building with answer set programming in the Gioia-Tauro seaport. Theor. Pract. Logic Program. **12**(3), 361–381 (2012)
13. Soininen, T., Niemelä, I.: Developing a declarative rule language for applications in product configuration. In: Gupta, G. (ed.) PADL 1999. LNCS, vol. 1551, pp. 305–319. Springer, Heidelberg (1999)
14. Tamura, N., Taga, A., Kitagawa, S., Banbara, M.: Compiling finite linear CSP into SAT. Constraints **14**(2), 254–272 (2009)

Advances in WASP

Mario Alviano, Carmine Dodaro, Nicola Leone, and Francesco Ricca(⊠)

Department of Mathematics and Computer Science, University of Calabria,
87036 Rende (CS), Italy
{alviano,dodaro,leone,ricca}@mat.unical.it

Abstract. ASP solvers address several reasoning tasks that go beyond
the mere computation of answer sets. Among them are cautious reasoning, for modeling query entailment, and optimum answer set computation, for supporting numerical optimization. This paper reports on
the recent improvements of the solver WASP, and details the algorithms
and the design choices for addressing several reasoning tasks in ASP. An
experimental analysis on publicly available benchmarks shows that the
new version of WASP outperforms the previous one. Comparing with the
state-of-the-art solver CLASP, the performance of WASP is competitive in
the overall for number of solved instances and average execution time.

1 Introduction

Answer Set Programming (ASP) [19] is a declarative programming paradigm
which has been proposed in the area of non-monotonic reasoning and logic programming. The idea of ASP is to represent a given computational problem by a
logic program whose answer sets correspond to solutions, and then use a solver
to find them. Answer set computation is a hard task, and is usually performed
by adapting techniques introduced for SAT solving, such as *learning*, *restarts*,
and *conflict-driven heuristics* [35].

Modern ASP solvers address several reasoning tasks that go beyond the mere
computation of answer sets. Among them, optimum answer set computation and
cautious reasoning have great impact in applications [1,10,14,26,33]. Optimum
answer set search amounts to finding an answer set that minimizes the violations
of the so-called weak constraints [13]. Cautious reasoning corresponds to the
computation of (a subset of) the certain answers, i.e., those atoms that belong
to all answer sets of a program. These tasks are computationally harder than
the answer set search [16], and require the implementation of specific evaluation
techniques.

This paper reports on the advances in the ASP solver WASP [2], focusing on
the latest extensions and improvements that we have implemented in the 2.1

This work was partially supported by MIUR within project "SI-LAB BA2KNOW –
Business Analitycs to Know", and by Regione Calabria, POR Calabria FESR
2007-2013, within project "ITravel PLUS" and project "KnowRex". Mario Alviano
was partly supported by the National Group for Scientific Computation (GNCS-
INDAM), and by Finanziamento Giovani Ricercatori UNICAL.

© Springer International Publishing Switzerland 2015
F. Calimeri et al. (Eds.): LPNMR 2015, LNAI 9345, pp. 40–54, 2015.
DOI: 10.1007/978-3-319-23264-5_5

version of the system. The new version is a natural evolution of the preliminary version presented in [5], bringing improvements in the implementation of the computation of answer sets as well as new algorithms for handling optimum answer set search and cautious reasoning.

Optimum answer set search is addressed in WASP 2.1 by combining answer set computation with MaxSAT algorithms [30] such as MGD [31], OPTSAT [34], PMRES [32], and BCD [20]. Another algorithm implemented in WASP 2.1 is OLL [8], introduced by the ASP solver CLASP, and successfully applied also to MaxSAT [29].

Cautious reasoning instead is addressed by implementing a new algorithm [4] inspired by backbone computation in propositional logic [28]. It is remarkable that WASP is the first ASP solver approaching this task by means of an *anytime* algorithm, i.e., an algorithm producing sound answers during its execution, and not just at the end. Being anytime is of particular relevance for real world applications, and in fact a large number of answers are often produced after a few seconds of computation, even when termination is not affordable in practice [4].

The efficiency of WASP 2.1 is assessed by comparing its performance with WASP 1.0 [2], the previous version of the solver, and with CLASP 3.1.1 [18], the latest version of the solver that won the 5th ASP Competition. The result highlights a substantial improvement w.r.t. the previous version, and a competitive performance w.r.t. CLASP in many cases, with an even superior performance in cautious reasoning.

2   ASP Language

This section recalls syntax and semantics of propositional ASP programs.

Syntax. Let $\mathcal{A}$ be a countable set of propositional atoms. A *literal* is either an atom (a positive literal), or an atom preceded by the *negation as failure* symbol $\sim$ (a negative literal). The complement of a literal ℓ is denoted $\overline{\ell}$, i.e., $\overline{a} = \sim a$ and $\overline{\sim a} = a$ for an atom a. This notation extends to sets of literals, i.e., $\overline{L} := \{\overline{\ell} \mid \ell \in L\}$ for a set of literals L. A *program* is a finite multiset of rules of the following form:

$$a_0 \leftarrow a_1, \ldots, a_m, \sim a_{m+1}, \ldots, \sim a_n \tag{1}$$

where $n \geq m \geq 0$ and each a_i $(i = 0, \ldots, n)$ is an atom. The atom a_0 is called head, and the conjunction $a_1, \ldots, a_m, \sim a_{m+1}, \ldots, \sim a_n$ is referred to as body. Rule r is said to be regular if $H(r) \neq \bot$, where $\bot$ is a fixed atom in $\mathcal{A}$, and a constraint otherwise. A constraint is possibly associated with a positive integer by the partial function *weight*. Constraints for which function *weight* is defined are called *weak constraints*, while the remaining constraints are referred to as *hard constraints*. The multisets of hard and weak constraints in Π are denoted $hard(\Pi)$ and $weak(\Pi)$, respectively, while the multiset of the remaining rules is denoted by $rules(\Pi)$. For a rule r of the form (1), the following notation is also used: $H(r)$ denotes the set containing the head atom a_0; $B(r)$ denotes the set $\{a_1, \ldots, a_m, \sim a_{m+1}, \ldots, \sim a_n\}$ of body literals; $B^+(r)$ and $B^-(r)$ denote the set of atoms appearing in positive and negative body literals, respectively, i.e., $B^+(r)$ is $\{a_1, \ldots, a_m\}$ and $B^-(r)$ is $\{a_{m+1}, \ldots, a_n\}$; $C(r) := H(r) \cup \overline{B(r)}$ is the clause representation of r.

Semantics. An *interpretation* I is a set of literals, i.e., $I \subseteq \mathcal{A} \cup \overline{\mathcal{A}}$. Intuitively, literals in I are true, literals whose complements are in I are false, and all other literals are undefined. I is total if there are no undefined literals, and I is inconsistent if $\perp \in I$ or there is $a \in \mathcal{A}$ such that $\{a, \sim a\} \subseteq I$. An interpretation I satisfies a rule r, denoted $I \models r$, if $C(r) \cap I \neq \emptyset$, while I violates r if $C(r) \subseteq \overline{I}$. A consistent, total interpretation I is a *model* of a program Π, denoted $I \models \Pi$, if $I \models r$ for all $r \in \Pi$. The semantics of a program Π is given by the set of its *answer sets*, or stable models [19]: Let Π^I denote the program reduct obtained by deleting from Π each rule r such that $B^-(r) \cap I \neq \emptyset$, and then by removing all the negative literals from the remaining rules. An interpretation I is an answer set for Π if $I \models \Pi$ and there is no total interpretation J such that $J \cap \mathcal{A} \subset I \cap \mathcal{A}$ and $J \models \Pi^I$. The set of all answer sets of a program Π is denoted $SM(\Pi)$.

Optimum Answer Sets. Let Π be a program with weak constraints. The cost of an interpretation I of Π is defined as follows:

$$cost_\Pi(I) := \sum_{r \in weak(\Pi):C(r) \subseteq I} weight(r). \tag{2}$$

An interpretation I is an *optimum answer set* of Π if $I \in SM(\Pi)$ and there is no $J \in SM(\Pi)$ such that $cost_\Pi(J) < cost_\Pi(I)$.

Cautious Consequences. Let Π be a program with no weak constraints. An atom $a \in \mathcal{A}$ is a cautious consequence of Π, denoted $\Pi \models_c a$, if $a \in I$ for all $I \in SM(\Pi)$.

3 Answer Set Computation in WASP 2.1

In this section we review the algorithms implemented in WASP 2.1 for the computation of an answer set. We first detail a transformation step of the input program performed at the beginning of the computation, and then we focus on the main strategies employed by WASP 2.1 for computing an answer set. The presentation is properly simplified to focus on the main principles.

3.1 Completion and Program Simplification

An important property of ASP programs is *supportedness*, i.e., all answer sets are supported models. A model I of a program Π is *supported* if each $a \in I \cap \mathcal{A}$ is supported, i.e., there exists a rule $r \in \Pi$ such that $H(r) = a$, and $B(r) \subseteq I$. Supportedness can be enforced according to different strategies [2,18]. WASP 2.1 implements a transformation step, called *Clark's completion*, which

---

Algorithm 1. ComputeAnswerSet

Input : A program Π and an interpretation I for Π
Output: An answer set for Π or $\bot$

1 **begin**
2 | **while** Propagate(I) **do**
3 | | **if** I *is total* **then** **return** I ;
4 | | $\ell :=$ ChooseUndefinedLiteral();
5 | | $I' :=$ ComputeAnswerSet($I \cup \{\ell\}, \Pi$);
6 | | **if** $I' \neq \bot$ **then** **return** I' ;
7 | | **if** *there are violated (learned) constraints* **then** **return** $\bot$;
8 | $\Pi := \Pi \cup$ AnalyzeConflictAndCreateConstraints(I);
9 | **return** $\bot$;

---

rewrites the input program Π into a new program $Comp(\Pi)$ whose classical models correspond to the supported models of Π [25]. In more detail, given a rule $r \in \Pi$, let aux_r denote a fresh atom, i.e., an atom not appearing elsewhere, the completion of Π consists of the following rules:

- $\leftarrow a, \sim aux_{r_1}, \ldots, \sim aux_{r_n}$, for each atom a occurring in Π, where $r_1, \ldots, r_n$ are the rules of Π whose head is a;
- $a_0 \leftarrow aux_r$ and $aux_r \leftarrow a_1, \ldots, a_m, \sim a_{m+1}, \ldots \sim a_n$, for each rule $r \in \Pi$ of the form (1);
- $\leftarrow aux_r, \overline{\ell}$, for each $r \in \Pi$ and $\ell \in B(r)$.

After the computation of Clark's completion, *simplification* techniques are applied in the style of SATELITE [15]. These consist of polynomial-time algorithms for strengthening and removing redundant rules, and also include atom elimination by means of rule rewriting.

3.2 Main Algorithm

In order to compute an answer set of a given program Π, WASP 2.1 first replaces Π with its completion $Comp(\Pi)$. After that, the program is passed to Algorithm 1, which is similar to the CDCL procedure in SAT solvers. The algorithm also receives a partial interpretation I, which is initially set to $\{\sim\bot\}$. Function Propagate (line 2) extends I with those literals that can be deterministically inferred. This function returns false if an inconsistency (or conflict) is detected, true otherwise. When no inconsistency is detected, interpretation I is returned if total (line 3). Otherwise, an undefined literal, say ℓ, is chosen according to some heuristic criterion (line 5). The computation then proceeds with a recursive call to ComputeAnswerSet on $I \cup \{\ell\}$ (line 5). In case the recursive call returns an answer set, the computation ends returning it (line 6). Otherwise, the algorithm unrolls choices until consistency of I is restored (line 7), and the computation resumes by propagating the consequences of the constraint learned by the conflict analysis. Conflicts detected during propagation are analyzed by procedure *AnalyzeConflictAndCreateConstraints*, which returns a new

Function Propagate(I)

```
1 while UnitPropagation(I) do
2   if not WellFoundedPropagation(I) then return true ;
3 return false;
```

constraint modeling the found conflict that is then added to Π (line 8). It is important to note that the main algorithm is usually complemented with some heuristic techniques that control the number of learned constraints, and possibly restart the computation to explore different branches of the search tree. Moreover, a crucial role is played by the heuristic criteria used for selecting branching literals. WASP 2.1 adopts the same branching, restart and deletion heuristics of the SAT solver GLUCOSE [11]. Propagation and constraint learning are described in more detail in the following.

Propagation. WASP 2.1 implements two deterministic inference rules for pruning the search space during answer set computation. These propagation rules are named *unit* and *well-founded*. Unit propagation is applied first (line 1 of function Propagate). It returns false if an inconsistency arises. Otherwise, well-founded propagation is applied (line 2). Function WellFoundedPropagation may infer the falsity of some atoms, in which case *true* is returned and unit propagation is applied. When no new atoms can be inferred by WellFoundedPropagation, function Propagate returns *true* to report that no inconsistency has been detected. In more details, unit propagation is as in SAT solvers: An undefined literal ℓ is inferred by unit propagation if there is a rule r that can be satisfied only by ℓ, i.e., r is such that $\ell \in C(r)$ and $C(r) \setminus \{\ell\} \subseteq \overline{I}$. Concerning well-founded propagation, we must first introduce the notion of unfounded sets. A set X of atoms is *unfounded* if for each rule r such that $H(r) \cap X \neq \emptyset$, at least one of the following conditions is satisfied: (i) $\overline{B(r)} \cap I \neq \emptyset$; (ii) $B^+(r) \cap X \neq \emptyset$. Intuitively, atoms in X can have support only by themselves. When an unfounded set X is found, function WellFoundedPropagation imposes the falsity of an atom in X. The falsity of other atoms in X will be imposed on subsequent calls to the function, unless an inconsistency arises during unit propagation. In case of inconsistencies, indeed, the unfounded set X is recomputed.

Conflict Analysis and Learning. Constraint learning acquires information from conflicts in order to avoid exploring the same search branch several times. WASP 2.1 adopts a learning schema based on the concept of the first Unique Implication Point (UIP) [35], which is computed by analyzing the so-called implication graph. Roughly, the implication graph contains a node for each literal in I, and arcs from ℓ_i to $\overline{\ell_0}$ ($i = 1, \ldots, n$; $n \geq 1$) if literal $\overline{\ell_0}$ is inferred by unit propagation on constraint $\leftarrow \ell_0, \ldots, \ell_n$. Each literal $\ell \in I$ is associated with a *decision level*, corresponding to the depth nesting level of the recursive call to ComputeAnswerSet on which ℓ is added to I. A node n in the implication graph is a UIP for a decision level d if all paths from the choice of level d to the conflict literals pass through n. Given a conflict at a level d, the first UIP is the UIP

Algorithm 2. Optimum Answer Set Search

    Input : A program Π
    Output: The optimum cost for Π
1 **begin**
2     $lower\_bound := 0;$ $wmax := \max\{weight(r) \mid r \in weak(\Pi)\};$
3     $R := \{r \in weak(\Pi) \mid weight(r) < wmax\};$
4     $lower\_bound := lower\_bound + \mathrm{CoreGuidedAlg}(\Pi, R);$
5     $wmax := \max\{weight(r) \mid r \in R\};$
6   **if** $R = \emptyset$ **then return** $lower\_bound;$
7   **goto** 3;

---

of level d that is closest to the conflict. The learning schema is as follows: Let u be the first UIP. Let L be the set of literals different form u occurring in a path from u to the conflict literals. The learned constraint comprises u and each literal ℓ such that the decision level of ℓ is lower than the one of u and there is an arc (ℓ, ℓ') in the implication graph for some $\ell' \in L$.

4 ASP Computational Tasks

WASP 2.1 addresses reasoning tasks that go beyond the mere computation of answer sets; and in particular optimum answer set computation and cautious reasoning, which are described in this section.

4.1 Optimum Answer Set Search

The computational problem analyzed in this section is referred to as *ASP optimization problem* and can be stated as follows: Given a coherent program Π, compute the *optimum cost* of Π, defined as the cost of an optimum answer set of Π. WASP 2.1 implements several strategies for computing an optimum model of a program, including model-guided and core-guided algorithms [30]. Intuitively, model-guided algorithms work by iteratively searching for an answer set until an optimum solution is found. Core-guided algorithms consider weak constraints as hard constraints and then work by iteratively relaxing some of them until an answer set is found, which eventually results in an optimum solution. In particular, WASP 2.1 implements the following algorithms:

- OPT, an algorithm inspired by OPTSAT [34] and its variant BASIC.
- MGD [31], a model-guided algorithm introduced for solving MaxSAT.
- OLL [8], a core-guided algorithm introduced in the context of ASP and then successfully ported to MaxSAT [29].
- PMRES, a core-guided algorithm implemented in the MaxSAT solver EVA [32].
- BCD, a core-guided algorithm implemented in the MaxSAT solver MSUN-CORE [20].
- INTERLEAVING, a strategy that combines the algorithms BASIC and OLL.

Algorithm 3. IterativeCoherenceTesting

   Input   : A program Π and a set of atoms Q

   Output: Atoms in Q that are cautious consequences of Π, or $\bot$

1 **begin**

2     $U := \emptyset;$ $O := Q;$

3     $I := \text{ComputeAnswerSet}(\Pi, \{\sim\bot\});$

4   **if** $I = \bot$ **then return** $\bot$;

5     $O := O \cap I;$

6   **while** $U \neq O$ **do**

7         $a := \text{OneOf}(O \setminus U);$

8         $I := \text{ComputeAnswerSet}(\Pi, \{\sim\bot, \sim a\});$

9   **if** $I = \bot$ **then** $U := U \cup \{a\}$;

10   **else** $O := O \cap I$;

11   **return** U;

Moreover, WASP 2.1 is the first ASP solver that implements the *stratification* technique [9]. Stratification refines core-guided algorithms on handling weighted instances. The idea is to focus at each iteration on weak constraints with higher weights by properly restricting the set of rules sent to the ASP solver. The effect is to improve the increment of the lower bound.

Algorithm 2 reports the procedure implemented by WASP 2.1 for addressing ASP optimization problems via core-guided algorithms [3], where CoreGuidedAlg denotes one of the core-guided algorithms available in WASP 2.1. The arguments of this function are passed by reference, so that any change to Π and R is preserved after the function returns. Algorithm 2 initially sets the lower bound to zero and variable $wmax$ to the greatest weight of any weak constraints of the program. Next, the set R of weak constraints having weight less than $wmax$ is computed (line 3). Intuitively, weak constraints in $weak(\Pi) \setminus R$ are processed, while the remaining weak constraints, i.e. those in R, are ignored. Algorithm CoreGuidedAlg is called on Π with the additional argument R and computes the cost of a solution considering only weak constraints which are not in R. The computed cost is then added to the lower bound (line 4), and $wmax$ is set to the greatest weight of any weak constraint in R (line 5). The process is then repeated until all weak constraints have been considered, i.e., until R is empty (line 6).

4.2 Cautious Reasoning

Cautious reasoning provides answers to the input query that are witnessed by all answer sets of the knowledge base. Cautious reasoning has been implemented by two ASP solvers, DLV [23] and CLASP [18], as a variant of their answer set search algorithms. In a nutshell, cautious reasoning can be obtained by reiterating the answer set search step according to a specific solving strategy. The procedure implemented by DLV searches for answer sets and computes their intersection, which eventually results in the set of cautious consequences of the input program.

At each step of the computation, the intersection of the identified answer sets represents an overestimate of the solution, which however is not provided as output by DLV. The procedure implemented by CLASP is similar, but the overestimate is showed and used to further constrain the computation. In fact, the overestimate is considered a constraint of the logic program, so that the next computed answer set is guaranteed to improve the current overestimate. The procedure implemented by WASP 2.1, whose pseudo-code is reported in Algorithm 3, is inspired by an algorithm for computing backbones of propositional formulas [28]. The algorithm receives as input a program Π, which is the completion of the input program, and a set of atoms Q representing answer candidates of a query, and produces as output either the largest subset of Q that only contains cautious consequences of Π, in case Π is coherent, or $\bot$ when Π is incoherent. Initially, the underestimate U and the overestimate O are set to $\emptyset$ and Q, respectively (line 2). A coherence test of Π is then performed (lines 3) by calling function ComputeAnswerSet, which actually implements answer set search as described in Sect. 3. If the program is incoherent then the computation stops returning $\bot$. Otherwise, the first answer set found improves the overestimate (line 5). At this point, estimates are improved until they are equal (line 6). Then, one cautious consequence candidate is selected by calling function OneOf (line 7). This candidate is then constrained to be false and an answer set is searched (line 8). If none is found then the underestimate can be increased (lines 9). Otherwise, the overestimate can be improved (lines 10).

An interesting property of the algorithm implemented by WASP 2.1 is that underestimates are produced during the computation of the complete solution. This is important since cautious reasoning is a resource demanding task, which is often not affordable to complete in reasonable time. The computation can thus be stopped either when a sufficient number of cautious consequences have been produced, or when no new answer is produced after a specified amount of time. Such algorithms are referred to as anytime in the literature. Moreover, WASP 2.1 also includes the algorithm implemented in CLASP, which has been modified in WASP 2.1 for producing cautious consequences during the computation [4]. As far as we know, WASP 2.1 is the only ASP solver which implements anytime algorithms for cautious reasoning. The empirical evaluation of these algorithms highlights that a large percentage (up to 90 % in some cases) of the sound answers can be provided after few seconds of the computation in several benchmarks [4].

5   Experiment

The performance of WASP 2.1 was compared with WASP 1.0 and CLASP 3.1.1. WASP 2.1 and CLASP use GRINGO [17] as grounder, while a custom version of the grounder DLV [7] is used by WASP 1.0. CLASP and WASP 1.0 have been executed using the same heuristic settings used in the 5th ASP Competition, while WASP 2.1 has been executed with its default heuristic which includes the algorithm OLL for programs weak constraints and the algorithm IterativeCoherenceTesting for cautious reasoning. We do not include WASP 2.0 presented in [5] since that executable was a preliminary version with no support for optimization constructs and for queries. Note that WASP 2.1 supports the ASP Core 2.0 [1] standard with limited support of disjunction. (The handling of additional languages constructs present in ASP Core 2.0 would have had no impact in the description of WASP 2.1 algorithms provided in this paper.) The experiment comprises two parts. The first part concerns a comparison of the solvers on instances taken from the 4th ASP Competition [1]. We refrain to use the instances from the 5th ASP Competition since it was a rerun of the 4th ASP Competition. The second part instead compares WASP 2.1 and CLASP on the task of cautious reasoning, and in particular in the context of Consistent Query Answering (CQA) [10]. CQA amounts to computing answers of a given query that are true in all repairs of an inconsistent database, where a repair is a revision of the original database that is maximal and satisfies its integrity constraints. The experiment was run on a four core Intel Xeon CPU X3430 2.4 GHz, with 16 GB of physical RAM, and operating system Debian Linux. Time and memory limits were set to 600 s and 15 GB, respectively. Performance was measured using the tools pyrunlim and pyrunner (https://github.com/alviano/python).

Part I: Instances of the 4th ASP Competition. Table 1 summarizes the number of solved instances and the average running time in seconds for each solver. In particular, the first two columns report the problems of the 4th ASP Competition and the total number of instances considered (#); the remaining columns report the number of solved instances within the time-out (solved), and the running time averaged over solved instances (time). The first observation is that WASP 2.1 outperforms WASP 1.0. In fact, WASP 2.1 solved 137 more instances than WASP 1.0 and it is faster in almost all benchmarks. Moreover, WASP 2.1 is comparable in performance with CLASP, even if the latter solves 7 instances more than the former. The advantage of CLASP with respect to WASP 2.1 is obtained in two benchmarks, namely GracefulGraphs and GraphColouring, whose instances show different symmetric solutions. Albeit CLASP does not implement any special technique for handling symmetric solutions, we observe that CLASP is considerably better than WASP 2.1 on those benchmarks. Our explanation is that the heuristic of WASP 2.1 seems to be ineffective on instances with different symmetric solutions. It is important to point out that even if CLASP and WASP 2.1 are comparable in the overall, there are nonetheless differences on specific instances. In particular, WASP 2.1 solves 11 instances which are not solved by CLASP.

For the sake of completeness, we also analyzed the behavior of the solvers by using a different memory setting. The performance of CLASP and WASP 2.1 are still comparable in case of a lower limit on the usage of memory. Nonetheless, on the considered benchmarks CLASP seems to use a larger amount of memory compared to WASP, which is due to a different implementation of program simplifications. Indeed, with a memory limit of 3 GB WASP 2.1 solves 359 instances, i.e., 3 instances more than CLASP.

Part II: Cautious Reasoning. Table 2 summarizes the number of solved instances and the average running time in seconds for each solver. We considered all co-NP

Table 1. Solved instances and average running time on benchmarks from 4th ASP Competition.

Problem	#	CLASP		WASP 1.0		WASP 2.1	
		Solved	Time	solved	Time	Solved	Time
BottleFillingProblem	30	30	7.5	30	64.8	30	6.8
CrossingMinimization	30	23	79.6	0	-	22	110.9
GracefulGraphs	30	15	103.8	9	169.4	10	73.4
GraphColouring	30	13	130.0	8	141.7	8	45.8
HanoiTower	30	28	53.2	15	142.0	30	48.2
IncrementalScheduling	30	3	232.4	0	-	4	262.9
Labyrinth	30	26	48.0	21	114.7	25	99.5
MaximalClique	30	30	64.6	2	549.8	29	151.7
NoMystery	30	9	74.6	5	275.3	7	108.1
PermutationPatternMatching	30	22	49.2	20	152.2	26	57.4
QualitativeSpatialReasoning	30	30	45.6	27	67.8	29	102.8
RicochetRobot	30	30	93.4	7	128.2	30	200.5
Sokoban	30	11	38.4	8	221.2	12	79.0
Solitaire	27	22	14.2	20	21.8	22	10.7
StableMarriage	30	29	282.6	27	189.7	29	220.7
StillLife	10	5	6.5	4	42.2	5	39.3
ValvesLocation	30	3	56.0	1	142.1	4	42.8
VisitAll	30	19	37.5	11	36.6	19	108.0
Weighted-Sequence Problem	30	25	66.1	14	193.0	25	125.3
Total	**547**	**373**	**75.3**	**229**	**122.8**	**366**	**102.5**

queries taken from [22] on 10 different databases. For each query, two different encodings have been used, namely BB for the encoding presented by Barceló and Bertossi in [12] and MRT for the encoding presented by Manna, Ricca, and Terracina in [27]. The number of solved instances within the time-out (solved), and the running time averaged over solved instances (time) are reported in the remaining columns. WASP 1.0 has been excluded from the comparison since it does not support the computation of cautious consequences. The first observation is that CLASP is competitive also on these benchmarks solving 85 % of the instances. Nonetheless, WASP 2.1 improves the performance of CLASP solving 94 % of the instances, which corresponds to 13 instances solved by WASP 2.1 that are not solved by CLASP. In those benchmarks, the algorithm implemented by WASP 2.1 is particularly effective compared to the one implemented in CLASP. Concerning the two different encodings, we observe that CLASP performs significantly better using the encoding presented in [27], solving 3 and 4 instances more than the encoding in [12] on the third and sixth queries, respectively. Similar considerations hold also for WASP 2.1 which is in general much faster when executed using the encoding presented in [27].

Table 2. Solved instances and average running time on consistent query answering [22].

Problem	#	CLASP		WASP 2.1	
		solved	time	solved	time
Query1-BB	10	10	26.0	10	31.0
Query2-BB	10	10	53.9	10	37.1
Query3-BB	10	6	231.0	10	78.5
Query4-BB	10	10	33.0	10	40.7
Query5-BB	10	10	67.2	10	53.1
Query6-BB	10	5	195.4	10	147.5
Query7-BB	10	5	245.2	6	248.6
Query1-MRT	10	10	13.9	10	16.6
Query2-MRT	10	10	31.3	10	24.2
Query3-MRT	10	9	192.9	10	66.7
Query4-MRT	10	10	15.2	10	18.3
Query5-MRT	10	10	15.6	10	18.8
Query6-MRT	10	9	194.8	10	68.8
Query7-MRT	10	5	194.2	6	241.9
Total	**140**	**119**	**89.2**	**132**	**67.8**

6   Related Work

During recent years, ASP has obtained growing interest since robust implementations were available. ASP solver can be classified into *based on translation* (or non-native) and *native* according to the evaluation strategies employed. Non-native solvers usually perform a translation from ASP to other theories and then use specific solvers for those theories as black boxes, while native solvers implement specific algorithms and data structures for dealing with ASP programs. Among non-native solvers, ASSAT [25], CMODELS [24], and LP2SAT [21] rewrite normal ASP programs into propositional formulas and then call an external SAT solver. WASP 2.1 is a native solver, thus, it adopts a very different solving approach with respect to non-native solvers. Nonetheless, some similarities exist. Indeed, WASP 2.1 uses Clark's completion as implemented by ASSAT [25]. Among the first effective native solvers that were proposed we mention DLV [23] and SMODELS [36], which implement a systematic backtracking without learning and adopt look-ahead heuristics, while WASP 2.1 is based on the CDCL algorithm [35]. WASP 2.1 shares with SMODELS the algorithm for the computation of unfounded sets based on source pointers [36], while DLV implements a pruning technique based on finding external supporting rules. Cautious reasoning is addressed in DLV only by implementing an algorithm based on the *enumeration of answer sets*, and DLV does not print any form of estimation of the result during the computation. We also note that DLV features brave reasoning, which is not currently supported by WASP 2.1. The native solvers that are more similar to WASP 2.1 are CLASP [18] and WASP 1.0 [2].

Comparison with CLASP. Both solvers are based on the CDCL algorithm using source pointers for detecting unfounded sets. There are nonetheless several technical differences with WASP 2.1 related to data structures and input simplification. Concerning optimization problems, both CLASP and WASP 2.1 implement algorithms BASIC, PMRES and OLL. The latter has been introduced in UNCLASP [8], an experimental branch of CLASP. However, WASP 2.1 implements several algorithms introduced for MaxSAT solving that are not implemented by CLASP. Moreover, WASP 2.1 improves the performance of the algorithm OLL on weighted benchmarks by implementing the *stratification* strategy. Concerning cautious reasoning, the algorithm implemented by CLASP is a smart variant of the algorithm that enumerates answer sets. In particular, an overestimate of the sound answers is considered as a constraint of the logic program, so that the next computed answer set is guaranteed to improve the current overestimate. WASP 2.1 differs from this solver because it is based on an algorithm checking the coherence of the input program with the complement of a cautious consequence candidate, for each candidate in the overestimate. A strength of the algorithm implemented by WASP 2.1 is being anytime, i.e., both underestimates and overestimates are produced during the computation, while this is not the case with the algorithm implemented by CLASP, which only prints overestimates during the computation. We also note that CLASP features brave reasoning, which is not currently supported by WASP 2.1.

Comparison with WASP 1.0. WASP 2.1 is a substantially revised version of WASP 1.0. Both solvers are based on the CDCL algorithm using source pointers for detecting unfounded sets. The most imporant difference is related to Clark's completion and program simplification in the style of SATELITE [15], which are not applied by WASP 1.0. Clark's completion brings advantages in terms of simplifying the implementation of the propagation procedure, which is more complex in WASP 1.0. In fact, WASP 1.0 implements a specific propagation procedure for handling supportedness, which requires complex data structures to achieve efficiency. Concerning optimization problems, both solvers implement several model-guided and core-guided algorithms, including OPT, MGD, OLL and BCD. However, WASP 2.1 implements also PMRES and a *stratification* technique, which are not implemented by WASP 1.0. Finally, WASP 1.0 has no front-end for dealing with cautious reasoning.

7   Conclusion

In this paper we reported on the recent improvement of the ASP solver WASP, and in particular we described the algorithms implemented in the system for addressing optimum answer set computation and cautious reasoning. The new solver was compared with both its predecessor and the latest version of CLASP.

In our experiment, WASP 2.1 outperforms WASP 1.0, and it is competitive with CLASP on publicly available benchmarks. Future work concerns the optimization of the internal data structures, and the reengineering of disjunctive rules in order to handle programs with unrestricted disjunction efficiently. We also plan to extend to ASP our recently proposed MaxSAT algorithms [6]. As a final remark, WASP 2.1 is freely available and the source can be download at https://github. com/alviano/wasp.git.

References

1. Alviano, M., Calimeri, F., Charwat, G., Dao-Tran, M., Dodaro, C., Ianni, G., Krennwallner, T., Kronegger, M., Oetsch, J., Pfandler, A., Pührer, J., Redl, C., Ricca, F., Schneider, P., Schwengerer, M., Spendier, L.K., Wallner, J.P., Xiao, G.: The fourth answer set programming competition: preliminary report. In: Cabalar, P., Son, T.C. (eds.) LPNMR 2013. LNCS, vol. 8148, pp. 42–53. Springer, Heidelberg (2013)
2. Alviano, M., Dodaro, C., Faber, W., Leone, N., Ricca, F.: WASP: a native ASP solver based on constraint learning. In: Cabalar, P., Son, T.C. (eds.) LPNMR 2013. LNCS, vol. 8148, pp. 54–66. Springer, Heidelberg (2013)
3. Alviano, M., Dodaro, C., Marques-Silva, J., Ricca, F.: Optimum stable model search: algorithms and implementation. J. Log. Comput. doi:10.1093/logcom/exv061
4. Alviano, M., Dodaro, C., Ricca, F.: Anytime computation of cautious consequences in answer set programming. TPLP 14(4–5), 755–770 (2014)
5. Alviano, M., Dodaro, C., Ricca, F.: Preliminary report on WASP 2.0 (2014). CoRR, abs/1404.6999
6. Alviano, M., Dodaro, C., Ricca, F.: A MaxSAT algorithm using cardinality constraints of bounded size. In: 24th International Joint Conference on Artificial Intelligence (To appear). IJCAI Organization, Buenos Aires, Argentina, July 2015
7. Alviano, M., Faber, W., Leone, N., Perri, S., Pfeifer, G., Terracina, G.: The disjunctive datalog system DLV. In: de Moor, O., Gottlob, G., Furche, T., Sellers, A. (eds.) Datalog 2010. LNCS, vol. 6702, pp. 282–301. Springer, Heidelberg (2011)
8. Andres, B., Kaufmann, B., Matheis, O., Schaub, T.: Unsatisfiability-based optimization in clasp. In: Dovier, A., Costa, V.S. (eds.) ICLP (Technical Communications). LIPIcs, vol. 17, pp. 211–221. Schloss Dagstuhl - Leibniz-Zentrum fuer Informatik (2012)
9. Ansótegui, C., Bonet, M.L., Levy, J.: SAT-based MaxSAT algorithms. Artif. Intell. 196, 77–105 (2013)
10. Arenas, M., Bertossi, L.E., Chomicki, J.: Answer sets for consistent query answering in inconsistent databases. TPLP 3(4–5), 393–424 (2003)
11. Audemard, G., Simon, L.: Predicting learnt clauses quality in modern SAT solvers. In: Boutilier, C. (ed.) IJCAI, pp. 399–404 (2009)
12. Barceló, P., Bertossi, L.: Logic programs for querying inconsistent databases. In: Dahl, V. (ed.) PADL 2003. LNCS, vol. 2562, pp. 208–222. Springer, Heidelberg (2002)
13. Buccafurri, F., Leone, N., Rullo, P.: Enhancing disjunctive datalog by constraints. IEEE Trans. Knowl. Data Eng. 12(5), 845–860 (2000)

14. Dodaro, C., Leone, N., Nardi, B., Ricca, F.: Allotment problem in travel industry: a solution based on ASP. In: ten Cate, B., Mileo, A. (eds.) RR 2015. LNCS, vol. 9209, pp. 77–92. Springer, Heidelberg (2015)

15. Eén, N., Biere, A.: Effective preprocessing in SAT through variable and clause elimination. In: Bacchus, F., Walsh, T. (eds.) SAT 2005. LNCS, vol. 3569, pp. 61–75. Springer, Heidelberg (2005)

16. Eiter, T., Gottlob, G., Mannila, H.: Disjunctive datalog. ACM TODS 22(3), 364–418 (1997)

17. Gebser, M., Kaminski, R., König, A., Schaub, T.: Advances in *gringo* series 3. In: Delgrande, J.P., Faber, W. (eds.) LPNMR 2011. LNCS, vol. 6645, pp. 345–351. Springer, Heidelberg (2011)

18. Gebser, M., Kaufmann, B., Schaub, T.: Conflict-driven answer set solving: from theory to practice. Artif. Intell. 187, 52–89 (2012)

19. Gelfond, M., Lifschitz, V.: Classical negation in logic programs and disjunctive databases. New Gener. Comput. 9(3/4), 365–386 (1991)

20. Heras, F., Morgado, A., Marques-Silva, J.: Core-guided binary search algorithms for maximum satisfiability. In: Burgard, W., Roth, D. (eds.) AAAI. AAAI Press (2011)

21. Janhunen, T.: Some (in)translatability results for normal logic programs and propositional theories. J. Appl. Non-Classical Logics 16(1–2), 35–86 (2006)

22. Kolaitis, P.G., Pema, E., Tan, W.: Efficient querying of inconsistent databases with binary integer programming. PVLDB 6(6), 397–408 (2013). http://www.vldb.org/pvldb/vol6/p397-tan.pdf

23. Leone, N., Pfeifer, G., Faber, W., Eiter, T., Gottlob, G., Perri, S., Scarcello, F.: The DLV system for knowledge representation and reasoning. ACM TOCL 7(3), 499–562 (2006)

24. Lierler, Y., Maratea, M.: Cmodels-2: SAT-based answer set solver enhanced to nontight programs. In: Lifschitz, V., Niemelä, I. (eds.) LPNMR 2004. LNCS (LNAI), vol. 2923, pp. 346–350. Springer, Heidelberg (2003)

25. Lin, F., Zhao, Y.: ASSAT: computing answer sets of a logic program by SAT solvers. Artif. Intell. 157(1–2), 115–137 (2004)

26. Manna, M., Oro, E., Ruffolo, M., Alviano, M., Leone, N.: The HɪLϵX system for semantic information extraction. Trans. Large-Scale Data Knowl. Centered Syst. 5, 91–125 (2012)

27. Manna, M., Ricca, F., Terracina, G.: Consistent query answering via ASP from different perspectives: theory and practice. TPLP 13(2), 227–252 (2013)

28. Marques-Silva, J., Janota, M., Lynce, I.: On computing backbones of propositional theories. In: Coelho, H., Studer, R., Wooldridge, M. (eds.) ECAI. FAIA, vol. 215, pp. 15–20. IOS Press (2010)

29. Morgado, A., Dodaro, C., Marques-Silva, J.: Core-guided MaxSAT with soft cardinality constraints. In: O'Sullivan, B. (ed.) CP 2014. LNCS, vol. 8656, pp. 564–573. Springer, Heidelberg (2014)

30. Morgado, A., Heras, F., Liffiton, M.H., Planes, J., Marques-Silva, J.: Iterative and core-guided MaxSAT solving: a survey and assessment. Constraints 18(4), 478–534 (2013)

31. Morgado, A., Heras, F., Marques-Silva, J.: Model-guided approaches for MaxSAT solving. In: ICTAI, pp. 931–938. IEEE (2013)

32. Narodytska, N., Bacchus, F.: Maximum satisfiability using core-guided MaxSAT resolution. In: Brodley, C.E., Stone, P. (eds.) AAAI, pp. 2717–2723. AAAI Press (2014)

33. Ricca, F., Grasso, G., Alviano, M., Manna, M., Lio, V., Iiritano, S., Leone, N.: Team-building with answer set programming in the gioia-tauro seaport. TPLP **12**(3), 361–381 (2012)
34. Rosa, E.D., Giunchiglia, E., Maratea, M.: A new approach for solving satisfiability problems with qualitative preferences. In: Ghallab, M., Spyropoulos, C.D., Fakotakis, N., Avouris, N.M. (eds.) ECAI. FAIA, vol. 178, pp. 510–514. IOS Press (2008)
35. Silva, J.P.M., Sakallah, K.A.: GRASP: a search algorithm for propositional satisfiability. IEEE Trans. Comput. **48**(5), 506–521 (1999)
36. Simons, P., Niemelä, I., Soininen, T.: Extending and implementing the stable model semantics. Artif. Intell. **138**(1–2), 181–234 (2002)

Improving Coordinated SMT-Based System Synthesis by Utilizing Domain-Specific Heuristics

Benjamin Andres[1], Alexander Biewer[2], Javier Romero[1],
Christian Haubelt[3], and Torsten Schaub[1,4(✉)]

[1] University of Potsdam, Potsdam, Germany
torsten@cs.uni-potsdam.de
[2] Robert Bosch GmbH, Stuttgart, Germany
[3] University of Rostock, Rostock, Germany
[4] INRIA Rennes, Rennes, France

Abstract. In hard real-time systems, where system complexity meets stringent timing constraints, the task of system-level synthesis has become more and more challenging. As a remedy, we introduce an SMT-based system synthesis approach where the Boolean solver determines a static binding of computational tasks to computing resources and a routing of messages over the interconnection network while the theory solver computes a global time-triggered schedule based on the Boolean solver's solution. The binding and routing is stated as an optimization problem in order to refine the solution found by the Boolean solver such that the theory solver is more likely to find a feasible schedule within a reasonable amount of time. In this paper, we enhance this approach by applying domain-specific heuristics to the optimization problem. Our experiments show that by utilizing domain knowledge we can increase the number of solved instances significantly.

1 Introduction

Embedded systems surround us in our daily life. Often, we interact or rely on them without noticing. Embedded systems are typically small application-specific computing systems that are part of a larger technical context. A computer program executed on a processor of an embedded system usually controls connected mechanical or electric components. For example, in a car, an embedded system manages the engine of the car while another one corrects over- and under-steering, e.g., when a driver goes into a bend too quickly.

Today, designing a new embedded system that serves a predefined purpose is becoming more and more challenging. With increasing system complexity due to regulatory requirements, customer demands, and migration to integrated architectures on massively parallel hardware, the task of system-level synthesis has become increasingly more challenging. During synthesis, the spatial binding of

---

T. Schaub—Affiliated with Simon Fraser University, Canada, and IIIS Griffith University, Australia.

© Springer International Publishing Switzerland 2015
F. Calimeri et al. (Eds.): LPNMR 2015, LNAI 9345, pp. 55–68, 2015.
DOI: 10.1007/978-3-319-23264-5_6

computational tasks to processing elements (PEs), the multi-hop routing of messages, and the scheduling of tasks and messages on shared resources has to be decided. Computational tasks can be viewed as small computer programs that realize control functions where messages between tasks are used to exchange information. In safety critical control application, e.g., the Electronic Stability Control (ESP) from the automotive domain, the correct functionality of an embedded system does not only depend on the correct result of a computation but also on the time, i.e. the schedule, when the result is available. Such systems are called (embedded) hard real-time systems.

With increasing complexity and interdependent decisions, system design demands for compact design space representations and highly efficient automatic decision engines, resulting in automatic system synthesis approaches. Previously presented approaches might fail due to the sheer system complexity in context of guaranteeing the timeliness of hard real-time systems. System synthesis based on satisfiability modulo theories (SMT) has been shown to solve this problem by splitting the work among a Boolean solver and a theory solver [1].

In the work at hand, we employed an answer set programming (ASP) solver to decide the static binding of computational tasks to the PEs of a given hardware architecture. Furthermore, the ASP solver determines for each message a static routing on the hardware architecture's network. Based on these decisions, a linear arithmetic solver ($\mathcal{T}$-solver) then computes global time-triggered schedules for all computational tasks as well as messages. Here, the $\mathcal{T}$-solver's solution guarantees the timing constraints of the system by construction. However, most often the time-triggered scheduling becomes too complex and the $\mathcal{T}$-solver cannot decide within a reasonable amount of time whether a binding and routing (B&R) provided by the ASP solver is schedulable or not. As a remedy, in [2] we introduced a *coordinated SMT-based system synthesis* approach that shows a significant better scalability compared to previous SMT-based synthesis approaches. We will present our proposed approach in detail in Sect. 3.

In [3] we presented a declarative framework for domain-specific heuristics in answer set solving. The approach allows modifying the heuristic of the solver directly from the ASP encoding. In this paper we show that the coordinated SMT-based system synthesis approach is considerably improved by applying domain-specific heuristics to the ASP solver `clasp`.

This paper is structured as follows. In Sect. 2 we provide a formal problem definition of the system synthesis problem for hard real-time systems. Furthermore, a formal system model is introduced. Section 3 explains our approach to solve the system synthesis problem in detail. In Sect. 4 we provide an introduction of the concepts enabling domain-specific heuristics in `clasp` along with the domain-specific heuristics we specified. The experimental results reported in Sect. 5 show the effectiveness of our heuristics. Section 6 concludes this paper.

2   Problem Formulation and System Model

In this section we introduce a formal system model that will be used in the remainder of this paper. Furthermore, on basis of our system model, we provide a problem formulation of the synthesis problem for hard real-time systems.

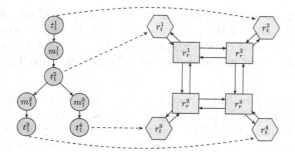

Fig. 1. An instance of our system synthesis model consisting of an application graph (left), a platform graph (right), and bindings of computational tasks to tiles.

In the paper at hand the platform or hardware architecture is modeled as a directed graph $g_P = (\mathbf{R}, \mathbf{E_R})$, the platform graph. Vertices $\mathbf{R}$ represent all possibly shared resources of a platform whereas the directed edges $\mathbf{E_R} \subseteq \mathbf{R} \times \mathbf{R}$ model interconnections between resources (cf. Fig. 1). The shared resources $\mathbf{R}$ can further be partitioned into tiles $\mathbf{R_t}$ and routers $\mathbf{R_r}$, with $\mathbf{R_t} \cup \mathbf{R_r} = \mathbf{R}$ and $\mathbf{R_t} \cap \mathbf{R_r} = \emptyset$. A tile $t \in \mathbf{R_t}$ implements exactly one processing element (PE) combined with local memory and is able to execute computational tasks. A router $r \in \mathbf{R_r}$ is able to transfer messages from one hardware resource to another. A mesh of interconnected routers form a so-called network-on-chip (NoC). NoCs are a preferred communication infrastructure as they scale well with a growing number of tiles that have to be interconnected in massive-parallel hardware architectures. For the sake of clarity, the remainder of this paper assumes a homogeneous hardware architecture, i.e., tiles possess the same processing power and routers forward messages within the same delay.

The applications that have to be executed on a hardware architecture are modeled by a set of applications $\mathbf{A}$. Each application $A^i \in \mathbf{A}$ is specified by the tuple

$$A^i = (g_A^i, P^i, D^i).$$

An application in hard real-time systems often controls a mechanical system and therefore requests to be executed periodically with the period P^i. In order to guarantee stability of a control algorithm, all computations and communication of an application have to be completed within the constrained relative deadline $D^i \leq P^i$.

Furthermore, each application $A^i \in \mathbf{A}$ is modeled as a directed acyclic graph $g_A^i = (\mathbf{T^i}, \mathbf{E_A^i})$, the application graph (cf. Fig. 1). The set of nodes $\mathbf{T^i} = \mathbf{T_t^i} \cup \mathbf{T_m^i}$ is the union of the set of computational tasks of an application $\mathbf{T_t^i}$ and the set of messages of an application $\mathbf{T_m^i}$. The directed edges $\mathbf{E_A^i} \subseteq (\mathbf{T_t^i} \times \mathbf{T_m^i}) \cup (\mathbf{T_m^i} \times \mathbf{T_t^i})$ of the graph g_A^i specify data dependencies between computational tasks and messages or vice versa.

With each computational task $t \in \mathbf{T_t^i}$ a worst-case execution time (WCET) C_t is associated that holds for all PEs. The workload W_t generated by a task

$t \in \mathbf{T_t^i}$ on a PE of tile is computed by $W_t = C_t/P_i$. Concerning messages $m \in \mathbf{T_m^i}$, the remainder of the paper assumes that a message equals an atomic entity that is transferred on the resources of the hardware architecture. With this definition, the worst-case transfer time of a message $m \in \mathbf{T_m^i}$ on a resource $r \in \mathbf{R_r}$ is defined by $C_{(m,r)}$.

Problem Formulation. In the design of time-triggered real-time systems, system synthesis is the task of computing a valid implementation $\mathbf{I} = (\mathbf{B}, \mathbf{R_m}, \mathbf{S})$ consisting of a binding $\mathbf{B} \subseteq (\bigcup_{A^i \in \mathbf{A}} \mathbf{T_t^i}) \times \mathbf{R_t}$, a routing of each message on a tree of resources $\mathbf{R_m} \subseteq \mathbf{R_r}$ and a feasible global time-triggered schedule $\mathbf{S}$. Here, the binding $\mathbf{B}$ contains for each computational task $t \in \mathbf{T_t^i}$ of an application $A^i \in \mathbf{A}$ exactly one tile it is executed on. A routing $\mathbf{R_m}$ defines a set of connected resources of the hardware architecture that are utilized during the transfer of a message. The time-triggered schedule

$$\mathbf{S} = \{\mathbf{s}_{(\lambda,\mathbf{r})} | \lambda \in (\mathbf{T_t^i} \cup \mathbf{T_m^i}), \mathbf{A^i} \in \mathbf{A}, \mathbf{r} \in \mathbf{R_m} \vee (\lambda, \mathbf{r}) \in \mathbf{B}\}$$

contains the start times $\mathbf{s}_{(\lambda,\mathbf{r})}$, generally in clock cycles, for each computational task that starts its execution on a tile and the start time of the transfer for each message on the resources on which the message is routed.

Example. Figure 1 presents an example of the system synthesis problem as described above. The platform graph $g_P = (\mathbf{R}, \mathbf{E_R})$ defines a regular 2×2 mesh with four tiles and four routers. All tiles are exclusively connected to exactly one router, while all routers are additionally connected to a set of two other routers. The example consists of only one application A_1 with four computational tasks and three messages. For the sake of clarity, we choose $P_1 = 10, D_1 = 9$, $C_{t_1^1} = C_{t_1^3} = C_{t_1^4} = 1, C_{t_1^2} = 2$ and $\forall m \in \mathbf{T_m^1}, \forall r \in \mathbf{R_r} : C_{(m,r)} = 1$. One possible implementation $\mathbf{I} = (\mathbf{B}, \mathbf{R_m}, \mathbf{S})$ is then given by the following sets:

$$\mathbf{B} = \{(t_1^1, r_t^2), (t_1^2, r_t^1), (t_1^3, r_t^4), (t_1^4, r_t^3)\}$$
$$\mathbf{R_{m_1^1}} = \{r_r^2, r_r^1\}$$
$$\mathbf{R_{m_1^2}} = \{r_r^1, r_r^3, r_r^4\}$$
$$\mathbf{R_{m_1^3}} = \{r_r^1, r_r^3\}$$
$$\mathbf{S} = \{ \ \mathbf{s}_{(t_1^1, r_t^2)} = 0, \ \mathbf{s}_{(m_1^1, r_r^2)} = 1, \ \mathbf{s}_{(m_1^1, r_r^1)} = 2, \ \mathbf{s}_{(t_1^2, r_t^1)} = 3,$$
$$\mathbf{s}_{(m_1^2, r_r^1)} = 5, \ \mathbf{s}_{(m_1^2, r_r^3)} = 6, \ \mathbf{s}_{(m_1^2, r_r^4)} = 7, \ \mathbf{s}_{(t_1^3, r_t^4)} = 8,$$
$$\mathbf{s}_{(m_1^3, r_r^1)} = 6, \ \mathbf{s}_{(m_1^3, r_r^3)} = 7, \ \mathbf{s}_{(t_1^4, r_t^3)} = 8 \ \}$$

3   SMT-Based System Synthesis

In the following section we present our SMT-based approach to solve the system synthesis problem introduced in the previous section. SMT-based system synthesis approaches, e.g., [1,4,5], gained a lot of attention in domains with complex system specifications and stringent timing constraints such as the automotive domain.

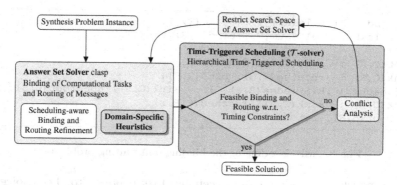

Fig. 2. Overview of the coordinated SMT-based synthesis approach introduced in this paper. We propose to utilize domain-specific heuristics in order to improve previous work [2].

In SMT-based system synthesis, the work is split between a Boolean solver and a theory solver ($\mathcal{T}$-solver) [1] (cf. Fig. 2). Here, the Boolean solver computes the static bindings of computational tasks to the PEs of a platform graph and the multi-hop routes of messages on a tree of resource of the platform. In contrast to related work [1,4,5], our approach implements the answer set solver `clasp` instead of a SAT solver, since answer set programming (ASP) has been shown to scale better in the determination of multi-hop routes for messages in the densely connected hardware architectures investigated in this paper [6].

3.1 Binding and Routing Using Answer Set Programming (ASP)

The ASP facts based on the formal model of the platform graph $g_P = (\mathbf{R_t} \cup \mathbf{R_r}, \mathbf{E_R})$ and the applications $A^i = (g_A^i = (\mathbf{T_t^i} \cup \mathbf{T_m^i}, \mathbf{E_A^i}), P^i, D^i) \in \mathbf{A}$ (cf. Sect. 2) for a system instance are defined as follows:

$$
\begin{aligned}
&\{task(t, W_t). \mid t \in \mathbf{T_t^i}, A^i \in \mathbf{A}, W_t = \lfloor 1000 \cdot C_t/P_i \rfloor\} \ \cup \\
&\{send(t, m). \mid (t, m) \in \mathbf{E_A^i}, t \in \mathbf{T_t^i}, m \in \mathbf{T_m^i}, A^i \in \mathbf{A}\} \ \cup \\
&\{receive(t, m). \mid (m, t) \in \mathbf{E_A^i}, t \in \mathbf{T_t^i}, m \in \mathbf{T_m^i}, A^i \in \mathbf{A}\} \ \cup \\
&\{tile(t). \mid t \in \mathbf{R_t}\} \ \cup \\
&\{router(r). \mid r \in \mathbf{R_r}\} \ \cup \\
&\{edge(r, \tilde{r}). \mid (r, \tilde{r}) \in \mathbf{E_R}\}.
\end{aligned} \tag{1}
$$

The ASP encoding of the binding and routing problem is depicted in Fig. 3. The rule in Line 1 specifies that every computational task provided in an instance must be mapped to exactly one tile (PE). Observe that the mapping of computational tasks t to a tile $r \in \mathbf{R_t}$ is represented by atoms `bind(`t`,`r`)` in an answer set. This provides the basis for specifying the routing of messages. The integrity constraint in Line 2 ensures that the workload of every tile does not exceed its maximal utilization. The routing is carried out by (recursively) constructing non-branching acyclic routes from resources of communication targets back to the resource of a sending task, where the routing stops. Line 4 (resp. 5)

```
1   1 {bind(T,R):tile(R)} 1 :- task(T,_).
2   :- tile(R), 1001 #sum{U,T:bind(T,R),task(T,U)}.

4   root(C,R) :- send(T,C), bind(T,R).
5   sink(C,R) :- receive(T,C), bind(T,R).
6   1 {reached(C,R,S):edge(R,S)} 1 :- sink(C,S), not root(C,S).
7   sink(C,R) :- sink(C,S), reached(C,R,S).
8   :- root(C,R), not sink(C,R).
```

Fig. 3. ASP encoding of the binding and routing problem.

identifies the tile the sending (resp. receiving) task is bound to. The choice rule in Line 6 connects each encountered target resource to exactly one predecessor, with the only exception that the target resource is not the resource the sender is bound to. Each connected resource is then identified as new target resource in Line 7. Finally, the integrity constraint in Line 8 requires that each resource with a sending task must be a target of the message.

3.2   Time-Triggered Scheduling

Based on the binding and routing decisions by the answer set solver, a time-triggered scheduling problem in linear arithmetic is formulated and solved by the $\mathcal{T}$-solver. The workload is splitted since large numbers are involved in time-triggered scheduling that do not scale well in Boolean solvers. The $\mathcal{T}$-solver computes the start times for each computational task on a PE and each message on the resources of the NoC such that all timing constraints are fulfilled. All constraints in the scheduling formulation are compositions of terms that are formulated in quantifier-free integer difference logic (QF_IDL), i.e., $\mathbf{s} - \tilde{\mathbf{s}} \leq k$, with $\mathbf{s}, \tilde{\mathbf{s}} \in \mathbb{N}$ being a start time variable of a computational task or message and a constant $k \in \mathbb{N}$. Constraints in the scheduling problem ensure for example that one resource is utilized at most by one computational task or message at the same time. Furthermore, constraints ensure the integrity of data flows between computational tasks and/or messages. Due to the limited space of the paper a detailed description of the scheduling problem and its formulation is omitted, but can be found in [7]. Our formulation of the scheduling problem is an adapted version of the one presented in [5].

3.3   Coupling of the Answer Set Solver and $\mathcal{T}$-solver

If the $\mathcal{T}$-solver is able to derive a time-triggered schedule based on the ASP solvers decision, the synthesis finishes with a valid solution. In contrast, if the $\mathcal{T}$-solver proves that a feasible schedule based on the binding and routing does not exist, a conflict analysis is started. Related work has shown that deriving a minimal reason (unsatisfiable core or irreducible inconsistent subset) why a schedule cannot be found can significantly increase scalability of SMT-based system synthesis [4].

In order to decrease the time for the $\mathcal{T}$-solver to prove that a binding and routing is not schedulable and to speed up a subsequent conflict analysis, we implemented a hierarchical scheduling scheme that has been established in previous work [5]. The $\mathcal{T}$-solver decides feasibility on subproblems before a schedule is derived for the complete system. Here, deciding feasibility, as well as the subsequent conflict analysis, of smaller subproblems tends to be much faster. Similar to [2], we apply the following hierarchical scheduling scheme:

1. Schedule the computational tasks on each tile independently (TS).
2. Schedule the computational tasks on each tile including incoming and outgoing messages from the tile independently (CS).
3. Schedule clusters of independent applications independently (AS).

If all scheduling problems in a hierarchical stage can be solved, the $\mathcal{T}$-solver starts to solve problems of the subsequent stage. In (AS), clusters of independent applications are specified such that no hardware resource is shared between two clusters. However, depending on the binding and routing, (AS) may be equivalent to the problem of scheduling the complete system.

Our approach implements a modified deletion filter [5] and forward filter [8] to compute a minimal reason why a schedule cannot be found. The deletion filter is applied on infeasible problems of (TS) whereas the forward filter is applied in conflict analysis in stages (CS) and (AS). Based on the first minimal reason from the conflict analysis in a hierarchical stage, the search space of the answer set solver is pruned via integrity constraints. As a result, further binding and routing solutions computed by the answer set solver do not lead to the same infeasibility that has already been observed. Note that the answer set solver is just halted while the $\mathcal{T}$-solver analyses a binding and routing. We do not restart the answer set solver repeatedly.

If a conflict has been found in scheduling hierarchy (TS), the deduced minimal reason only contains computational tasks. The constraint added to the context of the answer set solver ensures that the same set of conflicting computational tasks will not be bound to any PE again. In contrast, the result of a conflict analysis in stage (CS) and (AS) is a set of computational tasks and messages. Without breaking symmetries, the search space of the answer set solver is pruned by the concrete binding and routing of the conflicting computational tasks and messages. Once the search space of the answer set solver has been pruned, a new binding and routing is computed and subsequently analysed in the $\mathcal{T}$-solver. This iterative process stops once a feasible time-triggered schedule has been found or if the answer set solver returns that no further binding and routing can be derived. In the latter case the SMT-based system synthesis approach proved that no feasible solution to the synthesis problem exists for the provided problem specification.

3.4   Coordinated SMT-Based System Synthesis

Up to this point, we introduced our SMT-based system synthesis approach that is in general applicable to solve the synthesis problem presented in Sect. 2. However, a major drawback of the synthesis approach described so far lies in the

```
1  maxu(1000).
2  {maxu(MU-slice)} :- maxu(MU), omu(Mean), MU-slice>Mean.
3  :- tile(R), maxu(MU), MU+1 #sum {U,T:bind(T,R),task(T,U)}.
4  #maximize{1@2,MU:maxu(MU)}.

6  #minimize{1@1,C:reached(C,R,S)}.
```

Fig. 4. Encoding of the scheduling-aware binding and routing refinement realized as lexicographical optimization of the maximal utilization of all tiles and the number of routed messages.

often very time consuming scheduling in the T-solver. In extreme cases, deciding schedulability even of small subproblems can take up hours. This is especially the case if the workload of a PE is close to the maximum utilization of 100 %. As a remedy, in [2] we showed that the scalability of SMT-based synthesis can be improved considerably if the answer set solver and the T-solver are coordinated. The basic idea of the *coordinated SMT-based synthesis* is to let the answer set solver compute bindings and routings where schedulability can be decided within a reasonable time by the T-solver. This is realized by assigning a constant time budget to the answer set solver exclusively to refine (optimize) an initial binding and routing solution. With this *scheduling-aware binding and routing refinement* (cf. Fig. 2) the T-solver is expected to decide schedulability within a reasonable amount of time. Despite the additional time budget for the refinement, the coordinated SMT-based synthesis has been shown to scale significantly better than SMT-based approaches without coordination [2].

In the coordinated SMT-based synthesis approach of this paper, the scheduling-aware binding and routing refinement is realized by using a lexicographical optimization in clasp. With the highest priority, the load balancing of the PEs of the platform is optimized. Figure 4 depicts the encoding that realizes a simplified load balancing strategy. The basic idea of the strategy is to allow the solver to successively reduce the maximal utilization of PEs and then maximize the number of reductions. Beginning with a maximal utilization (maxu) of 1000 (Line 1) the choice rule in Line 2 allows to generate an additional maxu by reducing an existing one by a predefined amount slice, as long as the new maxu is greater then the optimal mean utilization omu(Mean). The integrity rule in Line 3 enforces all maximal utilizations and Line 4 maximizes the number of generated maxu. Note that we used this simplified strategy instead of a true min-max optimization due to performance reasons. With the second highest priority in the lexicographical optimization, we minimize the total number of routed messages in a system as shown in Line 6. While the coordinated SMT-based synthesis approach is part of the current state-of-the-art in symbolic system synthesis, the paper at hand aims on improving the coordinated approach by utilizing domain-specific heuristics in the answer set solver (cf. green box in Fig. 2). The following section describes in detail our formulated heuristics and the implementation details that enable the usage of domain-specific heuristics in clasp.

4   Heuristics

ASP provides a rich modelling language together with highly performant yet general-purpose solving techniques. Often these general-purpose solving capacities can be boosted by domain-specific heuristics. For this reason, we introduced a general declarative framework for incorporating domain-specific heuristics into ASP solving [3]. The rich modelling language is used to specify heuristic information, which is exploited by a dedicated Domain heuristic in clasp when it comes to non-deterministically assigning a truth value to an atom. Although this bears the risk of search degradation [9], it has already indicated great prospects by boosting optimization and planning in ASP [3]. In this framework, heuristic information is represented within a logic program by means of the dedicated predicate _heuristic. For expressing different types of heuristic information, the following basic modifiers are available: sign, level, init and factor. Here we explain the first two, that will be applied in our experiments, and refer the reader to [3] for further details. Modifier sign allows for controlling the truth value assigned to variables subject to a choice within the solver. With 1 representing true and -1 false, repectively. For example, given the program

```
{a}.       _heuristic(a,sign,1).
```

atom a is chosen with positive sign and the answer set {_heuristic(a,sign, 1),a} is produced, while replacing 1 by -1 we obtain {_heuristic(a,sign, -1)}. Modifier level establishes a ranking among atoms such that unassigned atoms of highest rank are chosen first, with the default rank of 0. Extending the previous program with

```
{b}.       _heuristic(b,sign, 1).
:- a, b.  _heuristic(a,level,1).
```

the solver chooses first a (because its level is 1) with a positive sign, and returns the answer set where a is true and b is false. Adding the fact _heuristic(b,level,2) atom b is chosen first with positive sign, and we obtain the answer set with b true and a false. Further extending the program with the fact _heuristic(a,level,3) yields once more the solution with a true and b false. This illustrates the fact that when an atom gets two values for the same modifier (in the example, 1 and 3 for the level of a), the one with higher absolute value takes precedence. The modifier true (false) is defined as the combination of a positive (negative) sign and a level. For instance, in the last example we could use _heuristic(a,true,3) instead of _heuristic(a,sign,1) and _heuristic(a,level,3). Note that domain heuristics are dynamic, in the sense that they depend on the changing partial assignment of the solver. As an example, consider the following program:

```
1  _heuristic(a,true, 1).    {a;b}.
2  _heuristic(b,sign, 1)  :-      a.
3  _heuristic(b,sign,-1)  :- not a.
```

The solver starts setting a to true, then the rule in Line 2 is fired, b is selected with a positive sign and the answer set with a and b is produced. On the other

```
1   % H1 - bindings first, then routing
2   _heuristic(bind(T,R),        level,2)  :- task(T,_), tile(R).
3   _heuristic(reached(C,R,S),level,1)  :- edge(R,S), send(_,C).
4   % H2 - discourage routing
5   _heuristic(reached(C,R,S),sign,-1)  :- edge(R,S), send(_,C).
6   % H3 - bind sender and receiver together
7   _heuristic(bind(T',R),sign,1)  :-
8        bind(T,R), send(T,C), receive(T',C).
9   % H4 - Clustering computational tasks
10  _heuristic(bind(T,R),true,A+2)  :-
11       bind(T',R), send(T',M), receive(T,M), belongs(A,T).
12  _heuristic(bind(T,R),true,A+1)  :-
13       bind(T',R'), send(T',M), receive(T,M),
14       neighbor(R',R), belongs(A,T).
15  _heuristic(bind(T,R),level,A)   :-
16       belongs(A,T), task(T,_), tile(R).
```

Fig. 5. Domain-specific heuristics used in our system synthesis approach.

hand, replacing modifier true by false we obtain instead a false and b false. This ability to represent dynamic heuristics is crucial to boost the performance of our system.

In order to refine the binding and routing provided by clasp to the $\mathcal{T}$-solver, we defined a number of different domain-specific heuristics. Figure 5 presents the heuristics that where most successful in increasing the number of solved instances and reducing the overall synthesis time. Except when one heuristic is overriden by another (e.g., H1 and H2 by H4) it makes sense to combine the different heuristics. In fact, our experiments in Sect. 5 show that this is highly advantageous.

The first heuristic H1 (Line 2–3) places a higher level on binding before routing, since the routing of messages is highly dependent on the binding of their corresponding tasks. Analysis of the answer sets provided by clasp shows that many messages are routed over the whole platform even though shorter paths exist. This is one of the major causes for failure during scheduling, since the $\mathcal{T}$-solver has to consider all resources a message is routed over. To reduce unnecessary detours, the heuristic H2 in Line 5 gives a negative sign to all reached/3 atoms.

Another technique to reduce the number of routed messages is to bind tasks exchanging a message with each other onto the same tile. This is stated by the heuristic H3 (Line 7), provided the sender is already bound to a tile.

The idea of H3 is extended in H4 (Line 10–16) where a task should be bound preferably onto the same tile as its corresponding communication partner or, to a lesser degree, to a neighboring tile. Note that only the rules for binding a receiving task to its sender are shown. The heuristics for encoding that a sending task should be bound to its receiver are analogous to the ones presented. Additionally, H4 tries to bind all task of one application before binding any other task. This allows to cluster tasks of one application onto the same tile before its full utilization is reached. The newly used fact `neighbor/2` identifies two neighboring tiles, while `belongs(A,T)` identifies tasks of the same application `A` and provides an ordering of applications. The identifier `A` is chosen in such a way that there are no conflicts with the offset of the heuristics in Lines 10 and 12. Note that both heuristics H3 and H4 are dynamic.

5   Experiments

In the following section we present the experimental results which quantify the improvement of our domain-specific heuristics from the previous section in coordinated SMT-based system synthesis introduced in Sect. 3.

All experiments were performed on a dual-processor Linux workstation containing two quad-core 2.4 GHz Intel Xeon E5620 and 24 GB RAM. All benchmark runs utilized only one CPU core and were carried out sequentially. We report average values over three independent runs per instance, to reduce the non-deterministic factors during the synthesis, e.g., through interrupting the process.

As answer set solver we implemented the python module `gringo`, containing `clasp` (version 3.1.0) with command-line switch `--configuration=auto`. In our initial tests unsatisfiable core based strategies (command-line switch `--opt-strategy=usc,4`) revealed to be highly efficient for this problem class and is taken as reference. Due to the algorithmic approach of unsatisfiable core based optimization, domain-specific heuristics are not working with unsatisfiable core based optimization strategies. We report results for the domain-specific heuristics with branch and bound based optimization (command-line switch `--opt-strategy=bb,2`). Both optimization strategies were selected as the best strategies by comparison of the results on test instances.

As $\mathcal{T}$-solver we implemented `yices` (version 2.3.0, [10]) with command-line arguments `--logic=QF_IDL` and `--arith-solver=floyd-warshall`. Additionally, the Z3 theorem prover (version 4.3.2, [11]) in logic `QF_IDL` was used in the forward filter in conflict analysis.

In our experiments, the overall time limit for a complete system synthesis run was set to 900 s. After this time, a still running benchmark was interrupted and documented as unsolved. The time for `clingo` to refine an initial binding and routing solution (optimization time) was set to 1 s. Note that in our tests more optimization time decreased the number of successful synthesised systems. We conjecture that the additional time spend on optimization reduces the time in the SMT-based synthesis approach to learn "just enough" infeasible binding and routing solutions within the overall time limit. However, the chosen optimization time is still sufficient to refine solutions such that the $\mathcal{T}$-solver can decide feasibility within a reasonable amount of time.

Concerning the instances of our experiments, the platform graphs $g_R = (\mathbf{R}, \mathbf{E_R})$ of all the system synthesis models were set to a regular 5×5-mesh composed of 25 tiles and 25 routers. The characteristics of an application $A_i = (g_A^i, P_i, D_i)$ in the system's application set were generated similar to the

Table 1. Experimental results of our benchmarks. First two lines are optimization only, while the lower ones include domain-specific heuristics and optimization (with −opt-strategy = bb,2).

Strategy/heuristic	Solved instances [of 60]			Successful synthesis	Couplings	Solving time		
	Completely	Partially	Unsolved			B&R	Scheduling	Total
Usc optimization	4	32	24	35 %	118	157.46	83.74	241.20
b&b optimization	0	20	40	13 %	199	280.03	154.76	434.79
Structural heuristic	34	8	18	62 %	16	241.91	7.05	248.96
H1	12	30	18	44 %	115	169.37	89.77	259.14
H1 + H2	17	28	15	53 %	102	150.86	51.14	202.00
H1 + H2 + H3	14	38	8	59 %	99	145.67	116.73	262.39
H4	34	25	1	79 %	40	54.75	56.16	110.90
H2 + H4	35	19	6	75 %	47	70.68	27.78	98.46
H4 + structural heuristics	38	17	5	79 %	36	48.00	43.48	91.48

ones reported in [2], representing relevant problem instances in the field of system synthesis. As a difference, the total system utilization in our instances is 70 % whereas the reported instances in [2] utilized only up to 40 %. In different tests we found that a system's utilization of nearly 70 % results in significant harder instances for the coordinated synthesis approach compared to utilization around approximately 60 %. Overall, we generated 60 different system instances with the average number of applications per instance being 88 ± 2, a total number of computational tasks of 391 ± 5 and a total number of messages of 303 ± 5.

Table 1 presents the selected results of our experiments. The first column presents the strategy and/or heuristic used in the experiment run. "Structural heuristics" references the commandline switch --dom-mod=5,8, automatically applying an heuristic with the false modifier to all atoms involved in a minimization statement. Note that all maximization statements are reduced to minimization before solving. The combination of multiple heuristics is depicted by "+". The next three columns depict the number of instances that where solved completely, partially or not at all, meaning that all three, some, or none of the three independent runs per instance were successful. "Successful synthesis" shows the amount of test runs that finished successfully in percent of overall 180 runs, "couplings" present the average number of couplings needed to solve one instance run with the respective approach. Note that both the resulting numbers are rounded up to integer values. The table concludes with the average time needed to solve the binding and routing ("B&R"), the scheduling and the complete system synthesis problem ("total") in seconds without instances that exceeded the time limit of 900 s. Note that "B&R" includes the time for the scheduling-aware binding and routing refinement (optimization time of 1 s per answer set solver call).

Although only a few (most interesting) heuristic combinations are shown, we conducted tests with over 70 different strategy/heuristic combinations in total. Many of these combinations yielded average or worse results. Test runs without optimization, i.e., without coordination of the answer set solver and the $\mathcal{T}$-solver, are omitted from the table. None of these synthesis runs were successful (even with the application of heuristics). Extensive data of our tests accompanied by the full encoding can be found in the Labs section of [13].

Comparing the result of the different heuristics reveals that dynamic heuristics are useful in practice and that the combination of different heuristics is most efficient. Note that the heuristic H4 is a stronger reformulation of H1 and H3.

While the structural-based heuristic solves as many instances as the best domain-specific heuristic H4 completely, a large number of instances were not solved at all. The structural-based heuristic also has a huge unbalance of workload between `clasp` and `yices`, with the pure B&R solving time without optimization over one order of magnitude larger than in the other approaches. The reason for this is that the structural-based heuristic is very aggressive in finding an optimal solution, with the first solution being very close to it. While this is very effective in terms of couplings needed, the overall runtime is worse than in the domain-specific heuristics. We suspect that structural-based heuristic does not scale well when the number of necessary routings increases. The combination of H4 and the structural heuristic is controversial. While it solves 4 more instances completely, the same number of instances were unsolved.

Our experiments show that the application of domain-specific heuristics yields a significant increase of successful synthesis runs with almost twice as many compared to the best strategy relying only on the scheduling-aware binding and routing refinement (using a `clasp`'s unsatisfiable core based optimization strategy). At the same time, both the average number of couplings and the total synthesis time is reduced. Furthermore, and more importantly, the number of completely solved instances were increased by a factor of 8.

6 Conclusion

In this paper we presented an SMT-based synthesis approach for hard real-time system synthesis. The approach at hand utilizes the answer set solver `clasp` for generating a binding and routing and the linear arithmetic solver `yices` ($\mathcal{T}$-solver) for time-triggered scheduling based on `clasp`'s decisions. We discussed that a scheduling-aware refinement of the solution provided by `clasp` is needed such that the $\mathcal{T}$-solver is able to solve the scheduling problem within a reasonable amount of time. Based on the optimization-based refinement of [2] we apply domain-specific heuristics, utilizing a novel technique [3] for specifying heuristics for `clasp`. A number of efficient domain-specific heuristics were proposed and benchmarked against optimization-only approaches, as well as structural heuristics provided by `clasp`. The results of the benchmark show that our domain-specific heuristics had a significant positive impact on both instances solved as well as on the runtime for solved instances. It is expected that the approach of utilizing domain-specific heuristics for the refinement of the solutions provided by the Boolean solver can be applied to other applications that combine Boolean with theory solving as well.

Future work includes the search for additional heuristics/approaches toward abolishing the need for optimization, currently consuming approximately half of the total synthesis time. We would also like to explore the possible advantages of a tighter integration of the Boolean and the theory solver as described in [8,12]. To the best of our knowledge none support domain-specific heuristics as utilized in clasp (3.1.0) yet.

Acknowledgments. This work was partly funded by DFG (550/9).

References

1. Reimann, F., Glaß, M., Haubelt, C., Eberl, M., Teich, J.: Improving platform-based system synthesis by satisfiability modulo theories solving. In: Proceedings of CODES+ISSS, pp. 135–144 (2010)
2. Biewer, A., Andres, B., Gladigau, J., Schaub, T., Haubelt, C.: A symbolic system synthesis approach for hard real-time systems based on coordinated SMT-solving. In: Proceedings of DATE, pp. 357–362 (2015)
3. Gebser, M., Kaufmann, B., Otero, R., Romero, J., Schaub, T., Wanko, P.: Domain-specific heuristics in answer set programming. In: Proceedings of AAAI, pp. 350–356 (2013)
4. Reimann, F., Lukasiewycz, M., Glaß, M., Haubelt, C., Teich, J.: Symbolic system synthesis in the presence of stringent real-time constraints. In: Proceedings of DAC, pp. 393–398 (2011)
5. Lukasiewycz, M., Chakraborty, S.: Concurrent architecture and schedule optimization of time-triggered automotive systems. In: Proceedings of CODES+ISSS, pp. 383–392 (2012)
6. Andres, B., Gebser, M., Schaub, T., Haubelt, C., Reimann, F., Glaß, M.: Symbolic system synthesis using answer set programming. In: Cabalar, P., Son, T.C. (eds.) LPNMR 2013. LNCS, vol. 8148, pp. 79–91. Springer, Heidelberg (2013)
7. Biewer, A., Munk, P., Gladigau, J., Haubelt, C.: On the influence of hardware design options on schedule synthesis in time-triggered real-time systems. In: Proceedings of MBMV, pp. 105–114 (2015)
8. Ostrowski, M., Schaub, T.: ASP modulo CSP: the clingcon system. Theory Pract. Logic Program. **12**(4–5), 485–503 (2012)
9. Järvisalo, M., Junttila, T., Niemelä, I.: Unrestricted vs restricted cut in a tableau method for boolean circuits. Ann. Math. Artif. Intell. **44**(4), 373–399 (2005)
10. Dutertre, B.: Yices 2.2. In: Biere, A., Bloem, R. (eds.) CAV 2014. LNCS, vol. 8559, pp. 737–744. Springer, Heidelberg (2014)
11. de Moura, L., Bjørner, N.S.: Z3: an efficient SMT solver. In: Ramakrishnan, C.R., Rehof, J. (eds.) TACAS 2008. LNCS, vol. 4963, pp. 337–340. Springer, Heidelberg (2008)
12. Janhunen, T., Liu, G., Niemelä, I.: Tight integration of non-ground answer set programming and satisfiability modulo theories. In: Proceedings of GTTV, pp. 1–13 (2011)
13. Potassco website. http://potassco.sourceforge.net

Integrating ASP into ROS for Reasoning in Robots

Benjamin Andres[1], David Rajaratnam[2], Orkunt Sabuncu[1], and Torsten Schaub[1,3]($\boxtimes$)

[1] University of Potsdam, Potsdam, Germany
torsten@cs.uni-potsdam.de
[2] University of New South Wales, Sydney, Australia
[3] INRIA Rennes, Rennes, France

Abstract. Knowledge representation and reasoning capacities are vital to cognitive robotics because they provide higher level functionalities for reasoning about actions, environments, goals, perception, etc. Although Answer Set Programming (ASP) is well suited for modelling such functions, there was so far no seamless way to use ASP in a robotic setting. We address this shortcoming and show how a recently developed ASP system can be harnessed to provide appropriate reasoning capacities within a robotic system. To be more precise, we furnish a package integrating the new version of the ASP solver *clingo* with the popular open-source robotic middleware Robot Operating System (ROS). The resulting system, *ROSoClingo*, provides a generic way by which an ASP program can be used to control the behaviour of a robot and to respond to the results of the robot's actions.

1 Introduction

Knowledge representation and reasoning capacities are vital to cognitive robotics because they provide higher level functionalities for reasoning about actions, environments, goals, perception, etc. While Answer Set Programming (ASP) is well suited for modelling high level functionalities, there was so far no seamless way to use ASP in a robotic setting. This is because ASP solvers were designed as one-shot problem solvers and thus lacked any reactive capabilities. So, for instance, each time new information arrived, the solving process had to be restarted from scratch.

In this paper, we address such shortcomings and show how a recently developed (multi-shot) ASP system [1] can be harnessed to provide knowledge representation and reasoning capabilities within a robotic system. We accomplish this by integrating a multi-shot ASP approach, where online information can be incorporated into an operative ASP solving process, into the popular open-source middleware ROS[1] (Robot Operating System; [2]).

T. Schaub—Affiliated with Simon Fraser University, Canada, and IIIS Griffith University, Australia.

[1] http://www.ros.org

© Springer International Publishing Switzerland 2015
F. Calimeri et al. (Eds.): LPNMR 2015, LNAI 9345, pp. 69–82, 2015.
DOI: 10.1007/978-3-319-23264-5_7

To be more precise, we furnish a ROS package integrating the ASP solver *clingo* 4 with the popular open-source ROS robotic middleware. The resulting system, called *ROSoClingo*, provides a generic method by which an ASP program can be used to control the behaviour of a robot and to respond to the results of the robot's actions. In this way, the *ROSoClingo* package plays the central role of fulfilling the need for high-level knowledge representation and reasoning in cognitive robotics by making details of integrating a reasoning framework within a ROS based system transparent to developers. As we detail below, the robotics developer can encode high-level planning tasks in ASP keeping only the interface requirements of the underlying behaviour nodes in mind and avoiding implementation details of their functionality (motion planning for example).

One crucial added value of our integration of reactive ASP framework into ROS is the facility of encoding adaptive behaviours directly in a declarative knowledge representation formalism. Additionally, the robot programmer can handle execution failures directly in the reasoning formalism. This paves the way for deducing new knowledge about the environment or diagnostic reasoning in the light of execution failures. The case study in Sect. 4 demonstrates these advantages of *ROSoClingo*.

Finally, it is worth mentioning a number of related approaches which utilize ASP or other declarative formalisms in cognitive robotics. In the work of [3,4] ASP is used for representing knowledge via a natural language based human robot interface. Additionally, action language formalisms and ASP have been used to plan and coordinate multiple robots for fulfilling an overall task [5,6]. ASP has also been used to integrate task and motion planning via external calls from action formalism to geometric reasoning modules [7]. However, all these implementations rely on one-shot ASP solvers and thus lack any reactive capabilities. Hence, they could greatly benefit from the reactive solving that comes from the usage of *ROSoClingo*.

In what follows, we provide the architecture and basic functionality of the *ROSoClingo* system. We then outline the ASP encoding for an example mail delivery robot. This example serves to highlight the features of the system but also serves as a guide for how an ASP encoding could be written for other application domains. The operations of the mail delivery robot are illustrated via a case-study conducted within a 3D simulation environment.[2] The features of *ROSoClingo* are discussed with reference to this case study and through comparisons to alternative approaches. Finally, it should be mentioned that the *ROSoClingo* system is publicly available [13] and we are committed to submitting the *ROSoClingo* package to the public ROS repository.

2   *ROSoClingo*

In this section, we describe the general architecture and functionality of the *ROSoClingo* system. With the help of the reactive ASP solver *clingo* (version 4), *ROSoClingo* provides high-level knowledge representation and reasoning

---

[2] http://gazebosim.org.

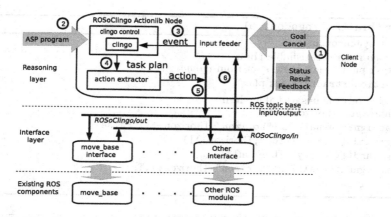

Fig. 1. The general architecture and main work flow of *ROSoClingo*.

capabilities to ROS based autonomous robots. Critically, *clingo* supports multi-shot reactive solving, where the solver does not simply terminate after an initial answer set computation, but instead enters a loop, incrementally incorporating new information into the solving process. For more extensive background to both *clingo* 4 and ROS the interested reader is referred to an extended version of this paper [8].

Figure 1 depicts the main components and workflow of the *ROSoClingo* system. It consists of a three layered architecture. The first layer consists of the core *ROSoClingo* component and the instantiation of a ROS *actionlib* API. In essence, this API simply exposes the services provided by *ROSoClingo* for use by other processes (i.e., ROS *nodes*). The package also defines the message structure for communication between the core *ROSoClingo* node and the various nodes of the interface layer. In contrast to the reasoning layer, the interface layer provides the data translations between what is required by the *ROSoClingo* node and any ROS components for which it needs to integrate. This architecture provides for a clean separation of duties, with the well-defined abstract reasoning tasks handled by the core node and the integration details handled by the interface nodes.

2.1 The *ROSoClingo* Core

The main *ROSoClingo* node is composed of a python module for the answer set solver *clingo* controlled by `clingoControl`, an `actionExtractor`, and an `inputFeeder`. Through its ROS *actionlib* API, it can receive goal and cancellation requests as well as send result, feedback, and status information back to a client node (marked by 1 in Fig. 1). The ASP program, encoding the high-level task planning problem, is given to the *ROSoClingo* node at system initialization (marked by 2). During initialization, *ROSoClingo* grounds the base subprogram of the ASP encoding and sets the current logical time point as well as the current horizon to 0. The logical time point identifies which actions of a task plan are to be executed next, while the horizon identifies the length of the task plan. The time point is incremented at the end of each cycle.

Algorithm 1. `clingoControl`

`solveAsynchronous`
if *clingo* returns satisfiable **then**
  | task plan ← get answer set from *clingo*
  |_ `actionExtractor`(task plan)

if *clingo* returns unsatisfiable **then**
  | horizon ← horizon + 1
  | `assignExternal(Fun("horizon",[horizon-1]),False)`
  | `ground([("transition",[horizon])])`
  | `ground([("query",[horizon])])`
  | `assignExternal(Fun("horizon",[horizon]),True)`
  |_ `clingoControl`

`occurs(Robot,Action,T)`	**Out**	Commanding the robot `Robot` to execute `Action` at time point T.
`event(Source,Event,T)`	**In**	Specifying an event `Event` from `Source` at time point T.
`event(request,(ID,Request),T)`	**In**	A special event, specifying a `Request` with id `ID` at time point T.
`event(request,(ID,cancel),T)`	**In**	A special event, canceling the request with id `ID` at time point T.

Fig. 2. Keywords used for communicating between *ROSoClingo* and *clingo*.

ROSoClingo's workflow starts with a goal arriving at the `inputFeeder` (marked by 1). If *clingo* is already in the process of searching for a task plan, the solving procedure is interrupted and the new goal is added to the solver. The goal request is transformed into an ASP fact and transmitted to *clingo* (marked by 3). Then `clingoControl` is called to resume the solving process with the additional goal.

Algorithm 1 presents the pseudo code representation of the `clingoControl` procedure. The *clingo* functions `assignExternal` as well as `ground` are explained in more detail in Sect. 3. It instructs the *clingo* solver to asynchronously find a task plan that satisfies all given goals. If *clingo* is able to find a valid task plan then the solution is forwarded to the `actionExtractor`. If no task plan is found for the current horizon, the horizon is incremented by one time step. This is realized by assigning `False` to the external atom that identifies the old horizon, followed by the grounding of the transition and query subprograms for the new horizon, and finally, the assignment of `True` to the external atom that identifies this new horizon. Note that the keyword `Fun` represents *clingo*'s data type for function terms, here applied to the external horizon atoms. Finally, `clingoControl` is called again to find a task plan with the new horizon. If an interrupt occurs, the solving process is stopped without *clingo* determining the (un-)satisfiability of the current program and `clingoControl` ends.

The `actionExtractor` identifies actions to be executed during the current logical time point and transforms them into *ROSoClingo* output messages (marked by 4). These messages are then transmitted via the `/ROSoClingo/out` *topic*[3] (marked by 5). It is then the task of the interface layer nodes to trans-

---

[3] *Topics* are a named publisher-subscriber communications mechanism for message passing between ROS nodes.

form them into goal requests for the underlying *actionlibs* and to compose a response once the action is executed. The response arrives at the `inputFeeder` component of the *ROSoClingo* node via the `/ROSoClingo/in` topic (marked by 6). The details of how the *ROSoClingo* interface layer interacts with existing ROS components are outlined in Sect. 2.2.

In contrast to goal requests, messages arriving at the `inputFeeder` component via `/ROSoClingo/out` are transformed into `event` predicates and then incorporated into the existing ASP program as external facts and processed by *clingo*. The keywords of Fig. 2 encode the protocol for this (internal) communication between *ROSoClingo* and *clingo*. The second column indicates whether the keyword is an input (in) or part of the output (out) of *clingo*. The (un)successful result of an action may generate new knowledge for the robot about the world (for example, the fact that a doorway is blocked or a new object is sensed).

Once all actions of the current time point report a result the cycle is completed and a new one is initiated, provided there are still actions left to be executed in the task plan. If the task plan is completed *ROSoClingo* waits for new goal requests to be issued.

Finally, it is worthwhile noting that the *ROSoClingo* package is able to support multiple goal requests at a time.

2.2   Integrating with Existing ROS Components

The core *ROSoClingo* node needs to issue commands to, and receive feedback from, existing ROS components. The complexity of this interaction is handled by the nodes at the interface layer (cf. Fig. 1). Unlike the components of the reasoning layer it is, unfortunately, not possible to define a single ROS interface to capture all interactions that may need to take place. Firstly, there is a need for data type conversions between the individual modules. Turning ROS messages into a suitable set of *clingo* statements therefore requires data type conversions that are specific for each action or service type.

A second complicating issue is that the level of abstraction of a ROS action may not be at the appropriate level required by the ASP program. For example, the *pose* goal for moving a robot consists of a Cartesian coordinate and orientation. However, reasoning about Cartesian coordinates may not be desirable when navigating between named locations such as corridors, rooms and offices. Instead one would hope to reason abstractly about these locations and the relationship between them; for example that the robot should navigate from the kitchen to the bedroom via the hallway.

While it is not possible to provide a single generic interface to all ROS components, it is however possible to outline a common pattern for such integration. For each existing component that needs to be integrated with *ROSoClingo* there must be a corresponding interface component. We therefore adopt a straightforward message type for messages sent by *ROSoClingo*. This type consists of an assigned name for the robot performing the action and the action to be executed. Note, the addition of robot names allows for the coordination of multiple robots, or multiple robot components, within a single ASP program and to identify the actions performed by each robot or component.

In a similar manner to the *ROSoClingo* output messages, the input messages also consist of a straightforward message type. These messages allow for an interface node to either respond with the success or failure of a *ROSoClingo* action, or alternatively to signal the result of some external or sensory input.

In the scope of the work presented in this paper we implemented an interface to the ROS move_base *actionlib*, a standard ROS component for driving a robot. The interface maps symbolic locations with specific coordinates in the environment, e.g. kitchen to (12.40,34.56,0.00), and vice versa. The interface node then tracks the navigation task and reports back to the *ROSoClingo* core the success or failure of its task.

3   ASP-based Task Planning in *ROSoClingo*

The methodology of *ROSoClingo*'s ASP-based approach to task planning is composed of two main activities, viz. *formalizing the dynamic domain* and *formalizing the task as a planning problem* in this domain. Each activity involves representing different types of knowledge related to the problem.

The basic principles of this methodology are similar to the general guidelines of representing dynamic domains and solving planning problems in ASP (either it is a direct ASP encoding [9] or an implementation of an action language via ASP [10]). However, since *ROSoClingo* relies upon the multi-shot solving capacities of the *clingo* 4 ASP system [1], the resulting encoding should meet the requirements of the incremental setting, where the whole program is structured as parametrizable subprograms. Multi-shot ASP solving is concerned with grounding and the integration of subprograms into the solving process, and is fully controllable from the procedural side, viz. the scripting language Python in our case. In explaining this process, we first concentrate on the methodology of representing various types of knowledge and later explain the way this knowledge is partitioned into subprograms.

For illustrating the methodology, consider the ASP encoding of a simplified mail delivery scenario, offering a well-known exemplary illustration of action formalisms in robotics [11,12]: A robot is given the task of picking up and delivering mail packages between offices. Whenever a mail delivery request is received, the robot has to navigate to the office requesting the delivery, pick up the mail package, and then navigate and deliver the item at the destination office. In addition, cancellation requests may happen. If the robot has already picked up the package, it must then return the package to the originating office. Additionally, some of the pathways in the environment may be blocked for some time.

We formalize the dynamic domain by representing the following types of knowledge. Due to space constraints, we provide only representative ASP snippets. One can find the full encoding at [13].

Static Knowledge. Time-independent parts of the domain constitute the static knowledge. In view of Sect. 4, we assume a world instance from the Willow

Garage office map and encode this map related information as static knowledge. The following is a snippet from the logic program declaring nodes of waypoints, which are composed of offices, corridors, and open areas, and connections among waypoints.

```
corridor(c1).  corridor(c2).  open(open1).  office(o4).
connection(c3,o4).  connection(c1,open1).  connection(c1,c2).

connection(X,Y) :- connection(Y,X).  waypoint(X) :- corridor(X).
waypoint(X) :- open(X).              waypoint(X) :- office(X).
```

In contrast to static knowledge, dynamic knowledge is time-dependent. In the following program snippets we use the parameter t to represent a time point. It is also used as an argument when declaring *clingo* 4's parameterizable subprograms (such as `#program transition(t)`). *ROSoClingo*'s control module incrementally grounds and integrates such programs with increasing integer values for t. For instance, the call `ground([("transition",[42])])` grounds the transition subprogram for planning horizon 42.

In order to specify a state of a dynamic domain, fluents (i.e., properties that change over time) are used. A state associated with a time point t is characterized by the fluents captured by atoms of the form `holds(F,t)` where F is an instance of a fluent. Figure 3 lists not only the fluents, but also the actions and exogenous events of the domain. While actions are performed by the robot, events may occur in the dynamic domain without the control of the robot. Actions and events occur within a state of the world and lead to some resulting state. We use the meta-predicates `occurs(A,t)` and `event(E,t)` for stating the occurrence of action A and event E respectively at time point t. We use the following choice rule to allow any action (extensions of `action` predicate includes all actions of the domain) to occur at time point t. The upper bound 1 concisely expresses that no concurrent task plans are permitted.

fluents	`at(W)`	the robot is at waypoint W
	`holding(O,P)`	the robot is holding the package (O,P)
	`received(request(O,P))`	the robot has received a delivery or cancellation
	`received(cancel(O,P))`	request for a package (O,P)
	`blocked(W,W')`	the path between waypoints W and W' is blocked
actions	`go(W)`	go to the waypoint W
	`pickup(O,P)`	pickup the package (O,P)
	`deliver(O,P)`	deliver the package (O,P)
events	`request(O,P)`	occurs on a request to delivering a package from office O to P
	`cancel(O,P)`	occurs whenever the request to delivering a package from office O to office P is cancelled
	`info(blocked(W,W'))` `info(unblocked(W,W'))`	occurs whenever a path between W and W' is blocked or unblocked
	`value(failure)`	occurs whenever an execution fails

Fig. 3. Fluents, actions, and events used to formalize the domain

```
{ occurs(A,t) : action(A) } 1.
```

Within the fluents, actions, or events of the domain, we identify each mail package delivery with the pair (O,P) consisting of its origin O and destination P. Although this leads to a simpler encoding, it does limit us to a single delivery from O to P at a time.

A crucial role in modeling exogenous events is played by *clingo*'s external directives [1]. An #external directive allows for, as yet, undefined atoms. To signal external events to the solver, *ROSoClingo* relies upon *clingo*'s library function assignExternal that allows for manipulating the truth values of external atoms. For instance, the following rules show how the goal request (based on the signature given in Fig. 2) is declared as an external atom and projected into exogenous event request(O,P).

```
#external event(request,(ID,bring(O,P)),t) :- office(O;P), id(ID).
event(request(O,P),t) :- event(request,(ID,bring(O,P)),t).
```

Recall that the first element of the occurs(Robot,Action,T) atom (Fig. 2) allows for reasoning with concurrent task plans for multi-robot scenarios or for robots with multiple actuators. However, we use occurs(A,t) in our case study, since we generate non-concurrent task plans for a single robot. The following rule adds the actuator name.

```
occurs(mailbot,A,t) :- occurs(A,t).
```

Static Causal Laws. This type of knowledge defines static relations among fluents. They play a role in representing indirect effects of actions. The following rule represents that blocked is symmetric and shows how one true blocked fluent can cause another blocked fluent to be true in a state.

```
holds(blocked(W,W'),t) :- holds(blocked(W',W),t).
```

Dynamic Causal Laws. Direct effects of actions and events are specified by dynamic causal laws. An action or event occurrence at time t can make its effect fluent hold at t. Additionally, the occurrence may cancel the perpetuation of fluents. To this end, we use atoms of the form abnormal(F,t) to express that fluent F must not persist to time point t. In robotics, however, action execution failures may occur. Whenever an underlying ROS node fails to perform an action, *ROSoClingo* triggers the value(failure) event to signal the execution failure to the encoding. We use atom executes(A,t) to decouple the occurrence of action A from its effects taking place.

```
executes(A,t) :- occurs(A,t), not event(value(failure),t).
```

This provides us with a concise way of blocking imaginary action effects and thus avoids inconsistencies between the actual world state and the robot's world view. Below are dynamic causal laws for action go(W) and event cancel(O,P).

```
holds(at(W),t)         :- executes(go(W),t).
abnormal(at(W'),t) :- executes(go(W),t), holds(at(W'),t-1).
holds(received(cancel(O,P)),t)      :- event(cancel(O,P),t).
abnormal(received(request(O,P)),t) :- event(cancel(O,P),t).
```

In addition, ASP's default reasoning capabilities, together with explicit executes and occurs statements, pave the way for reasoning with execution failures. For instance, the following rule enables the robot to conclude that the connection to a waypoint is blocked whenever the attempt to navigate to that waypoint fails. (See the third scenario in Sect. 4 for an illustration.)

```
holds(blocked(W',W),t) :- occurs(go(W),t), not executes(go(W),t),
                          holds(at(W'),t-1).
```

Action Preconditions. Action preconditions provide the executability conditions of an action in a state. We use atom poss(A,t) to state that action A is possible at t. Below are preconditions of action go(W). The integrity constraint makes sure that only actions take place whose preconditions are satisfied.

```
poss(go(W),t) :- holds(at(W'),t-1), connection(W',W),
                 not holds(blocked(W',W),t-1).
:- occurs(A,t), not poss(A,t).
```

Inertia. The following rule is a concise representation of the frame axiom.

```
holds(F,t) :- holds(F,t-1), not abnormal(F,t).
```

This completes the formalization of the dynamic domain. Next, we formalize the robot's task as a planning problem.

Initial Situation. The following rules represent the initial situation by stating the initial position of the robot.

```
init(at(open3)).
holds(F,0) :- init(F).
```

Goal Condition. The following snippet expresses the goal condition. This is the case whenever the robot has no pending delivery request and is not holding any package.

```
goal(t) :- not holds(received(request(_,_)),t),
           not holds(holding(_,_),t).
#external horizon(t).
:- not goal(t), horizon(t).
```

The integrity constraint makes the program unsatisfiable whenever the goal is not reached at the planning horizon. Clearly, this constraint must be removed whenever the horizon is incremented and a new instance with an incremented horizon is added. To this end, we take advantage of the external atom horizon(t) whose truth value can be controlled from *ROSoClingo* as shown in Algorithm 1. The manipulation of truth values of externals provides an easy mechanism to activate or deactivate ground rules on demand.

We have mentioned that *clingo* programs are structured into parametrizable subprograms. *ROSoClingo* relies on three subprograms, viz. base, transition(t), and query(t). The formalized knowledge is partitioned into these subprograms as follows: base contains the time-independent knowledge (static knowledge and initial situation), transition(t) contains the time-dependent knowledge (static and dynamic causal laws, action preconditions, and inertia), and finally query(t) contains the time-dependent volatile knowledge (goal condition). (See the full encoding at [13].)

4   Case Study

We now demonstrate the application of our *ROSoClingo* package in the mail delivery setting described in the previous section (Sect. 3). A robot is given the task of picking up and delivering mail packages between offices. Whenever a mail delivery request is received, the robot has to navigate to the office requesting the delivery, pick up the mail package, and then navigate and deliver the item at the destination office.

While the mailbot task is intrinsically dynamic in nature, a secondary source of dynamism is the external environment itself. Obstacles and obstructions are a natural part of a typical office environment, and it is in such cases that the need for high-level reasoning becomes apparent. Our scenario not only highlights the operations of a mail delivery robot in responding to new requests but also shows how such a robot can respond to a changing physical environment.

The office scenario is provided in simulation by the *Gazebo* 3D simulator using an openly accessible world model available for the Willow Garage[4] offices. The robot is a *TurtleBot* equipped with a Microsoft Kinect 3D scanner, which is a cost-effective and well supported robot suitable for small delivery tasks.

From the office environment a partial map has been generated using standard mapping software [14]. This static map is then used as the basis for navigation and robot localization. Furthermore, from this map a topological graph has been constructed to identify individual offices and waypoints that serve as a graph representation for logical reasoning and planning. While this graph has been hand-coded, topological graphs can also be generated through the use of automated techniques [15].

As previously outlined, *ROSoClingo* provides a simple mechanism for integration with other ROS components, including basic navigation. We further allow for external messages that can be sent to the robot informing it of paths that have been blocked and cleared. In an office environment this can correspond to public announcements, such as work being undertaken in a particular area. Such external messages can also be viewed in the context of the robot receiving additional sensor data.

Finally, as our robot was not equipped with a robot manipulator, item pickup and delivery functionality was simulated by a *ROSoClingo* interface that simply responds successfully to `pickup` and `deliver` action requests.

Scenarios. We consider three scenarios to highlight the behaviour of the mail-delivery robot when it detects and is informed of paths that have been blocked and cleared. In all three scenarios,[5] the robot is initially in the open area shown in Fig. 4.

In the first scenario, the robot is told that the corridor is blocked between points $C3$ and $C4$. It is then told to pick up an item from office $O9$ and deliver it to office $O14$. As *ROSoClingo* is able to plan at an abstract level it is able to know that it can move to $O9$ along the optimal route (i.e., via $C6$) but must return

---

[4] http://www.willowgarage.com.
[5] The videos of these scenarios are available at http://goo.gl/g8S5Ky.

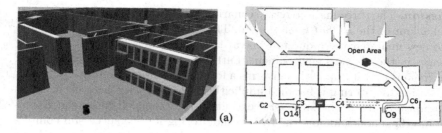

Fig. 4. (a) An office environment for a mail delivery robot, and (b) scenario showing delivery from $O9$ to $O14$, with a blockage dynamically appearing in the corridor between $C3$ and $C4$.

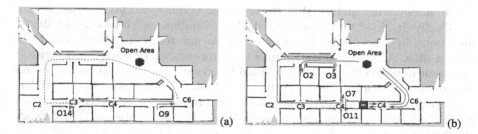

Fig. 5. (a) Scenario showing an obstruction being cleared allowing re-planning for a shorter path through $C4$ and $C3$, and (b) scenario showing adaptive behaviour where changes in the physical environment can affect the order in which tasks are performed.

through the open area and travel via the corridor point $C2$ in order to reach its destination $O14$. This path is indicated by the solid blue line in Fig. 4(b).

The second scenario (Fig. 5(a)) extends that of the first. From $O9$ the robot knows that the path between $C3$ and $C4$ is blocked so it starts to take the long way around as before. However, by the time it reaches $C6$ it has been informed that the blockage has been cleared. This triggers re-planning at the *ROSoClingo* level and the robot is turned around and the shorter path taken through $C4$ and $C3$ to the destination $O14$.

Finally the third scenario shows how dynamic changes to the physical environment can affect the order in which tasks are performed. In this scenario (Fig. 5(b)) the robot is first given a task to deliver an item from the office $O7$ to $O11$. While in the vicinity of $C6$ the robot is given a second task to take an item from $O2$ to $O3$. Since it reasons that it is already close to $O7$ the robot continues on with its first delivery task. However, as it progresses past $C4$ the robot detects that the path between $C4$ and $C5$ is blocked. Consequently, the robot has to turn around and take the longer route through the open area. But now offices $O2$ and $O3$ are closer to the robot than $O7$ and $O11$. This causes a change in the robot's task priorities and it swaps the order of tasks, performing the second delivery task first before continuing on with the original.

Discussion. The three mail-delivery scenarios outlined here showcase the adaptive behaviour of the *ROSoClingo* system. The robot is able to respond dynamically to new mail delivery requests while at the same time adapting intelligently to changes in the physical environment. Furthermore, an important property of *ROSoClingo* is that it implicitly performs a form of execution monitoring [16, 17].

Execution monitoring is handled implicitly by *ROSoClingo* because it makes no assumptions about the successful execution of actions. Rather, the *ROSoClingo* interface nodes handle the task of monitoring for the successful completion of actions. This information is then reported back to the reasoner and any failures are handled appropriately.

In fact, because execution monitoring is incorporated directly into the ASP reasoner, *ROSoClingo* can provide for much finer control than is allowed for by traditional systems such as [16]. In particular because execution monitors are specifically designed to deal with anomalous situations, such as action failures, they typically ignore external events that do not result in the failure of the current plan. At first glance, this may seem reasonable. However, in practice it can result in unintuitive and sub-optimal behaviour. For example, in the second mail delivery scenario (Fig. 5(a)) the robot replans on the announcement that a blockage has been cleared. Importantly, this re-planning is not triggered as a result of a failure of the current plan, but instead as a recognition of the existence of a better plan. In contrast, because the longer plan is still valid, a traditional execution monitoring based robot would ignore the positive information that the blockage has been cleared and the robot would simply follow the longer route.

Because of *ROSoClingo*'s ability to immediately adapt to new information it bears some resemblance to the Teleo-Reactive programming paradigm of [18]. This goal directed approach to reactive systems is based on *guarded* action rules which are being constantly monitored and triggered based on the satisfaction of rule conditions. However, while Teleo-Reactive systems can provide for highly dynamic behaviour, they typically do not incorporate the complex planning and reasoning functionality of traditional action languages. Hence, in the same way that action language formalisms are rarely applied to highly reactive problem domains, these reactive approaches are rarely applied in problems that require complex reasoning and planning.

However, in contrast to the dichotomy suggested by the difference between these two approaches, many practical real-world cognitive robotic problems do require both highly reactive behaviour and complex action planning. This is highlighted by our mail delivery scenarios where the robot has to undertake its mail deliver tasks while still operating in a dynamically changing physical environment. The successful application of *ROSoClingo* to this task shows that it can be seen as a step towards bridging these two approaches. A robot that incorporates complex reasoning and planning can at the same time adapt to a highly dynamic external environment.

5 Conclusion

We have developed a ROS package integrating *clingo* 4, an ASP solver featuring reactive reasoning, and the robotics middleware ROS. The resulting system, called *ROSoClingo*, fulfils the need for high-level knowledge representation and reasoning in cognitive robotics by providing a highly expressive and capable reasoning framework. *ROSoClingo* also makes details of integrating the ASP solver transparent for the developer, as it removes the need to deal with the mechanics of communicating between the solver and external (ROS) components.

Using reactive ASP and *ROSoClingo*, one can control the behaviour of a robot within a single framework in a fully declarative manner. This is particularly important when contrasted against Golog [11] based approaches where the developer must take care of the implementation (usually in Prolog) details of the control knowledge, and the underlying action formalism separately. We illustrated the usage of *ROSoClingo* via a three-fold case-study conducted with a ROS-based simulation of a robot delivering mail packages in the Willow Garage office environment using the *Gazebo* 3D simulator. We showed that ASP based robot control via *ROSoClingo* establishes a principled way of achieving adaptive behaviour in a highly dynamic environment.

This work on *ROSoClingo* opens up a number of avenues for future research. Here we concentrated on the use of *ROSoClingo* for high-level task planning. However *clingo* is a general reasoning tool with applications that extend to other areas of knowledge representation and reasoning such as diagnosis and hypothesis formation. Consequently, an important area for future research would be to consider the use of *ROSoClingo* in these contexts, such as a robot that makes and reasons about the causes of observations in its environment. Another line of future research is to utilize *clingo*'s optimization statements to find optimal task plans when costs of actions are not uniform [19].

Acknowledgments. This work was funded by ARC (DP150103034) and DFG (550/9).

References

1. Gebser, M., Kaminski, R., Kaufmann, B., Schaub, T.: Clingo = ASP + control: preliminary report. In: Leuschel, M., Schrijvers, T. (eds.) Technical Communications of the Thirtieth International Conference on Logic Programming (ICLP 2014). Theory and Practice of Logic Programming, Online Supplement (2014). http://arxiv.org/abs/1405.3694v1
2. Quigley, M., Gerkey, B., Conley, K., Faust, J., Foote, T., Leibs, J., Berger, E., Wheeler, R., Ng, A.: ROS: an open-source robot operating system. In: ICRA Workshop on OSS (2009)
3. Chen, X., Jiang, J., Ji, J., Jin, G., Wang, F.: Integrating NLP with reasoning about actions for autonomous agents communicating with humans. In: Proceedings of the IEEE/WIC/ACM International Conference on Intelligent Agent Technology (IAT 2009), pp. 137–140. IEEE (2009)
4. Chen, X., Ji, J., Jiang, J., Jin, G., Wang, F., Xie, J.: Developing high-level cognitive functions for service robots. In: van der Hoek, W., Kaminka, G., Lespérance, Y., Luck, M., Sen, S. (eds.) Proceedings of the Ninth International Conference on Autonomous Agents and Multiagent Systems (AAMAS 2010), pp. 989–996. IFAAMAS (2010)

5. Aker, E., Erdogan, A., Erdem, E., Patoglu, V.: Causal reasoning for planning and coordination of multiple housekeeping robots. In: Delgrande, J.P., Faber, W. (eds.) LPNMR 2011. LNCS, vol. 6645, pp. 311–316. Springer, Heidelberg (2011)
6. Erdem, E., Aker, E., Patoglu, V.: Answer set programming for collaborative house-keeping robotics: representation, reasoning, and execution. Intel. Serv. Robot. **5**(4), 275–291 (2012)
7. Erdem, E., Haspalamutgil, K., Palaz, C., Patoglu, V., Uras, T.: Combining high-level causal reasoning with low-level geometric reasoning and motion planning for robotic manipulation. In: Proceedings of the IEEE International Conference on Robotics and Automation (ICRA 2011), pp. 4575–4581. IEEE (2011)
8. Andres, B., Rajaratnam, D., Sabuncu, O., Schaub, T.: Integrating ASP into ROS for reasoning in robots: Extended version. Unpublished draft (2015). Available at [13]
9. Gelfond, M., Kahl, Y.: Knowledge Representation, Reasoning, and the Design of Intelligent Agents: The Answer-Set Programming Approach. Cambridge University Press, Cambridge (2014)
10. Baral, C., Gelfond, M.: Reasoning agents in dynamic domains. In: Minker, J. (ed.) Logic-Based Artificial Intelligence, pp. 257–279. Kluwer Academic, Dordrecht (2000)
11. Levesque, H., Reiter, R., Lespérance, Y., Lin, F., Scherl, R.B.: GOLOG: a logic programming language for dynamic domains. J. Logic Program. **31**(1–3), 59–83 (1997)
12. Thielscher, M.: Logic-based agents and the frame problem: a case for progression. In: Hendricks, V. (ed.) First-Order Logic Revisited: Proceedings of the Conference 75 Years of First Order Logic (FOL75), pp. 323–336. Logos, Berlin (2004)
13. Potassco website. http://potassco.sourceforge.net
14. Grisetti, G., Stachniss, C., Burgard, W.: Improved techniques for grid mapping with rao-blackwellized particle filters. IEEE Trans. Robot. **23**(1), 34–46 (2007)
15. Thrun, S., Bücken, A.: Integrating grid-based and topological maps for mobile robot navigation. In: Clancey, W., Weld, D. (eds.) Proceedings of the Thirteenth National Conference on Artificial Intelligence (AAAI 1996), pp. 944–950. AAAI/MIT Press, Portland (1996)
16. De Giacomo, G., Reiter, R., Soutchanski, M.: Execution monitoring of high-level robot programs. In: Cohn, A., Schubert, L., Shapiro, S. (eds.) Proceedings of the Sixth International Conference on Principles of Knowledge Representation and Reasoning (KR 1998), pp. 453–465. Morgan Kaufmann, Trento (1998)
17. Pettersson, O.: Execution monitoring in robotics: a survey. Robot. Autonom. Syst. **53**(2), 73–88 (2005)
18. Nilsson, N.: Teleo-reactive programs for agent control. J. Artif. Intell. Res. **1**, 139–158 (1994)
19. Khandelwal, P., Yang, F., Leonetti, M., Lifschitz, V., Stone, P.: Planning in action language BC while learning action costs for mobile robots. In: International Conference on Automated Planning and Scheduling (ICAPS) (2014)

Automated Inference of Rules with Exception from Past Legal Cases Using ASP

Duangtida Athakravi[1]([✉]), Ken Satoh[2], Mark Law[1],
Krysia Broda[1], and Alessandra Russo[1]

[1] Imperial College London, London, UK
{duangtida.athakravi07,mark.law09,k.broda,a.russo}@imperial.ac.uk
[2] National Institute of Informatics and Sokendai, Tokyo, Japan
ksatoh@nii.ac.jp

Abstract. In legal reasoning, different assumptions are often considered when reaching a final verdict and judgement outcomes strictly depend on these assumptions. In this paper, we propose an approach for generating a declarative model of judgements from past legal cases, that expresses a legal reasoning structure in terms of principle rules and exceptions. Using a logic-based reasoning technique, we are able to identify from given past cases different underlying defaults (legal assumptions) and compute judgements that cover all possible cases (including past cases) within a given set of relevant factors. The extracted declarative model of judgements can then be used to make deterministic automated inference on future judgements, as well as generate explanations of legal decisions.

1 Introduction

In legal reasoning, especially in continental laws, we use written rules to make a judgement in litigation. In these written rules, principle rules and exceptions are mentioned. It is related to proof of persuasion where conditions of principle rules must be proven by the side who claims holding the conclusion of principle rules, whereas exceptions must be proven by the side who denies the conclusion. Moreover, some rules are refined by the highest court in a country (the supreme court in Japan, for example) by adding some exceptions if the current principle rules or exceptions do not capture the conclusion of the current litigation.

In this paper, we consider the problem of generating a set of case-rules from previous cases judged by the court. Suppose that the following cases are found in the conclusion of *"depriving the other party of what he (or she) is entitled to expect under the contract"* in commercial litigation.

Case 1: The plaintiff (the buyer) showed that the goods were delivered on time but he failed to prove that there was a damage of goods. In this case, the judge decided that the seller did not deprive the buyer of what he expects.

Case 2: The plaintiff showed that the goods were delivered on time but the goods were damaged. The defendant failed to prove that it is repairable. In this case, the judge decided that the buyer was deprived of what he expects.

© Springer International Publishing Switzerland 2015
F. Calimeri et al. (Eds.): LPNMR 2015, LNAI 9345, pp. 83–96, 2015.
DOI: 10.1007/978-3-319-23264-5_8

Case 3: The plaintiff showed that the goods were delivered on time but the goods were damaged. Then, the defendant showed that the damage could be repaired and the buyer fixed an additional period of time for repair and the repair was completed in the additional period. In this case, the judge decided that the seller did not deprive the buyer of what he expects.

Case 4: The plaintiff showed that the goods were not delivered on time. Then, the defendant showed that the buyer fixed an additional period of time for the delivery but failed to prove that the goods were delivered in the period. In this case, the judge decided that the buyer was deprived of what he expects.

Case 5: The plaintiff showed that the goods were not delivered on time. Then, the defendant (the seller) showed that the buyer fixed an additional period of time and the goods were delivered in the period. In this case, the judge decided that the seller did not deprive the buyer of what he expects.

We would like to decide whether the following case satisfies the conclusion "depriving the other party of what he is entitled to expect under the contract":

New Case: *The plaintiff showed that the goods were delivered on time but were damaged. The defendant showed that it could be repaired and showed that the buyer fixed an additional period of time for repair, but failed to prove that the repair was not completed in the additional period.*

We can formalise all the factors mentioned in cases 1–5 as described below.

dot	The goods are delivered on time
fad	The buyer fixes an additional period for delivering the goods
dia	The goods are delivered in the above additional period
ooo	The goods are damaged
rpl	The goods are repairable
far	The buyer fixes an additional period for repair
ria	The goods are repaired in the above additional period

Past cases can then be expressed as pairs where the first argument denotes the factors mentioned in the case and the second argument the positive ($+dwe$) or negative ($-dwe$) conclusion "depriving the other party of what he is entitled to expect under the contract". We would like to deterministically infer what the conclusion would be for the new case, given the past cases:

Case 1	$\langle\{dot\}, -dwe\rangle$	**Case 4**	$\langle\{\neg dot, fad\}, +dwe\rangle$
Case 2	$\langle\{dot, ooo\}, +dwe\rangle$	**Case 5**	$\langle\{\neg dot, fad, dia\}, -dwe\rangle$
Case 3	$\langle\{dot, ooo, rpl, far, ria\}, -dwe\rangle$		
New Case	$\langle\{dot, ooo, rpl, far\}, ??\rangle$		

To make a judgement for a new case, the history of judgement revisions in past cases needs to be taken into account. In our example let us assume that if no factor is proven, the conclusion dwe will not be approved by judges ($-dwe$ is a conclusion in an empty case - a case with an empty set of factors). Then, Case 1 has the same conclusion as the initial empty case, so the factor dot is irrelevant to the judges. For Case 2 the conclusion (judgement) is reversed. Thus the combination of *dot* and *ooo* does affect the conclusion and therefore their simultaneous existence is an exceptional situation. For Case 3 the conclusion is reversed again into the original judgement. Then, we could say that the combination of *rpl*, *far*, *ria* is an exception to the exceptional situation in which *dot* and *ooo* exist. For Case 4, the conclusion differs from the initial empty case so the combination of ¬*dot* and *fad* represents an exceptional situation. Lastly for Case 5 the conclusion is reversed again into the original judgement. Therefore, *dia* constitutes an exception for exceptional situation in which ¬*dot* and *fad* exist. Finally, for the new case, given the factors {*dot*, *ooo*, *rpl*, *far*}, we should conclude +*dwe*, since *dot* and *ooo* hold in this case (and Case 2 was considered to be an exceptional situation with these factors), and it cannot be considered to be an exception for this exception since the only such exception known is the combination of *rpl*, *far*, *ria*, and in the new case *ria* does not hold.

In this paper, we aim to formalise the above reasoning and extract a set of rules that capture the past judgements and allow the automated inference of judgements for new cases. We view the cases and their judgements as outcomes of reasoning using an argumentation model [4]. Hence, attacks can be inferred from past cases that share some common factors but have opposing judgements, and consequently arguments are inferred from the factors that are uncommon between the two cases. Then, the purpose of relevant attacks is to identify only the necessary factors that affect the judgement of a new case. Consider the above cases. Case 1 does not contain any relevant information for deciding the outcome of future cases. Similarly, suppose Case 6 exists where the buyer was deprived of what he expects as the goods are delivered on time, but are damaged and repairable ⟨{*dot*, *ooo*, *rpl*}, +*dwe*⟩. This Case 6 will also be deemed irrelevant, due to the existence of Case 2 which tells us that {*dot*, *ooo*} are already sufficient for overturning the judgement. Thus, by extracting information about the relevant attacks and their arguments we can define rules reflecting the reasoning applied by the judge and use this reasoning to infer judgments for unseen cases. A meta-level representation of these rules is generated using the ASP solver Clingo [5].

The paper is structured as follows. Section 2 provides the formal definition of our computational model and introduces the notions of relevant attack and predicted judgement. Section 3 shows how the approach is implemented in Answer Set Programming (ASP) and illustrates its execution through the legal reasoning example described above. Sections 4 and 5 present, respectively, the correctness of the implementation and the evaluation of its scalability. Section 6 discusses related work and Sect. 7 concludes the paper.

2   Formalisation

In this section we give the formal definition of our computational model. We first define the formal representation of past cases and related judgements. Using this representation, we define the concepts of relevant attack and predicted judgements, and formalise the type of legal rules our approach is able to compute.

Definition 1 (Casebase). *Let F be a set of elements called* factors. *A case is a subset of F. A case with judgement is a pair $cj = \langle c, j \rangle$, where c is a case and $j \in \{+, -\}$. The set c is also referred to as the set of factors included in a case with judgement. Given a case with judgement cj, $case(cj)$ denotes the set of factors included in cj and $judgement(cj) = j$ denotes the judgement decision taken in the case. A* casebase, *denoted with CB, is a set of cases with judgements, namely a subset of $\mathcal{P}(F) \times \{+, -\}$, where $\mathcal{P}(F)$ is the powerset of F.*

Given a casebase CB we impose the restriction that for every $cj_1, cj_2 \in CB$, if $case(cj_1) = case(cj_2)$ then $judgement(cj_1) = judgement(cj_2)$. This avoids inconsistent casebases. We also assume that all casebases contain an element $\langle \emptyset, j_0 \rangle$ representing the empty case and *default judgement*, the assumed judgement in the absence of any factor, of the casebase.

Definition 2 (Raw Attack). *Let CB be a casebase. The* raw attack relation *is a set $RA \subseteq CB \times CB$ defined as the set of all pairs $\langle cj_1, cj_2 \rangle$ such that $\langle cj_1, cj_2 \rangle \in RA$ if and only if $case(cj_1) \supset case(cj_2)$ and $judgement(cj_1) \neq judgement(cj_2)$. For every pair $\langle cj_1, cj_2 \rangle \in RA$, we say cj_1 raw attacks cj_2 and we write $cj_1 \rightarrow_r cj_2$.*

Example 1. Let us consider the set of factors F given by $\{a, b, c, d, e, f\}$ and a casebase CB given by the named cases with judgements $\{c0 : \langle\{\}, -\rangle, c1 : \langle\{a\}, +\rangle, c2 : \langle\{c\}, +\rangle, c3 : \langle\{a, b\}, -\rangle, c4 : \langle\{a, b, c\}, +\rangle, c5 : \langle\{a, b, c, d\}, -\rangle\}$
Then, the raw attack relation over CB is given by the following set:
$$\{c1 \rightarrow_r c0, c2 \rightarrow_r c0, c4 \rightarrow_r c0, c3 \rightarrow_r c1, c5 \rightarrow_r c1, c5 \rightarrow_r c2, c4 \rightarrow_r c3, c5 \rightarrow_r c4\}.$$

Definition 3 (Relevant Attack). *Let CB be a casebase. The* relevant attack relation $AT \subseteq RA$ *is the set of pairs $\langle cj_1, cj_2 \rangle \in RA$ such that:*

- *$\langle cj_1, cj_2 \rangle \in AT$ if $case(cj_2) = \emptyset$ and there is no $cj_3 \rightarrow_r cj_2$ in RA such that $case(cj_1) \supset case(cj_3)$*
- *$\langle cj_1, cj_2 \rangle \in AT$ if there exists $\langle cj_2, cj_4 \rangle \in AT$ and there is no $cj_5 \rightarrow_r cj_2$ in RA such that $case(cj_1) \supset case(cj_5)$*
- *nothing else is in AT.*

Each element $\langle cj_1, cj_2 \rangle \in AT$ is denoted as $cj_1 \rightarrow cj_2$.

A relevant attack $cj_1 \rightarrow cj_2$ is between a case cj_1 that overturns the judgement of another case cj_2, with $case(cj_1) \supset case(cj_2)$, and such that either $judgement(cj_2) = j_0$, or cj_2 itself is an attacker in another relevant attack and there isn't another smaller attack against cj_2. Both scenarios imply that $case(cj_1)$ contains the relevant factors for overturning the judgement of cj_2. In summary, for a case to be relevant it must either be the default case, or it must be involved in a relevant attack against other relevant cases.

It can be observed from Definitions 2 and 3 that each casebase will have a unique set of raw attacks, and consequently a unique set of relevant attacks.

Example 2. Let us consider the set RA of raw attacks defined in Example 1, The set of relevant attacks is given by the following subset:
$$AT = \{c1 \rightarrow c0, c2 \rightarrow c0, c3 \rightarrow c1, c4 \rightarrow c3, c5 \rightarrow c4, c5 \rightarrow c2\}.$$

We can see that not all raw attacks are relevant attacks. Consider for instance the pair $\langle c5, c1 \rangle$. This is a raw attack but it is not a relevant attack as there exists a raw attack $c3 \rightarrow_r c1$ where $case(c3) \subset case(c5)$. For each relevant attack the set of factors, called *argument*, responsible for overturning the judgement can be deduced by comparing the factors in the two cases.

Definition 4 (Argument). *Let CB be a casebase and let AT be the set of relevant attacks with respect to CB. For each pair $\langle cj_1, cj_2 \rangle \in AT$, the set of factors representing the attack from cj_1 to cj_2 is given by $\alpha(cj_1, cj_2) = case(cj_1) - case(cj_2)$.*

When predicting the judgement of a new case, we consider past cases similar to it. Within the set of similar past cases, only some will be *active* with respect to the new case, meaning their ruling has not been overturned by other cases in the set. Furthermore, those that are overturned cannot themselves overturn the judgement of other cases. This is similar to the concept of alive and dead nodes in [14]. We define active case with judgement as follows.

Definition 5 (Active Case with Judgement). *Let CB be a casebase, AT be its corresponding set of relevant attacks, and c be a case. A case with judgement $cj \in CB$ is active with respect to c if and only if $case(cj) \subseteq c$, and for all $\langle cj_n, cj \rangle \in AT$, either $case(cj_n) \nsubseteq c$ or cj_n is not active with respect to c.*

We can infer the judgement of future cases from past cases' relevant attacks.

Definition 6 (Predicted judgement). *Let CB be a casebase, AT be the set of relevant attacks with respect to CB, and c be an unseen case (for all $cj \in CB$, $case(cj) \neq c$). The unique predicted judgement of c, denoted with $pj(c)$, is equal to the default judgement j_0 if and only if $\langle \emptyset, j_0 \rangle$ is active with respect to c.*

The aim of this work is to generate a judgement theory T from a given casebase CB and default judgement j_0 such that, given a new case c it is possible to predict the judgement of c.

3   Generating Case-Rules by ASP

In this section we describe how we can reason about past cases to generate the case-rules, namely legal reasoning structures expressed in terms of principle rules and exceptions. This is done by using Answer Set Programming (ASP) and the Clingo 3 ASP solver [5]. We begin by describing how relevant attacks can be inferred from examples of past cases with judgements. Using this we describe how case-rules can be generated in the form of a meta-level representation.

Our ASP computational framework uses a meta-level representation of the casebase and the judgement theory. Its reasoning process can be divided into three main steps: (i) extraction of relevant attacks from the given casebase; (ii) inference of the factors in the arguments of each relevant attack; (iii) generation of the judgement theory using the arguments. These steps are shown in Fig. 1, whose labelled programs are explained further in this section.

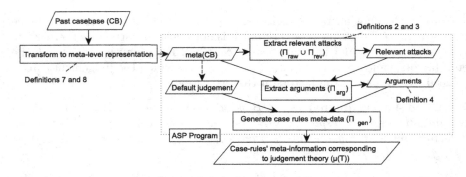

Fig. 1. Work flow for generating case-rules by ASP

3.1   Extracting Relevant Attacks from a Casebase

Each case with judgement can be seen as a definite clause.

Definition 7 (Rule representation of a case with judgement). *Let CB be a casebase. Each $cj \in CB$ can be expressed as a definite clause $r(cj)$ called a rule: $judgement(cj) : - f_1, \ldots, f_n$. where $f_i \in case(cj)$, for $1 \leq i \leq n$.*

Inferring relevant attacks and arguments means reasoning about the structure of the set of rules that express a given casebase. The meta-level representation of our cases with judgement describes the syntactic structure of each case in terms of the literals (i.e. factors) that appear in each case.

Definition 8 (Casebase meta-level representation). *Let CB be a casebase. Its meta-level representation, $meta(CB)$, is defined as:*

$$meta(CB) = \bigcup_{cj \in CB} \mu(r(cj)) \cup \tau(CB) \cup \delta(CB)$$

$\tau(CB) = \{\texttt{factor}(\texttt{f}_\texttt{i})|\texttt{f}_\texttt{i} \in \texttt{F}\}$, $\delta(CB)$ *is the meta-information about the default case,* $\delta(CB) = \{\texttt{default\_id(id(r(cj}_0)))\texttt{, default\_head(judgement(cj}_0))\}$, *and the function μ is defined as follows:*

$$\mu(r(cj)) = \begin{cases} \texttt{cb\_id(id(r(cj))).} \\ \texttt{is\_rule(id(r(cj)), judgement(cj)).} \\ \texttt{in\_rule(id(r(cj)), judgement(cj), f}_\texttt{i}\texttt{).} & \textit{for each } f_i \in case(cj) \end{cases}$$

This meta-level representation can be used to express the notion of raw attack ($\rightarrow_r$) given in Definition 2. The second and third rules below capture the condition $case(cj_1) \supset case(cj_2)$, where ID_1 and ID_2 are the IDs of cases with judgement cj_1 and cj_2 respectively; the condition $H_1 \neq H_2$ in the first rule below captures the condition $judgement(cj_1) \neq judgement(cj_2)$.

$$\Pi_{raw} = \begin{cases} \texttt{raw\_attack(ID}_1, \texttt{ID}_2) : -\texttt{factor\_subset(ID}_2, \texttt{ID}_1), \texttt{is\_rule(ID}_1, \texttt{H}_1), \\ \quad \texttt{is\_rule(ID}_2, \texttt{H}_2), \texttt{H}_1 \neq \texttt{H}_2. \\ \texttt{factor\_subset(ID}_1, \texttt{ID}_2) : -\texttt{cb\_id(ID}_1), \texttt{cb\_id(ID}_2), \\ \quad \texttt{not not\_factor\_subset(ID}_1, \texttt{ID}_2). \\ \texttt{not\_factor\_subset(ID}_1, \texttt{ID}_2) : -\texttt{cb\_id(ID}_1), \texttt{cb\_id(ID}_2), \texttt{factor(B)}, \\ \quad \texttt{in\_rule(ID}_1, \texttt{H}_1, \texttt{B}), \texttt{is\_rule(ID}_2, \texttt{H}_2), \texttt{not in\_rule(ID}_2, \texttt{H}_2, \texttt{B}). \end{cases}$$

The computation of the relevant attack relation (called *attack* in the program) uses ASP's choice operator (see first rule below) to select from the inferred raw attacks those relations that satisfy constraints (i) and (ii) given in Definition 3. They are captured by the second and third rules given below.

$$\Pi_{rev_1} = \begin{cases} \texttt{0 \{attack(ID}_1, \texttt{ID}_2)\} \ \texttt{1} : -\texttt{raw\_attack(ID}_1, \texttt{ID}_2). \\ : -\texttt{attack(ID}_1, \texttt{ID}_2), \texttt{not attackee(ID}_2). \\ : -\texttt{attack(ID}_1, \texttt{ID}_2), \texttt{raw\_attack(ID}_3, \texttt{ID}_2), \texttt{factor\_subset(ID}_3, \texttt{ID}_1). \\ \texttt{attackee(ID)} : -\texttt{default\_id(ID)}. \\ \texttt{attackee(ID}_2) : -\texttt{attack(ID}_2, \texttt{ID}_4). \end{cases}$$

To generate all relevant attacks in a given casebase, we use the following Clingo optimisation expression, which guarantees the maximum number of instances of the *attack* relation to be computed in a given solution.

$$\Pi_{rev_2} = \{\#\texttt{maximise}\{\texttt{attack(ID}_1, \texttt{ID}_2)\}.\}$$

Let $\Pi_{rev} = \Pi_{rev_1} \cup \Pi_{rev_2}$, and let $\Pi_{CB} = meta(CB) \cup \Pi_{raw} \cup \Pi_{rev}$. The answer set of $meta(CB) \cup \Pi_{raw}$ gives the subfactors and raw attacks between cases in the casebase. This is then used by Π_{rev} to generate all the relevant attacks AT. Thus the answer set of Π_{CB} contains the meta-level representation of the casebase, subfactors, raw attacks, relevant attacks and attackees.

3.2 Generating Meta-Level Information of Case-Rules

Using inferred relevant attacks and the meta-level representation of given cases, we can compute the set of arguments AR of the relevant attacks:

$$\Pi_{arg} = \begin{cases} \texttt{argument(ID}_1, \texttt{ID}_2, \texttt{Arg)} : -\texttt{attack(ID}_1, \texttt{ID}_2), \texttt{in\_rule(ID}_1, \texttt{H}_1, \texttt{Arg)}, \\ \quad\quad\quad \texttt{is\_rule(ID}_2, \texttt{H}_2), \texttt{not in\_rule(ID}_2, \texttt{H}_2, \texttt{Arg)}. \end{cases}$$

In order to predict the judgement of unseen cases, our judgement theory reflects the underlying reasoning applied throughout the past judged cases. The legal reasoning structure is normally composed of exceptions to a given default assumption (e.g. default judgement) and exceptions to exceptions.

Definition 9. *Given a casebase CB and its relevant attacks AT, the judgement theory T is the set of rules such that given a new case c, T derives the default judgement j_0 if and only if $pj(c) = j_0$. Let $ab(cj_i \rightarrow cj_j)$ be the reified atom of $cj_i \rightarrow cj_j$ in AT. The following rules are in the judgement theory[1]*

- *For empty case with judgement cj_0, and $cj_1 \rightarrow cj_0, \ldots, cj_n \rightarrow cj_0 \in AT$*
  $$judgement(cj_0) : -not\ ab(cj_1 \rightarrow cj_0), \ldots, not\ ab(cj_n \rightarrow cj_0).$$
- *For $f_1, \ldots, f_m \in \alpha(cj_x, cj_y)$, and $cj_{x+1} \rightarrow cj_x, \ldots, cj_{x+k} \rightarrow cj_x \in AT$*
  $$ab(cj_x \rightarrow cj_y) : -f_1, \ldots, f_m, not\ ab(cj_{x+1} \rightarrow cj_x), \ldots, not\ ab(cj_{x+k} \rightarrow cj_x).$$

Our computational approach abduces the above judgement theory in terms of its equivalent meta-level representation $\mu(T)$. The judgement theory itself is acquired by applying the inverse transformation $\mu^{-1}/1$. For example, μ^{-1} applied to the set of instances $\{\text{is\_rule}(id, h).\ \text{in\_rule}(id, h, b_1).\ \ldots\ \text{in\_rule}(id, h, b_n).\}$ gives the rule $h : -b_1, \ldots, b_n$.

To generate a legal reasoning structure each relevant attack inferred from the past cases has to be linked to a unique abnormality name. This Skolemisation is captured using the ASP choice rule given below, which associates the attack identifier AID of a given attack $a(ID_1, ID_2)$ with an abnormality name Ab. An integrity constraint guarantees that attack identifiers are unique.

$$\Pi_{gen_1} = \begin{cases} 1\ \{\text{id\_attack\_link}(\text{AID}, \text{a}(ID_1, ID_2)) : \text{abnormal}(\text{AID}, \text{Ab})\}\ 1 \\ \quad : -\text{attack}(ID_1, ID_2). \\ : -\text{id\_attack\_link}(\text{AID}, \text{At}), \text{id\_attack\_link}(\text{AID}_2, \text{At}), \text{AID}_1 \neq \text{AID}_2. \\ : -\text{id\_attack\_link}(\text{AID}, \text{At}_1), \text{id\_attack\_link}(\text{AID}, \text{At}_2), \text{At}_1 \neq \text{At}_2. \end{cases}$$

Abnormality names are defined by the basic types *abnormal* and *negated_abnormal* as illustrated below, where the number of relevant attacks can be obtained from the answer set of Π_{CB}:

$$\Pi_{gen_2} = \begin{cases} \text{gen\_id}(r0). \\ \text{gen\_id}(ri).\ \text{abnormal}(ri, abi).\ \text{negated\_abnormal}(ri, not\_abi). \\ \quad \textit{For } 1 \leq i \leq n, \textit{where } n \textit{ is the number of relevant attacks} \end{cases}$$

Once appropriate links between relevant attacks and abnormal identifiers are obtained through the choice rule given above, the meta-level representation of the rules defining these abnormalities can be inferred using the following two sets of rules. The first rule expresses the fact that the default judgement is the judgement of an empty factor set, thus it can only be applied on future cases if all attacks against it cannot be proven to hold. The second rule captures the meta-level representation of a principle rule of the form $j_0 \leftarrow not\ ab_1, \ldots, not\ ab_m$, defining the absence of exceptions to the default judgement.

$$\Pi_{gen_3} = \begin{cases} \text{is\_rule}(r0, \text{Def}) : -\text{default\_head}(\text{Def}). \\ \text{in\_rule}(r0, \text{Def}, \text{NAb}) : -\text{default\_head}(\text{Def}), \text{default\_id}(ID2), \\ \quad \text{id\_attack\_link}(\text{AID}, \text{a}(ID_1, ID_2)), \text{negated\_abnormal}(\text{AID}, \text{NAb}). \end{cases}$$

---

[1] where $\alpha/2$ represents arguments as described in Definition 4.

For each abduced link between relevant attack and abnormality names, the following set of rules allows the inference of the meta-level representation of rules of the form $ab_i \leftarrow f_{i1}, \ldots, f_{ik_i}, not\ ab_{mi1}, \ldots, not\ ab_p$ and $ab_q \leftarrow f_{q1}, \ldots, f_{qk_q}$.

$$\Pi_{gen_4} = \begin{cases} \texttt{is\_rule(AID, Ab)} : -\texttt{id\_attack\_link(AID, At)}, \texttt{abnormal(AID, Ab)}. \\ \texttt{in\_rule(AID, H, Arg)} : -\texttt{is\_rule(AID, H)}, \texttt{argument(ID1, ID2, Arg)}. \\ \quad \texttt{id\_attack\_link(AID, a(ID}_1, \texttt{ID}_2\texttt{))}, \\ \texttt{in\_rule(AID}_1, \texttt{H, NAb)} : -\texttt{is\_rule(AID}_1, \texttt{H)}, \texttt{negated\_abnormal(AID}_2, \texttt{NAb)}, \\ \quad \texttt{id\_attack\_link(AID}_1, \texttt{a(ID}_2, \texttt{ID}_1\texttt{))}, \texttt{id\_attack\_link(AID}_2, \texttt{a(ID}_3, \texttt{ID}_2\texttt{))}. \end{cases}$$

The first two rules allow the inference of the definition of predicate head abnormality name Ab that corresponds to the correct linked attack, for which the appropriate factors involved in the argument of this attack are inferred to be conditions in the body of such abnormality rule. The third rule captures the abnormality identifiers of subsequent attacks that invalidate the current exception.

Let $\Pi_{gen} = \Pi_{gen_1} \cup \Pi_{gen_2} \cup \Pi_{gen_3} \cup \Pi_{gen_4}$, and let $\Pi_{JT} = \Pi_{gen} \cup AR \cup \{default\_head(judgement(cj_0)), default\_id(id(r(cj_0)))\}$. The answer set of the program Π_{JT} projected over $is\_rule/2$ and $in\_rule/2$ corresponds to the meta-level representation of the judgement theory $\mu(T)$.

3.3   Application to Legal Reasoning Example

We show the result of our approach applied to the past cases with judgements described in Sect. 1^2. The following meta-level representation of the judgement theory is generated from Π_{gen}:

```
is_rule(r0,neg_dwe).          in_rule(r4,ab0,not_ab2). in_rule(r2,ab2,rpl).
in_rule(r0,neg_dwe,not_ab0).  is_rule(r5,ab1).          in_rule(r2,ab2,far).
in_rule(r0,neg_dwe,not_ab1).  in_rule(r5,ab1,neg_dot). in_rule(r2,ab2,ria).
is_rule(r4,ab0).              in_rule(r5,ab1,fad).      is_rule(r3,ab3).
in_rule(r4,ab0,dot).          in_rule(r5,ab1,not_ab3). in_rule(r3,ab3,dia).
in_rule(r4,ab0,ooo).          is_rule(r2,ab2).
```

This corresponds to the program:

```
neg_dwe :- not ab0, not ab1.    ab2 :- far, ria, rpl.
ab0 :- dot, ooo, not ab2.       ab3 :- dia.    ab1 :- fad, neg_dot, not ab3.
```

4   Correctness of the Generated Program

We provide the correctness of the program through two propositions. Proposition 1 ensures that all relevant attacks of the given casebase are computed, while Proposition 2 ensures the correct predicted judgement is computed. The proof of Proposition 1 is divided into four steps (see footnote 2). Lemma 1, establish the properties of an answer set of Π_{CB}. Lemma 2 shows that the union of

---

^{2} The full example with each ASP program's output, and proofs for the lemmas and propositions can be found at http://wp.doc.ic.ac.uk/spike/technical-reports/.

two answer sets of Π_{CB} is also an answer set of Π_{CB}. Lemma 3, shows that $ans(CB, AT) \in AS(\Pi_{CB})$, where $AS(P)$ is the set of answer sets of a given program P.

Definition 10. *Let* $\{S\} = AS(meta(CB) \cup \Pi_{raw})$, *$CB$ be a casebase, AT its relevant attacks, and* $Q = \{attack(id(r(cj_i)), id(r(cj_j))), attackee(id(r(cj_j))) \mid \langle cj_i, cj_j \rangle \in AT\}$. $ans(CB, AT)$ *is the interpretation* $S \cup Q \cup \{attackee(id(r(cj_0)))\}$.

Lemma 1. *Let S be the unique answer set of* $meta(CB) \cup \Pi_{raw}$ *and let* $\pi_p(A)$ *denote the projection of A over p. A set $A \in AS(\Pi_{CB})$ iff:*

(i) $\pi_{attack}(A) \subseteq \pi_{raw\_attack}(A)$
(ii) *For all* cj_1 *and* cj_2, *if* $attack(id(r(cj_1)), id(r(cj_2))) \in A$ *then* $attackee(id(r(cj_2))) \in A$
(iii) *For all cj_1 and cj_2, if $attack(id(r(cj_1)), id(r(cj_2))) \in A$ then there does not exists cj_3 such that $raw\_attack(id(r(cj_3)), id(r(cj_2))) \in A$ and $factor\_subset(id(r(cj_3)), id(r(cj_1))) \in A$)*
(iv) *For all cj, $attackee(id(r(cj)) \in A$ if and only if $default\_id(id(r(cj_3))$ is true or $attack(id(r(cj)), id(r(cj_x))) \in A$*
(v) *Let $\mathcal{L}$ be the language of $meta(CB) \cup \Pi_{raw}$, then for all s, $s \in \pi_{\mathcal{L}}(A)$ if and only if $s \in S$*

Lemma 2. *Given two answer sets A_1, $A_2 \in AS(\Pi_{CB})$, then $A_1 \cup A_2 \in AS(\Pi_{CB})$.*

Lemma 3. *Given a casebase CB, its raw attacks RA, and its relevant attack AT, $ans(CB, AT) \in AS(\Pi_{CB})$.*

Proposition 1. *Given a casebase CB with relevant attacks AT, $ans(CB, AT)$ is the unique optimal answer set of Π_{CB}.*

We assume that Π_{JT} will generate the correct judgement theory as described by Definition 9. The proof of Proposition 2 is divided into two parts. Lemma 6 shows that the judgement theory T can be partitioned into one part T_c responsible, and another T_{ex} irrelevant for the derivation of j_0. Thus T_c can be used to show that T derives the j_0 if and only if it is the predicted judgement.

Lemma 4. *Given a casebase CB and associated judgement theory T. Then for $A \in AS(\Pi_{JT})$, $\pi_{in\_rule, is\_rule}(A) = \mu(T)$.*

For rule $r \in T$, let $head(r)$ be its head literal, $body(r)$ be the set of its body literals, $fs(r)$ be the set of factors in its body literal, and def be a rule where $head(def) = j_0$. The following property can be derived from Definition 9.

Lemma 5. *Given a casebase CB with associated judgement theory T, let c be a new case, given as a set of factors. From Definition 9 for all abnormality rules ab in T there exists a sequence of rules not $head(ab) \in body(ab_{x_1}), \ldots$, not $head(ab_{x_n}) \in body(def)$ in T, where $n \geq 0$. The union of all its factors corresponds to a $cj \in CB$, and $cj \rightarrow cj_y$ for some $cj_y \in CB$. Abnormalities with a sequence such that $case(cj) \subseteq c$ is denoted by $seq(ab)$.*

Lemma 6. *Given a casebase CB with associated judgement theory T, and a new case c. Let $T_c = \{def\} \cup \{ab | ab \in T, seq(ab)\}$, and $T_{ex} = T \setminus T_c$. Then T derives j_0 iff T_c derives j_0.*

Proposition 2. *Given a casebase CB with associated judgement theory T, and a new case c. Let $\{A_T\} = AS(T \cup c)$. Then $j_0 \in A_T$ if and only if $pj(c) = j_0$.*

5 Evaluation

To test the performance of the approach we have applied it to randomly generated (consistent) casebases where the number of cases ranges from 20 to 100, and the number of factors ranges from 10 to 25. The time taken to generate the meta-level representation of the judgement theory for each casebase is presented in Fig. 2. To mitigate the grounding problem, the ASP program is split into two parts with the first program $\Pi_{CB} \cup \Pi_{arg}$ used for generating the relevant attacks and arguments (Fig. 2(a)), which are then added as facts to the second program Π_{gen} to generate the judgement theory (Fig. 2(b)). Both Figs. 2(a) and (b) show that the computational time increases with the number of cases, while the change in the number of factors is less important. It can be seen that the times in Fig. 2(a) are not significant when compared with those in Fig. 2(b). The reason for the much larger times in Fig. 2(b) is due to the use of Skolemisation in Π_{gen}[3]. However, the computational time of Π could be greatly decreased by using an external process to perform the Skolemisation, which would allow the generation of the judgement theory for casebases with greater than 100 cases.

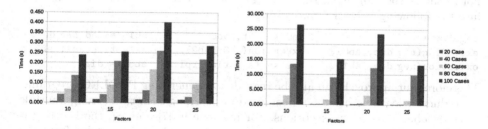

Fig. 2. Computational time for (a) $\Pi_{CB} \cup \Pi_{arg}$ to output $A_{atk,arg}$; (b) $\Pi_{gen} \cup \pi_{attack,argument,default\_id,default\_head}(A_{atk,arg})$ applied to randomly generated CB

6 Related Work

In this work we have shown how meta-level reasoning can be used for extracting information from past legal cases for generating case rules for deciding the

---

[3] While we were able to compute the relevant attacks and arguments for casebase with greater than 100 cases, we were unable to generate the judgement theory from them using Π_{gen}.

judgement of future cases. We have used notions of argumentation and ASP for computation, and while there have been many recent works [15] in representing argumentation frameworks and computing argumentation extensions using ASP, we are concerned with the extraction of information about the arguments and attacks from examples rather than computing extensions of a given framework.

Similar to legislators using past legal cases for creating or revising legislations, past work in legal reasoning has explored how this can be automated by formally reasoning reasoning from past cases [10,11], using argumentation [2], or using boolean function [12]. The system HYPO [1] also analyses factors in the form of dimensions, where a dimension is a structure containing a factor and the party it favours, to suggest the arguments and counter examples that plaintiff and defendant may use to further their objectives. It sorts case relevance using a claim lattice, a directed acyclic graph of dimensions and cases relevant to the set of dimensions. This differs from our approach where we sort cases using the relevant attacks, arranging them according to their ability to overturn judgements of other cases. Additionally, [9] shows how meta-level representation can be used for legal reasoning, however this concerns the representation and interpretation of the rules, and not the generation of rules.

The legal reasoning system PROLEG is used to represent rules and exceptions [13]. In fact, there is a correspondence between the representation used in PROLEG and the rules that we generate. PROLEG rules do not use negation as failure explicitly; exceptions are expressed by the form $exception(H, E)$ where H and E are atoms. The generated rule $C: -B_1, \ldots, B_n,$ not $E_1, \ldots,$ not E_m, where $n \geq 0$, $m \geq 0$, is represented in PROLEG as

$$C \Leftarrow B_1, \ldots, B_n. \qquad exception(C, E_1). \qquad \ldots \qquad exception(C, E_m).$$

For instance, the judgement theory we generated in in Sect. 3.3 can be translated into the following PROLEG program.

```
neg_dwe<=.                    ab0 <= dot, ooo.        ab1 <= fad,neg_dot.
exception(neg_dwe, ab0).      exception(ab0, ab2).    exception(ab1, ab3).
exception(neg_dwe, ab1).      ab2 <= far, ria, rpl.   ab3 <= dia.
```

Therefore, our approach can also be regarded as generating PROLEG programs.

Inductive learning has often been used for learning such rules, using multiple learning phases to learn exceptions. For instance in [7] and [8], the learning is split into two phases, the first phase learns the overly general rules from the examples, and the second phase specialises the general rules using exceptions. Our work is similar to the second phase, but with an assumed over-general rule (default judgement) to be given from the legal specification. Other similar work (e.g. [3]) uses prioritised logic to express preferences between the default rules.

7   Conclusion

We have presented a method for reasoning and extracting information from past cases to infer the arguments and attacks present in the decision for the judgement of a new case. We have defined the notion of minimal attacks for identifying the factors relevant to the judgements of cases, and describe how ASP can be used to generate these minimal attacks as well as how these attacks can be used for inferring rules for modelling the judgement using meta-level information. While not shown in this paper, PROLOG could be used instead of ASP, using its list structure to represent the rules.

For future work, it would be interesting to extend the meta-level representation and enhancing the reasoning approach for handling more complex casebases. For instance in [6], a different representation of the casebase is used where a factor favours either the defendant or plaintiff. Other extensions would be to consider a casebase with inconsistency and how the approach might handle conflicts, and how to make revision to an existing judgement theory.

Acknowledgment. This work was partially supported by JSPS KAKENHI Grant Numbers 26280091. The authors would like to thank Robert Kowalski for his valuable comments, and Kristijonas Cyras for his two examples, including Example 1, which indicated an error in the paper presented at the Eighth International Workshop on Juris-informatics (JURISIN 2014), Kanagawa, Japan.

References

1. Ashley, K.D.: Reasoning with cases and hypotheticals in hypo. Int. J. Man Mach. Stud. **34**, 753–796 (1991)
2. Bench-Capon, T., Prakken, H., Sartor, G.: Argumentation in legal reasoning. In: Simari, G., Rahwan, I. (eds.) Argumentation in Artificial Intelligence, pp. 363–382. Springer, US (2009)
3. Dimopoulos, Y., Kakas, A.: Learning non-monotonic logic programs: learning exceptions. In: Lavrač, N., Wrobel, S. (eds.) ECML 1995. LNCS, vol. 912, pp. 122–137. Springer, Heidelberg (1995)
4. Dung, P.M.: On the acceptability of arguments and its fundamental role in non-monotonic reasoning, logic programming and n-person games. Artif. Intell. **77**, 321–357 (1995)
5. Gebser, M., Kaufmann, B., Kaminski, R., Ostrowski, M., Schaub, T., Schneider, M.T.: Potassco: the potsdam answer set solving collection. AI Commun. **24**(2), 107–124 (2011)
6. Horty, J.F.: Reasons and precedent. In: Ashley, K.D., van Engers, T.M. (eds.) Proceedings of the Conference of the 13th International Conference on Artificial Intelligence and Law, Pittsburgh, USA, pp. 41–50. ACM (2011)
7. Inoue, K.: Learning extended logic programs. In: Proceedings of the 15th International Joint Conference on Artificial Intelligence, pp. 176–181 (1997)
8. Ohara, K., Taka, H., Babaguchi, N., Kitahashi, T.: Determination of general concept in learning default rules. In: Mizoguchi, R., Slaney, J.K. (eds.) PRICAI 2000. LNCS, vol. 1886, pp. 104–114. Springer, Heidelberg (2000)
9. Routen, T., Bench-Capon, T.J.M.: Hierarchical formalizations. Int. J. Man Mach. Stud. **35**(1), 69–93 (1991)
10. Satoh, K.: Translating case-based reasoning into abductive logic programming. In: Proceedings of 12th European Conference on Artificial Intelligence, Budapest, Hungary, pp. 142–146 (1996)
11. Satoh, K.: Statutory interpretation by case-based reasoning through abductive logic programming. JACIII **1**(2), 94–103 (1997)

12. Satoh, K.: Analysis of case-based representability of boolean functions by monotone theory. In: Richter, M.M., Smith, C.H., Wiehagen, R., Zeugmann, T. (eds.) ALT 1998. LNCS (LNAI), vol. 1501, pp. 179–190. Springer, Heidelberg (1998)
13. Satoh, K., Asai, K., Kogawa, T., Kubota, M., Nakamura, M., Nishigai, Y., Shirakawa, K., Takano, C.: PROLEG: an implementation of the presupposed ultimate fact theory of Japanese civil code by PROLOG technology. In: Bekki, D. (ed.) JSAI-isAI 2010. LNCS, vol. 6797, pp. 153–164. Springer, Heidelberg (2011)
14. Satoh, K., Kubota, M., Nishigai, Y., Takano, C.: Translating the Japanese presupposed ultimate fact theory into logic programming. In: JURIX 2009: The Twenty-Second Annual Conference on Legal Knowledge and Information Systems, The Netherlands, pp. 162–171. IOS Press (2009)
15. Toni, F., Sergot, M.: Argumentation and answer set programming. In: Balduccini, M., Son, T.C. (eds.) Logic Programming, Knowledge Representation, and Non-monotonic Reasoning. LNCS, vol. 6565, pp. 164–180. Springer, Heidelberg (2011)

Online Action Language $o\mathcal{BC}+$

Joseph Babb and Joohyung Lee[✉]

School of Computing, Informatics, and Decision Systems Engineering,
Arizona State University, Tempe, USA
{Joseph.Babb,joolee}@asu.edu

Abstract. We present an online action language called $o\mathcal{BC}+$, which extends action language $\mathcal{BC}+$ to handle external events arriving online. This is done by first extending the concept of online answer set solving to arbitrary propositional formulas, and then defining the semantics of $o\mathcal{BC}+$ based on this extension, similar to the way the offline $\mathcal{BC}+$ is defined. The design of $o\mathcal{BC}+$ ensures that any action description in $o\mathcal{BC}+$ satisfies the syntactic conditions required for the correct computation of online answer set solving, thereby alleviates the user's burden for checking the sophisticated conditions.

1 Introduction

While Answer Set Programming (ASP) is being widely applied to many challenging problems, most ASP applications are limited to offline usages. Continuous grounding and solving in view of possible yet unknown future events, such as the one required for the emerging applications in stream reasoning [1], is one of the main challenges in applying ASP to real-time dynamic systems.

Recently, there emerged the concept of *reactive answer set programming* [2], which is to incrementally ground and compose program slices taking into account external knowledge acquired asynchronously, thereby avoiding multiple unnecessary restarts of the grounding and solving process for each arrival of external inputs. For this, an online ASP program consists of multiple subprograms of different roles, and certain syntactic restrictions originating from the *module theorem* [3] are imposed to ensure the compositionality of their answer sets. The work led to an implementation OCLINGO,[1] which extends ASP grounder GRINGO and ASP solver CLASP in a monolithic way to handle external modules provided at runtime by a controller. However, checking the syntactic requirement for sound execution of online answer set solving is quite a complex task for the user, which significantly limits the usability of online answer set programming.

We address this challenge by introducing an online extension of high level action language $\mathcal{BC}+$ [4], which we call $o\mathcal{BC}+$. $\mathcal{BC}+$ is a recently proposed action language whose semantics is defined in terms of propositional formulas under the stable model semantics. It is shown in [4] that $\mathcal{BC}+$ is expressive enough to embed other action languages, such as $\mathcal{B}$, $\mathcal{C}$, $\mathcal{C}+$ [5] and $\mathcal{BC}$ [6]. Thus $o\mathcal{BC}+$ can be viewed as online extensions of these languages as well.

---

[1] http://www.cs.uni-potsdam.de/wv/oclingo/.

© Springer International Publishing Switzerland 2015
F. Calimeri et al. (Eds.): LPNMR 2015, LNAI 9345, pp. 97–111, 2015.
DOI: 10.1007/978-3-319-23264-5_9

Since the semantics of $\mathcal{BC}+$ is based on propositional formulas under the stable model semantics, we first generalize the result on online answer set solving to arbitrary propositional formulas, and define $o\mathcal{BC}+$ based on it. We demonstrate that $o\mathcal{BC}+$ provides a structured input language for online answer set solving, and thereby alleviates the user's burden for checking sophisticated conditions imposed on the input programs.

The paper is organized as follows. Section 2 reviews language $\mathcal{BC}+$ from [4] and the module theorem from [7]. Section 3 extends the concept of online answer set solving to propositional theories, based on which Sect. 4 defines the online extension of $\mathcal{BC}+$, and asserts that the design of the language ensures the syntactic conditions for applying online answer set solving.

2 Preliminaries

2.1 Review: Stable Models of Propositional Formulas

According to [8], stable models of a propositional formula are defined as follows. The *reduct* F^X of a propositional formula F relative to a set X of atoms is the formula obtained from F by replacing every maximal subformula that is not satisfied by X with $\perp$. Set X is called a *stable model* of F if it is a minimal set of atoms satisfying F^X. It is known that propositional logic programs can be identified with propositional formulas under the stable model semantics in the form of conjunctions of implications.

Throughout this paper, we consider propositional formulas whose signature σ consists of atoms of the form $c = v$,[2] where c is called a *constant* and is associated with a finite set $Dom(c)$ of cardinality ≥ 2, called the *domain*, and v is an element of its domain. If the domain of c is $\{\mathbf{f}, \mathbf{t}\}$ then we say that c is *Boolean*, and abbreviate $c = \mathbf{t}$ as c and $c = \mathbf{f}$ as $\sim c$.

2.2 Review: $\mathcal{BC}+$

Syntax. Language $\mathcal{BC}+$ includes two kinds of constants, *fluent constants* and *action constants*. Fluent constants are further divided into *regular* and *statically determined*.[3]

A *fluent formula* is a formula such that all constants occurring in it are fluent constants. An *action formula* is a formula that contains at least one action constant and no fluent constants.

A *static law* is an expression of the form

$$\textbf{caused } F \textbf{ if } G \tag{1}$$

---

[2] So $c = v$ is an atom in the propositional signature, and not an equality in first-order logic.

[3] Statically determined fluents are fluents whose values are completely determined by fluents in the same state, and not by direct effects of actions [5, Sect. 5.5].

where F and G are fluent formulas. An *action dynamic law* is an expression of the form (1) in which F is an action formula and G is a formula. A *fluent dynamic law* is an expression of the form

$$\textbf{caused } F \textbf{ if } G \textbf{ after } H \qquad (2)$$

where F and G are fluent formulas and H is a formula, provided that F does not contain statically determined constants. Static laws can be used to talk about causal dependencies between fluents in the same state. Action dynamic laws can be used to express causal dependencies between concurrently executed actions. A more common use of action dynamic laws is to express the assumption of an action being "exogenous" (the cause of the action is outside the domain description). Fluent dynamic laws can be used for describing direct effects of actions.

A *causal law* is a static law, an action dynamic law, or a fluent dynamic law. An *action description* is a finite set of causal laws.

The formula F in a causal law (1) or (2) is called the *head*, and G and H are called the *bodies*.

Semantics of $\mathcal{BC}+$. The semantics of $\mathcal{BC}+$ can be understood in terms of a "transition system"—a directed graph whose vertices are states of the world and edges represent transitions between states. For any action description D with a set σ^{fl} of fluent constants and a set σ^{act} of action constants, we define a sequence of propositional formulas $PF_0(D), PF_1(D), \ldots$ so that the stable models of $PF_m(D)$ represent paths of length m in the transition system corresponding to D. The signature of $PF_m(D)$ consists of atoms of the form $i{:}c{=}v$ such that

- for each fluent constant c of D, $i \in \{0, \ldots, m\}$ and $v \in Dom(c)$, and
- for each action constant c of D, $i \in \{0, \ldots, m-1\}$ and $v \in Dom(c)$.

By $i{:}F$ we denote the result of inserting $i{:}$ in front of every occurrence of every constant in formula F.

For any set $\mathbf{c}$ of symbols from σ^{fl} and σ^{act}, by $UEC_{\mathbf{c}}$ we denote the conjunction of

$$\bigwedge_{v \neq w \,\mid\, v,w \in Dom(c)} \neg(c = v \wedge c = w) \quad \wedge \quad \neg\neg \bigvee_{v \in Dom(c)} c = v, \qquad (3)$$

for all $c \in \mathbf{c}$, which represents the uniqueness and existence of values for the constants in $\mathbf{c}$.

For any atom $c{=}v$, "choice rule" $\{c{=}v\}^{\text{ch}}$ stands for $c{=}v \vee \neg(c{=}v)$, which, in the presence of (3), means that by default c is mapped to v [9].

The translation $PF_m(D)$ is the conjunction of

$j\!:\!F \leftarrow j\!:\!G$	for each static law (1) in D
$i\!:\!F \leftarrow i\!:\!G$	for each action dynamic law (1) in D
$(i{+}1)\!:\!F \leftarrow (i{+}1)\!:\!G \wedge i\!:\!H$	for each fluent dynamic law (2) in D
$\{0\!:\!c{=}v\}^{\mathrm{ch}}$	for each regular fluent c and every $v \in Dom(c)$
$j\!:\!UEC_{\sigma^{fl}} \qquad i\!:\!UEC_{\sigma^{act}}$	

$$(i = 0, \ldots, m{-}1, \quad j = 0, \ldots m).$$

We identify an interpretation I with the set of atoms that are satisfied by this interpretation. This allows us to represent any interpretation of the signature of $PF_m(D)$ in the form

$$(0 : s_0) \cup (0 : e_0) \cup (1 : s_1) \cup (1 : e_1) \cup \cdots \cup (m : s_m)$$

where $s_0, \ldots, s_m$ are interpretations of σ^{fl} and $e_0, \ldots, e_{m-1}$ are interpretations of σ^{act}.

States and transitions are defined in terms of stable models of $PF_0(D)$ and $PF_1(D)$ as follows.

Definition 1 (States and Transitions). *For any action description D of signature σ, a state of D is an interpretation s of σ^{fl} such that $0 : s$ is a stable model of $PF_0(D)$. A transition of D is a triple $\langle s, e, s' \rangle$ where s and s' are interpretations of σ^{fl} and e is an interpretation of σ^{act} such that $0\!:\!s \cup 0\!:\!e \cup 1\!:\!s'$ is a stable model of $PF_1(D)$.*

In view of the uniqueness and existence of value constraints for every state s and every fluent constant c, there exists exactly one v such that $c = v$ belongs to s; this v is considered the value of c in state s.

Given these definitions, we define the transition system $T(D)$ represented by an action description D as follows.

Definition 2 (Transition System). *A transition system $T(D)$ represented by an action description D is a labeled directed graph such that the vertices are the states of D, and the edges are obtained from the transitions of D as follows: for every transition $\langle s, e, s' \rangle$ of D, an edge labeled e goes from s to s'.*

Since the vertices and the edges of a transition system $T(D)$ are identified with the states and the transitions of D, we simply apply the definitions of a state and a transition to transition systems: A *state* of $T(D)$ is a state of D. A *transition* of $T(D)$ is a transition of D.

The stable models of $PF_m(D)$ represent the paths of length m in the transition system represented by D [4, Theorem 2].

2.3 Review: Module Theorem

We review the module theorem from [7] limited to the propositional case.

For any propositional formula F, by $At(F)$ we denote the set of all atoms occurring in F. The *head* atoms of F are defined to be the atoms that has an occurrence in F that is not in the antecedent of any implication (we understand $\neg F$ as an abbreviation of $F \to \bot$). By $Head(F)$ we denote the set of all head atoms of F.

A *module* $\mathbb{F}$ is a triple $(F, \mathcal{I}, \mathcal{O})$, where F is a propositional formula, and $\mathcal{I}$ and $\mathcal{O}$ are disjoint sets of atoms such that $At(F) \subseteq (\mathcal{I} \cup \mathcal{O})$.

Definition 3 (Module Stable Model). *We say that an interpretation I is a (module) stable model of a module $\mathbb{F} = (F, \mathcal{I}, \mathcal{O})$ if I is a stable model of $F \wedge \bigwedge_{A \in \mathcal{I}} \{A\}^{\mathrm{ch}}$.*

We refer the reader to [10] for the definition of a dependency graph of a propositional formula F relative to a set A of atoms, which we denote by $\mathrm{DG}[F;\ A]$.

Definition 4 (Joinability of Modules). *Two modules* $\mathbb{F}_1 = (F_1 \wedge H,\ \mathcal{I}_1,\ \mathcal{O}_1)$ *and* $\mathbb{F}_2 = (F_2 \wedge H,\ \mathcal{I}_2,\ \mathcal{O}_2)$ *are called* joinable *if*

- $\mathcal{O}_1 \cap \mathcal{O}_2 = \emptyset$,
- *each strongly connected component of* $\mathrm{DG}[F_1 \wedge F_2 \wedge H;\ \mathcal{O}_1 \cup \mathcal{O}_2]$ *is either a subset of* $\mathcal{O}_1$ *or a subset of* $\mathcal{O}_2$,
- $Head(F_1) \cap \mathcal{O}_2 = \emptyset$, *and* $Head(F_2) \cap \mathcal{O}_1 = \emptyset$.

Definition 5 (Join of Modules). *For any modules* $\mathbb{F}_1 = (F_1 \wedge H,\ \mathcal{I}_1,\ \mathcal{O}_1)$ *and* $\mathbb{F}_2 = (F_2 \wedge H,\ \mathcal{I}_2,\ \mathcal{O}_2)$ *that are joinable, the* join *of* $\mathbb{F}_1$ *and* $\mathbb{F}_2$, *denoted by* $\mathbb{F}_1 \sqcup \mathbb{F}_2$, *is defined to be the module* $(F_1 \wedge F_2 \wedge H,\ (\mathcal{I}_1 \cup \mathcal{I}_2) \setminus (\mathcal{O}_1 \cup \mathcal{O}_2),\ \mathcal{O}_1 \cup \mathcal{O}_2)$.

Given sets of atoms I_1, I_2, I_3, we say that I_1 and I_2 are I_3-compatible if $I_1 \cap I_3 = I_2 \cap I_3$.

Theorem 1 (Module Theorem [7]). *Let* $\mathbb{F}_1 = (F_1, \mathcal{I}_1, \mathcal{O}_1)$ *and* $\mathbb{F}_2 = (F_2, \mathcal{I}_2, \mathcal{O}_2)$ *be modules that are joinable, and let* I_i *($i = 1, 2$) be a subset of* $(\mathcal{I}_i \cup \mathcal{O}_i)$ *such that* I_1 *and* I_2 *are* $(\mathcal{I}_1 \cup \mathcal{O}_1) \cap (\mathcal{I}_2 \cup \mathcal{O}_2)$*-compatible. Then* $I_1 \cup I_2$ *is a stable model of* $\mathbb{F}_1 \sqcup \mathbb{F}_2$ *iff* I_1 *is a stable model of* $\mathbb{F}_1$ *and* I_2 *is a stable model of* $\mathbb{F}_2$.

3 Online Propositional Theories

We generalize the concept of online answer set solving to arbitrary propositional formulas as follows. This section inevitably has many notions, all of which are generalized from those in [2]. The generalization will be used in the next section in order to extend $\mathcal{BC}+$ to the online setting.

A *step-parametrized formula* $F[t]$ is a propositional formula which may contain *step-parameterized atoms* of the form $g(t) : a$, where t is a variable for nonnegative integers denoting a *step counter*, and $g(t)$ is some meta-level nonnegative integer valued arithmetic function whose only free variable is t. Given such a formula $F[t]$ and a nonnegative integer k, the *step-instantiated formula* $F[t/k]$ (or simply $F[k]$) is defined to be the propositional formula which is obtained from $F[t]$ by replacing every occurrence of every step-parametrized atom $g(t) : a$ with a standard atom $v : a$, where v is the value of $g(k)$. (Thus, $v : a$ is assumed to be in the underlying propositional signature.)

We define an *incremental theory* to be a triple $\langle B, P[t], Q[t] \rangle$ such that B is a propositional formula, and $P[t]$, $Q[t]$ are step-parametrized formulas. Informally, B is the *base component*, which describes static knowledge; $P[t]$ is the *cumulative component*, which contains information regarding every step that should be accumulated during execution; $Q[t]$ is the *volatile component*, which contains constraints or other information regarding the final step.

By an *online progression* $\langle E, F \rangle$ we denote some sequence of pairs of step-instantiated formulas $(E_i[e_i], F_i[f_i])$ for $i \geq 1$ with associated nonnegative integers e_i, f_i such that $e_i \leq f_i$. Intuitively, each $E_i[e_i]$ and $F_i[f_i]$ corresponds to stable and volatile knowledge acquired during execution, respectively. For each $(E_i[e_i], F_i[f_i])$, e_i and f_i denote the step for which they are relevant allowing knowledge to be acquired out of order. For example, $E_4[3]$ is the fourth piece of online input and contains information relevant to step 3.

Given an incremental theory $\langle B, P[t], Q[t] \rangle$, an online progression $\langle E, F \rangle$, and nonnegative integers j, k such that $e_1, \ldots, e_j, f_j \leq k$, the *incremental components* are

$$\{B, P[t/1], P[t/2], \ldots, P[t/k], Q[t/k], E_1[e_1], E_2[e_2], \ldots, E_j[e_j], F_j[f_j]\}. \quad (4)$$

As in [2], we define the *k-expanded propositional formula* $R_{j,k}$ of $\langle B, P[t], Q[t] \rangle$ w.r.t. $\langle E, F \rangle$ to be the conjunction of all formulas in (4).

Generalizing the notion of the simplification in [2], given a propositional formula F, we define the *simplification* of F onto a set A of atoms (denoted $Simplify(F, A)$) to be the formula obtained from F by replacing all occurrences of atoms p in F such that $p \notin (Head(F) \cup A)$ with $\bot$ and performing the following syntactic transformations recursively until no further transformations are possible:[4]

$$
\begin{array}{llll}
\neg\bot \;\mapsto\; \top & \neg\top \;\mapsto\; \bot & & \\
\bot \wedge F \;\mapsto\; \bot & F \wedge \bot \;\mapsto\; \bot & \top \wedge F \;\mapsto\; F & F \wedge \top \;\mapsto\; F \\
\bot \vee F \;\mapsto\; F & F \vee \bot \;\mapsto\; F & \top \vee F \;\mapsto\; \top & F \vee \top \;\mapsto\; \top \\
\bot \to F \;\mapsto\; \top & F \to \top \;\mapsto\; \top & \top \to F \;\mapsto\; F &
\end{array}
$$

We define the *modular instantiation* of F with respect to A, denoted $PM(F, A)$, to be the module $(Simplify(F, A), A, At(Simplify(F, A)) \backslash A)$.[5]

The idea of the simplification is to reduce the size of the input formulas by exploiting the fact that some atoms are known not to belong to any stable model.

---

[4] In [2], this process stops only at the second iteration.

[5] In practice when F is non-ground, we assume F is grounded first by substituting every variable with every element in the Herbrand universe.

However, unlike the offline solving, the values of external atoms in the online incremental computation are unknown at the time of simplifying the current module containing them. Thus, following [2], we associate each formula F in (4) with some designated set of external atoms $I(F)$ such that $Head(F) \cap I(F) = \emptyset$. Such atoms represent possible external inputs that may be introduced later by an online progression, and thus should be exempted from the current program simplification.

Given a module $\mathbb{F} = \langle F, \mathcal{I}, \mathcal{O} \rangle$, $Out(\mathbb{F})$ refers to $\mathcal{O}$.

Definition 6 (Modular Incremental Theories and Online Progression).
We say that an incremental theory $\langle B, P[t], Q[t] \rangle$ and an online progression $\langle E, F \rangle$ are modular if the following modules are well-defined for any nonnegative integers j, k such that $e_1, \ldots, e_j, f_j \leq k$.[6]

$$
\begin{aligned}
&\mathbb{P}_0 = PM(B,\ I(B)), &&\mathbb{E}_0 = \mathbb{F}_0 = \langle \top, \emptyset, \emptyset \rangle, \\
&\mathbb{P}_i = \mathbb{P}_{i-1} \sqcup PM(P[t/i],\ Out(\mathbb{P}_{i-1}) \cup I(P[t/i])), &&(i = 1, \ldots, k) \\
&\mathbb{Q}_i = PM(Q[t/i],\ Out(\mathbb{P}_i) \cup I(Q[t/i])), &&(i = 0, \ldots, k) \\
&\mathbb{E}_i = \mathbb{E}_{i-1} \sqcup PM(E_i[e_i],\ Out(\mathbb{P}_{e_i}) \cup Out(\mathbb{E}_{i-1}) \cup I(E_i[e_i])) &&(i = 1, \ldots, j) \\
&\mathbb{F}_i = PM(F_i[f_i],\ Out(\mathbb{P}_{f_i}) \cup Out(\mathbb{E}_i) \cup I(F_i[f_i])) &&(i = 1, \ldots, j) \\
&\mathbb{R}_{j,k} = \mathbb{P}_k \sqcup \mathbb{Q}_k \sqcup \mathbb{E}_j \sqcup \mathbb{F}_j.
\end{aligned}
$$

We refer to $\mathbb{R}_{j,k}$ as the *incremental composition* of the incremental theory $\langle B, P[t], Q[t] \rangle$ w.r.t. the online progression $\langle E, F \rangle$. Unlike the k-expanded propositional formula $R_{j,k}$, each component in the incrementally composed module $\mathbb{R}_{j,k}$ is simplified before being joined.

Given an incremental theory $\langle B, P[t], Q[t] \rangle$ and an online progression $\langle E, F \rangle$, we assume the precedence relation $\prec^*$ on the set

$$
\begin{aligned}
\{ &B, P[t/1], P[t/2], \ldots, P[t/k], E_1[e_1], E_2[e_2], \ldots, E_j[e_j], \\
&Q[t/0], Q[t/1], \ldots, Q[t/k], F_1[f_1], F_2[f_2], \ldots, F_j[f_j] \}
\end{aligned}
\tag{5}
$$

as the transitive closure of the following relation $\prec$:

$$
\begin{array}{lll}
B \prec P[t/1] \prec \cdots \prec P[t/k], & B \prec Q[t/0], & P[t/i] \prec Q[t/i] \quad (i \geq 1) \\
E_1[e_1] \prec \cdots \prec E_j[e_j], & E_i[e_i] \prec F_i[f_i] & (i \geq 1) \\
P[t/e_i] \prec E_i[e_i], & P[t/f_i] \prec F_i[f_i] & (i \geq 1).
\end{array}
$$

Additionally, we say that two formulas F and G in (5) *coexist* if they belong to (4) for some nonnegative integers j, k. Intuitively, F and G coexist if they are eventually composed together into some $\mathbb{R}_{j,k}$. For example, $P[t/1]$ and $Q[t/3]$ coexist as they are both present in $\mathbb{R}_{0,3}$, whereas $Q[t/1]$ and $Q[t/3]$ do not. In Fig. 1 the shaded blocks denote coexisting formulas for $j = 1, k = 2$.

---

[6] For notational simplicity, we define $E_0[e_0]$ and $F_0[f_0]$ to be $\top$, e_0, f_0 to be 0, and $I(E_0[e_0])$ and $I(F_0[f_0])$ to be $\emptyset$.

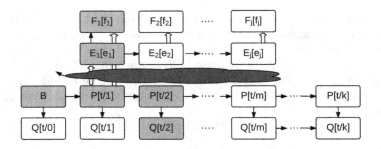

Fig. 1. Precedence graph of component formulas

Definition 7. *We say that an incremental theory $\langle B, P[t], Q[t] \rangle$ and an online progression $\langle E, F \rangle$ are* mutually revisable *if, for any distinct coexisting formulas G and H in (5), we have that $G \prec^* H$ whenever $Head(G) \cap (At(H) \setminus I(H)) \neq \emptyset$.*

Theorem 2 (Correctness of Incremental Composition). *Given an incremental theory $\langle B, P[t], Q[t] \rangle$ and an online progression $\langle E, F \rangle$ which are modular and mutually revisable, and nonnegative integers j, k such that $e_1, \ldots, e_j, f_j \leq k$, let $R_{j,k}$ be the k-expanded propositional formula w.r.t. $\langle E, F \rangle$, and let $\mathbb{R}_{j,k} = (H, \mathcal{I}, \mathcal{O})$ be the incremental composition of $\langle B, P[t], Q[t] \rangle$ w.r.t. $\langle E, F \rangle$. Then the stable models of $R_{j,k}$ coincide with the stable models of H, the formula of $\mathbb{R}_{j,k}$.*

It turns out that in the event that all explicit inputs (all $I(F)$ for $F \in (4)$) are empty, i.e., in the offline case, mutual revisability is a stronger condition than modularity. This means that in the offline case it is sufficient to just check that an incremental theory is mutually revisable.

Using Theorem 2, it is possible to incrementally ground, simplify, and solve a traditional ASP incremental theory in order to find the minimum k such that $R_{j,k}$ has an answer set without repeating previous work performed. In practice, this allows for a significant speedup when performing an iterative deepening search, such as when searching for a minimum length plan to accomplish a goal. In addition, the system is able to account for specific forms of online input in an equally efficient manner by allowing external information to be asserted in the online progression during execution.

Example 1 (Online ASP Solving). *Given an incremental theory*

$$\langle B, P[t], Q[t] \rangle = \langle \top, \quad \neg((t-1):q) \wedge \neg((t-1):p) \to t:p, \quad \neg(t:p) \to \bot \rangle$$

such that $I(B) = \emptyset$, $I(P[t/i]) = \{(i-1):q\}$, and $I(Q[t/i]) = \emptyset$ and an online progression $\langle E, F \rangle$.

Initially, $\mathbb{R}_{0,0}$ is constructed such that

$$\mathbb{P}_0 = \langle \top, \emptyset, \emptyset \rangle,$$
$$\mathbb{Q}_0 = PM(Q[t/0], Out(\mathbb{P}_0) \cup I(Q[t/0])) = \langle \top \to \bot, \emptyset, \emptyset \rangle,$$
$$\mathbb{R}_{0,0} = \mathbb{P}_0 \sqcup \mathbb{Q}_0 = \langle \top \to \bot, \emptyset, \emptyset \rangle.$$

Clearly, $\top \to \bot$ *has no stable models. As a result,* $\mathbb{R}_{0,1}$ *is attempted as follows:*

$$\mathbb{P}_1 = \mathbb{P}_0 \sqcup PM(P[t/1], Out(\mathbb{P}_0) \cup I(P[t/1])) = \mathbb{P}_0 \sqcup \langle \neg(0{:}q) \to 1{:}p, \{0{:}q\}, \{1{:}p\}\rangle$$
$$= \langle \neg(0{:}q) \to 1{:}p, \{0{:}q\}, \{1{:}p\}\rangle,$$
$$\mathbb{Q}_1 = PM(Q[t/1], Out(\mathbb{P}_1) \cup I(Q[t/1])) = \langle \neg(1{:}p) \to \bot, \{1{:}p\}, \emptyset\rangle,$$
$$\mathbb{R}_{0,1} = \mathbb{P}_1 \sqcup \mathbb{Q}_1 = \langle(\neg(0{:}q) \to 1{:}p) \wedge (\neg(1{:}p) \to \bot), \{0{:}q\}, \{1{:}p\}\rangle.$$

Solving is then halted as the formula in $\mathbb{R}_{0,1}$ *has one stable model* $\{1{:}p\}$. *However, with the arrival of the external event* $E_1[0] = 0{:}q$ *and* $F_1[0] = \top$ *such that* $I(E_1[0]) = I(F_1[0]) = \emptyset$, *we then must consider the construction of* $\mathbb{R}_{1,k}$ *rather than* $\mathbb{R}_{0,k}$. $\mathbb{R}_{1,1}$ *is constructed such that*

$$\mathbb{E}_1 = \langle \top, \emptyset, \emptyset \rangle \sqcup PM(E_1[e_1], Out(\mathbb{P}_{e_1}) \cup Out(\mathbb{E}_0) \cup I(E_1[e_1])) \qquad (e_1 \text{ is } 0)$$
$$= \langle \top, \emptyset, \emptyset \rangle \sqcup \langle 0{:}q, \emptyset, \{0{:}q\}\rangle = \langle 0{:}q, \emptyset, \{0{:}q\}\rangle,$$
$$\mathbb{F}_1 = PM(F_1[f_1], Out(\mathbb{P}_{f_1}) \cup Out(\mathbb{E}_1) \cup I(F_1[f_1])) \qquad (f_1 \text{ is } 0)$$
$$= \langle \top, \{0{:}q\}, \emptyset\rangle,$$
$$\mathbb{R}_{1,1} = \mathbb{P}_1 \sqcup \mathbb{Q}_1 \sqcup \mathbb{E}_1 \sqcup \mathbb{F}_1 = \langle(\neg(0{:}q) \to 1{:}p) \wedge 0{:}q \wedge (\neg(1{:}p) \to \bot), \emptyset, \{0{:}q, 1{:}p\}\rangle.$$

Once again, the formula of $\mathbb{R}_{1,1}$ *has no stable models, so the search is deepened to* $\mathbb{R}_{1,2}$ *as follows:*

$$\mathbb{P}_2 = \mathbb{P}_1 \sqcup PM(P[t/2], Out(\mathbb{P}_1) \cup I(P[t/2]))$$
$$= \mathbb{P}_1 \sqcup \langle \neg(1{:}q) \wedge \neg(1{:}p) \to 2{:}p, \{1{:}q, 1{:}p\}, \{2{:}p\}\rangle$$
$$= \langle(\neg(0{:}q) \to 1{:}p) \wedge (\neg(1{:}q) \wedge \neg(1{:}p) \to 2{:}p), \{0{:}q, 1{:}q\}, \{1{:}p, 2{:}p\}\rangle,$$
$$\mathbb{Q}_2 = PM(Q[t/2], Out(\mathbb{P}_2) \cup I(Q[t/2])) = \langle \neg(2{:}p) \to \bot, \{1{:}p, 2{:}p\}, \emptyset\rangle,$$
$$\mathbb{R}_{1,2} = \mathbb{P}_2 \sqcup \mathbb{Q}_2 \sqcup \mathbb{E}_1 \sqcup \mathbb{F}_1$$
$$= \langle(\neg(0{:}q) \to 1{:}p) \wedge (\neg(1{:}q) \wedge \neg(1{:}p) \to 2{:}p) \wedge 0{:}q \wedge (\neg(2{:}p) \to \bot),$$
$$\{1{:}q\}, \{0{:}q, 1{:}p, 2{:}p\}\rangle.$$

The formula of $\mathbb{R}_{1,2}$ *has a single stable model* $\{0{:}q, 2{:}p\}$.

4 Online Execution of $o\mathcal{BC}+$

Based on the concept of online propositional theories in the previous section, we define an online extension of $\mathcal{BC}+$, which provides a structured input language for online answer set solving that ensures the syntactic conditions of modularity and mutual revisability.

4.1 Syntax

The signature hierarchy of $o\mathcal{BC}+$ is extended from that of (offline) $\mathcal{BC}+$ by adding new sets of symbols called *external fluent constants* (denoted σ^{ef}) and *external action constants* (denoted σ^{ea}) such that $\sigma^{ef} \subseteq \sigma^{fl}$ and $\sigma^{ea} \subseteq \sigma^{act}$ (Fig. 2). We assume that the domain of each external fluent and action constant contains a special element u, which represents an *unknown* value. The syntax of causal laws is defined the same as in Sect. 2.2 except that external constants are allowed in the bodies but not in the heads.

An *observation* is an expression of the form

observed $c = v$ **at** m (6)

signature σ				
fluents (σ^{fl})			actions (σ^{act})	
internal		external (σ^{ef})	internal	external (σ^{ea})
regular	static. det.			

Fig. 2. Hierarchy of $o\mathcal{BC}+$ signature

where $c=v$ is an atom such that c is an external constant, v is a value other than u, and m is a nonnegative integer. An *observational constraint* is an expression of the form

$$\textbf{constraint } F \textbf{ at } m \qquad (7)$$

where F is a propositional formula containing no external constants and m is a nonnegative integer. We say that an observation (6) or observational constraint (7) is *dynamic* if it contains some action constant, otherwise we say it is *static*.

An *observation stream*, denoted $\mathcal{O}_{n,\widehat{m}}$, is a list $O_1, \ldots O_n$ such that

- for each $1 \leq i \leq n$, O_i is a finite set of observations (6) and observational constraints (7), and m_i is the maximum of each m among the static observations and constraints, and $m + 1$ among the dynamic observations and constraints;
- $\widehat{m}$ is the maximum of each m_i $(1 \leq i \leq n)$;
- for each external constant c and each m in $\{1, \ldots, \widehat{m}\}$, there is at most one observation (6) in $O_1 \cup \cdots \cup O_n$.

4.2   Semantics

Since any future external constants can take any values arbitrarily, the transition system in the presence of external constants can be defined straightforwardly by assigning arbitrary values to the external constants. That is, given an $o\mathcal{BC}+$ description and length m, we extend the propositional formula $PF_m(D)$ in Sect. 2.2 by adding the formulas $\{i : c = v\}^{\text{ch}}$ for every external constant c, every $v \in Dom(c)$, and every $i \in \{0, \ldots, m\}$ if c is a fluent constant, and $i \in \{0, \ldots, m-1\}$ if c is an action constant.

On the other hand, it is more meaningful to assume that the external input is "abnormal" to the system dynamics, and we want to "minimize" their effects. In other words, rather than arbitrary histories, we are interested in histories which are "normal" with respect to $\mathcal{O}_{n,\widehat{m}}$. Intuitively, in a normal history, the external constants are mapped to an unknown value unless the external observation asserts otherwise.

Formally, a history $\mathcal{H}_k$ of a transition system of length k is a sequence $\langle s_0, e_0, s_1, \ldots, e_{k-1}, s_k \rangle$ such that each $\langle s_i, e_i, s_{i+1} \rangle$ $(0 \leq i \leq k - 1)$ is a transition. We say that $\mathcal{H}_k$ satisfies $i : F$ where F is a fluent formula (action formula, respectively) if $s_i \models F$ ($e_i \models F$, respectively). Given an observation stream $\mathcal{O}_{n,\widehat{m}}$ and history $\mathcal{H}_k$ such that $k \geq \widehat{m}$, we say that $\mathcal{H}_k$ *observes* $\mathcal{O}_{n,\widehat{m}}$ if,

- for each observation (6) in $\mathcal{O}_{n,\widehat{m}}$, history $\mathcal{H}_k$ satisfies $m : c = v$, and
- for each observational constraint (7) in $\mathcal{O}_{n,\widehat{m}}$, history $\mathcal{H}_k$ satisfies $m : F$.

Named Sets:	Value:	
Status	$\{\mathtt{on}, \mathtt{off}\}$	
Boolean	$\{\mathtt{t}, \mathtt{f}\}$	
ExtBoolean	$\{\mathtt{t}, \mathtt{f}, \mathtt{u}\}$	

Notation: s ranges over elements in *Status*; v ranges over elements in *Boolean*.

Constants:	Type:	Domain:
Sw	regular fluent	*Status*
Light	statically determined fluent	*Status*
Flip	action	*Boolean*
Fault	regular fluent	*ExtBoolean*
ExtFault	external fluent	*ExtBoolean*
ReplaceBulb	action	*Boolean*

Causal laws:

inertial *Sw*   **inertial** *Fault* **after** *ReplaceBulb* $=\mathtt{f}$
exogenous *Flip*   **exogenous** *ReplaceBulb*

Flip **causes** $Sw = \mathtt{on}$ **if** $Sw = \mathtt{off}$ **nonexecutable** *ReplaceBulb* **if** $Flip = \mathtt{t}$
Flip **causes** $Sw = \mathtt{off}$ **if** $Sw = \mathtt{on}$ **caused** $Fault = v$ **if** $ExtFault = v$
default $Light = s$ **if** $Sw = s$ **caused** $Light = \mathtt{off}$ **if** $Fault = \mathtt{t}$
                                                  default $Fault = \mathtt{u}$ **after** *ReplaceBulb*

Fig. 3. Online faulty switch elaboration in $o\mathcal{BC}+$.

We say that $\mathcal{H}_k$ is *normal* with respect to $\mathcal{O}_{n,\widehat{m}}$, if it observes $\mathcal{O}_{n,\widehat{m}}$, and, for each external fluent constant (action constant, respectively) c and each $i \in \{0, \dots, k\}$ ($i \in \{0, \dots, k-1\}$, respectively), $\mathcal{H}_k$ satisfies $i : c = \mathtt{u}$ when there is no observation (6) in $\mathcal{O}_{n,\widehat{m}}$ such that $m = i$.

Intuitively, observations are non-monotonic observations the agent has made regarding the defined external actions and fluents. Meanwhile, the observational constraints serve to further limit past histories according to what the agent knows, such as what actions the agent has executed.

Example 2. *Consider a light switch problem where the light bulb may be burnt out. In the event this is the case, the light will not turn on until the bulb is replaced. This problem can be formalized in $o\mathcal{BC}+$ as shown in Fig. 3.[7] Intuitively, Fault is the agent's internal model of whether the light is burnt out, while ExtFault represents the agent's external observations.*

Normally, Fault is governed by inertia. However, in the event the agent gains additional information (i.e. observes whether there has been a fault) Fault is updated to reflect this. Performing ReplaceBulb will then reset the agent's internal model and the agent once again assumes that the fault has been fixed.

---

[7] It uses several abbreviations of causal laws as defined in [4].

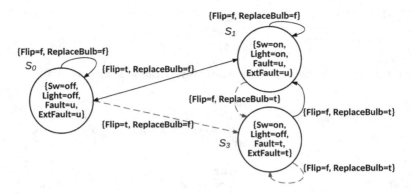

Fig. 4. A partial transition system of $\mathcal{D}^{oBC+}_{switch}$

Consider the transition system corresponding to the toggle switch elaboration. The minimum length history from

$$\mathcal{S}_0 = \{Switch = \mathtt{off}, \ Light = \mathtt{off}, \ Fault = \mathtt{u}, \ ExtFault = \mathtt{u}\}$$

to a state $\mathcal{S}$ such that $\mathcal{S} \models Light = \mathtt{on}$ are $\langle \mathcal{S}_0, \mathcal{E}_0, \mathcal{S}_1 \rangle$, and $\langle \mathcal{S}_0, \mathcal{E}_0, \mathcal{S}_2 \rangle$ where

$$\mathcal{E}_0 = \{Flip = \mathtt{t}, \ ReplaceBulb = \mathtt{f}\},$$
$$\mathcal{S}_1 = \{Switch = \mathtt{on}, \ Light = \mathtt{on}, \ Fault = \mathtt{u}, \ ExtFault = \mathtt{u}\}, and$$
$$\mathcal{S}_2 = \{Switch = \mathtt{on}, \ Light = \mathtt{on}, \ Fault = \mathtt{f}, \ ExtFault = \mathtt{f}\}.$$

Intuitively, the difference between $\mathcal{S}_1$ and $\mathcal{S}_2$ is that in $\mathcal{S}_1$ the agent has no knowledge as to whether a fault has occurred (i.e. the bulb has burnt out) whereas in $\mathcal{S}_2$ the agent knows that the light is fine. Of the two, only $\langle \mathcal{S}_0, \mathcal{E}_0, \mathcal{S}_1 \rangle$ is normal with respect to the online progression $\mathcal{O}_{0,0} = [\ ]$.

If, following the execution of Flip, the agent observes that a fault did occur, the knowledge can be added to the online progression producing

$$\mathcal{O}_{1,1} = [\{\textbf{observed } ExtFault = \mathtt{t} \textbf{ at } 1, \ \textbf{constraint } Flip = \mathtt{t} \textbf{ at } 0\}].$$

(The addition of the constraint enforces that the agent has executed $Flip = \mathtt{t}$ and prevents that action from being revised.) The new minimum length history from $\mathcal{S}_0$ to a state $\mathcal{S}$ such that $\mathcal{S} \models Light = \mathtt{on}$ and is normal w.r.t. $\mathcal{O}_{1,1}$ is $\langle \mathcal{S}_0, \mathcal{E}_0, \mathcal{S}_3, \mathcal{E}_1, \mathcal{S}_1 \rangle$ where

$$\mathcal{S}_3 = \{Switch = \mathtt{on}, Light = \mathtt{off}, Fault = \mathtt{t}, ExtFault = \mathtt{t}\}, and$$
$$\mathcal{E}_1 = \{Flip = \mathtt{f}, \ ReplaceBulb = \mathtt{t}\}.$$

This history essentially prescribes that the agent should replace the light bulb in order to attempt to fix the fault.

A partial specification of the transition system is shown in Fig. 4. The dashed edges depend on the assertion of an external constant and are not considered for transitions in normal histories.

Given an $o\mathcal{BC}+$ action description D, an observation stream $\mathcal{O}_{n,\widehat{m}}$, and some incrementally parametrized formula $\mathcal{Q}[t]$, we define the corresponding incremental theory $\langle B, P[t], Q[t] \rangle^{D,\mathcal{Q}[t]}$ and the online progression $\langle E, F \rangle^{\mathcal{O}_{n,\widehat{m}}}$ as follows.

$$B = \bigwedge \begin{cases} 0:F \leftarrow 0:G & \text{for each static law (1) in } D \\ 0:\{f=v\}^{\text{ch}} & \text{for each regular fluent } f \text{ and each } v \in Dom(f) \\ 0:\{f=u\}^{\text{ch}} & \text{for each external fluent } f \\ 0:UEC_{\sigma fl} & \end{cases}$$

$$P[t] = \bigwedge \begin{cases} t:F \leftarrow t:G & \text{for each static law (1) in } D \\ (t-1):F \leftarrow (t-1):G & \text{for each action dynamic law (1) in } D \\ t:F \leftarrow t:G \wedge (t-1):H & \text{for each fluent dynamic law (2) in } D \\ t:\{f=u\}^{\text{ch}} & \text{for each external fluent } f \\ (t-1):\{a=u\}^{\text{ch}} & \text{for each external action } a \\ t:UEC_{\sigma fl} & \\ (t-1):UEC_{\sigma act} & \end{cases}$$

$$Q[t] = \neg\neg\mathcal{Q}[t]$$

$$E_i[m_i] = \bigwedge \begin{cases} m_i:c=v & \text{for each observation (6) } \in O_i \\ \neg\neg m_i:F & \text{for each observational constraint (7) } \in O_i \end{cases}$$

$$F_i[m_i] = \top$$

The stable models of their incremental composition represent histories that are normal w.r.t. the observation.

Given a $o\mathcal{BC}+$ signature σ we define $At(\sigma)$ to be the set of atoms $c=v$ where $c \in \sigma$ and $v \in Dom(c)$. Furthermore, we define $At_u(\sigma)$ to be the set of all such atoms such that $v \neq u$.

We define the sets of explicit external inputs as follows:

- $I(B) = At_u(0:\sigma^{ef})$,
- $I(P[t/i]) = At_u(i:\sigma^{ef} \cup (i-1):\sigma^{ea})$,
- $I(Q[t/i]) = At_u(\bigcup_{0 \leq j < i}(j:\sigma^{ef} \cup j:\sigma^{ea}) \cup i:\sigma^{ef})$, and
- $I(E_i) = I(F_i) = \emptyset$.

The following proposition asserts that the translation of an $o\mathcal{BC}+$ description into propositional formulas ensures modularity and mutual revisability.

Theorem 3 (Modular and Mutually Revisable Construction). *Given an $o\mathcal{BC}+$ action description D and an observation stream $\mathcal{O}_{n,\widehat{m}}$, and a step-parameterized formula $\mathcal{Q}[t]$, the corresponding incremental theory $\langle B, P[t], Q[t] \rangle^{D,\mathcal{Q}[t]}$ and the corresponding online progression $\langle E, F \rangle^{\mathcal{O}_{n,\widehat{m}}}$ are modular and mutually revisable.*

The next theorem asserts that the stable models of the incremental assembly represents the histories in the transition system that are normal with respect to the online stream.

Theorem 4 (Correctness of Incremental Assembly). *Given an $o\mathcal{BC}+$ action description D, an observation stream $\mathcal{O}_{n,\widehat{m}}$, a step-parameterized $\mathcal{Q}[t]$, and some $k \geq \widehat{m}$, let $\mathbb{R}_{\widehat{m},k} = \langle H, \mathcal{I}, \mathcal{O} \rangle$ be the incremental composition of $\langle B, P[t], Q[t] \rangle^{D,\mathcal{Q}[t]}$ w.r.t. $\langle E, F \rangle^{\mathcal{O}_{n,\widehat{m}}}$. The stable models of H represent the histories of length k in the transition system described by D which (i) observe $\mathcal{O}_{n,\widehat{m}}$, (ii) are normal with respect to $\mathcal{O}_{n,\widehat{m}}$, and (iii) satisfy $\mathcal{Q}[t/k]$.*

5 Conclusion

We extended the concept of online answer set solving to propositional formulas under the stable model semantics, and based on this, designed a high level online action language $o\mathcal{BC}+$, whose structure ensures the syntactic conditions that are required for the correctness of online answer set solving.

Another high level language for OCLINGO ensuring the modularity condition of OCLINGO is Online Agent Logic Programming language from [11]. However, this is based on Agent Logic Programs, instead of action languages, and lacks negation.

$o\mathcal{BC}+$ is implemented in Version 3 of CPLUS2ASP. In addition to the static and the incremental mode already available in Version 2, which invoke CLINGO v3.0.5 and ICLINGO v3.0.5, respectively, newly introduced is the reactive mode, which invokes OCLINGO v3.0.92. The "reactive bridge" is a new software component, and acts as an intermediary between OCLINGO and a user-provided agent controller system. It allows the agent controller system to provide an $o\mathcal{BC}+$ observation stream during execution and receive updated solutions in the form of transition system histories. We refer the reader to the system homepage (http://reasoning.eas.asu.edu/cplus2asp) for more details and experiment results.

Acknowledgements. We are grateful to Michael Bartholomew, Yi Wang, and the anonymous referees for their useful comments on the draft. This work was partially supported by the National Science Foundation under Grant IIS-1319794 and South Korea IT R&D program MKE/KIAT 2010-TD-300404-001.

References

1. Valle, E.D., Ceri, S., van Harmelen, F., Fensel, D.: It's a streaming world! Reasoning upon rapidly changing information. IEEE Intell. Syst. **24**(6), 83–89 (2009)
2. Gebser, M., Grote, T., Kaminski, R., Schaub, T.: Reactive answer set programming. In: Delgrande, J.P., Faber, W. (eds.) LPNMR 2011. LNCS, vol. 6645, pp. 54–66. Springer, Heidelberg (2011)
3. Janhunen, T., Oikarinen, E., Tompits, H., Woltran, S.: Modularity aspects of disjunctive stable models. J. Artif. Intell. Res. **35**, 813–857 (2009)
4. Babb, J., Lee, J.: Action language $\mathcal{BC}+$: preliminary report. In: Proceedings of the AAAI Conference on Artificial Intelligence (AAAI) (2015)
5. Giunchiglia, E., Lee, J., Lifschitz, V., McCain, N., Turner, H.: Nonmonotonic causal theories. Artif. Intell. **153**(1–2), 49–104 (2004)
6. Lee, J., Lifschitz, V., Yang, F.: Action language $\mathcal{BC}$: preliminary report. In: Proceedings of International Joint Conference on Artificial Intelligence (IJCAI) (2013)
7. Babb, J., Lee, J.: Module theorem for the general theory of stable models. TPLP **12**(4–5), 719–735 (2012)
8. Ferraris, P.: Answer sets for propositional theories. In: Baral, C., Greco, G., Leone, N., Terracina, G. (eds.) LPNMR 2005. LNCS (LNAI), vol. 3662, pp. 119–131. Springer, Heidelberg (2005)
9. Bartholomew, M., Lee, J.: Stable models of multi-valued formulas: partial vs. total functions. In: Proceedings of International Conference on Principles of Knowledge Representation and Reasoning (KR), pp. 583–586 (2014)

10. Ferraris, P., Lee, J., Lifschitz, V., Palla, R.: Symmetric splitting in the general theory of stable models. In: Proceedings of International Joint Conference on Artificial Intelligence (IJCAI), pp. 797–803. AAAI Press (2009)
11. Cerexhe, T., Gebser, M., Thielscher, M.: Online agent logic programming with oClingo. In: Pham, D.-N., Park, S.-B. (eds.) PRICAI 2014. LNCS, vol. 8862, pp. 945–957. Springer, Heidelberg (2014)

aspartame: Solving Constraint Satisfaction Problems with Answer Set Programming

Mutsunori Banbara[3], Martin Gebser[1,6], Katsumi Inoue[4], Max Ostrowski[6], Andrea Peano[5], Torsten Schaub[2,6(✉)], Takehide Soh[3], Naoyuki Tamura[3], and Matthias Weise[6]

[1] Aalto University, HIIT, Greater Helsinki, Finland
[2] INRIA Rennes, Rennes, France
[3] Kobe University, Kobe, Japan
[4] NII Tokyo, Tokyo, Japan
[5] University of Ferrara, Ferrara, Italy
[6] University of Potsdam, Potsdam, Germany
torsten@cs.uni-potsdam.de

Abstract. Encoding finite linear CSPs as Boolean formulas and solving them by using modern SAT solvers has proven to be highly effective by the award-winning *sugar* system. We here develop an alternative approach based on ASP that serves two purposes. First, it provides a library for solving CSPs as part of an encompassing logic program. Second, it furnishes an ASP-based CP solver similar to *sugar*. Both tasks are addressed by using first-order ASP encodings that provide us with a high degree of flexibility, either for integration within ASP or for easy experimentation with different implementations. When used as a CP solver, the resulting system *aspartame* re-uses parts of *sugar* for parsing and normalizing CSPs. The obtained set of facts is then combined with an ASP encoding that can be grounded and solved by off-the-shelf ASP systems. We establish the competitiveness of our approach by empirically contrasting *aspartame* and *sugar*.

1 Introduction

Encoding finite linear Constraint Satisfaction Problems (CSPs; [2]) as propositional formulas and solving them by using modern solvers for Satisfiability Testing (SAT; [3]) has proven to be a highly effective approach by the award-winning *sugar*[1] system. The CP solver *sugar* reads a CSP instance and transforms it into a propositional formula in Conjunctive Normal Form (CNF). The translation relies on the order encoding [4,5], and the resulting CNF formula can be solved by an off-the-shelf SAT solver.

This paper is a greatly revised version of the workshop paper [1]. The work was funded by [1]AoF (251170), [6]DFG (SCHA 550/10-1), [5]5 × 1000 (UNIFE 2011), and [3]JSPS (KAKENHI 15K00099).

T. Schaub–Affiliated with Simon Fraser University, Canada, and IIIS Griffith University, Australia.

[1] http://bach.istc.kobe-u.ac.jp/sugar.

© Springer International Publishing Switzerland 2015
F. Calimeri et al. (Eds.): LPNMR 2015, LNAI 9345, pp. 112–126, 2015.
DOI: 10.1007/978-3-319-23264-5_10

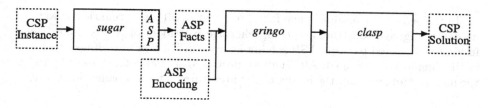

Fig. 1. Architecture of *aspartame*.

In what follows, we elaborate upon an alternative approach based on Answer Set Programming (ASP; [6]) and present the resulting *aspartame*[2] framework, serving two purposes. First, *aspartame* provides a library for solving CSPs as part of an encompassing logic program. Second, it constitutes an ASP-based CP solver similar to *sugar*. The major difference between *sugar* and *aspartame* rests upon the implementation of the translation of CSPs into Boolean constraint problems. While *sugar* implements a translation into CNF in Java, *aspartame* starts with a translation into a set of facts.[3] When used as a library, this set of facts (representing the CSP) must be supplied by the user.[4] In turn, these facts are combined with a general-purpose ASP encoding for CP solving (also based on the order encoding), which is subsequently instantiated by an off-the-shelf ASP grounder, in our case *gringo*. The resulting propositional logic program is then solved by an off-the-shelf ASP solver (here *clasp*). The architecture of *aspartame* is given in Fig. 1.

The high-level approach of ASP has obvious advantages. First, instantiation is done by general-purpose ASP grounders rather than dedicated implementations. Second, the elaboration tolerance of ASP allows for easy maintenance and modifications of encodings. And finally, it is easy to experiment with novel or heterogeneous encodings. However, the question is whether the high-level approach of *aspartame* matches the performance of the more dedicated *sugar* system. We empirically address this question by contrasting the performance of both CP solvers, while fixing the back-end solver to *clasp*, used as both a SAT and an ASP solver.

From an ASP perspective, we gain insights into advanced modeling techniques for solving CSPs. The ASP encoding implementing CP solving with *aspartame* has the following features:

- usage of function terms to abbreviate structural subsums
- avoidance of (artificial) intermediate Integer variables (to split sum expressions)
- a collection of encodings for the *alldifferent* constraint

---

[2] http://www.cs.uni-potsdam.de/aspartame.

[3] When used as CP solver, *aspartame* re-uses *sugar*'s front-end for parsing and normalizing (non-linear) CSPs. Also, we extended *sugar* to produce a fact-based representation.

[4] This will be integrated into *gringo*'s input language in the near future.

In the sequel, we assume some familiarity with ASP, its semantics as well as its basic language constructs. A comprehensive treatment of ASP can be found in [6], one oriented towards ASP solving is given in [7]. Our encodings are given in the language of *gringo* 4. Although we provide essential definitions of CSPs in the next section, we refer the reader to the literature [2] for a broader perspective.

2   Background

A *Constraint Satisfaction Problem* (CSP) is given by a pair $(\mathcal{V}, \mathcal{C})$ consisting of a set $\mathcal{V}$ of *variables* and a set $\mathcal{C}$ of *constraint clauses*. Every variable $x \in \mathcal{V}$ has an associated finite *domain* $D(x)$ such that either $D(x) = \{\top, \bot\}$ or $\emptyset \subset D(x) \subseteq \mathbb{Z}$; x is a *Boolean variable* if $D(x) = \{\top, \bot\}$, and an *Integer variable* otherwise. We denote the set of Boolean and Integer variables in $\mathcal{V}$ by $\mathcal{B}(\mathcal{V})$ and $\mathcal{I}(\mathcal{V})$, respectively. A constraint clause $C \in \mathcal{C}$ is a set of literals, and a *literal* is of the form e or $\bar{e}$, where e is either a Boolean variable in $\mathcal{B}(\mathcal{V})$, a linear inequality, or an *alldifferent* constraint. A *linear inequality* is an expression $\sum_{1 \leq i \leq n} a_i x_i \leq m$ in which m as well as all a_i for $1 \leq i \leq n$ are Integer constants and $x_1, \ldots, x_n$ are Integer variables in $\mathcal{I}(\mathcal{V})$. An *alldifferent* constraint (cf. [8]) applies if a subset $\{x_1, \ldots, x_n\}$ of Integer variables in $\mathcal{I}(\mathcal{V})$ is assigned to distinct values in their respective domains.[5]

Given a CSP $(\mathcal{V}, \mathcal{C})$, a *variable assignment* v is a (total) mapping $v : \mathcal{V} \to \bigcup_{x \in \mathcal{V}} D(x)$ such that $v(x) \in D(x)$ for every $x \in \mathcal{V}$. A Boolean variable $x \in \mathcal{B}(\mathcal{V})$ is *satisfied* wrt v if $v(x) = \top$. Likewise, a linear inequality $\sum_{1 \leq i \leq n} a_i x_i \leq m$ is *satisfied* wrt v if $\sum_{1 \leq i \leq n} a_i v(x_i) \leq m$ holds. An *alldifferent* constraint over subsets $\{x_1, \ldots, x_n\}$ of $\mathcal{I}(\mathcal{V})$ is *satisfied* wrt v if $v(x_i) \neq v(x_j)$ for all $1 \leq i < j \leq n$, respectively. Any Boolean variable, linear inequality, or *alldifferent* constraint that is not satisfied wrt v is *unsatisfied* wrt v. A constraint clause $C \in \mathcal{C}$ is *satisfied* wrt v if there is some literal $e \in C$ (or $\bar{e} \in C$) such that e is satisfied (or unsatisfied) wrt v. The assignment v is a *solution* for $(\mathcal{V}, \mathcal{C})$ if every $C \in \mathcal{C}$ is satisfied wrt v.

For illustration, consider a CSP $(\mathcal{V}, \mathcal{C})$ with Boolean and Integer variables $\mathcal{B}(\mathcal{V}) = \{b\}$ and $\mathcal{I}(\mathcal{V}) = \{x, y, z\}$, where $D(x) = D(y) = D(z) = \{1, 2, 3\}$, and constraint clauses $\mathcal{C} = \{C_1, C_2, C_3\}$ as follows:

$$C_1 = \{alldifferent(x, y, z)\} \tag{1}$$

$$C_2 = \{b, 4x - 3y + z \leq 0\} \tag{2}$$

$$C_3 = \{\bar{b}, -4x + 3y \leq -6\} \tag{3}$$

---

[5] Linear and non-linear inequalities relying on further comparison operators, such as $<, >, \geq, =$, and $\neq$, can be converted into the considered format via appropriate replacements [5]. Moreover, note that we here limit the constraints to the ones that are directly, i.e., without normalization by *sugar*, supported in our prototypical ASP encodings shipped with *aspartame*.

The *alldifferent* constraint in C_1 requires values assigned to x, y, and z to be mutually distinct. Assignments v satisfying the linear inequality $4x - 3y + z \leq 0$ in C_2 include $v(x) = 2$, $v(y) = 3$, and $v(z) = 1$ or $v(x) = 1$, $v(y) = 3$, and $v(z) = 2$, while assignments for the constraint $-4x + 3y \leq -6$ include $v(x) = 3$, $v(y) = 2$, and $v(z) = 1$ or $v(x) = 3$, $v(y) = 1$, and $v(z) = 2$. In view of the Boolean variable b, whose value allows for "switching" between the constraints in C_2 and C_3, we obtain the four solutions $v_1, \ldots, v_4$ for $(\mathcal{V}, \mathcal{C})$ shown alongside.

	b	x	y	z
v_1	$\bot$	2	3	1
v_2	$\bot$	1	3	2
v_3	$\top$	3	2	1
v_4	$\top$	3	1	2

3   The *aspartame* Approach

For using *aspartame* as a CP solver, we extended the front-end of the *sugar* system by an output component representing CSPs in terms of ASP facts. The latter also constitute the CSP instances when using *aspartame* as library. As usual, the resulting facts can then be combined with a first-order encoding processable with off-the-shelf ASP systems. In what follows, we describe *aspartame*'s fact format and we present dedicated ASP encodings utilizing function terms to capture substructures in CSP instances.

Fact Format. Facts express the variables and constraints of a CSP instance in the syntax of ASP grounders like *gringo*. Their format is easiest explained on the CSP from Sect. 2, whose fact representation is shown in Listing 1. While facts of the predicate `var/2` provide labels of Boolean variables, like b, the predicate `var/3` includes a third argument for declaring the domains of Integer variables, like x, y, and z. Domain declarations rely on function terms `range(l,u)`, standing for Integer intervals $[l, u]$. While one term, `range(1,3)`, suffices for the common domain $\{1, 2, 3\}$ of x, y, and z, in general, several intervals can be specified (via separate facts) to form non-continuous domains. Note that the interval format for Integer domains offers a compact fact representation of domains; e.g., the single term `range(1,10000)` captures a domain of 10000 elements. Furthermore, the usage of meaningful function terms avoids any need for artificial labels to refer to domains or parts thereof.

The literals of constraint clauses are also represented by means of function terms. In fact, the second argument of `constraint/2` in Line 3 of Listing 1 stands for *alldifferent*(x, y, z) from the constraint clause C_1 in (1), identified by the first argument of `constraint/2`. Since each fact of predicate `constraint/2` is supposed to describe a single literal only, constraint clause identifiers establish the connection between individual literals of a clause. The more complex term of the form `op(le,Σ,m)` in Line 5 stands for a linear inequality $\Sigma \leq m$. In particular, the inequality $4x - 3y + z \leq 0$ from C_2 is represented by nested `op(add,Σ,ax)` terms whose last argument ax and deepest Σ part are of the form `op(mul,a,x)`; such nesting corresponds to the precedence $(((4 * x) + (-3 * y)) + (1 * z)) \leq 0$. The representation by function terms captures linear inequalities of arbitrary arity and, as with Integer intervals, associates (sub)sums with canonical labels.

```
1   var(bool,b).        var(int,(x;y;z),range(1,3)).

3   constraint(1,global(alldifferent,arg(x,arg(y,arg(z,nil))))).
4   constraint(2,b).
5   constraint(2,op(le,op(add,op(add,op(mul,4,x),op(mul,-3,y)),op(mul,1,z)),0)).
6   constraint(3,op(neg,b)).
7   constraint(3,op(le,op(add,op(mul,-4,x),op(mul,3,y)),-6)).
```

Listing 1. Facts representing the CSP from Sect. 2.

The function term expressing the *alldifferent* constraint includes an argument list of the form $\texttt{arg}(x_1, \texttt{arg}(\ldots, \texttt{arg}(x_n, \texttt{nil}) \ldots))$, in which $x_1, \ldots, x_n$ refer to Integer variables. In Line 3 of Listing 1, an *alldifferent* constraint over arguments x is declared via $\texttt{global}(\texttt{alldifferent}, x)$; at present, $\texttt{alldifferent}$ is a fixed keyword in facts used by *aspartame*, but support for other kinds of global constraints can be added in the future.

First-Order Encoding. In addition to a dedicated output component of *sugar* for generating ASP facts, *aspartame* comes with several alternative first-order ASP encodings for solving CSP instances. In the following, we first describe a basic encoding that implements the order encoding techniques [5,9] concisely, and then present optimizations and extensions for the *alldifferent* constraints. We also show that these *aspartame* encodings can be used as library for solving CSPs and Constraint Optimization Problems. Due to lack of space, *aspartame* encodings presented here are restricted to unary constraint clauses and stripped off capacities for handling Boolean variables.

Basic Encoding. Listing 2 shows common auxiliary predicates shared by *aspartame* encodings. Given a CSP instance, for each Integer variable V, the domain values of the variables are kept in a *lua* table, and then the lower and upper bounds of each V are calculated via *lua* and captured in the second arguments of lb/2 and ub/2 respectively in Line 1–3. Each literal of a constraint clause is classified into a *alldifferent* constraint expressed by alldiff/1 or a linear inequality by wsum/1 in Line 5–7. The identifier in the first argument of constraint/2 is removed in Line 6–7, since the *aspartame* encoding presented here is restricted to unary constraint clauses. For each linear inequality $\sum_{i=1}^{n} a_i x_i \leq c$, in Line 6, Σ is sorted in descending order of $|D(x_i)|$ via *lua*, and then $\Sigma \leq c$ is captured in the argument of predicate wsum/1. Each sum Σ in $\texttt{op}(\_, \Sigma, \_)$ is decomposed into (sub)sums in Line 9–10, and then the lower and upper bounds of them are calculated and captured in the second arguments of inf/2 and sup/2 respectively in Line 13–19. In Line 11, a predicate $\texttt{unary\_exp}(e)$ is generated if e is an unary expression of the form $\texttt{op}(\texttt{mul}, a_i, x_i)$.

Listing 3 gives our encoding of Integer variables. For each variable V and each domain value A in $D(\texttt{V})$, we introduce a predicate $\texttt{p}(\texttt{V}, \texttt{A})$ expressing that $\texttt{V} \leq \texttt{A}$. The truth-assignments of the instances of p/2 are encoded by the choice rule in Line 1. The constraints in Line 2 and 3 ensure that each variable V has exactly one value in $D(\texttt{V})$. Note that the *lua* function $\texttt{getDom}(\texttt{V})$ returns $D(\texttt{V})$, and $\texttt{getSimpGT}(x, c) = \min\{d \in D(x) \mid d > c\}$. For illustration, consider an Integer variable $x \in \{2, 3, 4, 5, 6\}$ represented by $\texttt{var}(\texttt{int}, x, \texttt{range}(2, 6))$ as an

```
1   var(V) :- var(int,V,_).
2   lb(V,@getLB(V)) :- var(V).
3   ub(V,@getUB(V)) :- var(V).

5   global(I,global(Func,Args)) :- constraint(I,global(Func,Args)).
6   wsum(@sortWsum(L)) :- constraint(I,L), not global(_,L).
7   alldiff(Args) :- global(_,global(alldifferent,Args)).

9   exp(E)      :- wsum(op(_,E,_)).
10  exp(E1;E2) :- exp(op(add,E1,E2)).
11  unary_exp(op(mul,A,V)) :- exp(op(mul,A,V)).

13  inf(op(mul,A,V),A*LB) :- exp(op(mul,A,V)), A > 0, lb(V,LB).
14  inf(op(mul,A,V),A*UB) :- exp(op(mul,A,V)), A < 0, ub(V,UB).
15  inf(op(add,E1,E2),A+B) :- exp(op(add,E1,E2)), inf(E1,A), inf(E2,B).

17  sup(op(mul,A,V),A*UB) :- exp(op(mul,A,V)), A > 0, ub(V,UB).
18  sup(op(mul,A,V),A*LB) :- exp(op(mul,A,V)), A < 0, lb(V,LB).
19  sup(op(add,E1,E2),A+B) :- exp(op(add,E1,E2)), sup(E1,A), sup(E2,B).
```

Listing 2. Auxiliary predicates

```
1   { p(V,A) : A = @getDom(V) } :- var(V).
2   :- p(V,A), not p(V,B), B = @getSimpGT(V,A), A < UB, ub(V,UB).
3   :- not p(V,UB), var(V), ub(V,UB).
```

Listing 3. Encoding of Integer variables

ASP fact. The resulting propositional logic program and its answer sets obtained by *clasp* are as follows.

```
{p(x,2),p(x,3),p(x,4),p(x,5),p(x,6)}.
:- p(x,2),not p(x,3).
:- p(x,3),not p(x,4).
:- p(x,4),not p(x,5).
:- p(x,5),not p(x,6).
:- not p(x,6).
```

answer sets	interpretation
$\{p(x,6)\}$	$x = 6$
$\{p(x,5),p(x,6)\}$	$x = 5$
$\{p(x,4),p(x,5),p(x,6)\}$	$x = 4$
$\{p(x,3),p(x,4),p(x,5),p(x,6)\}$	$x = 3$
$\{p(x,2),p(x,3),p(x,4),p(x,5),p(x,6)\}$	$x = 2$

Constraints are encoded into clauses expressing conflict regions instead of conflict points. Especially, for any linear inequality $\sum_{i=1}^{n} a_i x_i \leq c$, the following holds [9].

$$\sum_{i=1}^{n} a_i x_i \leq c \Longleftrightarrow \begin{cases} (x_1 \leq \lfloor c/a_1 \rfloor) & (n = 1, a_1 > 0) \quad (4) \\ \neg(x_1 \leq \lceil c/a_1 \rceil - 1) & (n = 1, a_1 < 0) \quad (5) \\ \bigwedge_{d \in D(x_n)} \left((x_n \leq d - 1) \vee \sum_{i=1}^{n-1} a_i x_i \leq c - a_n d \right) & (n \geq 2, a_n > 0) \quad (6) \\ \bigwedge_{d \in D(x_n)} \left(\neg(x_n \leq d) \vee \sum_{i=1}^{n-1} a_i x_i \leq c - a_n d \right) & (n \geq 2, a_n < 0) \quad (7) \end{cases}$$

From this mathematical representation, we can define a recursive encoding procedure by considering $\sum_{i=1}^{n-1} a_i x_i \leq c - a_n d$ as the recursive call and also replacing the comparison of the form $x \leq c$ with the translation $\|x \leq c\|$. The translation $\|x \leq c\|$ is defined as $\top$ if $c \geq \max(D(x))$, $\bot$ else if $c < \min(D(x))$, and otherwise p(x,d) where $d = \max\{d' \in D(x) \mid d' \leq c\}$.

```
1   wsum(op(le,X,K-B*D)) :-
2          not unary(X) : opt_binary == 1;
3          wsum(op(le,op(add,X,op(mul,B,Y)),K)),
4          not p(Y,C) : B > 0, C = @getSimpLE(Y,D-1);
5          p(Y,D) : B < 0;
6          D = @getDomOpt(Y,B,K-Sup,K-Inf), inf(X,Inf), sup(X,Sup).

8   wsum(op(le,op(mul,B,Y),K-Inf)) :-
9          wsum(op(le,op(add,X,op(mul,B,Y)),K)),
10         K-Inf < Sup, inf(X,Inf), sup(op(mul,B,Y),Sup).
```

Listing 4. Encoding of $\sum_{i=1}^{n} a_i x_i \leq c$ $(n \geq 2)$

```
1   :- wsum(op(le,op(mul,A,X),K)), Inf <= K, K < Sup,
2          inf(op(mul,A,X),Inf), sup(op(mul,A,X),Sup),
3          not p(X,B) : A > 0; p(X,B) : A < 0;
4          B = @getLE(X,A,K).

6   :- wsum(op(le,op(mul,A,X),K)), Inf > K, inf(op(mul,A,X),Inf).
```

Listing 5. Encoding of $a_1 x_1 \leq c$

This encoding procedure can be optimized by considering the validity and inconsistency of the recursive part $\sum_{i=1}^{n-1} a_i x_i \leq c - a_n d$, which can reduce the number of iterations and clauses. The validity and inconsistency of the recursive part can be captured by inequalities $c - a_n d \geq \sup(\sum_{i=1}^{n-1} a_i x_i)$ and $c - a_n d < \inf(\sum_{i=1}^{n-1} a_i x_i)$ respectively, where $\inf(\Sigma)$ and $\sup(\Sigma)$ indicate the lower and upper bounds of linear expression Σ respectively. When the recursive part is valid, the clause containing it is unnecessary. When it is inconsistent, the literal of the recursive part can be removed, and moreover only one such clause is sufficient. Based on the observations above, we present a recursive ASP encoding of linear inequalities.

Encoding of $\sum_{i=1}^{n} a_i x_i \leq c$ $(n \geq 2)$ corresponding to (refoe:def:case:3) and (refoe:def:case:4) is presented in Listing 4. For a given linear inequality op(le, op(add, X, op(mul, B, Y)), K), in the first rule, a recursive part wsum(op(le, X, K − B ∗ D)) is generated for every domain value D in $D(Y)$ calculated via *lua* such that the recursive part becomes neither valid nor inconsistent, if "notp(Y, C)" holds when B > 0 (or p(Y, D) holds when B < 0). Intuitively, "notp(Y, C)" expresses $Y \geq D$ because of getSimpLE$(x,c) = \max\{d \in D(x) \mid d \leq c\}$. Note that the conditional literal in Line 2 is ignored, since the constant opt_binary is set to 0 initially. In the second rule, only one wsum(op(le, op(mul, B, Y), K − Inf)) corresponding to $a_n x_n \leq c - \inf(\sum_{i=1}^{n-1} a_i x_i)$ is generated if there exists at least one domain value in $D(Y)$ such that the recursive part becomes inconsistent.

Encoding of $a_1 x_1 \leq c$ corresponding to (6) and (7) is presented in Listing 5. For a given linear inequality op(le, op(mul, A, X), K), if it is neither valid nor inconsistent, the first rule calculates a bound B in $D(X)$ via *lua* and ensures that p(X, B) holds when A > 0 (or p(X, B) does not hold when A < 0). The *lua* function getLE(x,a,c) is defined as getSimpLE$(x,\lfloor c/a \rfloor)$ if $a > 0$, otherwise getSimpLE$(x,\lceil c/a \rceil - 1)$. If inconsistent, the second rule ensures that wsum(op(le, op(mul, A, X), K)) never holds.

```
1    :- wsum(op(le,op(add,op(mul,A,X),op(mul,B,Y)),K)),
2            not p(Y,C) : B > 0 , C = @getSimpLE(Y,D-1);
3               p(Y,D) : B < 0;
4            not p(X,E) : A > 0;
5               p(X,E) : A < 0;
6            D = @getDomOpt(Y,B,K-Sup,K-Inf), inf(op(mul,A,X),Inf), sup(op(mul,A,X),Sup),
7            E = @getLE(X,A,K-B*D).
```

Listing 6. Encoding of $\sum_{i=1}^{n} a_i x_i \leq c$ $(n = 2)$

We refer to the encoding of Listings 2–5 as *basic encoding*. This encoding can concisely implement CP solving based on the order encoding techniques by utilizing the feature of function terms. Moreover, it proposes an alternative approach to splitting sum expressions. In fact, the basic encoding splits them by generating the instances of predicate wsum/1 during recursive encoding, rather than by introducing intermediate Integer variables during preprocessing like a *sugar*'s CSP-to-CSP translation. It is noted that global constraints such as *alldifferent* and *cumulative* are first translated into linear inequalities by *sugar*'s front-end and then encoded by the basic encoding.

Optimizations and Extensions. The basic encoding generates some redundant clauses for linear inequalities of size two. Consider $x + y \leq 7$ represented by a function term op(le,op(add,op(mul,1,x),op(mul,1,y)),7), where $D(x) = D(y) = \{2, 3, 4, 5, 6\}$. The resulting propositional logic program is as follows.

```
:- not p(y,5). :- not p(x,5).
wsum(op(le,op(mul,1,x),2)) :- not p(y,4). :- wsum(op(le,op(mul,1,x),2)),not p(x,2).
wsum(op(le,op(mul,1,x),3)) :- not p(y,3). :- wsum(op(le,op(mul,1,x),3)),not p(x,3).
wsum(op(le,op(mul,1,x),4)) :- not p(y,2). :- wsum(op(le,op(mul,1,x),4)),not p(x,4).
```

The intermediate instances of wsum/1 are redundant and can be removed. This issue can be fixed by the *optimized encoding* which is an extension of the basic encoding by adding only the one rule of Listing 6 and by setting the constant opt_binary to 1. The rule of Listing 6 represents the special case of the first rule in Listing 4 for $\sum_{i=1}^{n} a_i x_i \leq c$ $(n = 2)$ and does not generate any intermediate instances of wsum/1. However, we keep generating such intermediate instances for $n > 2$ because they can be shared by different linear inequalities and can be effective in reducing the number of clauses. For the above example, the optimized encoding generates the following.

```
:- not p(y,5). :- not p(x,5).
:- not p(y,4), not p(x,2). :- not p(y,3), not p(x,3). :- not p(y,2), not p(x,4).
```

To extend our approach, we present four different encodings for the *alldifferent* constraints: *alldiffA*, *alldiffB*, *alldiffC*, and *alldiffD*. Listing 7 shows common auxiliary predicates shared by these encodings. For each variable V and domain value A in $D(V)$, we introduce a predicate val(V, A) that expresses V = A in Line 2. For each *alldifferent* constraint, its argument list is decomposed into a flat representation of the variables by alldiffArg/4 in Line 4–5, and then the lower and upper bounds of the variables are calculated and captured in the second and third arguments of alldiffRange/3 respectively. Note that, except in *alldiffD*, the *alldifferent* constraints are encoded by using the predicate val/2 [10].

```
1  alldiff :- alldiff(_).
2  val(V,A) :- var(int,V,_), p(V,A), not p(V,@getSimpLE(V,A-1)), alldiff.

4  alldiffArg(arg(F,A),1,F,A) :- alldiff(arg(arg(F,A),nil)).
5  alldiffArg(N,I+1,F,A) :- alldiffArg(N,I,_,arg(F,A)).

7  alldiffRange(A,LB,UB) :- alldiff(arg(A,nil)),
8       LB = #min {L,V : var(int,V,range(L,_)), alldiffArg(A,_,V,_)},
9       UB = #max {U,V : var(int,V,range(_,U)), alldiffArg(A,_,V,_)}.
```

Listing 7. Auxiliary predicates for the *alldifferent* constraints

```
1  :- alldiffRange(CI,LB,UB), X = LB..UB, val(V1,X), val(V2,X),
2     alldiffArg(CI,_,V1,_), alldiffArg(CI,_,V2,_), V1 < V2.
```

Listing 8. alldiffA encoding

The alldiffA encoding is the simplest one presented in listing 8. It consists of only one integrity constraint forbidding that two distinct variables have exactly the same domain value. The alldiffB encoding in listing 9 does exactly the same as alldiffA, but uses a cardinality constraint. The alldiffC encoding in listing 10 is more mature. We use a fixed ordering of the identifiers of the variables. In Line 1–2, whenever a variable V has been assigned to a value X, we derive a seen predicate for every variable whose identifier is less than that of V for the value X. In Line 3 we forbid that a variable V has the same value as a variable whose identifier is greater than that of V. The alldiffD encoding, not presented due to lack of space, uses hall intervals [11]. Therefore, for each variable V and each interval $[A, B]$, we introduce a Boolean variable $p(V, A, B)$ that is true iff V is in the interval $[A, B]$. Moreover, a linear inequality is added to constrain the number of variables having a value in this interval to $B - A$. We parametrized the alldiffD encoding with a maximal interval size.

Using *aspartame* Encodings as a Library. Let us consider the Two Dimensional Strip Packing (2sp) problems. Given a set of n rectangles and one large rectangle (called strip), the goal of the 2sp problem is to find the minimum strip height such that all rectangles are packed into the strip without overlapping.

A 2sp instance is given as a set of facts consisting of $\text{width}(W)$ and $\text{r}(i,w_i,h_i)$ for $1 \leq i \leq n$. The fact $\text{width}(W)$ expresses that the width of the strip is W, and $\text{r}(i,w_i,h_i)$ expresses the rectangle i with a width of w_i and a height of h_i. Using *aspartame* we introduce a pair of Integer variables (x_i, y_i) that express the position of lower left coordinates of each rectangle i, and then enforcing non-overlapping constraints $(x_i + w_i \leq x_j) \vee (x_j + w_j \leq x_i) \vee (y_i + h_i \leq y_j) \vee (y_j + h_j \leq y_i)$ for every two different rectangles i and j ($i < j$). For example, $x_i + w_i \leq x_j$ ensures that the rectangles i is located to the lefthand-side of j.

An ASP encoding for solving 2sp problems is shown in Listing 11. The predicates except $\text{le}/3$, $\text{width}/1$, and $\text{r}/3$ are defined in *aspartame* encodings. The Integer variable **height** with an initial range from **lb** to **ub**, in Line 3–4, is an objective variable that we want to minimize. The predicate $\text{le}(x,c,y)$ is intended

```
1   :- alldiffRange(CI,LB,UB),
2       X = LB..UB, 2{val(V,X) : alldiffArg(CI,_,V,_)}.
```

Listing 9. alldiffB encoding

```
1   seen(CI,I-1,X) :- alldiffArg(CI,I,V,_), 1 < I, val(V,X).
2   seen(CI,I-1,X) :- seen(CI,I,X), 1 < I.
3   :- alldiffArg(CI,I,V,_), val(V,X), seen(CI,I,X).
```

Listing 10. alldiffC encoding

to express $x + c \leq y$ where x and y are Integer variables and c is an Integer constant. The non-overlapping constraints can be concisely expressed by using cardinality constraints. In preliminary experiments (not presented in this paper), we confirmed that this encoding can be highly competitive in performance to a SAT-based approach for solving 2sp problems [12].

4   The *aspartame* System

As mentioned, *aspartame* re-uses *sugar*'s front-end for parsing and normalizing CSPs. Hence, it accepts the same input formats, viz. XCSP[6] and *sugar*'s native CSP format[7]. For this, we implemented an output hook for *sugar* that provides us with the resulting CSP instance in *aspartame*'s fact format. This format can also be used for directly representing linear arithmetic constraints within standard ASP encodings used for Constraint ASP (CASP; [13–16]). In both cases, the resulting facts are then used for grounding a dedicated ASP encoding (via the ASP grounder *gringo*). In turn, the resulting propositional logic program is passed to the ASP solver *clasp* that returns an assignment, representing a solution to the original CSP instance.

Our empirical analysis considers all instances of GLOBAL categories in the 2009 CSP Competition (see footnote 6). We ran them on a cluster of Linux machines equipped with dual Xeon E5520 quad-core 2.26 GHz processors and 48 GB RAM. We separated grounding and solving times, and imposed on each a limit of 1800 s and 16GB. While we count a grounding timeout as 1800s, we penalize unsuccessful solving with 1800 s if either solving or grounding does not finish in time.

At first, we analyze the difference between the *basic encoding* and its refinements from the previous section. To this end, Table 1 contrasts the results obtained from different ASP encodings as well as *sugar* (2.2.1). The name of the benchmark class and the number of instances is given in the first column. In each setting, the *trans* column shows the average time used for translating CSP problems into their final propositional format. For this purpose, *aspartame* uses

---

[6] http://www.cril.univ-artois.fr/CPAI09.

[7] http://bach.istc.kobe-u.ac.jp/sugar/package/current/docs/syntax.html.

```
1   var(int,    x(I), range(0,W-X))   :- r(I,X,Y), width(W).
2   var(int,    y(I), range(0,ub-Y))  :- r(I,X,Y).
3   var(int, height, range(lb,ub)).
4   objective(minimize, height).

6   1 { le(x(I),XI,x(J)) ; le(x(J),XJ,x(I)) ; le(y(I),YI,y(J)) ; le(y(J),YJ,y(I)) } :-
7           r(I,XI,YI), r(J,XJ,YJ), I < J.
8   le(y(I),Y,height) :- r(I,X,Y).

10  wsum(op(le,op(add,op(mul,1,X),op(mul,-1,Y)),-C)) :- le(X,C,Y).
```

Listing 11. Encoding of 2sp problems

gringo (4.5), while *sugar* uses a dedicated implementation resulting in a CNF in DIMACS format. Analogously, the *solve* column gives the average time for each benchmark class, showing the number of translation (to^t) and total timeouts (to). In all cases, we use *clasp*(3.1.1) as back-end ASP or SAT solver, respectively, in its ASP default configuration tweety. Comparing the *basic encoding* with the *optimized encoding*, we observe that the latter significantly reduces both solving and grounding timeouts (mainly due to the *CabinetStart1* class). Next, we want to investigate the impact of the recursive structure of our encodings. For this, we disabled splitting of linear constraints within *sugar*'s translation. This usually leads to an exponential increase in the number of clauses for *sugar*. The results are given in column *optimizednosplit* of Table 1. In fact, disabled splitting performs as good as the optimized encoding with splitting. In some cases, it even improves performance. As splitting constraints the right way usually depends heavily on heuristics, our recursive translation offers a heuristic-independent solution to this problem. Finally, although *aspartame* and *sugar* are at eye height regarding solving time and timeouts, *aspartame* falls short by an order of magnitude when it comes to translating CSPs into propositional format. Here the dedicated implementation of *sugar*[8] clearly outperforms the grounder-based approach of *aspartame*. On the other hand, our declarative approach allows us to easily modify and thus experiment with different encodings.

This flexibility was extremely useful when elaborating upon different encodings. While for the benchmarks used in Table 1 the *alldifferent* constraints are translated to linear constraints with the help of *sugar*, we now want to handle them by an encoding. To this end, Table 2 compares four alternative encodings for handling *alldifferent*. While variants *A*, *B*, *C* have already been presented above, variant *D* uses a more complex encoding using hall intervals of size three [11]. Our experiments show however that simple translations using binary inequalities like *A* and *B* are as good as more complex ones like *C* and even outperform more sophisticated ones as *D*. The last column shows the combination of non-splitting linear constraints (in *sugar*) and handling the *alldifferent* constraint with translation *B*. This is currently the best performing combination of encodings and constitutes the default setting of *aspartame* (2.0.0)[9].

---

[8] The timeouts of *sugar* during translation are always due to insufficient memory.

[9] The system is available at http://www.cs.uni-potsdam.de/aspartame/.

Table 1. Experiments comparing different encodings with sugar.

Benchmark Class	basic trans	solve(to^t,to)	optimized trans	solve(to^t,to)	optimizednosplit trans	solve(to^t,to)	sugar trans	solve(to^t,to)
CabinetStart1(40)	1800	1800 (40, 40)	53	20 (0, 0)	8	2 (0, 0)	4	12 (0, 0)
QG3(7)	2	515 (0, 2)	2	515 (0, 2)	2	515 (0, 2)	2	514 (0, 2)
QG4(7)	2	291 (0, 1)	2	278 (0, 1)	2	278 (0, 1)	2	269 (0, 1)
QG5(7)	1	168 (0, 0)	1	71 (0, 0)	1	71 (0, 0)	1	60 (0, 0)
QG6(7)	3	257 (0, 1)	4	257 (0, 1)	4	257 (0, 1)	2	257 (0, 1)
QG7	3	263 (0, 1)	4	259 (0, 1)	4	259 (0, 1)	2	258 (0, 1)
Allsquares(37)	49	385 (0, 5)	162	229 (0, 4)	33	278 (0, 4)	4	271 (0, 4)
AllsquaresUnsat(37)	49	745 (0, 15)	160	683 (0,13)	33	728 (0,13)	4	660 (0, 13)
Bibd1011(6)	19	5 (0, 0)	30	3 (0, 0)	15	6 (0, 0)	9	6 (0, 0)
Bibd1213(7)	28	13 (0, 0)	42	20 (0, 0)	20	15 (0, 0)	7	2 (0, 0)
Bibd6(10)	5	1 (0, 0)	8	1 (0, 0)	4	1 (0, 0)	3	1 (0, 0)
Bibd7(14)	6	1 (0, 0)	9	1 (0, 0)	5	1 (0, 0)	4	1 (0, 0)
Bibd8(7)	9	14 (0, 0)	14	3 (0, 0)	7	11 (0, 0)	4	2 (0, 0)
Bibd9(10)	13	3 (0, 0)	20	2 (0, 0)	9	3 (0, 0)	6	4 (0, 0)
BibdVariousK(29)	17	343 (0, 4)	23	298 (0, 4)	14	324 (0, 5)	6	266 (0, 3)
bqwh15106_glb(10)	0	0 (0, 0)	1	0 (0, 0)	1	0 (0, 0)	1	0 (0, 0)
bqwh18141_glb(10)	1	0 (0, 0)	1	0 (0, 0)	1	0 (0, 0)	1	0 (0, 0)
Cjss(10)	61	1084 (0, 6)	97	1085 (0, 6)	86	1091 (0, 6)	24	944 (0, 5)
Compet02(20)	199	67 (0, 0)	1165	200 (2, 2)	1164	200 (2, 2)	22	9 (0, 0)
Compet08(16)	17	146 (0, 1)	90	16 (0, 0)	91	16 (0, 0)	73	463 (0, 4)
CostasArray(11)	3	577 (0, 3)	15	381 (0, 2)	16	514 (0, 3)	2	362 (0, 2)
LatinSquare(10)	1	180 (0, 1)	2	180 (0, 1)	2	180 (0, 1)	1	180 (0, 1)
MagicSquare(18)	1057	1179 (10, 11)	1208	1103 (11, 11)	1444	1400 (14, 14)	629	756 (6, 7)
Medium(5)	305	117 (0, 0)	1717	1446 (4, 4)	1721	1446 (4, 4)	31	10 (0, 0)
Nengfa(3)	113	19 (0, 0)	777	5 (0, 0)	770	5 (0, 0)	4	6 (0, 0)
pigeons_glb(19)	0	0 (0, 0)	0	0 (0, 0)	0	0 (0, 0)	1	0 (0, 0)
PseudoGLB(100)	122	466 (5, 23)	142	482 (7,26)	12	382 (0,18)	95	488 (5, 26)
Rcpsp(39)	0	0 (0, 0)	1	0 (0, 0)	0	0 (0, 0)	1	0 (0, 0)
RcpspTighter(39)	0	0 (0, 0)	1	0 (0, 0)	0	0 (0, 0)	1	0 (0, 0)
Small(5)	14	2 (0, 0)	87	1 (0, 0)	87	1 (0, 0)	4	1 (0, 0)
Total	210	412 (55, 114)	163	273 (24, 78)	124	274 (20, 75)	44	252 (11, 70)

5   Discussion

CASP approaches [13–16] handle constraints in a lazy way by using off-the-shelf CP solvers either as back-end (*ezcsp*) or online propagator (*clingcon*). The approach in [15] (inca) uses a dedicated propagator to translate constraints (via the order encoding) during solving. Our approach can be seen as an "early" approach, translating all constraints before the solving process. As regards pure CP solving, *aspartame*'s approach can be seen as a first-order alternative to SAT-based approaches like *sugar* [5]. Although the performance of the underlying SAT solver is crucial, the SAT encoding plays an equally important role [17]. Among them, we find the direct [18,19], support [20,21], log [22,23], order [4,5], and compact order [24] encoding. The order encoding showed good performance for a wide range of CSPs [4,12,25–27]. In fact, the SAT-based CP solver *sugar* won the GLOBAL category at the 2008 and 2009 CP solver competitions [28]. Also, the SAT-based CP solver BEE [29] and the CLP system B-Prolog [30] use this encoding. In fact, the order encoding provides a compact translation of arithmetic constraints, while also maintaining bounds consistency by unit propagation.

We presented an alternative approach to solving finite linear CSPs based on ASP. The resulting system *aspartame* relies on high-level ASP encodings and

Table 2. Experiments comparing different encodings for alldifferent.

Benchmark Class	alldiffA trans	solve(to^t,to)	alldiffB trans	solve(to^t,to)	alldiffC trans	solve(to^t,to)	alldiffD trans	solve(to^t,to)	alldiffBnosplit trans	solve(to^t,to)
CabinetStart1(40)	54	20(0, 0)	53	20(0, 0)	53	20(0, 0)	65	23(0, 0)	8	2(0, 0)
QG3(7)	2	515(0, 2)	2	515(0, 2)	2	515(0, 2)	3	515(0, 2)	2	515(0, 2)
QG4(7)	2	276(0, 1)	2	278(0, 1)	2	283(0, 1)	3	290(0, 1)	2	278(0, 1)
QG5(7)	1	56(0, 0)	1	68(0, 0)	1	60(0, 0)	3	149(0, 0)	1	68(0, 0)
QG6(7)	4	257(0, 1)	4	257(0, 1)	4	257(0, 1)	6	258(0, 1)	4	257(0, 1)
QG7	4	259(0, 1)	4	260(0, 1)	4	259(0, 1)	4	261(0, 1)	4	260(0, 1)
Allsquares(37)	161	229(0, 4)	161	230(0, 4)	161	230(0, 4)	166	232(0, 4)	33	281(0, 4)
AllsquaresUnsat(37)	158	683(0,13)	158	682(0,13)	158	682(0,13)	168	701(0,13)	32	726(0,13)
Bibd1011(6)	30	3(0, 0)	30	3(0, 0)	31	3(0, 0)	32	3(0, 0)	15	6(0, 0)
Bibd1213(7)	42	20(0, 0)	42	20(0, 0)	43	20(0, 0)	43	24(0, 0)	20	15(0, 0)
Bibd6(10)	8	1(0, 0)	8	1(0, 0)	8	1(0, 0)	14	1(0, 0)	4	1(0, 0)
Bibd7(14)	9	1(0, 0)	9	1(0, 0)	9	1(0, 0)	9	1(0, 0)	5	1(0, 0)
Bibd8(7)	14	3(0, 0)	14	3(0, 0)	14	3(0, 0)	14	3(0, 0)	7	11(0, 0)
Bibd9(10)	20	2(0, 0)	20	2(0, 0)	20	2(0, 0)	21	3(0, 0)	9	3(0, 0)
BibdVariousK(29)	23	298(0, 4)	23	298(0, 4)	23	298(0, 4)	26	300(0, 4)	14	324(0, 5)
bqwh15106_glb(10)	0	0(0, 0)	0	0(0, 0)	0	0(0, 0)	4	1(0, 0)	0	0(0, 0)
bqwh18141_glb(10)	0	0(0, 0)	0	0(0, 0)	0	0(0, 0)	5	1(0, 0)	0	0(0, 0)
Cjss(10)	99	1085(0, 6)	99	1085(0, 6)	99	1085(0, 6)	97	1085(0, 6)	87	1091(0, 6)
Compet02(20)	834	22(0, 0)	836	18(0, 0)	840	21(0, 0)	1095	268(2, 2)	842	19(0, 0)
Compet08(16)	236	463(0, 4)	232	455(0, 4)	230	475(0, 4)	633	1498(0,13)	233	455(0, 4)
CostasArray(11)	5	503(0, 2)	5	349(0, 2)	5	494(0, 3)	10	517(0, 3)	6	343(0, 2)
LatinSquare(10)	0	180(0, 1)	0	180(0, 1)	0	180(0, 1)	2	180(0, 1)	0	180(0, 1)
MagicSquare(18)	1192	1107(11,11)	1192	1103(11,11)	1191	1104(11,11)	1196	1106(11,11)	1442	1401(14,14)
Medium(5)	1510	36(0, 0)	1499	34(0, 0)	1503	35(0, 0)	1700	1458(4, 4)	1507	34(0, 0)
Nengfa(3)	10	2(0, 0)	9	1(0, 0)	9	2(0, 0)	67	117(0, 0)	9	1(0, 0)
pigeons_glb(19)	0	0(0, 0)	0	0(0, 0)	0	0(0, 0)	0	0(0, 0)	0	0(0, 0)
PseudoGLB(100)	142	482(7,26)	142	482(7,26)	142	482(7,26)	142	483(7,26)	12	382(0,18)
Rcpsp(39)	1	0(0, 0)	1	0(0, 0)	1	0(0, 0)	1	0(0, 0)	0	0(0, 0)
RcpspTighter(39)	1	0(0, 0)	1	0(0, 0)	1	0(0, 0)	1	0(0, 0)	0	0(0, 0)
Small(5)	62	1(0, 0)	62	1(0, 0)	61	1(0, 0)	92	8(0, 0)	62	1(0, 0)
Total	148	269(18,76)	148	266(18,76)	148	269(18,77)	174	326(24,92)	110	264(14,72)

delegates both the grounding and solving tasks to general-purpose ASP systems. Furthermore, these encodings can be used as a library for solving CSPs as part of an encompassing logic program, as it is done in the framework of CASP. We have contrasted *aspartame* with its SAT-based ancestor *sugar*, which delegates only the solving task to off-the-shelf SAT solvers, while using dedicated algorithms for constraint preprocessing. Although *aspartame* does not fully match the performance of *sugar* from a global perspective, the picture is fragmented and leaves room for further improvements, especially for the translation process. Experience from *aspartame* will definitely help to build/improve the current experimental support for linear constraints in *gringo*. Despite all this, *aspartame* demonstrates that ASP's general-purpose technology allows to compete with state-of-the-art constraint solving techniques. In fact, the high-level approach of ASP facilitates extensions and variations of first-order encodings for dealing with particular types of constraints. In the future, we thus aim at investigating alternative encodings, e.g., regarding *alldifferent* constraints, as well as support of further global constraints.

References

1. Banbara, M., Gebser, M., Inoue, K., Schaub, T., Soh, T., Tamura, N., Weise, M.: Aspartame: solving CSPs with ASP. In: ASPOCP, abs/1312.6113, CoRR (2013)

2. Rossi, F., v Beek, P., Walsh, T. (eds.): Handbook of Constraint Programming. Elsevier, Melbourne (2006)
3. Biere, A., Heule, M., v Maaren, H., Walsh, T. (eds.): Handbook of Satisfiability. IOS, Amsterdam (2009)
4. Crawford, J., Baker, A.: Experimental results on the application of satisfiability algorithms to scheduling problems. In: AAAI, pp. 1092–1097. AAAI Press (1994)
5. Tamura, N., Taga, A., Kitagawa, S., Banbara, M.: Compiling finite linear CSP into SAT. Constraints **14**, 254–272 (2009)
6. Baral, C.: Knowledge Representation.Reasoning and Declarative Problem Solving. Cambridge University Press, Cambridge (2003)
7. Gebser, M., Kaminski, R., Kaufmann, B., Schaub, T.: Answer Set Solving in Practice. Morgan and Claypool Publishers, San Rafael (2012)
8. Beldiceanu, N., Simonis, H.: A constraint seeker: finding and ranking global constraints from examples. In: Lee, J. (ed.) CP 2011. LNCS, vol. 6876, pp. 12–26. Springer, Heidelberg (2011)
9. Tamura, N., Banbara, M., Soh, T.: Compiling pseudo-boolean constraints to SAT with order encoding. In: ICTAI, pp. 1020–1027. IEEE (2013)
10. Gent, I., Nightingale, P.: A new encoding of alldifferent into SAT. In: Workshop on Modelling and Reformulating Constraint Satisfaction Problems (2004)
11. Bessiere, C., Katsirelos, G., Narodytska, N., Quimper, C., Walsh, T.: Decompositions of all different, global cardinality and related constraints. In: IJCAI, pp. 419–424 (2009)
12. Soh, T., Inoue, K., Tamura, N., Banbara, M., Nabeshima, H.: A SAT-based method for solving the two-dimensional strip packing problem. Fund. Informaticae **102**, 467–487 (2010)
13. Gebser, M., Ostrowski, M., Schaub, T.: Constraint answer set solving. In: Hill, P.M., Warren, D.S. (eds.) ICLP 2009. LNCS, vol. 5649, pp. 235–249. Springer, Heidelberg (2009)
14. Balduccini, M.: Representing constraint satisfaction problems in answer set programming. In: ASPOCP, pp. 16–30 (2009)
15. Drescher, C., Walsh, T.: A translational approach to constraint answer set solving. Theor. Pract. Logic Program. **10**, 465–480 (2010)
16. Ostrowski, M., Schaub, T.: ASP modulo CSP: the clingcon system. Theor. Pract. Logic Program. **12**, 485–503 (2012)
17. Prestwich, S.: CNF encodings. In: [3], pp. 75–97
18. de Kleer, J.: A comparison of ATMS and CSP techniques. In: IJCAI, pp. 290–296 (1989)
19. Walsh, T.: SAT v CSP. In: CP, pp. 441–456 (2000)
20. Kasif, S.: On the parallel complexity of discrete relaxation in constraint satisfaction networks. Artif. Intell. **45**, 275–286 (1990)
21. Gent, I.: Arc consistency in SAT. In: ECAI, pp. 121–125 (2002)
22. Iwama, K., Miyazaki, S.: SAT-variable complexity of hard combinatorial problems. In: IFIP, pp. 253–258 (1994)
23. Van Gelder, A.: Another look at graph coloring via propositional satisfiability. Discrete Appl. Math. **156**, 230–243 (2008)
24. Tanjo, T., Tamura, N., Banbara, M.: Azucar: A SAT-based CSP solver using compact order encoding. In: Cimatti, A., Sebastiani, R. (eds.) SAT 2012. LNCS, vol. 7317, pp. 456–462. Springer, Heidelberg (2012)
25. Metodi, A., Codish, M., Stuckey, P.: Boolean equi-propagation for concise and efficient SAT encodings of combinatorial problems. J. Artif. Intell. Res. **46**, 303–341 (2013)

26. Ohrimenko, O., Stuckey, P., Codish, M.: Propagation via lazy clause generation. Constraints **14**, 357–391 (2009)
27. Banbara, M., Matsunaka, H., Tamura, N., Inoue, K.: Generating combinatorial test cases by efficient SAT encodings suitable for CDCL sat solvers. In: Fermüller, C.G., Voronkov, A. (eds.) LPAR-17. LNCS, vol. 6397, pp. 112–126. Springer, Heidelberg (2010)
28. Lecoutre, C., Roussel, O., van Dongen, M.: Promoting robust black-box solvers through competitions. Constraints **15**, 317–326 (2010)
29. Metodi, A., Codish, M.: Compiling finite domain constraints to SAT with BEE. Theor. Pract. Logic Program. **12**, 465–483 (2012)
30. Zhou, N.: The SAT compiler in B-prolog. The ALP Newsletter, March 2013

"Add Another Blue Stack of the Same Height!": ASP Based Planning and Plan Failure Analysis

Chitta Baral[1] and Tran Cao Son[2(✉)]

[1] Department of Computer Science and Engineering, Arizona State University,
Tempe, AZ, USA
[2] Department of Computer Science, New Mexico State University,
Las Cruces, NM, USA
tson@cs.nmsu.edu

Abstract. We discuss a challenge in developing intelligent agents (robots) that can collaborate with human in problem solving. Specifically, we consider situations in which a robot must use natural language in communicating with human and responding to the human's communication appropriately. In the process, we identify three main tasks. The first task requires the development of planners capable of dealing with *descriptive goals*. The second task, called *plan failure analysis*, demands the ability to analyze and determine the reason(s) why the planning system does not success. The third task focuses on the ability to understand communications via natural language. We show how the first two tasks can be accomplished in answer set programming.

1 Introduction

Human-Robot interaction is an important field where humans and robots collaborate to achieve tasks. Such interaction is needed, for example, in search and rescue scenarios where the human may direct the robot to do certain tasks and at times the robot may have to make its own plan. Although there has been many works on this topic, there has not been much research on interactive planning where the human and the robot collaborate in making plans. For such interactive planning, the human may communicate to the robot about some goals and the robot may make the plan, or when it is unable it may explain why it is unable and the human may make further suggestions to overcome the robot's problem and this interaction may continue until a plan is made and the robot executes it. The communication between the robot and the human happens in natural language as ordinary Search and Rescue officials may not be able to master the Robot's formal language to instruct it in situations of duress. Following is an abstract example of such an interactive planning using natural language communication.

Consider the block world domain in Fig. 1 (see, [1]). The robot has its own blocks and the human has some blocks as well. The two share a table and some other blocks as well. Suppose that the human communicates to the robot the sentence *"Add another blue stack of the same height!"* Even if we assume that the robot is able to recognize the color of the blocks, create, and execute plans

© Springer International Publishing Switzerland 2015
F. Calimeri et al. (Eds.): LPNMR 2015, LNAI 9345, pp. 127–133, 2015.
DOI: 10.1007/978-3-319-23264-5_11

HUMAN   ROBOT

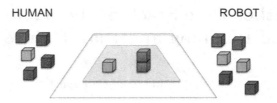

Fig. 1. A simplified version of the blocks world Example from the BAA (Color figure online)

for constructions of stack of blocks, such a communication presents several challenges to a robot. Specifically, it requires that the robot is capable of understanding natural language, i.e., it requires that the robot is able to identify that

- the human refers to stacks of only blue blocks (*blue stack*);
- the human refers to the height of a stack as the number of blocks on the stack;
- there is a blue stack of the height 2 on the table; and
- it should use its two blue blocks to build a new stack of two blue blocks.

It is easy to see that among the above points, the first three are clearly related to the research in natural language processing (NLP) and the last one to planning, i.e., an integration of NLP and planning is needed for solving this problem. Ideally, given the configuration in Fig. 1 and the reasoning above, the robot should devise a plan to create a new stack using its two blue blocks and execute the plan. On the other hand, if the robot has only one blue block, the robot should realize that it cannot satisfy the goal indicated by the human and it needs to respond to the human differently, for example, by informing the human that it does not have enough blue blocks or by asking the human for permission to use his blue blocks.

As planning in various settings has been considered in answer set programming and there have been attempts to translate NLP into ASP, we would like to explore the use of ASP and related tools in solving this type of problems. *Our focus in this paper is the interactive planning problem that the robot needs to solve.*

2   ASP Planning for "Add Another Blue Stack of the Same Height!"

What are the possible responses to the communication from the human *"Add another blue stack of the same height!"* for the robot? Assume that the robot can recognize that it needs to create and execute a plan in responding to the communication. We next describe an ASP implementation of an ASP planning module for the robot. Following the literature (e.g., [2]), this module should consist of the planning domain and the planning problem.

2.1 The Planning Domain

The block world domain D_b can be typically encoded by an ASP-program as follows:

- blocks are specified by the predicate bl (.);
- actions such as pont (X) or put (X,Y), where X and Y denote blocks, are defined using ASP-rules;
- properties of the world are described by fluents and encoded using predicates—similar to the use of predicates in Situation Calculus—whose last parameter represents the situation term or an time step in ASP. For simplicity, we use two fluents on(X,Y,T) and ont (X,T) which denote Y is on X or X is on the table, respectively, at the time point T;
- Rules encoding executability conditions and effects of actions are also included, for example, to encode the executable condition and effects of the action pont (X), we use the the rules

```
executable(pont(X),T):-time(T),action(pont(X)),clear(X,T).
ont(X,T+1):-time(T),occ(pont(X),T).
ont(X,T+1):-time(T),ont(X,T),#count{Z:occ(put(Z,X),T)}==0.
```

D_b can be used for planning [2]. To use D_b in planning, we need the following components: (*i*) rules for generating action occurrences; (*ii*) rules encoding the initial state; and (*iii*) rules checking for goal satisfaction.

2.2 What are we Planning for? and How do we Plan for it?

Having defined D_b, we now need to specify the planning problem. The initial state can be easily seen from the figure. The question is then what is the goal of the planning problem. Instead of specifying a formula that should be satisfied in the goal state, the sentence describes a desirable state. This desirable state has two properties: (*i*) it needs to have the second (or a new) blue stack; and (*ii*) the new blue stack has the same height as the existing blue stack. In other words, the goal is *descriptive*. We will call a planning problem whose goal is descriptive as a *descriptive planning problem*.

In the rest of this subsection, we will present an encoding of the planning problem for "*Add another blue stack of the same height!*." First, we need to represent the goal. Considering that a stack can be represented by the block at the top, the goal for the problem can be represented by a set of goal conditions in ASP as follows.

```
g_c(S,is,stack):-bl(S).    g_c(S,type,another):-bl(S).
g_c(S,color,blue):-bl(S). g_c(S,height,same):-bl(S).
```

These rules state that the goal is to build a stack represented by S, the stack is blue, has the same height, and it is a new stack. Note that these rules are still vague in the sense that they contain unspecified concepts, e.g., "*what is a stack?*," "*what does it mean to be the same height?*," etc.

The problem can be solved by adding rules to the block program D_b to define when a goal condition is satisfied. Afterwards, we need to make sure that all goal conditions are satisfied. As we only need to obtain one new stack, the following rules do that:

```
not_sat_goal(S,T):-bl(S),g_c(X,Y,Z), not sat(X,Y,Z,T).
sat_goal(S, T)    :-not not_sat_goal(T).
:- X = #count {S  : sat_goal(S, n)}, X ==0.
```

The first two rules define blocks that are considered to satisfy a goal condition. The last rule ensures that at least one of the blocks satisfies all four goal conditions.

We now need rules defining when different kinds of goal conditions are satisfied. For simplicity, let us consider stacks to have at least one block. In that case we have the following rule defining when a block is a stack.

```
sat(S,is,stack,T):-bl(S),time(T),clear(S,T).
```

We next define that the color of a stack at time point T is blue if all blocks in that stack at that time point are blue. The next rule simply says that a stack is blue if its top is blue and all blocks below the top are also blue.

```
sat(S,color,blue,T):-bl(S),time(T),color(S,blue),clear(S,T),
              #count{U:above(U,S,T), not color(U,blue)}==0.
```

Defining sat (S,height,same,T) is less straightforward. First, we define a predicate same_height (S,U,T) meaning that in the context of time step T, S and U have the same height (or two stacks represented by S and U have the same height). We then use this predicate to define sat (S,height,same,T). Note that the definition of the predicate sat (S,height,same,T) also enforces the constraint that the blue stack that is compared to the stack represented by S does not change.

```
same_height(S,U,T):-bl(S),bl(U),S!=U,ont(S,T),ont(U,T).
same_height(S,U,T):-bl(S),bl(U),S!=U,on(S1,S,T),on(U1,U,T),
              same_height(S1, U1, T).
sat(S,height,same,T):- sat(S,is,stack,T), S!=U,
      sat(U,is,stack,T),sat(U,color,blue,T),
      unchanged(U,T),clear(S,T),clear(U,T),same_height(S,U,T).
```

Similarly, defining sat(S, type, another,T) is also not straightforward. Intuitively, it means that there is another stack U different from S and U has not changed over time. We define it using a new predicate unchanged (U,T) which means that the stack with U at the top has not changed over time (from step 0 to step T).

```
sat(S,type,another,T):-bl(U),unchanged(U,T),same_height(S,U,T).
unchanged(U,T):- time(T),bl(U),not changed(U,T).
changed(U,T):- time(T),T>0,bl(U),above(V,U,0), not above(V,U,T).
changed(U,T):- time(T), T > 0, bl(U),ont(U,0), not ont(U,T).
changed(U,T):- time(T), T > 0, bl(U),not ont(U,0), ont(U,T).
changed(U,T):- time(T), T > 0, bl(U),not ont(U,0), on(X,U,T).
```

Let us denote with G_b the collection of rules developed in this section. To compute a plan, we will only need to add the initial state and the rules described above for enforcing that all goal conditions must be satisfied. Initializing the initial situation as 0 (Fig. 1), the description of who has which blocks and what is on the table can be expressed through the following facts, denoted with I_b:

```
has(human,b1,0).color(b1,red).has(human,b2,0). color(b2,brown).
has(human,b3,0).color(b3,brown).has(human,b4,0).color(b4,green).
has(human,b5,0).color(b5,blue). has(robot,b6,0). color(b6,red).
has(robot,b7,0). color(b7,red). has(robot,b8,0). color(b8,blue).
has(robot,b9,0).color(b9,blue).has(robot,b10,0).color(b12,red).
has(robot,b11,0).color(b11,green).ont(b12,0). color(b10,green).
ont(b13,0).  color(b13,blue).on(b13,b14,0).  color(b14,blue).
```

It is easy to see that the program $D_b \cup I_b \cup G_b$ with n=2 has two answer sets, one contains the atoms occ(pont(b8),0), occ(put(b8,b9),1) that corresponds to a possible plan pont(b8),put(b8,b9) for the robot; the other one corresponds to the plan pont(b9),put(b9,b8).

2.3   What if Planning Fails?

The previous subsection presents an initial configuration in which the robot can generate a plan satisfying the request from the human. Obviously, there are several configurations of the block world in which the planning generation phase can fail. What should be the robot's response? What is the ground for such response? Let us consider one of such situations.

Assuming a different initial configuration in which the robot has only one blue block, say b8 (or b9). Let I_{b_f} be the new initial situation. Furthermore, we require that the robot cannot use the block belonging to the human if it does not get the human's permission. In this case, the robot will not be able to add another blue stack of the same height to the table because there is no plan that can satisfy this goal. More specifically, it is because the robot does not have enough blue blocks for the task at hand. This is what the robot should respond back to the human user. We next discuss a possible way for the robot to arrive at this conclusion.

First, the encoding in the previous subsection must be extended to cover the requirement that the robot can only use its own blocks or the blocks on the table in the construction of the new stack. This can be achieved with the following rules:

```
available(X, T) :- time(T),bl(X),has(robot,X,T).
available(X, T) :- time(T),bl(X),ont(X, T),clear(X,T).
available(X, T) :- time(T),bl(X),above(Y,X,T),clear(X,T).
:-occ(pont(X),T),not available(X,T).
:-occ(put(Y,X),T),not available(X,T).
```

Let D_{b_f} be D_b unions with the above rules. It is easy to see that, for any constant n, the program $D_{b_f} \cup I_{b_f} \cup G_b$ does not have an answer set. When the planning fails, the robot should analyze the situation and come up with an appropriate response. A first reasonable step for the robot is to identify why the planning fails. As it turns out, the program described in the previous subsection only needs only minor changes to accomplish this task. First, we need to modify the goals as follows.

```
g_c(S,is,stack):-bl(S),ok(1).g_c(S,height,same):-bl(S),ok(3).
g_c(S,color,blue):-bl(S),ok(2).g_c(S,type,another):-bl(S),ok(4).
{ok(1..4)}4.ngoals(X):-X=#count{I:ok(I)}.#maximize {X:ngoals(X)}.
```

Let G_{b_f} be the new goal, obtained from G_b by replacing the rules defining g_c(.) with the above rules. It is easy to check that every answer set of the program $D_{b_f} \cup I_{b_f} \cup G_{b_f}$ does not contain the atom ok(2) which indicates that the goal condition g_c(S,color,blue) cannot be satisfied. As such, the robot should use the missing goal condition as a ground for its response. However, the robot could use this information in different ways. For example, it can tell the human that it does not have enough blue blocks or it could ask the human for permission to use the human's blue blocks to complete the goal. It means that the robot needs the ability to make assumptions and reason with them. This can be defined formally as follows.

Definition 1. *A* planning problem with assumptions *is a tuple* $\mathcal{P} = (D, I, G, AP, AF)$ *where* (D, I, G) *is a planning problem, AP is a set of actions, and AF is a set of fluents. We say that* $\mathcal{P}$ *needs a* plan failure analysis *if* (D, I, G) *has no solution.*

Intuitively, AP is the set of actions that the robot could execute and AF is a set of assumptions that the robot could assume. For example, for our running example, AP could contain the logic program encoding of an action ask(blue) whose effect is that the robot can use the blue block from the human; AF could be {has(robot,bx), color(bx,blue)}. So, a planning problem with assumption for the robot is $\mathcal{P}_{b_f} = (D_{b_f}, I_{b_f}, G_b, \{ask(blue)\}, \{has(robot, bx), color(bx, blue)\})$.

Definition 2. *A* plan failure analysis *for a planning problem with assumptions* $\mathcal{P} = (D, I, G, AP, AF)$ *is a pair* (A, F) *such that* $A \subseteq AP$, $F \subseteq AF$, *and the planning problem* $(D \cup A, I \cup F, G)$ *is solvable.* (A, F) *is a* preferred plan failure analysis *if there exists no analysis* (A', F') *such that* $A' \subsetneq A$ *or* $F' \subsetneq F$.

It is easy to see that $\mathcal{P}_{b_f} = (D_{b_f}, I_{b_f}, G_b, \{ask(blue)\}, \{has(robot, bx), color (bx, blue)\})$ has two preferred plan failure analyses; one, $(\emptyset, \{has(robot,bx), color(bx,blue)\})$, tells the robot that it does not have enough blue blocks; another one $(\{ask(blue)\}, \emptyset)$ tells the robot to ask for permission to use the human's blue blocks.

To compute a preferred plan failure analysis, we can apply the same method used in identifying a minimal set of satisfying goal conditions. We assume that for each action act in AP, AP consists of the rules defines its effects and a declaration of the form is_ap (act). For each fluent l in AF, we assume that AF contains a declaration is_af(1) as well as the rules for describing how l changes when actions are executed. Let D_a be the following set of rules:

```
{action(A) : is_ap(A)}.        a_assume(A):- action(A),is_ap(A).
{assume(L) : is_af(L)}.        L@0:-assume(L).
nl_assume(F):-F=#count{L:assume(L)}.
na_assume(Y):-Y=#coun {A:a_assume(A)}.
#minimize {1@1,F:nl_assume(F)}.  #minimize {1@1,A:na_assume(A)}.
```

The first rule says that the robot can assume any action in *AP*. Any action that is assumed will be characterized by the predicate a_assume (.) (the second rule). The third rule says that the robot can assume any fluent in *AF*. L@0 represents the fact that L is true at the time 0. The rest of the rules minimizes the number of actions and the number of assumed fluents, independent from each other. We can show that the new program returns possible preferred plan failure analyses of the problem.

3   Conclusions and Future Work

We describe two problems related to planning that need to be addressed for an effective collaboration between an intelligent system and human when communication via natural language is necessary. We show that ASP based approaches can be employed to deal with these two problems. Our discussion shows that ASP can play an important role in the development of intelligent systems that can interact with human via natural language. Our future goal is to develop further applications that integrate various aspects of AI including NLP, Ontologies and Reasoning by using KPARSER that has been used to address the Winograd Schema Challenge [3].

References

1. DARPA: Communicating with Computers (CwC) (2015)
2. Lifschitz, V.: Answer set programming and plan generation. AIJ **138**(1–2), 39–54 (2002)
3. Sharma, A., Aditya, S., Nguyen, V., Baral, C.: Towards addressing the winograd schema challenge - building and using needed tools. In: Proceedings of IJCAI (2015)

A Theory of Intentions for Intelligent Agents
(Extended Abstract)

Justin Blount[1], Michael Gelfond[2], and Marcello Balduccini[3]($^{(\boxtimes)}$)

[1] Southwest Research Institute, San Antonio, USA
[2] Texas Tech University, Lubbock, USA
michael.gelfond@ttu.edu
[3] Drexel University, Philadelphia, USA
{justin.blount,marcello.balduccini}@gmail.com

Abstract. We describe the $\mathcal{AIA}$ architecture for intelligent agents whose behavior is driven by their intentions and who reason about, and act in, changing environments. The description of the domain includes both the agent's environment and the general *Theory of Intentions*. The architecture is designed to enable agents to explain unexpected observations and determine which actions are *intended* at the present moment. Reasoning is reduced to computing answer sets of CR-Prolog programs constructed automatically from the agent's knowledge.

1 Introduction

This paper presents a new declarative methodology for the design and implementation of intelligent agents. We limit our attention to a single agent satisfying the following assumptions:

- the agent's environment, its mental state, and the effects of occurrences of actions can be represented by a transition diagram. States of the diagram contain physical properties of the world as well as mental attitudes of the agent. Transitions are caused by actions and reflect possible changes in these states;
- the agent is capable of making correct observations, remembering the domain history, and correctly recording the *results* of his *attempts* to perform actions;
- normally, the agent is capable of observing the occurrence of all relevant exogenous actions (actions not performed by the agent).

Our approach to agent design (referred to as $\mathcal{AIA}$) builds upon earlier work on the AAA architecture [1]. The main difference between AAA and $\mathcal{AIA}$ is in the organization of the control loop of the agent. In both cases the agent uses its knowledge about the domain to perform diagnostic and planning tasks. However, in our approach the loop is centered around the notion of intention, which is absent in AAA. The use of intentions simplifies and generalizes the loop and allows the agent to more naturally persist in its commitment to achieve its

© Springer International Publishing Switzerland 2015
F. Calimeri et al. (Eds.): LPNMR 2015, LNAI 9345, pp. 134–142, 2015.
DOI: 10.1007/978-3-319-23264-5_12

goals. Moreover, it allows an outside observer (including the agent designer) to reason about agent's behavior, e.g. to prove that the agent will never perform any unintended actions.

Our work is obviously related to the BDI agent model [7,9]. Space constraints prevent a thorough comparison, but we will mention some of the key differences. The BDI model is usually based on a rather complex logic, e.g. LORA [9], with multiple modalities that include beliefs, desires, intentions and time, as well as complex actions obtained from elementary actions by operators such as *if*, *while*, *choice*. By contrast, $\mathcal{AIA}$ is based on simpler, yet expressive logics that are directly executable. Revisions of beliefs, desires and intentions are also achieved quite differently in the two approaches, since BDI logics are monotonic, while $\mathcal{AIA}$ is based on non-monotonic logic. Finally, in BDI intentions are considered on par with beliefs and desires. In our work, on the other hand, intentions are precisely definable in terms of beliefs and desires. The hierarchical representation of activities in $\mathcal{AIA}$, which we introduce later in this paper, also paves the way towards establishing a connection between the flexibility of reasoning of the $\mathcal{AIA}$ architecture and the computational efficiency of HTN planning [8].

The main technical contributions of our work are the introduction of a formal theory of intentions ($\mathcal{TI}$) and the development of an algorithm which takes the agent's knowledge (including the theory of intentions), explains the unexpected observations and computes an action the agent will intend to perform. (Note, that when necessary, the second task can require planning).

The $\mathcal{TI}$ represents properties of intentions as a collection of statements of an action theory of $\mathcal{AL}$. This ensures declarativity and allows for seamless integration of $\mathcal{TI}$ with agent's knowledge about its domain and history. Existence of a reasoning algorithm ensures that the declarative specification of an agent can be actually executed. The algorithm is based on the reduction of the task of explaining observations and finding an intended action to the problem of computing answer sets of a program of CR-Prolog [3] automatically generated from the agent's knowledge. As usual answer sets are computed by a general purpose algorithm implemented by a comparatively efficient answer set solver [2]. A prototype implementation of a software called $\mathcal{AIA}$ *Agent Manager* allows to test this methodology. The following example informally describes the agent and illustrates its intended behavior by a number of simple (but non-trivial) scenarios.

Example 1. [Bob and John] Consider an environment that contains our agent Bob, his colleague John, and a row of four rooms, $r1, r2, r3, r4$ where consecutive rooms are connected by doorways, such that either agent may *move* between neighboring rooms. The door between $r3$ and $r4$ is special and can be *locked* and *unlocked* by both agents. If the door is *locked* then neither can *move* between those two rooms until it is *unlocked*. Bob and John *meet* if they are located in the same room.

Scenario 1: Planning to Achieve the Goal. Initially Bob knows that he is in $r1$, John is in $r3$, and the door between $r3$ and $r4$ is unlocked. Suppose that Bob's boss requests that he meet with John. This causes Bob to intend to meet with John. This type of intention is referred to as an *intention to achieve* a goal. Since Bob acts on his intentions, he uses his knowledge of the domain to choose a plan to achieve his goal. Of course Bob does not waste time and chooses the shortest plan that he expects to achieve his goal, that is to move from $r1$ to $r2$ and then to $r3$. A pair consisting of a goal and the plan aimed at achieving it is called an *activity*. To fulfill his intention, Bob *intends to execute* the activity consisting of the goal to meet John and the two step plan to move from $r1$ to $r3$. The process of executing an activity begins with a *mental*[1] action to *start* the activity. Assuming there are no interruptions, the process continues with the execution of each action in the plan (in this case, moving to $r2$, then to $r3$). After meeting John in $r3$ the process concludes with an action to *stop* the activity.

Scenario 2: Not Expected to Achieve Goal and Replanning. Suppose that as Bob is moving from $r1$ to $r2$ he observes John moving from $r3$ to $r4$. Bob should recognize that in light of this new observation the continued execution of his activity is not expected to achieve the goal, i.e. his activity is *futile*. As a result, he should *stop* executing his activity and *start* executing a new one (containing a plan to move to $r3$ and then to $r4$) that is expected to achieve the goal.

Scenario 3: Failure to Achieve, Diagnosis, and Replanning. Bob moved from $r1$ to $r2$ and then to $r3$, but observes that John is not there. Bob must recognize that his activity failed to achieve the goal. Further analysis should allow Bob to conclude that, while he was executing his activity, John must have moved to $r4$. Bob doesn't know exactly when John moved, but his intention will persist, and he will find a new activity (containing a plan to move to $r4$) to achieve his goal.

Scenario 4: Failure to Execute, Diagnosis, and Replanning. Believing that the door is unlocked, Bob attempts to move from $r3$ to $r4$, but is unable to perform the action. This is unexpected, but Bob realizes that John must have locked the door after moving to $r4$. Bob's new activity contains the same goal to meet John and a plan to unlock the door before moving to $r4$.   ◇

Despite the comparative simplicity of the tasks illustrated by these scenarios we are unaware of any systematic declarative methodology which will allow us to easily build an agent capable of the type of reasoning needed to perform them.

2   The Representation Language

The representation language adopted in this work is an extension of action language $\mathcal{AL}$ [4]. The language is parametrized by a sorted signature containing three special sorts *actions*, *fluents*, and *statics* (properties which can,

---

[1] Actions that directly affect an agent's mental state are referred to as *mental* actions, while those actions that directly affect the state of the environment are referred to as *physical* actions.

resp. cannot, be changed by actions). The fluents are partitioned into two sorts: *inertial* and *defined*. Together, statics and fluents are called *domain properties*. A *domain literal* is a domain property p or its negation $\neg p$. If domain literal l is formed by a fluent, we refer to it as a *fluent literal*; otherwise it is a *static literal*. Allowed statements are: *causal laws* (a **causes** l_{in} **if** $p_0, \ldots, p_m$), *state constraints* (l **if** $p_0, \ldots, p_m$), and *executability conditions* (**impossible** $a_0, \ldots, a_k$ **if** $p_0, \ldots, p_m$), where $k \geq 0$, a, a_i's are actions, l is a domain literal (the *head*), l_{in} is a literal formed by an inertial fluent, and p_i's are domain literals. Moreover, no negation of a defined fluent can occur in the heads of state constraints. The collection of state constraints whose head is a defined fluent f is referred to as the *definition of* f. As in logic programming, f is true if it follows from the truth of the body of at least one of its defining rules and is false otherwise. A *system description* of $\mathcal{AL}$ is a collection of statements of $\mathcal{AL}$ over some (implicitly defined) signature.

In this paper we expand the syntax of $\mathcal{AL}$ by requiring its signature to contain *activities*, consisting of a goal, a plan aimed at achieving the goal, and a name. We name activities by natural numbers. For instance we can denote Bob's activity from Scenario 1 by $\langle 1, [move(b, r1, r2), move(b, r2, r3)], meet(b, j) \rangle$. In $\mathcal{AL}$ this will be represented by the statics $activity(1)$, $comp(1, 1, move(b, r1, r2))$, $comp(1, 2, move(b, r2, r3))$, $length(1, 2)$, $goal(1, meet(b, j))$, where $comp(X, Y, A)$ states that A is the Yth element of the plan of activity X, and $length(X, N)$ says that the plan of activity X has length N. In this example both components of the activity's plan are actions. In general, they can be other, previously defined, activities.

Normally, a system description of $\mathcal{AL}$ is used together with a recorded history of the domain, i.e., a collection of agent's observations. In traditional action theories such histories determine past trajectories of the system, called *models*, the agent believes to be possible. If no such model exists the history is deemed inconsistent.

Compared to the AAA architecture, the present work also *expands the notion of domain history and modifies the notion of history's model*. The new domain history includes two more types of statements: $attempt(A, I)$ and $\neg hpd(A, I)$. The former indicates that the agent attempted to execute action A at step I. If at that point the preconditions for executability of A are satisfied, then action A is successful and, therefore, the domain history will contain $hpd(A, I)$; otherwise it will contain $\neg hpd(A, I)$. The notion of model is modified to allow the agent to explain unexpected observations by assuming the occurrence of a minimal collection of occurrences of exogenous actions missed by the agent.

3   Theory of Intentions

The agent's mental state is primarily described by the two inertial fluents $active\_goal(g)$ and $status(m, k)$. The latter holds when k is the index of the component of m that has most recently been executed, and $status(m, -1)$ holds when the agent does not intend to execute m^2. The inertial property of these two fluents elegantly captures the natural persistence of the agent's intentions.

The two mental actions $start(m)$ and $stop(m)$ directly affect the agent's mental state by initiating and terminating its intent to execute activity m. Special exogenous mental actions $select(g)$ and $abandon(g)$, which can be thought of as being performed by the agent's controller, initiate and terminate the agent's intent to achieve goal g. Special agent action $wait$, which has no executability conditions or effects (physical or mental), can be seen as doing nothing. Since action $wait$ has no effects, it is neither a mental nor physical action. All other agent and exogenous actions are said to be *physical*. While the agent's and exogenous mental actions do not affect the state of the physical environment, some physical actions may affect the agent's mental state. The properties of the above actions and fluents are expressed by a collection of axioms of $\mathcal{AL}^3$.

Defined fluent $active(M)$ is true when activity M has a status that is not equal to -1:

$$active(M) \quad \textbf{if} \; \neg status(M, -1). \tag{1}$$

Action $start$ sets the value of $status$ to 0 and an agent cannot $start$ an active activity. Similarly action $stop$ deactivates, and an agent cannot $stop$ an inactive activity.

Defined fluent $child(M1, M)$ is true when $M1$ is the current component of M:

$$child(M1, M) \quad \textbf{if} \; comp(M, K + 1, M1), status(M, K). \tag{2}$$

Similarly, $child\_goal(G1, G)$ is true when G and $G1$ are the goals of M and $M1$, and $descendant(M1, M)$ is defined recursively in terms of $child$. Sub-activities and sub-goals are represented by defined fluent $minor(\cdot)$. We refer to activities and goals that are not $minor$ as *top-level*. Special exogenous action $select$ activates a goal, and $abandon$ deactivates a goal. A state constraint is also included, which ensures that top-level goals are no longer active once they have been achieved.

The next axioms describe the propagation of the intent to achieve a goal to its child goal. Of course, the parent goal may be a top-level or minor goal.

The first axiom in (3) says that an unachieved minor goal $G1$ of an activity $M1$ becomes $active$ when $M1$ is the next component of an ongoing activity M. The second says that a minor goal $G1$ is no longer active when it is achieved:

$$
\begin{aligned}
active\_goal(G1) \quad &\textbf{if} \; \neg G1, \; minor(G1), child\_goal(G1, G), active\_goal(G), \\
& goal(M1, G1), status(M1, -1). \\
\neg active\_goal(G1) \quad &\textbf{if} \; G1, \; minor(G1), child\_goal(G1, G), active\_goal(G).
\end{aligned}
\tag{3}
$$

Not shown here are the third and forth axioms, which say that a minor goal $G1$ is no longer active when its parent is no longer active, and that a minor goal $G1$ of $M1$ is no longer active when $M1$ has been executed (i.e. its status is equal to its length). Defined fluents $in\_progress(M)$ and $in\_progress(G)$ are true when

---

[2] The mental state includes statics, which describe activities.

[3] For space reasons we omit formal representations of some axioms.

M and its goal G are both active. Defined fluent $next\_act(M, A)$ is true if agent action A is the next action of the ongoing execution of M. For a physical agent action of M, the axiom is:

$$next\_act(M, A) \text{ if } phys\_agent\_act(A), status(M, K),$$
$$comp(M, K + 1, A), in\_progress(M). \tag{4}$$

Executing the next physical action of M increments the status of M:

$$A \textbf{ causes } status(M, K + 1) \text{ if } next\_act(M, A), status(M, K),$$
$$comp(M, K + 1, A), phys\_agent\_act(A). \tag{5}$$

Along the same lines, stopping an activity causes its descendants to be inactive:

$$stop(M) \textbf{ causes } status(M1, -1) \text{ if } descendant(M1, M). \tag{6}$$

An *intentional system description* $\mathcal{D}$ consists of a description of the agent's physical environment, a collection of activities, and the theory of intentions. Paths in the transition diagram $\mathcal{T}(\mathcal{D})$ correspond to physically possible *trajectories* of the domain. A state of the trajectory is divided into two parts: *physical* and *mental* consisting of all physical and mental fluent literals respectively.

4   The $\mathcal{AIA}$ Control Strategy

In our architecture the agent's behavior is specified by the following $\mathcal{AIA}$ *control loop*:

1. interpret observations;
2. find an intended action e;
3. attempt to perform e and update history with a record of the attempt;
4. observe the world, update history with observations, and go to step **1**.

In step **1** the agent uses diagnostic reasoning to explain unexpected observations. The agent explains these observations by hypothesizing that some exogenous actions occurred unobserved in the past. In step **2** the agent finds an *intended action*, i.e.: to continue executing an ongoing activity that is expected to achieve its goal; to *stop* an ongoing activity whose goal is no longer active (either achieved or abandoned); to *stop* an activity that is no longer expected to achieve its goal; or to *start* a chosen activity that is expected to achieve his goal.

In general, a history Γ of the domain defines trajectories in the transition diagram satisfying Γ. These trajectories define possible pasts of the domain compatible with observations in Γ and the assumptions about the agent's observation strategy and ability. This however does not mean that every action in such a model is intended by an agent. This is the case, for example, if Bob procrastinates and *waits* instead of performing the intended action $start(1)$. It can be shown, however, that this is impossible for histories produced by an agent executing the $\mathcal{AIA}$ control loop. Every agent's action in every model of such a history is intended. Such histories are called *intentional*, and this is exactly what we require from an intentional agent.

5   The Reasoning Algorithms

In this section we present a refinement of the $\mathcal{AIA}$ control loop in which reasoning tasks are reduced to computing answer sets of a CR-Prolog program
constructed from the intentional system description and the domain history.

The program, denoted by $\Pi(\mathcal{D}, \Gamma_n)$, consists of: the translation of $\mathcal{D}$ into
ASP rules $(\Pi(\mathcal{D}))$; rules for computing models of Γ_n $(\Pi(\Gamma_n))$; and rules for
determining intended actions at n $(IA(n))$. Construction of $\Pi(\mathcal{D})$ is based on the
diagnostic module of AAA [6]. In addition to standard axioms, it contains axioms
encoding our domain assumptions and the effects of a record $attempt(A, I)$ and
$\neg hpd(A, I)$. A consistency-restoring rule of CR-Prolog allows us to compute
minimal explanations of unexpected observations:

$$occurs(A, I2) \xleftarrow{+} phys\_exog\_act(A), curr\_step(I1),\ I2 < I1.$$
$$unobs(A, I) \leftarrow I < I1, phys\_exog\_act(A), occurs(A, I), not\ hpd(A, I).$$
$$number\_unobs(N, I) \leftarrow curr\_step(I), N = \#count\{unobs(EX, IX)\}.$$

The following lemma links the first step and the answer sets of $\Pi(\mathcal{D}) \cup \Pi(\Gamma_n)$.

Lemma 1. *If Γ_n is an intentional history of $\mathcal{D}$, then P_n is a model of Γ_n iff
P_n is defined by some answer set A of $\Pi = \Pi(\mathcal{D}) \cup \Pi(\Gamma_n)$. Moreover, for
every answer set A of Π, $number\_unobs(x, n) \in A$ iff there are x unobserved
occurrences of exogenous actions in A.*

To perform the second step – finding an intended action – we will need program
$IA(n)$. It consists of an atom $interpret(x, n)$ where x is the number of unobserved
exogenous actions in the models of Γ_n and the collection of rules needed to
compute an intended action. A constraint requires the agent to adhere to the
outcome of the reasoning completed in step 1 by preventing the agent from
assuming additional occurrences of exogenous actions. Next we notice that the
collection of possible histories can be divided in four categories. The categories,
which are uniquely determined by a mental state of the agent, are used in the
rules for computing intended actions.

For example, a history belongs to category 1 if the agent has neither goal
nor activity to commit to. In this case the intended action is to *wait*. This is
defined by a rule in which literal $active\_goal\_or\_activity(I)$ is true when there is
an active goal or activity at the current step I. Similarly a history is of category
2 if the agent's top-level activity is active but its goal is not, and in this case the
intended action is to *stop* the activity. A history is of category 3 if the agent's
top-level activity and goal are both active. In this situation the intended action
will be the next action of activity M. But there is an exception to this rule —
the agent needs to check that this activity still has a chance achieve his goal.
Other rules cause an atom $proj\_success(M, I)$ to be part of an answer set if the
next action is intended. If no such answer set exists, then the activity is futile,
and the intended action is to *stop*. This is achieved by a cr-rule:

$$futile(M, I) \xleftarrow{+} interpret(N, I), category\_3(M, I), \neg proj\_success(M, I).$$

Finally, category 4 corresponds to the case in which there is an active goal but the agent does not yet have a plan to achieve it. In this case an intended action will begin executing either an activity containing a shortest plan for achieving this goal or *wait* if such activity does not exist. The planning task uses cr-rules similar to those in [5]. The resulting program shows that, by mixing regular ASP rules with consistency restoring rules, CR-Prolog is capable of expressing rather non-trivial forms of reasoning. The following lemma ensures that step **2** of the $\mathcal{AIA}$ control loop – finding an intended action – is reduced to computing answer sets of $\Pi(\mathcal{D}, \Gamma_n)$.

Lemma 2. *Let Γ_n be an intentional history and x be the number of unobserved occurrences of exogenous actions in a model of Γ_n. Action e is an intended action of Γ_n iff some answer set A of $\Pi(\mathcal{D}, \Gamma_n) \cup \{interpret(x, n).\}$ contains the atom $intended\_act(e, n)$.*

6   Conclusions

This paper describes the $\mathcal{AIA}$ architecture for intelligent agents whose behavior is driven by their intentions and who reason about, and act in, changing environments. We presented a formal model of an intentional agent and its environment that includes the theory of intentions $\mathcal{TI}$. Such a model was capable of representing activities, goals, and intentions. We presented an algorithm that takes the agent's knowledge (including $\mathcal{TI}$), explains unexpected observations, and computes the agent's intended action. Both reasoning tasks are reduced to computing answer sets of CR-prolog programs. A prototype can be found at http://www.depts.ttu.edu/cs/research/krlab/software-aia.php.

References

1. Balduccini, M., Gelfond, M.: The AAA architecture: an overview. In: AAAI Spring Symposium on Architecture of Intelligent Theory-Based Agents (2008)
2. Balduccini, M.: CR-MODELS: an inference engine for CR-prolog. In: Baral, C., Brewka, G., Schlipf, J. (eds.) LPNMR 2007. LNCS (LNAI), vol. 4483, pp. 18–30. Springer, Heidelberg (2007)
3. Balduccini, M., Gelfond, M.: Logic programs with consistency-restoring rules. In: Doherty, P., McCarthy, J., Williams, M.A. (eds.) International Symposium on Logical Formalization of Commonsense Reasoning. pp. 9–18. AAAI 2003 Spring Symposium Series, March 2003
4. Baral, C., Gelfond, M.: Reasoning agents in dynamic domains. In: Workshop on Logic-Based Artificial Intelligence. Kluwer Academic Publishers, June 2000
5. Blount, J., Gelfond, M.: Reasoning about the intentions of agents. In: Artikis, A., Craven, R., Kesim Çiçekli, N., Sadighi, B., Stathis, K. (eds.) Logic Programs, Norms and Action. LNCS, vol. 7360, pp. 147–171. Springer, Heidelberg (2012)
6. Gelfond, M., Kahl, Y.: Knowledge Representation, Reasoning, and the Design of Intelligent Agents: The Answer-Set Programming Approach. Cambridge University Press, Cambridge (2014)

7. Rao, A.S., Georgeff, M.P.: Modeling rational agents within a BDI-architecture. In: Proceedings of the 2nd International Conference on Principles of Knowledge Representation and Reasoning. pp. 473–484. Morgan Kaufmann publishers Inc., San Mateo, CA, USA (1991)
8. Sacerdoti, E.: The nonlinear nature of plans. In: Proceedings of IJCAI-1975 (1975)
9. Wooldridge, M.: Reasoning about Rational Agents. The MIT Press, Cambridge (2000)

Answer Set Programming Modulo Acyclicity

Jori Bomanson[1], Martin Gebser[1,2], Tomi Janhunen[1], Benjamin Kaufmann[2], and Torsten Schaub[2,3(✉)]

[1] Aalto University, HIIT, Espoo, Finland
{jori.bomanson,tomi.janhunen}@aalto.fi
[2] University of Potsdam, Potsdam, Germany
{gebser,kaufmann,torsten}@cs.uni-potsdam.de
[3] INRIA Rennes, Rennes, France

Abstract. Acyclicity constraints are prevalent in knowledge representation and, in particular, applications where acyclic data structures such as DAGs and trees play a role. Recently, such constraints have been considered in the satisfiability modulo theories (SMT) framework, and in this paper we carry out an analogous extension to the answer set programming (ASP) paradigm. The resulting formalism, ASP modulo acyclicity, offers a rich set of primitives to express constraints related with recursive structures. The implementation, obtained as an extension to the state-of-the-art answer set solver CLASP, provides a unique combination of traditional unfounded set checking with acyclicity propagation.

1 Introduction

Acyclic data structures such as DAGs and trees occur frequently in applications. For instance, Bayesian [1] and Markov [2] network learning as well as Circuit layout [3] are based on respective conditions. When logical formalisms are used for the specification of such structures, dedicated *acyclicity constraints* are called for. Recently, such constraints have been introduced in the *satisfiability modulo theories* (SMT) framework [4] for extending Boolean satisfiability in terms of graph-theoretic properties [5,6]. The idea of *satisfiability modulo acyclicity* [7] is to view Boolean variables as conditionalized edges of a graph and to require that the graph remains acyclic under variable assignments. Moreover, the respective theory propagators for acyclicity have been implemented in contemporary CDCL-based SAT solvers, MINISAT and GLUCOSE, which offer a promising machinery for solving applications involving acyclicity constraints.

In this paper, we consider acyclicity constraints in the context of *answer set programming* (ASP) [8], featuring a rule-based language for knowledge representation. While SAT solvers with explicit acyclicity constraints offer an alternative

---

This work was funded by AoF (251170), DFG (SCHA 550/8 and 550/9), as well as DAAD and AoF (57071677/279121). An extended draft with additional elaborations and experiments is available at http://www.cs.uni-potsdam.de/wv/publications/.
T. Schaub—Affiliated with Simon Fraser University, Canada, and IIIS Griffith University, Australia.

© Springer International Publishing Switzerland 2015
F. Calimeri et al. (Eds.): LPNMR 2015, LNAI 9345, pp. 143–150, 2015.
DOI: 10.1007/978-3-319-23264-5_13

mechanism to implement ASP via appropriate translations [7], the goal of this paper is different: the idea is to incorporate acyclicity constraints into ASP, thus accounting for extended rule types as well as reasoning tasks like enumeration and optimization. The resulting formalism, *ASP modulo acyclicity*, offers a rich set of primitives to express constraints related with recursive structures. The implementation, obtained as an extension to the state-of-the-art answer set solver CLASP [9], provides a unique combination of traditional unfounded set [10] checking and acyclicity propagation [5].

2   Background

We consider logic programs built from rules of the following forms:

$$a \leftarrow b_1, \ldots, b_n, \text{not } c_1, \ldots, \text{not } c_m. \tag{1}$$

$$\{a\} \leftarrow b_1, \ldots, b_n, \text{not } c_1, \ldots, \text{not } c_m. \tag{2}$$

$$a \leftarrow k \leq [b_1 = w_1, \ldots, b_n = w_n, \text{not } c_1 = w_{n+1}, \ldots, \text{not } c_m = w_{n+m}]. \tag{3}$$

Symbols $a, b_1, \ldots, b_n, c_1, \ldots, c_m$ stand for (propositional) *atoms*, $k, w_1, \ldots, w_{n+m}$ for non-negative integers, and not for (default) *negation*. Atoms like b_i and negated atoms like not c_i are called *positive* and *negative literals*, respectively. For a *normal* (1), *choice* (2), or *weight* (3) *rule* r, we denote its *head* atom by $\text{head}(r) = a$ and its *body* by $\text{B}(r)$. By $\text{B}(r)^+ = \{b_1, \ldots, b_n\}$ and $\text{B}(r)^- = \{c_1, \ldots, c_m\}$, we refer to the *positive* and *negative body* atoms of r. When r is a weight rule, the respective sequence of *weighted literals* is denoted by $\text{WL}(r)$, and its restrictions to positive or negative literals by $\text{WL}(r)^+$ and $\text{WL}(r)^-$. A normal rule r such that $\text{head}(r) \in \text{B}(r)^-$ is called an *integrity constraint*, and we below skip $\text{head}(r)$ and not $\text{head}(r)$ for brevity, where $\text{head}(r)$ is an arbitrary atom occurring in r only. A *weight constraint program* P, or simply a *program*, is a finite set of rules; P is a *choice program* if it consists of normal and choice rules only, and a *positive program* if it involves neither negation nor choice rules.

Given a program P, let $\text{head}(P) = \{\text{head}(r) \mid r \in P\}$ and $\text{At}(P) = \text{head}(P) \cup \bigcup_{r \in P}(\text{B}(r)^+ \cup \text{B}(r)^-)$ denote the sets of head atoms or all atoms, respectively, occurring in P. The *defining rules* of an atom $a \in \text{At}(P)$ are $\text{Def}_P(a) = \{r \in P \mid \text{head}(r) = a\}$. An *interpretation* $I \subseteq \text{At}(P)$ satisfies $\text{B}(r)$ for a normal or choice rule r iff $\text{B}(r)^+ \subseteq I$ and $\text{B}(r)^- \cap I = \emptyset$. The weighted literals of a weight rule r evaluate to $\text{v}_I(\text{WL}(r)) = \sum_{1 \leq i \leq n, b_i \in I} w_i + \sum_{1 \leq i \leq m, c_i \notin I} w_{n+i}$; when r is a weight rule, I satisfies $\text{B}(r)$ iff $k \leq \text{v}_I(\text{WL}(r))$. For any rule r, we write $I \models \text{B}(r)$ iff I satisfies $\text{B}(r)$, and $I \models r$ iff $I \models \text{B}(r)$ implies $\text{head}(r) \in I$. The *supporting rules* of P with respect to I are $\text{SR}_P(I) = \{r \in P \mid \text{head}(r) \in I, I \models \text{B}(r)\}$. Moreover, I is a *model* of P, denoted by $I \models P$, iff $I \models r$ for every $r \in P$ such that r is a normal or weight rule. A model I of P is a *supported model* of P when $\text{head}(\text{SR}_P(I)) = I$. Note that any positive program P possesses a unique *least model*, denoted by $\text{LM}(P)$.

For a normal or choice rule r, $\text{B}(r)^I = \text{B}(r)^+$ denotes the reduct of $\text{B}(r)$ with respect to an interpretation I, and $\text{B}(r)^I = (\max\{0, k - \text{v}_I(\text{WL}(r)^-)\} \leq$

$WL(r)^+)$ is the reduct of $B(r)$ for a weight rule r. The *reduct* of a program P with respect to an interpretation I is $P^I = \{\text{head}(r) \leftarrow B(r)^I \mid r \in SR_P(I)\}$. Then, I is a *stable model* of P iff $I \models P$ and $LM(P^I) = I$. While any stable model of P is a supported model of P as well, the converse does not hold in general. However, the following concept provides a tighter notion of support achieving such a correspondence.

Definition 1. *A model I of a program P is* well-supported *by a set $R \subseteq SR_P(I)$ of rules iff* $\text{head}(R) = I$ *and there is some ordering $r_1, \ldots, r_n$ of R such that, for each $1 \le i \le n$,* $\text{head}(\{r_1, \ldots, r_{i-1}\}) \models B(r_i)^I$.

In fact, a (supported) model I of a program P is stable iff I is well-supported by some subset of $SR_P(I)$, and several such subsets may exist. The notion of well-support counteracts circularity in the *positive dependency graph* $DG^+(P) = \langle At(P), \succeq \rangle$ of P, whose edge relation $a \succeq b$ holds for all $a, b \in At(P)$ such that $\text{head}(r) = a$ and $b \in B(r)^+$ for some rule $r \in P$. If $a \succeq b$, we also write $\langle a, b \rangle \in DG^+(P)$.

3   Acyclicity Constraints

In [5], the SAT problem has been extended by explicit acyclicity constraints. The basic idea is to label edges of a directed graph with dedicated Boolean variables. While satisfying the clauses of a SAT instance referring to these labeling variables, also the directed graph consisting of edges whose labeling variables are true must be kept acyclic. Thus, the graph behind the labeling variables imposes an additional constraint on satisfying assignments. In what follows, we propose a similar extension of logic programs subject to stable model semantics.

Definition 2. *The* acyclicity extension *of a logic program P is a pair $\langle V, e \rangle$, where*

1. *V is a set of nodes and*
2. *$e : At(P) \to V \times V$ is a partial injection that maps atoms of P to edges.*

In the sequel, a program P is called an *acyclicity program* if it has an acyclicity extension $\langle V, e \rangle$. To define the semantics of acyclicity programs, we identify the graph of the acyclicity check as follows. Given an interpretation $I \subseteq At(P)$, we write $e(I)$ for the set of edges $e(a)$ induced by atoms $a \in I$ for which $e(a)$ is *defined*. For a given acyclicity extension $\langle V, e \rangle$, the graph $e(At(P))$ is the maximal one that can be obtained under any interpretation and is likely to contain cycles. If not, then the extension can be neglected altogether as no cycles can arise. To be precise about the acyclicity condition being imposed, we recall that a graph $\langle V, E \rangle$ with the set $E \subseteq V^2$ of edges has a *cycle* iff there is a non-trivial directed path from any node $v \in V$ back to itself via the edges in E. An *acyclic* graph $\langle V, E \rangle$ has no cycles of this kind.

Definition 3. *Let P be an acyclicity program with an acyclicity extension $\langle V, e \rangle$. An interpretation $M \subseteq \mathrm{At}(P)$ is a stable (or supported) model of P subject to $\langle V, e \rangle$ iff M is a stable (or supported) model of P such that the graph $\langle V, e(M) \rangle$ is acyclic.*

Example 1. Consider a directed graph $\langle V, E \rangle$ and the task to find a Hamiltonian cycle through the graph, i.e., a cycle that visits each node of the graph exactly once. Let us encode the graph by introducing the fact $\mathtt{node}(v)$ for each $v \in V$ and the fact $\mathtt{edge}(v, u)$ for each $\langle v, u \rangle \in E$. Then, it is sufficient (i) to pick beforehand an arbitrary initial node, say v_0, for the cycle, (ii) to select for each node exactly one outgoing and one incoming edge to be on the cycle, and (iii) to check that the cycle is not completed before the path spanning along the selected edges returns to v_0. Assuming that a predicate $\mathtt{hc}$ is used to represent selected edges, the following (first-order) rules express (ii):

$$1\{\mathtt{hc}(v, u) : \mathtt{edge}(v, u)\}1 \leftarrow \mathtt{node}(v).$$
$$1\{\mathtt{hc}(v, u) : \mathtt{edge}(v, u)\}1 \leftarrow \mathtt{node}(u).$$

To enforce (iii), we introduce an acyclicity extension $\langle V, e \rangle$, where e maps an atom $\mathtt{hc}(v, u)$ to an edge $\langle v, u \rangle$ whenever v and u are different from v_0. ∎

Our next objective is to relate acyclicity programs with ordinary logic programs in terms of translations. It is well-known that logic programs subject to stable model semantics can express reachability in graphs, which implies that also acyclicity is expressible. To this end, we present a translation based on *elimination orderings* [11].

Definition 4. *Let P be an acyclicity program with an acyclicity extension $\langle V, e \rangle$. The translation $\mathrm{Tr}_{\mathrm{EL}}(P, V, e)$ extends P as follows.*

1. *For each atom $a \in \mathrm{At}(P)$ such that $e(a) = \langle v, u \rangle$, the rules:*

$$\mathtt{el}(v, u) \leftarrow \mathtt{not}\ a. \tag{4}$$
$$\mathtt{el}(v, u) \leftarrow \mathtt{el}(u). \tag{5}$$

2. *For each node $v \in V$ such that $\langle v, u_1 \rangle, \ldots, \langle v, u_k \rangle$ are the edges in $e(\mathrm{At}(P))$ starting from v:*

$$\mathtt{el}(v) \leftarrow \mathtt{el}(v, u_1), \ldots, \mathtt{el}(v, u_k). \tag{6}$$
$$\leftarrow \mathtt{not}\ \mathtt{el}(v). \tag{7}$$

The intuitive reading of the new atom $\mathtt{el}(v, u)$ is that the edge $\langle v, u \rangle \in e(\mathrm{At}(P))$ has been eliminated, meaning that it cannot belong to any cycle. Analogously, the atom $\mathtt{el}(v)$ denotes the elimination of a node $v \in V$. By the rule (4), an edge $\langle v, u \rangle$ is eliminated when the atom a such that $e(a) = \langle v, u \rangle$ is false, while the rule (5) is applicable once the end node u is eliminated. Then, the node v gets eliminated by the rule (6) if all edges starting from it are eliminated. Finally, the constraint (7) ensures that all nodes are eliminated. That is, the success of the acyclicity test presumes that $\mathtt{el}(v, u)$ or $\mathtt{el}(v)$, respectively, is derivable for each edge $\langle v, u \rangle \in e(\mathrm{At}(P))$ and each node $v \in V$.

Theorem 1. *Let P be an acyclicity program with an acyclicity extension $\langle V, e \rangle$ and $\text{Tr}_{\text{EL}}(P, V, e)$ its translation into an ordinary logic program.*

1. *If M is a stable model of P subject to $\langle V, e \rangle$, then $M' = M \cup \{\text{el}(v, u) \mid \langle v, u \rangle \in e(\text{At}(P))\} \cup \{\text{el}(v) \mid v \in V\}$ is a stable model of $\text{Tr}_{\text{EL}}(P, V, e)$.*
2. *If M' is a stable model of $\text{Tr}_{\text{EL}}(P, V, e)$, then $M = M' \cap \text{At}(P)$ is a stable model of P subject to $\langle V, e \rangle$.*

Transformations in the other direction are of interest as well, i.e., the goal is to capture stable models by exploiting the acyclicity constraint. While the existing translation from ASP into SAT modulo acyclicity [7] provides a starting point for such a transformation, the target syntax is given by rules rather than clauses.

Definition 5. *Let P be a weight constraint program. The acyclicity translation of P consists of $\text{Tr}_{\text{ACYC}}(P) = \bigcup_{a \in \text{At}(P)} \text{Tr}_{\text{ACYC}}(P, a)$ with an acyclicity extension $\langle \text{At}(P), e \rangle$ such that $e(\text{dep}(a, b)) = \langle a, b \rangle$ for each edge $\langle a, b \rangle \in \text{DG}^+(P)$, where $\text{Tr}_{\text{ACYC}}(P, a)$ extends $\text{Def}_P(a)$ for each atom $a \in \text{At}(P)$ as follows.*

1. *For each edge $\langle a, b \rangle \in \text{DG}^+(P)$, the choice rule:*

$$\{\text{dep}(a, b)\} \leftarrow b. \tag{8}$$

2. *For each defining rule (1) or (2) of a, the rule:*

$$\text{ws}(r) \leftarrow \text{dep}(a, b_1), \ldots, \text{dep}(a, b_n), \text{not } c_1, \ldots, \text{not } c_m. \tag{9}$$

3. *For each defining rule (3) of a, the rule:*

$$\text{ws}(r) \leftarrow k \leq [\text{dep}(a, b_1) = w_1, \ldots, \text{dep}(a, b_n) = w_n,$$
$$\text{not } c_1 = w_{n+1}, \ldots, \text{not } c_m = w_{n+m}]. \tag{10}$$

4. *For $\text{Def}_P(a) = \{r_1, \ldots, r_k\}$, the constraint:*

$$\leftarrow a, \text{not } \text{ws}(r_1), \ldots, \text{not } \text{ws}(r_k). \tag{11}$$

The rules (9) and (10) specify when r provides well-support for a, i.e., the head atom a non-circularly depends on $B(r)^+ = \{b_1, \ldots, b_n\}$. The constraint (11) expresses that $a \in \text{At}(P)$ must have a well-supporting rule $r \in \text{Def}_P(a)$ whenever a is true. To this end, respective dependencies have to be established in terms of the choice rules (8).

Theorem 2. *Let P be a weight constraint program and $\text{Tr}_{\text{ACYC}}(P)$ its translation into an acyclicity program with an acyclicity extension $\langle \text{At}(P), e \rangle$.*

1. *If M is a stable model of P, then there is an ordering $r_1, \ldots, r_n$ of some $R \subseteq \text{SR}_P(M)$ such that $M' = M \cup \{\text{ws}(r) \mid r \in R\} \cup \{\text{dep}(\text{head}(r_i), b) \mid 1 \leq i \leq n, b \in B_i\}$, where $B_i \subseteq B(r_i)^+ \cap \text{head}(\{r_1, \ldots, r_{i-1}\})$ for each $1 \leq i \leq n$, is a supported model of $\text{Tr}_{\text{ACYC}}(P)$ subject to $\langle \text{At}(P), e \rangle$.*

2. *If M' is a supported model of $\mathrm{Tr_{ACYC}}(P)$ subject to $\langle \mathrm{At}(P), e \rangle$, then $M = M' \cap \mathrm{At}(P)$ is a stable model of P and M is well-supported by $R = \{r \mid \mathbf{ws}(r) \in M'\}$.*

It is well-known that supported and stable models coincide for *tight* logic programs [12,13]. The following theorem shows that translations produced by $\mathrm{Tr_{ACYC}}$ possess an analogous property subject to the acyclicity extension $\langle \mathrm{At}(P), e \rangle$. This opens up an interesting avenue for investigating the efficiency of stable model computation—either using unfounded set checking or the acyclicity constraint, or both.

Theorem 3. *Let P be a weight constraint program and $\mathrm{Tr_{ACYC}}(P)$ its translation into an acyclicity program with an acyclicity extension $\langle \mathrm{At}(P), e \rangle$. Then, M is a supported model of $\mathrm{Tr_{ACYC}}(P)$ subject to $\langle \mathrm{At}(P), e \rangle$ iff M is a stable model of $\mathrm{Tr_{ACYC}}(P)$ subject to $\langle \mathrm{At}(P), e \rangle$.*

As witnessed by Theorems 2 and 3, the translation $\mathrm{Tr_{ACYC}}$ provides means to capture stability in terms of the acyclicity constraint. However, the computational efficiency of the translation can be improved when additional constraints governing $\mathrm{dep}(v, u)$ atoms are introduced. The purpose of these constraints is to falsify dependencies in settings where they are not truly needed. We first concentrate on choice programs and will then extend the consideration to weight rules below. The following definition adopts the cases from [7] but reformulates them in terms of rules rather than clauses.

Definition 6. *Let P be a choice program. The strong acyclicity translation of P, denoted by $\mathrm{Tr_{ACYC+}}(P)$, extends $\mathrm{Tr_{ACYC}}(P)$ as follows.*

1. *For each $\langle a, b \rangle \in \mathrm{DG}^+(P)$, the constraint:*

$$\leftarrow \mathrm{dep}(a, b), \mathbf{not}\ a. \tag{12}$$

2. *For each $\langle a, b \rangle \in \mathrm{DG}^+(P)$ and $r \in \mathrm{Def}_P(a)$ such that $b \notin \mathrm{B}(r)^+$, the constraint:*

$$\leftarrow \mathrm{dep}(a, b), \mathbf{ws}(r). \tag{13}$$

Intuitively, dependencies from a are not needed if a is false (12). Quite similarly, a particular dependency may be safely preempted (13) if the well-support for a is provided by a rule r not involving this dependency.

The strong acyclicity translation for weight rules includes additional subprograms.

Definition 7. *Let P be a weight constraint program and $r \in P$ a weight rule of the form (3), where $\mathrm{head}(r) = a$, $|\{b_1, \ldots, b_n\}| = n$, and $w_1, \ldots, w_n$ are ordered such that $w_{i-1} \leq w_i$ for each $1 < i \leq n$. The strong acyclicity translation $\mathrm{Tr_{ACYC+}}(P)$ of P is fortified as follows.*

1. For $1 < i \leq n$, the rules:

$$\mathtt{nxt}(r, i) \leftarrow \mathtt{dep}(a, b_{i-1}). \tag{14}$$
$$\mathtt{nxt}(r, i) \leftarrow \mathtt{nxt}(r, i-1). \tag{15}$$
$$\mathtt{chk}(r, i) \leftarrow \mathtt{nxt}(r, i), \mathtt{dep}(a, b_i). \tag{16}$$

2. The weight rule:

$$\mathtt{red}(r) \leftarrow k \leq [\mathtt{chk}(r, 2) = w_2, \ldots, \mathtt{chk}(r, n) = w_n,$$
$$\mathtt{not}\ c_1 = w_{n+1}, \ldots, \mathtt{not}\ c_m = w_{n+m}]. \tag{17}$$

3. For each $\langle a, b \rangle \in DG^+(P)$ such that $b \in B(r)^+$, the constraint:

$$\leftarrow \mathtt{dep}(a, b), \mathtt{red}(r). \tag{18}$$

The idea is to cancel dependencies $\langle a, b \rangle \in DG^+(P)$ by the constraint (18) when the well-support obtained though r can be deemed redundant by the rule (17). To this end, the rules of the forms (14) and (15) identify an atom among $b_1, \ldots, b_n$ of smallest weight having an active dependency from a, i.e., $\mathtt{dep}(a, b_i)$ is true, provided such an i exists. By the rules of the form (16), any further dependencies are extracted, and (17) checks whether the remaining literals are sufficient to reach the bound k. If so, all dependencies from a are viewed as redundant. This check covers also cases where, e.g., negative literals suffice to satisfy the body and positive dependencies play no role.

4   Discussion

In this paper, we propose a novel SMT-style extension of ASP by explicit acyclicity constraints in analogy to [5]. These kinds of constraints have not been directly addressed in previous SMT-style extensions of ASP [14–16]. The new extension, herein coined ASP modulo acyclicity, offers a unique set of primitives for applications involving DAGs or tree structures. One interesting application is the embedding of ASP itself, given that unfounded set checking can be captured (Theorem 2). The utilized notion of well-supporting rules resembles *source pointers* [17], used in native answer set solvers to record rules justifying true atoms. In fact, a major contribution of this work is the implementation of new translations and principles in tools. For instance, CLASP [9] features enumeration and optimization, which are not supported by ACYCMINISAT and ACYCGLUCOSE [5]. Thereby, a replication of supported (and stable) models under translations can be avoided by using the projection capabilities of CLASP [18]. Last but not least, acyclicity programs enrich the variety of modeling primitives available to users.

References

1. Cussens, J.: Bayesian network learning with cutting planes. In: Proceeding UAI 2011, pp. 153–160. AUAI Press (2011)

2. Corander, J., Janhunen, T., Rintanen, J., Nyman, H., Pensar, J.: Learning chordal Markov networks by constraint satisfaction. In: Proceeding NIPS 2013, NIPS Foundation, pp. 1349–1357 (2013)
3. Erdem, E., Lifschitz, V., Wong, M.D.F.: Wire routing and satisfiability planning. In: Palamidessi, C., Moniz Pereira, L., Lloyd, J.W., Dahl, V., Furbach, U., Kerber, M., Lau, K.-K., Sagiv, Y., Stuckey, P.J. (eds.) CL 2000. LNCS (LNAI), vol. 1861, pp. 822–836. Springer, Heidelberg (2000)
4. Barrett, C., Sebastiani, R., Seshia, S., Tinelli, C.: Satisfiability modulo theories. In: Handbook of Satisfiability, pp. 825–885. IOS Press (2009)
5. Gebser, M., Janhunen, T., Rintanen, J.: SAT modulo graphs: Acyclicity. In: Fermé, E., Leite, J. (eds.) JELIA 2014. LNCS, vol. 8761, pp. 137–151. Springer, Heidelberg (2014)
6. Bayless, S., Bayless, N., Hoos, H., Hu, A.: SAT modulo monotonic theories. In: Proceeding AAAI 2015, pp. 3702–3709. AAAI Press (2015)
7. Gebser, M., Janhunen, T., Rintanen, J.: Answer set programming as SAT modulo Acyclicity. In: Proceeding ECAI 2014, pp. 351–356. IOS Press (2014)
8. Baral, C.: Knowledge Representation, Reasoning and Declarative Problem Solving. Cambridge University Press, Cambridge (2003)
9. Gebser, M., Kaufmann, B., Schaub, T.: Conflict-driven answer set solving: from theory to practice. Artif. Intell. 187–188, 52–89 (2012)
10. Van Gelder, A., Ross, K., Schlipf, J.: The well-founded semantics for general logic programs. J. ACM 38(3), 620–650 (1991)
11. Gebser, M., Janhunen, T., Rintanen, J.: ASP encodings of Acyclicity properties. In: Proceeding KR 2014. AAAI Press (2014)
12. Fages, F.: Consistency of Clark's completion and the existence of stable models. J. Methods Logic Comput. Sci. 1, 51–60 (1994)
13. Erdem, E., Lifschitz, V.: Tight logic programs. Theor. Pract. Logic Program. 3(4–5), 499–518 (2003)
14. Gebser, M., Ostrowski, M., Schaub, T.: Constraint answer set solving. In: Hill, P.M., Warren, D.S. (eds.) ICLP 2009. LNCS, vol. 5649, pp. 235–249. Springer, Heidelberg (2009)
15. Liu, G., Janhunen, T., Niemelä, I.: Answer set programming via mixed integer programming. In: Proceeding KR 2012, pp. 32–42. AAAI Press (2012)
16. Lee, J., Meng, Y.: Answer set programming modulo theories and reasoning about continuous changes. In: Proceeding IJCAI 2013, pp. 990–996. IJCAI/AAAI Press (2013)
17. Simons, P., Niemelä, I., Soininen, T.: Extending and implementing the stable model semantics. Artif. Intell. 138(1–2), 181–234 (2002)
18. Gebser, M., Kaufmann, B., Schaub, T.: Solution enumeration for projected Boolean search problems. In: van Hoeve, W.-J., Hooker, J.N. (eds.) CPAIOR 2009. LNCS, vol. 5547, pp. 71–86. Springer, Heidelberg (2009)

A Framework for Goal-Directed Query Evaluation with Negation

Stefan Brass[✉]

Institut für Informatik, Martin-Luther-Universität Halle-Wittenberg,
Von-Seckendorff-Platz 1, 06099 Halle (Saale), Germany
`brass@informatik.uni-halle.de`

Abstract. This paper contains a proposal how goal-directed query evaluation for the well-founded semantics WFS (and other negation semantics) can be done based on elementary program transformations. It also gives a new look at the author's SLDMagic method, which has several advantages over the standard magic set method (e.g., for tail recursions).

1 Introduction

The efficient evaluation of queries to logic programs with nonmonotic negation remains an everlasting problem. Of course, big achievements have been made, but at the same time problem size and complexity grows, so that any further progress can increase the practical applicability of logic-based, declarative programming.

In this paper, we consider non-disjunctive Datalog, i.e. pure Prolog without function symbols, but with unrestricted negation in the body. All rules must be range-restricted (allowed), i.e. variables appearing in the head or a negative body literal must also appear in a positive body literal.

We are mainly interested in the well-founded semantics WFS, but since our method is based on elementary program transformations, it can also be used as a pre-computation step for other semantics. Of course, it is well-known that because of odd loops over negation, for the stable model semantics it does not suffice to follow only the predicate calls from a given goal. But see, e.g. [2,11].

The magic set transformation [1] is the best known method for goal-directed query evaluation in deductive databases. However, it has a number of problems:

- For tail recursions, magic sets are significantly slower than SLD-resolution (e.g. a quadratic number of facts derived compared with an SLD-tree containing only a linear number of nodes/literals). This problem applies to all methods which store literals implicitly proven in the SLD-tree ("lemmas").
- It sometimes transforms non-recursive programs into recursive ones. In the same way, a stratified program can be transformed into a non-stratified one.
- It can only pass values for arguments to called predicates, not more general conditions. Furthermore, optimizations based on the evaluation sequence of called literals are restricted to the bodies of single rules.
- Quite often, variable bindings are projected away in order to call a predicate, and are later recovered by a costly join when the predicate succeeded.

© Springer International Publishing Switzerland 2015
F. Calimeri et al. (Eds.): LPNMR 2015, LNAI 9345, pp. 151–157, 2015.
DOI: 10.1007/978-3-319-23264-5_14

While some of these problems have been solved with specific optimizations, our "SLDMagic"-method [3] solved all these problems by simulating SLD-resolution bottom-up.

Of course, sometimes magic sets are better: If the same literal is called repeatedly in different contexts, magic sets proves it only once, whereas SLD resolution proves it again each time. Therefore, SLDMagic had the possibility to mark positive body literals with call(...), in which case they were proven separately in a different SLD-tree. I.e. we assume that the programmer or an automatic optimizer marks some of the positive body literals of the rules with the special keyword "call". In this way, the best of both methods can be combined. Furthermore, when we want to compute the structure of occurring goals already at "compile-time" (when the facts/relations for database predicates are not yet known), we needed to introduce at least one "call" in recursions which are not tail-recursions. Since call-literals are proven in a subproof (much like negation as failure), the length of the occurring goals in the SLD-tree became bounded, and we could encode entire goals as single facts (where the arguments represented the data values only known at "run time").

However, SLDMagic could not handle negation. It is the purpose of this paper to remedy this problem, and to give a different look at the method, which opens perspectives for further improvements. Whereas earlier, we described the method by means of partial evaluation of a meta-interpreter, we now combine it with our work on elementary program transformations.

Together with Jürgen Dix, we investigated ways to characterize non-monotonic semantics of (disjunctive) logic programs by means of elementary program transformations [4,5]. We used the notion of conditional facts, introduced by Bry [7] and Dung/Kanchansut [10]. A conditional fact is a ground rule with only negative body literals. If a nonmonotonic semantics permits certain simple transformations, in particular the "Generalized Principle of Partial Evaluation" [9,12] (which is simply unfolding for non-disjunctive programs), and the elimination of tautologies, any (ground) program can be equivalently transformed into a set of conditional facts. If we also permit the evaluation of negative body literals in obvious cases, we can compute a unique normal form of the program, called the "residual program". From this, one can directly read the well-founded model: If A is given as (unconditional) fact, it is true, if there is no rule with A in the head, it is false, and in all other cases, it is undefined. The residual program is also equivalent to the original program under the stable model semantics.

Because the residual program can grow to exponential size in rare cases, we restricted unfolding and "delayed" also positive body literals until their truth value became obvious (as for the negative body literals before) [6]. Combined with magic sets, this resulted in a competitive evaluation procedure for WFS. But there is further optimization potential if we use ideas from SLDMagic. It is also nice if the framework is based entirely on elementary program transformations.

This paper contains only some preliminary ideas (it is a "short paper" about work in progress). A more complete version is being prepared. Progress will be reported at: http://www.informatik.uni-halle.de/~brass/negeval/.

There are some obvious similarities to SLG-resolution [8,13]. However, when a tail-recursive predicate is tabled in order to ensure termination, SLG-resolution has the same problem as magic sets. Furthermore, our approach is explained based only on rules, whereas SLG-resolution needs more complex data structures. But much more work has been done on the efficient implementation of SLG-resolution in the XSB system, whereas our approach is still at the beginning.

2   Goal-Directed Query Evaluation Based on Program Transformations

We assume that the query is given as a rule $\mathsf{answer}(\mathsf{X}_1, \ldots, \mathsf{X}_m) \leftarrow \mathsf{B}_1 \wedge \cdots \wedge \mathsf{B}_n$, where the special predicate answer does not otherwise appear in the program. The body of this query (rule) is the classical query. In this way, we do not have to track substitutions for the variables from the query while the proof proceeds. This is automatically done if we compute the instances of the special predicate answer which are derivable from the program plus this rule.

The method generates rules which must be considered (starting with this query rule). The rule body is a goal, as would appear in an SLD-tree. The rule head is the literal which has to be proven. First this is $\mathsf{answer}(\mathsf{X}_1, \ldots, \mathsf{X}_m)$, but as the proof progresses, the answer variables X_i are instantiated. Furthermore, negative body literals and call-literals cause subqueries to appear.

The occurring rules can be seen as being generated from the given program (extended by the query rule) using elementary transformations. However, in contrast to our earlier work, there are two important differences:

- The occurring rules are often not ground. In the theoretical part of our work on negation semantics, we started with the full ground instantiation of the given program. Of course, if we now want to use program transformations as a practical means of computation, we must work with non-ground rules. They can be understood as a compact representation of the set of their ground instances. Most modern semantics including WFS do not distinguish between a rule with variables and its set of ground instances.
- For goal-directed query evaluation, we do not want to consider the entire program. The "relevance" property of a nonmonotonic semantics [9] ensures that it is sufficient to look only at ground literals which are reachable from the query via the call-graph. WFS has the relevance property, the stable model semantics does not, but see [11]. Note that relevance is applied repeatedly during the evaluation. When we found that a rule instance is not applicable, it vanishes from the call graph and might remove large parts of the program.

Of course, we do not first take the entire program, and then delete non-relevant parts. Instead, we have a "working set" of rules we consider. We apply transformations on this set until a normal form is reached, i.e. until no further transformation is applicable. We only have to ensure that as long as there still is a relevant rule instance in the given program, it is considered, i.e. a transformation remains applicable.

In order to improve termination, we need to avoid the generation of rules which differ from an already generated rule only by a renaming of variables.

Definition 1 (Variable Normalization). *Let an infinite sequence $V_1, V_2, \ldots$ of variables be given. A rule $A \leftarrow B_1 \wedge \cdots \wedge B_n$ is variable-normalized iff it only contains variables from the set $\{V_1, V_2, \ldots\}$ and for each occurrence of V_i, all variables $V_1, \ldots, V_{i-1}$ occur to the left. The function* std *renames the variables of a rule to $V_1, V_2, \ldots$ in the order of occurrence, i.e. produces a variant of the given rule which is variable-normalized.*

Our method works with a set of rules which is initialized with the query. We write R for the current rule set to distinguish it from the given logic program P. We use a second set D of rules for the "deleted" rules. In this way, both sets R and D can only grow monotonically during the computation, which improves termination: It is not possible that a rule is added, deleted, and then added again. However, only the non-deleted rules in R really participate in the computation.

Definition 2 (Computation State). *A computation state is a pair (R, D) of sets of variable-normalized rules such that $D \subseteq R$. A rule in $R - D$ is called active, a rule in D is called deleted.*

Let the query Q be answer$(X_1, \ldots, X_m) \leftarrow B_1 \wedge \cdots \wedge B_n$. *The initial computation state is (R_0, D_0) with $R_0 := \{std(Q)\}$ and $D_0 := \emptyset$.*

2.1 Positive Body Literals

In the following, we use the term "positive body literal" for a body literal without negation and without "call". Literals with "call" are "call-literals" (they are never negated because negation already implies a subproof like "call" does).

Positive body literals are solved with unfolding (an SLD resolution step). If a rule contains several positive body literals, a selection function restricts unfolding to one of these. We require that a recursive positive body literal can only be selected last (when there are no other positive body literals). This implies that there can be only one recursive positive body literal, but this is no restriction: Other such literals can be made call-literals. The purpose of this condition is to make the length of the occurring rules bounded (to ensure termination). Negative body literals which are added during the tail-recursion are no problem if we eliminate duplicates: They are ground, so the total number is still bounded (although the bound depends on the data, whereas the bound for the positive body literals depends only on the program rules). Non-ground call-literals which are added during the repeated unfolding of a tail-recursive rule cannot be permitted. But again, this is no restriction because we can make the tail-recursive literal a call-literal, too (at the cost of losing the tail-recursion optimization).

The selection function only restricts the unfolding of positive body literals. Work on negative body literals and call-literals is not restricted, although an implementation is free to decide in each step which of several applicable transformations it uses. The reason why we cannot prescribe a single "active" literal in each rule is that because of rules like p←¬p, the evaluation of a negative body literal (in this case, ¬p) can "block". But there might be another body literal which would fail (this is similar to the "fairness" requirement of SLD resolution). The same can happen with call-literals.

The computation steps are relations between computation states. An implementation can follow any path until no further transformation is possible.

Definition 3 (Unfolding). *Let* $A \leftarrow B_1 \land \cdots \land B_n$ *be a rule in* $R - D$, *where* B_i *is a positive literal selected by the selection function. Let further* $A' \leftarrow B'_1 \land \cdots \land B'_m$ *be a variant of a rule in* P *with fresh variables (i.e. the variables renamed such that they are disjoint from those occurring in* $A \leftarrow B_1 \land \cdots \land B_n$), *such that* B_i *and* A' *are unifiable with most general unifier* θ. *Let*

$$R' := R \cup \{\text{std}(\theta(A \leftarrow B_1 \land \cdots \land B_{i-1} \land B'_1 \land \cdots \land B'_m \land B_{i+1} \land \cdots \land B_n))\}$$

If $R' \neq R$, *we write* $(R, D) \mapsto_U (R', D)$.

Note that because R is a set, unfolding with a rule like $p(X) \leftarrow p(X)$ does not result in a new rule and therefore cannot lead to non-termination. We need to assume that the negation semantics permits the deletion of tautologies.

Definition 4 (Deletion After Complete Unfolding). *Let* $A \leftarrow B_1 \land \cdots \land B_n$ *be a rule in* $R - D$, *where* B_i *is a positive literal selected by the selection function. Let all rules which can be generated from the given rule by unfolding be already contained in* R, *and let* $D' := D \cup \{A \leftarrow B_1 \land \cdots \land B_n\}$. *If* $D' \neq D$, *the transformation step* $(R, D) \mapsto_D (R, D')$ *is permitted.*

If there is only a single matching rule, or one immediately unfolds with all matching rules (e.g. in case of set-oriented evaluation with a database predicate), one can "delete" the rule with the unfolded call immediately. But by separating the two steps, other evaluation orders are possible, e.g. doing a depth-first search.

2.2 Negative Body Literals

Definition 5 (Complement Call). *Let* $A \leftarrow B_1 \land \cdots \land B_n$ *be a rule in* $R - D$, *and* B_i *be a negative ground literal. Let* A' *be the corresponding positive literal, i.e.* $B_i = \neg A'$. *Let* $R' := R \cup \{A' \leftarrow A'\}$. *If* $R' \neq R$, *we write* $(R, D) \mapsto_C (R', D)$.

Of course, $A' \leftarrow A'$ is a tautology. But it is important because it sets up a new query. So when we want to work on a negative literal, we try to prove the corresponding positive literal. This is the same as SLDNF-resolution would do.

The next transformation handles the case where a negative literal is proven by failure to prove the corresponding positive literal.

Definition 6 (Positive Reduction). *Let* $A \leftarrow B_1 \land \cdots \land B_n$ *be a rule in* $R - D$, *where* B_i *is a negative ground literal. Let* A' *be the corresponding positive literal, i.e.* $B_i = \neg A'$. *If* R *contains* $A' \leftarrow A'$, *but* $R - D$ *does not contain any rule with head* A', *then* $(R, D) \mapsto_P (R', D')$ *with*

$$R' := R \cup \{A \leftarrow B_1 \wedge \cdots \wedge B_{i-1} \wedge B_{i+1} \wedge \cdots \wedge B_n\}$$
$$D' := D \cup \{A \leftarrow B_1 \wedge \cdots \wedge B_n\}.$$

(The new rule is variable-normalized since B_i was ground.)

The next transformation handles the case that the selected negative body literal is obviously false, because the corresponding positive literal was proven:

Definition 7 (Negative Reduction). *Let $A \leftarrow B_1 \wedge \cdots \wedge B_n$ be a rule in $R - D$, where B_i is a negative ground literal. Let A' be the corresponding positive literal, i.e. $B_i = \neg A'$. If R contains A' (as a rule with empty body, i.e. a fact), then $(R, D) \mapsto_N (R, D')$ with $D' := D \cup \{A \leftarrow B_1 \wedge \cdots \wedge B_n\}$.*

2.3   Call Literals

The specially marked call-literals are semantically positive literals, but they are not solved by unfolding. Instead, a subproof is set up (as for negative literals):

Definition 8 (Start of Subproof). *Let $A \leftarrow B_1 \wedge \cdots \wedge B_n$ be a rule in $R - D$, and the literal B_i be of the form $\mathrm{call}(A')$. Let $R' := R \cup \{A' \leftarrow A'\}$. If $R' \neq R$, we write $(R, D) \mapsto_S (R', D)$.*

The following transformation is similar to positive reduction combined with a very special case of unfolding:

Definition 9 (Return). *Let $A \leftarrow B_1 \wedge \cdots \wedge B_n$ be a rule in $R - D$, and the literal B_i be of the form $\mathrm{call}(A')$. Suppose further that there is a rule $A'' \leftarrow B'_1 \wedge \cdots \wedge B'_m$ where all body literals are negative (i.e. a conditional fact), such that A' and A'' are unifiable with most general unifier θ. Let*

$$R' := R \cup \{\mathrm{std}(\theta(A \leftarrow B_1 \wedge \cdots \wedge B_{i-1} \wedge B'_1 \wedge \cdots \wedge B'_m \wedge B_{i+1} \wedge \cdots \wedge B_n))\}$$

If $R' \neq R$, we write $(R, D) \mapsto_R (R', D)$.

A call is complete when a kind of small fixpoint is reached in the larger set of rules constructed. Negative literals can be evaluated later, but for positive body literals and call-literals, all possible derivations must be done:

Definition 10 (End of Subproof). *Let $R_0 \subseteq R_1 \subseteq R$ be rule sets, such that*

- *Each rule in R_0 contains a call-literal,*
- *the transformations $\mapsto_U$ (Unfolding), $\mapsto_S$ (Start of Subproof), $\mapsto_R$ (Return) are not applicable in R_1 (i.e. everything derivable is already contained in R_1).*

If $R_0 \not\subseteq D$, the "End of Subproof" transformation is applicable:

$$(R, D) \mapsto_E (R, D \cup R_0).$$

This transformation is relatively complicated because it includes a kind of loop detection. It must be able to handle cases like:

$$p(X) \leftarrow call(q(X)).$$
$$q(X) \leftarrow call(p(X)).$$

An alternative for the "return" operation is not to unfold, but just ground the call if it matches the head of an "extended conditional fact", which is a ground rule with only negative and call-literals in the body. The call-literal would be removed only if it matches a fact (without condition). This operation "success" of [6] (and the converse "failure") help to avoid a possible exponential blowup.

References

1. Beeri, C., Ramakrishnan, R.: On the power of magic. In: Proceedings of Sixth ACM Symposium on Principles of Database Systems (PODS 1987), pp. 269–284. ACM (1987)
2. Bonatti, P.A., Pontelli, E., Son, T.C.: Credulous resolution for answer set programming. In: Proceedings of the 23rd National Conference on Artificial Intelligence (AAAI 2008), vol. 1, pp. 418–423. AAAI Press (2008)
3. Brass, S.: SLDMagic — the real magic (with applications to web queries). In: Lloyd, W., et al. (eds.) CL 2000. LNCS, vol. 1861, pp. 1063–1077. Springer, Heidelberg (2000)
4. Brass, S., Dix, J.: Characterizations of the stable semantics by partial evaluation. In: Marek, V.W., Nerode, A., Truszczyński, M. (eds.) LPNMR 1995. LNCS (LNAI), vol. 928, pp. 85–98. Springer, Heidelberg (1995)
5. Brass, S., Dix, J.: A general approach to bottom-up computation of disjunctive semantics. In: Dix, J., Pereira, L.M., Przymusinski, T.C. (eds.) Nonmonotonic Extensions of Logic Programming. LNCS (LNAI), vol. 927, pp. 127–155. Springer, Heidelberg (1995)
6. Brass, S., Dix, J., Freitag, B., Zukowski, U.: Transformation-based bottom-up computation of the well-founded model. Theor. Pract. Logic Program. **1**(5), 497–538 (2001)
7. Bry, F.: Logic programming as constructivism: a formalization and its application to databases. In: Proceedings of the 8th ACM SIGACT-SIGMOD-SIGART Symposium on Principles of Database Systems (PODS 1989), pp. 34–50 (1989)
8. Chen, W., Warren, D.S.: A goal-oriented approach to computing the well-founded semantics. J. Logic Program. **17**, 279–300 (1993)
9. Dix, J., Müller, M.: Partial evaluation and relevance for approximations of the stable semantics. In: Raś, Z.W., Zemankova, M. (eds.) MIS 1994. LNCS (LNAI), vol. 869, pp. 511–520. Springer, Heidelberg (1994)
10. Dung, P.M., Kanchansut, K.: A fixpoint approach to declarative semantics of logic programs. In: Proceedings of North American Conference on Logic Programming (NACLP 1989), pp. 604–625 (1989)
11. Marple, K., Gupta, G.: Galliwasp: a goal-directed answer set solver. In: Albert, E. (ed.) LOPSTR 2012. LNCS, vol. 7844, pp. 122–136. Springer, Heidelberg (2013)
12. Sakama, C., Seki, H.: Partial deduction of disjunctive logic programs: a declarative approach. In: Fribourg, L., Seki, H. (eds.) LOPSTR 1994 and META 1994. LNCS, vol. 883, pp. 170–182. Springer, Heidelberg (1994)
13. Swift, T.: Tabling for non-monotonic programming. Ann. Math. Artif. Intell. **25**, 201–240 (1999)

Implementing Preferences with *asprin*

Gerhard Brewka[3], James Delgrande[2], Javier Romero[4],
and Torsten Schaub[1,2,4,5 (✉)]

[1] INRIA Rennes, Rennes, France
[2] Simon Fraser University, Burnaby, Canada
[3] Universität Leipzig, Leipzig, Germany
[4] Universität Potsdam, Potsdam, Germany
`torsten@cs.uni-postsdam.de`
[5] IIIS Griffith University, Brisbane, Australia

Abstract. *asprin* offers a framework for expressing and evaluating combinations of quantitative and qualitative preferences among the stable models of a logic program. In this paper, we demonstrate the generality and flexibility of the methodology by showing how easily existing preference relations can be implemented in *asprin*. Moreover, we show how the computation of optimal stable models can be improved by using declarative heuristics. We empirically evaluate our contributions and contrast them with dedicated implementations. Finally, we detail key aspects of *asprin*'s implementation.

1 Introduction

Preferences are pervasive and often are a key factor in solving real-world applications. This was realized quite early in Answer Set Programming (ASP; [1]), where solvers offer optimization statements representing ranked, sum-based objective functions (viz. *#minimize* statements or weak constraints [2,3]). On the other hand, such quantitative ways of optimization are often insufficient for applications and in stark contrast to the vast literature on qualitative and hybrid means of optimization [4–8].

This gulf is bridged by the *asprin* system, which offers a flexible and general framework for implementing complex combinations of quantitative and qualitative preferences. The primary contribution of this paper is to substantiate this claim by showing how easily selected approaches from the literature can be realized with *asprin*. In particular, we detail how answer set optimization [5], minimization directives [2], strict partially ordered sets [8], and the non-temporal part of the preference language in [7] can be implemented with *asprin*. Moreover, we sketch how the implementations of ordered disjunctions [4] and penalty-based answer set optimization [6] are obtained. In fact, *asprin*'s simple interface and straightforward methodology reduces the implementation of customized preferences to defining whether one model is preferred to another. This also lays

---

This work was funded by DFG (BR 1817/5; SCHA 550/9) and NSERC.

F. Calimeri et al. (Eds.): LPNMR 2015, LNAI 9345, pp. 158–172, 2015.
DOI: 10.1007/978-3-319-23264-5_15

bare *asprin*'s expressive power, which is delineated by the ability to express a preference's decision problem within ASP. In view of the practical relevance of preferences, we also investigate the use of ASP's declarative heuristics for boosting the search for optimal stable models. In particular, we are interested how the combination of ASP's two general-purpose frameworks for preferences and heuristics compares empirically with dedicated implementations.

The paper [9] introduced *asprin*'s approach and focused on fundamental aspects. Apart from a formal elaboration of *asprin*'s propositional language, it provided semantics and encodings for basic preferences from *asprin*'s library (like `subset` or `less(weight)`). As well, it empirically contrasted the implementation of such basic preferences with the dedicated one in *clasp* and analyzed *asprin*'s scalability in view of increasingly nested preference structures. Here, we build upon this work and focus on engineering aspects. First, we introduce the actual first-order preference modeling language of *asprin*, including its safety properties, and carve out its simple interfaces and easy methodology. Second, we demonstrate how existing preferences from the literature can be implemented via *asprin*. In doing so, we provide best practice examples for preference engineers. Third, we show how declarative heuristics can be used for boosting the computation of optimal models. And last but not least, we detail aspects of *asprin*'s implementation and contrast it with dedicated ones.

In what follows, we rely upon a basic acquaintance with ASP [1,10].

2 *asprin*'s Approach at a Glance

asprin allows for declaring and evaluating preference relations among the stable models of a logic program. Preferences are declared by *preference statements*, composed of an identifier s, a type t, and an argument set: '`#preference(s,t){`e_1`;`...`;`e_n`}` : B.' The identifier names the preference relation, whereas its type and arguments define the relation; the set B of built-in or domain predicates is used for instantiation.[1] Identifiers and types are represented by terms, while each argument e_j is a *preference element*. For safety, variables appearing in s or t must occur in a positive atom of B. Let us consider an example before delving into further details:

```
#preference(costs,less(weight)){C :: activity(A) : cost(A,C)}.
```

This statement declares a preference relation `costs` with type `less(weight)`. Given atoms `cost(sauna,40)` and `cost(dive,70)`, grounding results in one of the simplest form of preference elements, namely the weighted literals `40::activity(sauna)` and `70::activity(dive)`. Informally, the resulting preference relation prefers stable models whose atoms induce the minimum sum of weights. Hence, models with neither `sauna` nor `dive` are preferred over those with only `sauna`. Stable models with only `dive` are still less preferred, while those with both `sauna` and `dive` are least preferred. We refer the reader to [9] on how preference statements induce preference relations by applying preference types to preference elements. And we focus in what follows on *asprin*'s syntactic features.

---

[1] Just as with bodies, we drop curly braces from such sets.

Preference elements can be more complex than in the example. In the most general case, we even admit conditional elements, which are used to capture conditional preferences. Moreover, preference types may refer to other preferences in their arguments, which is used for preference composition. For instance, the statement

`#preference(all,pareto){name(costs),name(temps)}.`

defines a preference relation `all`, which is the Pareto ordering of preference relations `costs` and `temps`.

More formally, a *preference element* is of the form '$F_1 > \ldots > F_m \mid\mid F : B$' where each F_r is a set of weighted formulas, F is a non-weighted Boolean formula, and B is as above. We drop '>' if $m = 1$, and '$\mid\mid F$' and '$: B$' whenever F and/or B are empty, respectively. Intuitively, r gives the rank of the respective set of weighted formulas. This can be made subject to condition F by using the conditional '$\mid\mid$'. Preference elements provide a (possible) structure to a set of weighted formulas by giving a means of conditionalization and a symbolic way of defining pre-orders.

A set of weighted formulas F_r is represented as '$F_1 ; \ldots ; F_m$'. We drop the curly braces if $m = 1$. And finally, a *weighted formula* is of the form '$t :: F$' where t is a term tuple and F is a either a Boolean formula or a naming atom. We may drop $::$ and simply write F whenever t is empty. Boolean formulas are formed from atoms, possibly preceded by strong negation ('-'), using the connectives **not** (negation), **&** (conjunction) and **|** (disjunction). Parentheses can be written as usual, and when omitted, negation has precedence over conjunction, and conjunction over disjunction. Naming atoms of form `name(s)` refer to the preference associated with preference statement `s` (cf [9]). For safety, variables appearing in a weighted formula must occur in a positive atom of the set B from either the corresponding preference element or preference statement. Examples of preference elements include '`a(X)`', '`42::b(X)`', '`{1::name(p);2::name(q)}`', '`{a(X);b(X)} > {c(X);d(X)}`', and '`a(X) > b(X) || c(X): dom(X)`'.

Since preference statements may only be auxiliary, a preference relation must be distinguished for optimization. This is done via an optimization statement of form '`#optimize(s).`' with the name of the respective preference statement as argument.

Finally, a *preference specification* is a set of preference statements along with an optimization directive. It is valid if grounding results in acyclic and closed naming dependencies along with a single optimization directive (see [9] for details).

Once a preference specification is given, the computation of preferred stable models is done via a branch-and-bound process relying on *preference programs*. Such programs, which need to be defined for each preference type, take two reified stable models and decide whether one is preferred to the other. An optimal one is computed iteratively by repeated calls to an ASP solver. First, an arbitrary stable model of the underlying program is generated; then, this stable model is "fed" to the preference program to produce a better one, etc. Once the preference program becomes unsatisfiable, the last stable model obtained is an optimal one.

The basic algorithm is described in [9]; it is implemented via *clingo 4*'s Python library, providing operative grounder and solver objects.

asprin also provides a *library* containing a number of predefined, common, preference types along with the necessary preference programs. Users happy with what is available in the library can thus use the available types without having to bother with preference programs at all. However, if the predefined preference types are insufficient, users may define their own relations. In this case, they also have to provide the preference programs *asprin* needs to cope with the new preference relations.

3 Embedding Existing Approaches

The implementation of customized preference types in *asprin* boils down to furnishing a preference program for the preference that is subject to optimization. For the sake of generality, this is usually done for the preference type, which then gets instantiated to the specific preference relation of interest.

The purpose of a preference program is to decide whether one stable model is strictly preferred to another wrt the corresponding preference relation. To this end, we reify stable models and represent them via the unary predicates `holds/1` and `holds'/1`. More formally, we define for a set X of atoms, the following sets of facts:

$$H(X) = \{\texttt{holds}(a)\,.\mid a \in X\} \text{ and } H'(X) = \{\texttt{holds'}(a)\,.\mid a \in X\}$$

Then, given a preference statement identified by **s**, the program P_s is a preference program implementing preference relation $\succ_\mathbf{s}$, if for sets X, Y of atoms, we have

$$X \succ_\mathbf{s} Y \text{ iff } P_\mathbf{s} \cup H(X) \cup H'(Y) \text{ is satisfiable.} \qquad (1)$$

See [9] for a formal elaboration of preference programs.

In what follows, we explain *asprin*'s interfaces and methodology for implementing preference programs.

To begin with, *asprin* represents preference specifications in a dedicated fact format. Each optimization directive '`#optimize(s).`' is represented as a fact

   `optimize(s).`

Next, each preference statement '`#preference(s,t)` $\{e_1;\dots;e_n\}$: B.' gives rise to n rules encoding preference elements along with one rule of form

   `preference(s,t):- ` B.

In turn, preference elements are represented by several facts, each representing a comprised weighted formula. Recall that a weighted formula F_k of form '$t::F$' occurs in some set $\boldsymbol{F}_i$ of form '$F_1;\dots;F_m$' (or equals F_0) of a preference element e_j of form '$\boldsymbol{F}_1 > \dots > \boldsymbol{F}_n \mid\mid F_0 : B_j$' that belongs itself to a preference statement **s** as given above. Given this, the weighted formula F_k is translated into a rule of the form

   `preference(s,(`j`,`v`),`i`,for(`t_F`),t):- ` B_j`, ` B.

where j and i are the indices of e_j and $\mathbf{F}_i$, respectively, v is a term tuple containing all variables appearing in the rule, and t_F is a term representing the Boolean formula F by using function symbols _not/1, _and/2, and _or/2 in prefix notation. For example, the formula (not a(X) | b(X)) & c(X) is translated into _and(c(X),_or(_not(a(X)),b(X))). For representing condition F_0, we set i to 0. A naming atom name(s) is represented analogously, except that for(t_F) is replaced by name(s).

For instance, the earlier preference statement costs is translated as follows.

```
preference(costs,(1,(A,C)),1,for(activity(A)),(C)) :- cost(A,C).
preference(costs,less(weight)).
```

Grounding the first rule in the presence of cost(sauna,40) and cost(dive,70) yields two facts, representing the weighted literals 40::activity(sauna) and 70::activity(dive).

Second, *asprin* extends the basic truth assignment to atoms captured by holds/1 and holds'/1 to all Boolean formulas occurring in the preference specification at hand. To this end, formulas are represented as terms as described above. Hence, for any formula F occurring in the preference specification, *asprin* warrants that holds(t_F) is true whenever F is entailed by the stable model X captured in $H(X)$, where t_F is the term representation of F. This is analogous for holds'/1.

Third, in *asprin*'s methodology a preference program is defined generically for the preference type, and consecutively instantiated to the specific preference in view of its preference elements. Concretely, *asprin* stipulates that preference programs define the unary predicate better/1, taking preference identifiers as arguments. The user's implementation is required to yield better(s) whenever the stable model captured by $H(X)$ is strictly better than that comprised in $H'(X)$. For illustration, consider the preference program for *asprin*'s prefabricated preference type less(weight).

```
1   better(P) :- preference(P,less(weight)),
2        1 #sum {-W,X: holds(X),  preference(P,_,_,for(X),(W));
3            W,X: holds'(X), preference(P,_,_,for(X),(W))}.
```

asprin complements this by the generic integrity constraint

    :- not better(P), optimize(P).

ensuring that better(P) holds whenever P is subject to optimization and enforces the fundamental property of preference programs in (1).

All in all, a preference program thus consists of (i) facts representing preference and optimization statements, (ii) auxiliary rules, extending predicates holds/1 and holds'/1 to Boolean formulas as well as the above integrity constraint, and finally (iii) the definition of the preference type(s). While parts (i) and (ii) are provided by *asprin*, only part (iii) must be provided by the "preference engineer". Our methodology accounts for this by defining predicate better/1. However, this is not strictly necessary as long as all three parts constitute a preference program by fulfilling (1).

Additionally, the customization of preferences can draw upon *asprin*'s library containing various pre-defined preference types. This includes the primitive types subset and superset, less(cardinality) and more(cardinality), less(weight) and more(weight), along with the composite types neg, and pareto, and lexico. In fact, for these types, *asprin* not only provides definitions of better(s) but also its non-strict, and equal counterparts, namely, bettereq/1, equal/1, worse/1, and worseeq/1. Such definitions are very useful in defining aggregating preference types such as pareto (see below).

Answer set optimization. For capturing answer set optimization (ASO; [5]), we consider ASO rules of form

$$\phi_1 > \cdots > \phi_m \leftarrow B \qquad (2)$$

where each ϕ_i is a propositional formula for $1 \le i \le m$ and B is a rule body.

The semantics of ASO is based on satisfaction degrees for rules as in (2). The satisfaction degree of such a rule r in a set of atoms X, written $v_X(r)$, is 1 if $X \not\models b$ for some $b \in B$, or if $X \models b$ for some $\sim b \in B$, or if $X \not\models \phi_i$ for every $1 \le i \le m$, and it is $\min\{k \mid X \models \phi_k, 1 \le k \le m\}$ otherwise. Then, for sets X, Y of atoms and a set O of rules of form (2), $X \succeq_O Y$ if for all rules $r \in O$, $v_X(r) \le v_Y(r)$, and $X \succ_O Y$ is defined as $X \succeq_O Y$ but $Y \not\succeq_O X$.

In *asprin*, we can represent an ASO rule r as in (2) as preference statement of form

#preference(s_r,aso){$\phi_1 > \ldots > \phi_m$ || B}.

A set $\{r_1, \ldots, r_n\}$ of ASO rules is represented by corresponding preference statements s_{r_1} to s_{r_n} along with an aggregating pareto preference subject to optimization.

#preference(paraso,pareto){name(s_{r_1}), ... name(s_{r_n})}.
#optimize(paraso).

Note that aggregating preferences other than pareto could be used just as well.

The core implementation of preference type aso is given in Lines 1–23 below. Predicate one/1 is true whenever an ASO rule has satisfaction degree 1 wrt the stable model captured by $H(X)$. The same applies to one'/1 but wrt $H'(Y)$.

```
1   one(P) :- preference(P,aso),
2              not holds(F) : preference(P,_,R,for(F),_), R>1.
3   one(P) :- preference(P,aso),
4              holds(F), preference(P,_,1,for(F),_).
5   one(P) :- preference(P,aso),
6              not holds(F), preference(P,_,0,for(F),_).

8   one'(P) :- preference(P,aso),
9               not holds'(F) : preference(P,_,R,for(F),_), R>1.
10  one'(P) :- preference(P,aso),
11              holds'(F), preference(P,_,1,for(F),_).
12  one'(P) :- preference(P,aso),
13              not holds'(F), preference(P,_,0,for(F),_).
```

With these rules, we derive better(s_r) in Line 15 whenever some ASO rule r has satisfaction degree 1 in X and one greater than 1 in Y. Otherwise, better(s_r) is derivable in Line 16 whenever r has satisfaction degree R in X but none of the formulas ϕ_1 to ϕ_R are true in Y. This is analogous for bettereq/1 in lines 20–23.

```
15  better(P) :- preference(P,aso), one(P), not one'(P).
16  better(P) :- preference(P,aso),
17      preference(P,_,R,for(F),_), holds(F), R > 1, not one'(P),
18      not holds'(G) : preference(P,_,R',for(G),_), 1 < R',R' <= R.

20  bettereq(P) :- preference(P,aso), one(P).
21  bettereq(P) :- preference(P,aso),
22      preference(P,_,R,for(F),_), holds(F), R > 1, not one'(P),
23      not holds'(G) : preference(P,_,R',for(G),_), 1 < R',R' < R.
```

The remaining rules implement the composite preference type pareto.

```
25  better(P) :- preference(P,pareto),
26              better(R),       preference(P,_,_,name(R),_),
27              bettereq(Q) : preference(P,_,_,name(Q),_).
```

Note how pareto makes use of both the strict and non-strict preference types, viz. better/1 and bettereq/1. In fact, we only list this here for completeness since the definition could be imported from *asprin*'s library.

Altogether, the rules in Line 1–27 capture the semantics of ASO. To see this, consider a set O of ASO rules and the program P_O consisting of Line 1–27 along with the facts for the preference and optimization statements corresponding to O and the auxiliary rules in (ii) mentioned above. Then, we can show that $X \succ_O Y$ holds iff $P_O \cup H(X) \cup H'(Y)$ is satisfiable.

asprin also includes an implementation of the ASO extension with penalties introduced in [6]. Here, each formula ϕ_i in (2) is extended with a weight and further weight-oriented cardinality- and inclusion-based composite preference types are defined. The implementation in *asprin* extends the one presented above by complex weight handling and is thus omitted for brevity. Similarly, logic programs with ordered disjunction [4] are expressible in *asprin* via the translation to ASO described in [5].

Partially Ordered Sets. In [8], qualitative preferences are modeled as a strict partially ordered set $(\Phi, <)$ of literals. The literals in Φ represent propositions that are preferably satisfied and the strict partial order $<$ on Φ gives their relative importance. We (slightly) generalize this to sets of Boolean formulas. Then, for sets X, Y of atoms and a strict partially ordered set $(\Phi, <)$, $X \succ_{(\Phi,<)} Y$ if there exists a formula $\phi \in \Phi$ such that $X \models \phi$ and $Y \not\models \phi$, and for every formula $\phi \in \Phi$ such that $Y \models \phi$ and $X \not\models \phi$, there is a formula $\phi' \in \Phi$ such that $\phi' < \phi$ and $X \models \phi'$ but $Y \not\models \phi'$.

We represent a partially ordered set $(\Phi, <)$ by a preference statement $s_{(\Phi,<)}$ of form:

$$\texttt{\#preference(}s_{(\Phi,<)}\texttt{,poset)} \; \Phi \cup \{\phi' > \phi \mid \phi' < \phi \}.$$

The preference type poset captures all preference relations $\succ_{(\Phi,<)}$ for all strict partially ordered sets $(\Phi, <)$.

The core implementation of preference type poset is given in Lines 1–13 below. In fact, Line 1 to 4 are only given for convenience to project the components of $(\Phi, <)$.

```
1  poset(P,F)    :- preference(P,poset),
2             preference(P,_,_,for(F),_).
```

```
3   poset(P,F,G)  :- preference(P,poset),
4                   preference(P,I,1,for(F),_), preference(P,I,2,for(G),_).

6   better(P,F)   :- preference(P,poset),
7                    poset(P,F), holds(F), not holds'(F).
8   notbetter(P)  :- preference(P,poset),
9                    poset(P,F), not holds(F), holds'(F),
10                   not better(P,G) : poset(P,G,F).

12  better(P) :- preference(P,poset),
13               better(P,_), not notbetter(P).
```

Given the reification of two sets X, Y in terms of holds/1 and holds'/1, we derive an instance of better(P,F) whenever $X \models \phi_F$ but $Y \not\models \phi_F$ (and F is the representation of ϕ_F). Similarly, we derive notbetter(P) whenever there is a formula ϕ_F such that $Y \models \phi_F$ and $X \not\models \phi_F$ but better(P,G) fails to hold for all ϕ_G preferred to ϕ_F by the strict partial order $<$. Finally, these two auxiliary predicates are combined in Line 12 and 13 to define the preference type poset.

Finally, we sketch how these rules capture the intended semantics. For this, given a strict partially ordered set $(\Phi, <)$, we consider the program $P_{(\Phi,<)}$ consisting of the rules in Line 1–13, the facts for the preference and optimize statements corresponding to $(\Phi, <)$, and the auxiliary rules (ii) described above. Then, we can show that for two sets of atoms X and Y, $X \succ_{(\Phi,<)} Y$ holds iff $P_{(\Phi,<)} \cup H(X) \cup H'(Y)$ is satisfiable.

Son and Pontelli [7] propose a language for specifying preferences in planning that distinguishes three types of preferences: basic, atomic, and general preferences. A basic preference is originally expressed by a propositional formula using Boolean as well as temporal connectives. Given that our focus does not lie on planning, we restrict basic preferences to Boolean formulas. Then, for sets X, Y of atoms and a formula ϕ, [7] defines $X \succ_\phi Y$ by if $X \models \phi$ and $Y \not\models \phi$.

In *asprin*, such a basic preference is declared by a preference statement s_ϕ of form

```
#preference(s_φ,basic){ φ }.
```

And the preference type basic is implemented by the following rule.

```
better(P) :- preference(P,basic), preference(P,_,_,for(F),_),
             holds(F), not holds'(F).
```

Interestingly, atomic and general preferences can be captured by composite preferences pre-defined in *asprin*'s library. That is, the language constructs !, &, |, and ◁ directly correspond to neg, and, pareto, and lexico. For brevity, we refrain from further details and refer the reader to [7,9] for formal definitions.

Optimization Statements. Finally, it is instructive to see how common optimization statements are expressed in *asprin*.[2] A #*minimize* directive is of the form

$$\#minimize\{w_1@k_1, t_1 : \ell_1, \ldots, w_n@k_n, t_n : \ell_n\}$$

where each w_i and k_i is an integer, and $\ell_i = \ell_{i_1}, \ldots, \ell_{i_k}$ and $t_i = t_{i_1}, \ldots, t_{i_m}$ are tuples of literals and terms, respectively. For a set X of atoms and an integer k, let

---

[2] The decomposition of weak constraints is analogous, and is omitted for brevity.

Σ_k^X denote the sum of weights w_i over all occurrences of elements $(w_i@k_i, t_i : \ell_i)$ in M such that $X \models \ell_i$. Then, for sets X, Y of atoms and minimize statement M as above, $X \succ_M Y$ if $\Sigma_k^X < \Sigma_k^Y$ and $\Sigma_{k'}^X = \Sigma_{k'}^Y$ for all $k' > k$.

In *asprin*, a minimize statement M as above can be represented by the following preference specification.

```
#preference(s_M,lexico){-k :: name(s_k) | (w@k,t:ℓ) ∈ M}.
#preference(s_k,less(weight)){w,(t) :: ℓ | (w@k,t:ℓ) ∈ M}.
#optimize(s_M).
```

The preference type less(weight) is defined as follows.

```
better(P) :- preference(P,less(weight)),
  1 #sum {-W,T,F : holds(F),   preference(P,_,_,for(F),(W,T));
          W,T,F : holds'(F),  preference(P,_,_,for(F),(W,T))}.
```

Note that by wrapping tuples t into (t), we only deal with pairs $w, (t)$ rather than tuples of varying length.

asprin's separation of preference declarations from optimization directives not only illustrate how standard optimization statements conflate both concepts but it also explicates the interaction of preference types lexico and less(weight).

4   Heuristic Support in *asprin*

Optimization problems are clearly more difficult than decision problems, since they involve the identification of optimal solutions among all feasible ones. To this end, it seems advantageous to direct the solving process towards putative optimal solutions by supplying heuristic information. Although this runs the risk of search degradation [11], it has already indicated great prospects by improving regular optimization in ASP [12] as well as qualitative preferences [8]. While the latter had to be realized by modifications to a SAT solver, in *asprin* we draw upon the integration with *clingo 4*'s declarative heuristic framework [12]. Heuristic information is represented in a logic program by means of the dedicated predicate _heuristic. Different types of heuristic information can be controlled with *clingo 4*'s domain heuristic along with the basic modifiers sign, level, init, and factor. In brief, sign allows for controlling the truth value assigned to variables subject to a choice within the solver, while level establishes a ranking among atoms such that unassigned atoms of highest rank are chosen first. With init, a value is added to the initial heuristic score of an atom. The whole search is biased with factor by multiplying heuristic scores by a given value. Furthermore, modifiers true and false are defined as the combination of a positive sign and a level, and a negative sign and a level, respectively. See [12] for a details.

This framework seamlessly integrates into *asprin* by means of so-called *heuristic programs*, where heuristics for concrete preference types may be specified. For example, consider the following heuristic program for less(weight):

```
_heuristic(holds(X),false,1) :- preference(P,less(weight)),
                                preference(P,_,_,for(X),_).
```

This tells the solver to decide first on formulas appearing in preference statements of type `less(weight)` and to assign *false* to them. As another example, we can replicate the modification of the sign heuristic proposed in [8] for `poset` as follows:

```
_heuristic(holds(X),sign,1)  :-  preference(P,poset),
                                  preference(P,_,_,for(X),_).
```

The idea is to assign *true* when deciding on formulas of a `poset` preference statement. In general, the goal of these heuristic programs is to direct the search towards optimal solutions in such a way that fewer intermediate solutions have to be computed.

For activating the domain heuristics, the option `--heuristic=Domain` must be supplied. In addition, *asprin* provides an easy way to modify it from the command line via option `--domain-heuristic=<m>[,<v>]`; it turns on the domain heuristic and applies heuristic modifier `m` with value `v` (1 by default) to the formulas occurring in preference statements. For example, instead of adding the previous heuristic program for `poset`, we could have issued the option `--domain-heuristic=sign`.

As put forward in [8,13], even domain heuristics alone may be used to compute optimal models by a single call to a solver. In other words, preference types may actually be implemented by domain heuristics. For example, the preference type `subset` can alternatively be implemented by the following heuristic program

```
_heuristic(holds(X),false,1)  :-  preference(P,subset),
                                  preference(P,_,_,for(X),_).
```

which guarantees that the first answer set computed is (already) optimal. Similarly, the following heuristic program implements a more sophisticated version of `poset`:

```
_heuristic(holds(F),true,1)  :-  preference(P,poset),
                                 preference(P,_,_,for(F),_),
                                 assigned(P,G) : poset(P,G,F).
```

```
assigned(P,G)  :-  poset(P,G,_),       holds(G).
assigned(P,G)  :-  poset(P,G,_),  not holds(G).
```

With `poset/3` defined as in Sect. 3, predicate `assign/2` represents that a formula is assigned by the solver. Given this heuristic program, the solver prefers to satisfy formulas whose dominating formulas are already assigned. As shown in [8], such a heuristic guarantees that the first optimal model computed is optimal. The *asprin* library includes such heuristic program also for `aso`. Using them, there is no need for checking the optimality of a solution. For this case, option `--mode=heuristic` tells *asprin* to avoid the check and activate the domain heuristic.

5   Using the *asprin* System

asprin is implemented in Python (2.7) and consists of a parser along with a solver that uses *clingo 4*'s Python library. This library provides *clingo* objects maintaining a logic program and supporting methods for adding, deleting and grounding rules, as well as for solving the current logic program. This approach allows for continuously changing the logic program at hand without any need for re-grounding rules. Also, it benefits from information learned in earlier solving steps.

The input of *asprin* consists of a set of ASP files structured by means of *clingo 4*'s #program directives into base, preference, and heuristic programs. Base programs consist typically of a problem instance and encoding, and may contain a preference specification (just as with #*minimize* statements).[3] Rules common to all types of preference programs are grouped under program blocks headed by '#program preference.', while type-specific ones use '#program preference(t).' where t is the preference type. Similarly for '#program heuristic.' and '#program heuristic(t).'. Among all the type-specific preference and heuristic programs in the input files, *asprin* only loads those for the preference types appearing in the preference specification of the base program. On the other hand, for every preference type t of the preference specification, *asprin* requires a corresponding preference program 'preference(t)', and when using option --mode=heuristic there must also be a corresponding heuristic program 'heuristic(t)'.[4] *asprin*'s implementation relies on the correctness of preference and heuristic programs. In other words, if the preference (or heuristic) programs implement correctly the corresponding preference types, then *asprin* also functions correctly, as shown in [9].

asprin's parser starts translating preference and optimization statements as explained in Sect. 3. Then every atom a appearing in a weighted formula is reified into _holds$(a,0)$ adding a rule of form '_holds$(a,0)$:- a.' to the base program. Similar auxiliary rules are added for handling Boolean formulas. The successive answer sets computed by *asprin* are reified into atoms of the form _holds(t_F,n) where n takes successively increasing integer values starting with 1. Next, preference programs are slightly modified for comparing answer sets numbered m1 and m2. Atoms of the form holds(t) and holds'(t) are translated into _holds(t,m1) and _holds(t,m2), respectively. After parsing, the base program generated by the parser is solved by *clingo*. If the program is unsatisfiable, then *asprin* terminates and returns UNSAT. Otherwise, *asprin* enters a loop, where the last generated answer set is reified into facts of form _holds(t_F,n); the preference program is grounded setting m1 to 0 and m2 to n; and *clingo* solves the resulting program. If a new answer set is found, *asprin* returns to the beginning of the loop, and otherwise it returns the last answer set found. By construction, this last answer set is optimal wrt the preference in focus.

asprin can be configured by several command line options. As with standard ASP solvers, a natural number n tells *asprin* how many optimal models should be computed (where 0 initiates the computation of all optimal models). Option --project allows for projecting the optimal models on the atoms occurring in

---

[3] If no preference specification is given, *asprin* computes answer sets of the base program.

[4] In this case, for computing a single optimal model no preference program is needed.

the preference specification. Options for modifying the underlying *clingo* solver can be directly issued from the command line. More options and details are obtained with *asprin*'s `--help` option.

6 Empirical Evaluation of *asprin*

In [9], we contrasted *asprin* (1.0) 's performance with that of *clingo 4.4* on basic weight- and subset-based preferences. We begin with extending this study to investigate the impact of heuristic information on *asprin*'s performance. To this end, we consider in Table 1 the benchmarks from [9] and solve them by increasing the heuristic influence on *asprin*'s search. We ran all benchmarks with *asprin 1.1* using *clingo 4.5* on a Linux machine with an Intel Dual-Core Xeon 3.4 GHz processor, imposing a limit of 900 s and 4 GB of memory per run. A timeout is counted as 900 s. Each entry in Table 1 gives average time and in parentheses the number of enumerated models and timeouts. The number of enumerated models reflects how well *asprin* converges to the optimum. Each group of four data columns contains results from running *asprin* in its default setting and heuristics modifying `sign`, `level` and both, viz. `false`. The first group deals with *w*eight-based optimization and the second with *s*ubset-based optimization. Overall each heuristic modification improves over the standard in terms of runtime and convergence. This somewhat holds for timeouts as well, though certain heuristic settings degenerate on specific classes. Generally, we observe that the stronger the heuristic influence, the better *asprin* converges to the optimum. Interestingly, the best runtime is however obtained with the least interfering strategy, simply preferring negative signs for preference elements. On the other hand, classes like *Puzzle* are resistant to heuristic manipulations and weight-based optimization is even worse in this case. Here, convergence is immediate and cannot be improved by heuristic means. The bad performance can thus be explained by the interference of the heuristics with the final UNSAT problem needed for establishing optimality. Just modifying the `sign` thus appears as the best overall compromise, boosting convergence without overly hindering the final UNSAT proof. Otherwise, the best heuristic modification must be decided case-by-case.

Table 1. Comparing *asprin* with different heuristic settings

Benchmark \ System	$asprin_w$	$asprin_w$+s	$asprin_w$+l	$asprin_w$+f	$asprin_s$	$asprin_s$+s	$asprin_s$+l	$asprin_s$+f
Ricochet (30) 20.00	432(8, 4)	407(7,4)	68(1, 0)	71(1, 0)	365(8,3)	461(7,10)	69(1, 0)	71(1, 0)
Timetabling(12)23687.75	345(285, 3)	255(202,2)	900(4,12)	6(1, 0)	217(144,2)	21(18, 0)	900(2,12)	5(1, 0)
Puzzle (7) 580.57	82(2, 0)	112(2,0)	136(2, 0)	416(2, 1)	31(1,0)	32(1, 0)	21(1, 0)	51(1, 0)
Crossing (24) 211.92	104(42, 1)	98(35,0)	805(19,20)	387(6, 6)	0(6,0)	1(6, 0)	7(9, 0)	3(1, 0)
Valves (30) 56.63	69(7, 0)	65(6,0)	460(8,11)	715(0, 22)	38(4,0)	39(4, 0)	339(4, 6)	673(0,21)
Expansion (30) 7501.87	216(299, 0)	10(15,0)	38(7, 0)	12(3, 0)	64(295,0)	14(54, 0)	4(4, 0)	3(1, 0)
Repair (30) 6750.73	76(48, 0)	15(47,0)	71(3, 2)	8(2, 0)	8(43,0)	3(11, 0)	1(1, 0)	1(1, 0)
Diagnosis (30) 1669.00	196(341, 3)	76(66,0)	43(4, 0)	118(3, 2)	19(338,0)	2(39, 0)	0(1, 0)	0(1, 0)
$\varnothing(\varnothing, \Sigma)$	190(129,11)	130(48,6)	315(6,45)	217(2, 31)	93(105,5)	72(18,10)	168(3, 18)	101(1,21)

Our next series of experiments aims at comparing the general-purpose approach of *asprin* with dedicated implementations of `aso` [14] and `poset` [8] preferences. In both cases, we use their benchmark generators and sets to contrast the approaches. The experimental settings are the same as above.

First, we compare *asprin* with the system for `aso` preferences [14]; it implements a branch-and-bound approach in C++ and calls *clingo* each time from scratch via a system call. We refer to it as *aso*. We also used the benchmark generator from [14] to generate random 3CNF formulas with n variables and $4n$ clauses. For each formula of n variables, it randomly generates $3n$ preference rules with $a > \neg a$ or $\neg a > a$ for some a in the head, and 0 to 2 literals in the body. In addition, the approach handles ranked `aso` preferences (aso_l), which amounts to an aggregation of `aso` preferences with `lexico` in $asprin_{l+a}$. The gen-

Table 2. Comparing *asprin* with *aso*

n	*aso*	aso_l	$asprin_a$	$asprin_{l+a}$
350	9(0)	17(0)	4(0)	5(0)
360	14(0)	22(0)	48(0)	50(0)
370	15(0)	25(0)	38(0)	39(0)
380	10(0)	23(0)	8(0)	9(0)
390	59(0)	72(0)	50(1)	52(1)
400	22(0)	33(0)	28(0)	30(0)
410	87(1)	96(1)	124(2)	125(2)
420	97(1)	108(1)	60(0)	62(0)
430	68(0)	79(0)	144(0)	147(0)
440	165(3)	175(3)	165(2)	167(2)
450	45(0)	61(0)	52(0)	54(0)
460	112(1)	125(1)	117(2)	120(2)
470	201(4)	210(4)	161(2)	162(2)
480	152(2)	165(2)	70(1)	72(1)
490	206(2)	218(2)	265(4)	267(4)
$\varnothing(\Sigma)$	84(14)	95(14)	89(14)	91(14)

erator accounts for this by assigning a higher rank to half of the `aso` rules. The results of comparing both systems on both sets of random benchmarks are shown in Table 2. Each cell gives average runtime and number of timeouts. We see that the general-purpose approach of *asprin* is comparable with the dedicated approach of [14] on their benchmark set. On the other hand, we observed with *asprin* a very fast convergence, so that no real difference can be expected on these benchmarks.

Table 3. Comparing *satpref* and *asprin* under different heuristic settings

Benchmark\System	*satpref*	*satpref+s*	*satpref+H*	$asprin_p$	$asprin_p+s$	$asprin_p+H$
0.0	0(29, 0)	0(1, 0)	0(1, 0)	1(16, 0)	0(2, 0)	0(1, 0)
0.00621	0(35, 0)	0(1, 0)	90(1, 6)	1(17, 0)	1(2, 0)	1(1, 0)
0.01243	1(75, 0)	1(3, 0)	118(1, 7)	6(26, 0)	2(3, 0)	3(1, 0)
0.02486	8(388, 0)	6(10, 0)	635(1, 38)	55(74, 0)	9(8, 0)	64(1, 4)
0.04972	67(1463, 2)	16(36, 0)	900(0,100)	318(203, 16)	26(17, 0)	176(1, 14)
1.0	850(10315,88)	243(590,10)	177(1, 12)	856(323, 92)	174(96, 0)	280(1, 24)
$\varnothing(\varnothing,\Sigma)$	154(2051,90)	44(107,10)	320(1,163)	206(110,108)	35(21, 0)	88(1, 42)
MAXSAT	54(8849, 0)	9(7, 0)	62(1, 0)	835(957, 31)	109(31, 3)	171(1, 6)
PBO/pbo-mqc-nencdr	5(267, 0)	2(2, 0)	664(1, 88)	150(207, 14)	9(2, 0)	244(1, 20)
PBO/pbo-mqc-nlogencdr	3(228, 0)	1(2, 0)	237(1, 21)	110(214, 3)	5(2, 0)	141(1, 15)
PSEUDO/primes	110(396,18)	110(1,18)	110(1, 18)	215(334, 27)	106(5,17)	110(1, 17)
PSEUDO/routing	346(409, 4)	49(1, 0)	50(1, 0)	85(475, 0)	4(1, 0)	86(1, 1)
Partial-MINONE	14(2, 0)	14(2, 0)	7(1, 0)	24(2, 0)	24(1, 0)	25(1, 0)
$\varnothing(\varnothing,\Sigma)$	88(1692,22)	31(2,18)	188(1,127)	236(365, 75)	43(7,20)	129(1, 59)

Next, we compare *asprin* with the system *satpref* for `poset` preferences [15]. Interestingly, *satpref* not only extends the SAT solver *minisat* with branch-and-bound-based optimization but also uses heuristic support for boosting optimization. Table 3 contains the results of our comparison on benchmarks from [15]. The first six lines of data stem from 600 random instances (each with 500 variables and 1750 clauses), in which an order $a > b$ between variables a and b is generated with the probabilities given in the left column. The second six lines of data stem from instances taken from various competitions (cf. [15]). As above, we compare both systems in their basic setting and with sign-based heuristics. In addition, we contrast the declarative heuristics from the end of Sect. 4 ($asprin_p$+H) with its hard-coded counterpart in *satpref*+H. Such a heuristic ensures that the first found model is optimal. In fact, as above, the best results with both systems are obtained with a light sign-based heuristics. The slight edge of *satpref* over $asprin_p$ is due to additional grounding efforts (given that problems are expressed in ASP). Despite this, the experiments show that the general-purpose approach of *asprin* is overall comparable with the dedicated approach of *satpref*.

7   Discussion

We have presented *asprin*, a general and flexible ASP-based system for representing and evaluating combinations of quantitative and qualitative preferences. We presented *asprin*'s first-order modeling language and showed how existing (and future) preferences can be expressed in *asprin*. We showed that our general-purpose approach matches the performance of dedicated systems for aso and poset preferences. Moreover, we demonstrated how well-chosen heuristics can boost the optimization process.

References

1. Baral, C.: Knowledge Representation. Reasoning and Declarative Problem Solving. Cambridge University Press, Cambridge (2003)
2. Simons, P., Niemelä, I., Soininen, T.: Extending and implementing the stable model semantics. Artif. Intell. **138**(1–2), 181–234 (2002)
3. Leone, N., Pfeifer, G., Faber, W., Eiter, T., Gottlob, G., Perri, S., Scarcello, F.: The DLV system for knowledge representation and reasoning. ACM TOCL **7**(3), 499–562 (2006)
4. Brewka, G.: Logic programming with ordered disjunction. In: Proceedings of AAAI, pp. 100–105. AAAI Press (2002)
5. Brewka, G., Niemelä, I., Truszczyński, M.: Answer set optimization. In: Proceedings of IJCAI, pp. 867–872. Morgan Kaufmann (2003)
6. Brewka, G.: Complex preferences for answer set optimization. In: Proceedings of KR, pp. 213–223. AAAI Press (2004)
7. Son, T., Pontelli, E.: Planning with preferences using logic programming. Theor. Pract. Logic Program. **6**(5), 559–608 (2006)
8. Di Rosa, E., Giunchiglia, E., Maratea, M.: Solving satisfiability problems with preferences. Constraints **15**(4), 485–515 (2010)
9. Brewka, G., Delgrande, J., Romero, J., Schaub, T.: asprin: Customizing answer set preferences without a headache. In: Proceedings of AAAI, pp. 1467–1474. AAAI Press (2015)
10. Gebser, M., Kaminski, R., Kaufmann, B., Schaub, T.: Answer Set Solving in Practice. Morgan and Claypool Publishers, San Rafael (2012)

11. Järvisalo, M., Junttila, T., Niemelä, I.: Unrestricted vs restricted cut in a tableau method for boolean circuits. Ann. Math. Artif. Intell. **44**(4), 373–399 (2005)
12. Gebser, M., Kaufmann, B., Otero, R., Romero, J., Schaub, T., Wanko, P.: Domain-specific heuristics in answer set programming. In: Proceedings of AAAI, pp. 350–356. AAAI Press (2013)
13. Castell, T., Cayrol, C., Cayrol, M., Le Berre, D.: Using the Davis-Putnam procedure for an efficient computation of preferred models. In: Proceedings of ECAI, pp. 350–354. Wiley (1996)
14. Zhu, Y., Truszczynski, M.: On optimal solutions of answer set optimization problems. In: Cabalar, P., Son, T.C. (eds.) LPNMR 2013. LNCS, vol. 8148, pp. 556–568. Springer, Heidelberg (2013)
15. Di Rosa, E., Giunchiglia, E.: Combining approaches for solving satisfiability problems with qualitative preferences. AI Commun. **26**(4), 395–408 (2013)

Diagnosing Automatic Whitelisting for Dynamic Remarketing Ads Using Hybrid ASP

Alex Brik[1]([⊠]) and Jeffrey Remmel[2]

[1] Google Inc., Mountain View, USA
abrik@google.com
[2] Department of Mathematics, UC San Diego, San Diego, USA

Abstract. Hybrid ASP (H-ASP) is an extension of ASP that allows users to combine ASP type rules and numerical algorithms. Dynamic Remarketing Ads is Google's platform for serving customized ads based on past interactions with a user. In this paper we will describe the use of H-ASP to diagnose failures of the automatic whitelisting system for Dynamic Remarketing Ads. We will show that the diagnosing task is an instance of a computational pattern that we call the Branching Computational Pattern (BCP). We will then describe a Python H-ASP library (H-ASP PL) that allows to perform computations using a BCP, and we will describe a H-ASP PL program that solves the diagnosing problem.

Past research has demonstrated that logic programming with the answer-set semantics, known as *answer-set programming* or *ASP*, for short, is an expressive knowledge-representation formalism [2,9–13]. The availability of the nonclassical negation operator *not* allows the user to model incomplete information, frame axioms, and default assumptions such as normality assumptions and the closed-world assumption efficiently. Modeling these concepts in classical propositional logic is less direct [9] and typically requires much larger representations.

A fundamental methodological principle behind ASP, which was identified in [12], is that to model a problem, one designs a program so that its answer sets *encode* or *represent* problem solutions. Niemelä [13] has argued that logic programming with the stable-model semantics should be thought of as a language for representing constraint satisfaction problems. Thought of from this point of view, ASP systems are ideal logic-based systems to reason about a variety of types of data and integrate quantitative and qualitative reasoning. ASP systems allow the users to describe solutions by giving a series of constraints and letting an ASP solver search for solutions.

To solve many of the real world problems one needs to have an ASP programming environment where one can perform external data searches and bring back information that can be used in the program. Extensions of ASP that allow such external data searches include DLV^{DB} system [15] for querying relational databases, VI programs [6] for importing knowledge from external sources, HEX programs [7] which allow access to external data sources via external atoms, $GRINGO$ grounder that provides an interface for calling function written in Lua during the grounding process [8], and Hybrid ASP (H-ASP) introduced by the authors in [4].

© Springer International Publishing Switzerland 2015
F. Calimeri et al. (Eds.): LPNMR 2015, LNAI 9345, pp. 173–185, 2015.
DOI: 10.1007/978-3-319-23264-5_16

In this paper, we will discuss applications of H-ASP which can perform external data searches. In particular, we shall discuss an example of problems that can be solved by processing a connected directed acyclic graph (cDAG for short) where each vertex of the cDAG contains both logical and non-logical information in the form of parameters. The cDAG can be generated by following a multi-step pattern of computation which we will call the Branching Computational Pattern (BCP for short). At any stage in the computation we are given a set of vertices. These vertices can either be an initial set of vertices or a set of vertices produced at the previous step. Then the BCP instance creates multiple new branches emanating from a particular vertex. For each new branch, the BCP instance performs a computation using the data from the vertex and possibly auxiliary data from external repositories to derive new logical information and parameters at that vertex as well as pass relevant logical information and new parameters to its children. The result of such computation defines new vertices of the cDAG. Then new edges from each parent vertex to its children are added to the cDAG. The resulting cDAG is called a BCP cDAG.

The focus of this paper is the problem of diagnosing failures of the automatic whitelisting system for Dynamic Remarketing Ads (automatic whitelisting, for short). Dynamic Remarketing Ads is Google's platform for delivering ads which are customized to an individual user based on the user's past interactions with the advertiser such as the user's previously viewed items or abandoned shopping carts. In order for Google to start serving dynamic remarketing ads for a particular advertiser that advertiser needs to be whitelisted, i.e. the advertiser has to have been added to a list of advertisers that are known to use Dynamic Remarketing Ads. Whitelisting is done automatically by a system that detects whether an advertiser is ready to serve dynamic remarketing ads based on the logs and the content of ads databases.

There are nine cases when an advertiser can be automatically whitelisted. In each case there is a set of constraints that need to be satisfied in order for that case to apply. The technical challenge in using ASP for diagnosing automatic whitelisting is that in order to check the constraints it is necessary to search data stored in Google's various data repositories. The fact that in our application, the amount of data is quite large and the repository contents change in real time makes pre-computing impractical. Moreover, data searches in repositories often depend on the data obtained in the previous steps. Under these circumstances what is required is an extension of ASP that allows the following: (1) conclusions to be derived conditional on the results of the external data searches, and (2) parameter passing between the algorithms that perform data searches. H-ASP provides this functionality. To solve the problem of diagnosing failures of the automatic whitelisting system we have implemented a Python library, which we call H-ASP PL for running H-ASP programs that use a certain subset of H-ASP rules. We have then created a H-ASP program that runs using H-ASP PL library. The program was successfully used for several months in the cases of many advertisers.

Another problem that can be solved by processing a cDAG is computing an optimal strategy for an agent acting in a dynamic domain. In [5], the authors showed how H-ASP programs can be used to combine logical reasoning, continuous parameters, and probabilistic reasoning in the context of computing optimal strategy using Markov Decision Processes. A feature of the solution in [5] was that one started with a basic H-ASP program and performed a series of program transformations so that one could compute a maximal stable model which contains all of the stable models of the original program. That is, according to the H-ASP semantics, which we will define in a subsequent section, a H-ASP program can have multiple stable models where some of the stable models can be subsets of other stable models. In the context of computing an optimal strategy, such stable models would describe only a part of the evolution tree of the dynamic domain. Such dynamic domains can be represented as BCP cDAGs. Hence, each stable model represents a part of the BCP cDAG. However, in order to compute an optimal strategy, one needs to compute the entire evolution tree of a dynamic domain. Thus the full BCP cDAG is required.

We have a similar situation in the case of diagnosing automatic whitelisting. We are also interested in the full BCP cDAG, which in this case will encode all of the possible whitelisting paths.

There already exists a literature discussing the use of ASP for diagnosing malfunctioning devices. In [1] Balduccini and Gelfond describe an approach for diagnosing a malfunctioning device based on the theory of action language $\mathcal{AL}$. In their approach the underlying diagnostic program explicitly describes the laws that govern the behavior of the dynamic domain, and the non-monotonicity of ASP is used to compute all the possible scenarios under which a malfunction could occur. In the case of diagnosing automatic whitelisting, we use the non-monotonicity of ASP to compute all the possible scenarios for a malfunction, however we do not explicitly describe the laws that govern the behavior of the dynamic domain. The latter is motivated by the relative simplicity of the domain for the automatic whitelisting, and by the time constraints of the project.

There are two main advantages of using H-ASP rather than a common programming language such as Python directly. The first advantage is the efficiency of representation. Our H-ASP program specifies how automatic whitelisting occurs and lets the solver report back failures when automatic whitelisting fails. An equivalent Python program will have to specify both how the automatic whitelisting occurs and the details of the diagnostic logic. Hence, H-ASP program is smaller than one would expect the equivalent Python program to be. The second advantage is the robustness of the H-ASP program. Because H-ASP program describes mostly the problem domain, it is easier to update the program when changes to the automatic whitelisting logic occur. This is important since the requirements for automatic whitelisting are continually being modified.

The outline of this paper is the following. In Sect. 1, we formally define the Branching Computational Pattern (BCP). Our problem of diagnosing automatic whitelisting is a special case of the BCP. In Sect. 2, we give an overview of H-ASP. In Sect. 3, we discuss the computational pattern as it relates to H-ASP and we briefly describe the H-ASP PL library. In Sect. 4, we present a toy example to illustrate how the problem of diagnosing automatic whitelisting is solved. In Sect. 5, we describe the semantics of H-ASP PL, and the Local Algorithm which is used in H-ASP PL. In Sect. 6, we discuss some of the related work and give the closing comments.

1   The Branching Computational Pattern (BCP)

Let $\langle V, E \rangle$ be a connected directed acyclic graph (cDAG). Let $R(V, E)$ be the set of vertices with no in-edges. If $|R(V, E)| = 1$, then we will refer to the unique vertex $r(V, E) \in R(V, E)$ as the root node. If $(v, w) \in E$, we will say that v is a *parent* of w and that w is a *child* of v. If there exists $v_1, v_2, ..., v_n$ such that for all $i \in \{2, ..., n\}$, $(v_{i-1}, v_i) \in E$ and $v_1 = v$ and $v_n = w$, then we say that v is an *ancestor* of w and w is a *descendant* of v. It is easy to see that for all $v \in V$ such that $v \notin R(V, E)$, there exists a vertex in $R(V, E)$ which is an ancestor of v.

Our branching computational patterns allow the user to compute a cDAG $\langle V, E \rangle$ where each vertex $v \in V$ is a pair $(A, \mathbf{p})$ where A is a set of propositional atoms and $\mathbf{p}$ is a vector of parameter values representable by a computer. We will refer to such a cDAG as a *computational cDAG*. If all $(A, \mathbf{p}) \in V$, $A \subseteq At$ and $\mathbf{p} \in S$, then we will say that $\langle V, E \rangle$ is a *computational cDAG over At and S*. At each cDAG vertex $(A, \mathbf{p})$, the computation consists of the two steps:

1. use A and $\mathbf{p}$ to choose algorithms (that will possibly access external data repositories and/or perform computations) to produce the set of next parameter value vectors $\mathbf{q}_1, ..., \mathbf{q}_k$ and
2. for each $\mathbf{q}_i$ produced in step 1, derive atoms $B_{i,1}, B_{i,2}, ..., B_{i,m_i}$.

The set of children of $(A, \mathbf{p})$ will be the pairs $(\{B_{i,1}, B_{i,2}, ..., B_{i,m_i}\}, \mathbf{q}_i)$ for $i = 1, ..., k$. To produce the root nodes of the cDAGs, step 2 is applied to the initial set of parameter values specified as an input. The computation can then be repeated at each child node.

This computational process is illustrated in Fig. 1. At the node *C0*, we obtain *data1* from database *Database1* and *data2* from database *Database2*. *C0* then creates three new children: $(\{A_i\}, fi(data1, data2))$ for $i = 1, 2, 3$ where fi is one of the computational algorithms associated with an H-ASP rule that can

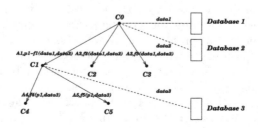

Fig. 1. Computational pattern illustration

be applied at $C0$. At node $C1$, we obtain *data3* from database *Database3*. Then $C1$, creates two new children: $(\{A_i\}, fi(p1, data3))$ for $i = 4, 5$. This figure illustrates the main aspects of the BCP. At each node, external data sources can be accessed. This new data, the value of the parameters stored at the node, and the logical information stored at the node are then used to create new children by passing new parameters and logical information to each child as well as updating the logic information stored at the node.

The problem of diagnosing failures for automatic whitelisting fits the general BCP paradigm. There are nine cases for the automatic whitelisting. In order to help in understanding the types of criteria used, we shall describe one of these cases.

Case One: An advertiser is whitelisted if the advertiser has installed a Javascript tag containing the id of one of the advertiser's products, a user visits advertiser's website, and the advertiser has created a dynamic remarketing ad.

Here the initial set of candidate advertisers for whitelisting is obtained from the table, which we will call T_init, containing the information about advertisers who have installed a Javascript tag. Whether a user has visited advertiser's website can be determined by examining a log called L_userevents. The id of each advertiser from the candidate set will be used to search L_userevents to determine whether a user has visited advertiser's website or not. The remaining advertiser ids will be used to determine whether the advertiser has created a dynamic remarketing ad by using an external function GetCreatedAdIds().

The cDAG representing the automatic whitelisting system can be constructed by making a vertex represent a whitelisting condition that needs to be satisfied. An edge from a vertex x to a vertex y will be added to indicate that the condition for y needs to be checked immediately after checking the condition for x. This may be necessary, for instance if data derived when checking the condition for x needs to be used to check the condition for y. A root node will have as its children the first conditions for each of the nine cases. Thus the cDAG will be a tree with 9 branches.

2 Hybrid ASP

In this section we shall give a brief overview of H-ASP. A H-ASP program P has an underlying parameter space S and a set of atoms At. Elements of S are of the form $\mathbf{p} = (t, x_1, \ldots, x_m)$ where t is time and x_i are parameter values. We shall let $t(\mathbf{p})$ denote t and $x_i(\mathbf{p})$ denote x_i for $i = 1, \ldots, m$. We refer to the elements of S as *generalized positions*. The universe of P is $At \times S$. For ease of notation, we will often identify an atom and the string representing an atom.

Let $M \subseteq At \times S$. Define $\widehat{M} = \{\mathbf{p} \in S : (\exists a \in At)((a, \mathbf{p}) \in M)\}$. For a generalized position $\mathbf{p} \in S$, define $W_M(\mathbf{p}) = \{a \in At : (a, \mathbf{p}) \in M\}$. A *hybrid state* at generalized position $\mathbf{p} \in S$ is a pair $(W_M(\mathbf{p}), \mathbf{p})$. In general, a pair $(A, \mathbf{p})$ where $A \subseteq At$ and $\mathbf{p} \in S$ will be referred to as a *hybrid state*. For a hybrid state $(A, \mathbf{p})$, we write $(A, \mathbf{p}) \in M$ if $\mathbf{p} \in \widehat{M}$ and $W_M(\mathbf{p}) = A$.

A *block* B is an object of the form $B = a_1, \ldots, a_n, not\ b_1, \ldots, not\ b_m$ where $a_1, \ldots, a_n, b_1, \ldots, b_m \in At$. We let $B^- = not\ b_1, \ldots, not\ b_m$. Given $M \subseteq At \times S$, $B = a_1, \ldots, a_n, not\ b_1, \ldots, not\ b_m$, and $\mathbf{p} \in S$, we say that M satisfies B at the generalized position $\mathbf{p}$, written $M \models (B, \mathbf{p})$, if $(a_i, \mathbf{p}) \in M$ for $i = 1, \ldots, n$ and $(b_j, \mathbf{p}) \notin M$ for $j = 1, \ldots, m$. If B is empty, then $M \models (B, \mathbf{p})$ automatically holds.

There are two types of rules in H-ASP.

Advancing rules are of the form $a \leftarrow B_1; B_2; \ldots; B_r : A, O$ where A is an algorithm, each B_i is a block, and $O \subseteq S^r$ is such that if $(\mathbf{p}_1, \ldots, \mathbf{p}_r) \in O$, then $t(\mathbf{p}_1) < \cdots < t(\mathbf{p}_r)$, $A(\mathbf{p}_1, \ldots, \mathbf{p}_r) \subseteq S$, and for all $\mathbf{q} \in A(\mathbf{p}_1, \ldots, \mathbf{p}_r)$, $t(\mathbf{q}) > t(\mathbf{p}_r)$. Here and in the next rule, we allow n or m to be equal to 0 for any given i. Moreover, if $n = m = 0$, then B_i is empty and we automatically assume that B_i is satisfied by any $M \subseteq At \times S$. We shall refer to O as the *constraint set* of the rule and the algorithm A as the *advancing algorithm* of the rule. The idea is that if $(\mathbf{p}_1, \ldots, \mathbf{p}_r) \in O$ and for each i, B_i is satisfied at the generalized position $\mathbf{p}_i$, then the algorithm A can be applied to $(\mathbf{p}_1, \ldots, \mathbf{p}_r)$ to produce a set of generalized positions O' such that if $\mathbf{q} \in O'$, then $t(\mathbf{q}) > t(\mathbf{p}_r)$ and $(a, \mathbf{q})$ holds.

Stationary rules are of the form $a \leftarrow B_1; B_2; \ldots; B_r : H, O$ where each B_i is a block, $O \subseteq S^r$ is such that if $(\mathbf{p}_1, \ldots, \mathbf{p}_r) \in O$, then $t(\mathbf{p}_1) < \cdots < t(\mathbf{p}_r)$, and H is a Boolean algorithm defined on O. We shall refer to O as the *constraint set* of the rule and the algorithm H as the *Boolean algorithm* of the rule. The idea is that if $(\mathbf{p}_1, \ldots, \mathbf{p}_r) \in O$ and for each i, B_i is satisfied at the generalized position $\mathbf{p}_i$, and $H(\mathbf{p}_1, \ldots, \mathbf{p}_r)$ is true, then $(a, \mathbf{p}_r)$ holds.

A *H-ASP Horn program* is a H-ASP program which does not contain any negated atoms in At. Let P be a Horn H-ASP program, let $I \in S$ be an initial condition. Then the one-step provability operator $T_{P,I}$ is defined so that given $M \subseteq At \times S$, $T_{P,I}(M)$ consists of M together with the set of all $(a, J) \in At \times S$ such that

(1) there exists a stationary rule $C = a \leftarrow B_1; B_2; \ldots; B_r : H, O$ and $(\mathbf{p}_1, \ldots, \mathbf{p}_r) \in O \cap \left(\widehat{M} \cup \{I\}\right)^r$ such that $(a, J) = (a, \mathbf{p}_r)$, $M \models (B_i, \mathbf{p}_i)$ for $i = 1, \ldots, r$, and $H(\mathbf{p}_1, \ldots, \mathbf{p}_r) = 1$ or

(2) there exists an advancing rule $C = a \leftarrow B_1; B_2; \ldots; B_r : A, O$ and $(\mathbf{p}_1, \ldots, \mathbf{p}_r) \in O \cap \left(\widehat{M} \cup \{I\}\right)^r$ such that $J \in A(\mathbf{p}_1, \ldots, \mathbf{p}_r)$ and $M \models (B_i, \mathbf{p}_i)$ for $i = 1, \ldots, r$.

The stable model semantics for H-ASP programs is defined as follows. Let $M \subseteq At \times S$ and $I \in S$. An H-ASP rule $C = a \leftarrow B_1; \ldots, B_r : A, O$ is *inconsistent* with (M, I) if for all $(\mathbf{p}_1, \ldots, \mathbf{p}_r) \in O \cap \left(\widehat{M} \cup \{I\}\right)^r$, either (i) there is an i such that $M \not\models (B_i^-, \mathbf{p}_i)$, (ii) $A(\mathbf{p}_1, \ldots, \mathbf{p}_r) \cap \widehat{M} = \emptyset$ if A is an advancing algorithm, or (iii) $A(\mathbf{p}_1, \ldots, \mathbf{p}_r) = 0$ if A is a Boolean algorithm. Then we form the Gelfond-Lifschitz reduct of P over M and I, $P^{M,I}$ as follows.

(1) Eliminate all rules that are inconsistent with (M, I).

(2) If the advancing rule $C = a \leftarrow B_1; \ldots, B_r : A, O$ is not eliminated by (1), then replace it by $a \leftarrow B_1^+; \ldots, B_r^+ : A^+, O^+$ where for each i, B_i^+ is the result of removing all the negated atoms from B_i, O^+ is equal to the set of all $(\mathbf{p}_1, \ldots, \mathbf{p}_r)$ in $O \cap \left(\widehat{M} \cup \{I\}\right)^r$ such that $M \models (B_i^-, \mathbf{p}_i)$ for $i = 1, \ldots, r$ and $A(\mathbf{p}_1, \ldots, \mathbf{p}_r) \cap \widehat{M} \neq \emptyset$, and $A^+(\mathbf{p}_1, \ldots, \mathbf{p}_r)$ is defined to be $A(\mathbf{p}_1, \ldots, \mathbf{p}_r) \cap \widehat{M}$.

(3) If the stationary rule $C = a \leftarrow B_1; \ldots, B_r : H, O$ is not eliminated by (1), then replace it by $a \leftarrow B_1^+; \ldots, B_r^+ : H|_{O^+}, O^+$ where for each i, B_i^+ is the result of removing all the negated atoms from B_i, O^+ is equal to the set of all $(\mathbf{p}_1, \ldots, \mathbf{p}_r)$ in $O \cap \left(\widehat{M} \cup \{I\}\right)^r$ such that $M \models (B_i^-, \mathbf{p}_i)$ for $i = 1, \ldots, r$ and $H(\mathbf{p}_1, \ldots, \mathbf{p}_r) = 1$.

Then M is a *stable model* of P with initial condition I if $\bigcup_{k=0}^{\infty} T_{PM,I,I}^k (\emptyset) = M$.

We say that M is a *single trajectory stable model* of P with initial condition I if M is a stable model of P with initial condition I and for each $t \in \{t(\mathbf{p}) | \mathbf{p} \in S\}$, there exists at most one $\mathbf{p} \in \widehat{M} \cup \{I\}$ such that $t(\mathbf{p}) = t$.

We say that an advancing algorithm A lets a parameter y be *free* if the domain of y is Y and for all generalized positions $\mathbf{p}$ and $\mathbf{q}$ and all $y' \in Y$, whenever $\mathbf{q} \in A(\mathbf{p})$, then there exist $\mathbf{q}' \in A(\mathbf{p})$ such that $y(\mathbf{q}') = y'$ and $\mathbf{q}$ and $\mathbf{q}'$ are identical in all the parameter values except possibly y. We say that an advancing algorithm A *fixes* a parameter y if A does not let y be free.

The reason for introducing the last two definitions is that we often want to limit the effects of algorithms to specifying only a subsets of the parameters to make programs easier to understand. The exact mechanism for doing so will be discussed in the next section. For now, however we will note that the parameters that the advancing algorithm will be responsible for producing will correspond to the fixed parameters of the algorithm. The rest of the parameters will correspond to the free parameters.

3   H-ASP Library

H-ASP programs can be used to perform the computations for a BCP. In fact, BCP computations can be carried by by H-ASP programs which use only a restricted set of H-ASP which we call H-ASP program of order 1. A H-ASP program P is of order 1 if all its advancing rules are of the form $a \leftarrow B : A, O$, and all its stationary rules of the form $a \leftarrow B : H, O$. If P is of order 1, then we will say that $\mathbf{p}$ is a *child* of $\mathbf{q}$ (and $\mathbf{q}$ is a parent of $\mathbf{p}$) under P, I if there exists a stable model M of P with the initial condition I and there exists an advancing rule $a \leftarrow B : A, O \in P$ such that $M \models (B, \mathbf{q})$ and $\mathbf{q} \in O$ and $\mathbf{p} \in A(\mathbf{q})$.

Given an H-ASP program P of order 1 and the initial condition I, we define the computational cDAG *induced* by P, I, $comp(P, I) = \langle V, E \rangle$, as follows.

(1) V is the set of all the hybrid states $(A, \mathbf{p})$ such that there exists a stable model M of P with initial condition I and $(A, \mathbf{p}) \in M$.

(2) E is the set of all pairs $((A, \mathbf{p}), (B, \mathbf{q})) \in V^2$ such that there exists a stable model M of P initial condition I and $(A, \mathbf{p}) \in M$ and $(B, \mathbf{q}) \in M$ and $(A, \mathbf{p})$ is a parent of $(B, \mathbf{q})$ under P, I.

We can prove the following theorem.

Theorem 1. *Let At be a set of propositional atoms and let S be a set of parameter values. Let $\langle V, E \rangle$ be a computational cDAG over At and S. Then there exists a H-ASP program P of order 1 and an initial condition I for P such that $comp(P, I) = \langle W, U \rangle$ is isomorphic to $\langle V, E \rangle$. Moreover, P can be chosen to have a maximal stable model M and a parameter space X such that there exists a map $\pi : X \to S$ so that the isomorphism g from $comp(P, I)$ to $\langle V, E \rangle$ is defined by setting $g((A, \mathbf{p})) = (A \cap At, \pi(\mathbf{p}))$ for $(A, \mathbf{p}) \in W$ where W is the set of all the hybrid states of M.*

The proof of the theorem consists of construction of a H-ASP program P of order 1, initial condition I and a simple isomorphism π that satisfy the conditions of the theorem. The idea is that for every edge $((A, \mathbf{p}), (B, \mathbf{q})) \in E$, P contains a set of rules with constraint sets that are satisfied only by $\pi(\mathbf{p})$, and that generate $(B \cap At, \pi(\mathbf{q}))$.

It is also easy to see that for a H-ASP program P of order 1 and the initial condition I, the computation of a stable model is performed according to the BCP along the computational cDAG induced by P, I.

Practical applications of BCP's require either a computer language or a library. Due to the time constraints of our project, we created a Python library which allows us to compute the stable models for H-ASP programs of order 1. However, to make the task of programming with the library easier, we have added one more type of rule beyond those allowed in H-ASP programs of order 1. We will now briefly describe some of the key features of the library and the programs called H-ASP PL programs that it processes.

Let At be the set of atoms and let S be the parameter space.

1. H-ASP PL programs consists of three types of H-ASP rules: advancing rules of the form $a \leftarrow B : A, O$, stationary rules of the form $a \leftarrow B : H, O$, and stationary rules of the form $a \leftarrow B_1; B_2 : G, \Theta$.

2. The parameters in the parameter space S are named. If $\mathbf{p}$ is a generalized position and Q is a parameter, we denote the value of Q at $\mathbf{p}$ by $\mathbf{p}[Q]$.

3. The time parameter is named TIME and it is assumed that every advancing algorithm increments the value of TIME by 1. That is, if A is an advancing algorithm, $\mathbf{p}$ is a generalized position, then for all $\mathbf{q} \in A(\mathbf{p})$, $\mathbf{q}[\text{TIME}] = \mathbf{p}[\text{TIME}] + 1$. Because of this assumption, the advancing algorithms are not required to specify the value of the parameter TIME.

4. In [4], the authors suggested an indirect approach by which the advancing algorithms can specify the values for only some of the parameters. The approach requires extending the Herbrand base of a program P by a set of new

atoms S_1, ..., S_n one for each parameter. Suppose that there is an advancing algorithm A in a rule $a \leftarrow B : A, O$ that specifies parameters with indexes $i_1, i_2, \ldots, i_k$ and lets other parameters be free. Then we add to P rules of the form $S_{i_j} \leftarrow B : A, O$ for each j from 1 to k. This is, repeated for every advancing rule of P. Then if M is a stable model of P and $\mathbf{p} \in \widehat{M}$, we will require that $\{S_1, \ldots, S_n\} \subseteq W_M(\mathbf{p})$. This will ensure that every parameter at $\mathbf{p}$ is set by some advancing rule. For our library, we assume that this mechanism is used by any H-ASP program that it will process. This allows us to implement it implicitly without requiring the H-ASP user to specify the additional rules.

5. For our application, we would like to have the ability to apply a constraint to two hybrid states belonging to the same single trajectory stable model of P. In order to do that in our library, for a stationary rule of the form $a \leftarrow B_1; B_2 : G, \Theta$, we assume that if $G(\mathbf{p}, \mathbf{q}) = 1$, then $t(\mathbf{p}) + 1 = t(\mathbf{q})$ and there exists a single trajectory stable model M of P such that $\{\mathbf{p}, \mathbf{q}\} \subseteq \widehat{M}$.

4   Example

The following example illustrates a typical step of processing performed by the program for diagnosing automatic whitelisting. The complete diagnosis requires many steps of this type.

Suppose that a decision is to be made based on a decision tree containing two branches. In the first branch data repository 1 is searched for the data D1 relevant for condition C1. If condition C1 is satisfied based on D1, then data repository 2 is searched for the data D2[D1] which is dependent on D1. Condition C2 is then evaluated based on D2[D1]. In the second branch, data repository 3 is searched for the data D3. Condition C3 is then evaluated based on D3. If in neither branch 1 nor branch 2 all of the corresponding constraints are satisfied, then a negative decision is made (see Fig. 2).

An explanation of a negative decision would have to describe which condition in each of the two decision branches has failed. The idea is to create a H-ASP PL program P that will produce a stable model whose computational cDAG will model the decision tree in that the nodes of the computational cDAG will correspond to the conditions of the decision tree, the edges will correspond to the successor relations of the tree.

We will need the following H-ASP PL parameters: DATA - to pass the data from the state corresponding to C1 to the state corresponding to C2, EXPLA-NATION - to record the description of the conditions that were not satisfied.

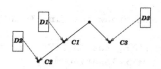

Fig. 2. Decision tree diagram

The advancing algorithm SetData{i} for $i \in \{1, 2, 3\}$ gets data D{i} and sets the output parameter DATA. Moreover, the advancing algorithm SetData2 uses the value of DATA parameter in order to get data D2[D1]. In order to check condition Ci where $i \in \{1, 2, 3\}$, the Boolean algorithms CheckCondition{i} for $i \in \{1, 2, 3\}$ are implemented and return TRUE iff Ci is satisfied. EXPLANATION parameter is set by the advancing algorithms Condition{i}Fails to a string "condition {i} is not satisfied" for $i \in \{1, 2, 3\}$. The advancing algorithm SetExplanationEmpty sets the value of EXPLANATION to the empty string.

This gives us the following H-ASP PL rules (rule label is in the brackets in the following format R{branch#}{time that the rule will affect}.{rule index}):

```
# Initial branching. IsTime0 is a boolean algorithm that returns TRUE iff
# the time of the input generalized position is 0.
  [R1.0.1] BRANCH1 :- not BRANCH2: IsTime0
  [R2.0.1] BRANCH2 :- not BRANCH1: IsTime0
# Produce generalized position for making decision C1.
  [R1.1.1] CHECK_C1 :- BRANCH1: SetData1
  [R1.1.2] CHECK_C1 :- BRANCH1: SetExplanationEmpty
# Check condition C1.
  [R1.1.3] C1_SAT :- CHECK_C1: CheckCondition1
  [R1.1.4] C1_DONE :- CHECK_C1
# If condition C1 is not satisfied then generate an explanation and set
# DATA to empty - branch 1 negative decision is explained.
  [R1.2.1] EXPLAINED :- C1_DONE, not C1_SAT: Condition1Fails
  [R1.2.2] END :- C1_DONE, not C1_SAT: SetDataEmpty
# If condition C1 is satisfied, then proceed with getting D2[D1]
# Data D1 is the value of DATA for the appropriate generalized position.
  [R1.2.3] CHECK_C2 :- C1_SAT: SetData2
  [R1.2.4] CHECK_C2 :- C1_SAT: SetExplanationEmpty
# Check condition C2.
  [R1.2.5] C2_SAT :- CHECK_C2: CheckCondition2
  [R1.2.6] C2_DONE :- CHECK_C2
# If condition C2 is not satisfied then produce an explanation.
  [R1.3.1] EXPLAINED :- C2_DONE, not C2_SAT: Condition2Fails
# If condition C2 is satisfied, set EXPLANATION to empty.
  [R1.3.2] END :- C2_DONE, C2_SAT: SetExplanationEmpty
# In both cases set DATA to empty.
  [R1.3.3] END :- C2_DONE: SetDataEmpty
# Now, for branch 2 - get D3.
  [R2.1.1] CHECK_C3 :- BRANCH2: SetData3
  [R2.1.2] CHECK_C3 :- BRANCH2: SetExplanationEmpty
# Check condition C3.
  [R2.1.3] C3_SAT :- CHECK_C3: CheckCondition3
  [R2.1.4] C3_DONE :- CHECK_C3
# If C3 is not satisfied then state that in EXPLANATION, otherwise set
# EXPLANATION to empty.
  [R2.2.1] EXPLAINED :- C3_DONE, not C3_SAT: Condition3Fails
  [R2.2.2] END :- C3_SAT: SetExplanationEmpty
# In both cases set DATA to empty.
  [R2.2.3] END :- C3_DONE: SetDataEmpty
```

5   Semantics of H-ASP PL

The semantics for H-ASP PL programs is a variant of the H-ASP stable model semantics. In order to diagnose failures of automatic whitelisting system, all the cases for whitelisting will need to be examined. Our H-ASP PL program will describe all of the whitelisting cases. We will be interested in a stable model of the underlying H-ASP program that will describe the results of examining each case. We would like this to be the unique maximal stable model. We will thus construct the semantics of H-ASP PL so that for a valid H-ASP PL program, its underlying H-ASP program has a unique maximal stable model, which after an appropriate transform, will be the stable model of the H-ASP PL program.

The semantics of H-ASP PL programs are defined in two steps. For a H-ASP PL program W, a transform $Tr[PL]$ is used to produce a H-ASP program $Tr[PL](W)$. Then the transform Tr introduced in [3] is used to produce a H-ASP program $Tr(Tr[PL](W))$. $Tr(Tr[PL](W))$ has the following properties for an initial condition I of $Tr[PL](W)$:

1. There is a bijection between the set of stable models of $Tr[PL](W)$ with the initial condition I and the set of stable models of $Tr(Tr[PL](W))$ with the corresponding initial condition $J(I)$.
2. For the initial condition $J(I)$, $Tr(Tr[PL](W))$ has a unique maximal stable model M'_{max}. M'_{max} is maximal in a sense that it contains all the stable models of $Tr(Tr[PL](W))$ with the initial condition $J(I)$.

Then we set the stable model of W with initial condition I to be the unique maximal stable model M'_{max} of $Tr(Tr[PL](W))$ with initial condition $J(I)$.

The transform $Tr[PL]$ is defined similarly to the transform $Tr\#$ introduced in [3] for transforming valid H-ASP# programs. The definitions of both transforms are omitted due to the space constraints. The following new theorem states that any computational cDAG representable by a computer can be computed by a H-ASP PL program.

Theorem 2. *Let At be a set of propositional atoms, and let S be a set whose elements are representable by a computer. Let $\langle V, E \rangle$ be a computational cDAG over At and S. Then there exists a H-ASP PL program P and an initial condition I for P such that $comp(Tr(Tr[PL](P)), J(I)) = \langle W, U \rangle$ is isomorphic to $\langle V, E \rangle$.*

The theorem is proved by constructing an isomorphism based on a H-ASP PL program P and initial condition I, chosen so that $Tr(Tr[PL](P))$ and $J(I)$ are the H-ASP program and the corresponding initial condition constructed in the proof of Theorem 1.

For a valid H-ASP PL program W and initial condition I of $Tr[PL](W)$, the maximal stable model M'_{max} of $Tr(Tr[PL](W))$ with the initial condition $J(I)$ can be computed by the Local Algorithm [3].

An informal description of the Local Algorithm is as follows. The Local Algorithm is a multi-stage process where the hybrid states derived at stage n are used to derive hybrid states at stage $n+1$. Suppose that a hybrid state $(V, \mathbf{p})$ was derived by the Local Algorithm at stage n. The Local Algorithm first uses all the advancing rules applicable at $(V, \mathbf{p})$ to derive a set of the candidate next hybrid states $(Z_1, \mathbf{q}_1)$, ..., $(Z_k, \mathbf{q}_k)$. For each $(Z_i, \mathbf{q}_i)$ the stationary rules applicable at $(Z_i, \mathbf{q}_i)$ are then used to form an ASP program $D(Z_i, \mathbf{q}_i)$. Suppose that the stable models of $D(Z_i, \mathbf{q}_i)$ are Y_1, ..., Y_m (for each Y_j we have that $Z_i \subseteq Y_j$). Then the set of the next hybrid states with the generalized position $\mathbf{q}_i$ is $(Y_1, \mathbf{q}_i)$, ..., $(Y_m, \mathbf{q}_i)$.

Theorem 3. *(Based on theorem 82, [3]) For a valid H-ASP PL program W and initial condition I of $Tr[PL](W)$, the result of the Local Algorithm applied to W produces $M'_{\max}$, which is the unique maximal stable model of $Tr(Tr[PL](W))$ with the initial condition $J(I)$.*

6 Conclusion

The extensions of ASP that allow external data searches include DLV^{DB} system [15], VI programs [6], GRINGO grounder [8]. In [14], however Redl notes that HEX programs [7] can be viewed as a generalization of these formalisms. We will thus only describe the relation of our work to HEX programs.

HEX programs are an extension of ASP programs that allow accessing external data sources via external atoms. The external atoms admit input and output variables, which after grounding, take predicate or constant values for the input variables, and constant values for the output variables. Through the external atoms and under the relaxed safety conditions, HEX programs can produce constants that don't appear in the original program. The main similarities with this approach and our approach are that both H-ASP PL and HEX programs allow the use of external data sources, and both support mechanisms for passing the information between external algorithms. The main differences are the following: (1) in H-ASP PL the information processed by the external algorithms represents a type of information that is different from the information contained in the logical atoms, (2) the H-ASP PL programs have a built-in support for producing BCP cDAGs, and (3) a H-ASP program underlying the H-ASP PL definitions has a unique maximal stable model under H-ASP stable model semantics. The latter two properties make the H-ASP PL very convenient for the problem of diagnosing automatic whitelisting.

A relation of our approach for solving diagnostic problems to that of Balduccini and Gelfond (see [1]) was discussed in the beginning of this paper. We think that developing an approach similar to that of Balduccini and Gelfond for H-ASP is an interesting problem for future work.

In this paper we have discussed the use of H-ASP to diagnose failures of the automatic whitelisting system for Google's Dynamic Remarketing Ads. The software, which we discuss in this paper was used to diagnose and fix failures of the automatic whitelisting for many advertisers over a time interval of several months. Whereas the time needed to diagnose a single failure without the software was 30–60 min, the time needed to diagnose a single failure using the software was 1–3 min. The declarative nature of the H-ASP PL program made it easy to update the software so as to reflect multiple changes to automatic whitelisting system that occurred over time.

References

1. Balduccini, M., Gelfond, M.: Diagnostic reasoning with a-prolog. TPLP **3**(4–5), 425–461 (2003)
2. Baral, C.: Knowledge Representation, Reasoning and Declarative Problem Solving. Cambridge University Press, Cambridge (2003)
3. Brik, A.: Extensions of Answer Set Programming. Ph.D. thesis, UC San Diego (2012)
4. Brik, A., Remmel, J.B.: Hybrid ASP. In: Gallagher, J.P., Gelfond, M. (eds.) ICLP (Technical Communications). LIPIcs, vol. 11, pp. 40–50. Schloss Dagstuhl - Leibniz-Zentrum fuer Informatik (2011)
5. Brik, A., Remmel, J.B.: Computing a finite horizon optimal strategy using hybrid ASP. In: NMR (2012)
6. Calimeri, F., Cozza, S., Ianni, G.: External sources of knowledge and value invention in logic programming. Ann. Math. Artif. Intell. **50**(3–4), 333–361 (2007)
7. Eiter, T., Ianni, G., Schindlauer, R., Tompits, H.: A uniform integration of higher-order reasoning and external evaluations in answer-set programming. In: Kaelbling, L.P., Saffiotti, A. (eds.) Proceedings of the Nineteenth International Joint Conference on Artificial Intelligence, IJCAI 2005, Edinburgh, Scotland, UK, 30 July–5 August 2005, pp. 90–96. Professional Book Center (2005)
8. Gebser, M., Kaufmann, B., Kaminski, R., Ostrowski, M., Schaub, T., Schneider, M.T.: Potassco: the potsdam answer set solving collection. AI Commun. **24**(2), 107–124 (2011)
9. Gelfond, M., Leone, N.: Logic programming and knowledge representation - the a-prolog perspective. Artif. Intell. **138**(1–2), 3–38 (2002)
10. Lifschitz, V.: Action languages, answer sets and planning. In: Apt, K.R., Marek, V.W., Truszczynski, M., Warren, D.S. (eds.) The Logic Programming Paradigm: A 25-Year Perspective, pp. 357–373. Springer, Heidelberg (1999)
11. Marek, V.W., Remmel, J.B.: Set constraints in logic programming. In: Lifschitz, V., Niemelä, I. (eds.) LPNMR 2004. LNCS (LNAI), vol. 2923, pp. 167–179. Springer, Heidelberg (2003)
12. Marek, V.W., Truszczynski, M.: Stable models and an alternative logic programming paradigm. In: Apt, K.R., Marek, V.W., Truszczynski, M., Warren, D.S. (eds.) The Logic Programming Paradigm: A 25-Year Perspective, pp. 375–398. Springer, Heidelberg (1999)
13. Niemelä, I.: Logic programs with stable model semantics as a constraint programming paradigm. Ann. Math. Artif. Intell. **25**(3–4), 241–273 (1999)
14. Redl, C.: Answer Set Programming with External Sources: Algorithms and Efficient Evaluation. Ph.D. thesis, Vienna University of Technology (2015)
15. Terracina, G., Leone, N., Lio, V., Panetta, C.: Experimenting with recursive queries in database and logic programming systems. TPLP **8**(2), 129–165 (2008)

Performance Tuning in Answer Set Programming

Matthew Buddenhagen[✉] and Yuliya Lierler

University of Nebraska at Omaha, Omaha, Nebraska, USA
{mbuddenhagen,ylierler}@unomaha.edu

Abstract. Performance analysis and tuning are well established software engineering processes in the realm of imperative programming. This work is a step towards establishing the standards of performance analysis in the realm of answer set programming – a prominent constraint programming paradigm. We present and study the roles of human tuning and automatic configuration tools in this process. The case study takes place in the realm of a real-world answer set programming application that required several hundred lines of code. Experimental results suggest that human-tuning of the logic programming encoding and automatic tuning of the answer set solver are orthogonal (complementary) issues.

1 Introduction

Performance analysis, profiling, and tuning are well established software engineering processes in the realm of imperative programming. Performance analysis tools – profilers – collect and analyze memory usage, utilization of particular instructions, or frequency and duration of function calls. This information aids programmers in the performance optimization of code. Profilers for imperative programming languages have existed since the early 1970s, and the methodology of their design as well as their usage is well understood. The situation changes when we face constraint programming paradigms.

Answer set programming (ASP) [12,13] is a prominent representative of constraint programming. In ASP, the tools for processing problem specifications, or *encodings*, are called (answer set) *solvers*. The crucial difference between the imperative and constraint programming paradigms exemplified by ASP, is that, in the latter, the connection between the encoding and solver's execution is very subtle. Consequently, performance analysis methods that matured within imperative programming are not applicable to constraint programming. In addition, the following observations apply: (i) specified problems in constraint programming paradigms are often NP complete and commonly result in significant computational effort by solvers, (ii) there are typically a variety of ways to encode the same problem, (iii) solvers offer different heuristics, expose numerous parameters, and their running time is sensitive to the configuration used.

We would like to thank Joshua Irvin, Marius Lindauer, Peter Schüller, Benjamin Susman, Miroslaw Truszczynski, and Victor Winter for valuable discussions related to this paper as well as anonymous reviewers for their comments.

F. Calimeri et al. (Eds.): LPNMR 2015, LNAI 9345, pp. 186–198, 2015.
DOI: 10.1007/978-3-319-23264-5_17

In this work, we undertake a case study towards outlining methodology of performance analysis in constraint programming. The case study takes place in the realm of a real-world answer set programming application that required several hundred lines of code. To the best of our knowledge, this is the first effort of its kind. Earlier efforts include the work by Gebser et al. [5] and [3], who present a careful analysis of performance tuning for the *n-queens* and *ricochet robots* problems, respectively. These problems are typically modeled within a page in ASP. Parsing is one of the important tasks in natural language processing. Lierler and Schüller [11] developed an ASP-based natural language parser called ASPCCGTK. The focus of this work is the performance tuning process during the development of ASPCCGTK. The original design of the parser was based on the observation that the construction of a parse tree for a given English sentence can be seen as an instance of a planning problem. System ASPCCGTK version 0.1 (ASPCCGTK-0.1) and ASPCCGTK version 0.2 (ASPCCGTK-0.2) vary only in how specifications of the planning problem are stated, while the constraints of the problem remain the same. Yet, the performance of ASPCCGTK-0.1 and ASPCCGTK-0.2 differs significantly for longer sentences. The way from ASPCCGTK-0.1 to ASPCCGTK-0.2 comprised 20 encodings, and along that way, grounding size and solving time were the primary measures directing the changes in the encodings. Rewriting suggestions by Gebser et al. [5] guided the ASPCCGTK encodings tuning.

The goal of present paper is threefold. First, this is an effort to reconstruct and document the "20-encodings" way from ASPCCGTK-0.1 to ASPCCGTK-0.2. Second, by undertaking this effort we will make a solid step toward outlining a performance analysis methodology for constraint programming. Third, we study the question of how tuning solver parameters by means of automatic configuration tools [10] effects the performance of the studied encodings. The last question helps us understand the placement of such tools on the performance analysis map in constraint programming. Despite the fact that changing a solver's settings may substantially influence its performance, it is common to only consider the performance of a solver's default configuration. Yet, it is unclear whether the best performing encoding when using a solver's default configuration would remain the best with respect to a tuned solver configuration. Silverthorn et al. [14] performed a case study that estimated the effect of parameter tuning as well as portfolio solving approach exemplified by CLASPFOLIO [6] on performance of solvers in context of three applications. A part of the current study is a logical continuation of that effort. In summary, this paper provides experimental evidence to support the validity of a performance tuning approach that first relies on the default solver settings while browsing the encodings and second tunes the solver's parameters on the best encoding to gain a better performing solution.

The outline of the paper follows: We start with a review of basic answer set programming and modeling concepts. We then present the process of performance tuning undertaken in ASPCCGTK. We review automatic configuration and present the details of the experimental analysis performed. Last, we provide the conclusions based on the experimental and analytic findings of this work.

2   Answer Set Programming and Modeling Guidelines

Answer set programming [12,13] is a declarative programming formalism based on the answer set semantics of logic programs [8]. The concept of ASP is to first represent a given problem by a program whose answer sets correspond to solutions. Second, a solver is used to generate answer sets for this program. Unlike imperative programming, where programs specify how to find a solution from given inputs, an ASP program encodes a specification of the problem itself. The ASP system comprises two tools: grounder and solver. In this work we use solver CLASP[1] [7] and its front-end grounder GRINGO [4].

Atoms and rules are basic elements of the ASP language, and a typical logic programming rule has the form of a Prolog rule. For instance, the program

$$p.$$
$$q \leftarrow p, \; not \; r.$$

is composed of such rules. This program has one answer set $\{p, q\}$. In a rule, the right hand side of an arrow is called the *body* of a rule, the left hand side is called the *head*. A rule whose body is empty is called a *fact*. The first rule of the program above is a fact. Intuitively, facts are always part of any program's answer set. In addition to Prolog rules, GRINGO also accepts rules of other kinds – "choices", "constraints" and "aggregates". For example, rule

$$\{p, q, r\}.$$

is a choice rule. Answer sets of this one-rule program are arbitrary subsets of the atoms p, q, r. A *constraint* is a rule with an empty head that encodes a condition on answer sets. For instance, the constraint $\leftarrow p, \; not \; q.$ eliminates answer sets that include p and do not include q.

The grounder GRINGO allows the user to specify large programs in a compact way, using rules with schematic variables and other abbreviations. GRINGO takes a program "with abbreviations" as an input and produces its propositional (ground) counterpart by using an "intelligent instantiation" procedure to produce propositional program that preserves the answer sets of original program. The program is then processed by the solver CLASP, which finds its answer sets. The inference mechanism of CLASP is related to propositional satisfiability (SAT) solvers [7].

We do not expect the reader to be familiar with the concept of an answer set. For the purpose of this paper, it is sufficient to know that answer sets are special ground atom subsets of the given logic program.

A common ASP practice is to devise a generic problem encoding that can be coupled with a specific problem instance to produce a solution. A problem instance typically consists of facts built from atoms of a particular predicate signature that we call an *input* signature. Dedicated predicate symbols in a generic encoding are meant to encode the solution, and we call the set composed of these predicate symbols an *output* signature. Sometimes it is important

---

[1] http://potassco.sourceforge.net/.

to distinguish between logic programs that encode problem specifications and those that encode a problem instance. In these cases, we refer to the former as *e-programs* and the latter as *i-programs*. To illustrate these ASP concepts, consider sample *graph coloring* problem:

> *A 3-coloring of a graph is a labeling of its vertexes with at most 3 colors such that no two vertexes sharing the same edge have the same color.*

An ASP e-program

$$\Pi_{color} \left| \begin{array}{l} color(1). \quad color(2). \quad color(3). \\ \{c(V,I)\} \leftarrow vtx(V), \; color(I). \\ \leftarrow c(V,I), \; c(V,J), \; I < J, \; vtx(V), \; color(I), \; color(J). \\ \leftarrow c(V,I), \; c(W,I), \; vtx(V), \; vtx(W), \; color(I), \; edge(V,W). \\ \leftarrow not\; c(V,1), \; not\; c(V,2), \; not\; c(V,3), vtx(V). \end{array} \right.$$

encodes a generic solution to this problem. The first three facts of the encoding specify that there are three distinct colors: 1, 2 and 3. A choice rule in line two states that each vertex V may be assigned some colors. The third line says it is impossible for a vertex to be assigned two colors. The fourth line says that two adjacent vertexes may not be assigned the same color. The last line states that every vertex must be assigned a color. Predicate signature $\{c\}$ is an output signature of program Π_{color}. Predicate signature $\{edge, vtx\}$ is an input signature so that an i-program has the following form for a given graph (V, E)

$$\begin{array}{ll} vtx(v). & (v \in V) \\ edge(v,w). & (\{v,w\} \in E) \end{array}$$

The union of any problem instance and program Π_{color} will result in a program whose answer sets encode 3-coloring of a graph.

Gebser et al. [5] outline the "hints on modeling" in ASP that follow:

1. Keep the grounding compact:
   (i) If possible, use aggregates; (ii) Try to avoid combinatorial blow-up; (iii) Project out unused variables; (iv) But don't remove too many inferences!
2. Add additional constraints to prune the search space:
   (i) Consider special cases; (ii) Break symmetries; (iii) Test whether the additional constraints really help
3. Try different approaches to model the problem
4. It (still) helps to know the systems:
   (i) GRINGO offers options to trace the grounding process; (ii) CLASP offers many options to configure the search

To the best of our knowledge, this is the prime account of guidelines for performance tuning in ASP. We call this list *Performance Guidelines*.

3   ASPCCGTK and Human-Driven ASP Performance Tuning

Lierler and Schüller [11] describe parts of the ASP-based natural language parser ASPCCGTK encoding. The ASPCCGTK website – http://www.kr.tuwien.ac.at/staff/former_staff/ps/aspccgtk/ – contains the complete application code. Versions ASPCCGTK-0.1 and ASPCCGTK-0.2 differ only in how specifications of the parsing task are stated, but the difference in performance of these encodings is significant. The way from ASPCCGTK-0.1 to ASPCCGTK-0.2 is comprised of 20 manually generated versions. The Performance Guidelines items 1 and 2 guided the way in considering the various encodings.

We now enumerate the program rewriting techniques that were used to tune ASPCCGTK. We start by introducing a concept of "output-equivalent" programs, which provides an important semantic property to capture a broad class of useful rewriting techniques. We conjecture that most of the ASPCCGTK encodings are output-equivalent. We believe that a future study of output-equivalent rewriting techniques will allow the rewriting-based tuning process (stemming from items 1 and 2 of Performance Guidelines) to be automated to a large extent. We conclude this section by presenting the historical ASPCCGTK encoding tree and the details of the tuning methodology used in the process. The encoding tree presents the details on the evolution of the ASPCCGTK.

Programs Π_1 and Π_2 are called *strongly equivalent* if for any program Π, answer sets of $\Pi \cup \Pi_1$ and $\Pi \cup \Pi_2$ coincide [2]. Strong equivalence was introduced to formalize the semantic properties of techniques that could be used in optimizing ASP code. In practical settings, the concept of strong equivalence is rather restrictive. For example, transformations on programs often involve changing the predicate signature, and strong equivalence is inadequate to capture such transformations.

We introduce the notion of "output-equivalent" programs to cope with the shortcomings of strong equivalence. Given a logic program Π, by $i(\Pi)$ and $o(\Pi)$ we denote their input and output signatures respectively. For a set X of atoms and a set of predicate symbols P, by $X_{|P}$ we denote the subset of X that contains all atoms in X whose predicate symbol is in P. For instances, $\{q(a,b),p(a),p(b),r(X)\}_{\{r\}} = \{r(X)\}$. We say that e-programs Π and Π' are output-equivalent if (i) their input and output signatures coincide and (ii) for any i-program I in their input signature, any answer set X of $I \cup \Pi$ is such that there is an answer set X' of $I \cup \Pi'$ and $X_{|o(\Pi)} = X'_{|o(\Pi)}$, and vice versa. In other words, both e-programs "agree" on the atoms in the output signature with respect to the same input. Output-equivalence relates to *uniform* equivalence [2].

We now present the ASP "code-change" classification that is then used to construct the ASPCCGTK encoding tree. In ASPCCGTK tree, each transition is marked by the kind of rewrite applied to the parent encoding. We conjecture that all rewriting techniques but one, called "output signature change", result in output-equivalent programs. It is a direction of future work to generally describe the presented rewriting techniques and formally claim that such rewritings are output-equivalence preserving.

Concretion ($\mathcal{C}$) replaces overly general rules by their effectively used, partial instantiations. For example, consider e-program

$$q(X, Y) \leftarrow p(X),\ p(Y)$$
$$u(X) \leftarrow q(X, X), \tag{1}$$

whose input signature is $\{p\}$ and output signature is $\{u\}$. Using concretion on (1) will result in program

$$q(X, X) \leftarrow p(X),\ p(X).$$
$$u(X) \leftarrow q(X, X).$$

The latter program will normally result in a smaller grounding.

Projection[2] ($\mathcal{P}$) reduces the number of schematic variables in a rule so that a fewer number of ground instances is produced. Consider e-program

$$u(X) \leftarrow p(X, V),\ q(X, Y, Z, 0), r(Z, W), \tag{2}$$

whose input signature is $\{p, q, r\}$ and output signature is $\{u\}$. One way to apply projection to this program results in

$$u(X) \leftarrow p(X, W),\ q\_new(X, Z), r(Z, W).$$
$$q\_new(X, Z) \leftarrow q(X, Y, Z, 0). \tag{3}$$

Simplification ($\mathcal{S}$) The idea of this technique is to reduce the number of rules, particularly constraints, by eliminating the rules that are "entailed" by the rest of a program. For instance, consider e-program

$$\{u(X)\} \leftarrow p(X).$$
$$\{v(X)\} \leftarrow q(X).$$
$$\leftarrow p(X),\ q(X).$$
$$\leftarrow u(X),\ v(X),$$

whose input signature is $\{p, q\}$ and output signature is $\{u, v\}$. By simplification we may eliminate the last rule of this program.

Equivalence ($\mathcal{E}$) replaces some rules of the program by strongly equivalent rules. For instance, a program

$$\{u(X, Y)\} \leftarrow p(X),\ q(Y)$$
$$\leftarrow u(X, Y),\ u(X, Y'),\ Y \neq Y'$$

is strongly equivalent to program $\{u(X, Y) : q(Y)\}1 \leftarrow p(X)$.

Auxiliary Signature Reduction ($\mathcal{A}$) reduces the program's signature by reformulating problem specifications by means of fewer predicates. For instance, reformulating program (3) as (2) will give us such effect.

Output Signature Change ($\mathcal{O}$) changes the output signature of a program to allow different sets of predicates to encode the solution.

---

[2] Terms Concretion and Projection were coined by Gebser et al. [5].

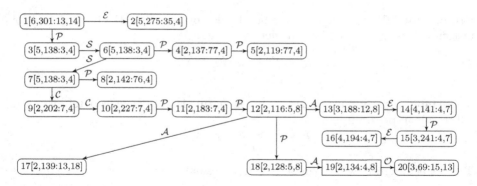

Fig. 1. ASPCCGTK encodings tree.

Figure 1 presents the relations between the 20 encodings considered on the way from ASPCCGTK-0.1 to ASPCCGTK-0.2. Each node in this tree represents an ASPCCGTK encoding and is annotated by five numbers. The first number is the encoding id and the others are discussed later in the section along with the tuning methodology used to transition from one encoding to another. An arrow in the tree suggests that an encoding of a "child" node is a modification of its "parent" node encoding. For instance, encodings 2 and 3 are both modifications of encoding 1. Each arrow is annotated by a tag corresponding to the technique used to obtain the new encoding. We followed the practice of making the smallest possible change per revision. For example, when technique $\mathcal{A}$ was used then no more than one auxiliary predicate was eliminated from the encoding. ASPCCGTK-0.1 comprises encoding 1. Encoding 19 was identified as the "winner" and is the designated encoding ASPCCGTK-0.2.

A set of 30 problem instances, randomly selected from the Penn Treebank[3], was used to benchmark each ASPCCGTK encoding. Following parameters were used to evaluate the quality of each encoding: (i) number of time or memory outs (3000 sec. timeout), (ii) average ground size, (iii) average solving time (default configuration of CLASP v 2.0.2), (iv) average grounding time (default configuration of GRINGO). In Fig. 1, each encoding id is annotated by four numbers $[o,s{:}g,z]$, where o is the total number of timeouts/memory outs, s is the average solving time (in seconds; on instances that did not timeout/memoryout), g is the average grounding time (in seconds; on instances that did not timeout/memoryout)), and z and 10^5 are factors relating to the average number of ground rules reported by CLASP. The last number provides the relative size of ground instances produced by GRINGO. These numbers were obtained in experiments using a Xeon X5355 @ 2.66GHz CPU.

The rules of thumb used in evaluating which encoding is better follow:

1. if number of time or memory outs of encoding E exceeds these of encoding E' then E' is a better encoding, otherwise

---

[3] http://www.cis.upenn.edu/~treebank/.

2. if cumulative average grounding and solving time of E exceeds that of E' then E' is a better encoding, otherwise
3. if grounding size of E exceeds that of E' then E' is a better encoding.

These rules were followed "softly" during the tuning process. For instance, encoding 19 is deemed to be the best, based on solver performance, even though the rules above suggest that 12 is the better encoding.

4   Automatic Algorithm Configuration and Tuning

Performance of answer set solvers greatly depends on their parameters-settings. In automatic algorithm configuration, the tuner evaluates the various parameter settings of the system in question and suggests an optimized configuration. Formally, the algorithm configuration problem can be formulated as follows: given a parametrized (target) algorithm $\mathcal{A}$, a set of problem instances (inputs) I, and a cost metric c, find parameter settings of $\mathcal{A}$ that minimize c on I. A *parameter-setting* is a name-value pair (p, v), where p is a parameter name and v is a value. A *configuration* is a set of parameter-settings. By $\mathcal{A}(\mathcal{P}, I)$, we denote an execution of algorithm $\mathcal{A}$ on instance I given parameter-settings $\mathcal{P}$. The cost metric c is often the runtime required to solve a problem instance, yet other factors such as solution quality maybe included. Various tools for solving the algorithm configuration problem have been proposed in the literature. System SMAC[4] [9] is a representative of such tools, and is based on the sequential model-based algorithm configuration method. Other such systems include PARAMILS [10] (precursor of SMAC) and GGA [1].

The rules of thumb listed in the end of Sect. 3 intuitively make sense, but given the disjointness of problem specifications from solving technology, there is no reason to believe that these rules achieve the best result in practice. Lierler and Schüller [11] and Silverthorn et al. [14] report that after applying automatic configuration tool PARAMILS to CLASP on the best encoding, the tuned version of CLASP outperformed the default version by a factor of 5. This observation raises the question: if CLASP were tuned on each encoding, would we still find 19 to be the best performing encoding as we described in Sect. 3? This is the question that we analyze in the rest of this paper.

We start by using the automatic configuration system SMAC version 2.06.01 to tune CLASP for each ASPCCGTK encoding. SMAC is susceptible to over-tuning. To account for this possibility, SMAC accepts both a training set of instances and a validation set. Upon reaching the end of the user-specified training time limit, SMAC uses the learned parameterization to execute a solver with found parameters on each instance of the validation set, and reports slower of the two execution metrics (one on the training set and another on the validation set) as its final result. To make final comparison of the performance of tuned versions of CLASP versus its default settings, we used a so-called held-out set of instances.

---

[4] http://www.cs.ubc.ca/labs/beta/Projects/SMAC/.

To create our pool of problem instances for SMAC, we classified the Penn Treebank instances (sentences) by word count into five word intervals, and restricted our selections to sentences having between 6 and 25 words. Our choice of boundaries was based on the previous analysis of ASPCCGTK. The time spent by ASPCCGTK parsing sentences with less than 6 words was negligible while there was marked increase in the number of solver timeouts for sentences with more than 25 words. To ensure an even distribution across the instance classes, we randomly selected an equal number of sentences from each class when creating our three disjoint test sets: a held-out set of 60 instances, a training set of 300 instances, and a validation set of 100 instances.

We used SMAC with its default setting for all but four parameters, whose values and snippets follow:

- *deterministic* is set to True. This parameter governs whether or not the target algorithm $\mathcal{A}$ is treated as deterministic. When set to True, SMAC will never execute $\mathcal{A}(\mathcal{P}, I)$ twice for any configuration $\mathcal{P}$ and instance I.
- *cutoffTime* is set to 300 seconds. Thus CPU time limit is 300 seconds for an individual target algorithm run $\mathcal{A}(\mathcal{P}, I)$.
- *wallclock-limit* is set to 480000 seconds (5.56 days). It instructs SMAC to terminate after using up a given amount of wall-clock time.
- *run-obj* is set to RUNTIME. It specifies to SMAC that the objective type that we are optimizing for is runtime.

Each execution of SMAC is non-deterministic. To account for this, performing several parallel runs is recommended by its developers. For each encoding, we executed ten instances of the SMAC tuning process and chose the best-performing configuration. The ten instances were run in parallel on independent CPU cores of a local resource cluster.

When using SMAC, the target algorithm is typically executed by way of a wrapper application. At a minimum, the wrapper implements the SMAC interface contract and calls the target algorithm with the specified parameter set, but may include other useful features such as coordinating parallel executions of SMAC. For our experiment we utilized PICLASP 1.0[5], a Python-based, SMAC-compatible wrapper for CLASP, developed by Marius Lindauer. PICLASP is explicitly compatible with the CLASP 2.1.x series, and for our experiment we used CLASP-2.1.3. To execute SMAC against the target algorithm, the algorithm's configurable parameters and their domains must be specified in parameter configuration space file. Lindauer provides a parameter configuration file for CLASP 2.1.x in PICLASP distribution, which we were able to use without modification. We implemented a small modification to PICLASP that allowed the use of separate training and validation sets, and we also created a benchmarking tool based on the Lindauer's CLASP wrapper class.

Our benchmarking tool, BENCHER, uses the CLASP wrapper class to conveniently invoke CLASP for each member of a benchmark set. When appended to the BENCHER command line, a CLASP parameter string is passed through

---

[5] http://www.cs.uni-potsdam.de/piclasp/.

to the solver, providing an easy way to test SMAC resultant parameterizations. If no additional parameters are provided, the solver operates in default mode. The CLASP result for each instance and the average performance is output as a JSON file to facilitate additional analysis if desired. Our modified version of PICLASP, BENCHER, the twenty encodings of ASPCCGTK, and our three instance test sets, can be downloaded from the University of Nebraska at Omaha web server: http://faculty.ist.unomaha.edu/ylierler/projects/smac-aspccg.zip.

Automatic tuning was conducted on a high-performance cluster node, powered by dual, 6-core, Intel Xeon X5660 2.8 GHz HT processors. Each CPU had 6 physical, hyper-threaded cores providing a total of 24 virtual cores. The node had a total of 256 GB memory and a 500 GB SAN partition allocated to the experiment. We had dedicated access to the node during our experiment and used a local resource management queue to execute parallel SMAC instances exclusively on the experimental node. For each of the twenty ASPCCGTK encodings we tested, we initiated a parallel execution of ten SMAC instances with each instance executing on a separate core, with 2GB of allocated memory. Each execution of the CLASP solver was allowed 300 seconds (5 minutes) of CPU time to complete, and executions exceeding 300 seconds were reported as Timeouts. This cutoff value was selected based on previous analysis of ASPCCGTK that showed the solver was typically able to complete in less than 300 seconds for sentences having 25 or fewer words. We selected a value that would allow adequate time for the solver to complete, but would not diminish too greatly the time SMAC spent probing the parameter space and formulating solutions.

The SMAC automatic configuration phase timeout was configured at 5.56 days. We chose this value based on preliminary executions of SMAC over increasing lengths of time and comparing the benchmark times of the resulting parameterizations. We chose encoding with id 8 for the initial trials because its default benchmark time was adjacent to the median default benchmark time. Initially, speedup was significant but degraded to marginal improvements over time in what approximated a logarithmic rate. We chose a time that was clearly within the region of diminishing returns to allow for variability in the encodings, and yield more consistent results. In practice, we spent 22 weeks to tune all of the ASPCCGTK encodings.

5   Experimental Results

Figure 2 graphs the default and auto-tuned solver execution times of each ASPC-CGTK encoding on the held-out set. The **Default** series represents average runtime using the default CLASP parameter values, and the **SMAC** series times were achieved using the optimized parameter configurations yielded by SMAC. Recall that the runtime variations in the default scenario are attributable to human-tuning efforts. Figure 2 reveals an observable relationship between human-tuned performance and auto-tuned performance. The results suggest that the performance optimization rules of thumb applied along the way from ASPCCGTK-0.1 to ASPCCGTK-0.2 remain valid, and automatic configuration of the solver compliments the human efforts as opposed to nullifying or subsuming their effects.

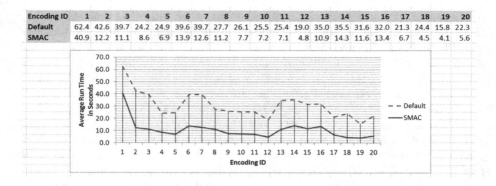

Encoding ID	1	2	3	4	5	6	7	8	9	10	11	12	13	14	15	16	17	18	19	20
Default	62.4	42.6	39.7	24.2	24.9	39.6	39.7	27.7	26.1	25.5	25.4	19.0	35.0	35.5	31.6	32.0	21.3	24.4	15.8	22.3
SMAC	40.9	12.2	11.1	8.6	6.9	13.9	12.6	11.2	7.7	7.2	7.1	4.8	10.9	14.3	11.6	13.4	6.7	4.5	4.1	5.6

Fig. 2. Default and SMAC benchmarks

We note that speedup in auto-tuning ranged from 1.53 to 5.40 and averaged 3.26. Generally speedup deviates around the average but remains relatively consistent except in extreme cases. The worst performing encoding resulted in the least speedup and three of the best performing encodings had above average speedup. Encoding 18 stands out as a significant outlier, having only the sixth best Default benchmark but the greatest speedup of 5.4.

Figures 3 and 4 present the results on the following inquiry. We reconsidered 30 problem instances that played the key role in human-tuning described in Sect. 3. Recall that they were randomly selected without regard to the complexity of these instances, and substantially differ from the instances in held-out set. This set of instances includes two sentences of length 42 and 52 words; six and eleven sentences comprised of 30 and 20 words respectively; and eleven sentences that range between 9 and 19 words. We collected the following statistics on Intel(R) Core(TM) i7-3770 CPU @ 3.40 GHz using the 30 afore mentioned instances: (i) runtime with a default parameterization of CLASP-2.1.3, (ii) runtime with the parameterization of CLASP-2.1.3 reported best for the encoding in question, (iii) runtime of the parameterization of CLASP-2.1.3 reported best for encoding 1. The timeout was set at 3600 seconds.

Figure 3 presents average run times (that also include time spent on grounding) for instances that did not time or memory out on any of the encodings given any CLASP configuration. Figure 4 presents the cumulative number of time and memory outs. Row **Original** presents the data stemming from the original human-tuning process, repeating some of the information presented in Fig. 1. Row **Rerun** presents the newly obtained numbers for the default parameterization of CLASP (note that the machine and CLASP version differ from **Original**). Row **SMAC** presents the data for the version of CLASP deemed to be best by SMAC for the respective encoding. Row **SMAC (Enc 1)** presents the data for the version of CLASP deemed to be best by SMAC for encoding 1. Presented data supports two major observations: (i) 30 random instances versus the instances of held-out set do not seem to change the outlook on which encoding is the "winner"; (ii) the parameter settings suggested by SMAC for the encoding 1 perform nearly as well as the encoding-specific SMAC parameterizations. The latter

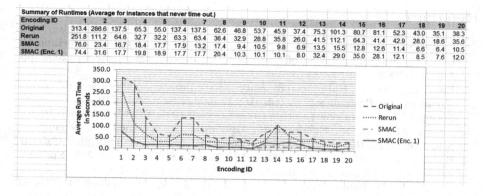

Summary of Runtimes (Average for instances that never time out.)

Encoding ID	1	2	3	4	5	6	7	8	9	10	11	12	13	14	15	16	17	18	19	20
Original	313.4	286.6	137.5	65.3	55.0	137.4	137.5	62.6	46.8	53.7	45.9	37.4	75.3	101.3	80.7	81.1	52.3	43.0	35.1	38.3
Rerun	251.8	111.2	64.6	32.7	32.2	63.3	63.4	36.4	32.9	28.8	35.8	26.0	41.5	112.1	64.3	41.4	42.9	28.0	18.6	35.6
SMAC	76.0	23.4	16.7	18.4	17.7	17.9	13.2	17.4	9.4	10.5	9.8	6.9	13.5	15.5	12.8	12.6	11.4	6.6	6.4	10.5
SMAC (Enc. 1)	74.4	31.6	17.7	19.8	18.9	17.7	17.7	20.4	10.3	10.1	10.1	8.0	32.4	29.0	35.0	28.1	12.1	8.5	7.6	12.0

Fig. 3. Original test set runtimes

Summary of Timeouts

Encoding ID	1	2	3	4	5	6	7	8	9	10	11	12	13	14	15	16	17	18	19	20
Original	6	5	5	2	2	5	5	2	2	2	2	3	4	3	4	2	2	2	2	3
Rerun	5	5	5	5	2	5	5	3	2	2	2	3	2	2	3	2	3	2	2	2
SMAC	3	2	2	2	1	2	2	2	2	1	2	1	2	2	2	3	2	1	1	2
SMAC (Enc. 1)	3	2	2	2	2	2	2	2	2	2	1	2	2	2	2	2	2	2	3	1

Fig. 4. Original test set timeouts

observation suggests that it is meaningful to use automatic configuration tuning early in the human-tuning process as a means to speed up the tuning process. It also makes sense to perform automatic configuration of parameters on the "winner", since the resulting solver optimization is presumably unique to the encoding in question.

6 Conclusions

Returning to our three stated objectives, we satisfied the first one by reconstructing and documenting the human effort to optimize the ASPCCGTK parser described in Sect. 3. The benchmark results clearly illustrate the effects due to the progressive application of output-equivalent rewriting techniques along the way from ASPCCGTK 0.1 to ASPCCGTK 0.2. Secondly, by achieving our first objective, we have validated the principles of ASP performance tuning as suggested by Gebser et al. [5], and established such a methodology within the context of a real world application. We believe that this provides a concrete basis for future work and the development of generally applicable automated ASP code rewriting-optimization tools. Finally, our efforts help clarify the role of automatic configuration tools within the context of constraint programming and performance optimization. Our results lead us to conclude that human-tuning of the ASP implementation and automatic tuning of the solver appear to be orthogonal issues, with auto-tuning having a linear affect on performance. Further, code-based optimization principles seem to take precedence over automatic configuration.

References

1. Ansótegui, C., Sellmann, M., Tierney, K.: A gender-based genetic algorithm for the automatic configuration of algorithms. In: Proceedings of the CP 2009, pp. 142–157 (2009)
2. Eiter, T., Fink, M.: Uniform equivalence of logic programs under the stable model semantics. In: Palamidessi, C. (ed.) ICLP 2003. LNCS, vol. 2916, pp. 224–238. Springer, Heidelberg (2003). http://dx.doi.org/10.1007/978-3-540-24599-5_16
3. Gebser, M., Jost, H., Kaminski, R., Obermeier, P., Sabuncu, O., Schaub, T., Schneider, M.: Ricochet robots: a transverse ASP benchmark. In: Cabalar, P., Son, T.C. (eds.) LPNMR 2013. LNCS, vol. 8148, pp. 348–360. Springer, Heidelberg (2013). http://dx.doi.org/10.1007/978-3-642-40564-8_35
4. Gebser, M., Kaminski, R., Kaufmann, B., Ostrowski, M., Schaub, T., Thiele, S.: A User's Guide to gringo, clasp, clingo, and iclingo (2010). http://potassco. sourceforge.net
5. Gebser, M., Kaminski, R., Kaufmann, B., Schaub, T.: Challenges in answer set solving. In: Balduccini, M., Son, T.C. (eds.) Logic Programming, Knowledge Representation, and Nonmonotonic Reasoning. LNCS, vol. 6565, pp. 74–90. Springer, Heidelberg (2011)
6. Gebser, M., Kaminski, R., Kaufmann, B., Schaub, T., Schneider, M.T., Ziller, S.: A portfolio solver for answer set programming: preliminary report. In: Delgrande, J.P., Faber, W. (eds.) LPNMR 2011. LNCS, vol. 6645, pp. 352–357. Springer, Heidelberg (2011)
7. Gebser, M., Kaufmann, B., Neumann, A., Schaub, T.: Conflict-driven answer set solving. In: Proceedings of 20th International Joint Conference on Artificial Intelligence (IJCAI 2007), pp. 386–392. MIT Press (2007)
8. Gelfond, M., Lifschitz, V.: The stable model semantics for logic programming. In: Proceedings of International Logic Programming Conference and Symposium, pp. 1070–1080 (1988)
9. Hutter, F., Hoos, H.H., Leyton-Brown, K.: Sequential model-based optimization for general algorithm configuration. In: Coello, C.A.C. (ed.) LION 2011. LNCS, vol. 6683, pp. 507–523. Springer, Heidelberg (2011)
10. Hutter, F., Hoos, H., Leyton-Brown, K., Stützle, T.: ParamILS: an automatic algorithm configuration framework. J. Artif. Intell. Res. (JAIR) **36**, 267–306 (2009)
11. Lierler, Y., Schüller, P.: Parsing combinatory categorial grammar via planning in answer set programming. In: Erdem, E., Lee, J., Lierler, Y., Pearce, D. (eds.) Correct Reasoning. LNCS, vol. 7265, pp. 436–453. Springer, Heidelberg (2012)
12. Marek, V., Truszczyński, M.: Stable models and an alternative logic programming paradigm. In: Apt, K.R., Marek, V.W., Truszczynski, M., Warren, D.S. (eds.) The Logic Programming Paradigm: a 25-Year Perspective, pp. 375–398. Springer Verlag, Berlin (1999)
13. Niemelä, I.: Logic programs with stable model semantics as a constraint programming paradigm. Ann. Math. Artif. Intell. **25**, 241–273 (1999)
14. Silverthorn, B., Lierler, Y., Schneider, M.: Surviving solver sensitivity: an asp practitioner's guide. In: International Conference on Logic Programming (ICLP) (2012). http://www.cs.utexas.edu/users/ai-lab/pub-view.php?PubID=127153

Enablers and Inhibitors in Causal Justifications of Logic Programs

Pedro Cabalar and Jorge Fandinno[(✉)]

Department of Computer Science, University of Corunna, A Corunna, Spain
{cabalar,jorge.fandino}@udc.es

Abstract. In this paper we propose an extension of logic programming (LP) where each default literal derived from the well-founded model is associated a justification represented as an algebraic expression. This expression contains both causal explanations (in the form of proof graphs built with rule labels) and terms under the scope of negation that stand for conditions that enable or disable the application of causal rules. Using some examples, we discuss how these new conditions, we respectively call *enablers* and *inhibitors*, are intimately related to default negation and have an essentially different nature from regular cause-effect relations. The most important result is a formal comparison to the recent algebraic approaches for justifications in LP: *Why-not Provenance* (WnP) and *Causal Graphs* (CG). We show that the current approach extends both WnP and CG justifications under the Well-Founded Semantics and, as a byproduct, we also establish a formal relation between these two approaches.

1 Introduction

The strong connection between Non-Monotonic Reasoning (NMR) and Logic Programming (LP) semantics for default negation has made possible that LP tools became nowadays an important paradigm for Knowledge Representation (KR) and problem-solving in Artificial Intelligence (AI). In particular, *Answer Set Programming* (ASP) [1,2] has raised as a preeminent LP paradigm for practical NMR with applications in diverse areas of AI including planning, reasoning about actions, diagnosis, abduction and beyond. The ASP semantics is based on *stable models* [3] and is also closely related to the other mainly accepted interpretation for default negation, *well-founded* semantics (WFS) [4]. One interesting difference between these two LP semantics and classical models (or even other NMR approaches) is that true atoms in LP must be founded or justified by a given derivation. These *justifications* are not provided in the semantics itself, but can be syntactically built in some way in terms of the program rules, as studied in several approaches [5–11].

Rather than manipulating justifications as syntactic objects, two recent approaches have considered multi-valued extensions of LP where justifications are treated as *algebraic* constructions: *Why-not Provenance* (WnP) [12] and *Causal Graphs* (CG) [13]. Although these two approaches present formal

© Springer International Publishing Switzerland 2015
F. Calimeri et al. (Eds.): LPNMR 2015, LNAI 9345, pp. 199–212, 2015.
DOI: 10.1007/978-3-319-23264-5_18

similarities, they start from different understandings of the idea of justification. On the one hand, WnP answers the query of why some literal L might hold by providing conjunctions of "hypothetical modifications" on the program that would allow deriving L. These modifications include rule labels, expressions like $not(A)$ with A an atom, or negations '¬' of the two previous cases. As an example, a justification for L like $r_1 \wedge not(p) \wedge \neg r_2 \wedge \neg not(q)$ means that the presence of rule r_1 and the absence of atom p would allow deriving L if both rule r_2 were removed and atom q were added to the program. If we want to explain why L *actually* holds, we have to restrict to justifications without '¬', that is, those without program modifications (which will be the focus of this paper).

On the other hand, CG-justifications start from identifying program rules as *causal laws* so that, for instance, $(p \leftarrow q)$ can be read as "event q *causes* effect p." Under this viewpoint, (positive) rules offer a natural way for capturing the concept of *causal production*, i.e. a continuous chain of events that has helped to cause or produce an effect [14,15]. The explanation of a true atom is made in terms of graphs formed by rule labels that reflect the ordered rule applications required for deriving that atom. These graphs are obtained by algebraic operations exclusively applied on the positive part of the program. Default negation in CG is understood as absence of cause and, consequently, a false atom has no justification.

The explanation of an atom A in CG is more detailed than in WnP, since the former contains graphs that correspond to all relevant proofs of A whereas in WnP we just get conjunctions that do not reflect any particular ordering among rule applications. However, as explained before, CG does not reflect the effect of default negation in a given derivation and, sometimes, this information is very valuable, especially if we want to answer questions of the form "why not."

To understand the kind of problems we are interested in, consider the following example. A drug d in James Bond's drink causes his paralysis p provided that he was not given an antidote a that day. We know that Bond's enemy, Dr. No, poured the drug:

$$p \leftarrow d, not\ a \qquad (1)$$

$$d \qquad (2)$$

In this case it is obvious that d causes p, whereas the absence of a just *enables* the application of the rule. Now, suppose we are said that Bond is daily administered an antidote by the MI6, unless it is a holiday h:

$$a \leftarrow not\ h \qquad (3)$$

Adding this rule makes a to become an *inhibitor* that prevents d to cause p. But suppose now that we are in a holiday, that is, fact h is added to the program (1)–(3). Then, the inhibitor a is *disabled* and d causes p again. However, we do not consider that the holiday h is a (productive) cause for Bond's paralysis p although, indeed, the latter counterfactually depends on the former: "had not been a holiday h, Bond would have not been paralysed." We will say that the fact h, which disables an inhibitor of d, is an *enabler* of d.

In this work we propose dealing with these concepts of enablers and inhibitors by augmenting CG justifications with a new negation operator '$\sim$' in the CG causal algebra. We show that this new approach, we call *Extended Causal Justifications* (ECJ) captures WnP justifications under the Well-founded Semantics and, as a byproduct, we establish a formal relation between WnP and CG.

The rest of the paper is structured as follows. The next section defines the new approach. Sections 3 and 4 explain through a running example the formal relations to CG and WnP, respectively. The next section discusses some related work and, finally, Sect. 6 concludes the paper.

2   Extended Causal Justifications (ECJ)

A *signature* is a pair $\langle At, Lb \rangle$ of sets that respectively represent *atoms* (or *propositions*) and *labels*. Intuitively, each atom in At will be assigned justifications built with rule labels from Lb. These justifications will be expressions that combine four different algebraic operators: a product '$*$' representing conjunction or joint causation; a sum '$+$' representing alternative causes; a non-commutative product '$\cdot$' that captures the sequential order that follows from rule applications; and a non-classical negation '$\sim$' which will precede inhibitors (negated labels) and enablers (doubly negated labels).

Definition 1 (Term). *Given a set of labels Lb, a* term, *t is recursively defined as one of the following expressions $t ::= l \mid \prod S \mid \sum S \mid t_1 \cdot t_2 \mid \sim t_1$ where $l \in Lb$, t_1, t_2 are in their turn terms and S is a (possibly empty and possibly infinite) set of terms. A term is* elementary *if it has the form $\sim\sim l$, $\sim l$ or l with $l \in Lb$ being a label.*   □

When $S = \{t_1, \ldots, t_n\}$ is finite we simply write $\prod S$ as $t_1 * \cdots * t_n$ and $\sum S$ as $t_1 + \cdots + t_n$. Moreover, when $S = \emptyset$, we denote $\prod S$ by 1 and $\sum S$ by 0, as usual, and these will be the identities of the product '$*$' and the addition '$+$', respectively. We assume that '$\cdot$' has higher priority than '$*$' and, in its turn, '$*$' has higher priority than '$+$'.

Definition 2 (Value). *A* (causal) value *is each equivalence class of terms under axioms for a completely distributive (complete) lattice with meet '$*$' and join '$+$' plus the axioms of Fig. 1. The set of (causal) values is denoted by $\mathbf{V}_{Lb}$.*   □

Note that $\langle \mathbf{V}_{Lb}, +, *, \sim, 0, 1 \rangle$ forms a pseudo-complemented, completely distributive (complete) lattice whose meet and join are, as usual, the product '$*$' and the addition '$+$'. Note also that all three operations, '$*$', '$+$' and '$\cdot$' are associative. Product '$*$' and addition '$+$' are also commutative, and they hold the usual absorption and distributive laws with respect to infinite sums and products of a completely distributive lattice. We say that a term is in *negation normal form* (NNF) if no operators are in the scope of negation '$\sim$'. Without loss of generality, we assume from now on that all terms are in NNF. The lattice order relation is defined as usual in the following way:

$$ t \leq u \qquad \text{iff} \qquad (t * u = t) \qquad \text{iff} \qquad (t + u = u) $$

pseudo-complement	De Morgan	excluded middle	appl. negation
$t * {\sim}t = 0$	${\sim}(t{+}u){=}({\sim}t * {\sim}u)$	${\sim}t + {\sim}{\sim}t{=}1$	${\sim}(t \cdot u){=}{\sim}(t * u)$
${\sim}{\sim}{\sim}t{=}{\sim}t$	${\sim}(t * u){=}({\sim}t{+}{\sim}u)$		

Associativity	Absorption	Identity	Annihilator
$t \cdot (u{\cdot}w) = (t{\cdot}u) \cdot w$	$t \quad = t + u \cdot t \cdot w$	$t = 1 \cdot t$	$0 = t \cdot 0$
	$u \cdot t \cdot w = t * u \cdot t \cdot w$	$t = t \cdot 1$	$0 = 0 \cdot t$

Indempotence	Addition distributivity	Product distributivity
$l \cdot l = l$	$t \cdot (u{+}w) = (t{\cdot}u) + (t{\cdot}w)$	$c \cdot d \cdot e = (c \cdot d) * (d \cdot e)$ with $d \neq 1$
	$(t + u) \cdot w = (t{\cdot}w) + (u{\cdot}w)$	$c \cdot (d * e) = (c \cdot d) * (c \cdot e)$
		$(c * d) \cdot e = (c \cdot e) * (d \cdot e)$

Fig. 1. Properties of the '$\sim$' and '$\cdot$'operators (c, d, e are terms without '+' and l is a label).

Consequently 1 and 0 are respectively the top and bottom elements with respect to the $\leq$ order relation.

Definition 3 (Labelled logic program). *Given a signature* $\langle At, Lb \rangle$, *a (labelled logic) program* P *is a set of rules of the form:*

$$r_i : \quad H \leftarrow B_1, \ldots, B_m, \text{ not } C_1, \ldots, \text{ not } C_n \qquad (4)$$

where $r_i \in Lb$ *is a label or* $r_i = 1$, H *(the* head *of the rule) is an atom and* B_i's *and* C_i's *(the* body *of the rule) are either atoms or terms.* □

When $n = 0$ we say that the rule is *positive*, furthermore, if $m = 0$ we say that the rule is a *fact* and omit the symbol '$\leftarrow$.' When $r_i \in Lb$ we say that the rule is *labelled*; otherwise $r_i = 1$ and we omit both r_i and ':'. By these conventions, for instance, an unlabelled fact A is actually an abbreviation of $(1 : A \leftarrow)$. A program P is *positive* when all its rules are positive, i.e. it contains no default negation. It is *uniquely labelled* when each rule has a different label or no label at all. In this paper, we will we assume that programs are uniquely labelled. Furthermore, for clarity sake, we also assume that, for every atom $A \in At$, there is an homonymous label $A \in Lb$, and that each fact A in the program actually stands for the labelled rule $(A : A \leftarrow)$. For instance, following these conventions, a possible labelled version for the James Bond's program could be program P_1 below:

$$r_1 : \quad p \leftarrow d, not\, a \qquad\qquad\qquad d$$
$$r_2 : \quad a \leftarrow not\, h \qquad\qquad\qquad\qquad h$$

where facts d and h stand for rules $(d : d \leftarrow)$ and $(h : h \leftarrow)$, respectively.

A *CP-interpretation* is a mapping $I : At \longrightarrow \mathbf{V}_{Lb}$ assigning a value to each atom. For interpretations I and J we say that $I \leq J$ when $I(A) \leq J(A)$ for each atom $A \in At$. Hence, there is a $\leq$-bottom interpretation $\mathbf{0}$ (resp. a $\leq$-top interpretation $\mathbf{1}$) that stands for the interpretation mapping each atom A to 0 (resp. 1). The value assigned to a negative literal $not\, A$ by an interpretation I, denoted as $I(not\ A)$, is defined as: $I(not\ A) \stackrel{\text{def}}{=} \sim I(A)$. Similarly, for a term t, $I(t) \stackrel{\text{def}}{=} [t]$ is the equivalence class of t.

Definition 4 (Reduct). *Given a program P and an interpretation I we denote by P^I the positive program containing a rule like*

$$r_i : H \leftarrow B_1, \ldots, B_m,\ I(not\ C_1), \ldots,\ I(not\ C_n) \tag{5}$$

per each rule of the form (4) in P.

Definition 5 (Model). *An interpretation J is a (causal) model of a rule like (5) iff*

$$\big(\ J(B_1) * \ldots * J(B_m) * I(not\ C_1) * \ldots * I(not\ C_n)\ \big) \cdot r_i\ \leq\ J(H)$$

and it is a model of P^I, written $J \models P^I$, iff it is a model of all rules in P^I. The operator $\Gamma_P(I)$ returns the least model of the positive program P^I.

Program P^I is positive and, as happens in standard logic programming, it also has a *least* causal model. Furthermore the operator Γ_P is anti-monotonic, and therefore Γ_P^2 is monotonic having a least fixpoint $\mathbb{L}_P$ and a greatest fixpoint $\mathbb{U}_P \stackrel{\text{def}}{=} \Gamma_P(\mathbb{L}_P)$ that respectively correspond to the justifications for true and for non-false atoms in the (standard) well-founded model (WFM), we denote W_P. A *query literal (q-literal)* L is either an atom A, its default negation '*not A*' or the expression '*undef A*' meaning that A is undefined.

Definition 6 (Causal well-founded model). *Given a program P, the* causal well-founded model $\mathbb{W}_P$ *is a mapping from q-literals to values s.t.*

$$\mathbb{W}_P(A) \stackrel{\text{def}}{=} \mathbb{L}_P(A) \quad \mathbb{W}_P(not\ A) \stackrel{\text{def}}{=} \sim \mathbb{U}_P(A) \quad \mathbb{W}_P(undef\ A) \stackrel{\text{def}}{=} \sim \mathbb{W}_P(A) * \sim \mathbb{W}_P(not\ A) \quad \square$$

As we will formalise below, when A is undefined in the standard well-founded model, $\mathbb{L}_P(A) \neq \mathbb{U}_P(A)$ and, thus, $\mathbb{W}_P(undef\ A) \neq 0$. Continuing with our running example, the causal WFM of program P_1 corresponds to $\mathbb{W}_{P_1}(d) = d$, $\mathbb{W}_{P_1}(h) = h$, $\mathbb{W}_{P_1}(a) = \sim h \cdot r_2$ and $\mathbb{W}_{P_1}(p) = (\sim\sim h * d) \cdot r_1 + (\sim r_2 * d) \cdot r_1$. Intuitively $(\sim\sim h * d) \cdot r_1$ means that the fact h (double negated label $\sim\sim h$) has enabled d (non negated label) to produce p by means of rule r_1. In its turn, $(\sim r_2 * d) \cdot r_1$ means that $d \cdot r_1$ would have been sufficient, had not been present r_2. Furthermore, $\mathbb{W}_{P_1}(a)$ means that a does not hold because the fact h (negated label $\sim h$) has inhibited rule r_2 to produce it. The following definitions formalise these concepts.

Let l be a label occurrence in a term t in the scope of $n \geq 0$ negation $\sim$ operators. We say that l is a *odd* or an *even* occurrence if n is odd or even, respectively. We further say that it is *strictly even* if it is even and $n > 0$.

Definition 7 (Justification). *Given a program P and a q-literal L we say that a term with no sums E is a* (sufficient causal) justification *for L iff $E \leq \mathbb{W}_P(L)$.*

Odd (resp. strictly even) labels[1] in E are called inhibitors *(resp.* enablers*) of E. A justification is said to be* inhibited *if it contains some inhibitor and it is said to be* enabled *otherwise.*   □

For instance, in our previous example, there are two justifications, $E_1 = (\sim\sim h * d)\cdot r_1$ and $E_2 = (\sim r_2 * d)\cdot r_1$, for atom p. Justification E_1 is enabled because it contains no inhibitors (in fact, E_1 is the unique real support for p). Moreover, h is an enabler in E_1 because it is strictly even (it is in the scope of double negation). On the contrary, E_2 is disabled because it contains the inhibitor r_2 (because it occurs here in the scope of one negation). Intuitively, r_2 has prevented $d\cdot r_1$ to become a justification of p. The next theorem shows that the literals satisfied by the standard WFM are precisely those ones containing at least one enabled justification in the causal WFM.

Theorem 1. *Let P be a program and W_P its (standard) well-founded model. A q-literal L holds with respect to W_P if and only if there is some enabled justification E of L, that is $E \leq \mathbb{W}_P(L)$ and E does not contain (odd) negative labels.*   □

Back to our example program P_1, as we had seen, atom p had an enabled justification $(\sim\sim h * d)\cdot r_1$. The same happens for atoms d and h whose respective justifications are just their own atom labels. Therefore, these three atoms hold in the standard WFM, W_{P_1}. On the contrary, as we discussed before, the only justification for a is inhibited by h, and thus, a does not hold in W_{P_1}. We can further check that a is false in W_{P_1} (it is not undefined) because literal *not a* holds, since $\mathbb{W}_{P_1}(not\,a) = \sim\sim h + \sim r_2$ provides two justifications, being the first one, $\sim\sim h$, enabled (it contains no inhibitors). The interest of an inhibited justification for a literal is to point out "potential" causes that have been prevented by some abnormal situation. In our case, the presence of $\sim h$ in $\mathbb{W}_{P_1}(a) = \sim h\cdot r_2$ points out that an exception h has prevented r_2 to cause a. When the exception is removed, the inhibited justification (after removing the inhibitors) becomes an enabled justification r_2 for a.

Theorem 2. *Let E be an inhibited justification of some atom A with respect to program P. Let Q be the result of removing from P all rules r_i whose labels are inhibitors in E. Similarly, let F be the result of removing those inhibitors $\sim r_i$ from E. Then F is an enabled justification of A with respect to Q.*   □

3   Relation to Causal Graph Justifications

We discuss now the relation between ECJ and CG approaches. Formally, ECJ extends CG causal terms by the introduction of the new negation operator '$\sim$'. Semantically, however, there are more differences than a simple syntactic extension. A first minor difference is that ECJ is defined in terms of a WFM, whereas

---

[1] We just mention labels, and not their occurrences because terms are in NNF and E contains no sums: having odd and even occurrences of a same label would mean that $E = 0$.

CG defines (possibly) several causal stable models. In the case of stratified programs, this difference is irrelevant, since the WFM is complete and coincides with the unique stable model. A second, more important difference is that CG exclusively considers productive causes in the justifications, disregarding additional information like the inhibitors or enablers from ECJ. As a result, a false atom in CG has *no justification* – its causal value is 0 because there was no way to derive the atom. For instance, in program P_1, the only CG stable model I just makes $I(a) = 0$ and we lose the inhibited justification $\sim h \cdot r_2$ (default r_2 could not be applied). True atoms like p also lose any information about enablers: $I(p) = d \cdot r_1$ and nothing is said about $\sim \sim h$. Another consequence of the CG orientation is that negative literals *not A* are never assigned a cause (different from 0 or 1), since they cannot be "derived" or produced by rules. In the example, we simply get $I(not\ a) = 1$ and $I(not\ p) = 0$.

To further illustrate the similitudes and differences between ECJ and CG, consider the following program P_2 capturing a variation of the Yale Shooting Scenario.

$$d_{t+1} : dead_{t+1} \leftarrow shoot_t,\ loaded_t,\ not\ ab_t \qquad \overline{loaded_0} \qquad load_1$$
$$l_{t+1} : loaded_{t+1} \leftarrow load_t \qquad\qquad\qquad\quad \overline{dead_0} \qquad water_3$$
$$a_{t+1} : ab_{t+1} \leftarrow water_t \qquad\qquad\qquad\qquad \overline{ab_0} \qquad shoot_8$$

plus the following rules corresponding inertia axioms

$$F_{t+1} \leftarrow F_t,\ not\ \overline{F}_{t+1} \qquad\qquad \overline{F}_{t+1} \leftarrow \overline{F}_t,\ not\ F_{t+1}$$

for $F \in \{loaded, ab, dead\}$. Atoms of the form $\overline{A}$ represent the strong negation of A and we assume we disregard models satisfying both A and $\overline{A}$. Atom $dead_9$ does not hold in the standard WFM of P_2, and so there is no CG-justification for it. Note here the importance of default reasoning. On the one hand, the default flow of events is that the turkey, Fred, continues to be alive when nothing threats him. Hence, we do not need a cause to explain why Fred is alive. On the other hand, shooting a loaded gun would normally kill Fred, being this a cause of its death. But, in this example, another exceptional situation – *water* spilled out – has *inhibited* this existing threat and allowed the world to flow as if nothing had happened (that is, following its default behaviour).

In the CG-approach, $dead_9$ is simply false by default and no justification is provided. However, a gun shooter could be "disappointed" since another conflicting default (shooting a loaded gun *normally* kills) has not worked. Thus, an expected answer for the shooter's question "why *not dead_9*?" is that $water_3$ broke the default, disabling d_9. In fact, ECJ yields the following inhibited justification for $dead_9$:

$$\mathbb{W}_{P_2}(dead_9) = (\sim water_3 * shoot_8 * load_1 \cdot l_2) \cdot d_9 \qquad\qquad (6)$$

meaning that $dead_9$ could not be derived because of inhibitor $water_3$ prevented the application of r_1 to cause the death of Fred. Moreover, according to

Theorem 2, if we remove fact $water_3$ (the inhibitor) from P_2, then for the new program P_3 we get:

$$\mathbb{W}_{P_3}(dead_9) = (shoot_8 * load_1 \cdot l_2) \cdot d_9 \qquad (7)$$

which is nothing else but the result of removing $\sim water_3$ from (6). In fact, the only CG stable model of P_3 makes this same assignment (7) which also corresponds to the causal graph depicted in Fig. 2. In the general case: CG-justifications intuitively correspond to enabled justifications after forgetting all the enablers. We formalise next the correspondence between CG and ECJ justifications.

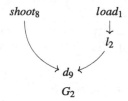

Fig. 2. Cause of $dead_9$ in program P_2.

Definition 8 (Causal values). *A CG term, t is a term with no negation '$\sim$'. CG values are the equivalence classes of CG terms under the axioms of Definition 2. The set of CG values is denoted by $\mathbf{C}_{Lb}$. We also define a mapping $\lambda^c : \mathbf{V}_{Lb} \longrightarrow \mathbf{C}_{Lb}$ from values into CG values in the following recursive way:*

$$\lambda^c(t) \stackrel{\text{def}}{=} \begin{cases} \lambda^c(u) \odot \lambda^c(w) & \text{if } t = u \odot v \text{ with } \odot \in \{+, *, \cdot\} \\ 1 & \text{if } t = \sim\sim l \text{ with } l \in Lb \\ 0 & \text{if } t = \sim l \text{ with } l \in Lb \\ l & \text{if } t = l \text{ with } l \in Lb \end{cases}$$

Note that we have assumed that t is in negation normal form. Otherwise $\lambda^c(t) \stackrel{\text{def}}{=} \lambda^c(u)$ where u is the equivalent term in negation normal form. □

Function λ^c maps every negated label $\sim l$ to 0 (which is the annihilator of both product '$*$' and application '$\cdot$' and the identity of addition '$+$'). Hence λ^c removes all the inhibited justifications. Furthermore λ^c maps every doubly negated label $\sim\sim l$ to 1 (which is the identity of both product '$*$' and application '$\cdot$'). Therefore λ^c removes all the enablers (i.e. doubly negated labels $\sim\sim l$) for the remaining (i.e. enabled) justifications.

Definition 9 (CG stable models). *Given a program P, a CG stable model is a mapping $\tilde{I} : At \longrightarrow \mathbf{C}_{Lb}$ from atoms to CG values such that there exists a fixpoint I of the operator Γ_P^2 satisfying $\lambda^c(I(A)) = \lambda^c(\Gamma_P(I(A)))$ and $\tilde{I}(A) \stackrel{\text{def}}{=} \lambda^c(I(A))$ for every atom A.* □

Theorem 3. *For any program P, the CG values (Definition 8) and the CG stable models (Definition 9) are exactly the causal values and causal stable models defined in [13].*   □

Theorem 3 shows that Definition 9 is an alternative definition of CG causal stable models. Furthermore, it settles that every causal model corresponds to some fixpoint of the operator Γ_P^2. Therefore, for every enabled justification there is a corresponding CG-justification common to all stable models. In order to formalise this idea we just take the definition of causal explanation from [16]. A graph of labels is a *causal graph* (c-graph) if it is a directed graph, transitively and reflexively closed. Furthermore we also define a one-to-one correspondence between c-graphs and causal values.

$$value(G) \overset{\text{def}}{=} \prod\{ \, v_1 \cdot v_2 \mid (v_1, v_2) \text{ is an edge of } G \, \}$$

Definition 10 (CG-justification). *Given an interpretation I we say that a c-graph G is a* (sufficient) *CG-justification for an atom A iff $value(G) \leq \tilde{I}(A)$.*   □

Note that mapping $value(\cdot)$ is a one-to-one correspondence and, thus, we can define $graph(v) \overset{\text{def}}{=} value^{-1}(v)$ for all $v \in \mathbf{C}_{Lb}$.

Theorem 4. *Let P be a program. For any enabled justification E of some atom A w.r.t. $\mathbb{W}_P$, i.e. $E \leq \mathbb{W}_P(A)$, there is a CG-justification $G \overset{\text{def}}{=} graph(\lambda^c(E))$ of A with respect to any stable model $\tilde{I}$ of P.*   □

As happens between the (standard) Well-founded and Stable Model semantics, the converse of Theorem 4 does not hold in general. For instance, let P_4 be the program consisting on the following rules:

$$r_1 : a \leftarrow not\, b \qquad r_2 : b \leftarrow not\, a, not\, c \qquad c \qquad r_3 : c \leftarrow a \qquad r_4 : d \leftarrow b, not\, d$$

The (standard) WFM of P_4 is two-valued and corresponds to the unique (standard) stable model $\{a, c\}$. Furthermore there are two causal explanations of c with respect to this unique stable model: the fact c and the pair of rules $r_1 \cdot r_3$. Note that when c is removed $\{a, c\}$ is still the unique stable model, but all atoms are undefined in the WFM. Hence, $r_1 \cdot r_3$ is a justification with respect to the unique stable model of the program, but not with respect to is WFM.

4   Relation to Why-Not Provenance

An evident similarity between ECJ and WnP is the use of an alternating fixpoint operator [17] which has been actually borrowed from WnP. However there are some slight differences. A first one is that we have incorporated from CG the non-commutative operator '·' which allows capturing, not only which rules justify a given atom, but also the dependencies among these rules. The second is the use of a *non-classical* negation '$\sim$' that is crucial to distinguish between productive causes and enablers, something that cannot be represented with the classical negation '¬' in WnP since double negation can always be removed. Apart from the interpretation of negation in both formalisms, there are other differences too. As an example, let us compare the justifications we obtain for $dead_9$ in program P_3. While for ECJ we obtained (7) (or graph G_2 in Fig. 2), the corresponding WnP justification has the form:

$$l_2 \wedge d_9 \wedge load_1 \wedge shoot_8 \wedge not(water_0) \wedge not(water_1) \wedge \ldots \wedge not(water_7) \quad (8)$$

A first observation is that the subexpression $l_2 \wedge d_9 \wedge load_1 \wedge shoot_8$ constitutes, informally speaking, a "flattening" of (7) (or graph G_2) where the ordering among rules has been lost. We get, however, new labels in the form of $not(A)$ meaning that atom A is required not to be a program fact, something that is not present in CG-justifications. For instance, (8) points out that *water* can not be spilt on the gun along situations $0, \ldots, 7$. Although this information can be useful for debugging (the original purpose of WnP) its inclusion in a causal explanation is obviously inconvenient from a Knowledge Representation perspective, since it explicitly *enumerates all the defaults* that were applied (no water was spilt at any situation) something that may easily blow up the (causally) irrelevant information in a justification.

An analogous effect happens with the enumeration of exceptions to defaults, like inertia. Take program P_5 obtained from P_2 by removing all the performed actions, i.e., facts $load_1$, $water_3$, and $shoot_7$. As expected, Fred will be alive, $\overline{dead_t}$, at any situation t by inertia. ECJ will assign no cause for $dead_t$, not even any inhibited one, i.e. $\mathbb{W}_P(\overline{dead_t}) = 1$ and $\mathbb{W}_P(dead_t) = 0$ for any t. However, there are many WnP justifications of $dead_t$ corresponding to *all the plans* for killing Fred in t steps. For instance, among others, all the following:

$$d_9 \wedge \neg not(load_0) \wedge r_2 \wedge \neg not(shoot_1) \wedge not(water_0)$$
$$d_9 \wedge \neg not(load_0) \wedge r_2 \wedge \neg not(shoot_2) \wedge not(water_0) \wedge not(water_1)$$
$$d_9 \wedge \neg not(load_1) \wedge r_2 \wedge \neg not(shoot_3) \wedge not(water_0) \wedge not(water_1) \wedge not(water_2)$$
$$\ldots$$

are WnP-justifications for $dead_9$. The intuitive meaning of expressions of the form $\neg not(A)$ is that $dead_9$ can be justified by adding A as a fact to the program. For instance, the first conjunction means that it is possible to justify $dead_9$ by adding the facts $load_0$ and $shoot_1$ and not adding the fact $water_0$. We will call these justifications, which contain a subterm of the form $\neg not(A)$, *hypothetical* in the sense that they involve some hypothetical program modification.

Definition 11 (Provenance values) *A* provenance term, *t is recursively defined as one of the following expressions* $t:: = l \mid \prod S \mid \sum S \mid \neg t_1$ *where* $l \in Lb$, t_1, t_2 *are in their turn provenance terms and S is a (possibly empty and possible infinite) set of provenance terms. Provenance values are the equivalence classes of provenance terms under the equivalences of the Boolean algebra. We denote by* $\mathbf{B}_{Lb}$ *the set of provenance values over Lb.* ☐

Informally speaking, with respect to ECJ, we have removed the application '·' operator, whereas product '∗' and addition '+' hold the same equivalences as in Definition 2 and negation '$\sim$' has been replaced by '$\neg$' from Boolean algebra. Thus, '$\neg$' is classical and satisfies all the axioms of '$\sim$' plus $\neg\neg t = t$. Note also that, we have followed the convention from [12] of using the symbols '$\wedge$' instead of '$*$' to represent the meet and '$\vee$' instead of '$+$' to represent the join when we write provenance formulae. We define a mapping $\lambda^p : \mathbf{V}_{Lb} \longrightarrow \mathbf{B}_{Lb}$ in the following recursive way:

$$
\lambda^p(t) \stackrel{\text{def}}{=} \begin{cases}
\lambda^p(u) \odot \lambda^p(w) & \text{if } t = u \odot v \text{ with } \odot \in \{+, *\} \\
\lambda^p(u) * \lambda^p(w) & \text{if } t = u \cdot v \\
\neg \lambda^p(u) & \text{if } t = \sim u \\
l & \text{if } t = l \text{ with } l \in Lb
\end{cases}
$$

Definition 12 (Provenance). *Given a program P the why-not provenance program $\mathfrak{P}(P) \stackrel{\text{def}}{=} P \cup P'$ where P' contains a labelled fact of the form of $(\sim not(A) : A)$ for each atom $A \in At$ not occurring in P as a fact. We will write $\mathfrak{P}$ instead of $\mathfrak{P}(P)$ when the program P is clear by the context. We denote by $Why_P(L) \stackrel{\text{def}}{=} \lambda^p(\mathbb{W}_{\mathfrak{P}}(L))$ the why-not provenance of a q-literal L.* □

Theorem 5. *For any program P, the provenance of a literal according to Definition 12 is equivalent to the provenance defined in [12].* □

Theorem 5 shows that the provenance of a literal can obtained from replacing the negation '$\sim$' by '$\neg$' and '$\cdot$' by '$*$' in the causal WFM of the augmented program $\mathfrak{P}$.

Theorem 6. *For any program P, a conjunction of literals D is a non-hypothetical WnP justification of some q-literal L, i.e. $D \leq Why_P(L)$ iff there is a justification E, i.e. $E \leq \mathbb{W}_P(L)$ s. t. $\lambda^p(E)$ is the result of removing labels of the form 'not(A)' from D.* □

Theorem 6 establishes a correspondence between non-hypothetical WnP-justifications and (flattened) ECJ justifications. In the case of hypothetical justifications, they are not directly captured by ECJ, but can be obtained using the augmented program $\mathfrak{P}$ as stated by Theorem 5. As a byproduct we establish a formal relation between WnP and CG.

Theorem 7. *Let P be a program and D be a non-hypothetical and enabled WnP-justification of some atom A, i.e. $D \leq Why_P(A)$. Then, for all CG stable model $\tilde{I}$ of P, there is some CG-justification G w.r.t. $\tilde{I}$ s.t. D contains all the vertices of G.* □

5   Related Work

There exists a vast literature on causal reasoning in Artificial Intelligence (AI). Papers on reasoning about actions and change [18–20] have been traditionally focused on using causal inference to solve representational problems (mostly, the frame, ramification and qualification problems) without paying much attention to the derivation of cause-effect relations. Perhaps the most established AI approach for causality is relying on *causal networks* [21] (See [22] for an updated version). In this approach, it is possible to conclude cause-effect relations like "*A* has been an actual cause of *B*" from the behaviour of structural equations by applying, under some *contingency* (an alternative model in which some values are fixed) the *counterfactual dependence* interpretation from [23]: "had *A* not happened, *B* would not have happened." Causal networks and ECJ differ in their final goals. While the former focuses on revealing a *unique* everyday-concept of causation (*actual causation*), the latter tries to provide precise definitions of *different concepts of causation*, leaving the choice of which concept corresponds to a particular scenario to the programmer. The approach of *actual causation* has also been followed in LP by [24,25].

As has been slightly discussed in the introduction, ECJ is also related to work by Hall [14,15], who has emphasized the difference between two types of causal relations: *dependence* and *production*. The former relies on the idea "that counterfactual dependence between wholly distinct events is sufficient for causation." The latter is characterised by being *transitive*, *intrinsic* (two processes following the same laws must be both or neither causal) and *local* (causes must be connected to their effects via sequences of causal intermediates). In this sense, WnP is more oriented to *dependence* while CG is mostly related to *production*. ECJ is a combination of the dependence-oriented approach from WnP and the production-oriented behaviour from CG.

Focusing on LP, our work obviously relates to explanations obtained from ASP debugging approaches [5–11]. The most important difference of these works with respect to ECJ, and also WnP and CG, is that the last three provide fully algebraic semantics in which justifications are embedded into program models. A formal relation between [11] and WnP was established in [26] and so, using Theorems 5 and 6, it can be directly extended to ECJ, but at the cost of flattening the graph information (i.e. losing the order among rules).

6   Conclusions

In this paper we have introduced a unifying approach that combines causal production with enablers and inhibitors. We formally capture inhibited justifications by introducing a "non-classical" negation '$\sim$' in the algebra of causal graphs (CG). A inhibited justification is nothing else but an expression containing some negated label. We have also distinguished productive causes from enabling conditions (counterfactual dependences that are not productive causes) by using a double negation '$\sim \sim$' for the latter. The existence of enabled justifications is a sufficient and necessary condition for the truth of a literal. Furthermore, our justifications capture, under the Well-founded semantics, both Causal Graph and Why-not Provenance justifications. As a byproduct we established a formal relation between these two approaches.

Using an example, we have also shown how to capture causal knowledge in the presence of dynamic defaults – those whose behaviour are not predetermined, but rely on some program condition – as for instance the inertia axioms. As pointed out by [27], causal knowledge is structured by a combination of *inertial laws* – how the world would evolve if nothing intervened – and *deviations* from these inertial laws. The importance of default knowledge has been widely recognised as a cornerstone of the problem of actual causation in [28, 29] among others.

Interesting issues for future study are incorporating enabled and inhibited justifications to the stable model semantics and replacing the syntactic definition in favour of a logical treatment of default negation, as done for instance with the Equilibrium Logic [30] characterisation of stable models. Other natural steps would be the consideration of syntactic operators for capturing more specific knowledge about causal information, like the influence of a particular event or label in a conclusion, and the representation of non-deterministic causal laws, by means of disjunctive programs and the incorporation of probabilistic knowledge. From a KR point of view, another interesting future line of study is to apply our semantics to other traditional examples of the actual causation literature like *short-circuits* and *switches* [15].

Acknowledgements. We are thankful to Carlos Damasio for his suggestions and comments on earlier versions of this work. We also thank the anonymous reviewers for their help to improve the paper. This research was partially supported by Spanish Project TIN2013-42149-P.

References

1. Niemelä, I.: Logic programs with stable model semantics as a constraint programming paradigm. Ann. Math. Artif. Intell. **25**, 241–273 (1999)
2. Marek, V.W., Truszczyński, M.: Stable models and an alternative logic programming paradigm. In: Apt, K.R., Marek, V.W., Truszczyński, M., Warren, D.S. (eds.) The Logic Programming Paradigm. Artificial Intelligence, pp. 375–398. Springer, Heidelberg (1999)
3. Gelfond, M., Lifschitz, V.: The stable model semantics for logic programming. In: Logic Programming: Proceedings of the Fifth International Conference and Symposium, vol. 2 (1988)
4. Van Gelder, A., Ross, K.A., Schlipf, J.S.: The well-founded semantics for general logic programs. J. ACM (JACM) **38**, 619–649 (1991)
5. Specht, G.: Generating explanation trees even for negations in deductive database systems. In: LPE 1993, pp. 8–13 (1993)
6. Pemmasani, G., Guo, H.-F., Dong, Y., Ramakrishnan, C.R., Ramakrishnan, I.V.: Online justification for tabled logic programs. In: Kameyama, Y., Stuckey, P.J. (eds.) FLOPS 2004. LNCS, vol. 2998, pp. 24–38. Springer, Heidelberg (2004)
7. Gebser, M., Pührer, J., Schaub, T., Tompits, H.: Meta-programming technique for debugging answer-set programs. In: Proceedings of the 23rd Conference on Artificial Inteligence (AAAI 2008) (2008)
8. Oetsch, J., Pührer, J., Tompits, H.: Catching the ouroboros: on debugging non-ground answer-set programs. Theory Pract. Logic Program. **10**, 513–529 (2010)
9. Schulz, C., Toni, F.: ABA-based answer set justification. In: TPLP 13(4-5-Online-Supplement) (2013)

10. Denecker, M., De Schreye, D.: Justification semantics: a unifiying framework for the semantics of logic programs. In: Proceedings of the LPNMR Workshop (1993)
11. Pontelli, E., Son, T.C., El-Khatib, O.: Justifications for logic programs under answer set semantics. Theory Pract. Logic Program. (TPLP) **9**, 1–56 (2009)
12. Viegas Damásio, C., Analyti, A., Antoniou, G.: Justifications for logic programming. In: Cabalar, P., Son, T.C. (eds.) LPNMR 2013. LNCS, vol. 8148, pp. 530–542. Springer, Heidelberg (2013)
13. Cabalar, P., Fandinno, J., Fink, M.: Causal graph justifications of logic programs. TPLP **14**, 603–618 (2014)
14. Hall, N.: Two concepts of causation. In: Collins, J., Hall, E.J., Paul, L.A. (eds.) Causation and Counterfactuals, pp. 225–276. MIT Press, Cambridge (2004)
15. Hall, N.: Structural equations and causation. Philos. Stud. **132**, 109–136 (2007)
16. Cabalar, P., Fandinno, J., Fink, M.: A complexity assessment for queries involving sufficient and necessary causes. In: Fermé, E., Leite, J. (eds.) JELIA 2014. LNCS, vol. 8761, pp. 297–310. Springer, Heidelberg (2014)
17. Van Gelder, A.: The alternating fixpoint of logic programs with negation. In: Proceedings of the eighth ACM SIGACT-SIGMOD-SIGART Symposium on Principles of Database Systems, pp. 1–10. ACM (1989)
18. Lin, F.: Embracing causality in specifying the indirect effects of actions. In: Mellish, C.S. (ed.)Proceedings of the International Joint Conference on Artificial Intelligence (IJCAI). Morgan Kaufmann, Montreal (1995)
19. McCain, N., Turner, H.: Causal theories of action and change. In: Proceedings of the AAAI 1997, pp. 460–465 (1997)
20. Thielscher, M.: Ramification and causality. Artif. Intell. J. **1–2**, 317–364 (1997)
21. Pearl, J.: Causality: Models, Reasoning, and Inference. Cambridge University Press, New York (2000)
22. Halpern, J.Y., Hitchcock, C.: Actual causation and the art of modeling. arXiv preprint arXiv:1106.2652 (2011)
23. Hume, D.: An enquiry concerning human understanding (1748) Reprinted by Open Court Press. LaSalle, IL (1958)
24. Meliou, A., Gatterbauer, W., Halpern, J.Y., Koch, C., Moore, K.F., Suciu, D.: Causality in databases. IEEE Data Eng. Bull. **33**, 59–67 (2010)
25. Vennekens, J.: Actual causation in CP-logic. TPLP **11**, 647–662 (2011)
26. Viegas Damásio, C., Analyti, A., Antoniou, G.: Justifications for logic programming. In: Cabalar, P., Son, T.C. (eds.) LPNMR 2013. LNCS, vol. 8148, pp. 530–542. Springer, Heidelberg (2013)
27. Maudlin, T.: Causation, counterfactuals, and the third factor. In: Collins, J., Hall, E.J., Paul, L.A. (eds.) Causation and Counterfactuals. MIT Press, Cambridge (2004)
28. Halpern, J.Y.: Defaults and normality in causal structures. In: Brewka, G., Lang, J. (eds.) Proceedings of the Eleventh International Conference on Principles of Knowledge Representation and Reasoning (KR 2008), pp. 198–208. AAAI Press (2008)
29. Hitchcock, C., Knobe, J.: Cause and norm. J. Philos. **11**, 587–612 (2009)
30. Pearce, D.: A new logical characterisation of stable models and answer sets. In: Dix, J., Przymusinski, T.C., Moniz Pereira, L. (eds.) NMELP 1996. LNCS, vol. 1216. Springer, Heidelberg (1997)

Efficient Problem Solving on Tree Decompositions Using Binary Decision Diagrams

Günther Charwat[✉] and Stefan Woltran

Institute of Information Systems, TU Wien, Wien, Austria
{gcharwat,woltran}@dbai.tuwien.ac.at

Abstract. Dynamic programming (DP) on tree decompositions is a well studied approach for solving hard problems efficiently. Usually, implementations rely on tables for storing information, and algorithms specify how tuples are manipulated during traversal of the decomposition. However, a bottleneck of such table-based algorithms is relatively high memory consumption. Binary Decision Diagrams (BDDs) and related concepts have been shown to be very well suited to store information efficiently. While several techniques have been proposed that combine DP with efficient BDD-based storage for some particular problems, in this work we present a general approach where DP algorithms are specified on a logical level in form of set-based formula manipulation operations that are executed directly on the BDD data structure. In the paper, we provide several case studies in order to illustrate the method at work, and report on preliminary experiments. These show promising results, both with respect to memory and run-time.

1 Introduction

For problems that are known to be intractable, one approach is to exploit structural properties of the given input. An important parameter of graph-based instances is "tree-width", which, roughly speaking, measures the tree-likeness of the input. Tree-width is defined on so-called tree decompositions [30], where the instance is split into smaller parts, thereby taking into account its structure. The problem at hand can then be solved by dynamic programming (DP). Many problems are fixed-parameter tractable (fpt) with respect to tree-width, i.e., solvable in time $f(k) \cdot n^{\mathcal{O}(1)}$ where k is the tree-width, n is the input size and f is some computable function. Note that here the explosion in run-time is confined to k instead of the input size. Courcelle showed that every problem that is definable in monadic second-order logic (MSO) is fixed-parameter tractable with respect to tree-width [12]. There, the problem is solved via translation to a finite tree automaton (FTA). However, the algorithms resulting from such "MSO-to-FTA" translation are oftentimes impractical due to large constants [29]. One approach

---

This work has been supported by the Austrian Science Fund (FWF): Y698, P25607, P25518.

F. Calimeri et al. (Eds.): LPNMR 2015, LNAI 9345, pp. 213–227, 2015.
DOI: 10.1007/978-3-319-23264-5_19

to overcome this problem is to develop dedicated DP algorithms for the problems at hand (e.g., [6,21]). Such algorithms typically rely on tables for storing information, resulting in a large memory footprint. This problem has been addressed, e.g., by proposing heuristics [4] or reducing the number of simultaneously stored tables [2].

In this work we mitigate the problem by developing DP algorithms with *native* support for efficient storage. In our approach, Binary Decision Diagrams (BDDs) [9] serve as the data structure. BDDs have undergone decades of research and are a well-established concept used, e.g., in model-checking [27], planning [22] and software verification [5]. Our approach is in line with recent research that studies the effectiveness of exploiting tree-width by applying decomposition techniques in combination with decision diagrams. In the area of knowledge compilation, so-called "Tree-of-BDDs" [19,33] are constructed in an offline phase from a given CNF, and queried in the online phase to answer questions on this data structure in linear time. Furthermore, Algebraic Decision Diagrams (ADDs) [3] are used for compiling Bayesian networks in such a way that the structure of the network can be exploited in order to compute inference efficiently [11]. Combining DP and decision diagrams has been proven well-suited also for Constraint Optimization Problems (COPs) [31]. The key idea is to employ ADDs to store the set of possible solutions, and the branch-and-bound algorithm is executed on a decomposition of the COP instance. This was shown to be superior to earlier approaches in [8], where additionally (no)good recording is applied during computation.

In this work we continue this promising branch of research. However, from a conceptual perspective, our algorithms are specified on a logical level as formulae. This gives a compact and exact specification of algorithms, which are executed directly on the BDDs in form of BDD manipulation operations. In contrast to table-based DP algorithms, we do not manipulate tuples directly, but modify the *set* of models. Furthermore, in the course of this work we develop two different DP algorithm design paradigms, which we call *early decision method* (EDM) and *late decision method* (LDM). In EDM, information is incorporated in the BDD as soon as it becomes available when traversing the tree decomposition and is thus similar to the approach usually employed in standard table-based implementations. As we will see, LDM gives rise to novel DP algorithms where the BDD manipulation operations are delayed until just before the information is removed. We illustrate these concepts by providing several case studies that exemplarily show how DP algorithms can be implemented following our approach. These prototypical problems differ in that only fixed information, also changing information or even connectedness has to be handled appropriately. While we focus here on problems that are NP-complete, we plan to apply our method also to problems beyond NP (thus covering applications from the AI and LPNMR domain) with the long-term goal to extend our way of DP algorithm specification to all MSO-definable problems. To summarize, the main contributions of this paper are as follows:

- An approach for specifying DP algorithms on tree decompositions via formula manipulation, and two design patterns called early/late decision method (EDM/LDM).

- Case studies of 3-COLORABILITY, STABLE EXTENSION (from the field of argumentation) and HAMILTONIAN CYCLE to illustrate our method at work.
- A performance analysis that compares memory and time requirements of our approach with available DP implementations, indicating that our approach significantly reduces memory requirements and gives advantages in performance.[1]

2   Background

Tree Decompositions. Tree decompositions, introduced in [30], are defined as follows.

Definition 1. *A* tree decomposition *of an undirected graph $G = (V, E)$ is a pair $(\mathcal{T}, \mathcal{X})$ where $\mathcal{T} = (V_\mathcal{T}, E_\mathcal{T})$ is a tree and $\mathcal{X} : V_\mathcal{T} \to 2^V$ assigns to every node $V_\mathcal{T}$ of the tree a set of vertices V from the original graph. The sets of vertices $\mathcal{X} = (X_t)_{t \in V_\mathcal{T}}$ have to satisfy the following conditions: (a) $\bigcup_{t \in V_\mathcal{T}} X_t = V$. (b) $\{x, y\} \in E \Rightarrow \exists t \in V_\mathcal{T} : \{x, y\} \subseteq X_t$. (c) $x \in X_{t'} \wedge x \in X_{t''} \wedge t''' \in path(t', t'') \Rightarrow x \in X_{t'''}$.*

X_t is also called the bag *for the vertex $t \in V_\mathcal{T}$. The* width *w of the decomposition is $max_{t \in V_\mathcal{T}} |X_t| - 1$. The* tree-width *$k$ of a graph is the minimum width over all its tree decompositions.*

Intuitively, this definition guarantees that every vertex of the graph is contained in some bag of the tree decomposition, adjacent vertices appear together in some bag, and that nodes that contain the same vertex are connected. For problems on directed graphs, Definition 1 can be naturally extended. We will denote an edge between two vertices x, y by $\{x, y\}$, and directed arcs by (x, y). Furthermore, for a decomposition node t, we denote by $E_t = \{\{x, y\} \in E \mid x, y \in X_t\}$ the edges of G induced by the vertices X_t, and analogously by A_t the arcs in t. It is well-known that obtaining an optimal decomposition (with respect to width) is NP-hard [1], but there are heuristics that provide a "good" decomposition in polynomial time [7,13,14]. For the ease of representation, we consider a special type of tree decomposition throughout this work.

Definition 2. *A tree decomposition $\mathcal{T} = (V_\mathcal{T}, E_\mathcal{T})$ is called* normalized *if each $t \in V_\mathcal{T}$ is of one of the following types: (1)* Leaf *node: t has no child nodes. (2)* Introduction *node: t has exactly one child node t' with $X_{t'} \subset X_t$ and $|X_{t'}| = |X_t| - 1$. (3)* Removal *node: t has exactly one child node t' with $X_t \subset X_{t'}$ and $|X_{t'}| = |X_t| + 1$. (4)* Join *node: t has exactly two child nodes t' and t'' with $X_t = X_{t'} = X_{t''}$.*

Furthermore, without loss of generality, we assume that $X_r = \emptyset$ for the root node r of $\mathcal{T}$. Note that such a normalized decomposition can be obtained in linear time from an arbitrary one without increasing the tree-width [23].

---

[1] A prototype system, called *dynBDD*, which is built on top of the BDD library CUDD [32] is available under http://dbai.tuwien.ac.at/proj/decodyn/dynbdd.

Example 1. Figure 1 shows an example graph G and a possible (normalized) tree decomposition T (T_n) of width 2.The tree decompositions are optimal w.r.t. width.

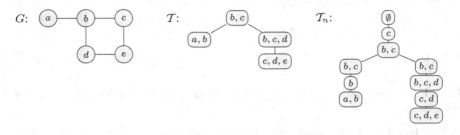

Fig. 1. Graph G and possible (normalized) tree decomposition T (T_n) of G.

(Reduced Ordered) Binary Decision Diagrams. In our approach, Reduced Ordered Binary Decision Diagrams (ROBDDs) [9] serve as the data structure for storing information during the traversal of the decomposition.

Definition 3. *An* Ordered Binary Decision Diagram $B = (V_B, A_B)$ *is a rooted, connected, directed acyclic graph where $V_B = V_T \cup V_N$ and $A_B = A_\top \cup A_\bot$. The following conditions have to be satisfied:*

1. *V_T may contain the terminal nodes $\top$ and $\bot$.*
2. *V_N contains the internal nodes, where each $v \in V_N$ represents a variable v.*
3. *Each $v \in V_N$ has exactly one outgoing arc in $A_\top$ and one in $A_\bot$, represented by a solid and a dashed arc respectively.*
4. *For every path from the root to a terminal node, each variable occurs at most once and in the same order (i.e., we have a strict total order over the variables).*

In Reduced OBDDs *(ROBBD)*, isomorphic nodes are merged into a single node with several incoming edges. Furthermore, nodes $v \in V_N$ where both outgoing arcs reach the same node $v' \in V_B$, are removed.

Given an OBDD B, propositional variables V_N and an assignment A to V_N, the *corresponding path* in B is the unique path from the root node to a terminal node, such that for every $v \in V_N$ it includes the outgoing arc in $A_\top$ $(A_\bot)$ iff A gets assigned true (false) for v. A is a satisfying assignment of the function represented by B iff the path ends in $\top$. With slight abuse of notation, in the following we will specify BDDs by giving the function in form of a logic formula.

Example 2. Figure 2 shows an OBBD B and the corresponding ROBBD B_{red} for formula $(a \land b \land c) \lor (a \land \neg b \land c) \lor (\neg a \land b \land c)$. Nodes c_1, c_2 and c_3 represent the same variable c and have arcs to the same terminal nodes. Hence, these isomorphic nodes are merged to a single node c. Then, both outgoing arcs of b_1 reach c, and b_1 is removed. Furthermore, c_4 is removed.

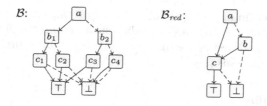

Fig. 2. OBBD and ROBBD of formula $(a \wedge b \wedge c) \vee (a \wedge \neg b \wedge c) \vee (\neg a \wedge b \wedge c)$.

BDDs support standard logical operators *conjunction* ($\wedge$), *disjunction* ($\vee$), *negation* ($\neg$) and *equivalence* ($\leftrightarrow$). Furthermore, for a BDD $\mathcal{B}$, *existential quantification* over a set of variables V, $V \subseteq V_N$ is denoted by $\exists V \mathcal{B}$. *Restriction* of a variable $v \in V_N$ to true ($\top$) or false ($\bot$) and *renaming* to a variable v' is denoted by $\mathcal{B}[v/\cdot]$ where $\cdot \in \{\top, \bot, v'\}$. For sets of variables $V \subseteq V_N$, $\mathcal{B}[V/\cdot]$ with $\cdot \in \{\top, \bot, V'\}$ and $V' = \{v' \mid v \in V\}$, denotes restriction or renaming of each $v \in V$ by applying $\mathcal{B}[v/\cdot]$.

3   DP on Tree Decompositions with BDDs

Our algorithms follow a general pattern of how the solution is constructed. First, the graph representation of the input instance is decomposed. Next, the decomposition is normalized in linear time. The resulting tree decomposition $\mathcal{T}$ is traversed in bottom-up order, and at each node $t \in V_{\mathcal{T}}$ the associated BDD, denoted by $\mathcal{B}_t$, is manipulated according to the problem at hand. In the root node r of the decomposition (where $X_r = \emptyset$), either $\mathcal{B}_r = \top$ or $\mathcal{B}_r = \bot$ holds, representing the solution to the problem.

We present two algorithm design choices, which we call the *early decision method* (EDM), where information is incorporated within introduction nodes, and the *late decision method* (LDM), where BDD manipulation is delayed until removal of vertices. For unsatisfiable instances, EDM potentially detects conflicts earlier during the traversal of the decomposition. However, LDM gives advantages when specifying more involved algorithms (see Sect. 4). Note that EDM is similar to the approach employed in standard table-based implementations, while LDM is usually harder to implement on tables.

For the case studies presented in the following, we will specify the manipulation operations on $\mathcal{B}_t$ based on the node type of t, with $\mathcal{B}_t^l$ representing the BDD resulting from a leaf node operation, $\mathcal{B}_t^i$ (introduction node), $\mathcal{B}_t^r$ (removal node), and $\mathcal{B}_t^j$ (join node). Nodes t', t'' denote child nodes, $\mathcal{B}_{t'}, \mathcal{B}_{t''}$ the BDDs constructed in the child nodes, and u the introduced or removed vertex (if any). All $\mathcal{B}_t$ for $t \in V_{\mathcal{T}}$ are required to share the same global variable ordering for efficiency during manipulation. In general, the size of the stored BDDs (i.e., the number of nodes in $\mathcal{B}_t$) is bounded by $\mathcal{O}(2^{wl})$ where w is the width of $\mathcal{T}$ and l the number of variables stored per bag element (i.e., vertex of the original input graph). However, in practice the size may be exponentially smaller, in particular in case a "good" variable ordering is applied [20]. Since finding an optimal variable ordering is in general NP-hard [9], we rely on BDD-internal heuristics for finding such a good ordering [32]. With this, the BDDs require much less space than an equivalent table representation as used in state-of-the-art systems (see Sect. 4).

3.1   3-Colorability

The 3-COLORABILITY problem (*"Given a graph G, is G 3-colorable?"*) is very well-suited to illustrate how DP algorithms for problems that are FPT with respect to tree-width can be specified following our approach. As input, the algorithms expect a simple graph $G = (V, E)$. Furthermore, we define the set of colors $C = \{r, g, b\}$. The following variables are to be used within the BDDs: For all $c \in C$ and $x \in V$, the truth value of variable c_x denotes whether vertex x gets assigned color c.

EDM. The BDD manipulation operations given below are applied at the respective decomposition nodes. We have to guarantee that every vertex gets assigned exactly one color, and adjacent vertices do not have the same color. Intuitively, $\mathcal{B}_t^l$ and $\mathcal{B}_t^i$ are constructed by adding the respective constraints for introduced vertices. In $\mathcal{B}_t^r$, due to the definition of tree decompositions, we know that all constraints related to removed vertex u were already taken into account. Hence, we can abstract away the variables associated with u, thereby keeping the size of the BDD bound by the width of the decomposition. In join nodes, $\mathcal{B}_t^j$ combines the intermediate results obtained in the child nodes of the decomposition.

$$\mathcal{B}_t^l = \bigwedge_{c \in C} \bigwedge_{\{x,y\} \in E_t} \neg(c_x \wedge c_y) \wedge \bigwedge_{x \in X_t} (r_x \vee g_x \vee b_x) \wedge$$
$$\bigwedge_{x \in X_t} \left(\neg(r_x \wedge g_x) \wedge \neg(r_x \wedge b_x) \wedge \neg(g_x \wedge b_x) \right)$$

$$\mathcal{B}_t^i = \mathcal{B}_{t'} \wedge \bigwedge_{c \in C} \bigwedge_{\{x,u\} \in E_t} \neg(c_x \wedge c_u) \wedge (r_u \vee g_u \vee b_u) \wedge$$
$$\neg(r_u \wedge g_u) \wedge \neg(r_u \wedge b_u) \wedge \neg(g_u \wedge b_u)$$

$$\mathcal{B}_t^r = \exists r_u g_u b_u [\mathcal{B}_{t'}] \qquad\qquad \mathcal{B}_t^j = \mathcal{B}_{t'} \wedge \mathcal{B}_{t''}$$

LDM. Another possibility for specifying the algorithm is to incorporate information as late as possible, that is, when a vertex is removed from the decomposition. In leaf nodes the BDD $\mathcal{B}_t^l$ is initialized with $\top$, and in introduction nodes the BDD $\mathcal{B}_t^i$ corresponds to that of the child nodes. When a vertex u is removed, one variable of r_u, g_u, b_u is set to true, thereby assigning to the vertex exactly one color $c \in C$. Furthermore, neighboring vertices x with $\{x, u\} \in E_{t'}$ must not get assigned the same color, which is achieved by adding $\neg c_x$ to the formula. $\mathcal{B}_t^r$ is simply the disjunction over the three BDDs resulting from the choice of the color. As in EDM, it is sufficient to construct $\mathcal{B}_t^j$ via conjunction of the child BDDs.

$$\mathcal{B}_t^l = \top \qquad\qquad \mathcal{B}_t^i = \mathcal{B}_{t'} \qquad\qquad \mathcal{B}_t^j = \mathcal{B}_{t'} \wedge \mathcal{B}_{t''}$$

$$\mathcal{B}_t^r = \Big(\mathcal{B}_{t'}[r_u/\top, g_u/\bot, b_u/\bot] \wedge \bigwedge_{\{x,u\}\in E_{t'}} \neg r_x\Big) \vee$$
$$\Big(\mathcal{B}_{t'}[r_u/\bot, g_u/\top, b_u/\bot] \wedge \bigwedge_{\{x,u\}\in E_{t'}} \neg g_x\Big) \vee$$
$$\Big(\mathcal{B}_{t'}[r_u/\bot, g_u/\bot, b_u/\top] \wedge \bigwedge_{\{x,u\}\in E_{t'}} \neg b_x\Big)$$

3.2 Stable Extension

The STABLE EXTENSION problem ("*Given an argumentation framework AF, does there exist a stable extension in AF?*") is a well-known problem from the area of abstract argumentation [16]. An argumentation framework $AF = (V, A)$ is a directed graph where the vertices V represent arguments and the arcs A the so-called attack relation between arguments. A stable extension $\mathcal{E}$ of an AF is a set $\mathcal{E} \subseteq V$ that is (i) conflict-free, i.e., for all $x, y \in \mathcal{E} : (x, y) \notin A$ holds, and (ii) all arguments are either in the set or defeated, i.e., for all $x \in V : x \in \mathcal{E} \vee (\exists(y, x) \in A : y \in \mathcal{E})$ holds. In the area of argumentation, DP algorithms for various semantics have been studied in [17].

In our BDD-based approach we specify the following variables. For all $x \in V$, the truth value of variable i_x denotes whether argument x is in some $\mathcal{E}$. Furthermore, the assignment of true to variable d_x represents that x is defeated. Additionally, for a node t of the tree decomposition, we denote by $D_t = \{d_x \mid x \in X_t\}$ the defeated arguments in t. Here, i_x for $x \in V$ are variables with *fixed* truth assignment (i.e., containment in an extension is *guessed once*), while all d_x have a truth assignment that *changes* during the traversal of the tree decomposition (an argument may *become* defeated in a decomposition node).

$$\mathcal{B}_t^l = \bigwedge_{(x,y)\in A_t} (\neg i_x \vee \neg i_y) \wedge \bigwedge_{y\in X_t} \Big(d_y \leftrightarrow \bigvee_{(x,y)\in A_t} i_x\Big)$$

$$\mathcal{B}_t^i = \exists D_{t'}' \Big[\mathcal{B}_{t'}[D_{t'}/D_{t'}'] \wedge \bigwedge_{\{u,y\}\in E_t} (\neg i_u \vee \neg i_y) \wedge \Big(d_u \leftrightarrow \bigvee_{(x,u)\in A_t} i_x\Big) \wedge$$
$$\bigwedge_{\substack{(u,y)\in A_t \wedge \\ u\neq y}} (d_y \leftrightarrow d_y' \vee i_u) \wedge \bigwedge_{y\in X_t \wedge (u,y)\notin A_t} (d_y \leftrightarrow d_y') \Big]$$

$$\mathcal{B}_t^r = \mathcal{B}_{t'}[i_u/\top, d_u/\bot] \vee \mathcal{B}_{t'}[i_u/\bot, d_u/\top]$$

$$\mathcal{B}_t^j = \exists D_t' \exists D_t'' \Big[\mathcal{B}_{t'}[D_t/D_t'] \wedge \mathcal{B}_{t''}[D_t/D_t''] \wedge \bigwedge_{x\in X_t} (d_x \leftrightarrow d_x' \vee d_x'') \Big]$$

EDM. In leaf nodes, variable d_y for arguments $y \in X_t$ is true (i.e., defeated) iff one of its attacking arguments is in the stable extension, and adjacent arguments can not be both in the extension. In $\mathcal{B}_t^i$, for u the formula is constructed as in leaf nodes. In order to update the truth value of defeat variables, for any argument y we apply a general pattern of $\exists y'[\mathcal{B}_{t'}[y/y'] \wedge (y \leftrightarrow (y' \vee cond))]$, that is, renaming, potentially adding conditions ($cond$), and removing the renamed variable y'. Here, $cond$ contains i_u in case u is an incoming neighbor of y. By this, the size of the BDDs remains bounded by the width of the decomposition. In removal nodes, u must either be contained in the extension, or it is defeated. Note that the conflict-free property would be violated in case u is both in the extension and defeated. In $\mathcal{B}_t^j$, the defeat information is propagated via renaming, equivalence, and existential quantification.

$$\mathcal{B}_t^l = \bigwedge_{x \in X_t} \neg d_x \qquad\qquad \mathcal{B}_t^i = \mathcal{B}_{t'} \wedge \neg d_u$$

$$\mathcal{B}_t^r = \phi_t^r[i_u/\top, d_u/\bot] \vee \phi_t^r[i_u/\bot, d_u/\top] \text{ with}$$

$$\phi_t^r = \exists D_{t'}' \Big[\mathcal{B}_{t'}[D_{t'}/D_{t'}'] \wedge \bigwedge_{\{u,y\} \in E_{t'}} (\neg i_u \vee \neg i_y) \wedge$$
$$\bigwedge_{y \in X_t} (d_y \leftrightarrow d_y' \vee_{(u,y) \in A_{t'}} i_u) \wedge (d_u \leftrightarrow d_u' \vee \bigvee_{(x,u) \in A_{t'}} i_x) \Big]$$

$$\mathcal{B}_t^j = \exists D_t' D_t'' \Big[\mathcal{B}_{t'}[D_t/D_t'] \wedge \mathcal{B}_{t''}[D_t/D_t''] \wedge \bigwedge_{x \in X_t} (d_x \leftrightarrow d_x' \vee d_x'') \Big]$$

LDM. Here, all information is considered when a vertex is removed. Hence, introduced vertices cannot become defeated in leaf or introduction nodes, and the corresponding variables are initialized with $\bot$. In $\mathcal{B}_t^r$, we guess whether the removed vertex u is in the extension or defeated. Furthermore, we guarantee conflict-freeness with vertices adjacent to u. A vertex y becomes defeated if it is attacked by u and u is in the extension, and u is defeated if it is attacked by some vertex that was already removed, or by an in-vertex on an arc in $A_{t'}$. Note that we use a small disjunction symbol with condition whenever there is at most one disjunction in the instantiated formula, and a large symbol otherwise. $\mathcal{B}_t^j$ is specified as in EDM.

3.3   Hamiltonian Cycle

The HAMILTONIAN CYCLE problem (*"Given a graph $G = (V, E)$, does there exist a Hamiltonian Cycle in G?"*) requires a more involved algorithm specification. Monolithic propositional encodings (where the whole instance is available at once) allow one to assign a global order over the variables that specifies the ordering over the vertices in the cycle. However, in our DP-based approach, we are restricted to information that is available in the current decomposition node. Hence, we consider a *relative* ordering as follows. The idea is to first specify exactly one incoming and one outgoing edge for each vertex. For $x \in V$, the truth value of variable i_x (o_x) denotes that it has an outgoing (incoming) edge. A selected edge $\{x, y\} \in E$ is represented by variable t_{xy}. Second, we have to guarantee that we have a *single* cycle that covers all vertices. Therefore we select a fixed vertex $f \in V$ that denotes where the cycle starts and ends. Variable a_{xy} for $x, y \in V$ denotes that x lies after y on the path from f to f. For a tree decomposition node t we have $S_t = \{i_x, o_x, a_{xy} \mid x, y \in X_t\}$. Furthermore, for a vertex $u \in X_t$, let $T_{t,u} = \{t_{xu}, t_{ux} \mid \{x, u\} \in E_t\}$. Due to space limitations in the following we only present the LDM version.

LDM. In leaf and introduction nodes all changing variables are initialized with $\bot$. In removal nodes, at least one incoming edge for removed vertex u is selected. Here, i'_u is true iff the incoming neighbor of u was already removed from the bag. Furthermore, at most one incoming edge from $E_{t'}$ is selected. Finally, if i'_u is true, we cannot select an additional incoming edge, and the incoming and outgoing edges for u have to be different. The same construction is used to guarantee exactly one outgoing edge for u. For vertices $x \in X_t$, i_x and o_x are updated in case u was a neighbor of x. Again, at most one incoming (outgoing) edge must be selected. For $x, z \in X_t$, a_{xz} becomes true if $u \neq f$ and u lies on the path between x and z. With this, we keep information on the path (from f to f), restricted to X_t, where the truth value of t_{xy}-variables represents selected edges in X_t and a_{xy}-variables denote that x is before y on the path where intermediate vertices were already removed. Finally, in case a_{xx} for $x \neq f$ is true, we know that there is a cycle that does not cover f, and is therefore no Hamiltonian cycle. In join nodes, i_x, o_x and a_{xy} variables are propagated as usual. Here, whenever both i'_x and i''_x are true, due to the connectedness condition of tree decompositions and the fact that these variables are updated when a vertex is removed, x has different incoming edges and cannot be a solution. The same holds for outgoing edges.

$$\mathcal{B}^l_t = \bigwedge_{x \in X_t} (\neg i_x \wedge \neg o_x) \wedge \bigwedge_{x,y \in X_t} \neg a_{xy}$$

$$\mathcal{B}^i_t = \mathcal{B}_{t'} \wedge \neg i_u \wedge \neg o_u \wedge \bigwedge_{x \in X_t} (\neg a_{xu} \wedge \neg a_{ux})$$

$$\mathcal{B}^r_t = \exists T_{t',u} S'_{t'} \Big[\mathcal{B}_{t'}[S_t/S'_{t'}] \wedge (i'_u \vee \bigvee_{\{x,u\} \in E_{t'}} t_{xu}) \wedge (o'_u \vee \bigvee_{\{u,y\} \in E_{t'}} t_{uy}) \wedge$$

$$\bigwedge_{\substack{\{x',u\} \in E'_t \wedge \\ \{x'',u\} \in E'_t \wedge x' \neq x''}} \big(\neg(t_{x'u} \wedge t_{x''u}) \wedge \neg(t_{ux'} \wedge t_{ux''})\big) \wedge$$

$$\bigwedge_{\{x,u\} \in E_{t'}} \big(\neg(i'_u \wedge t_{xu}) \wedge \neg(o'_u \wedge t_{ux}) \wedge \neg(t_{xu} \wedge t_{ux})\big) \wedge$$

$$\bigwedge_{x \in X_t} \Big(\big(i_x \leftrightarrow (i'_x \vee_{\{u,x\} \in E_{t'}} t_{ux})\big) \wedge \big(o_x \leftrightarrow (o'_x \vee_{\{x,u\} \in E_{t'}} t_{xu})\big) \Big) \wedge$$

$$\bigwedge_{\{x,u\} \in E_{t'}} \big(\neg(i'_x \wedge t_{ux}) \wedge \neg(o'_x \wedge t_{xu})\big) \wedge$$

$$\bigwedge_{x,z \in X_t} \Big(a_{xz} \leftrightarrow a'_{xz} \vee_{u \neq f} \big((a'_{xu} \vee_{\{x,u\} \in E_{t'}} t_{xu}) \wedge (a'_{uz} \vee_{\{u,z\} \in E_{t'}} t_{uz}) \big) \Big) \wedge$$

$$\bigwedge_{x \in X_t \wedge x \neq f} \neg a_{xx} \Big]$$

$$\mathcal{B}_t^j = \exists S'_t S''_t \Big[\mathcal{B}_{t'}[S_t/S'_t] \wedge \mathcal{B}_{t''}[S_t/S''_t] \wedge$$

$$\bigwedge_{x \in X_t} \Big(\big((i_x \leftrightarrow (i'_x \vee i''_x)) \wedge (o_x \leftrightarrow (o'_x \vee o''_x)) \big) \wedge \neg(i'_x \wedge i''_x) \wedge \neg(o'_x \wedge o''_x) \Big) \wedge$$

$$\bigwedge_{x,y \in X_t} \big(a_{xy} \leftrightarrow (a'_{xy} \vee a''_{xy}) \big) \Big]$$

4   Experimental Analysis

The aforementioned algorithms were implemented in the prototype system *dynBDD* that utilizes the library CUDD [32] for efficient BDD management and the HTDECOMP library [14] for constructing the tree decompositions by applying the "min-degree" heuristics. We compare run-time and memory requirements to freely available implementations that also utilize the concept of DP on tree decompositions, namely SEQUOIA [25] (version 0.9) and D-FLAT [6] (version 1.0.0). Furthermore, for the area of abstract argumentation, the DP-based dynPARTIX system [10] (version 2.0) is available. SEQUOIA implements a game-theoretic approach [24]. As input, SEQUOIA expects the problem to be formulated as an MSO formula. The instance is decomposed and the DP algorithm automatically generated and executed. D-FLAT combines DP with answer-set programming (ASP). In contrast to SEQUOIA, the user specifies an ASP encoding that is executed at each node of the decomposition, thereby defining the DP algorithm explicitly. dynPARTIX comprises of implementations of reasoning tasks relevant to the field of argumentation.

All experiments were performed on a single core of an AMD Opteron 6308 (3.5 GHz) processor running Debian GNU/Linux 7 (kernel 3.2.0-4-amd64). Each run was limited to 10 min (Timeout) and 4 GB of memory (Memout). Instances were generated using the random graph model due to Erdös and Rényi [18]. This allows us to compare the implementations on various instances that cover a broad range of different widths. Below, we denote by n the number of vertices and by p the edge probability of an instance. Since run-time depends on the heuristically obtained tree decomposition, we run the algorithms on the same normalized tree decompositions (if not stated otherwise).

3-Colorability. To analyze the performance, the EDM and LDM versions of the algorithms were implemented in dynBDD and D-FLAT. For SEQUOIA, the performance of the MSO-based algorithm is measured. Overall, 252 instances with n between 10 and 1000 and p between 0.001 (very sparse) and 0.2 (dense) were tested. Figure 3 (upper left) shows the number of solved instances per system and implementation variant. SEQUOIA implements a pre-check to tell whether it is capable of solving the instance. *Error* denotes the number of times this check failed. Results show that EDM is by far superior to LDM. Here, we observed that EDM detects conflicts in unsatisfiable instances earlier than LDM. In fact, additional analysis showed that for satisfiable instances dynBDD (EDM) performed only marginally better than dynBDD (LDM). In total, 161 instances were solved by both dynBDD (EDM) and D-FLAT (EDM). Figure 3 (lower left) gives details on the accumulated run-time and memory usage of the best-performing systems over these instances. The figure shows how many instances were solved after a certain amount of time. For example, D-FLAT (EDM) solved the first 100 instances in approx. 1500 seconds with a total of 4700 MB of memory usage, whereas dynBDD (EDM) required only 30 seconds and 500 MB. In total, dynBDD (EDM) required approx. 18 % less time and 47 % less memory on solved instances, and solved 36 instances more than D-FLAT (EDM). Note that we omitted dynBDD (LDM) since it solves significantly less instances than the best systems.

Regarding the width w of the tree decompositions, dynBDD (EDM) solved satisfiable instances up to $w = 48$ and unsatisfiable instances up to $w = 944$. While unsatisfiable instances of high width may be easily solvable due to early

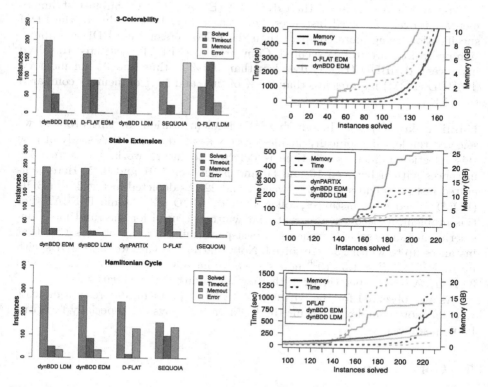

Fig. 3. Result Overview: System comparison and detailed results for best-performing systems.

conflict detection, the measured width for solved satisfiable instances is quite large. Recall that the size of the BDD (or an equivalent table representation) may be up to $\mathcal{O}(2^{wl})$ which corresponds to $2^{48 \cdot 3} \approx 2.2 \cdot 10^{43}$ for $w = 48$ and 3 variables per bag element. This indicates that BDDs are indeed memory-efficient.

Stable Extension. We compare dynBDD with D-FLAT as well as dynPAR-TIX. Note that dynPARTIX constructs decompositions on the same heuristics as used for the other systems, but obtaining the same decomposition can not be guaranteed. Furthermore, we compare dynBDD to SEQUOIA. However, to the best of our knowledge, SEQUOIA is incapable of handling directed graphs. To give an impression on its performance, we decided to show results for the related INDEPENDENT DOMINATING SET problem.

260 instances with n between 10 and 100 and p between 0.001 and 0.1 were tested. Figure 3 (middle column) illustrates the overall number of solved instances per system. The run-time of the problem-tailored implementation in dynPARTIX (that implements an approach similar to EDM) is almost the same as that of dynBDD (EDM). However, the advantage of BDDs becomes evident with respect to memory, where dynBDD (EDM) requires less than 8 % of that of dynPARTIX. Although dynBDD (LDM) solves slightly less instances and requires more memory than dynBDD (EDM), it is by far fastest implementation over all solved instances. One reason may be that due to the LDM specification being more compact than the EDM version, less BDD manipulation operations are to be executed. Compared to D-FLAT (restricted to solved instances), dynBDD (EDM) uses less than 0.6 % of time and 9 % of memory. dynBDD (LDM) requires less than 0.5 % of time and 14 % of memory compared to D-FLAT.

Hamiltonian Cycle. In our dynBDD implementation for this problem we selected the lexicographically smallest vertex as fixed vertex f. A study of how this selection influences run-time is deferred to future work. We tested 390 instances with n between 10 and 50 and p between 0.01 and 0.25. Here, generated instances had a width between 1 and 22. As depicted in Fig. 3, dynBDD (LDM) solved most instances, followed by dynBDD (EDM) and D-FLAT. For this problem, it becomes apparent that width is crucial for the run-time. Considering instances solved by the three best-performing systems, we observed that instances up to width 12 were solved. Note that in this case we have BDDs with up to $2^{12 \cdot (6+4 \cdot 12)} \approx 1.3 \cdot 10^{36}$ nodes since we have $3 \cdot 1$ i_x, $3 \cdot 1$ o_x, 12 t_{xy}, and $3 \cdot 12$ a_{xy}-variables per vertex in the bag (including renamed variables).

As also observed for the STABLE EXTENSION problem, our results indicate that for more complex problems the LDM variant pays off, especially for satisfiable instances.

5   Conclusion

In this work we showed how classical DP algorithms on tree decompositions can be reformulated in order to be executed on Binary Decision Diagrams. This gives rise to algorithms that are specified on a logical level where - opposed to manipulation of tuples in a table-based specification - the set of models is modified by executing operations directly on the BDD. Furthermore, we studied two algorithm design patterns, namely early (EDM) and late decision method (LDM), and illustrated the concepts by providing several case studies. The case studies are exemplary for NP-complete problems that are tractable w.r.t. tree-width. The corresponding algorithms are specified solely on fixed variables (3-COLORABILITY), additionally require changing variables (STABLE EXTENSION) and handling of connectedness within the DP algorithm (HAMILTONIAN CYCLE). From a practical perspective, our work is in line with the freely available systems D-FLAT [6], SEQUOIA [25] and dynPARTIX [10]. Our preliminary experiments showed that the implementation of dynBDD indeed mitigates performance and memory shortcomings of these systems. In particular, results indicate that for problems which require a more involved algorithm, LDM is superior to EDM. Note that our system currently does not implement any problem-specific short-cuts and that the libraries have been employed as black-box tools.

In the future, we want to tighten the integration of BDD handling and the tree decomposition (in particular, to obtain a good ordering of variables in the BDD from the structure of the decomposition) and to study how problem-specific shortcuts can be incorporated. Additionally, our approach natively supports parallel problem solving (over decomposition branches), which would be a complementary approach to recent developments on parallel BDD implementations [15,26]. Finally, our approach can directly be extended to problems that involve optimization. Here, ADDs as well as Multi-valued Decision Diagrams (MDDs) (see, e.g., [28]) and related data structures can serve as appropriate tools. Most importantly, we want to study our approach in the context of problems that are hard for the second level of the polynomial hierarchy (e.g., Circumscription, Abduction) and ultimately provide a tool-set for all MSO-definable problems.

References

1. Arnborg, S., Corneil, D.G., Proskurowski, A.: Complexity of finding embeddings in a k-tree. SIAM J. Algebraic Discrete Methods **8**, 277–284 (1987)
2. Aspvall, B., Telle, J.A., Proskurowski, A.: Memory requirements for table computations in partial k-tree algorithms. Algorithmica **27**(3), 382–394 (2000)
3. Bahar, R., Frohm, E., Gaona, C., Hachtel, G., Macii, E., Pardo, A., Somenzi, F.: Algebric decision diagrams and their applications. Formal Meth. Syst. Des. **10**(2–3), 171–206 (1997)
4. Betzler, N., Niedermeier, R., Uhlmann, J.: Tree decompositions of graphs: saving memory in dynamic programming. Discrete Optim. **3**(3), 220–229 (2006)
5. Beyer, D., Stahlbauer, A.: BDD-based software verification - applications to event-condition-action systems. STTT **16**(5), 507–518 (2014)
6. Bliem, B., Morak, M., Woltran, S.: D-FLAT: declarative problem solving using tree decompositions and answer-set programming. TPLP **12**(4–5), 445–464 (2012)
7. Bodlaender, H.L., Koster, A.M.C.A.: Treewidth computations I. Upper Bounds. Inf. Comput. **208**(3), 259–275 (2010)

8. Boutaleb, K., Jégou, P., Terrioux, C.: (No)good recording and ROBDDs for solving structured (V)CSPs. In: Proceedings of the ICTAI, pp. 297–304. IEEE Computer Society (2006)

9. Bryant, R.E.: Graph-based algorithms for Boolean function manipulation. IEEE Trans. Comput. **100**(8), 677–691 (1986)

10. Charwat, G., Dvořák, W.: dynPARTIX 2.0 - Dynamic programming argumentation reasoning tool. In: Proceedings of the COMMA, FAIA, vol. 245, pp. 507–508. IOS Press (2012)

11. Chavira, M., Darwiche, A.: Compiling Bayesian networks using variable elimination. In: Proceedings of the IJCAI, pp. 2443–2449 (2007)

12. Courcelle, B.: The monadic second-order logic of graphs. I. Recognizable sets of finite graphs. Inf. Comput. **85**(1), 12–75 (1990)

13. Dechter, R.: Constraint Processing. Morgan Kaufmann, San Francisco (2003)

14. Dermaku, A., Ganzow, T., Gottlob, G., McMahan, B., Musliu, N., Samer, M.: Heuristic methods for hypertree decomposition. In: Gelbukh, A., Morales, E.F. (eds.) MICAI 2008. LNCS (LNAI), vol. 5317, pp. 1–11. Springer, Heidelberg (2008)

15. van Dijk, T., Laarman, A., van de Pol, J.: Multi-core BDD operations for symbolic reachability. Electr. Notes Theor. Comput. Sci. **296**, 127–143 (2013)

16. Dung, P.M.: On the acceptability of arguments and its fundamental role in non-monotonic reasoning, logic programming and n-person games. Artif. Intell. **77**(2), 321–357 (1995)

17. Dvořák, W., Pichler, R., Woltran, S.: Towards fixed-parameter tractable algorithms for abstract argumentation. Artif. Intell. **186**, 1–37 (2012)

18. Erdös, P., Rényi, A.: On random graphs. I. Publicationes Mathematicae (Debrecen) **6**, 290–297 (1959)

19. Fargier, H., Marquis, P.: Knowledge compilation properties of Trees-of-BDDs, revisited. In: Proceedings of the IJCAI, pp. 772–777 (2009)

20. Friedman, S.J., Supowit, K.J.: Finding the optimal variable ordering for binary decision diagrams. In: Proceedings of the IEEE Design Automation Conference, pp. 348–356. ACM (1987)

21. Groër, C., Sullivan, B.D., Weerapurage, D.: INDDGO: Integrated network decomposition & dynamic programming for graph optimization. Technical reports ORNL/TM-2012/176 (2012)

22. Kissmann, P., Hoffmann, J.: BDD ordering heuristics for classical planning. J. Artif. Intell. Res. (JAIR) **51**, 779–804 (2014)

23. Kloks, T.: Treewidth, Computations and Approximations. LNCS, vol. 842. Springer, Heidelberg (1994)

24. Kneis, J., Langer, A., Rossmanith, P.: Courcelle's theorem - A game-theoretic approach. Discrete Optimization **8**(4), 568–594 (2011)

25. Langer, A., Reidl, F., Rossmanith, P., Sikdar, S.: Evaluation of an MSO-solver. In: Proceedings of the ALENEX, pp. 55–63 (2012)

26. Lovato, A., Macedonio, D., Spoto, F.: A thread-safe library for binary decision diagrams. In: Giannakopoulou, D., Salaün, G. (eds.) SEFM 2014. LNCS, vol. 8702, pp. 35–49. Springer, Heidelberg (2014)

27. Męski, A., Penczek, W., Szreter, M., Woźna-Szcześniak, B., Zbrzezny, A.: BDD-versus SAT-based bounded model checking for the existential fragment of linear temporal logic with knowledge: algorithms and their performance. Auton. Agent. Multi-Agent Syst. **28**(4), 558–604 (2014)

28. Miller, D.M.: Multiple-valued logic design tools. In: Proceedings of the MVL, pp. 2–11 (1993)

29. Niedermeier, R.: Invitation to fixed-parameter algorithms. Oxford Lecture Series in Mathematics and its Applications, vol. 31. OUP, Oxford (2006)
30. Robertson, N., Seymour, P.D.: Graph minors. III. Planar tree-width. J. Comb. Theory, Ser. B **36**(1), 49–64 (1984)
31. Sachenbacher, M., Williams, B.C.: Bounded search and symbolic inference for constraint optimization. In: Proceedings of the IJCAI, pp. 286–291. PBC (2005)
32. Somenzi, F.: CU Decision Diagram package release 2.5.0. Department of Electrical and Computer Engineering, University of Colorado at Boulder (2012)
33. Subbarayan, S.: Integrating CSP decomposition techniques and BDDs for compiling configuration problems. In: Barták, R., Milano, M. (eds.) CPAIOR 2005. LNCS, vol. 3524, pp. 351–365. Springer, Heidelberg (2005)

Knowledge Acquisition via Non-monotonic Reasoning in Distributed Heterogeneous Environments

Stefania Costantini[(✉)]

Dipartimento di Ingegneria e Scienze dell'Informazione e Matematica (DISIM),
Universitá degli Studi dell'Aquila, L'Aquila, Italy
`stefania.costantini@univaq.it`

Abstract. The role of data and knowledge exchange is becoming increasingly important. The approach of DACMAS [1] proposes a quite general modeling of Multi-Agent Systems (MAS), including data representation in a MAS via DRL-Lite Ontologies. Yet, data/knowledge acquisition from heterogeneous sources which are not agents and which are external to the MAS is not provided. In the Knowledge Representation and Reasoning field, this topic is coped with by mMCSs (Managed Multi-Context Systems). In this paper, we propose an integration of the two approaches into DACMACSs. The aim is to obtain an enhanced integrated flexible framework where non-monotonicity is present: in the modalities for defining knowledge acquisition; in the conditions for triggering the acquisition and for knowledge exploitation.

1 Introduction

The importance of data and knowledge exchange in Artificial Intelligence applications is constantly increasing. In many application fields it is particularly important to comprise and elaborate information provided by multiple sources. Distributed autonomous evolving applications are the realm of intelligent software agents and Multi-Agent Systems (MAS). Many approaches to MAS are based upon computational logic, where logical agents are able to reason, and to perform non-monotonic reasoning when needed (cf. the Proceedings of the "Computational Logic in Multi-Agent Systems" (CLIMA) Workshop Series). Logic-based data management and exchange are therefore important issues in logical agents. Such agents are required to perform non-monotonic reasoning both in defining and executing patterns for knowledge exchange and in the modalities for knowledge exploitation.

DACMAS (Data-Aware Commitment-based Multi-Agent Systems) [1] is a quite general model of multi-agent systems, which explicitly introduces elements of knowledge representation in the specification of such systems. Knowledge and data in DACMASs are in fact represented according to the DRL-Lite Description Logic [2]. A DACMAS always includes an *institutional* agent which owns a "global" TBox, specifying the domain in which the MAS operates, whereas each

© Springer International Publishing Switzerland 2015
F. Calimeri et al. (Eds.): LPNMR 2015, LNAI 9345, pp. 228–241, 2015.
DOI: 10.1007/978-3-319-23264-5_20

participating agent is equipped with its local ABox. The DACMAS approach is particularly interesting because, apart from a general specification of data management and communicative features, it remains very general about an agent program's definition, and can be thus specialize to many existing agent-oriented logic languages.

However, the possibility for agents to interact with components of a different nature which are part of the agents' environment is an important aspect which is lacking in DACMASs. As a matter of fact, even the most comprehensive software engineering approaches such as, e.g., "Agents and Artifacts" (cf. [3] and the references therein) assume that external sources can be either "agentified" or "wrapped", as in fact, citing the proposers, "An artifact is a computational, programmable system resource, that can be manipulated by agents, residing at the same abstraction level of the agent abstraction class". In a real setting this may be too strong an assumption, at least concerning external knowledge bases. What must be necessarily known about such external sources is that they can be queried and should return certain kinds of information: however, modification is in general not allowed, and availability of a better description cannot be taken for granted. Moreover, wrapping a source according to a certain specification would imply that all agents in a MAS have a uniform view of that source. This assumption may be reductive under the perspective of knowledge representation and reasoning: in fact, each agent might in principle interpret, elaborate, reason upon and incorporate the acquired knowledge in a way which is especially tailored to its tasks and objectives.

In the Artificial Intelligence and Knowledge Representation field, the Multi-Context Systems (MCS) approach has been proposed to model information exchange among heterogeneous sources [4–6]. MCSs are defined so as to drop the assumption of making such sources in some sense homogeneous: rather, the approach deals explicitly with their different representation languages and semantics. Heterogeneous sources are called "contexts" and in the MCS understanding they are fundamentally different from agents as they do not have reactive, proactive and social capabilities, but can simply be queried and updated. MCSs have evolved from the simplest form [4] to managed MCS (mMCS) [7], and reactive mMCS [6] for dealing with external inputs such as a stream of sensor data.

In this paper we propose to combine DACMAS with mMCS. The aim is to obtain the formalization of multi-agent systems which are able to flexibly interact with heterogeneous external information sources, by means of suitable agent-oriented modalities. In mMCSs, communication among contexts occurs via special non-monotonic "bridge rules", which are assumed to be automatically applied whenever applicable. We allow also agents to be equipped with bridge rules. However, we devise an entirely new mechanism for bridge-rule activation. In the proposed approach, that we call DACMACS (Data-Aware Commitment-based managed Multi-Agent-Context Systems) agents do not mandatorily apply bridge rules. Rather, being autonomous entities, DACMACS agents are able to apply bridge rules proactively and non-monotonically, depending upon requirements which are specific for each agent.

The main features of the proposal are the following. (1) Agents can query (sets of) contexts, but contexts cannot query agents. (2) Agents are equipped with bridge rules, whose application is however activated via special *trigger rules*, which allow a bridge rule to be invoked upon certain conditions and/or according to a certain timing. (3) The result of a bridge rule is interpreted as an agent-generated internal event, which is captured by reactive rules which may determine modifications to the agent's ABox according to suitable reasoning techniques for the analysis and the incorporation of the newly acquired knowledge.

The integration among DACMASs and mMCSs has been purposedly devised so as to be smooth, thus requiring as few modifications to the formal machinery as possible. DACMACS therefore: (i) is based upon the combination of well-established existing technologies; (ii) provides a uniform representation and a formal semantics for the resulting class of systems; (iii) preserves most of the interesting formal properties of both mMCSs and DACMASs that therefore extend to DACMACS; (iv) introduces however novel agent-oriented interaction patterns; (v) introduces semantic notions that constitute an enhancement over both DACMASs and reactive mMCSs.

The paper is organized as follows: in Sects. 2 and 3 we provide the necessary background notions about mMCSs and DACMASs. In Sect. 4 we present and illustrate the new approach of DACMACSs, and discuss its properties. In Sect. 5 we present (in a nutshell, for lack of space) an example that we consider as representative of a potential real-world application domain of DACMACSs.

2   Background: mMCS

Managed Multi-Context systems (mMCS) [5–7]) model the information flow among multiple possibly heterogeneous data sources. The device for doing so is constituted by "bridge rules", which are similar to datalog rules (cf., e.g., [8] for a survey about datalog and the references therein for more information) but allow for knowledge acquisition from external sources, as in each element of their "body" the "context", i.e. the source, from which information is to be obtained is explicitly indicated. In the short summary of mMCS provided below we basically adopt the formulation of [6], which is simplified w.r.t. [7].

Reporting from [5], a logic L is a triple $(KB_L; Cn_L; ACC_L)$, where KB_L is the set of admissible knowledge bases of L, which are sets of KB-elements ("formulas"); Cn_L is the set of acceptable sets of consequences, whose elements are data items or "facts" (in [5] these sets are called "belief sets"; we adopt the more neutral terminology of "data sets"); $ACC_L : KB_L \to 2^{Cn_L}$ is a function which defines the semantics of L by assigning to each knowledge-base a set of "acceptable" sets of consequences. A managed Multi-Context System (mMCS) $M = (C_1, \ldots, C_n)$ is a heterogeneous collection of contexts where $C_i = (L_i; kb_i; br_i)$ and L_i is a logic, $kb_i \in KB_{L_i}$ is a knowledge base (below "knowledge base") and br_i is a set of bridge rules. Each such rule is of the following form, where the left-hand side $o(s)$ is called the *head*, also denoted as $hd(\rho)$, the right-hand side is called the *body*, also denoted as $body(\rho)$, and the comma stand for conjunction.

$o(s) \leftarrow (c_1 : p_1), \ldots, (c_j : p_j), not\,(c_{j+1} : p_{j+1}), \ldots, not\,(c_m : p_m).$

For each bridge rule included in a context C_i, it is required that $kb_i \cup o(s)$ belongs to KB_{Li} and, for every $k \leq m$, c_k is a context included in M, and each p_k belongs to some set in KB_{L_k}. The meaning is that $o(s)$ is added to the consequences of kb_i whenever each p_r, $r \leq j$, belongs to the consequences of context c_r, while instead each p_w, $j < w \leq m$, does not belong to the consequences of context c_w. While in standard MCSs the head s of a bridge rule is simply added to the "destination" context's knowledge base kb, in managed MCS the kb is subjected to an elaboration w.r.t. s according to a specific operator o and to its intended semantics. Formula s itself can be elaborated by o, for instance with the aim of making it compatible with kb's format, or by performing more involved elaboration.

If $M = (C_1, \ldots, C_n)$ is an MCS, a data state (or, equivalently, belief/knowledge state), is an n-uple $S = (S_1, \ldots, S_n)$ such that each S_i is an element of Cn_i. Desirable data states are those where each S_i is acceptable according to ACC_i. A bridge rule ρ is applicable in a knowledge state iff for all $1 \leq i \leq j : p_i \in S_i$ and for all $j + 1 \leq k \leq m : p_k \notin S_k$. Let $app(S)$ be the set of bridge rules which are applicable in data state S.

For a logic L, $F_L = \{s \in kb \,|\, kb \in KB_L\}$ is the set of formulas occurring in one of its knowledge bases. A *management base* is a set of operation names (briefly, operations) OP, defining elaborations that can be performed on formulas, e.g., addition of, revision with, etc. For a logic L and a management base OP, the set of operational statements that can be built from OP and F_L is $F_L^{OP} = \{o(s) \,|\, o \in OP, s \in F_L\}$. The semantics of such statements is given by a management function, which maps a set of operational statements and a knowledge base into a set of modified knowledge bases. In particular, a management function over a logic L and a management base OP is a function $mng : 2^{F_L^{OP}} \times KB_L \to 2^{KB_L} \setminus \emptyset$.

Semantics of an mMCS is in terms of *equilibria*. A data state $S = (S_1, \ldots, S_n)$ is an equilibrium for an MCS $M = (C_1, \ldots, C_n)$ iff, for $1 \leq i \leq n$, $kb_i' = S_i \in ACC_i(mng_i(app(S), kb_i))$.

Thus, an equilibrium is a global data state composed of acceptable data states, one for each context, encompassing inter-context communication determined by bridge rules and the elaboration resulting from the operational statements and the management functions. Equilibria may not exist (where conditions for existence have been studied, and basically require the avoidance of cyclic bridge-rules application), or may contain inconsistent data sets (local inconsistency, w.r.t. *local consistency*). A management function is called *local consistency (lc-) preserving* iff, for every given management base, kb' is consistent. It can be proved that an mMCS where all management functions are lc-preserving is locally consistent.

Notice that bridge rules are intended to be applied whenever they are applicable, so inter-context communication automatically occurs, though mediated via the management functions.

3   Background: DACMAS

We remind the reader that in DLR-Lite Ontologies [2] we have the following. (1) A TBox is a finite set of assertions specifying: concepts and relations; inclusion and disjunction among concepts/relations; key assertions for relations. (2) An ABox is a finite set of assertions concerning concept and relation membership, defined in accordance with the TBox. In essence, a TBox describes the structure of the data/knowledge, and the ABox specifies the actual data/knowledge instance. (3) In DLR-Lite, data can be queried via UCQs (Union of Conjunctive Queries) and ECQs (Existential Conjunctive Queries): the latter are FOL (First-Order Logic) queries involving negation, conjunction and the existential quantifier, whose atoms are UCQs.

Knowledge and data in agents composing a DACMAS are in fact represented in DRL-Lite: a DACMAS always includes an *institutional* agent which owns a "global" TBox, specifying properties of the domain in which the MAS operates, whereas each participating agent is equipped with its local ABox.

Communication among DACMAS agents may occur through simple event-based reaction, or via *commitments*, which are a relatively recent though very well-established general paradigm for agent interaction (cf. [9] and the references therein). A commitment $C_{x,y}(ant, csq)$ in particular relates a *debtor agent* x to a *creditor agent* y where x commits to bring about *csq* whenever *ant* holds. Commitment lifecycle (they can be created, fulfilled, canceled, etc.) is managed by a so-called "commitment machine".

Formally, a DACMAS (Data-Aware Commitment-based Multi-Agent System) is (from [1]) a tuple $\langle \mathcal{T}, \mathcal{E}, \mathcal{X}, \mathcal{I}, \mathcal{C}, \mathcal{B} \rangle$ where: (i) $\mathcal{X}$ is a finite set of agent specifications; (ii) $\mathcal{T}$ is a global DLR-Lite TBox, which is common to all agents participating in the system; (iii) $\mathcal{I}$ is a specification for the "institutional" agent; (iv) $\mathcal{E}$ is a set of predicates denoting events (where the predicate name is the event type, and the arity determines the content/payload of the event); (v) $\mathcal{C}$ is a contractual specification; (vi) and $\mathcal{B}$ is a Commitment Box (CBox). The global TBox includes the list of the names of all participating agents in connection to their specifications. Each agent is equipped with a local ABox, consistent with the global TBox, where however the ABoxes of the various agents are not required to be mutually consistent. The set consisting of the union of the global TBox and an agent's local ABox constitutes the agent's knowledge base. The institutional agent is a special agent who is aware of every message exchanged in the system, and can query all the ABoxes. In addition, it is responsible of the management of commitments, whose concrete instances are maintained in the Commitment Box $\mathcal{B}$, and it does so based on the *Commitment Rules* in $\mathcal{C}$, which define the commitment machine. An execution semantics for DACMASs is provided in [1], in terms of a transition system constructed by means of a suitable algorithm.

Each agent's specification includes a (possibly empty set of): *communicative rules*, which proactively determine events to be sent to other agents; *update rules*, which are internal reactive rules that update the local ABox upon sending/receiving an event to/from another agent.

A communicative rule has the form $Q(r, \hat{x})$ **enables** $EV(\hat{x})$ **to** r where: Q is an ECQ_l query, which is an ECQ with *location argument* l of the form $@Ag$ to specify the agent to which the query is directed (if omitted, then the agent queries its own ABox); $\hat{x}$ is a set of tuples representing the results of the query; $EV(\hat{x})$ is an event supported by the system, i.e., predicate EV belongs to $\mathcal{E}$; r is a variable, denoting an agent's name. Whenever the rule is proactively applied, if query Q evaluates to true (i.e., if the query succeeds) then $EV(\hat{x})$ and r are instantiated via one among the answers returned by the query, according to the agent's own choice. For instance, an agent can make an enquiry about the name r of the provider of some service: if several names are returned, only one is chosen. Then it sends to the selected provider a subscription request (instantiated with the necessary information $\hat{x}$) in order to be able to access the service.

Update rules are ECA-like rules[1] of the following form, where α is an action, the other elements are as before, and each rule is to be applied whenever an event is either sent or received, as specified in the rule itself:

on $EV(\hat{x})$ **to** r **if** $Q(r, \hat{x})$ **then** $\alpha(r, \hat{x})$(**on** − **send**/**on** − **receive**)

Update rules may imply the insertion in the agent's ABox of new data items not previously present in the system, taken from a countably infinite domain Δ. For instance, after subscription to a service an agent can receive offers and issue orders, the latter case determining the creation of a commitment (managed by the institutional agent).

An agent specification is a tuple $\langle sn, \mathcal{A}_{sn}, \Gamma \rangle$, where sn is the agent specification name, $\mathcal{A}_{sn}$ the local ABox, and Γ is the set of communicative and update rules characterizing the agent.

4 DACMACS

Let a logic, a management base and management functions be as specified in Sect. 2. The definition of a DACMACS (Data-Aware Commitment-based managed Multi-Agent-Contexts Systems) extends that of DACMASs as the set of participating agents is augmented with a set of contexts, to be understood as external data/knowledge sources which can be consulted by agents.

The global TBox specifies, as in DACMASs, the ontological specification shared by all agents in the system. We introduce explicit lists of agents' and context's names. We also introduce a global ABox, not present in DACMAS, which: links agents' names with their specification, and contexts' names with their *roles*, where a role specifies the function that a context assumes for agents. E.g., a context's name *studoff* may correspond to context role *student_office*, and context name *poldept* to *police_department*. For simplicity, we assume here that roles are specified as constants. We also assume that context names include all the information needed for actually posing queries (e.g., context names might coincide with their URIs). Each DACMACS agent may specify in its local ABox more information about context roles, or also list additional contexts (with corresponding roles) which are specifically known to that agent.

---

[1] As it is well-known, "ECA" rules stands for "Event-Condition-Action" rules, and specify reaction to events.

In DACMACS, agents' specification (seen below) is richer than in DACMAS, as it may include bridge rules, similar to those of mMCSs, whose activation is however proactive rather than mandatory. Each agent is equipped with local management functions.

Definition 1. *A DACMACS (Data-Aware Commitment-based managed Multi-Agent-Context System) is a tuple $\langle \mathcal{T}, \mathcal{E}, \mathcal{N}, \mathcal{X}, \mathcal{Y}, \mathcal{I}, \mathcal{A}, \mathcal{C}, \mathcal{B} \rangle$ where $\mathcal{T}, \mathcal{E}, \mathcal{X}, \mathcal{I}, \mathcal{C}$ and $\mathcal{B}$ are the same as for DACMASs, and: (i) $\mathcal{N}$ is a set of agents' names, listing the agents (beyond the institutional agent) which compose the MAS; (ii) $\mathcal{Y}$ is a set of contexts' names, listing the contexts which are globally known to the MAS; (iii) $\mathcal{A}$ is a global ABox, which is consistent with the global TBox and with all agents' local ABoxes.*

Context names in the global and local ABoxes might also be linked to the information about the related query language. However, again for the sake of simplicity though without loss of generality, we assume that all contexts accept datalog queries, in particular of the following form.

Definition 2. *An agent-to-context datalog query is defined as follows:*

$$Q :\text{-} A_1, \ldots, A_n, not\, B_1, \ldots, not\, B_m \ \ with\ n + m > 0$$

where the left-hand-side Q can stand in place of the right-hand-side. The comma stand for conjunction, and each of the A_is is either an atom or a binary expression involving connectives such as equality, inequality, comparison, applied to variables occurring in atoms and to constants. Each atom has a (possibly empty) tuple of arguments and can be either ground, i.e., all arguments are constants, or non-ground, i.e., arguments include both constants and variables, to be instantiated to constants in the query results. All variables which occur either in Q or in the B_is also occur in the A_is.

Intuitively, the conjunction of the A_is selects a set of tuples and the B_is rule some of them out. Q is essentially a placeholder for the whole query, but also projects over the wished-for elements of the resulting tuples.

Each context may include bridge rules, of the form specified in Sect. 2, where however the body refers to contexts only, i.e., contexts cannot query agents. The novelty of our approach is that also agents may be equipped with bridge rules, for extracting data from contexts. Agents' bridge rules are more general than those in mMCSs, as a bridge rule combines information extracted from a number of contexts, where each context returns data/knowledge according to a query of the form defined in Definition 2. As in DACMASs, we assume that the new data items possibly added to the agent's ABox belong to the same countably infinite domain Δ. However, as seen below bridge rules in agents are not automatically applied as in mMCSs: rather, they are proactively activated by agents upon need.

Definition 3. *A bridge rule occurring in an agent's specification has the following form.*

$A(\hat{x})$ **determinedby** $E_1, \ldots, E_k, not\, G_{k+1}, \ldots, not\, G_r$

where $A(\hat{x})$, *called the* conclusion *of the rule, is an atom over tuple of arguments* $\hat{x}$. *The right-hand-side is called the* body *of the rule, and is a conjunction of queries on external contexts. Precisely, each of the* E_i*s and each of the* G_i*s (where* $k > 0$ *and* $r \geq 0$*) can be either of the form* $DQ_i(\hat{x}_i) : c_i$ *or of the form* $DQ_i(\hat{x}_i) : q_i$ *where:* DQ_i *is a datalog query (defined according to Definition 2) over tuple of arguments* $\hat{x}_i$*;* c_i *is a context listed in the local ABox with its role, and thus locally known to the agent;* $q_i = Role@inst(role_i)$ *is a context name obtained by means of a standard query* $Role@inst$ *to the institutional agent* inst *(notation '@' is borrowed from standard DACMASs), performed by providing the context role* $role_i$*. We assume that all variables occurring in* $A(\hat{x})$ *and in each of the* G_i*s also occur in the* E_i*s. The comma stands for conjunction. Assuming (without loss of generality) that all the* $\hat{x}_i$*s have the same arity, when the rule is* triggered *(see Definition 4 below) then the* E_i*s may produce a set of tuples, some of which are discarded by the* G_i*s. Finally,* $A(\hat{x})$ *is obtained as a suitable projection. Within an agent, different bridge rules have distinct conclusions. The management operations and function are defined separately (see Definition 5 below).*

E.g., $Role@inst(student\_office)$ returns the name of the context corresponding to the student office. There is, as anticipated before, an important difference between bridge rules in contexts and bridge rules in agents. Each bridge rule occurring in a context is meant to be automatically applied whenever the present data state of the entire DACMACS entails the rule body. The new knowledge corresponding to the rule head is added (via the management function) to the context's knowledge base. Instead, bridge rules in agents are meant to be proactively activated by the agent itself. To this aim, we introduce new kinds of rules not present in DACMAS.

Definition 4. *A (timed)* trigger rule *is a proactive rule of the form*
    $Q(\hat{x})$ **enables** $A(\hat{y})$ $[Time \mid Frequency]$
where: Q *is an* ECQ_l *query, and* $\hat{x}$ *a set of tuples representing the results of the query;* $A(\hat{y})$ *is the conclusion of exactly one of the agent's bridge rules. If query* Q *evaluates to true, then* $A(\hat{y})$ *is (partially) instantiated via one among the answers returned by the query, according to the agent's own choice, and the corresponding bridge rule is triggered. This means, for a bridge rule of the form* $A(\hat{x})$ **determinedby** $E_1, \ldots, E_k, not\ G_{k+1}, \ldots, not\ G_r$, *that the conclusion* $A(\hat{x})$ *is allowed to be derived if the rule is applicable, i.e., if system's present data state entails its body. The options in square brackets, if specified, state that the bridge rule with conclusion* $A(\hat{y})$ *should be triggered either at time instant* $Time$ *(according to the system's own internal measurement), or at a frequency* $Frequency$, *expressed in terms of time instants.*

Since agents' bridge rules are executed neither automatically nor simultaneously, we have to revise the definition of management function with respect to the original definition of Sect. 2. First, notice that for each agent included in a DACMACS the underlying logic $(KB_L; Cn_L; ACC_L)$ is such that: KB_L is composed of the global TBox plus the global ABox and the agent's local ABox;

ACC_L is determined by the DRL-Lite semantics, according to which elements of Cn_L are computed. If an agent is equipped with n bridge rules then there will be n operators in the agent's management base, one for each bridge rule, i.e., $OP = \{op_1, \ldots, op_n\}$. Each of them may perform any form of reasoning, though it will at least make the acquired knowledge compatible with the global TBox, by means of techniques that we do not consider here. F_L^{OP} is defined as in Sect. 2, but instead of a single management function there will now be n management functions $mng_{b_1}, \ldots, mng_{b_n}$, one per each bridge rule. They can however be factorized within a single agent's management function with signature (as in MCSs), for agent i,

$$mng_i : 2^{F_L^{OP}} \times KB^L \rightarrow 2^{KB_L} \setminus \emptyset$$

which specializes into the mng_{b_i}s according to the bridge-rule head. Whenever a bridge rule is triggered, its result is interpreted as an agent's generated event and is reacted to via a special ECA rule:

Definition 5. *A* bridge-update rule *has the form* upon $A(\hat{x})$ **then** $\beta(\hat{x})$ *where:* $A(\hat{x})$ *is the conclusion of exactly one bridge rule, and $\hat{x}$ a set of tuples representing the results of the application of the bridge rule; $\beta(\hat{x})$ specifies the operator, management function and actions to be applied to $\hat{x}$.*

The definition of β may imply querying the ABoxes of the agent and of the institutional agent, performing knowledge format conversion, belief revision, etc., so as to re-elaborate the agent's ABox. The minimal requirement is that of keeping the ABox consistent: i.e., in mMCS terms, the management function specified in β is assumed to be lc-preserving. DACMACS agents' specification is therefore augmented w.r.t. that of DACMAS ones:

Definition 6. *An agent specification is a tuple $\langle sn, \mathcal{A}_{sn}, \Gamma \rangle$, where sn is the agent specification name, $\mathcal{A}_{sn}$ the local ABox, and Γ is the set of rules characterizing the agent. In particular, $\Gamma = \Gamma_{cu} \cup \Gamma_{btu} \cup \Gamma_{aux}$, where Γ_{cu} is, like in DACMAS, the set of communicative and update rules; in addition Γ_{btu} is the set of bridge, trigger and bridge-update rules, and Γ_{aux} the set of the necessary auxiliary rules (defining the predicates occurring in the body of the former rules).*

The definition of data state and of equilibria must be extended with respect to those provided in Sect. 2, and not only because a data state now includes both contexts' and agents' sets of consequences. As mentioned, in MCSs a bridge rule is applied whenever it is applicable. This however does not in general imply that it is applied only once, and that an equilibrium, once reached, lasts forever. In fact, contexts are in general able to incorporate new data items from the external environment (which may include, as discussed in [6], the input provided by sensors). Therefore, a bridge rule is in principle re-evaluated whenever a new result can be obtained, thus leading to evolving equilibria. In DACMACSs, there is the additional issue that for a bridge rule to be applied it is not sufficient that it is applicable in the MCS sense (i.e.-, when its body is true), but it must also be triggered by a corresponding timed trigger rule.

Similarly to what is done in Linear Time Logic (LTL) we assume a discrete, linear model of time where each state/time instant can be represented by an integer number. States $t_0, t_1, \ldots$ can be seen as time instants in abstract terms, as we have $t_{i+1} - t_i = \delta$, where δ is the actual interval of time after which we assume a given system to have evolved. In DACMACSs, both agents' and contexts' knowledge base contents may change not only in consequence of communication, but also due to interaction with an external environment. Thus, each agent's or context's knowledge base can be subjected to *updates*, where an update is understood as a finite set composed of new facts asserted to be true, and/or new facts asserted to be false, and/or already-known facts changing their truth value. Updates are not in general just "blindly" accepted, rather they are incorporated into the existing knowledge by means of some kind of *update operator* which performs knowledge base revision, for which several techniques exist that for lack of space we are not able to mention here.

In the definitions below, given a DACMACS M, we assume that $(C_1, \ldots, C_j)$ are the composing contexts and $(A_{j+1}, \ldots, A_m)$ the composing agents, $j \geq 0, m > 0$; let $n = j + m$.

Definition 7. *Let M be a DACMACS including $n = j + m$ contexts and agents. Let $\Pi = \Pi_1, \Pi_2, \ldots$ be a sequence of finite updates performed to agents' and contexts' private knowledge bases at time instants $t_1, t_2, \ldots$, respectively. Thus, for $i > 0$, $\Pi_i = \langle \Pi_{iC_1}, \ldots \Pi_{iC_j}, \Pi_{iA_{j+1}}, \ldots \Pi_{iA_m} \rangle$ is a tuple composed of the updates performed to each context and agent. Let $\mathcal{U}_E$, $E \in \{C_1, \ldots, C_j, A_{j+1}, \ldots, A_m\}$, be the* update operator *that each agent or context employs for incorporating the new information, and let $\mathcal{U}$ be the tuple composed of all these operators.*

Definition 8. *Let M be a DACMACS including $n = j + m$ contexts and agents. A* timed data state *of M at time T is a tuple $S^T = (S_1^T, \ldots, S_n^T)$ such that each S_i^T is an element of Cn_i at time T. The timed data state S^0 is an equilibrium according to an analogous definition as for mMCSs, i.e. iff, for $1 \leq i \leq n$, there exists $S_i^0 \in ACC_i(mng_i(app(S^0), kb_i))$.*

Each transition from a timed data state to the next one, and consequently the definition of an equilibrium, is determined both by the update operators and by the application of bridge rules, where in a DACMACS a bridge rule is applied only when it has been triggered, according to the specified timing/frequency.

Definition 9. *Let M be a DACMACS, and let S^T be a timed data state of M at time T. A bridge rule ρ occurring in each composing context or agent is potentially applicable in S^T iff S^T entails its body. For contexts, entailment is the same as in MCSs. For agents, entailment implies that all queries in the rule body succeed w.r.t. S^T. A bridge rule is applicable in a context whenever it is potentially applicable. A bridge rule with head $A(\hat{y})$ is applicable in an agent A_j whenever it is potentially applicable and there exists a trigger rule of the form $Q(\hat{x})$* enables *$A(\hat{y})$ in the specification of A_j such that $Cn_j \models Q(\hat{x})$. If the trigger rule is timed and specifies an application time $\hat{T}$, then it must be $T = \hat{T}$. If the trigger rule is timed and specifies a frequency $\hat{F}$, then it must be $T = \hat{F} * N$, for some N^2. Let $app(S^T)$ be the set of bridge rules which are applicable in data state S^T.*

---

[2] To be more precise, T is required to be the nearest approximation for $\hat{T}$ or respectively $\hat{F} * N$.

In DACMACSs, bridge rules are applied after performing the update of each component's knowledge base. So, data states following the initial one S^0 are equilibria if they are composed of acceptable data sets, considering however knowledge base update, which is performed before inter-agent-context communication determined by bridge rules.

Definition 10. *Let M be a DACMACS, let E_i, $1 \leq E \leq n$, be any composing agent/context, $\mathcal{U}_i$ be the corresponding update operator, Π be a given update sequence and Π_T^i be the update performed upon E_i at time T. Let S^T be a timed data state of M at time $T \leq 0$. A timed data state of M at time $T + 1$ a is an equilibrium iff, for $1 \leq i \leq n$, $S_i^{T+1} \in ACC_i(mng_i(app(S^T), \mathcal{U}_i(kb_i, \Pi_T^i)))$.*

Definition 11. *Let M be a DACMACS. A safe evolution trajectory of M w.r.t. a sequence Π of updates is a sequence $S^0, S^1, \ldots$ where the S^is, $i \geq 0$ are timed data states of M such that $\forall T \geq 0$, S^T is an equilibrium, and S^{T+1} is obtained from S^T and Π as specified in Definition 10.*

The above-introduced notions of equilibrium and safe evolution trajectory generalize those defined in [6] for reactive mMCSs, where input from sensor data is considered, and is coped with by extending bridge rules applicability. In our case, the introduction of an explicit update operator allows for sensor data incorporation, as well as for any other knowledge base manipulation. Updates and bridge rules application are independent, where at each state either the update or the set of applicable bridge rules or both can be empty or not. Bridge rules application is however performed upon the contents of the updated knowledge bases.

4.1   Properties of DACMACSs

In the terminology of [5], we require all management functions (both those related to agents and those related to contexts) to be *local consistency (lc-)preserving*. We also require the update operators to preserve consistency of agents' and contexts' knowledge bases. We thus obtain the following, which is the analogous to Proposition 2 stated in [5] for mMCSs:

Proposition 1. *Let D be a DACMACS such that all update operators and management functions associated to the composing agents and contexts are lc-preserving. Then D is locally consistent.*

Global consistency of all data sets in a DACMACS is not required here, as it is not required in DACMAS in the first place: in fact, incoherences may reflect agents' local views. By adopting the A-ILTL (Agent-Oriented Interval LTL) interval linear temporal logic [10,11], several interesting properties of a DACMACS can be defined and verified. For instance, for proposition φ, it can be checked whether φ holds for agent A included in a DACMACS (i.e., in the above terminology, $A \in \{A_{j+1}, \ldots, A_n\}$) in some equilibrium reached at a certain time or within some time interval, given sequence Π of updates and a resulting safe evolution trajectory.

A simple property that might be interesting to check can be for instance $F_n \varphi_{A_v}$ which specifies that φ becomes true in agent A_v at some time less than or equal to n (F standing for "eventually", or "finally"). Given the above definitions, this accounts to check whether

$$(\varphi \in S_v^0) \vee (\varphi \in S_v^n) \vee \exists m \text{ where } 0 < m < n \text{ such that } F_m \varphi_{A_v}$$

For finite updates this and many other properties are decidable, where the complexity of check depends upon the complexity of the update operators and of the management functions. However, to deal with complexity an approach is proposed in [11] for run-time rather than a-priori checking of such properties.

5   Medical Example

We will now introduce an example (necessarily simple due to lack of space) inspired to a real-life medical problem. Let us model the situation in DAC-MACS's terms, assuming to have an agent called *Paula* like the patient we imagine it is designed to care for. Let us assume the agent's ABox to be defined as follows (we basically adopt the datalog notation, because it is widely known). The patient, Paula, is a 78 years old lady. Being 78 implies being elderly, which is not a disease but nevertheless it is, from the sanitary point of view, a special condition.

> *person*(*paula*).   *age*(*paula*, *78*).
> *special_condition*(*X*, *elderly*) :- *person*(*X*), *age*(*X*, *A*), *A* > 70.

Let us now assume that this elderly lady has certain health conditions, and takes certain medicaments in relation to these conditions. In particular, she takes an anticoagulant (and other medicaments not listed here) for coping with a cardiac insufficiency.

> *has_disease*(*paula*, *cardiac_insufficiency*).
> *takes_medicament*(*P*, *anticoagulant*) :- *has_disease*(*P*, *cardiac_insufficiency*).

Unfortunately, Paula presents some disquieting symptoms, namely extreme weakness, that can be attributed to low hemoglobin level. Then, it is stated that hemoglobin level is obtained within a blood test, and that its value is worrying (cannot be explained in a trivial way) if it is less than 10.

> *has_symptom*(*paula*, *extreme_weakness*).
> *symptom_cause*(*extreme_weakness*, *hemoglobin_level*).
> *blood_test*(*hemoglobin_level*).
> *anomalous_value*(*P*, *hemoglobin_level*) :- *person*(*P*), *value*(*P*, *hemoglobin*, *V*), *V* ≤ *10*.

The following trigger rule states that, if a person P has a symptom S which has a possible cause T detectable as a blood test with value V, then: the agent has to obtain from the institutional agent *inst* the specification name of the context corresponding to the blood test records via the query *Spec@inst*(*Rec*, *blood_tests_records*), which returns its result in the variable *Rec*.

So, the bridge rule for testing Paula's hemoglobin level V by querying this context can be proactively activated.

$person(P)$ AND $has\_symptom(P, S)$ AND
$symptom\_cause(S, T)$ AND $blood\_test(T)$ AND
$Spec@inst(Rec, blood\_tests\_records)$ **enables** $blood\_test\_result(Paula, T, V, Rec)$.

The bridge rule simply queries Rec for the required value V of parameter T for person P (assuming that the blood test record context by default returns the most recent value).

$blood\_test\_result(P, T, V, Rec)$ **determinedby** $value(P, T, V) : Rec$.

he result is then simply added to the agent's ABox by a bridge-update rule:

upon $value(P, T, V)$ **then** $add(value(P, T, V))$.

The following trigger rule states that if a person which has a certain disease and some special condition has an anomalous value of some factor V, then the medical diagnosis context $Diag$ (whose name is obtained as before from the institutional agent) must be consulted in order to obtain the list C of possible causes.

$person(P)$ AND $has\_disease(P, D)$ AND
$special\_condition(P, S)$ AND $anomalous\_value(P, V)$ AND
$Spec@inst(Diag, medical\_diagnosis)$ **enables** $poss\_causes(P, D, S, V, C, Diag)$.

A corresponding sample bridge rule can be the following (more realistically however, several diagnostic knowledge bases might be consulted via more involved queries).

$poss\_causes(P, D, S, V, C, Diag)$ **determinedby** $poss\_causes(P, D, S, V, C) : Diag$.

The related bridge-update rule starts a reasoning process to infer from given parameters and possible causes C the most plausible cause R.

upon $poss\_causes(P, D, S, V, C, Diag)$ **then** $infer\_plausible(P, D, S, V, C, R)$.

Actually R was, for Paula, an internal hemorrhage due to the long-termed use of anticoagulants. So, her life was saved and her relatively good health and satisfactory quality of life have been restored.

6   Concluding Remarks

In this paper we proposed DACMACS, which is a flexible framework augmenting agents with controlled interaction and knowledge exchange with external heterogeneous data sources. We have defined notions of timed data states and equilibria. Similarly to what argued for DACMAS, instances of DACMACS are implementable via publicly available technologies, taking as basis any of the existing logical agent-oriented programming languages. The execution semantics of a DACMACS can be defined by extending the *transition system* defined for DACMAS in [1]. Thus, results provided in [1] for DACMASs about verification using the μ-calculus can be extended to DACMACSs, though we defer discussion to a future paper.

Among the future directions of this work we may include the possibility to equip also contexts with ontological descriptions, and cope with ontology conversions in bridge rules application. We might also consider the aspect, not tackled in mMCSs either, of actual applicability of bridge rules, which occurs only if needed contents are delivered within a deadline. The issue of verification and of the balance among a-priori and run-time verification deserves further exploration.

References

1. Montali, M., Calvanese, D., De Giacomo, G.: Verification of data-aware commitment-based multiagent system. In: Proceedings of AAMAS (2014)
2. Baader, F., Calvanese, D., McGuinness, D.L., Nardi, D., Patel-Schneider, P.F.: The Description Logic Handbook: Theory, Implementation, and Applications. Cambridge University Press, Cambridge (2003)
3. Omicini, A., Ricci, A., Viroli, M.: Artifacts in the a&a meta-model for multi-agent systems. Auton. Agent. Multi-Agent Syst. **17**(3), 432–456 (2008)
4. Brewka, G., Eiter, T.: Equilibria in heterogeneous nonmonotonic multi-context systems. In: Proceedings of the 22nd AAAI Conference on Artificial Intelligence, pp. 385–390. AAAI Press (2007)
5. Brewka, G., Eiter, T., Fink, M.: Nonmonotonic multi-context systems: a flexible approach for integrating heterogeneous knowledge sources. In: Balduccini, M., Son, T.C. (eds.) Logic Programming, Knowledge Representation, and Nonmonotonic Reasoning. LNCS, vol. 6565, pp. 233–258. Springer, Heidelberg (2011)
6. Brewka, G., Ellmauthaler, S., Pührer, J.: Multi-context systems for reactive reasoning in dynamic environments. In: ECAI 2014, Proceedings of the 21st European Conference on Artificial Intelligence. IJCAI/AAAI (2014)
7. Brewka, G., Eiter, T., Fink, M., Weinzierl, A.: Managed multi-context systems. In: IJCAI 2011, Proceedings of the 22nd International Joint Conference on Artificial Intelligence, IJCAI/AAAI, pp. 786–791 (2011)
8. Apt, K.R., Bol, R.: Logic programming and negation: a survey. J. Log. Program. **19–20**, 9–71 (1994)
9. Singh, M.P.: Commitments in multiagent systems: some history, some confusions, some controversies, some prospects. In: Paglieri, F., Tummolini, L., Falcone, R., Miceli, M. (eds.) The Goals of Cognition. Essays in Honor of Cristiano Castelfranchi, pp. 601–626. College Publications, London (2012)
10. Costantini, S.: Self-checking logical agents. In: Proceedings of the Eighth Latin American Workshop on Logic, Languages, Algorithms and New Methods of Reasoning LA-NMR 2012. CEUR Workshop Proceedings of the vol. 911, pp. 3–30 CEUR-WS.org (2012). Invited Paper, Extended Abstract in Proceedings of the AAMAS 2013
11. Costantini, S., De Gasperis, G.: Runtime self-checking via temporal (meta-)axioms for assurance of logical agent systems. In: Proceedings of the LAMAS 2014, 7th Workshop on Logical Aspects of Multi-Agent Systems, held at AAMAS 2014, pp. 241–255 (2014)

Digital Forensics Evidence Analysis: An Answer Set Programming Approach for Generating Investigation Hypotheses

Stefania Costantini[1]([✉]), Giovanni De Gasperis[1], and Raffaele Olivieri[1,2]

[1] Dipartimento di Ingegneria e Scienze dell'Informazione e Matematica,
Universitá degli Studi dell'Aquila, Via Vetoio 1, 67100 L'Aquila, Italy
{stefania.costantini,giovanni.degasperis}@univaq.it
[2] The Italian Department of Scientific Investigations of Carabinieri,
Raggruppamento Carabinieri Investigazioni Scientifiche (Ra.C.I.S.),
Viale di Tor di Quinto 119, 00191 Rome, Italy
raffaele.olivieri@gmail.com

Abstract. The results of the evidence analysis phase in Digital Forensics (DF) provide objective data which however require further elaboration by the investigators: in fact, they must contextualize analysis results within an investigative environment so as to provide possible hypotheses that can be proposed as proofs in court, to be evaluated by lawyers and judges. Aim of our research has been that of exploring the applicability of Answer Set Programming (ASP) to the automatization of evidence analysis. This brings many advantages, among which that of making different possible investigative hypotheses explicit, whereas different human experts working on the case often devise and select, relying on intuition, discordant interpretations. Very complex investigations for which human experts can hardly find solutions turn out in fact to be reducible to optimization problems in classes P or NP or not far beyond, that can thus be expressed in ASP. As a proof of concept, in this paper we present the formulation of some real investigative cases via simple ASP programs, and discuss how this leads to the formulation of concrete investigative hypotheses.

1 Introduction

Digital Forensics (DF) is a branch of criminalistics which deals with the identification, acquisition, preservation, analysis and presentation of the information content of computer systems, or in general of digital devices [1,2]. The aim is to identify digital sources of proofs, and to organize such proofs in order to make them robust in view of their discussion in court, either in civil or penal trials. Digital forensics is concerned with the analysis of potential elements of proof after a crime has been committed ("post-mortem"). Clearly, the development of digital forensics is highly related to the development of Information and Communication Technologies in the last decades, and to the widespread diffusion of electronic devices and infrastructures. It involves several disciplines such as

F. Calimeri et al. (Eds.): LPNMR 2015, LNAI 9345, pp. 242–249, 2015.
DOI: 10.1007/978-3-319-23264-5_21

computer science, electronic engineering, various branches of law, investigation techniques and criminological sciences. Rough evidence must be in fact used to elicit hypotheses concerning events, actions and facts (or sequences of them) with the goal to present them in court. Evidence analysis involves examining fragmented incomplete knowledge, and defining complex scenarios by aggregation, likely involving time, uncertainty, causality, and alternative possibilities. No single methodology exists today for digital evidence analysis. The scientific investigation experts usually proceed by means of their expertise, which results from a mix of experience and intuition.

As a matter of fact, evidence acquisition is supported by a number of hardware and software tools, either closed- or open- source. These tools are continuously evolving to follow the evolution of the involved technologies and devices. Evidence analysis is instead much less supported. In evidence analysis, technicians and experts perform the following tasks: (i) collect, categorize and revise the evidence items retrieved from electronic devices; (ii) examine them so as to hypothesize the possible existence of a crime and potential crime perpetrators; (iii) elicit from the evidence possible proofs that support the hypotheses; (iv) organize and present the proofs in a form which is acceptable by the involved parties, namely lawyers and judges, which may include to exhibit explicit supporting arguments. Few software tools exist that cover only some partial aspects. Furthermore, all of them are "black box" tools, i.e., they provide results without motivation or explanation, and without any possibility of verification. Thus, such results can hardly be presented as reliable proofs to the involved parties. Moreover, the absence of decision support systems leads to undesirable uncertainty about the outcome of evidence analysis. Often, different technicians analyzing the same case reach different conclusions, and this may determine different judge's decisions in court.

Formal and verifiable artificial intelligence and automated reasoning methods and techniques for evidence analysis would be very useful for the elicitation of sources of proof. Several aspects should to be taken into account such as timing of events and actions, possible (causal) correlations, context in which suspicious actions occurred, skills of the involved suspects, validity of alibis, etc. Moreover, given available evidence, different possible underlying scenarios may exist, that should be identified, examined and evaluated. All the above should be performed via "white box" techniques, meaning that such techniques should be verifiable with respect to the results they provide, how such results are generated, and how the results can be explained. The new wished-for software tools should be reliable and provide a high level of assurance, in the sense of confidence in the system's correct behavior. Computational logic is a suitable candidate to definition and implementation of such tools, and non-monotonic reasoning is clearly extensively required.

The long-term objective of this research is to provide law enforcement, investigators, intelligence agencies, criminologists, public prosecutors, lawyers and judges with decision-support-systems that can effectively aid them in their activities. The adoption of such systems can contribute to making legal proceedings clearer and faster, and also under some respects more reliable. In fact, the choice of computational logic as a basis guarantees transparency and verifiability of tools and results. The objective of the present paper is to provide a proof-of-concept of the applicability of computational logic and non-monotonic reasoning to such tasks. To this aim we adopted Answer Set Programming (ASP, cf., among others, [3–7]), rather than other equally suitable non-monotonic reasoning tools, because ASP programs are declarative and readable even by the non-expert.

In fact, in order to convince the several parties involved, whatever limited their computer science expertise might be, we have considered fragments of some investigative cases and have transposed them into existing and popular combinatorial problems. We have then represented such problems via simple self-explanatory answer set programs which provide results which are easy to understand. However, we have considered fragments of real cases which are presently being investigated by the Italian Department of Scientific Investigations of Carabinieri[1] as in fact one of the authors of this paper as a member officer of a DF laboratory. Even though the encodings of the combinatorial problems are known (and very simple), the mapping of the DF cases to such problems is novel. The reader may refer to [8] for an extended version of this paper, including the answer set programming encoding of all the examples, and the answer sets resulting from the experiments. In the next sections we present three sample cases, and then conclude.

2   Case 1: Data Recovery and File Sharing Hypotheses

The Investigative Case. The judicial authority requested the digital forensics laboratory to analyze the contents of an hard disk, in order to check for the presence of illegal contents files. If so, they requested to check for potential activities of sharing on Internet of illegal materials.

The hard disk under analysis was physically damaged (as often done by criminals if they suspect capture). Therefore, after a head replacement, the evidence acquisition phase recovered a large amount of files (of various types: images, videos, documents, etc.), however without their original name. This because the damage present on the disk plates disallowed the recovery the information of the MFT[2]. For this reason, an arbitrary name has been provisionally assigned to all the recovered files. Information about the real file names and their original location in the file system is thus missing.

Elements. By analyzing the recovered files, technicians detected the occurrence of: (i) Files with illegal contents. (ii) Various "INDX files", corresponding in the NTFS file system to directory files, which contain the follow META-DATA: filename; physical and logical size of the file; created, accessed, modified and changed timestamps. (iii) Index related to eMule (which is a widely-used file-exchange application), including a file containing sharing statistics, whose original name is "known.met".

---

[1] The Police branch of the Italian Army http://www.carabinieri.it.

[2] Master File Table: structured block table containing the attributes of all files in the volume of NTFS file systems, which are those used in Windows operating systems.

Starting from the elements described above, we have been able to reply to the judicial authority's question with: reasonably reliable hypotheses of association of the recovered files to the respective original names; a reasonable certainty that illegal files were actually exchanged on the Internet. This has been obtained by modeling the given problem by means of a very simple well-known ASP example.

The Stable Marriage Problem. The Marriage Problem (or SMP - Stable Marriage Problem) is an NP-hard optimization problem which finds a stable matching between two sets of elements S_1 and S_2 (say *men* and *women*) given a set of preferences for each element. A matching is a mapping from the elements of one set to the elements of the other set which thus creates a set of pairs (A, B) where $A \in S_1$ and $B \in S_2$. A matching is stable whenever it is not the case that: (i) some element $\hat{A}$ of the first matched set prefers some given element $\hat{B}$ of the second matched set over the element to which $\hat{A}$ is already matched, and (ii) $\hat{B}$ also prefers $\hat{A}$ over the element to which $\hat{B}$ is already matched.

Reduction. The given problem is reducible to SMP as follows. *men*: defined as the set of names extracted from directory files "INDX files"; *women*: defined as the set of recovered files which have been provisionally assigned arbitrary names. The *preferences list* (or relation order) between the *men* and *women* sets is derived from the comparison of the properties of the individual recovered files (file type, size, etc.) with those identified in "INDX files". From the answer sets resulting from the encoding reported in [8], it has been possible to formulate hypotheses about the original names of the recovered files. Moreover, by comparing the file names indexed in the file *known.met*[3], it has also been possible to make reasonable assumptions about the effective sharing of files with illegal content.

3 Case 2: Path Verification

The Investigative Case. After a heinous crime, an allegedly suspect has been arrested. The police sequestered all his mobile devices (smartphone, route navigator, tablet, etc.). The judicial authority requested the DF laboratory to analyze the digital contents of the mobile devices in order to determine their position with respect to the crime site during an interval of time which includes the estimated time when the crime was perpetrated.

Elements. From the analysis of the mobile devices, a set of geographical GPS coordinates have been extracted, some of them related to the the time interval under investigation. There are however some gaps, one of them certainly due to a proven switch off of few minutes around the crime time. To start with, a list called GPS-LIST is generated, collecting all the positions extracted from the various devices, grouped and ordered by *time unit* of interest (seconds, multiple

---

[3] As mentioned, known.met is a file of the widely-used *eMule* file-exchange application that stores the statistics of all files that the software shared, all files present in the download list and downloaded in the past.

of seconds, minutes, etc.). The objective is that of establishing whether the known GPS coordinates are compatible with some path which locates the given mobile devices at the crime site during the given time interval. If no such path exists, then the suspect must be discharged. If some compatible path is found, then the investigation about the potential perpetrator can proceed. The objective has been reached via reduction to the following simple game.

Hidato Puzzle (Hidoku). Hidato is a logical puzzle (also known as "Hidoku") invented by the Israeli mathematician Dr. Gyora Benedek. The aim of Hidato is to fill a matrix of numbers, partially filled a priori, using consecutive numbers connected over a horizontal, vertical or diagonal ideal line. Below we show, as a simple example, a 6 × 6 initial matrix.

18	0	0	0	26	0
19	0	0	27	0	0
0	14	0	0	23	31
1	0	0	8	33	0
0	0	5	0	0	0
0	0	10	0	36	35

Reduction. It has been possible to perform the reduction of the given investigation problem to the "Hidato Puzzle" problem, by creating a matrix representing the geographical area of interest where each element of the matrix represents a physical zone crossable in a unit of time. The physical size of the individual cell of the matrix (grid) on the map will be proportionate to both the time unit to be considered and the hypothetical transfer speed. The matrix has been populated with the elements of the previously-created LIST-GPS, i.e., with known positions of the suspect.

Considering the above matrix, assume that the crime has been committed at location 34, at a time included in the interval with lower bound corresponding to when the suspect was at location 1 and upper bound corresponding to when the suspect was at location 36. All devices have been provably switched off between locations 5 and 10.

The two answer sets resulting from the encoding reported in [8] for the given example, and shown in the picture below, actually correspond to paths which are compatible with the hypothesis of the suspect committing the crime. In fact, location 34 consistently occurs between locations 5 and 10.

4   Case 3: Alibi Verification

The Investigative Case. During an investigation concerning a bloody murder, it is necessary to check the alibi provided by a suspect. In the questioning, the suspect has been rather vague about the timing of his movements. However, he declared what follows: to have left home (place X) at a certain time; to have

18	20	28	29	26	25
19	17	21	27	30	24
13	14	16	22	23	31
1	12	15	8	33	32
2	11	5	9	7	34
3	4	10	6	**36**	35

18	20	28	29	26	25
19	17	21	27	30	24
16	14	13	22	23	31
1	15	12	8	33	32
2	11	5	9	7	34
3	4	10	6	**36**	35

reached the office at place Y where he worked on the computer for a certain time; to have subsequently reached place Z where, soon after opening the entrance door, he discovered the body and raised the alarm. In order to verify the suspect's alibi, the judicial authority requested the DF laboratory to analyze: the contents of the smartphone owned by the suspect; the computer confiscated in place Y, where the suspect says to have worked; a video-surveillance equipment installed at a post office situated near place Z, as its video-camera surveys the street that provides access to Z.

Elements. The coroner's analysis on the body has established the temporal interval including the time of death. From the forensic analysis of the smarphone it has been possible to compile a list of GPS positions related to a time interval including the time of death, denoted by GPS-LIST. The analysis of the computer allowed the experts to extract the list of accesses on the day of the crime, denoted by LOGON-LIST. The analysis of the video-surveillance equipment allowed the experts to isolate some sequences, denoted by VIDEO-LIST, that show a male subject whose somatic features are compatible with the suspect. All the above lists have been ordered according to the temporal sequence of their elements. The investigation case at hand can be modeled as a planning problem where time is a fundamental element in order to establish whether a sequence of actions exist that may allow to reach a certain objective within a certain time. Several approaches to causal and temporal reasoning in ASP exist, that could be usefully exploited for this kind of problem[4]. Here, for lack of space and for the sake of simplicity we model the problem by means of the very famous "Monkey & Banana" problem, which is the archetype of such kind of problems in artificial intelligence.

Monkey and Banana. The monkey and banana problem is a well-known toy problem in artificial intelligence, particularly in logic programming and planning. The specification of "Monkey & Banana" is the following: A monkey is in a room. Suspended from the ceiling is a banana, beyond the monkey's reach. In the room there is also a chair (in some versions there is a stick, that we do not consider). The ceiling is just the right height so that a monkey standing on a chair could knock the banana down (in the more general version by using the stick, in our version just by hand). The monkey knows how to move around, carry other things around, reach for the banana. What is the best sequence of actions for the monkey? The initial conditions are that: the chair is not just

---

[4] For lack of space we cannot provide the pertinent bibliography: please refer to [9] and to the references therein.

below the bananas, rather it is in a different location in the room; the monkey is in a different location with respect to the chair and the bananas.

Reduction. The reduction of the case at hand to the "Monkey & Banana" problem is the following. Monkey = Suspect. Banana = Body. Eats Banana = Raise Alarm. Initial Position Monkey = X. Initial Position Chair = Y. Below Banana = Z. Walks = Walks. Move Chair = Motion to Z. Ascend = Open the Door. Idle = Unknown Action. Notice the reduction of the "idle" state of the monkey to unknown actions that the suspect may have performed at that time.

Problem's constraint are that, at any time, the monkey: may perform only one action at each time instant among walk, move chair, stand on chair, or stay idle; if the monkey stands on the chair, it cannot walk, and it cannot climb further; if the chair is not moved then it stays where it is, and vice versa if it is moved it changes its position; the monkey is somewhere in the room, where it remains unless it walks, which implies changing position; the monkey may climb or move the chair only if it is in the chair's location; the monkey can reach the banana only if it has climbed the chair, and the chair is under the banana.

Among the answer sets resulting from the encoding reported in [8] there are many which suggest suspicious behavior. In particular, one outlines a scenario where the initial suspect's actions are unknown. Then he moves to the crime site where however he has the time and opportunity to commit the crime at step 4. Even worse is another answer set, where the suspect moves to the crime site, than moves back to the office, moves a second time to the crime site where again he has the time and opportunity to commit the crime at step 4. As the suspect's presence at the crime site is confirmed by the video-surveillance equipment records, this behavior is suggestive of, e.g., going to meet the victim and having a discussion, going back to the office (maybe to get a weapon) and then actually committing the crime.

5   Conclusions

In this paper we have demonstrated the applicability of non-monotonic reasoning techniques to evidence analysis in digital forensics by mapping some fragments of real cases to existing simple answer set programs. The application of artificial intelligence and in particular of non-monotonic reasoning techniques to evidence analysis is a novelty: in fact, even very influential publications in digital forensics such as [1,2] are basically a guide for human experts about how to better understand and exploit digital data. Therefore the present work, though preliminary, opens significant new perspectives.

Future developments include building a toolkit exploiting not only ASP but also other non-monotonic-reasoning techniques such as abduction, temporal reasoning, causal reasoning and others, as elements of decision-support-systems that can effectively aid investigation activities and support of the production of evidence to be examined in trial. The multidisciplinary future challenge is that of making such tools formally accepted in court proceedings. From the technical point of view, for making such tools acceptable and perceived as reliable it is crucial to develop verification, certification, assurance and explanation techniques.

References

1. Casey, E.: Handbook of Digital Forensics and Investigation. Elsevier, California (2009)
2. Casey, E.: Digital Evidence and Computer Crime: Forensic Science, Computers, and the Internet. Elsevier, London (2011). books.google.com
3. Gelfond, M., Lifschitz, V.: The stable model semantics for logic programming. In: Kowalski, R., Bowen, K. (eds.) Proceedings of the 5th International Conference and Symposium on Logic Programming, pp. 1070–1080. MIT Press (1988)
4. Gelfond, M., Lifschitz, V.: Classical negation in logic programs and disjunctive databases. New Gener. Comput. **9**, 365–385 (1991)
5. Baral, C.: Knowledge Representation, Reasoning and Declarative Problem Solving. Cambridge University Press, Cambridge (2003)
6. Leone, N.: Logic programming and nonmonotonic reasoning: from theory to systems and applications. In: Baral, C., Brewka, G., Schlipf, J. (eds.) LPNMR 2007. LNCS (LNAI), vol. 4483, p. 1. Springer, Heidelberg (2007)
7. Truszczyński, M.: Logic programming for knowledge representation. In: Dahl, V., Niemelä, I. (eds.) ICLP 2007. LNCS, vol. 4670, pp. 76–88. Springer, Heidelberg (2007)
8. Costantini, S., DeGasperis, G., Olivieri, R.: How answer set programming can help in digital forensic investigation. In: Ancona, D., Maratea, M., Mascardi, V. (eds.) 30th Convegno Italiano di Logica Computazionale (Italian Conference on Computational Logic), CILC2015, Proceedings, University of Genova (2015). To appear on CEUR Workshop Proceedings. http://cilc2015.dibris.unige.it
9. Cabalar, P.: Causal logic programming. In: Erdem, E., Lee, J., Lierler, Y., Pearce, D. (eds.) Correct Reasoning. LNCS, vol. 7265, pp. 102–116. Springer, Heidelberg (2012)

A Formal Theory of Justifications

Marc Denecker[1]([✉]), Gerhard Brewka[2], and Hannes Strass[2]

[1] Department of Computer Science, K.U. Leuven, 3001 Heverlee, Belgium
marcd@cs.kuleuven.be
[2] Computer Science Institute, Leipzig University, Leipzig, Germany

Abstract. We develop an abstract theory of justifications suitable for describing the semantics of a range of logics in knowledge representation, computational and mathematical logic. A theory or program in one of these logics induces a semantical structure called a justification frame. Such a justification frame defines a class of justifications each of which embodies a potential *reason* why its facts are true. By defining various evaluation functions for these justifications, a range of different semantics are obtained. By allowing nesting of justification frames, various language constructs can be integrated in a seamless way. The theory provides elegant and compact formalisations of existing and new semantics in logics of various areas, showing unexpected commonalities and interrelations, and creating opportunities for new expressive knowledge representation formalisms.

1 Introduction

In this paper we introduce a new semantical framework suitable for describing semantics of a broad class of existing and new (nonmonotonic) logics. These logics are from the area of knowledge representation, nonmonotonic reasoning and mathematical logic. The purpose of the framework is four-fold: (1) By providing a unified semantical account of different semantical principles, it highlights the differences between different semantics within the same formalism (e.g., various semantics of logic programs) and (2) it highlights common semantical principles of different formalisms (e.g. logic programs vs. argumentation frameworks); (3) for existing formalisms it gives rise to new semantics nobody has thought of before; (4) last but not least it provides ways for seamless integration of various expressive language constructs in a single logic.

As for (4), it is a fundamental goal of the field of knowledge representation to build expressive languages, that is, ones that provide a range of different language constructs. It is well-known that (certainly for nonmonotonic logics) extending a logic with new language constructs can be very difficult. An illustration is the saga of extending logic and answer set programming with aggregates [1,2], a topic on which many effort-years, several PhD theses and countless papers have been devoted. Clearly, it would be of great value to have a clean, modular way to compose (nonmonotonic) logics from existing language constructs. A powerful such principle is presented here, in the form of *nesting*.

© Springer International Publishing Switzerland 2015
F. Calimeri et al. (Eds.): LPNMR 2015, LNAI 9345, pp. 250–264, 2015.
DOI: 10.1007/978-3-319-23264-5_22

The framework we present is based on so-called justification frames. They specify a class of defined and parameter facts and include a set of semantic rules $x \leftarrow S$ (x a defined fact, S a set of facts) that provide potential reasons for defined facts x. Justifications J are graphs obtained by concatenation of such reasons. In the context of an interpretation $\mathfrak{A}$, a justification J may justify a fact x or not, depending on the branches below x in J. The value of these branches in $\mathfrak{A}$ is specified by a mapping from sequences of facts to $\{t, f\}$, called a *branch evaluation*. A justification system consists of a justification frame together with a branch evaluation. A *supported interpretation* of such a system is one in which true facts x are those with a justification J in which all branches from x evaluate to t. A very useful feature of the framework is the nesting of justification systems. This yields a powerful way for meaning-preserving integration of different language constructs by nesting the justification system of one construct in that of another.

Several semantical frameworks of a seemingly similar kind already exist. In particular, our framework is a (substantial) extension of a justification framework for logic programming proposed in [3]. Another framework that comes to mind is approximation fixpoint theory (AFT) developed by Denecker, Marek and Truszczyński [4]. We believe that the framework presented here has some clear advantages over AFT:

- AFT is a more abstract algebraical operator-based approach, whereas the justification framework rests on a logical notion: *a justification as a reason for a fact to hold*. The advantage of AFT's abstract approach is that it is applicable to a broader range of logics, (e.g., logic programming, AEL, default logic). The advantage of the logical approach here is that it is much more intuitive.
- (Direct) application of AFT induces non-standard *ultimate* variants of stable and well-founded semantics [5], whereas the justification framework formalizes standard versions.
- The justification framework identifies a single source of difference in various semantics, namely the way infinite branches are evaluated in justification graphs.
- The nesting of justification systems is a new technique for seamless integration of complex language constructs (e.g., aggregates, rule sets) in one logic. Nesting is a feature derived from μ-calculus [6] and nested fixpoint logics [7]. This feature is not supported by AFT nor in any other semantic frameworks that we know of.
- Last but not least, justifications are used in the implementation of practical systems such as clasp [8] and IDP [9].

As a consequence, to the best of our knowledge, no other framework (including AFT) is currently capable of formalizing and integrating the logics treated in this paper. These logics are from mathematical logic and formal methods, knowledge representation and nonmonotonic reasoning, and include logic programs and answer set programs under various semantics [10], abstract argumentation [11], inductive definitions, co-inductive logic programming [12] and nested inductive/coinductive definitions [13].

2   Justification Frames

Let $\mathcal{F}$ be a set and $\sim: \mathcal{F} \to \mathcal{F}$ an involution, that is, a bijection that is its own inverse (hence for all $x \in \mathcal{F}$, $\sim\sim x = x$). We assume $\mathcal{F}$ has a partition $\{\mathcal{F}^p, \mathcal{F}^n\}$ such that $\sim\mathcal{F}^p = \mathcal{F}^n$ (and vice versa). We call elements of $\mathcal{F}$ facts, of $\mathcal{F}^p$ positive facts, and of $\mathcal{F}^n$ negative facts. We call $\sim x$ the complement of x. The operator $\sim$ is called the complementation operator. A good intuition for the complementation operator is as (classical) negation. We frequently call $\mathcal{F}$ a *fact space*.

As will become clear, it can be useful to have facts $t, u, i \in \mathcal{F}^p$ and their complements $\sim t, \sim u, \sim i \in \mathcal{F}^n$. We call them *logical* facts. For convenience, we denote $\sim t$ as f. They are interpreted facts. They stand respectively for *true, unknown (positive), inconsistent (positive), false, unknown (negative) and inconsistent (negative)*.

Definition 1. *An interpretation $\mathfrak{A}$ is a subset of $\mathcal{F}$ such that $t, i, \sim i \in \mathfrak{A}$ and $f, u, \sim u \notin \mathfrak{A}$ when present in $\mathcal{F}$. The set of interpretations is denoted $\mathbb{I}$. An interpretation $\mathfrak{A}$ is* consistent *if for every non-logical fact $x \in \mathcal{F}$, $\mathfrak{A}$ contains at most one of x and $\sim x$. $\mathfrak{A}$ is* anti-consistent *if for every non-logical fact $x \in \mathcal{F}$, $\mathfrak{A}$ contains at least one of x and $\sim x$. $\mathfrak{A}$ is* exact *if consistent and anti-consistent.*

Interpretations as defined above can be viewed as 4-valued interpretations, consistent interpretations correspond to 3-valued interpretations, exact interpretations to standard 2-valued interpretations. In particular, x is true in $\mathfrak{A}$ if $x \in \mathfrak{A}$ and $\sim x \notin \mathfrak{A}$; false if $x \notin \mathfrak{A}$ and $\sim x \in \mathfrak{A}$; unknown if $x, \sim x \notin \mathfrak{A}$ and inconsistent if $x, \sim x \in \mathfrak{A}$. Thus, interpretations assign each of t, f, u, i its standard truth value (if present in $\mathcal{F}$).

Example 1. Let $\mathcal{F}$ be the set of literals of a propositional vocabulary Σ. The positive facts are the atoms (i.e., $\mathcal{F}^p = \Sigma$), the negative facts are the negative literals (classical negation) (i.e., $\mathcal{F}^n = \{\neg p \mid p \in \Sigma\}$). The complementation operator $\sim$ is the obvious one. For $a, b, c, d \in \Sigma$, the set $\mathfrak{A} = \{a, \neg a, b, \neg c\}$ is a 4-valued interpretation where a is inconsistent, b is true, c is false and d is unknown.

Using the vocabulary of facts and their negations, we can formulate rules that can be used to justify the truth of facts.

Definition 2. *A justification frame $\mathcal{JF}$ is a structure $(\mathcal{F}, \mathcal{F}_d, R)$ such that:*

- *$\mathcal{F}_d \subseteq \mathcal{F}$ is closed under $\sim$, that is, $\sim\mathcal{F}_d = \mathcal{F}_d$;*
- *$t, f, u, \sim u, i, \sim i \notin \mathcal{F}_d$;*
- *$R \subseteq \mathcal{F}_d \times 2^{\mathcal{F}}$.*

We view a tuple $(x, S) \in R$ as a rule and present rules as $x \leftarrow S$. We call S a case of x in $\mathcal{JF}$ if $(x, S) \in R$. The set of cases of x in $\mathcal{JF}$ is denoted $\mathcal{JF}(x)$. We define the set of parameter facts of $\mathcal{JF}$ as $\mathcal{F} \backslash \mathcal{F}_d$ and denote it as $\mathcal{F}_o$; we also sometimes write $x \leftarrow S \in \mathcal{JF}$ and mean $(x, S) \in R$.

A justification frame contains a set of rules. The interpretation varies. We may view $\mathcal{F}_d$ as a set of facts defined by their rules, while $\mathcal{F}_o$ is a set of parameter symbols. Or we might view $\mathcal{F}_d$ as endogenous facts in a causal system, while $\mathcal{F}_o$ are exogenous facts. The idea is that the endogenous facts are governed by causal relationships described by the rules, while exogenous facts are governed by the external environment (e.g., a human agent or external system). More interpretations are possible.

It is possible that for some $x \in \mathcal{F}_d$, there are no rules with x in the head ($\mathcal{JF}(x) = \emptyset$). Then x is never "justified". It is also possible that $x \leftarrow \emptyset \in R$, then x is always "justified". The role of the parameters is illustrated below.

Example 2. We construct a justification frame for defining the transitive closure of a graph. Consider a set V of nodes and define $\mathcal{F}_o = \{E(a,b), \sim E(a,b) \mid a, b \in V\}$. Each exact interpretation of $\mathcal{F}_o$ determines a graph $G = (V, E)$. Set the defined facts to $\mathcal{F}_d = \{Path(a,b), \sim Path(a,b) \mid a, b \in V\}$ and all facts to $\mathcal{F} = \mathcal{F}_d \cup \mathcal{F}_o$. The intended interpretation of fact $Path(a,b)$ is that there is a path from a to b in graph G. The rules R correspond to those of the monotone inductive definition that $Path$ is the transitive closure of E:

$$R = \{Path(a,b) \leftarrow \{E(a,b)\}, Path(a,b) \leftarrow \{Path(a,c), Path(c,b)\} \mid a, b, c \in V\}$$

Later we will see how to derive the rules for negative facts $\sim Path(a,b)$ and how to determine the interpretation for the defined facts given an arbitrary interpretation of the parameter facts. Note that the rules define $Path$ for each choice of edges E but do not constrain G in any way. It is in this respect that $E(a,b)$ literals are called parameters.

We next associate an operator on interpretations with each justification frame. This operator is – in essence – like (Fittings 4-valued version of) the immediate consequence operator T_P for logic programs [14,15]. It takes as input an interpretation and returns an interpretation containing exactly those defined facts that are justified by the input.

Definition 3. *We define the derivation operator of $\mathcal{JF}$ as the mapping $T_{\mathcal{JF}}$: $\mathbb{I}_{\mathcal{F}} \rightarrow \mathbb{I}_{\mathcal{F}_d}$ that maps an interpretation $\mathfrak{A}$ of $\mathcal{F}$ to an interpretation $T_{\mathcal{JF}}(\mathfrak{A})$ of $\mathcal{F}_d$ such that for each $x \in \mathcal{F}_d$, we set $x \in T_{\mathcal{JF}}(\mathfrak{A})$ iff there exists $x \leftarrow S \in \mathcal{JF}$ such that $S \subseteq \mathfrak{A}$.*

The framework below will be geared towards systems where each case of fact x provides a sufficient condition for x while the set of cases of x represent a necessary condition for x in the sense that if x is true, then at least one case must apply. Stated more concisely, the framework is geared towards fixpoints of the operator. However, this is still quite vague (an operator may have many sorts of fixpoints) and will be refined later (when we define various "branch evaluations"). This operator can be used to define when two given justification frames are equivalent.

Definition 4. *Two justification frames $\mathcal{JF}, \mathcal{JF}'$ are equivalent (denoted $\mathcal{JF} \equiv \mathcal{JF}'$) if and only if $T_{\mathcal{JF}} = T_{\mathcal{JF}'}$.*

We call a rule $x \leftarrow S$ redundant in R if there is a more general rule $x \leftarrow S' \in R$ such that $S' \subsetneq S$. Redundant rules may always be deleted from the rule set of a justification frame, as long as the more general rule stays. Formally, let $\mathcal{JF} = \langle \mathcal{F}, \mathcal{F}_d, R \rangle$ and $Re \subseteq R$ be a set of rules each of which is redundant in $R \backslash Re$. Then it is easy to see that $\mathcal{JF}' = \langle \mathcal{F}, \mathcal{F}_d, R \backslash Re \rangle$ is equivalent to $\mathcal{JF}$. However, it is not always possible to remove all redundant rules from a justification frame.

Example 3. Take the (infinite) justification frame $\mathcal{JF}$ with $\mathcal{F} = \{p, {\sim}p, q_i, {\sim}q_i \mid i \in \mathbb{N}\}$, $\mathcal{F}_d = \{p, {\sim}p\}$ and $R = \{p \leftarrow \{q_n, q_{n+1}, \ldots\} \mid n \in \mathbb{N}\}$. Every rule of R is redundant in R. Deleting all of them leads to a justification frame that is not equivalent to $\mathcal{JF}$.

We will only be concerned with justification frames where every defined fact has at least one rule, and rule bodies are not empty. We call them *proper* justification frames.

Definition 5. *A justification frame $\mathcal{JF}$ is* proper *if for all $x \in \mathcal{F}_d$, we have $\mathcal{JF}(x) \neq \emptyset$ and $x \leftarrow \emptyset \notin \mathcal{JF}$.*

Each justification frame can be translated to an equivalent proper one. For each $\mathcal{JF}$, we define its proper justification frame as $\mathcal{JF}'$ with identical sets of parameter facts and defined facts and the following rules (for $x \in \mathcal{F}_d$):

- all rules $x \leftarrow S \in \mathcal{JF}$ such that $S \neq \emptyset$;
- rule $x \leftarrow \{t\}$ if $x \leftarrow \emptyset \in \mathcal{JF}$;
- rule $x \leftarrow \{f\}$ if $x \in \mathcal{F}_d$ and $\mathcal{JF}(x) = \emptyset$.

Proposition 1. $\mathcal{JF} \equiv \mathcal{JF}'$, *that is, $T_{\mathcal{JF}}$ and $T_{\mathcal{JF}'}$ are identical on all interpretations.*

We will – for readability – sometimes present justification frames that are not proper and take them to mean their equivalent proper justification frames.

3   Justifications and Branch Evaluations

Next, we define the central concept of the paper, a justification for a given justification frame. This will be our first step in defining the semantics of justification frames.

Definition 6. *Let $\mathcal{JF} = \langle \mathcal{F}, \mathcal{F}_d, R \rangle$ be a justification frame. A $\mathcal{JF}$-justification J is a subset of R containing at most one rule $x \leftarrow S$ for each $x \in \mathcal{F}_d$. J is called* complete *if for each $x \in \mathcal{F}_d$ there is some $x \leftarrow S$ in J. If $x \leftarrow S \in J$, we denote $J(x) = S$. If J is not complete we call it* partial.

Alternatively, a justification J can be seen as a partial function from $\mathcal{F}_d$ to $2^{\mathcal{F}}$ such that $x \leftarrow J(x) \in R$ if J is defined in $x \in \mathcal{F}_d$. If some $x \in \mathcal{F}_d$ has no rules $x \leftarrow S$ (i.e., if $\mathcal{JF}(x) = \emptyset$), then no complete justification exists for $\mathcal{JF}$.

Proper justification systems $\mathcal{JF}$ have the interesting property that complete justifications exist and that justifications J are equivalent to a special class of directed graphs G on domain $\mathcal{F}$ such that if a fact x has children in G, then $x \in \mathcal{F}_d$ and the set S of its children is a case of x in $\mathcal{JF}$ (that is, $x \leftarrow S$ is a rule of $\mathcal{JF}$). It is easy to see that if $\mathcal{JF}$ is proper, there is a one-to-one correspondence between justifications J and such graphs G. Indeed, given J, we derive $G = \{(x,y) \mid y \in J(x)\}$; vice versa, given G, J is defined in x if x has children, and then $J(x) = \{y \mid (x,y) \in G\}$. Notice that this correspondence is not one-to-one in case $\mathcal{JF}$ is not proper. For example, if $x \leftarrow \emptyset \in \mathcal{JF}$, then if x has no children in G, it is unclear whether J is undefined in x or whether $J(x) = \emptyset$. By Proposition 1, we may impose (w.l.o.g.) the condition that justification frames are proper. In examples we will represent and treat justifications as graphs.

Proposition 2. *For any proper $\mathcal{JF}$, a complete justification corresponds to a graph whose leaves are $\mathcal{F}_o$ and for each non-leaf x with children S exists a rule $x \leftarrow S$ in $\mathcal{JF}$.*

A complete justification J contains for each fact $x \in \mathcal{F}_d$ a potential reason (or a justification, or argument, or cause, etc.) for x to be true: this reason is expressed in the subtree of J below x. Of course, not every such reason is good. It can be flawed for external reasons (e.g., it is based on a parameter fact y that is false in the world) or because of intrinsic reasons (e.g., there is a cyclic argument). In the framework defined here, the support given by J to x in an interpretation $\mathfrak{A}$ is determined by evaluating the branches below x in J. With each branch B we can associate a unique fact, denoted $\mathcal{B}(B)$, so that B evaluates positively in $\mathfrak{A}$ iff $\mathcal{B}(B) \in \mathfrak{A}$. Thus, J justifies x iff $\mathcal{B}(B) \in \mathfrak{A}$, for all branches leaving x. Below, we formalize these concepts.

Definition 7. *An $\mathcal{F}_d$-branch B (briefly, a branch) of $\mathcal{JF}$ for $x_0 \in \mathcal{F}_d$ is an infinite sequence $x_0 \to x_1 \to \dots$ such that $x_i \in \mathcal{F}_d$ or a finite sequence $x_0 \to \dots \to x_n$ such that $x_i \in \mathcal{F}_d$ for $i < n$ and $x_n \in \mathcal{F}_o$. An infinite branch (compactly, an ∞-branch) is positive (negative) if it has a tail of positive (negative) facts. It is mixed if neither positive nor negative. A branch evaluation $\mathcal{B}$ is a mapping from branches to facts.*

A branch contains at least two facts. In a mixed ∞-branch, each tail contains infinitely many positive and negative facts.

Definition 8. *A branch of a complete justification J from defined fact x_0 is a maximally long path $x_0 \to x_1 \to \dots$ in J. (Hence, $x_{i+1} \in J(x_i)$ for all $i \geq 0$.)*

It is obvious that a branch of a complete J from x is a branch in the sense of Definition 7. (This property holds for proper justification frames but not in general).

Example 4. We define four branch evaluations that later will be shown to induce well-known logic programming semantics. For every branch $B = x_0 \to x_1 \to \dots$, we define $\mathcal{B}_{sp}(B) = x_1$. (The subscript refers to "supported" semantics.) Next, we define three more branch evaluations all of which map B to its leaf x_n if B is a finite branch $x_0 \to \dots \to x_n$. If B is an ∞-branch, we have:

- (Kripke-Kleene) $\mathcal{B}_{KK}(B) = u$ if $x_0 \in \mathcal{F}^p$ and $\mathcal{B}_{KK}(B) = \sim u$ if $x_0 \in \mathcal{F}^n$.
- (stable) $\mathcal{B}_{st}(B) = t$ if B consists of negative facts only; $\mathcal{B}_{st}(B) = f$ if B consists of positive facts only; otherwise, $\mathcal{B}_{st}(B) = x_i$ if x_i is the first fact in B with another sign than x_0.
- (well-founded) $\mathcal{B}_{wf}(B) = t$ if B is a negative ∞-branch; $\mathcal{B}_{wf}(B) = f$ if B is a positive ∞-branch; $\mathcal{B}_{wf}(B) = u$ if B is mixed and $x_0 \in \mathcal{F}^p$; $\mathcal{B}_{wf}(B) = \sim u$ if B is mixed and $x_0 \in \mathcal{F}^n$.

The names suggest a connection to different semantics of logic programs. This will be explored below.

Definition 9. *Given some branch evaluation $\mathcal{B}$, we say that $x \in \mathcal{F}_d$ is supported by J in $\mathfrak{A}$ (under $\mathcal{B}$) if for all branches $B = x \rightarrow \ldots$ in J, we find $\mathcal{B}(B) \in \mathfrak{A}$. We say that x is supported by $\mathcal{JF}$ in $\mathfrak{A}$ (under $\mathcal{B}$) if there exists a complete justification J of $\mathcal{JF}$ such that x is supported by J in $\mathfrak{A}$. We denote the set of supported facts by $\mathcal{S}_{\mathcal{JF}}^{\mathcal{B}}(\mathfrak{A})$.*

Using the specific branch evaluation $\mathcal{B}_{sp}$ allows to express the derivation operator associated to a justification frame.

Proposition 3. *For $\mathcal{B}_{sp}$ we have $\mathcal{S}_{\mathcal{JF}}^{\mathcal{B}_{sp}}(\mathfrak{A}) = T_{\mathcal{JF}}(\mathfrak{A})$.*

This property does not hold for other branch evaluations. Each combination of justification frame $\mathcal{JF}$ and branch evaluation $\mathcal{B}$ induces an operator $\mathcal{S}_{\mathcal{JF}}^{\mathcal{B}}(\cdot)$ from interpretations of $\mathcal{F}$ to interpretations of $\mathcal{F}_d$.

Proposition 4. *If $\mathcal{JF}, \mathcal{JF}'$ are equivalent (i.e., they induce the same operator) then for each branch evaluation $\mathcal{B}$, the operators $\mathcal{S}_{\mathcal{JF}}^{\mathcal{B}}(\cdot)$ and $\mathcal{S}_{\mathcal{JF}'}^{\mathcal{B}}(\cdot)$ are identical.*

The meaning of a justification frame is not only specified by the set of its rules but also by the selected branch evaluation. The same set of rules may have a different meaning for a different branch evaluation. This is captured in the next definition, another central concept of the paper: a *justification system* is a justification frame extended with a branch evaluation.

Definition 10. *A justification system is a structure $\langle \mathcal{F}, \mathcal{F}_d, R, \mathcal{B} \rangle$ with $\langle \mathcal{F}, \mathcal{F}_d, R \rangle$ a justification frame and $\mathcal{B}$ a branch evaluation on that frame.*

Again, we can associate an operator with a justification system just like we did for justification frames. We only have to additionally take into account the specific branch evaluation at hand.

Definition 11. *Let $\mathcal{JS} = \langle \mathcal{F}, \mathcal{F}_d, R, \mathcal{B} \rangle$ be a justification system and let $\mathcal{JF} = \langle \mathcal{F}, \mathcal{F}_d, R \rangle$ be its included justification frame. With $\mathcal{JS}$ we associate the operator $\mathcal{S}_{\mathcal{JF}}^{\mathcal{B}}(\cdot) : \mathbb{I}_{\mathcal{F}} \rightarrow \mathbb{I}_{\mathcal{F}_d}$ and denote the mapping of an interpretation $\mathfrak{A}$ under this operator as $\mathcal{JS}(\mathfrak{A})$. The justified interpretations of a justification system $\mathcal{JS}$ are the fixpoints of $\mathcal{S}_{\mathcal{JF}}^{\mathcal{B}}(\cdot)$.*

The operator $\mathcal{S}_{\mathcal{JF}}^{\mathcal{B}}(\cdot) : \mathbb{I}_{\mathcal{F}} \rightarrow \mathbb{I}_{\mathcal{F}_d}$ can only be iterated if its domain and co-domain are identical, that is, if $\mathcal{F}_o = \emptyset$. There is a simple way to fix this: each operator O from $\mathbb{I}_{\mathcal{F}}$ to $\mathbb{I}_{\mathcal{F}_d}$ has a canonical extension $O' : \mathbb{I}_{\mathcal{F}} \rightarrow \mathbb{I}_{\mathcal{F}}$ defined as $O'(\mathfrak{A}) = O(\mathfrak{A}) \cup (\mathfrak{A} \cap \mathcal{F}_o)$. The extended operator copies the interpretation of the parameters and can be iterated.

4 Reconstructions

This section shows how several established knowledge representation formalisms can be reconstructed within our theory of justifications. Often, formalisms make implicit semantic assumptions – e.g. in logic programs any atom without a rule is considered to be false. The next definitions show how our theory makes such assumptions explicit.

Take a justification frame $\mathcal{JF} = \langle \mathcal{F}, \mathcal{F}_d, R \rangle$ and a fact $x \in \mathcal{F}_d$. We can view the body of a rule $x \leftarrow S \in R$ as a logical conjunction of literals, $\delta_S = \bigwedge_{y \in S} y$. The set of all cases for x then can be thought of as a (possibly infinite) disjunction of such conjunctions, $\gamma_x = \bigvee_{x \leftarrow S \in R} \delta_S$. In a sense, γ_x characterizes the truth value of x in any given interpretation. Intuitively, our definition of complement closure below aims to construct the rules that are needed to characterize the negation of x, the fact $\sim x$. To obtain these rules we use the negation of γ_x, that is, $\neg \gamma_x = \neg \bigvee_{x \leftarrow S \in R} \delta_S \equiv \bigwedge_{x \leftarrow S \in R} \neg \delta_S$. To get actual rules according to the definition of a justification system, we consider the DNF of $\neg \gamma_x$, that is, all possible ways of making all possible cases for x inapplicable.

Definition 12. *A selection function for $x \in \mathcal{F}_d$ is a function $\mathbf{S}$ from the set $\mathcal{JF}(x)$ of cases of x to $\mathcal{F}$ such that $\mathbf{S}(S) \in S$ for each $S \in \mathcal{JF}(x)$. A complement selection of $x \in \mathcal{F}_d$ in $\mathcal{JF}$ is a set $\{\sim \mathbf{S}(S) \mid S \in \mathcal{JF}(x)\}$ for some selection function $\mathbf{S}$ for x.*

The complement selections of x correspond to the disjuncts in the DNF of $\neg \gamma_x$. Each selects at least one element from each case of x and adds the negations of all selected elements in one set. It can be seen that if all elements of this set are true, then the bodies of all rules of x are false. Vice versa, if the bodies of all rules of x are false then all elements of at least one complement selection are true.

Definition 13. *Let $\mathcal{JF} = \langle \mathcal{F}, \mathcal{F}_d, R \rangle$ be a justification frame such that either one of: (1) $x \leftarrow S \in R$ implies that $x \in \mathcal{F}^p$ (there are no rules for negative facts); or (2) $x \leftarrow S \in R$ implies that $x \in \mathcal{F}^n$ (there are no rules for positive facts). The complement closure of $\mathcal{JF}$ is the justification frame $\langle \mathcal{F}, \mathcal{F}_d, R \cup R^c \rangle$ where R^c consists of all rules $\sim x \leftarrow S$ with $x \in \mathcal{F}_d$ and S a complement selection of x in $\mathcal{JF}$.*

Example 5 (Continuation of Example 2). The complement closure of the justification frame for transitive closure contains all possible rules of the form $\sim Path(a, b) \leftarrow S$, where S is any subset of $\mathcal{F}$ that contains at least $\sim E(a, b)$ and for every $c \in V$ either the literal $\sim Path(a, c)$ or the literal $\sim Path(c, b)$.

We now look at how existing semantics of existing formalisms can be reconstructed using justification systems.

Argumentation. Our first reconstruction shows how Dung-style argumentation frameworks (AFs) [11] give rise to justification frames. Argumentation frameworks are a simple, popular formalism for representing arguments and attacks between these arguments. More precisely, an AF is a pair $F = (A, X)$ where A is a set of (atomic) arguments and $X \subseteq A \times A$ is a binary relation ("attack") on arguments. Intuitively, if $(a, b) \in X$, then argument a attacks argument b.

Definition 14. *Given an AF $F = (A, X)$, the justification frame associated with F is $\mathcal{JF}_F = (\mathcal{F}, \mathcal{F}_d, R)$ such that $\mathcal{F} = \mathcal{F}_d = A \cup \sim A$ and $R = \{\sim a \leftarrow \{b\} \mid (b, a) \in X\}$.*

Thus in the resulting fact space $\mathcal{F}$ there is a fact a for each argument $a \in A$, and a fact $\sim a$ for the opposite of each argument $a \in A$. ll of these facts are (going to be) defined. The rules of $\mathcal{JF}_F$ for negative $\sim a \in \mathcal{F}$ encode the meaning of "attack": an argument a is rejected (that is, its opposite $\sim a$ is true) if one of its attackers b is accepted (b is true). The complement closure $\mathcal{JF}_F^c$ will additionally contain the rules $R^c = \{a \leftarrow \{\sim b \mid (b, a) \in X\} \mid a \in A\}$. Intuitively, these derived rules express that for an argument a to be accepted, all its attackers b must be rejected (that is, their opposites $\sim b$ must be true). Using one and the same branch evaluation, namely $\mathcal{B}_{sp}$, we can show that justification systems allow us to reconstruct the major semantics of abstract argumentation frameworks.

Proposition 5. *Let $F = (A, X)$ be an argumentation framework and $\mathcal{JF}_F^c$ be the complement closure of its associated justification frame $\mathcal{JF}_F$. A consistent interpretation $\mathfrak{A}$*

- *is an exact fixpoint of $T_{\mathcal{JF}_F^c}$ iff it is stable for F;*
- *is a fixpoint of $T_{\mathcal{JF}_F^c}$ iff it is complete for F;*
- *is a $\subseteq$-maximal fixpoint of $T_{\mathcal{JF}_F^c}$ iff it is preferred for F;*
- *is the $\subseteq$-least fixpoint of $T_{\mathcal{JF}_F^c}$ iff it is grounded for F;*
- *satisfies $\mathfrak{A} \subseteq T_{\mathcal{JF}_F^c}(\mathfrak{A})$ iff it is admissible for F.*

Proof (Sketch). Consistent interpretations $\mathfrak{A}$ can be seen as three-valued interpretations on the set A. It can be shown that $T_{\mathcal{JF}_F}$ is (isomorphic to) the three-valued characteristic operator of the AF F. The claims then follow from Propositions 4.4 to 4.9 in [16]. □

Four out of five of these types of fact sets correspond to specific sorts of fixpoints of $T_{\mathcal{JF}_F^c}$. Thus an argument (or its opposite) belongs to such a set iff it is justified by one of its cases. The exceptions are the admissible sets, which are only postfixpoints: facts in an admissible set have to be justified by it but not all facts justified by such a set must belong to the set. Although we technically use sets, these semantics are three-valued and thus closer to the notion of a labelling than to that of a set-based extension.

Logic Programming. Justification frames differ from propositional logic programs in two ways: (a) the presence of a set $\mathcal{F}_o$ of parameter literals (whose interpretation is not defined by the rules), and (b) the presence of rules with negation in the head (which via complement closure can be derived from those for positive literals).[1]

Definition 15. *Let Π be a (propositional) logic program over atoms Σ. The justification frame associated with Π is the structure $\mathcal{JF}_\Pi = (\mathcal{F}, \mathcal{F}_d, \Pi)$ where $\mathcal{F}_d$ is the set of all literals over Σ, and $\mathcal{F} \backslash \mathcal{F}_d = \mathcal{F}_o$ is the set of logical facts.*

We assume without loss of generality that $\mathcal{JF}_\Pi$ is proper, i.e. it contains at least one rule per non-logical fact (possibly $p \leftarrow \{f\}$) and rule bodies are non-empty. While the above definition is about propositional logic programs, the approach easily generalizes to the predicate case by simply instantiating the rules using first-order interpretations. We now establish the connection between branch evaluations and various semantics of logic programs. Since Π contains only rules for atoms, we apply complement closure.

Theorem 1. *Let Π be a logic program and $\mathcal{JF}$ be the complement closure of $\mathcal{JF}_\Pi$.*

- *An exact interpretation $\mathfrak{A}$ is an exact fixpoint of $\mathcal{S}_{\mathcal{JF}}^{\mathcal{B}_{sp}}(\cdot)$ iff $\mathfrak{A}$ is a supported model of Π.*
- *An interpretation $\mathfrak{A}$ is a fixpoint of $\mathcal{S}_{\mathcal{JF}}^{\mathcal{B}_{KK}}(\cdot)$ iff $\mathfrak{A}$ is the Kripke Kleene model of Π.*
- *An interpretation $\mathfrak{A}$ is an exact fixpoint of $\mathcal{S}_{\mathcal{JF}}^{\mathcal{B}_{st}}(\cdot)$ iff $\mathfrak{A}$ is a stable model of Π.*
- *An interpretation $\mathfrak{A}$ is a fixpoint of $\mathcal{S}_{\mathcal{JF}}^{\mathcal{B}_{wf}}(\cdot)$ iff $\mathfrak{A}$ is the well-founded model of Π.*

For some branch evaluations $\mathcal{B}$, the value of $\mathcal{S}_{\mathcal{JF}}^{\mathcal{B}}(\mathfrak{A})$ depends entirely on the value of parameter facts in $\mathfrak{A}$. This is the case if the branch evaluation maps every branch to a parameter fact.

Definition 16. *A branch evaluation $\mathcal{B}$ is parametric if for every branch B, $\mathcal{B}(B)$ is a parameter fact.*

Proposition 6. *If $\mathcal{B}$ is parametric, then for every parameter interpretation $\mathfrak{A}_p \subseteq \mathcal{F}_o$, $\mathfrak{A} = \mathcal{S}_{\mathcal{JF}}^{\mathcal{B}}(\mathfrak{A}_p)$ is the unique fixpoint of $\mathcal{S}_{\mathcal{JF}}^{\mathcal{B}}(\cdot)$ such that $\mathfrak{A} \cap \mathcal{F}_o = \mathfrak{A}_p$.*

The branch evaluations $\mathcal{B}_{KK}$ and $\mathcal{B}_{wf}$ are parametric. They are used in logic programming semantics that have a unique model. $\mathcal{B}_{sp}$ and $\mathcal{B}_{st}$ are not parametric, and thus, supported and stable semantics admit for any number of models.[2]

---

[1] Negation in the head of (extended) answer set programs is different from the negation studied here, and the justification semantics defined below is not directly suitable to compute answer sets of programs with explicit negation. We focus on systems where the rules for facts and their negations are complementary, hence negation is classical. In contrast, rules of ASP for negative literals are independent of those for positive literals.

[2] By dropping the constraint that $\mathcal{F}_o$ consists of logical facts only, we obtain extensions of all main semantics for a parameterized variant of logic programming.

Example 6. Consider the branch evaluation $\mathcal{B}_{cowf}(B)$ defined like $\mathcal{B}_{wf}$ except that positive ∞-branches are evaluated to t and negative ones to f. The semantics induced by $\mathcal{B}_{cowf}$ is a sort of well-founded semantics that "prefers" maximal models. This induces a coinductive semantics as in, for example, μ-calculus [6] and coinductive logic programming [12]. For an illustration, consider the set D of finite and infinite lists over $\{A, B\}$ and $\mathcal{JF} = \langle \mathcal{F}, \mathcal{F}_d, R \rangle$ where $\mathcal{F}_d = \{P(s) \mid s \in D\}$ and $\mathcal{F}$ the extension of $\mathcal{F}_d$ with logical facts and, using Prolog notation $[H|T]$ for lists with head H and tail T, the rule set $R = \{ P([A, B|s]) \leftarrow \{P(s)\} \mid s \in D \}$. After taking the complement closure, an interpretation $\mathfrak{A}$ is a fixpoint of $\mathcal{S}_{\mathcal{JF}}^{\mathcal{B}_{cowf}}(\cdot)$ iff $\mathfrak{A} = \{P([A, B, A, B, A, B, \ldots])\}$. Had we used $\mathcal{B}_{wf}$ ("preferring" minimal models), the fixpoint would have been $\emptyset$.

5   Nested Justification Systems

Modularity and composition are key properties of knowledge representation languages. We compose (parametric) justification systems by *nesting* them. It is important to note that nesting as presented in this section is restricted to parametric justification systems for reasons that will become clear soon.

Definition 17. *Let $\mathcal{F}$ be a set of facts. A nested justification system on $\mathcal{F}$ is a tuple $\mathcal{JS} = \langle \mathcal{F}, \mathcal{F}_{dg}, \mathcal{F}_{dl}, R, \mathcal{B}, \{\mathcal{JS}^1, \ldots, \mathcal{JS}^k\} \rangle$ where:*

- *$\langle \mathcal{F}, \mathcal{F}_{dl}, R, \mathcal{B} \rangle$ is a parametric justification system.*
- *Each $\mathcal{JS}^i$ is a nested justification system on fact space $\mathcal{F}^i = (\mathcal{F} \backslash \mathcal{F}_{dg}) \cup \mathcal{F}_{dg}^i$ with (globally) defined facts $\mathcal{F}_{dg}^i$.*
- *$\mathcal{F}_{dg}$ is the disjoint union of $\mathcal{F}_{dl}$ and $\mathcal{F}_{dg}^1, \ldots, \mathcal{F}_{dg}^k$.*

A nested justification system is a tree-like definition that defines the set $\mathcal{F}_{dg}$ of globally defined facts. This set is partitioned into $k+1$ subsets. One subset, $\mathcal{F}_{dl}$, consists of facts that are *locally* defined in the root of the tree by R. The rest of the facts are defined in one of the k nested subdefinitions $\mathcal{JS}^i$ of $\mathcal{JS}$. The branch evaluation of $\mathcal{JS}$ is defined for branches of locally defined facts only. The parameters of subdefinitions $\mathcal{JS}^i$ are those of $\mathcal{JS}$ augmented with the locally defined facts. In particular, for each $\mathcal{JS}^i$, facts defined in siblings $\mathcal{JS}^j$ with $j \neq i$ are not to appear as parameters of $\mathcal{JS}^i$. In leaves of the tree, we have $k = 0$ and $\mathcal{F}_{dg} = \mathcal{F}_{dl}$.

The semantics of nested justification systems is based on two notions: compression and unfolding. We start explaining the latter. Let R_1 be a set of rules defining facts $\mathcal{F}_{d1} \subseteq \mathcal{F}$, R a second set of rules in fact space $\mathcal{F}$. The unfolding of R_1 on $\mathcal{F}_{d1}$ in R, denoted $UNF_{(\mathcal{F}_{d1}, R_1)}(R)$, is the set of rules that can be obtained from any $x \leftarrow S \in R$ by replacing each fact $y \in S$ defined in R_1, in an arbitrary way, by the body facts of a rule $y \leftarrow S' \in R_1$. E.g., suppose R contains rule $a \leftarrow \{g, b\}$ and R_1 contains the rules $b \leftarrow \{c, d\}$, $b \leftarrow \{f\}$ for b. Then unfolding R_1 on $\{b\}$ in R replaces that rule of R by two rules, $a \leftarrow \{g, c, d\}$ and $a \leftarrow \{g, f\}$. Compression turns a nested definition into an (equivalent) unnested one.

Definition 18. *Let* $\mathcal{JS} = \langle \mathcal{F}, \mathcal{F}_{dg}, \mathcal{F}_{dl}, R, \mathcal{B}, \{\mathcal{JS}^1, \ldots, \mathcal{JS}^k\}\rangle$ *be a justification system. Its compression* $\mathcal{C}(\mathcal{JS})$ *is defined inductively:* $\mathcal{C}(\mathcal{JS}) = \langle \mathcal{F}, \mathcal{F}_{dg}, R_c, \mathcal{B}\rangle$ *where*

$$R_c = R^s \cup UNF_{(\mathcal{F}_{dg} \setminus \mathcal{F}_{dl}, R^s)}(R)$$

with $R^s = R_c^1 \cup \cdots \cup R_c^k$ *and* R_c^i *is the set of rules* $x \leftarrow S$ *such that* $x \in \mathcal{F}_{dg}^i$ *and* $S = \{\mathcal{B}^i(B) \mid B \text{ is a branch of } x \text{ in } J\}$ *for some complete justification* J *of* $\mathcal{JS}^i$.

Notice that in the base case $k = 0$, justification system $\mathcal{JS}$ and its compression $\mathcal{C}(\mathcal{JS})$ are essentially the same. Now we see why all branch evaluations $\mathcal{B}$ used in different nodes must be parametric, and why definitions cannot use facts defined in nodes not on the path from the root to the current node. Under these conditions, subdefinitions $\mathcal{JS}^i$ are translated in an equivalence preserving way in a set of flat rules $x \leftarrow S$ where S contains only parameter facts and locally defined facts of $\mathcal{JS}$. With the set R^s of all these rules, we eliminate non-locally defined facts in bodies of the local definition R by unfolding R^s on $\mathcal{F}_{dg} \setminus \mathcal{F}_{dl}$ in R, thus producing rules that contain only parameter and locally defined bodies of $\mathcal{JS}$. For the resulting definition R_c, we use $\mathcal{B}_{sp}$ for branches of facts of $\mathcal{F}_{dg} \setminus \mathcal{F}_{dl}$ and the locally given $\mathcal{B}$ for branches of locally defined facts. Hence, the operator $\mathcal{S}_{\mathcal{C}(\mathcal{JS})}^{\mathcal{B}}(\cdot)$ is well-defined and its fixpoints define the semantics of $\mathcal{JS}$.

Example 7. We define a nested justification system with $\mathcal{JS}^2$ nested in $\mathcal{JS}^1$. Take the list domain D as in Example 6, $\mathcal{F}_{dl}^1 = \{P(s) \mid s \in D\}$ and $\mathcal{F}_{dl}^2 = \{Q(s) \mid s \in D\}$, $\mathcal{B}^1 = \mathcal{B}_{cowf}$ and $\mathcal{B}^2 = \mathcal{B}_{wf}$ and finally,

$$R_1 = \{P(s) \leftarrow \{Q(s)\} \mid s \in D\}$$
$$R_2 = \{Q([A|s]) \leftarrow \{P(s)\} \mid s \in D\} \cup \{Q([B|s]) \leftarrow \{Q(s)\} \mid s \in D\}$$

After taking the complement closure of $\mathcal{JS}^1$ and its compression, the rules defining positive facts are, for each $s \in D$:

$$P([B, \ldots, B, A|s]) \leftarrow \{P(s)\} \quad \text{and} \quad Q([B, \ldots, B, A|s]) \leftarrow \{P(s)\}$$

In the unique supported interpretation of the compression, both P and Q are the set of all lists with infinitely many occurrences of A.

In our final example, we show how our justification framework can treat logic programs with aggregates in rule bodies. In particular, this illustrates the power of nesting.

Example 8. Consider a logic program rule with a weight constraint, that is,

$$p \leftarrow i \leq \{l_1, \ldots, l_n\} \leq j \tag{1}$$

with $1 \leq i \leq j \leq n$, meaning that p is true if at least i and at most j literals from the set $L = \{l_1, \ldots, l_n\}$ are true. An LP rule (1) is translated into the set of $\mathcal{JF}$ rules

$$R_p = \{p \leftarrow L^+ \cup {\sim}L^- \mid L^+, L^- \subseteq L \text{ and } |L^+| = i \text{ and } |L^-| = n - j\}$$

Alternatively, we can use nested definitions. A weight-constraint rule (1) appears in the top-level definition with $q_{i \leq \{l_1,...,l_n\} \leq j}$ a single new atom in the body. Such new atoms are then defined by a nested definition $q_{i \leq \{l_1,...,l_n\} \leq j} \leftarrow L^+ \cup \sim L^-$ with L^+ and L^- as in R_p above. It is not difficult to see that the compression of the second approach yields the first. This application further clarifies aggregates, nesting and compression.

6   Discussion

Justifications as mathematical semantical constructs have appeared in different ways in different areas. In [17,18], stable and answer set semantics are defined for programs using justifications similar to ours. Phrased in the terms of our paper, atoms x are justified by sets $S = \{\mathcal{B}_{st}(B) \mid B$ is a branch of x in $J\}$ for some complete justification J of the program. Tree-shaped justifications were used in [3] to build a semantical framework for (abductive) logic programming. [19] present an algebra of tree-shaped justifications for atomic positive facts for logic programming. [20] propose justification graphs for justifying the truth value of atoms in answer sets of logic programs. Our study differs on the technical level and generalizes these works in several dimensions, e.g. by considering parameters, alternative branch evaluations (e.g. coinduction), nesting and novel applications (e.g. to argumentation frameworks). Justifications as datastructures are used in the ASP solver clasp [8] and in the FO(ID) model expander IDP3 [9]. Justifications are underlying provenance systems in the context of databases [21].

The justification framework defined above is of great amplitude and much uncharted territory lies in front. It covers a remarkable amount of existing semantics in different areas. Here we showed this for argumentation frameworks and logic programming. The framework also induces new and more general versions of these formalisms. For example it comprises nested logic programs with negation and feedback, a new formalism that remains to be studied.[3] Alternative branch evaluations can be introduced. For example, some have argued that in the logic program $\{P \leftarrow \neg P\}$, P should be inconsistent while in $\{P \leftarrow \neg Q, Q \leftarrow \neg P\}$, P and Q should be undefined [23]. A refinement of $\mathcal{B}_{wf}$ that would distinguish between these cases would be the one that assigns i to branches B with complementary facts $x, \sim x$. This remains to be explored. The justification framework may be applicable to many other logics as well. We already mentioned coinductive logic programming [12] (as illustrated by Example 6); we expect the framework to cover other mathematical and knowledge representation logics of nested induction/coinduction such as μ-calculus [6], FO(LFP) with nesting [7] and FO(FD) [13]. We believe that our justification theory can also be applied to assumption-based argumentation [24] and abstract dialectical frameworks [25]. The approach is promising as well for logics of causality such as FO(C) [26]. All these connections still need to be investigated. As mentioned in Sect. 1, we know

---

[3] This should not be confused with the nested logic programs of [22], where nesting refers to the expressions inside a logic program rule, and not sets of rules being nested altogether.

of only one other approach with a comparable coverage: Approximation Fixpoint Theory (AFT) [4]. While AFT is defined in a different way (as an algebraical fixpoint theory of lattice operators), it was used to characterize about the same semantics for the same logics and the question is if there is a relationship between both frameworks. Such a relationship would further broaden the application of our justification framework, for example to autoepistemic logic and default logic.

References

1. Pelov, N., Denecker, M., Bruynooghe, M.: Well-founded and stable semantics of logic programs with aggregates. TPLP **7**(3), 301–353 (2007)
2. Son, T.C., Pontelli, E.: A constructive semantic characterization of aggregates in answer set programming. TPLP **7**(3), 355–375 (2007)
3. Denecker, M., De Schreye, D.: Justification semantics: A unifying framework for the semantics of logic programs. In: LPNMR, pp. 365–379. MIT Press (1993)
4. Denecker, M., Marek, V., Truszczyński, M.: Approximations, stable operators, well-founded fixpoints and applications in nonmonotonic reasoning. In: Minker, J. (ed.) Logic-Based Artificial Intelligence, pp. 127–144. Springer, New York (2000)
5. Denecker, M., Marek, V., Truszczyński, M.: Ultimate approximation and its application in nonmonotonic knowledge representation systems. Inf. Comput. **192**(1), 84–121 (2004)
6. Kozen, D.: Results on the propositional μ-calculus. Theoret. Comput. Sci. **27**(1), 333–354 (1983)
7. Park, D.: Fixpoint induction and proofs of program properties. Mach. Intell. **5**, 59–78 (1969)
8. Gebser, M., Kaufmann, B., Schaub, T.: Conflict-driven answer set solving: from theory to practice. Artif. Intell. **187**, 52–89 (2012)
9. Mariën, M., Wittocx, J., Denecker, M., Bruynooghe, M.: SAT(ID): satisfiability of propositional logic extended with inductive definitions. In: Kleine Büning, H., Zhao, X. (eds.) SAT 2008. LNCS, vol. 4996, pp. 211–224. Springer, Heidelberg (2008)
10. Gelfond, M., Lifschitz, V.: Classical negation in logic programs and disjunctive databases. New Gener. Comput. **9**, 365–385 (1991)
11. Dung, P.M.: On the acceptability of arguments and its fundamental role in non-monotonic reasoning, logic programming and n-person games. Artif. Intell. **77**, 321–358 (1995)
12. Gupta, G., Bansal, A., Min, R., Simon, L., Mallya, A.: Coinductive logic programming and its applications. In: Dahl, V., Niemelä, I. (eds.) ICLP 2007. LNCS, vol. 4670, pp. 27–44. Springer, Heidelberg (2007)
13. Hou, P., De Cat, B., Denecker, M.: FO(FD): extending classical logic with rule-based fixpoint definitions. Theor. Pract. Log. Program. **10**(4–6), 581–596 (2010)
14. van Emden, M.H., Kowalski, R.A.: The semantics of predicate logic as a programming language. J. ACM **23**(4), 733–742 (1976)
15. Fitting, M.: Fixpoint semantics for logic programming: a survey. Theoret. Comput. Sci. **278**(1–2), 25–51 (2002)
16. Strass, H.: Approximating operators and semantics for abstract dialectical frameworks. Artif. Intell. **205**, 39–70 (2013)
17. Fages, F.: A new fixpoint semantis for general logic programs compared with the well-founded and the stable model semantics. In: ICLP, p. 443. MIT Press (1990)

18. Schulz, C., Toni, F.: ABA-based answer set justification. Theor. Pract. Log. Program. **13**(4-5-Online-Supplement) (2013)
19. Cabalar, P., Fandinno, J., Fink, M.: Causal graph justifications of logic programs. Theor. Pract. Log. Program. **14**(4–5), 603–618 (2014)
20. Pontelli, E., Son, T.C., Elkhatib, O.: Justifications for logic programs under answer set semantics. Theor. Pract. Log. Program. **9**(1), 1–56 (2009)
21. Viegas Damásio, C., Analyti, A., Antoniou, G.: Justifications for logic programming. In: Cabalar, P., Son, T.C. (eds.) LPNMR 2013. LNCS, vol. 8148, pp. 530–542. Springer, Heidelberg (2013)
22. Lifschitz, V., Tang, L.R., Turner, H.: Nested expressions in logic programs. Ann. Math. Artif. Intell. **25**(3–4), 369–389 (1999)
23. Hallnäs, L.: Partial inductive definitions. Theor. Comp. Sci. **87**(1), 115–142 (1991)
24. Bondarenko, A., Dung, P.M., Kowalski, R.A., Toni, F.: An abstract, argumentation-theoretic approach to default reasoning. Artif. Intell. **93**, 63–101 (1997)
25. Brewka, G., Woltran, S.: Abstract dialectical frameworks. In: KR, pp. 102–111 (2010)
26. Bogaerts, B., Vennekens, J., Denecker, M., Van den Bussche, J.: FO(C): a knowledge representation language of causality. TPLP **14**((4–5–Online–Supplement)), 60–69 (2014)

A New Computational Logic Approach to Reason with Conditionals

Emmanuelle-Anna Dietz[(✉)] and Steffen Hölldobler

International Center for Computational Logic, TU Dresden,
01062 Dresden, Germany
{dietz,sh}@iccl.tu-dresden.de

Abstract. We present a new approach to evaluate conditionals in human reasoning. This approach is based on the weak completion semantics which has been successfully applied to adequately model various other human reasoning tasks in the past. The main idea is to explicitly consider the case, where the condition of a conditional is *unknown* with respect to some background knowledge, and to evaluate it with minimal revision followed by abduction. We formally compare our approach to a recent approach by Schulz and demonstrate that our proposal is superior in that it can handle more human reasoning tasks.

1 Introduction

Conditionals are statements of the form *if condition then consequence*. *Indicative conditionals* are conditionals whose condition may or may not be *true* and, consequently, whose consequence also may or may not be *true*; however, the consequence is asserted to be *true* if the condition is *true*. On the contrary, the condition of a *subjunctive* or *counterfactual conditional* needs to be *false* [21];[1] however, in the counterfactual circumstance of the condition being *true*, the consequence is asserted to be *true*. We will distinguish between both types by expressing them in their indicative or subjunctive mood, respectively. It is generally accepted that conditionals in natural language do not have the same interpretation as material (or truth functional) conditionals [10]. A lot of approaches have been proposed but it seems that there is no agreement on a general theory [11]. We briefly discuss a few of them.

Ramsey [32] proposed to test conditionals by assuming the condition hypothetically and verify whether the consequence follows. This approach is problematic in case the current state is not consistent with the condition. Stalnaker [38] extended Ramsey's approach and suggested minimal revision. Lewis [22] showed that Stalnaker and Thomason's counterfactual theory of possible worlds [39] had some technical problems and developed an approach of maximal world-similarity [21,23]. Ginsberg's possible worlds approach [13] towards

---

The authors are mentioned in alphabetical order.

[1] This and other definitions are controversially discussed within the fields of philosophy and psychology [14]. Some require, that counterfactuals must be in the subjunctive mood or can only be evaluated in a state that is different wrt the current one [43].

© Springer International Publishing Switzerland 2015
F. Calimeri et al. (Eds.): LPNMR 2015, LNAI 9345, pp. 265–278, 2015.
DOI: 10.1007/978-3-319-23264-5_23

counterfactuals might be one of the first in the field of AI. It has been improved by requiring relevancy [28]. Other early approaches have been proposed in [3,35]. The logic programming approaches presented in [1,2,31,41,42] are inspired by Pearl's structural theory of counterfactuals in Bayesian networks [26,27]. The distinction between causal and counterfactual reasoning, based on Pearl's theory, is been extensively discussed in [37]. Rescher [33,34] presented a systematic reconstruction of the belief system using principles of saliency and prioritization, which only requires to consider immediately relevant beliefs.

The question which we shall be discussing in this paper is how to automate reasoning such that conditionals are evaluated by an automated deduction system like humans do. This will be done in a context of logic programming (cf. [24]), weak completion [16], abduction [19], Stenning and van Lambalgen's representation of implications as well as their semantic operator [40] and three-valued Łukasiewicz logic [25], which has been put together in [9,15–18]. This approach–which we call WCS for *weak completion semantics*–has been applied to adequately model the suppression [7] and the selection tasks [8], the belief-bias effect [29] as well as contextual abductive reasoning with side-effects [30].

The methodology of the WCS approach applied to reasoning about conditionals differs significantly from methods and techniques applied in well-known approaches. It is inspired by [27] and we agree with Rescher's view to concentrate on the relevant knowledge and minimally revising the current state. We evaluate conditionals with respect to some background knowledge and explicitly treat the case where the conditions of a conditional are *unknown*. In this case we apply a monotonic form of minimal revision followed by abduction in order to satisfy the conditions. We apply the new method to indicative as well as counterfactual conditionals. As a very similar approach has been proposed by Schulz recently [36], we formally show that the WCS approach is more general. Finally, we discuss some open questions and point to future research.

2   Preliminaries

We assume the reader to be familiar with logic and logic programming, but recall basic notions and notations. A *(logic) program* is a finite set of (program) clauses of the form $A \leftarrow \top$, $A \leftarrow \bot$ or $A \leftarrow B_1 \wedge \ldots \wedge B_n$, $n > 0$, where A is an atom, B_i, $1 \leq i \leq n$, are literals and $\top$ and $\bot$ denote truth and falsehood, resp. A is called *head* and $\top$, $\bot$ as well as $B_1 \wedge \ldots \wedge B_n$ are called *body* of the corresponding clause. Clauses of the form $A \leftarrow \top$ and $A \leftarrow \bot$[2] are called *positive* and *negative facts*, respectively. We restrict terms to be constants and variables only, i.e., we consider *data logic programs*. Throughout this paper, $\mathcal{P}$ denotes a program. We assume for each $\mathcal{P}$ that the alphabet consists precisely of the symbols occurring in $\mathcal{P}$ and that non-propositional programs contain at least one constant. When writing sets of literals we will omit curly brackets if the set has only one element.

---

[2] We consider weak completion semantics and, hence, a clause of the form $A \leftarrow \bot$ is turned into $A \leftrightarrow \bot$ provided that this is the only clause in the definition of A.

$g\mathcal{P}$ denotes the set of all ground instances of clauses occurring in $\mathcal{P}$. A ground atom A is *defined* in $g\mathcal{P}$ iff $g\mathcal{P}$ contains a clause whose head is A; otherwise A is said to be *undefined*. $def(\mathcal{S}, \mathcal{P}) = \{A \leftarrow body \in g\mathcal{P} \mid A \in \mathcal{S} \vee \neg A \in \mathcal{S}\}$ is called *definition* of $\mathcal{S}$ in $\mathcal{P}$, where $\mathcal{S}$ is a set of ground literals. $\mathcal{S}$ is said to be *consistent* iff it does not contain a pair of complementary literals.

For a given $\mathcal{P}$, consider the following transformation: (1) For each defined atom A, replace all clauses of the form $A \leftarrow body_1, \ldots, A \leftarrow body_m$ occurring in $g\mathcal{P}$ by $A \leftarrow body_1 \vee \ldots \vee body_m$. (2) Replace all occurrences of $\leftarrow$ by $\leftrightarrow$. The obtained set is called *weak completion* of $\mathcal{P}$ or $wc\mathcal{P}$.[3]

We consider the three-valued Łukasiewicz (or Ł-) logic [25] and represent each interpretation I by $\langle I^\top, I^\bot \rangle$, where $I^\top = \{A \mid I(A) = \top\}$, $I^\bot = \{A \mid I(A) = \bot\}$, $I^\top \cap I^\bot = \emptyset$, and each ground atom $A \notin I^\top \cup I^\bot$ is mapped to U (*unknown*). Let $\langle I^\top, I^\bot \rangle$ and $\langle J^\top, J^\bot \rangle$ be two interpretations. We define:

$$\langle I^\top, I^\bot \rangle \subseteq \langle J^\top, J^\bot \rangle \text{ iff } I^\top \subseteq J^\top \text{ and } I^\bot \subseteq J^\bot \text{ and}$$
$$\langle I^\top, I^\bot \rangle \cup \langle J^\top, J^\bot \rangle = \langle I^\top \cup J^\top, I^\bot \cup J^\bot \rangle.$$

It has been shown in [16] that logic programs as well as their weak completions admit a least model under Ł-logic. Moreover, the least Ł-model of $wc\mathcal{P}$ can be obtained as the least fixed point of the following semantic operator, which is due to Stenning and van Lambalgen [40]: $\Phi_\mathcal{P}(\langle I^\top, I^\bot \rangle) = \langle J^\top, J^\bot \rangle$, where

$$J^\top = \{A \mid A \leftarrow body \in g\mathcal{P} \text{ and } body \text{ is } true \text{ under } \langle I^\top, I^\bot \rangle\},$$
$$J^\bot = \{A \mid def(A, \mathcal{P}) \neq \emptyset \text{ and}$$
$$body \text{ is } false \text{ under } \langle I^\top, I^\bot \rangle \text{ for all } A \leftarrow body \in def(A, \mathcal{P})\}.$$

Weak completion semantics (WCS) is the approach to consider weakly completed logic programs and to reason with respect to the least Ł-models of these programs. We write $\mathcal{P} \models_{wcs} F$ iff formula F holds in the least Ł-model of $wc\mathcal{P}$. In the remainder of this paper, $\mathcal{M}_\mathcal{P}$ denotes the least Ł-model of $wc\mathcal{P}$.

The $\Phi_\mathcal{P}$ operator differs from the semantic operator defined by Fitting in [12] in that the additional condition $def(A, \mathcal{P}) \neq \emptyset$ is required in the definition of $J^\bot$. This condition states that A must be defined in order to be mapped to *false*, whereas in the Kripke-Kleene-semantics considered by Fitting an atom is mapped to *false* if it is undefined. This reflects precisely the difference between the weak completion and the completion semantics. The Kripke-Kleene-semantics was also applied in [40]. However, as shown in [16] this semantics is not only the cause for a technical bug in one theorem of [40], but it does also lead to a non-adequate model of some human reasoning tasks. Both, the technical bug as well as the non-adequate modeling, can be avoided by using WCS.

As shown in [9], WCS is related to the well-founded semantics (WFS) as follows: Let $\mathcal{P}$ be a program which does not contain a positive loop and let $\mathcal{P}^+ = \mathcal{P} \setminus \{A \leftarrow \bot \mid A \leftarrow \bot \in \mathcal{P}\}$. Let u be a new nullary relation symbol not occurring in $\mathcal{P}$ in $\mathcal{P}^* = \mathcal{P}^+ \cup \{B \leftarrow u \mid def(B, \mathcal{P}) = \emptyset\} \cup \{u \leftarrow \neg u\}$. Then, the least Ł-model of $wc\mathcal{P}$ and the well-founded model for $\mathcal{P}^*$ coincide. The programs specified in [7,8], which model the suppression and the selection task as well as

---

[3] Note that undefined atoms are not identified with $\bot$ as in the completion of $\mathcal{P}$ [5].

the programs presented in the sequel of this paper are acyclic and, thus, tight. Therefore, our results hold for both, WCS and WFS.

An *abductive framework* consists of a logic program $\mathcal{P}$, a set of *abducibles* $\mathcal{A}_\mathcal{P} = \{A \leftarrow \top \mid def(A, \mathcal{P}) = \emptyset\} \cup \{A \leftarrow \bot \mid def(A, \mathcal{P}) = \emptyset\}$, a set of *integrity constraints* $\mathcal{IC}$, i.e., expressions of the form $\bot \leftarrow B_1 \wedge \ldots \wedge B_n$, and the entailment relation $\models_{wcs}$. An abductive framework is denoted by $\langle \mathcal{P}, \mathcal{A}_\mathcal{P}, \mathcal{IC}, \models_{wcs} \rangle$.

One should observe that each $\mathcal{P}$ and, in particular, each finite set of positive and negative ground facts has an L-model. For the latter, this can be obtained by mapping all heads occurring in this set to *true*. Thus, in the following definition, explanations as well as the union of a program and an explanation are satisfiable.

An *observation* $\mathcal{O}$ is a set of ground literals; it is *explainable* in the framework $\langle \mathcal{P}, \mathcal{A}_\mathcal{P}, \mathcal{IC}, \models_{wcs} \rangle$ iff there exists a (minimal) $\mathcal{E} \subseteq \mathcal{A}_\mathcal{P}$ called *explanation* such that $\mathcal{M}_{\mathcal{P} \cup \mathcal{E}}$ satisfies $\mathcal{IC}$ and $\mathcal{P} \cup \mathcal{E} \models_{wcs} L$ for each $L \in \mathcal{O}$.

3　The Semantic Operator Revisited

Before looking into conditionals, we need to reconsider the $\Phi_\mathcal{P}$ operator and establish some of its properties. The least fixed point of $\Phi_\mathcal{P}$ can be computed by iterating the operator starting with the empty interpretation: $\Phi_\mathcal{P} \uparrow 0 = \langle \emptyset, \emptyset \rangle$, $\Phi_\mathcal{P} \uparrow (n+1) = \Phi_\mathcal{P}(\Phi_\mathcal{P} \uparrow n)$ for all $n \in \mathbb{N}$.

Proposition 1

1. $\Phi_\mathcal{P}$ *is monotonic.*
2. *For all* $n \geq 0$ *we find* $\Phi_\mathcal{P}(\Phi_\mathcal{P} \uparrow n) \supseteq \Phi_\mathcal{P} \uparrow n$.
3. *For all* $n \geq 0$ *we find* $\Phi_\mathcal{P} \uparrow (n+1) = \Phi \uparrow n \cup \langle J^\top, J^\bot \rangle$, *where*

$$J^\top = \{A \mid \Phi \uparrow n(A) = U \text{ and } A \leftarrow body \in g\mathcal{P} \text{ and } \Phi \uparrow n(body) = \top\},$$
$$J^\bot = \{A \mid \Phi \uparrow n(A) = U \text{ and } def(A, \mathcal{P}) \neq \emptyset \text{ and }$$
$$\Phi \uparrow n(body) = \bot \text{ for all } A \leftarrow body \in def(A, \mathcal{P})\}.$$

Proof

1. Has been shown as Proposition 3.21 in [20].
2. The proof is by induction on n: The case $n = 0$ holds because

$$\Phi_\mathcal{P}(\Phi_\mathcal{P} \uparrow 0) = \Phi_\mathcal{P}(\langle \emptyset, \emptyset \rangle) \supseteq \langle \emptyset, \emptyset \rangle.$$

From the induction hypothesis $\Phi_\mathcal{P}(\Phi_\mathcal{P} \uparrow n) \supseteq \Phi_\mathcal{P} \uparrow n$ we conclude by the monotonicity of $\Phi_\mathcal{P}$ that $\Phi_\mathcal{P}(\Phi_\mathcal{P}(\Phi_\mathcal{P} \uparrow n)) \supseteq \Phi_\mathcal{P}(\Phi_\mathcal{P} \uparrow n)$, which is equal to $\Phi_\mathcal{P}(\Phi_\mathcal{P} \uparrow (n+1)) \supseteq \Phi_\mathcal{P} \uparrow (n+1)$.
3. $\Phi \uparrow (n+1) = \Phi \uparrow n \cup \Phi \uparrow (n+1) = \Phi \uparrow n \cup (\Phi \uparrow (n+1) \backslash \Phi \uparrow n) = \Phi \uparrow n \cup \langle J^\top, J^\bot \rangle$.

4　Revision

The first new concept introduced in this paper is a revision operator which revises a given program with respect to a set of literals. Revision will be needed when evaluating counterfactual conditionals in order to revise background knowledge in the form of a logic program such that a previously *false* condition is mapped to *true* under the revised program. Somehow surprisingly it will turn out that revision is also needed in the evaluation of indicative conditionals.

Let S be a finite and consistent set of ground literals in

$$rev(\mathcal{P}, S) = (\mathcal{P} \setminus def(S, \mathcal{P})) \cup \{A \leftarrow \top \mid A \in S\} \cup \{A \leftarrow \bot \mid \neg A \in S\},$$

where A denotes an atom. $rev(\mathcal{P}, S)$ is called the *revision of $\mathcal{P}$ with respect to S*.

Proposition 2

1. *rev is non-monotonic, i.e., there exist $\mathcal{P}$, S and F such that $\mathcal{P} \models_{wcs} F$ and $rev(\mathcal{P}, S) \not\models_{wcs} F$.*
2. *rev is monotonic, i.e., $\mathcal{M}_{\mathcal{P}} \subseteq \mathcal{M}_{rev(\mathcal{P},S)}$, if $\mathcal{M}_{\mathcal{P}}(L) = U$ for all $L \in S$.*
3. $\mathcal{M}_{rev(\mathcal{P},S)}(S) = \top$.

Proof

1. Let $\mathcal{P} = \{a \leftarrow \top\}$, $S = \{\neg a\}$ and $F = a$. We find $\mathcal{M}_{\mathcal{P}} = \langle a, \emptyset \rangle$, $rev(\mathcal{P}, S) = \{a \leftarrow \bot\}$, $\mathcal{M}_{rev(\mathcal{P},S)} = \langle \emptyset, a \rangle$, $\mathcal{P} \models_{wcs} a$ and $rev(\mathcal{P}, S) \not\models_{wcs} a$.
2. $\mathcal{M}_{\mathcal{P}}$ and $\mathcal{M}_{rev(\mathcal{P},S)}$ can be computed by iterating $\Phi_{\mathcal{P}}$ and $\Phi_{rev(\mathcal{P},S)}$, respectively. By induction on n we can show that for all $n \in \mathbb{N}$ the relationship $\Phi_{\mathcal{P}} \uparrow n \subseteq \Phi_{rev(\mathcal{P},S)} \uparrow n$ holds. In case $n = 0$ we find $\Phi_{\mathcal{P}} \uparrow 0 = \langle \emptyset, \emptyset \rangle = \Phi_{rev(\mathcal{P},S)} \uparrow 0$. We assume that the result holds for n and turn to the induction step:

$$\Phi_{\mathcal{P}} \uparrow (n+1) = \Phi_{\mathcal{P}}(\Phi_{\mathcal{P}} \uparrow n) = \langle I^{\top}, I^{\bot} \rangle, \tag{1}$$

where

$$I^{\top} = \{A \mid A \leftarrow body \in g\mathcal{P}, \Phi_{\mathcal{P}} \uparrow n(body) = \top\}$$
$$I^{\bot} = \{A \mid def(A, \mathcal{P}) \neq \emptyset, \Phi_{\mathcal{P}} \uparrow n(body) = \bot \text{ for all } A \leftarrow body \in def(A, \mathcal{P})\}$$

As $\mathcal{M}_{\mathcal{P}}(L) = U$ for all $L \in S$, we find that $atom(L)$ is neither in $I^{\top}$ nor in $I^{\bot}$, where $atom(L) = L$ if L is an atom and $atom(L) = A$ if $L = \neg A$. By the definition of revision, however, $atom(L)$ is either in $J^{\top}$ or in $J^{\bot}$, where

$$J^{\top} = \{A \mid A \leftarrow body \in g\,rev(\mathcal{P}, S), \Phi_{rev(\mathcal{P},S)} \uparrow n(body) = \top\}$$
$$J^{\bot} = \{A \mid def(A, rev(\mathcal{P}, S)) \neq \emptyset,$$
$$\Phi_{rev(\mathcal{P},S)} \uparrow n(body) = \bot \text{ for all } A \leftarrow body \in def(A, rev(\mathcal{P}, S))\}$$

As $\mathcal{P}$ and $rev(\mathcal{P}, S)$ contain identical definitions for atoms not occurring in S we conclude by the induction hypothesis that $I^{\top} \subseteq J^{\top}$, $I^{\bot} \subseteq J^{\bot}$ and

$$\langle I^{\top}, I^{\bot} \rangle \subseteq \langle J^{\top}, J^{\bot} \rangle = \Phi_{rev(\mathcal{P},S)}(\Phi_{rev(\mathcal{P},S)} \uparrow n) = \Phi_{rev(\mathcal{P},S)} \uparrow (n+1). \tag{2}$$

The result follows by combining (1) and (2) and the induction theorem.
3. Follows immediately from the definition of revision and Proposition 1.

5 Indicative Conditionals

Conditions as well as consequences of conditionals are assumed to be finite sets (or conjunctions) of ground literals. When parsing conditionals we assume that information concerning the mood of the conditionals has been extracted. In this section we focus on the indicative mood. In the sequel, let $cond(C, D)$ be a conditional with condition C and consequence D, both of which are assumed to be finite and consistent sets of literals.

Conditionals are evaluated with respect to some background information specified as a program and a set of integrity constraints. More specifically, as the weak completion of each program admits a least Ł-model, conditionals are evaluated under the least Ł-model of a program. In the reminder of this section let P be a program, IC be a finite set of integrity constraints, and M_P be the least Ł-model of wcP such that M_P satisfies IC.

In this setting we propose to evaluate indicative conditionals as follows:

1. If $M_P(C) = \top$ and $M_P(D) = \top$, then $cond(C, D)$ is *true*.
2. If $M_P(C) = \top$ and $M_P(D) = \bot$, then $cond(C, D)$ is *false*.
3. If $M_P(C) = \top$ and $M_P(D) = U$, then $cond(C, D)$ is *unknown*.
4. If $M_P(C) = \bot$, then $cond(C, D)$ is *vacuous*.
5. If $M_P(C) = U$, then evaluate $cond(C, D)$ with respect to $M_{P'}$, where
   - $M_{P'}$ is the least Ł-model of wcP',
   - $P' = rev(P, S) \cup \mathcal{E}$,
   - S is a smallest subset of C and $\mathcal{E} \subseteq \mathcal{A}_{rev(P,S)}$ is a minimal explanation for $C \backslash S$ such that $M_{P'}(C) = \top$ and $M_{P'}$ satisfies IC.

In words, if the condition of a conditional is *true*, then the conditional is evaluated as implication in Ł-logic. If the condition is *false*, then Ł-logic would assign *true* to the conditional. However, we believe that such a conditional is in fact not an indicative but a subjunctive one. For the time being, we assign *vacuous* to it and intend to investigate this case more thoroughly in the future.

The novel contribution concerns the case that the condition C of a conditional is *unknown*. In this case we propose to split C into two disjoint subsets S and $C \backslash S$, where the former is treated by revision and the latter by abduction. In case C contains some literals which are *true* and some which are *unknown* under M_P, then the former will be part of $C \backslash S$ because the empty explanation explains them. As we assume S to be minimal this approach is called *minimal revision followed by abduction* (MRFA). Furthermore, because revision as well as abduction are only applied to literals which are assigned to *unknown*, the approach is monotonic.

6 Examples

We will focus on the novel and interesting case in our approach, i.e., the one where the condition of a conditional is *unknown*. In particular, we consider the *firing squad example* discussed in [26]: *If the court orders an execution (exec), then the captain will give the signal (sig) upon which riflemen A (rmA) and B (rmB) will shoot the prisoner. Consequently, the prisoner will be dead (dead).* We assume that the court's decision is *unknown*, that both riflemen are accurate, alert and law-abiding, and that the prisoner is unlikely to die from any other cause. This background information can be captured in the following program

$$\mathcal{P}_1 = \{ \ sig \leftarrow exec \wedge \neg ab_1, \ rmA \leftarrow sig \wedge \neg ab_2, \ rmB \leftarrow sig \wedge \neg ab_3,$$
$$dead \leftarrow rmA \wedge \neg ab_4, \ dead \leftarrow rmB \wedge \neg ab_5, \ alive \leftarrow \neg dead \wedge \neg ab_6 \ \}$$
$$\cup \ \{ \ ab_i \leftarrow \bot \mid i \in [1,6] \ \},$$

where we have presented conditionals by licenses for implications as proposed in [40].[4] We obtain $\mathcal{M}_{\mathcal{P}_1} = \langle \emptyset, \{ab_i \mid i \in [1,6]\}\rangle$. The set $\mathcal{A}_{\mathcal{P}_1}$ of abducibles in the abductive framework $\langle \mathcal{P}_1, \mathcal{A}_{\mathcal{P}_1}, \emptyset, \models_{wcs}\rangle$ is $\{exec \leftarrow \top, exec \leftarrow \bot\}$. The explanation $\mathcal{E}_1^\top = \{exec \leftarrow \top\}$ explains $\{sig, rmA, rmB, dead, \neg alive\}$, whereas $\mathcal{E}_1^\bot = \{exec \leftarrow \bot\}$ explains $\{\neg sig, \neg rmA, \neg rmB, \neg dead, alive\}$. The observation $\{\neg sig, rmA\}$ cannot be explained at all because $wc\mathcal{A}_{\mathcal{P}_1} = \{exec \leftrightarrow \top \vee \bot\} \equiv \{exec \leftrightarrow \top\}$, where $\equiv$ denotes semantic equivalence. We will now evaluate three indicative conditionals with respect to $\mathcal{M}_{\mathcal{P}_1}$:

1. *If the prisoner is alive, then the captain did not signal*, i.e. $cond(alive, \neg sig)$: As *alive* can be explained by $\mathcal{E}_1^\bot$ and $\mathcal{M}_{\mathcal{P}_1 \cup \mathcal{E}_1^\bot}(\neg sig) = \top$, the conditional is *true*. In this case revision is not needed.
2. *If rifleman A shot, then rifleman B shot as well*, i.e., $cond(rmA, rmB)$: As rmA can be explained by $\mathcal{E}_1^\top$ and $\mathcal{M}_{\mathcal{P}_1 \cup \mathcal{E}_1^\top}(rmB) = \top$, the conditional is *true*. Again, revision is not needed.
3. *If the captain gave no signal and rifleman A decides to shoot, then the court did not order an execution*, i.e. $cond(\{\neg sig, rmA\}, \neg exec)$: As $\{\neg sig, rmA\}$ cannot be explained, $\mathcal{P}_1$ must be revised. Let $\mathcal{P}_2 = rev(\mathcal{P}_1, rmA)$. We obtain $\mathcal{M}_{\mathcal{P}_2} = \langle \{rmA, dead\}, \{alive\} \cup \{ab_i \mid i \in [1,6]\}\rangle$. $\neg sig$ is still mapped to *unknown* under this model, but it can now be explained by $\{exec \leftarrow \bot\}$. Hence, the conditional is *true*. In [6] an abstract reduction system for the evaluation of indicative conditionals has been specified, where separate rules allow the application of revision and abduction in any order, and it was shown that there is no other minimal revision followed by abduction for $cond(\{\neg sig, rmA\}, \neg exec)$.

Note that the results would be different under the well-founded semantics: The well-founded model of $\mathcal{P}_1$ is $\langle \{alive\}, \{ab_i \mid i \in [1,6]\} \cup \{exec, sig, rmA, rmB, dead\}\rangle$. As no atom is *unknown* in this model, no conditional evaluates to *unknown*. Accordingly, only revision can be applied for the evaluation of each conditional.

---

[4] In this section, the abnormality predicates are not needed. We have kept them to be in line with our general approach to model human reasoning episodes (see e.g. [7]) and to be able to extend the example in the future by, for example, considering the case that the captain is not law-abiding or that a rifle is malfunctioning.

7   The Approach by Schulz

Schulz [36] presents another computational logic approach based on L-logic, where the $\Phi_{\mathcal{P}}$ operator is modified such that it allows to evaluate conditionals. In this section, let $\mathcal{S}$ be a finite and consistent set of ground literals. $\mathcal{S}$ can be identified with the interpretation $\langle \mathcal{S}^\top, \mathcal{S}^\perp \rangle$, where $\mathcal{S}^\top = \{A \mid A \in \mathcal{S}\}$ and $\mathcal{S}^\perp = \{A \mid \neg A \in \mathcal{S}\}$ and A denotes a ground atom.

Let $I = \langle I^\top, I^\perp \rangle$ be an interpretation. Schulz defines

$$\Psi_{\mathcal{P}}(\langle I^\top, I^\perp \rangle) = \langle I^\top, I^\perp \rangle \cup \langle J^\top, J^\perp \rangle,$$

where

$$J^\top = \{A \mid I(A) = \mathsf{U},\ A \leftarrow body \in g\mathcal{P} \text{ and } I(body) = \top\},$$
$$J^\perp = \{A \mid I(A) = \mathsf{U},\ def(A, \mathcal{P}) \neq \emptyset \text{ and } I(body) = \perp \text{ for all } A \leftarrow body \in \mathcal{P}\}.$$

In contrast to the $\Phi_{\mathcal{P}}$ operator, which is iterated starting with the empty interpretation, the $\Psi_{\mathcal{P}}$ operator is iterated as follows: $\Psi_{\mathcal{P}} \uparrow 0 = \langle \mathcal{S}^\top, \mathcal{S}^\perp \rangle$ and $\Psi_{\mathcal{P}} \uparrow (n+1) = \Psi_{\mathcal{P}}(\Psi_{\mathcal{P}} \uparrow n)$. As shown in [36], the Ψ operator admits a least fixed point which shall be denoted by $\mathsf{lfp}\ \Psi_{\mathcal{P},\mathcal{S}}$. Moreover, in [36] reasoning is performed with respect to this fixed point, i.e. $\mathcal{P}, \mathcal{S} \models_s F$ iff $\mathsf{lfp}\ \Psi_{\mathcal{P},\mathcal{S}}(F) = \top$.

7.1   Some Properties

Let us identify some general properties of the operators $\Phi_{\mathcal{P}}$ and $\Psi_{\mathcal{P}}$.

Proposition 3. $\mathsf{lfp}\ \Phi_{\mathcal{P}}$ and $\mathsf{lfp}\ \Psi_{\mathcal{P},\mathcal{S}}$ exist.

The existence of $\mathsf{lfp}\ \Phi_{\mathcal{P}}$ and $\mathsf{lfp}\ \Psi_{\mathcal{P},\mathcal{S}}$ was established in [16] and [36], respectively. The following proposition is an immediate consequence of the definition of the $\Psi_{\mathcal{P}}$ operator and corresponds to Proposition 2(3.).

Proposition 4. For all $L \in \mathcal{S}$ we find $\mathcal{P}, \mathcal{S} \models_s L$.

7.2   The Correspondence

We show the correspondence between the approach by Schulz and our approach.

Lemma 5. For all $n \in \mathbb{N}$, we find $\Phi_{rev(\mathcal{P},\mathcal{S})} \uparrow n \subseteq \Psi_{\mathcal{P}} \uparrow n \subseteq \Phi_{rev(\mathcal{P},\mathcal{S})} \uparrow (n+1)$.

Proof. To simplify the presentation we will omit the indices of the operators $\Psi_{\mathcal{P}}$ and $\Phi_{rev(\mathcal{P},\mathcal{S})}$ in this proof. The proof is by induction on n. In case $n = 0$ we find

$$\Phi{\uparrow}0 = \langle \emptyset, \emptyset \rangle \subseteq \langle \mathcal{S}^\top, \mathcal{S}^\perp \rangle = \Psi{\uparrow}0 \subseteq \langle I^\top, I^\perp \rangle = \Phi{\uparrow}1,$$

where

$$I^\top = \{A \mid A \leftarrow \top \in rev(\mathcal{P}, \mathcal{S})\} \qquad\qquad \supseteq \mathcal{S}^\top,$$
$$I^\perp = \{A \mid def(A, rev(\mathcal{P}, \mathcal{S})) = \{A \leftarrow \perp\}\} \supseteq \mathcal{S}^\perp.$$

As induction hypothesis we assume that the result holds for n, i.e.,

$$\Phi \uparrow n \subseteq \Psi \uparrow n \subseteq \Phi \uparrow (n+1). \tag{3}$$

In the induction step we need to show that the result holds for $n + 1$. We start by showing that

$$\Phi \uparrow (n+1) \subseteq \Psi \uparrow (n+1). \tag{4}$$

By Proposition 1(3.) and the definition of Ψ this corresponds to

$$\Phi \uparrow n \cup \langle I^\top, I^\perp \rangle \subseteq \Psi \uparrow n \cup \langle J^\top, J^\perp \rangle,$$

where

$$
\begin{aligned}
I^\top &= \{A \mid \Phi \uparrow n(A) = \mathrm{U} \text{ and } A \leftarrow body \in g\mathcal{P} \text{ and } \Phi \uparrow n(body) = \top\}, \\
I^\perp &= \{A \mid \Phi \uparrow n(A) = \mathrm{U} \text{ and } def(A, \mathcal{P}) \neq \emptyset \text{ and} \\
&\qquad \Phi \uparrow n(body) = \perp \text{ for all } A \leftarrow body \in g\mathcal{P}\}, \\
J^\top &= \{A \mid \Psi \uparrow n(A) = \mathrm{U} \text{ and } A \leftarrow body \in g\mathcal{P} \text{ and } \Psi \uparrow n(body) = \top\}, \\
J^\perp &= \{A \mid \Psi \uparrow n(A) = \mathrm{U} \text{ and } def(A, \mathcal{P}) \neq \emptyset \text{ and} \\
&\qquad \Psi \uparrow n(body) = \perp \text{ for all } A \leftarrow body \in g\mathcal{P}\}.
\end{aligned}
$$

From the induction hypothesis (3) we conclude that

$$\Phi \uparrow n \subseteq \Psi \uparrow n \cup \langle J^\top, J^\perp \rangle. \tag{5}$$

Now suppose that $A \in I^\top$. Then, $\Phi \uparrow n(A) = \mathrm{U}$ and we distinguish two cases:

1. If $\Psi \uparrow n(A) = \mathrm{U}$, then $A \in J^\top$ because of the induction hypothesis (3).
2. If $\Psi \uparrow n(A) \neq \mathrm{U}$, then A must already been assigned to either *true* or *false* under $\Psi \uparrow n$. By (3), $\Psi \uparrow n \subseteq \Phi \uparrow (n+1)$ and, hence, $\Psi \uparrow n(A) = \top$.

Likewise, we find for $A \in I^\perp$ that either $A \in J^\perp$ or $\Psi \uparrow n(A) = \perp$. Therefore,

$$\langle I^\top, I^\perp \rangle \subseteq \Psi \uparrow n \cup \langle J^\top, J^\perp \rangle, \tag{6}$$

and (4) follows immediately from (5) and (6).

We turn to the proof of

$$\Psi \uparrow (n+1) \subseteq \Phi \uparrow (n+2). \tag{7}$$

By the definition for Ψ and Proposition 1(3.), this corresponds to

$$\Psi \uparrow n \cup \langle J^\top, J^\perp \rangle \subseteq \Phi \uparrow (n+1) \cup \langle I^\top, I^\perp \rangle,$$

where

$$
\begin{aligned}
J^\top &= \{A \mid \Psi \uparrow n(A) = \mathrm{U} \text{ and } A \leftarrow body \in g\mathcal{P} \text{ and } \Psi \uparrow n(body) = \top\}, \\
J^\perp &= \{A \mid \Psi \uparrow n(A) = \mathrm{U} \text{ and } def(A, \mathcal{P}) \neq \emptyset \text{ and} \\
&\qquad \Psi \uparrow n(body) = \perp \text{ for all } A \leftarrow body \in g\mathcal{P}\}, \\
I^\top &= \{A \mid \Phi \uparrow (n+1)(A) = \mathrm{U} \text{ and } A \leftarrow body \in g\mathcal{P} \text{ and } \Phi \uparrow (n+1)(body) = \top\}, \\
I^\perp &= \{A \mid \Phi \uparrow (n+1)(A) = \mathrm{U} \text{ and } def(A, \mathcal{P}) \neq \emptyset \text{ and} \\
&\qquad \Phi \uparrow (n+1)(body) = \perp \text{ for all } A \leftarrow body \in g\mathcal{P}\}.
\end{aligned}
$$

By the induction hypothesis (3) we find

$$\Psi \uparrow n \subseteq \Phi \uparrow (n+1) \cup \langle I^\top, I^\perp \rangle. \tag{8}$$

Now suppose that $A \in J^\top$. Then, $\Psi \uparrow n(A) = \mathrm{U}$ and we distinguish two cases.

1. If $\Phi \uparrow (n+1)(A) = \mathrm{U}$, then $A \in I^\top$ because of the induction hypothesis (3).
2. If $\Phi \uparrow (n+1)(A) \neq \mathrm{U}$, then A must already be assigned to either *true* or *false* under $\Phi \uparrow (n+1)$. By (4), $\Phi \uparrow (n+1)(A) = \top$.

Likewise, we find for $A \in J^\perp$ that either $A \in I^\perp$ or $\Phi \uparrow (n+1)(A) = \perp$. Therefore,

$$\langle J^\top, J^\perp \rangle \subseteq \Phi \uparrow (n+1) \cup \langle I^\top, I^\perp \rangle. \tag{9}$$

and (7) follows immediately from (8) and (9).

We can now prove our main result, the correspondence of the two operators.

Theorem 6. lfp $\Phi_\mathcal{P} = $ lfp $\Psi_{\mathcal{P},\mathcal{S}}$.

Proof. The result follows immediately from Proposition 3 and and Lemma 5.

Let us consider an example from [36]: *If you drop a wine glass (drop) then the wine glass breaks (broken). She drops the wine glass.* This scenario can be represented by the program $\mathcal{P}_3 = \{broken \leftarrow drop \wedge \neg ab, \; ab \leftarrow \perp, \; drop \leftarrow \top\}$. Now, consider $\mathcal{S} = \{\neg broken\}$. Then, $rev(\mathcal{P}_3, \mathcal{S}) = \{ab \leftarrow \perp, drop \leftarrow \top, broken \leftarrow \perp\}$, $\mathcal{S}^\top = \emptyset$, $\mathcal{S}^\perp = \{broken\}$, and the two fixed points are computed as follows:

	$\Psi_{\mathcal{P}_3}$	$\Phi_{rev(\mathcal{P}_3,\mathcal{S})}$
$\uparrow 0$	$\langle \emptyset, \{broken\} \rangle$	$\langle \emptyset, \emptyset \rangle$
$\uparrow 1$	$\langle \{drop\}, \{ab, broken\} \rangle$	$\langle \{drop\}, \{ab, broken\} \rangle$

Coming back to the examples discussed in Sect. 6 we observe that they can be modeled by Schulz' approach only if the appropriate initial set $\mathcal{S}$ is given. Schulz does not provide any means to obtain these sets. One should note that these sets are not simply the unknown conditions of the given conditionals. We compute the additional assignments by MRFA (as explained in Sect. 5).

8   Counterfactual Conditionals

We now consider counterfactual conditionals, i.e., conditionals whose condition is false. The following *forest fire* example is from [4]. The conditional *if there had not been so many dry leaves on the forest floor* ($\neg dl$), *then the forest fire would not have occurred* ($\neg ff$), $cond(\neg dl, \neg ff)$, is to be evaluated with respect to

$$\mathcal{P}_4 = \{ff \leftarrow l \wedge \neg ab, \; l \leftarrow \top, \; ab \leftarrow \neg dl, \; dl \leftarrow \top\},$$

which states that *lightning* (l) *causes fire* (ff) *if nothing abnormal is taking place* ($\neg ab$), *lightning happened, the absence of dry leaves* ($\neg dl$) *is an abnormality* (*ab*), *and dry leaves* (*dl*) *are present.* We obtain $\mathcal{M}_{\mathcal{P}_4} = \langle\{dl, l, f\!f\}, \{ab\}\rangle$, where condition $\neg dl$ of the counterfactual is *false.* Let $\mathcal{S} = \{\neg dl\}$, $\mathcal{S}^\top = \emptyset$, $\mathcal{S}^\perp = \{dl\}$,

$$rev(\mathcal{P}_4, \mathcal{S}) = \{f\!f \leftarrow l \wedge \neg ab, \ l \leftarrow \top, \ ab \leftarrow \neg dl, \ dl \leftarrow \perp\},$$

and the fixed points can be computed as follows:

	$\Psi_{\mathcal{P}_4}$	$\Phi_{rev(\mathcal{P}_4,\mathcal{S})}$
↑0	$\langle\emptyset, \{dl\}\rangle$	$\langle\emptyset, \emptyset\rangle$
↑1	$\langle\{l, ab\}, \{dl\}\rangle$	$\langle\{l\}, \{dl\}\rangle$
↑2	$\langle\{l, ab\}, \{dl, f\!f\}\rangle$	$\langle\{l, ab\}, \{dl\}\rangle$
↑3	$\langle\{l, ab\}, \{dl, f\!f\}\rangle$	$\langle\{l, ab\}, \{dl, f\!f\}\rangle$

As $\neg f\!f$ is *true* in the least Ł-model, we conclude that the counterfactual is *true.*

Let us extend the example by adding *arson* (*a*) *causes a forest fire* (*f\!f*):

$$\mathcal{P}_5 = \{f\!f \leftarrow l \wedge \neg ab_1, \ f\!f \leftarrow a \wedge \neg ab_2, \ l \leftarrow \top, \ ab_1 \leftarrow \neg dl, \ dl \leftarrow \top, \ ab_2 \leftarrow \perp\}.$$

We find $\mathcal{M}_{\mathcal{P}_5} = \langle\{dl, l, f\!f\}, \{ab_1, ab_2\}\rangle$ and $\mathcal{M}_{rev(\mathcal{P}_5, \neg dl)} = \langle\{l, ab_1\}, \{dl, ab_2\}\rangle$. $cond(\neg dl, \neg f\!f)$ is now *unknown*, because $f\!f$ is *unknown* in the least model. The results are different under the well-founded semantics: The well-founded model of $rev(\mathcal{P}_5, \neg dl)$ is $\langle\{l, ab_1\}, \{dl, ab_2, a, f\!f\}\rangle$, accordingly $cond(\neg dl, \neg f\!f)$ is *true*. We assume that the evaluation according to the weak completion semantics is more appropriate: The answer to whether $cond(\neg dl, \neg f\!f)$ is valid, should be *it is unknown* because arson could have been the cause for the forest fire.

The two reasoning episodes exemplify how we intend to evaluate counterfactuals. Let $\mathcal{C}$ be the condition of a counterfactual. If $\mathcal{C}$ is *true*, then the counterfactual is *vacuous* in that it is not a subjunctive but rather an indicative conditional. If $\mathcal{C}$ is *false*, then (non-monotonic) revision is applied to force $\mathcal{C}$ to be *true* and the counterfactual is evaluated with respect to the revised program. Both examples are treated this way. The most interesting case is again where $\mathcal{C}$ is *unknown*. We have two options:

1. We can apply MRFA to map $\mathcal{C}$ to *true* and evaluate the counterfactual with respect to the modified program. Note that this program is a monotonic extension of the original program.
2. We can use MRFA to map $\mathcal{C}$ to *false* and thereafter (non-monotonically) revise the modified program to force $\mathcal{C}$ to be *true.*

Obviously, both options do not necessarily lead to the same result. Consider

$$\mathcal{P}_6 = \{f\!f \leftarrow l \wedge \neg ab \wedge \neg rain, \ l \leftarrow \top, \ ab \leftarrow \neg dl, \ dl \leftarrow \neg rain\}$$

The last clause states *if it doesn't rain, then the leaves are dry.* $\mathcal{M}_{\mathcal{P}_6} = \langle\{l\}, \emptyset\rangle$, where $\mathcal{C} = \{\neg dl\}$ is mapped to *unknown*. For option 1., $\mathcal{C}$ is explained by $\mathcal{E}^\top = \{rain \leftarrow \top\}$, and leads to $\mathcal{M}_{\mathcal{P}_6 \cup \mathcal{E}^\top} = \langle\{l, ab, rain\}, \{f\!f, dl\}\rangle$. For option 2., we need to make $\mathcal{C}$ *false* first, which is done by abducing $\mathcal{E}^\perp = \{rain \leftarrow \perp\}$. We find that $\mathcal{M}_{\mathcal{P}_6 \cup \mathcal{E}^\perp} = \langle\{l, dl, f\!f\}, \{ab, rain\}\rangle$, and revise with respect to $\mathcal{C}$:

$$rev(\mathcal{P}_6 \cup \mathcal{E}^\perp, \mathcal{C}) = \{f\!f \leftarrow l \wedge \neg ab \wedge \neg rain, \ l \leftarrow \top, \ ab \leftarrow \neg dl, \ dl \leftarrow \perp, rain \leftarrow \perp\}$$

We obtain $\mathcal{M}_{rev(\mathcal{P}_6 \cup \mathcal{E}^\perp, \mathcal{S})} = \langle\{l, ab\}, \{f\!f, dl, rain\}\rangle$ which differs from $\mathcal{M}_{\mathcal{P}_6 \cup \mathcal{E}^\perp}$ with respect to *rain*.

9   Conclusion

This paper presents a novel computational logic approach for the evaluation of conditionals. In the case where the condition of a conditional is *unknown*, we propose to explain as many literals as possible by abduction, and if necessary, revise the remaining ones. We formally show the correspondence to Schulz' approach and observe that we can handle more human reasoning tasks. In fact, we are unaware of any computational logic approach which can handle as many human reasoning episodes as our approach based on the weak completion semantics. However, there are still many open and interesting questions, some of which will be mentioned in the sequel.

Similarly to an abductive procedure for the condition of a conditional, we can extend our approach by introducing abduction for the consequence, in case the consequence is still *unknown*. For instance, consider again program $\mathcal{P}_5$ in Sect. 8: After the revision step for $\mathcal{C} = \neg dl$, $\mathcal{D} = \neg ff$ is still *unknown*. There is a minimal explanation for $\mathcal{D}$, namely $\mathcal{E} = \{a \leftarrow \bot\}$. Accordingly, we can now conclude that the counterfactual, $cond(\mathcal{C}, \mathcal{D})$ is *true*, if *there is no arson*.

Another issue that we need to investigate–and already proposed in [6]–is to carry out psychological experiments which verify whether our assumption of MRFA is indeed adequate for human reasoning. Furthermore, as discussed in the last part of Sect. 8, we need to clarify which of the two options is more adequate, in case the condition of a counterfactual is *unknown*.

We also need to look into the revision operator. So far, the operator is quite simple and straightforward. On the other hand, revision has been intensely studied in the the field of computational logic and we need to investigate whether and how these approaches can be adapted to a setting which treats *unknown* conditions explicitly and considers the weak completion semantics.

Acknowledgements. We like to thank Luís Moniz Pereira, Bob Kowalski and Marco Ragni for many discussions and comments on earlier drafts of our work.

References

1. Baral, C., Gelfond, M., Rushton, J.N.: Probabilistic reasoning with answer sets. TPLP **9**(1), 57–144 (2009)
2. Baral, C., Hunsaker, M.: Using the probabilistic logic programming language p-log for causal and counterfactual reasoning and non-naive conditioning. In: Veloso, M.M. (ed.) IJCAI, pp. 243–249 (2007)
3. Bench-Capon, T.J.M.: Representing counterfactual conditionals. In: Cohn, A.G. (ed.) Proceedings of the Artificial Intelligence and Simulation of Behaviour, pp. 51–60. Pitman and Kaufmann, Brighton, England (1989)
4. Byrne, R.M.J.: The Rational Imagination: How People Create Alternatives to Reality. MIT Press, Cambridge (2007)
5. Clark, K.L.: Negation as failure. In: Gallaire, H., Minker, J. (eds.) Logic and Data Bases, vol. 1, pp. 293–322. Plenum Press, New York (1978)

6. Dietz, E.-A., Hölldobler, S., Pereira, L.M.: On indicative conditionals. In: Hölldobler, S., Liang, Y. (eds.) Proceedings of the 1st International Workshop on Semantic Technologies, CEUR Workshop Proceedings, vol. 1339, pp. 19–30 (2015)
7. Dietz, E.-A., Hölldobler, S., Ragni, M.: A computational logic approach to the suppression task. In: Miyake, N., Peebles, D., Cooper, R.P. (eds.) Proceedings of the 34th Annual Conference of the Cognitive Science Society, pp. 1500–1505. Cognitive Science Society, Austin, TX (2012)
8. Dietz, E.-A., Hölldobler, S., Ragni, M.: A computational logic approach to the abstract and the social case of the selection task. In: 11th International Symposium on Logical Formalizations of Commonsense Reasoning (2013)
9. Dietz, E.-A., Hölldobler, S., Wernhard, C.: Modeling the suppression task under weak completion and well-founded semantics. J. Appl. Non-Class. Logics 24(1–2), 61–85 (2014)
10. Edgington, D.: On conditionals. Mind 104(414), 235–329 (1995)
11. Evans, J., Over, D.: If. Oxford cognitive science series. Oxford University Press, Oxford (2004)
12. Fitting, M.: A Kripke-Kleene semantics for logic programs. J. Logic Program. 2(4), 295–312 (1985)
13. Ginsberg, M.L.: Counterfactuals. Artif. Intell. 30(1), 35–79 (1986)
14. Hoerl, C., McCormack, T., Beck, S.R.: Understanding counterfactuals and causation. In: Hoerl, C., McCormack, T., Beck, S.R. (eds.) Consciousness and Self-consciousness. Oxford University Press, Oxford (2011)
15. Hölldobler, S., Kencana Ramli, C.D.P.: Logic programs under three-valued Łukasiewicz semantics. In: Hill, P.M., Warren, D.S. (eds.) ICLP 2009. LNCS, vol. 5649, pp. 464–478. Springer, Heidelberg (2009)
16. Hölldobler, S., Kencana Ramli, C.D.P.: Logics and networks for human reasoning. In: Alippi, C., Polycarpou, M., Panayiotou, C., Ellinas, G. (eds.) ICANN 2009, Part II. LNCS, vol. 5769, pp. 85–94. Springer, Heidelberg (2009)
17. Hölldobler, S., Philipp, T., Wernhard, C.: An abductive model for human reasoning. In: Logical Formalizations of Commonsense Reasoning, Papers from the AAAI 2011 Spring Symposium, AAAI Spring Symposium Series, Technical Reports, pp. 135–138. AAAI Press, Cambridge, MA (2011)
18. Hölldobler, S., Ramli, C.K.: Contraction properties of a semantic operator for human reasoning. In: Li, L., Yen, K.K. (eds.) Proceedings of the Fifth International Conference on Information, pp. 228–231. International Information Institute (2009)
19. Kakas, A.C., Kowalski, R.A., Toni, F.: Abductive logic programming. J. Logic Comput. 2(6), 719–770 (1993)
20. Kencana Ramli, C.D.: Logic programs and three-valued consequence operators. Master's thesis, Institute for Artificial Intelligence, TU Dresden, 2009
21. Lewis, D.: Counterfactuals. Blackwell Publishers, Oxford (1973)
22. Lewis, D.: Probabilities of conditionals and conditional probabilities. Philos. Rev. 95, 581–589 (1976)
23. Lewis, D.: On the Plurality of Worlds. Blackwell Publishers, Oxford (1986)
24. Lloyd, J.W.: Foundations of Logic Programming. Springer, Heidelberg (1984)
25. Łukasiewicz, J.: O logice trójwartościowej. Ruch Filozoficzny, 5:169–171, 1920: English translation: on three-valued logic. In: Łukasiewicz, J., Borkowski, L. (eds.) Selected Works, pp. 87–88. North Holland, Amsterdam (1990)
26. Pearl, J.: Causality: Models, Reasoning, and Inference. Cambridge University Press, New York (2000)
27. Pearl, J.: The algorithmization of counterfactuals. Ann. Math. Artif. Intell. 61(1), 29–39 (2011)

28. Pereira, L.M., Aparício, J.N.: Relevant counterfactuals. EPIA 1989. LNCS, vol. 390, pp. 107–118. Springer, Heidelberg (1989)
29. Pereira, L.M., Dietz, E.-A., Hölldobler, S.: A computational logic approach to the belief bias effect. In: Proceedings of the 14th International Conference on Principles of Knowledge Representation and Reasoning (2014)
30. Pereira, L.M., Dietz, E.-A., Hölldobler, S.: Contextual abductive reasoning with side-effects. TPLP **14**, 633–648 (2014)
31. Pereira, L.M., Saptawijaya, A.: Abduction and beyond in logic programming with application to morality. IfCoLog J. Logics Appl. Special issue on "Frontiers of Abduction" (2015, accepted)
32. Ramsey, F.: The Foundations of Mathematics and Other Logical Essays. Harcourt, Brace and Company, New York (1931)
33. Rescher, N.: What If?: Thought Experimentation In Philosophy. Transaction Publishers, New Brunswick (2005)
34. Rescher, N.: Conditionals. MIT Press, Cambridge (2007)
35. Routen, T., Bench-Capon, T.J.M.: Hierarchical formalizations. Int. J. Man-Mach. Stud. **35**(1), 69–93 (1991)
36. Schulz, K.: Minimal models vs. logic programming: the case of counterfactual conditionals. J. Appl. Non-Class. Logics **24**(1–2), 153–168 (2014)
37. Sloman, S.: Causal Models How People Think about the World and Its Alternatives. Oxford University Press, Oxford (2005)
38. Stalnaker, R.C.: A theory of conditionals. In: Rescher, N. (ed.) Studies in Logical Theory, pp. 98–112. Blackwell, Oxford (1968)
39. Stalnaker, R.C., Thomason, R.H.: A semantic analysis of conditional logic. Theoria **36**, 23–42 (1970)
40. Stenning, K., van Lambalgen, M.: Human Reasoning and Cognitive Science. A Bradford Book. MIT Press, Cambridge (2008)
41. Vennekens, J., Bruynooghe, M., Denecker, M.: Embracing events in causal modelling: interventions and counterfactuals in CP-logic. In: Janhunen, T., Niemelä, I. (eds.) JELIA 2010. LNCS, vol. 6341, pp. 313–325. Springer, Heidelberg (2010)
42. Vennekens, J., Denecker, M., Bruynooghe, M.: CP-logic: a language of causal probabilistic events and its relation to logic programming. CoRR, abs/0904.1672 (2009)
43. Woodward, J.: Psychological studies of causal and counterfactual reasoning. In: Hoerl, C., McCormack, T., Beck, S.R. (eds.) Understanding Counterfactuals, Understanding Causation: Issues in Philosophy and Psychology, Consciousness and Self-consciousness. Oxford University Press, Oxford (2011)

Interactive Debugging of Non-ground ASP Programs

Carmine Dodaro[1]([✉]), Philip Gasteiger[2], Benjamin Musitsch[2],
Francesco Ricca[1], and Kostyantyn Shchekotykhin[2]

[1] Department of Mathematics and Computer Science, University of Calabria,
Rende (CS), Italy
{dodaro,ricca}@mat.unical.it

[2] Alpen-Adria-Universität Klagenfurt, Klagenfurt, Austria
{philip.gasteiger,benjamin.musitsch,
kostyantyn.shchekotykhin}@gmail.com

Abstract. Answer Set Programming (ASP) is an expressive paradigm
for problem solving. Although the basic syntax of ASP is not particularly difficult, the identification of (even trivial) mistakes may be painful
and absorb a lot of time. The development of programs can be made
faster and comfortable by resorting to an effective program debugger. In
this paper we present a new interactive debugging method for ASP. The
method points to a buggy non-ground rule identified by asking the programmer a sequence of questions on an expected answer set. The method
has been implemented on top of the WASP solver. The tight integration
with the solver allows to avoid efficiency problems due to the grounding
blowup induced by modern reification-based debuggers.

1 Introduction

Answer Set Programming (ASP) [5] is a declarative problem solving paradigm
proposed in the area of logic programming and non-monotonic reasoning. Computational problems of comparatively high complexity [7] can be solved in ASP,
which provides a clear separation between the specification of a problem and
the computation of its solutions by an ASP solver. The suitability of the ASP
framework for problem solving is witnessed by the large number of applications
that have been developed [2,6,13]. The applications of ASP to real-world problems outlined several advantages of this paradigm from a software engineering
viewpoint. Namely, ASP programs are flexible, intuitive, extensible and easy to
maintain [13].

Although the basic syntax of ASP is not particularly difficult, the identification of (even trivial) faults can be tedious and time consuming. Techniques
and tools, called *debuggers*, can help the programmer to deal with faults in ASP

This research was partially funded by the Carinthian Science Fund (KWF-
3520/26767/38701). This work was partially supported by MIUR within project "SI-
LAB BA2KNOW – Business Analitycs to Know", and by Regione Calabria, POR
Calabria FESR 2007-2013, within project "ITravel PLUS" and project "KnowRex".

© Springer International Publishing Switzerland 2015
F. Calimeri et al. (Eds.): LPNMR 2015, LNAI 9345, pp. 279–293, 2015.
DOI: 10.1007/978-3-319-23264-5_24

programs, thus, making the development process faster and more comfortable. In the last few years, a number of tools for debugging ASP programs were proposed [4,11,18,19,22]. Among the most prominent approaches are the ones based on the notion of meta-programming [11,19] that work by applying ASP itself to debug a faulty ASP program. The idea is to generate a factual representation of the faulty program – also called a reification of the program – and combine it with another ASP program – the debugging program – which models the possible causes of a fault. Reification-based debuggers have some issues that may make them either difficult to apply or even inapplicable in some cases. The first issue was already observed in [19], and consists of the fact that the generated meta-programs might be unsolvable by the best ASP systems available. The reason is that prior to computation of answer sets ASP systems compute the ground instantiation of the input by means of a grounder. The reification used in meta-programming approach may cause the ground debugging program to be so large that either the grounder is not able to produce it in a reasonable time or the solver is not able to elaborate it. This issue is due to the very nature of the meta-programming approach which requires a ground debugging program to comprise all possible explanations of all possible faults of the input program.

In this paper we present a new interactive debugging method for non-ground ASP programs that allows the programmer to find the cause of a problem. The new method builds a diagnosis identifying a set of non-ground rules that include the ones causing a bug. The user, upon request, can improve the diagnosis by interacting with the debugger, which automatically builds queries on the truth of literals in an expected answer set. The additional knowledge injected in the system by the user, who answers queries during the debugging session, allows the debugger to refine the diagnosis up to a point in which the (non-ground) rules causing a fault are easily identified. The debugging method presented in this paper is the first one –to the best of our knowledge– that can take profit of the literal assumption interface and conflict analysis services provided by modern ASP solvers [1,10]. We have implemented our method on top of the grounder GRINGO [10] and the solver WASP [1]. Moreover, we compared our solution with related approaches, and we report on the results of an experiment assessing our implementation on benchmarks already used in the literature for this purpose [21].

Summarizing, the main contributions of this paper are: (i) a new interactive debugging method for non-ground ASP programs; (ii) a description of the implementation of the new approach; (iii) the analysis of the result of an experiment evidencing that our debugger can handle instances that are pragmatically out of reach for state-of-the-art reification-based debuggers. Indeed, these latter suffer for the intrinsic grounding blowup that is avoided in our approach that can be integrated within the solver.

2   Preliminaries

In this section we recall Answer Set Programming (ASP) syntax and semantics as well as some properties of ASP programs. In the following we assume the reader familiar with basic logic-programming notions.

Syntax. A *disjunctive logic program* (DLP) Π is a finite set of rules of the form

$$a_1 \vee \cdots \vee a_m \leftarrow l_1, \ldots, l_n \qquad (1)$$

where all $a_1, \ldots, a_m$ are atoms and $l_1, \ldots, l_n$ are literals, $m, n \geq 0$. A *literal* is an atom a_i or its negation $\sim a_i$, where $\sim$ denotes *negation as failure*. The complement of a literal l (set of literals L) is denoted $\bar{l}$ ($\bar{L}$), and $\bar{L} := \{\bar{l} \mid l \in L\}$. Each *atom* is an expression of the form $p(t_1, \ldots, t_k)$, where p is a predicate symbol and $t_1, \ldots, t_k$ are *terms*, i.e. are either variables or constants. An atom (literal, rule) is *ground* if it is variable-free. Rules are required to be *safe*, i.e., variables must occur in at least one positive literal of the body. In each rule r of the form (1) the set of atoms $H(r) = \{a_1, \ldots, a_m\}$ is called *head* and the set of literals $B(r) = \{l_1, \ldots, l_n\}$ is called *body*. Moreover, we differentiate between the disjoint sets $B^+(r), B^-(r) \subseteq B(r)$ comprising *positive* and *negative* body literals respectively. A rule is called *fact* if $|H(r)| = 1$ and $B(r) = \emptyset$, *constraint* if $H(r) = \emptyset$ and *normal rule* if $|H(r)| = 1$ and $B(r) \neq \emptyset$. The set of all atoms occurring in Π is denoted $At(\Pi)$.

Semantics. Let Π be an ASP program. The *Herbrand universe* U_Π and the *Herbrand base* B_Π of Π are defined as usual. The semantics of Π is given in terms of its ground instantiation Π_G. Π_G is the set of all the ground instances of rules of Π that can be obtained by substituting variables with constants from U_Π.

A set L of ground literals is said to be *consistent* if, for every atom $l \in L$, its complementary literal $\bar{l}$ is not contained in L. An interpretation I for Π_G is a consistent set of ground literals over atoms in B_Π. A ground literal l is *true* w.r.t. I if $l \in I$; l is *false* w.r.t. I if its complementary literal is in I; l is *undefined* if it is neither true nor false. An interpretation I is *total* if, for each atom a in B_Π, either a or $\sim a$ is in I. A total interpretation M is a *model* for Π_G if, for every $r \in \Pi_G$, at least one atom in the head is true w.r.t. M whenever all literals in the body are true w.r.t. M. Given two interpretations I_1 and I_2, $I_1 \leq^+ I_2$ iff for each positive literal $a \in I_1$ it also holds that $a \in I_2$.

Let Π_G be a ground program and I an interpretation. The *reduct* or *Gelfond-Lifschitz transform* [12] of Π_G w.r.t. I is the ground program Π_G^I, obtained from Π_G by (i) deleting all rules $r \in \Pi_G$ whose negative body is false w.r.t. I and (ii) deleting the negative body from the remaining rules. An *answer set* of a program Π is a model M of Π_G that is $\leq^+$-minimal model of Π_G^M, i.e., there is no other model M' of Π_G^M s.t. $M' \leq^+ M$. The set of all answer sets of a program Π is denoted $AS(\Pi)$. The program Π is *incoherent*, if the set of all answer sets $AS(\Pi) = \emptyset$, and *coherent*, otherwise.

Properties. We now recall some definition and properties of answer sets that are useful in the remainder of the paper. Given an interpretation I for a ground program Π, we say that a ground atom $a \in I$ is *supported* w.r.t. I if there exists a *supporting* rule $r \in \Pi_G$ such that $B(r) \subseteq I$, and $(H(r) \setminus \{a\}) \cap I = \emptyset$, i.e., all body literals are true and a is the only true atom in the head. An important property of answer sets is the so-called supportedness, i.e., all atoms in an answer set are supported. Moreover, answer sets are unfounded free, i.e., an answer set contains no unfounded sets. A set $A \subseteq B_\Pi$ is *unfounded* w.r.t. an interpretation I if for each rule $r \in \Pi_G$ such that $\underline{H(r)} \cap A \neq \emptyset$ it holds that (a) $I \cap (H(r) \setminus A) \neq \emptyset$ or (b) $B^+(r) \cap A \neq \emptyset$ or (c) $\overline{B(r)} \cap I \neq \emptyset$. Intuitively, atoms in A can have support only by themselves. Given a ground program Π_G, an *odd loop* is a fragment of a program in which an atom $a \in At(\Pi_G)$ depends recursively on itself through an odd number of negative arcs in the dependency graph $G(\Pi_G) = (At(\Pi_G), E)$, where the set of edges E comprises positive $\{(h, b)^+ \mid r \in \Pi_G, h \in H(r), b \in B^+(r)\}$ and negative $\{(h, b)^- \mid r \in \Pi_G, h \in H(r), b \in B^-(r)\}$ arcs. Note that, constraints can be viewed as simple odd loops, since a constraint $r_c \in \Pi$ can be rewritten as a normal rule with an odd loop $idr_c \leftarrow B(r_c), \sim idr_c$, where idr_c is a fresh atom. A ground program Π_G is incoherent only if it contains an *odd loop*.

3   Query-Based ASP Debugging

Debugging is the process of finding bugs in a program, where a bug commonly indicates an incorrect behavior of a program. In a nutshell, during the development of a program, the programmer tests it running a (usually small) set of problem instances for which he/she already knows an expected solution. Intuitively, a bug is revealed in two possible cases: (i) the answer sets computed by the solver are not compliant with the expected/intended solution; (ii) the tested program is incoherent but at least one answer set is expected. The combination of a program with an expected output to be verified is called in jargon a *test case*. The practice of defining test cases is common in the development independently from the programming paradigm, and can also be intended as part of the development process if one resorts to a systematic test-driven programming methodology. In ASP, test cases may be defined by using specific testing languages such as the one supported by ASPIDE [8,9], or can be generated automatically using existing methods [14]. In order to simplify the presentation a test case is defined as the combination of a program Π (under test) and a set of literals O. Intuitively, literals in O are expected to be true in an intended answer set of Π. A test case fails if $\Pi \cup \{\leftarrow \bar{l} \mid l \in O\}$ is incoherent. Note that both cases of bugs mentioned above can be captured by properly specifying a failing test case. In fact, given a failing test case the debugging process may identify the rules of the program causing the failure. For this reason, in the following we assume w.l.o.g. that a debugger takes as input an incoherent program (originating from a failing test case) and aims at pointing the user the source of the incorrect behavior. In case the set of rules provided to the user is either too large then the debugger interacts with the user querying for additional information about the expected behavior in terms of expected truth of literals. This information is used to further narrow the diagnosed set of buggy rules. In the following we first present the core of the approach that works on an inconsistent debug program, and then we describe the behavior of our method in specific scenarios in which some atom expected to be true misses a supporting rule, a case that may requires an additional step before being possibly re-conducted to the first case.

3.1 Debugging Inconsistency

Debugging an incoherent ASP program is a complex computational task. Therefore, prior to invoking a debugger, a user may add some of the rules of an incoherent program Π to the *background knowledge*, i.e. a set of rules $\mathcal{B}$ that must be considered as correct by a debugger. Defining the background knowledge can simplify the debugging process. In the following examples, we consider all facts as correct, i.e. $\mathcal{B} = \{r \in \Pi \mid r \text{ is a fact}\}$.

Definition 1 (Debugging Program). *Let Π be an incoherent program $\mathcal{B} \subseteq \Pi$ be a background knowledge. Then, a* debugging program Δ *is such that $\Pi \cap \Delta = \mathcal{B}$ and for each rule $r \in \Pi \setminus \mathcal{B}$ there is a rule $r' \in \Delta$ such that all the following conditions hold: (i) $h(r') = h(r)$; (ii) $B^+(r') = B^+(r) \cup \{\_debug(id_r, \boldsymbol{vars})\}$; and (iii) $B^-(r') = B^-(r)$. Here atoms of the form $\_debug(id_r, \boldsymbol{vars})$ are called debugging atoms, where the tuple of variables $\boldsymbol{vars}$ comprises all variables occurring in $B^+(r)$ and id_r is an identifier of the rule $r \in \Pi$.*

In other words given program Π to debug, its associated debugging program Δ is obtained by Π by adding to each rule that is not a fact an additional positive body atom that is called debug atom. Given a debugging program Δ a grounder generates a ground program Δ^G. Each rule r_i in Δ^G is identified by a unique ground debugging atom, say d_i. All d_i appearing in Δ^G are collected in the *set of literals Asm.*

Basically, the idea used in the following is to solve under the assumption that literals in Asm belong to answer sets. In practice this is done via the so-called solving-under-assumptions feature of modern ASP solvers [1,10] that produces, as a byproduct of conflict analysis [1,10], an *unsatisfiable core* comprising the assumptions that cause the inconsistency. In our case the debug atoms $d_i \in Asm$ corresponding to rules causing the inconsistency are detected.

Definition 2 (Unsatisfiable Core). *Let Δ^G be a ground debugging program and $\mathbf{D} \subseteq At(\Delta^G)$ be a set of ground debugging atoms of Δ^G. Then an* unsatisfiable core *is a set of atoms $C \subseteq \mathbf{D}$ such that if every atom $a \in C$ is assumed to be true, i.e. $C \subseteq Asm$, the program Δ^G is incoherent. An unsatisfiable core C is minimal iff there is not unsatisfiable core C' such that $C' \subset C$.*

In our setting, the set Asm comprises only debugging atoms of the debugging program Δ^G, thus an unsatisfiable core is a set of debugging atoms $\{d_1, \ldots, d_k\} \subseteq \mathbf{D}$. Note that, the implemented approaches to core computation do not guarantee the minimality of returned cores. The minimization of an unsatisfiable core can be done by algorithms like QUICKXPLAIN [15]. Such divide-and-conquer algorithms allow a solver to compute a minimal unsatisfiable core using $O(|C| \log \frac{|C'|}{|C|} + 2|C|)$ calls to the solving algorithm, where $|C'|$ and $|C|$ are the corresponding cardinality of input and minimal unsatisfiable cores, respectively.

Example 1 ([19]). A conference system is designed to assign papers to reviewers according to their bids. The latter have values 0 – conflict; 1 – indifference and

2 – would like to review. In case a bid is not known the default value 1 must be assumed. The test program L_1 defines two papers and two reviewers

$$L_1 = \{pc(m1), pc(m2), paper(p1), bid(m1, p1, 2),$$
$$r_1 : some\_bid(M, P) \leftarrow bid(M, P, X),$$
$$r_2 : bid(M, P, 1) \leftarrow \sim some\_bid(M, P), pc(M), paper(P)\}$$

The program L_1 is incoherent because the ground rules $\{r_1', r_2'\} \subseteq \Delta^G$ contain an odd loop w.r.t. the atom $some\_bid(m2, p1)$ under assumption that both debug atoms are true, i.e. $\{\_debug(r1, m2, p1, 1), \_debug(r2, m2, p1)\} \subseteq Asm$.

$$r_1' : some\_bid(m2, p1) \leftarrow bid(m2, p1, 1), \_debug(r1, m2, p1, 1),$$
$$r_2' : bid(m2, p1, 1) \leftarrow \sim some\_bid(m2, p1), \_debug(r2, m2, p1).$$

The unsatisfiable core $C = \{\_debug(r1, m2, p1, 1), \_debug(r2, m2, p1)\}$ returned by an ASP solver is minimal and allows a user to consider two possible assumptions. Assuming that r_1' is faulty, i.e. replace $\_debug(r1, m2, p1, 1)$ with $\overline{\_debug(r1, m2, p1, 1)}$ in the set Asm, the solver finds an answer set A_1 comprising atoms $\{some\_bid(m1, p1), bid(m2, p1, 1)\}$. If the user assumes that r_2' is faulty, i.e. inverts $\_debug(r2, m2, p1)$ in Asm, then the answer set A_2 comprising $\{some\_bid(m1, p1)\}$ is returned. Analysis of these two answer sets shows that they assign different truth values to the atom $bid(m2, p1, 1)$. If the atom $bid(m2, p1, 1)$ is required in the missing answer set, then the first assumption is correct and the user has to correct r_1. Otherwise, the second assumption must be selected and the user has to modify the rule r_2. Since in our case the atom $bid(m2, p1, 1)$ is expected the rule r_1 must be corrected. One of the possible modifications would be $r_1 : some\_bid(M, P) \leftarrow bid(M, P, X), X \neq 1$. The addition of an $X \neq 1$ condition to the body allows to exclude the rule $some\_bid(m2, p1) \leftarrow bid(m2, p1, 1)$ from Π_G and thus solves the problem.

The fault localization in the program L_1 can be done automatically by the debugger. Given an unsatisfiable core C, the debugger can *relax* the core by updating the set of assumptions as $Asm' = (Asm \setminus D) \cup \{\overline{d_i} \mid d_i \in D\}$, where $D = \{d_1, \ldots, d_k\} \subseteq C$ is the set of updated debugging atoms. However, for the minimal core C the update of one debugging atom, i.e. $D = \{d_i\}$ for some $d_i \in C$, is sufficient to obtain at least one answer set of L_1. The answer sets obtained by two relaxations of L_1, i.e. $D_1 = \{\_debug(r1, m2, p1, 1)\}$ and $D_2 = \{\_debug(r2, m2, p1)\}$, can be compared by a debugger to find discrepancies between them. Namely, in Example 1 the atom $bid(m2, p1, 1)$ represents this difference between answer sets A_1 and A_2. If the debugger gets the information that the atom $bid(m2, p1, 1)$ must be in the missing answer set of the program, it will return rule $r_1 \in \Pi$ and its grounded version as an explanation of the fault. Otherwise, the rule $r_2 \in \Pi$ is the correct explanation.

Definition 3 (Diagnosis). *Let Δ^G be a ground debugging program and $\mathbf{D} \subseteq At(\Delta^G)$ be a set of ground debugging atoms of Δ^G. Then a diagnosis is a set of atoms $D \subseteq \mathbf{D}$ such that if the set of literals $\{\overline{d_i} \mid d_i \in D\}$ is assumed to be true,* *i.e.* $\{\overline{d_i} \mid d_i \in D\} \subseteq Asm$, *then the program* Δ^G *is coherent. A diagnosis* D *is minimal iff there is no* $D' \subset D$ *such that* D' *is a diagnosis.*

In Example 1 there are two diagnoses $D_1 = \{\_debug(r1, m2, p1, 1)\}$ and $D_2 = \{\_debug(r2, m2, p1)\}$ of cardinality 1, i.e. two single faults. However, if a program comprises multiple faults, i.e. two or more (non-)intersecting unsatisfiable cores, then computation of diagnoses of higher cardinality is required. Given an algorithm for computation of minimal unsatisfiable cores the minimal diagnoses of arbitrary cardinality can be computed using HS-TREE algorithm [23]. This algorithm uses the property that a minimal diagnosis is a minimal hitting set of all unsatisfiable cores (see Theorem 4.4 [23]). To find minimal hitting sets HS-TREE generates a labeled tree in a breadth-first order and applies a pruning strategy that guarantees the minimality of found diagnoses.

Differentiation between any two minimal diagnoses D_1 and D_2 can be done by comparing answer sets of the debugging program Δ^G under assumptions $\{\overline{d} \mid d \in D_1\}$ and $\{\overline{d} \mid d \in D_2\}$. Given any of these assumptions the solver relaxes the corresponding rules of Δ^G thus removing the odd loops. Since a relaxed program comprises no odd loops, it is coherent and the solver returns at least one answer set. The focus of the answer set analysis lies on the identification of an atom q such that q is in every answer set of one relaxed program and is not in any answer set of another program. Clearly, the atom q allows us to discriminate between the diagnoses. E.g. let q be in all answer sets of a relaxed program w.r.t. D_1 and not in any answer of a relaxed program w.r.t. D_2. If some oracle, like a programmer, provides an information that the answer sets of a correct program must comprise q, then the debugger can accept D_1 and reject D_2.

Generally, given a set of diagnoses $\mathcal{D}$, an atom q can be used to a partition elements of $\mathcal{D}$ into three sets $\mathcal{D}^{\mathcal{P}}, \mathcal{D}^{\mathcal{N}}$ and $\mathcal{D}^{\emptyset}$, where (a) answer sets of every relaxed program w.r.t. a diagnosis of the set $\mathcal{D}^{\mathcal{P}}$ comprise q; (b) the diagnoses $\mathcal{D}^{\mathcal{N}}$ result in relaxed programs which answer sets do not comprise q; and (c) the set $\mathcal{D}^{\emptyset}$ comprises all diagnoses that cannot be classified neither to $\mathcal{D}^{\mathcal{P}}$ nor to $\mathcal{D}^{\mathcal{N}}$.

Definition 4 (Query). *Let* Δ^G *be a ground debugging program,* $\mathbf{D} \subseteq At(\Delta^G)$ *be a set of debugging atoms and* $\mathcal{D}$ *be a set of diagnoses for* Δ^G. *A query is an atom* $q \in At(\Delta^G) \setminus \mathbf{D}$ *such that the set of diagnoses can be partitioned into three sets* $\mathcal{D}^{\mathcal{P}}, \mathcal{D}^{\mathcal{N}}$ *and* $\mathcal{D}^{\emptyset}$, *where*

- $\forall D \in \mathcal{D}^{\mathcal{P}}$ *it holds that* $\forall A \in AS(\Delta^G) : q \in A$, *i.e.* $\Delta^G \cup \{\leftarrow \sim q\}$ *is coherent, under assumption* $\{\overline{d_i} \mid d_i \in D\}$
- $\forall D \in \mathcal{D}^{\mathcal{N}}$ *it holds that* $\forall A \in AS(\Delta^G) : q \notin A$, *i.e.* $\Delta^G \cup \{\leftarrow q\}$ *is coherent, under assumption* $\{\overline{d_i} \mid d_i \in D\}$
- $\mathcal{D}^{\emptyset} = \mathcal{D} \setminus (\mathcal{D}^{\mathcal{P}} \cup \mathcal{D}^{\mathcal{N}})$

Given a set of diagnoses $\mathcal{D} = \{D_1, \ldots, D_m\}$ a query can be computed as follows:

1. Generate a power set $\mathcal{P}(\mathcal{D})$ of the set of diagnoses $\mathcal{D}$.
2. Assign to a set $\mathcal{D}^\mathcal{P}$ an element of $\mathcal{P}(\mathcal{D})$ and find an intersection V of all answer sets obtained for Δ^G under each of the assumptions $\{\overline{d_i} \mid d_i \in D\}$, $\forall D \in \mathcal{D}^\mathcal{P}$.
3. If $V = \emptyset$, then goto Step 1; otherwise select an atom $q \in V$ as a query.
4. Classify the remaining diagnoses $D_j \in \mathcal{D} \setminus \mathcal{D}^\mathcal{P}$ to the sets $\mathcal{D}^\mathcal{P}, \mathcal{D}^\mathcal{N}$ and $\mathcal{D}^\emptyset$ by verifying for each D_j conditions given in Definition 4.
5. If $\mathcal{D}^\mathcal{N} \neq \emptyset$ then return a query q. Otherwise, remove the set $\mathcal{D}^\mathcal{P}$ selected in the first step from $\mathcal{P}(\mathcal{D})$ and goto Step 1.

Note that in practice the set of all diagnoses is often approximated by a set of k diagnoses, where k is some small predefined constant. The experimental results obtained for diagnosis of description logics knowledge bases [25,26] show that $k = 9$ allows a debugger to achieve a good balance between query computation time and the number of queries required to identify the source of a fault.

The query generation algorithm can generate multiple queries for a set of diagnoses. Therefore, multiple query selection strategies are suggested in the literature [16]. These strategies can be roughly classified into myopic and look-ahead ones. The myopic strategies select a query depending on a partition $\mathcal{D}^\mathcal{P}, \mathcal{D}^\mathcal{N}, \mathcal{D}^\emptyset$ and do not need any additional information. For instance, a popular *Split-in-half* measure prefers queries that partition the set of diagnoses such that answer sets of a half of diagnoses comprise q and the other half is not, i.e., $q = \arg\min_q ||\mathcal{D}^\mathcal{P}| - |\mathcal{D}^\mathcal{N}|| + |\mathcal{D}^\emptyset|$. The look-ahead strategies [24,25] use additional information about possible cause of a fault to select the most informative query. One of the measures that showed good performance is *Entropy*. Given a probability $p(D_i)$ for each diagnosis $D_i \in \mathcal{D}$, it selects a query that minimizes the expected entropy of a set of diagnoses after a user answers a query, i.e. it selects a query that maximizes the information gain.

3.2 Debugging Unsupported and Unfounded Atoms

An ASP program can also be faulty if an atom is expected to be true but either there is no supporting rule or it is in an unfounded set. In our setting such cases are detected when an unsatisfiable core comprises only debug atoms related to the original test cases or to the previous queries. Let a be one of such atoms, and $Supp(a) = \{r \mid a \in H(r)\}$ be the set of rules having a in the head. The query q in case of unsupported and unfounded atoms is selected from the set of atoms occurring in some rule having a in the head, i.e., on $\bigcup_{r \in Supp(a)} (B^+(r) \cup (H(r) \setminus \{a\}) \cup \{a' \mid \sim a' \in B^-(r)\})$.

Example 2 ([19]). Consider a similar problem as in Example 1 in which PC members can give bids from 0 – conflict to 2 – want to review. According to conference rules assignments of papers to PC members must not be conflicting (rule r_1). A conflict can occur in two situations: (i) a PC member provided a bid 0 for a paper (rule r_2) or (ii) a PC member authored a paper (rule r_3). In the latter case the conflict of interest, i.e. bid 0, must be derived automatically (rule r_4).

$$L_2 = \{pc(m1), paper(p1), bid(m1, p1, 2), assigned(p1, m1), author(p1, m1),$$
$$r_1 : \leftarrow assigned(P, M), bid(M, P, 0),$$
$$r_2 : conflict(M, P) \leftarrow bid(M, P, 0),$$
$$r_3 : conflict(M, P) \leftarrow pc(M), paper(P), author(M, P),$$
$$r_4 : bid(M, P, 0) \leftarrow pc(M), paper(P), conflict(M, P)\}$$

Given the fact that $p1$ authored $m1$ and the fact that $m1$ is assigned to $p1$, the program L_2 is expected to be incoherent. However, a solver finds and returns an answer set in which $p1$ gets $m1$ to review and no conflict is detected. In order to identify the problem the user adds a test case, which requires L_2 to have an answer set comprising atom $conflict(m1, p1)$. Processed by a debugger the test case is added to the debugging program Δ_2^G as rule r_5: $L_2' = \Delta_2^G \cup \{r5 :\leftarrow \sim conflict(m1, p1), \_debug(r5, m1, p1)\}$. Given a grounded debugging program L_2' debugger finds an unsatisfiable core $C = \{\_debug(r5, m1, p1)\}$. Therefore, the only diagnosis $D = \{\_debug(r5, m1, p1)\}$ suggests that the test case added by the user is faulty and the atom $conflict(m1, p1)$ must be false in all answer sets. Since the diagnosis comprises no other rules, it is possible to infer that atom $conflict(m1, p1)$ has no supporting rules. Therefore, the debugger to cope with this scenario determines all rules that are relevant to this atom, i.e. rules having $conflict(m1, p1)$ in their heads, and offer them to the user as a possible cause of a bug, since these are not able to support $conflict(m1, p1)$ that was expected to be true. In particular, rule r_2 is not supporting because atom $bid(m1, p1, 0)$ is in the positive body and it cyclically depends on atom $conflict(m1, p1)$, while r_3 is not supporting $conflict(m1, p1)$ because $author(m1, p1)$ is false. In case the user is not satisfied by this answer (i.e., supporting rules are correct), the analysis continues and, in particular, the debugger considers atoms $bid(m1, p1, 0)$ and $author(m1, p1)$ as query, because they cause missing support of $conflict(m1, p1)$. For example the query-based debugger may ask for the truth value of the atom $author(m1, p1)$ (from missing support of rule r_3). Now, it is easy to understand that r_3 is faulty because of a different order of variables in the predicate $author$.

4   Implementation

Our implementation of the debugger consists of two components: the debugging grounder GRINGO-WRAPPER and a modified version of the ASP solver WASP [1] called DWASP. The tools are available under https://github.com/gaste/gringo-wrapper and https://github.com/gaste/dwasp. Figure 1 illustrates the interaction of both components to debug a program Π. First, the program Π is read by GRINGO-WRAPPER from either the standard input or several input files. The debugging grounder internally transforms Π to Δ, passes the result to GRINGO [10] and outputs the ground debugging program Δ^G to the standard output. DWASP reads Δ^G and starts the interactive debugging session.

Grounding with GRINGO-WRAPPER. The task of GRINGO-WRAPPER is to obtain the grounded debugging program Δ^G, given an input program Π. First, Π is translated to the debugging program Δ, as described in Definition 1. All facts of

Fig. 1. Interaction of GRINGO-WRAPPER and DWASP in debugging mode.

Π are assumed to be correct, i.e. the background knowledge $\mathcal{B}$ comprises all facts of Π. After this transformation, GRINGO is used to obtain the ground version Δ^G of Δ. However, GRINGO performs several optimizations during grounding, such as deriving new facts from normal rules [10]. Although these optimizations potentially decrease the time required by the solver, they are counter-productive when debugging a logic program because wrong facts could be derived from faulty rules. Moreover, GRINGO might remove entire non-ground rules that are missing support. In this case, GRINGO-WRAPPER issues a warning message that highlights the rules that were removed by GRINGO.

In order to avoid the removal of atoms or simplification of rules done by GRINGO during the grounding, rules of the form $a(t_1, \ldots, t_m) \vee na(t_1, \ldots, t_m) \leftarrow l_1(t_1), \ldots, l_m(t_m)$ are initially added to the program for each atom $a(t_1, \ldots, t_m)$ occurring in the head of a rule whose body is $l_1(t_1), \ldots, l_m(t_m)$ where $na(t_1, \ldots, t_m)$ is a fresh atom. In the postprocessing step, the ground instantiations of the added rules are then removed. The same implementation trick is also applied for preserving debug atoms, but disjunctive rules are not removed in order to make those atoms irrelevant for the coherence of the processed program (roughly, so that they can have a supporting rule).

Debugging with DWASP. The query-based debugging approach presented in Sect. 3 has been implemented in DWASP, which exploits the assumptions interface of the ASP solver WASP [1]. The pseudo-code of DWASP is reported in

---

Algorithm 1. Debug mode of DWASP

input: An incoherent, ground debugging program Δ^G, maximum number of diagnoses k

1 **begin**
2 　$Asm \leftarrow \{d \mid d \text{ is a debugging atom of } \Delta^G\}$;
3 　**while** *user continues debugging session* **do**
4 　　print non-ground rules inside the minimal core;
5 　　$\mathcal{D} \leftarrow \text{ComputeDiagnoses}(Asm, k)$;
6 　　$q \leftarrow \text{DetermineQuery}(Asm, \mathcal{D})$;
7 　　**if** *user answers that the atom q should be inside the model* **then**
8 　　　$Asm \leftarrow Asm \cup \{q\}$;
9 　　**else**
10 　　　$Asm \leftarrow Asm \cup \{\overline{q}\}$;

Algorithm 1. First, all debug atoms of Δ^G are added to the set Asm (line 2), which is later used to compute a set of diagnoses $\mathcal{D}$ comprising at most k elements (line 5). The computation diagnoses is done by a HS-TREE algorithm which, in turn, uses DWASP and QUICKXPLAIN [15] to find minimal unsatisfiable cores. Given a set of diagnoses, the query atom q is computed in a way that maximizes the *Split-in-half* measure as described in Sect. 3. That is, we start from partitions $\mathcal{D}^P$ such that $2|\mathcal{D}^P| = |\mathcal{D}|$ and continue until a query is found. Finally, DWASP asks the user whether q should be true (false) in an expected answer set and adds the corresponding literal q ($\bar{q}$) to the set of assumptions (lines 7–10). The debugging session continues either until only one diagnosis remains or a user spots the error.

Example 3 (Debugging of a faulty program). Consider the following program encoding the graph coloring problem:

$r_1 : arc(1,2) \leftarrow \qquad r_2 : arc(2,3) \leftarrow$ % *Define graph by arc predicate*

$r_3 : node(X) \leftarrow arc(X,Y)$ %*Compute nodes from arcs*

$r_4 : node(X) \leftarrow arc(Y,X)$

$r_5 : col(X,b) \vee col(X,r) \vee col(X,y) \leftarrow node(X)$ % *Assign exactly one color to each node*

$r_6 : \leftarrow col(X,C1), col(Y,C2), arc(X,Y)$ % *Different colors to adjacent nodes*

The program is grounded using GRINGO-WRAPPER and the result is saved in the file `coloring.dbg`. DWASP is then started by typing the following command:

    wasp --debug=coloring.dbg

The solver then computes the unsatisfiable core and waits until a command is typed. For instance, in order to view the computed core the following command is typed:

    WDB> show core ground

and the results are the ground rules contained in the unsatisfiable core. Another option is to use the following command:

    WDB> ask

which generates a query on an atom of the input program. The query proposed by the solver is the following:

    Should 'col(2,b)' be in the model? (y/n/u)

Since node 2 can be colored with blue, the user types yes. Then, DWASP computes another set of diagnoses and asks the following query:

    WDB> ask
    Should 'col(1,r)' be in the model? (y/n/u)

Since node 1 can be colored with red, the user types again yes. Thus, the process ends up with only one rule inside the core, which is shown by typing the following command:

    WDB> show core nonground
    <- col(X, C1), col(Y, C2), arc(X,Y).
       { C1/r, C2/b, X/1, Y/2 }

At this point, it is easy to see that rule r_6 is faulty because the check C1=C2 is missing.

It is worth mentioning that for simplifying the presentation the example shown above only focuses on the main commands. Nonetheless, DWASP implements several other commands that give to the user the full control on a debug sessioning, e.g. commands for saving/restoring a debugging session, for retracting previous answers, etc.

Performance Analysis. We have assessed the performance of our implementation by comparing it with OUROBOROS [19,21] debugger, which is the only maintained solution able to cope with non-ground programs. In particular we have employed the same ASP encodings and instances taken from ASP competitions that have been used in [21] for analyzing the performance of a debugger. For grounding we use GRINGO (v4.4.0) in both methods. We have measured both the increase in grounding size w.r.t. the mere execution of the GRINGO grounder on the instances (detailed results are reported at https://github.com/gaste/gringo-wrapper). In our approach the increase in grounding size is due to the fact that the GRINGO-WRAPPER disables the optimizations performed by GRINGO, whereas in OUROBOROS the grounding of an ASP program modeling debugging is required. Considering the instances that were groundable with GRINGO within 5 min by our Intel Core i7-3667U machine with 8 GB of RAM, we report that in our approach the size of the instantiation of the debugging program is from 1.5 to 3 times the size of grounding the original program, whereas the debugging program of OUROBOROS generates groundings that are from 50 times up to 9382 times larger than the original program. Note that, the performance of our approach is only limited by the performance of the underlying solver, whereas in the case of OUROBOROS the limit is in the grounding of the debugging program, which may not be feasible.

5   Related Work

Modern ASP debugging approaches can be separated into integrated and declarative approaches. The first approaches are based on a tight integration with the solver, whereas the second ones are solver independent and are based on *meta-programming*.

The DLV debugger developed in [20] is an example of an integrated approach. It uses the reason calculus to detect and store the choices made by the solver during the backtracking phases in a *reasons table*. The table can be queried to justify the presence/absence of a literal in an answer set or to explain the incoherency of the program. This debugging system is however very limited, since it uses specific features of the DLV system and can only provide a partial interpretation justifying the lack of a model. IDEAS [4] is another procedural approach aiming at two types of problems: (a) why a set of atoms S is in an answer set A and (b) why S is not in any answer set. Both IDEAS algorithms are similar to the ones implemented in ASP solvers and try to decide which rules are responsible for derivation or non-derivation of atoms in S. The interactivity of IDEAS, as well as of all other modern debuggers allows a programmer: (1) to query a system for an explanation of an observed fault, (2) analyze the obtained results and (3) reformulate the query to make it more precise. In our approach we reuse the algorithms implemented in a solver and are able to find required refinements automatically, thus, making the steps (2) and (3) obsolete.

The declarative debuggers use a program over a meta language – a kind of ASP solver simulation – to manipulate a program over an object language – the faulty program. Each answer set of a meta-program comprises a *diagnosis*, which is a set of meta-atoms describing the cause why some interpretation of the faulty program is not its answer set. An approach used in SMDEBUG [27] addresses debugging of incoherent non-disjunctive ASP programs by adaption of model-based diagnosis [23]. Similarly to our approach the debugger focuses on detection of odd loops, but cannot detect problems arising due to unfounded sets. The SPOCK [11] and OUROBOROS [19,21] debuggers extend SMDEBUG by enabling identification of problems connected with unfounded sets. Both approaches represent the input program in a reified form allowing application of a debugging meta-program. In case of SPOCK the debugging can be applied only to grounded programs, whereas OUROBOROS can tackle non-grounded programs as well. The main problem of meta-programming approaches is that often the grounding of the debugging meta-program explodes. This is due to the fact that the ground debugging program has to comprise all atoms explaining all possible faults in an input faulty program, which is not the case in our approach. Moreover, our approach generalizes the interactive query-based method built on top of SPOCK [24] by enabling its application to non-ground programs.

There are other approaches enabling faults localization in ASP, but not directly comparable with DWASP, include Consistency-Restoring Prolog [3], translation of ASP programs to natural language [17], visualization of justifications for an answer set [22] as well as stepping thought an ASP program [18]. Combination of these approaches with ideas implemented in DWASP is a part of our future work.

6   Summary and Future Work

In this paper we presented an interactive debugging method for non-ground ASP programs, that can be efficiently implemented into any ASP solver supporting conflict analysis and literal assumption interface. We implemented our debugging method by extending the ASP solver WASP, and we reported an experiment demonstrating its applicability on ASP programs that cannot be handled using alternative debuggers.

As far as future work is concerned, we are working on a graphical debugger that will be integrated in the ASPIDE [9] IDE for ASP. Moreover we are planning to extend our methodology to programs featuring optimization constructs.

References

1. Alviano, M., Dodaro, C., Leone, N., Ricca, F.: Advances in wasp. In: Calimeri, F., Ianni, G., Truszczynski, M. (eds.) LPNMR. LNCS, vol. 9345, pp. 40–54. Springer, Heidelberg (2015)

2. Aschinger, M., Drescher, C., Friedrich, G., Gottlob, G., Jeavons, P., Ryabokon, A., Thorstensen, E.: Optimization methods for the partner units problem. In: Achterberg, T., Beck, J.C. (eds.) CPAIOR 2011. LNCS, vol. 6697, pp. 4–19. Springer, Heidelberg (2011)

3. Balduccini, M., Gelfond, M.: Logic programs with consistency-restoring rules. In: AAAI Spring Symposiumm, pp. 9–18 (2003)

4. Brain, M., Vos, M.D.: Debugging logic programs under the answer set semantics. In: Workshop on ASP, pp. 141–152 (2005)

5. Brewka, G., Eiter, T., Truszczynski, M.: Answer set programming at a glance. Commun. ACM **54**(12), 92–103 (2011)

6. Calimeri, F., Ianni, G., Ricca, F.: The third open answer set programming competition. TPLP **14**(1), 117–135 (2014)

7. Eiter, T., Gottlob, G., Mannila, H.: Disjunctive datalog. ACM TODS **22**(3), 364–418 (1997)

8. Febbraro, O., Leone, N., Reale, K., Ricca, F.: Unit testing in ASPIDE. In: INAP/WLP, pp. 345–364 (2011)

9. Febbraro, O., Reale, K., Ricca, F.: ASPIDE: integrated development environment for answer set programming. In: Delgrande, J.P., Faber, W. (eds.) LPNMR 2011. LNCS, vol. 6645, pp. 317–330. Springer, Heidelberg (2011)

10. Gebser, M., Kaminski, R., Kaufmann, B., Schaub, T.: Answer Set Solving in Practice. Morgan & Claypool Publishers, California (2012)

11. Gebser, M., Pührer, J., Schaub, T., Tompits, H.: A meta-programming technique for debugging answer-set programs. In: AAAI, pp. 448–453 (2008)

12. Gelfond, M., Lifschitz, V.: Classical negation in logic programs and disjunctive databases. New Gener. Comput. **9**(3–4), 365–386 (1991)

13. Grasso, G., Leone, N., Manna, M., Ricca, F.: ASP at work: spin-off and applications of the DLV system. In: Balduccini, M., Son, T.C. (eds.) Logic Programming, Knowledge Representation, and Nonmonotonic Reasoning. LNCS, vol. 6565, pp. 432–451. Springer, Heidelberg (2011)

14. Janhunen, T., Niemelä, I., Oetsch, J., Pührer, J., Tompits, H.: Random vs. structure-based testing of answer-set programs: an experimental comparison. In: Delgrande, J.P., Faber, W. (eds.) LPNMR 2011. LNCS, vol. 6645, pp. 242–247. Springer, Heidelberg (2011)

15. Junker, U.: QuickXplain: Preferred explanations and relaxations for over-constrained problems. In: AAAI, pp. 167–172 (2004)

16. de Kleer, J., Williams, B.C.: Diagnosing multiple faults. Artif. Intell. **32**(1), 97–130 (1987)

17. Mikitiuk, A., Moseley, E., Truszczynski, M.: Towards debugging of answer-set programs in the language PSpb. In: IC-AI, pp. 635–640 (2007)

18. Oetsch, J., Pührer, J., Tompits, H.: Stepping through an answer-set program. In: Delgrande, J.P., Faber, W. (eds.) LPNMR 2011. LNCS, vol. 6645, pp. 134–147. Springer, Heidelberg (2011)

19. Oetsch, J., Pührer, J., Tompits, H.: Catching the Ouroboros: on debugging non-ground answer-set programs. TPLP **10**(4–6), 2010 (2010)

20. Perri, S., Ricca, F., Terracina, G., Cianni, D., Veltri, P.: An integrated graphic tool for developing and testing dlv programs. In: SEA Workshop, pp. 86–100 (2007)

21. Polleres, A., Frühstück, M., Schenner, G., Friedrich, G.: Debugging non-ground ASP programs with choice rules, cardinality and weight constraints. In: Cabalar, P., Son, T.C. (eds.) LPNMR 2013. LNCS, vol. 8148, pp. 452–464. Springer, Heidelberg (2013)

22. Pontelli, E., Son, T.C., El-Khatib, O.: Justifications for logic programs under answer set semantics. TPLP **9**(1), 1–56 (2009)
23. Reiter, R.: A theory of diagnosis from first principles. Artif. Intell. **32**(1), 57–95 (1987)
24. Shchekotykhin, K.: Interactive query-based debugging of ASP programs. In: AAAI, pp. 1597–1603 (2015)
25. Shchekotykhin, K., Friedrich, G., Fleiss, P., Rodler, P.: Interactive ontology debugging: two query strategies for efficient fault localization. J. Web Semant. **12–13**, 88–103 (2012)
26. Shchekotykhin, K., Friedrich, G., Rodler, P., Fleiss, P.: Sequential diagnosis of high cardinality faults in knowledge-bases by direct diagnosis generation. In: ECAI, pp. 813–818 (2014)
27. Syrjänen, T.: Debugging inconsistent answer set programs. In: NMR, pp. 77–84 (2006)

Linking Open-World Knowledge Bases Using Nonmonotonic Rules

Thomas Eiter and Mantas Šimkus$^{(\boxtimes)}$

Institute of Information Systems, TU Wien, Vienna, Austria
simkus@dbai.tuwien.ac.at

Abstract. Integrating knowledge from various sources is a recurring problem in Artificial Intelligence, often addressed by *multi-context systems (MCSs)*. Existing MCSs however have limited support for the open-world semantics of *knowledge bases (KBs)* expressed in knowledge representation languages based on first-order logic. To address this problem we introduce *knowledge base networks (KBNs)*, which consist of open-world KBs linked by non-monotonic bridge rules under a stable model semantics. Basic entailment in KBNs is decidable whenever it is in the individual KBs. This is due to a fundamental representation theorem, which allows us to derive complexity results, and also gives a perspective for implementation. In particular, for networks of KBs in well-known *Description Logics (DLs)*, reasoning is reducible to reasoning in nonmonotonic *dl-programs*. As a by product, we obtain an embedding of a core fragment of Motik and Rosati's hybrid MKNF KBs, which amount to a special case of KBNs, to dl-programs. We also show that reasoning in networks of ontologies in lightweight DLs is not harder than in answer set programming.

1 Introduction

Integrating information from various *knowledge bases (KBs)* is a recurring problem in Artificial Intelligence and a major issue for Knowledge Representation and Reasoning. *Multi-context systems (MCSs)* [2,5,12,21] are a well-known approach to address this challenge. MCSs interlink individual KBs (called *contexts*) with *bridge rules* that enable making inferences across KBs, in a way such that a global system semantics emerges from the local KBs. While a variety of MCSs are available with different expressivity of bridge rules and knowledge base languages, they provide limited support for KBs with model-based semantics under an open-world view.

Interlinking KBs with open-world semantics is a relevant problem, as such KBs are becoming increasingly popular, e.g. in the form of various Description Logic (DL) ontologies or geographic information systems. However, combining such individual KBs into a more complex hybrid KB is not trivial, e.g. the naive union of DL KBs may lead to an inconsistency or fall into an undecidable fragment of first-order logic. Employing such open-world KBs in MCSs is not obvious, especially due to the significant semantic differences and the computational

© Springer International Publishing Switzerland 2015
F. Calimeri et al. (Eds.): LPNMR 2015, LNAI 9345, pp. 294–308, 2015.
DOI: 10.1007/978-3-319-23264-5_25

challenges. Specifically, the authors of [21] considered the open-world setting, but for KBs expressed in propositional logic only. The powerful MCS framework in [5], being very abstract, is not well-suited for an appealing model-based and decidable integration of open-world KBs expressed in decidable languages (see Sect. 5 for a discussion).

To get a feeling of our motivation, consider the following example:

Example 1. Assume we have a pair κ_1, κ_2 of contexts, where κ_1 is an ontology that describes restaurants and food, and κ_2 is a geospatial database. The contexts make the open-world assumption and may include disjunctive information such as: every wine must be assigned a country of origin, which can be arbitrarily chosen from the set of all countries. The open-world setting intuitively means that reasoning involves a case-by-case analysis of the possibilities (the *possible worlds*) that emerge from the knowledge in contexts. Suppose we would like to add to the knowledge in κ_1 the following. A user should be advised to reserve the restaurant x in case x is located in Manhattan and is not known to be a fast food restaurant. In our approach such information would be expressed by a *bridge rule* as follows:

$$1 : \mathsf{BookingAdvised}(x) \leftarrow 1 : \mathsf{Restaurant}(x),$$
$$2 : \mathsf{inside}(x, \text{``Manhattan''}),$$
$$1 : \textit{not}\, \mathsf{type}(x, \text{``Fast Food''}).$$

A natural question to ask the above multi-context system is to retrieve all restaurants for which reservation is advisable.

Motivated by the above challenges, we introduce a novel kind of MCS called *knowledge base networks (KBNs)*. A KBN consists of a collection of KBs with open-world semantics in terms of first-order structures, interlinked by nonmonotonic bridge rules. Such a network is equipped, similar as in [21], with a stable model semantics inspired by [11], in a way such that the local impact on KBs is minimal, i.e., the open-world perspective is kept as much as possible.

The new formalism has the following attractive properties.

- KBNs allow one to connect KBs expressed in a wide range of languages. The only requirement is that they have a model-based semantics which assigns each KB a set of first-order structures (i.e. the possible worlds). In particular, all ontology languages based on first-order logic are supported by KBNs.
- Reasoning in KBNs is decidable whenever it is in the comprising KBs. In particular, this holds for entailment of ground facts in the stable models of a given KBN. This is due to a fundamental result (Theorem 2) by which stable models can be represented as *knowledge states* (finite sets of ground atoms).
- Furthermore, relying on the finite representation of stable models, we obtain generic algorithms and complexity results for reasoning in KBNs. We instantiate the latter for the case where KBs are formulated in well-known DLs (e.g. DL-Lite and $\mathcal{SHIQ}$) and show that entailment of facts is often not harder than in answer set programming (ASP) or in DLs. In particular, for the so-called lightweight DLs it coincides with the complexity of ASP.

- Finally, it has an implementation perspective that leverages on existing solvers. To wit we provide a translation of KBNs with DL KBs into *dl-programs* [10], such that the stable models of the KBN correspond to the strong answer sets of the resulting dl-program. As a bonus, we obtain an embedding of a core fragment of hybrid MKNF [20] (corresponding to singleton KBNs) into dl-programs. As a converse embedding was known, this establishes a strong connection between hybrid MKNF and dl-programs.

2 Knowledge Base Networks

In this section, we formally define knowledge base networks. To this end, we first describe KBs that may take part in a network. We aim to be as general as possible, and thus consider all KBs with semantics in terms of first-order structures (*interpretations*). After formally defining interpretations, we introduce the syntax of bridge rules and the stable model semantics for KBNs.

We assume disjoint sets Const and Rel of *constants* and *relation symbols* of *arity* ≥ 0, respectively. Usually interpretations in first-order logic (FOL) are defined for a *signature*, i.e. a collection of symbols. We consider in this paper structures that interpret a given finite set σ of constants and, for simplicity, *all* relation symbols. Since KBs will be linked by rules based on epistemic queries, we need to make some assumptions. We assume that σ is part of the interpretation domain and all $c \in \sigma$ are interpreted as themselves. This is a variant of the *standard name assumption (SNA)*, which is a common assumption for epistemic logics based on FOL (see, e.g., [7,20]).

Definition 1. *Assume a finite set σ of constants. A σ-interpretation is a tuple $\mathcal{I} = (\Delta^{\mathcal{I}}, \cdot^{\mathcal{I}})$, where (i) $\Delta^{\mathcal{I}} \neq \emptyset$ is a set such that $\sigma \subseteq \Delta^{\mathcal{I}}$, called the domain of $\mathcal{I}$, and (ii) $\cdot^{\mathcal{I}}$ is a function that assigns to each n-ary relation symbol $R \in$ Rel an n-ary relation $R^{\mathcal{I}} \subseteq (\Delta^{\mathcal{I}})^n$. If σ is irrelevant or clear from the context, we call $\mathcal{I}$ an interpretation. We further assume a nullary relation $\perp$ such that $(\perp)^{\mathcal{I}} = \emptyset$ for all interpretations $\mathcal{I}$.*

We next give a general notion of KBs. The semantics to each KB is given by associating to it a possibly infinite set of interpretations (called *models* or *possible worlds*).

Definition 2. *Let $\mathcal{K}$ denote an infinite set of objects, called* knowledge bases *(KBs). We further assume a binary "models" relation $\models$ between interpretations and KBs. If $\mathcal{I} \models \kappa$, then $\mathcal{I}$ is called a* model *of the KB $\kappa \in \mathcal{K}$.*

In our approach, KBs communicate via rules that involve first-order atoms over the constants in Const and an additional countably infinite set V of *variables*. *Positive literals* (or, *atoms*) have the form $R(\vec{t})$, where $R \in$ Rel is n-ary and $\vec{t}$ is an n-tuple of *terms*, i.e., of elements from V $\cup$ Const. Instead of writing $\perp()$ we simply write $\perp$. *Negative literals* are expressions *not* B, where B is an atom. A (positive or negative) literal is *ground*, if it has no variables.

Definition 3. *Assume a KB $\kappa \in \mathcal{K}$, a σ-interpretation $\mathcal{I}$, a ground atom $B = R(\vec{t})$, and a set I of ground atoms. We extend the $\models$ relation as follow s; we write*

- *$\mathcal{I} \models R(\vec{t})$ if $\vec{t} \in R^{\mathcal{I}}$ and $t_i \in \sigma$ for every term t_i in $\vec{t}$;*
- *$\mathcal{I} \models I$ if $\mathcal{I} \models B$ holds for every $B \in I$;*
- *$\kappa, I \models B$ if $\mathcal{I} \models B$ for every $\mathcal{I}$ such that $\mathcal{I} \models \kappa$ and $\mathcal{I} \models I$;*
- *$\kappa \models B$ if $\kappa, \emptyset \models B$.*

We will use *"conditional entailment"* to refer to the problem of deciding $\kappa, I \models B$, where κ, I and B are as above. Note that conditional entailment is a generalized version of classical entailment from a KB, i.e. the case when $I = \emptyset$.

We next define *bridge rules*. They are syntactically similar to the rules in standard ASP, but employ indexed literals to be able to refer to different KBs of a KBN.

Definition 4. *An* indexed literal *is an expression $k : L$, where $k \geq 1$ is an integer and L is a literal; it is* positive, *if L is positive, and* negative *otherwise. A* (bridge) rule *ρ is an expression of the form*

$$L_0 \leftarrow L_1, \ldots, L_n, \tag{1}$$

where $L_0, \ldots, L_n$ are indexed literals. head$(\rho) = L_0$ *is the* head *of ρ, while* body$(\rho) = \{L_1, \ldots, L_n\}$ *is the* body *of ρ. We denote by $\mathsf{V}(\rho)$ the set of variables that occur in ρ. We assume rules are* safe, *i.e. each $x \in \mathsf{V}(\rho)$ occurs in a positive body literal of ρ. A rule ρ is* ground, *if $\mathsf{V}(\rho) = \emptyset$.*

We now formally define KBNs. In addition to a specification of KBs and interconnecting bridge rules, a KBN also specifies a finite set of "shared" constants. Intuitively, knowledge exchange between KBs will be limited to information about these constants.

Definition 5. *A* knowledge base network (KBN) *is a tuple $\mathcal{N} = (\vec{\kappa}, \mathcal{R}, \sigma)$, where*

(i) *$\vec{\kappa} = \langle \kappa_1, \ldots, \kappa_n \rangle$ is a tuple of KBs,*
(ii) *$\mathcal{R}$ is a finite set of bridge rules such that $k \in \{1, \ldots, n\}$ for every literal $k : L$ occurring in $\mathcal{R}$, and*
(iii) *$\sigma \subseteq \mathsf{Const}$ is a finite set of constants.*

We require $c \in \sigma$ for every constant c that occurs in $\mathcal{R}$. If "not" does not appear in $\mathcal{R}$, then $\mathcal{R}$ and $\mathcal{N}$ are called positive.

Example 2. Consider the following example of a KBN with two KBs, two bridge rules and two shared constants. More precisely, take a set $\sigma = \{c_1, c_2\}$ of constants, and a pair κ_1, κ_2 of KBs. We build the following KBN $\mathcal{N} = (\langle \kappa_1, \kappa_2 \rangle, \{\rho_1, \rho_2\}, \sigma)$, where ρ_1, ρ_2 are bridge rule with a single variable x as follows:

$$\rho_1 = 1 : A(x) \leftarrow 2 : D(x), 2 : not\, E(x),$$
$$\rho_2 = 2 : F(x) \leftarrow 1 : B(x).$$

Intuitively, the rules ensure that the following holds for every shared constant c in σ:

– if $D(c)$ is known in κ_2 but $E(c)$ is not known in κ_2, then $A(c)$ is known in κ_1;
– if $B(c)$ is known in κ_1, then $F(c)$ is known in κ_2.

Semantics. To formalize the semantics of KBNs we will work on sets of interpretations. We use $\mathbb{I}, \mathbb{I}', \mathbb{I}_1, \mathbb{I}_2$ and so forth to denote such sets. For two tuples $\vec{\mathbb{I}} = (\mathbb{I}_1, \ldots, \mathbb{I}_n)$ and $\vec{\mathbb{I}}' = (\mathbb{I}'_1, \ldots, \mathbb{I}'_n)$ of sets of interpretations, we write $\vec{\mathbb{I}} \subseteq \vec{\mathbb{I}}'$ if $\mathbb{I}_1 \subseteq \mathbb{I}'_1, \ldots, \mathbb{I}_n \subseteq \mathbb{I}'_n$. We write $\vec{\mathbb{I}} \subset \vec{\mathbb{I}}'$ if $\vec{\mathbb{I}} \subseteq \vec{\mathbb{I}}'$ and $\vec{\mathbb{I}}' \not\subseteq \vec{\mathbb{I}}$.

We give KBNs a semantics in terms of *epistemic models*. Roughly speaking, the latter assign every κ_i in a KBN $\mathcal{N}$ a set $\mathbb{I}_i$ of models of κ_i so that each rule ρ in $\mathcal{N}$ is satisfied, where rule satisfaction is defined via epistemic evaluation of rule literals. Intuitively, such a literal $i : R(\vec{t})$ (resp., $i : not\, R(\vec{t})$) has value *true*, if it is true in each (resp., false in some) interpretation in $\mathbb{I}_i$. In this way, rules transfer *knowledge* between KBs. Intuitively, rules should have minimal effect on the local KB models, i.e. preserve as much knowledge of individual KBs as possible. This minimality requirement will eventually lead to *stable models*.

Definition 6. *An* epistemic interpretation *for a KBN* $\mathcal{N} = (\vec{\kappa}, \mathcal{R}, \sigma)$ *is a* $|\vec{\kappa}|$-*tuple* $\vec{\mathbb{I}} = (\mathbb{I}_1, \ldots, \mathbb{I}_{|\vec{\kappa}|})$, *where each* $\mathbb{I}_i$ *of* $\vec{\mathbb{I}}$ *is a nonempty set of* σ-*interpretations.*

Our next goal is to formally define epistemic models. We start from the satisfaction of ground literals and ground rules.

Definition 7. *Assume a set* $\mathbb{I}$ *of interpretations and a ground atom* B. *We write* $\mathbb{I} \models B$ *if* $\mathcal{I} \models B$ *for all* $\mathcal{I} \in \mathbb{I}$. *We write* $\mathbb{I} \models not\, B$ *if* $\mathbb{I} \not\models B$, *i.e.* $\mathcal{I} \not\models B$ *for some* $\mathcal{I} \in \mathbb{I}$. *For an epistemic interpretation* $\vec{\mathbb{I}} = (\mathbb{I}_1, \ldots, \mathbb{I}_n)$ *and a literal* $k : L$, *we write* $\vec{\mathbb{I}} \models k : L$ *if* $\mathbb{I}_k \models L$. *For a ground rule (1), we let* $\vec{\mathbb{I}} \models \rho$ *if* $\vec{\mathbb{I}} \models L_1, \ldots, \vec{\mathbb{I}} \models L_n$ *implies* $\vec{\mathbb{I}} \models L_0$.

In particular, for each epistemic interpretation $\vec{\mathbb{I}} = (\mathbb{I}_1, \ldots, \mathbb{I}_n)$ and $1 \leq k \leq n$, we have $\vec{\mathbb{I}} \not\models k{:}\bot$, as $\mathbb{I}_k \neq \emptyset$, i.e., epistemic interpretations are locally consistent at each KB.

The satisfaction of rules with variables is defined via *grounding*.

Definition 8. *Given a set* $\sigma \subseteq \mathsf{Const}$, *a* σ-*substitution is any function* $\delta : \mathsf{V} \to \sigma$. *By* $\delta(\rho)$ *we denote the rule obtained from a rule* ρ *by replacing each variable* x *of* ρ *with* $\delta(x)$. *For a set* $\mathcal{R}$ *of rules, we let* $\mathsf{ground}(\mathcal{R}, \sigma) = \{\delta(\rho) \mid \rho \in \mathcal{R}, \delta \text{ is a } \sigma\text{-substitution}\}$.

We can now finalize the definition of epistemic models.

Definition 9. *Assume a KBN* $\mathcal{N} = (\vec{\kappa}, \mathcal{R}, \sigma)$ *and an epistemic interpretation* $\vec{\mathbb{I}}$ *for* $\mathcal{N}$. *We write* $\vec{\mathbb{I}} \models \mathcal{R}$, *if* $\vec{\mathbb{I}} \models \rho$ *for every* $\rho \in \mathsf{ground}(\mathcal{R}, \sigma)$. *We write* $\vec{\mathbb{I}} \models \mathcal{N}$ *if* $\vec{\mathbb{I}} \models \mathcal{R}$ *and* $\mathcal{I} \models \kappa_i$ *for every* $1 \leq i \leq |\vec{\kappa}|$ *and* $\mathcal{I} \in \mathbb{I}_i$. *If* $\vec{\mathbb{I}} \models \mathcal{N}$, *then* $\vec{\mathbb{I}}$ *is an* epistemic model *of* $\mathcal{N}$.

Stable Models. We next define *stable models*, which intuitively minimize the knowledge loss by maximally preserving models of individual KBs. In addition, stable models minimize the truth of atoms that occur negatively in rule bodies, thus implementing default negation. To this end, we adapt for our purposes the Gelfond-Lifschitz reduct [11].

Definition 10. *Given a KBN* $\mathcal{N} = (\vec{\kappa}, \mathcal{R}, \sigma)$ *and an epistemic interpretation* $\vec{\mathbb{I}}$ *for* $\mathcal{N}$, *we denote by* $\mathcal{R}^{\vec{\mathbb{I}}}$ *the set of rules obtained from* ground$(\mathcal{R}, \sigma)$ *by deleting*

(i) every rule with a body literal $k : not\ B$ *s.t.* $\vec{\mathbb{I}} \models k : B$;

(ii) every negative literal in the remaining rules.

Then, $\mathcal{N}^{\vec{\mathbb{I}}} = (\vec{\kappa}, \mathcal{R}^{\vec{\mathbb{I}}}, \sigma)$ *is called the* reduct *of* $\mathcal{N}$ *w.r.t.* $\vec{\mathbb{I}}$.

We are ready to define stable models of KBNs. Due to the adopted epistemic approach we *maximize* sets of models, which contrasts but is equivalent to model minimization in standard ASP (see the seminal paper [18] for this observation).

Definition 11. *An epistemic interpretation* $\vec{\mathbb{I}}$ *for a KBN* $\mathcal{N}$ *is a* stable model *of* $\mathcal{N}$, *if (1)* $\vec{\mathbb{I}} \models \mathcal{N}$ *and (2) there is no* $\vec{\mathbb{I}}'$ *such that* $\vec{\mathbb{I}} \subset \vec{\mathbb{I}}'$ *and* $\vec{\mathbb{I}}' \models \mathcal{N}^{\vec{\mathbb{I}}}$. *We say* $\mathcal{N}$ *is* consistent, *if* $\mathcal{N}$ *has a stable model.*

Example 3 (Cont'd). We illustrate the above concepts by an example. Let κ_1 have the set of models $\mathbb{I} = \{\mathcal{I}_1, \mathcal{I}_2, \mathcal{I}_3\}$ and κ_2 the set of models $\mathbb{J} = \{\mathcal{J}_1, \mathcal{J}_2, \mathcal{J}_3\}$ with

$$\mathcal{I}_1 = \{A(c_1), B(c_1), C(c_1)\}, \quad \mathcal{J}_1 = \{D(c_1), E(c_1), F(c_1)\},$$
$$\mathcal{I}_2 = \{A(c_1), B(c_1)\}, \quad \mathcal{J}_2 = \{D(c_1), E(c_1)\},$$
$$\mathcal{I}_3 = \{A(c_2)\}, \quad \mathcal{J}_3 = \{D(c_1), F(c_1)\}.$$

With a slight abuse of notation each model is given by the ground atoms it satisfies. Assume the case where $\mathbb{I}' = \{\mathcal{I}_1, \mathcal{I}_2, \mathcal{I}_3\}$ and $\mathbb{J}' = \{\mathcal{J}_1, \mathcal{J}_2\}$. It can be easily verified that $(\mathbb{I}', \mathbb{J}')$ is an epistemic model of $\mathcal{N}$. However, it is not a stable model of $\mathcal{N}$ because $(\mathbb{I}', \mathbb{J}' \cup \{\mathcal{J}_3\})$ is an epistemic model of $\mathcal{N}^{(\mathbb{I}', \mathbb{J}')}$. An intuitive reason to discard $(\mathbb{I}', \mathbb{J}')$ as a stable model is that $E(t)$ is epistemically true in $\mathbb{J}'$, but this is not "justified" by the bridge rules and the models of κ_1 and κ_2.

Consider the case where $\mathbb{I}' = \{\mathcal{I}_1, \mathcal{I}_2\}$ and $\mathbb{J}' = \{\mathcal{J}_1, \mathcal{J}_3\}$. One can easily see that such $(\mathbb{I}', \mathbb{J}')$ is a stable model of $\mathcal{N}$.

In standard ASP, a consistent *positive* program, i.e. program without an occurrence of "*not*", has a unique stable model. The same holds for any consistent positive KBN $\mathcal{N}$. This follows from the fact that the component-wise union of two epistemic models of $\mathcal{N}$ is again an epistemic model of $\mathcal{N}$.

Theorem 1. *Every positive KBN* $\mathcal{N}$ *has at most one stable model.*

Singleton KBNs, i.e. KBNs $\mathcal{N} = (\vec{\kappa}, \mathcal{R}, \sigma)$ with $|\vec{\kappa}| = 1$, remain interesting as they allow for closed-world reasoning about a KB (we shall see later that hybrid MKNF KBs can be seen as singleton KBNs). For such KBNs, we drop the unique prefix "1:" from rule literals, and also identify epistemic interpretations $\vec{\mathbb{I}} = (\mathbb{I}_1)$ of $\mathcal{N}$ with simply $\mathbb{I}_1$. In this case, $\mathcal{R}$ is syntactically a regular ASP program; the connection between singleton KBNs and ASP can be elaborated as follows.

Proposition 1. *Assume a regular ASP program* P. *Let* σ *be the set of constants that appear in* P, *and let* $\top$ *be a KB such that* $\mathcal{I} \models \top$ *for any interpretation* $\mathcal{I}$. *Then,* I *is a stable model of* P *iff the singleton KBN* $\mathcal{N} = (\top, P, \sigma)$ *has a stable model* $\mathbb{I}$ *such that* $I = \{R(\vec{t}) \mid \mathbb{I} \models R(\vec{t})$ *and* $R(\vec{t})$ *is ground*$\}$.

3   Reasoning in Knowledge Base Networks

We show in this section that basic inference in KBNs is decidable, provided reasoning in individual KBs is decidable. In particular, we consider the task of deciding, given a KBN $\mathcal{N}$ and a ground indexed atom $k : R(\vec{t})$, whether $\vec{\mathbb{I}} \models k : R(\vec{t})$ holds for every stable model $\vec{\mathbb{I}}$ of $\mathcal{N}$. We will refer to this problem as "*ground entailment*". We show that this problem is decidable if for each individual KB of $\mathcal{N}$ there is a procedure to check conditional (non-)entailment.

As a stable model may contain infinitely many interpretations of infinite size, we resort to a *finite representation* that is similar in spirit to those in [7,20, 22,23]. In particular, stable models will be represented using *stable knowledge states*, which are finite sets of ground indexed atoms satisfying certain stability conditions. To this end, given a KBN $\mathcal{N} = (\vec{\kappa}, \mathcal{R}, \sigma)$, we use base($\mathcal{N}$) to denote the set of all indexed atoms that can be built from constants in σ, relations and indices that occur in $\mathcal{R}$.

Definition 12. *Given an epistemic interpretation $\vec{\mathbb{I}}$ for a KBN $\mathcal{N}$, we let $K_{\mathcal{N}}(\vec{\mathbb{I}})$ denote the set of atoms $B \in$ base($\mathcal{N}$) such that $\vec{\mathbb{I}} \models B$.*

Observe that $K_{\mathcal{N}}(\vec{\mathbb{I}})$ is always finite, even if $\vec{\mathbb{I}}$ is infinite.

For rules over indexed literals, we define the notions of a classical model and a variation of the Gelfond-Lifschitz reduct.

Definition 13. *For a set I of ground indexed atoms and a ground atom B, we write $I \models k : B$, if $k : B \in I$ and $I \models k : not\, B$, if $k : B \notin I$. Assuming ρ is a ground rule $L_0 \leftarrow L_1, \ldots, L_n$, we write $I \models \rho$ in case $I \models L_1, \ldots, I \models L_n$ implies $I \models L_0$. Given a set $\mathcal{R}$ of ground rules, the rule set $\mathcal{R}^I$ is obtained from $\mathcal{R}$ by deleting*

(a) every rule with a body literal $k : not\, B$ s.t. $k : B \in I$, and
(b) every negative literal in the remaining rules.

We can now present stable knowledge states; informally, they are knowledge states that can be justified by the KBs and the bridge rules of the given KBN. For a set I of ground indexed atoms and $i \geq 1$, let $I_{|i} = \{B \mid i : B \in I\}$.

Definition 14. *Let $\mathcal{N} = (\vec{\kappa}, \mathcal{R}, \sigma)$ be a KBN. A set $I \subseteq$ base($\mathcal{N}$) is closed w.r.t. $\mathcal{N}$, if the following holds for every $1 \leq i \leq |\vec{\kappa}|$:*

(i) $\kappa_i, I_{|i} \not\models \bot$, and
(ii) $\kappa_i, I_{|i} \models B$ implies $i : B \in I$ for every $i : B \in$ base($\mathcal{N}$).

A set $I \subseteq$ base($\mathcal{N}$) is a stable knowledge state of $\mathcal{N}$, if

(A) I is closed w.r.t. $\mathcal{N}$,
(B) $I \models$ ground($\mathcal{R}, \sigma$), and
(C) there exists no $J \subset I$ such that (i) J is closed w.r.t. $\mathcal{N}$ and (II) $J \models$ (ground($\mathcal{R}, \sigma$))^I.

Algorithm 1. NONENTAILS

Input: KBN $\mathcal{N} = (\vec{\kappa}, \mathcal{R}, \sigma)$, ground indexed atom $i : B$
Output: TRUE iff $\mathcal{N}$ has a stable model $\vec{\mathbb{I}}$ s.t. $\vec{\mathbb{I}} \not\models i : B$

1 Guess a set $I \subseteq \mathsf{base}(\mathcal{N})$
2 Compute $\mathcal{N}^I := (\vec{\kappa}, \mathcal{R}^I, \sigma)$
3 Let $J = \text{LEASTKNOWLEDGESTATE}(\mathcal{N}^I)$
4 **return** TRUE iff $J = I$ and $\kappa_i, I|_i \not\models B$

The next theorem draws a strong connection between stable models and stable knowledge states of a KBN.

Theorem 2 (representation). *Assume a KBN $\mathcal{N}$. If $\vec{\mathbb{I}}$ is a stable model of $\mathcal{N}$, then $K_{\mathcal{N}}(\vec{\mathbb{I}})$ is a stable knowledge state of $\mathcal{N}$. Conversely, if I is a stable knowledge state of $\mathcal{N}$, then $\mathcal{N}$ has a stable model $\vec{\mathbb{I}}$ such that $K_{\mathcal{N}}(\vec{\mathbb{I}}) = I$.*

This theorem implies that stable models of a KBN $\mathcal{N}$ can be represented by finite sets of indexed atoms from $\mathsf{base}(\mathcal{N})$. Observe that for finite $\mathcal{R}$ and σ, there are only finitely many candidate sets $I \subseteq \mathsf{base}(\mathcal{N})$, and for each such I it is decidable whether it satisfies (B) of Definition 14 and likewise, whether J satisfies (C.ii). However, checking whether I satisfies (A) or J satisfies (C.i) relies on the decidability of conditional entailment in the individual KBs of $\mathcal{N}$.

Based on these observations, we give in Algorithm 1 a nondeterministic procedure for checking nonentailment of ground atoms from KBNs, which uses an oracle for conditional entailment in individual KBs. It uses Algorithm 2 as a subroutine that ensures via fix-point computation that I is the least $J \subseteq \mathsf{base}(\mathcal{N})$ that is closed w.r.t. $\mathcal{N}$ (note that $\mathcal{N}$ and $\mathcal{N}^I$ share the KBs), which implies that (A)–(C) hold.

Technically, we say a set $\mathcal{L} \subseteq \mathcal{K}$ of KBs is *decidable*, if given $\kappa \in \mathcal{L}$, a set of atoms I, and a ground atom B, testing $\kappa, I \not\models B$ is decidable. We say a KBN $\mathcal{N}$ is *over* $\mathcal{L}$, if $\kappa \in \mathcal{L}$ for each KB of $\mathcal{N}$. In addition, we say $\mathcal{L}$ is in a complexity class $\mathcal{C}$ if testing $\kappa, I \not\models B$ is in $\mathcal{C}$. We say that the *data complexity* of $\mathcal{L}$ is in a complexity class $\mathcal{C}$ if the following problem is in $\mathcal{C}$ for every $\kappa \in \mathcal{L}$ and ground atom B: given a set of atoms I, check whether $\kappa, I \not\models B$. I.e. for data complexity, the size of KBs and queries is assumed to be fixed.

Theorem 3 (decidability). *Assume a decidable set $\mathcal{L} \subseteq \mathcal{K}$ of KBs. Then ground entailment from KBNs over $\mathcal{L}$ is decidable.*

3.1 Complexity

As we have seen, the complexity of ground entailment depends on the complexity of conditional entailment in the individual KBs. For a set $\mathcal{L} \subseteq \mathcal{K}$ in $\mathcal{C}$, it is easy to see that non-entailment from KBNs over $\mathcal{L}$ is in the class NEXP$^{\mathcal{C}}$. We next consider this problem more for the important case of bounded predicate arities in rules.

Algorithm 2. LEASTKNOWLEDGESTATE

Input: A positive KBN $\mathcal{N} = (\vec{\kappa}, \mathcal{R}, \sigma)$
Output: $K_{\mathcal{N}}(\mathbb{I})$ for the unique stable model $\mathbb{I}$ of $\mathcal{N}$ if $\mathcal{N}$ is consistent.

1 $I \leftarrow \emptyset$
2 **repeat**
3 $\quad$ $I' \leftarrow I$
4 $\quad$ **foreach** $\rho \in \mathsf{ground}(\mathcal{R}, \sigma)$ *s.t.* $\mathsf{body}(\rho) \subseteq I$ **do**
5 $\quad\quad$ $\lfloor$ $I \leftarrow I \cup \{\mathsf{head}(\rho)\}$
6 $\quad$ **foreach** $i : B \in \mathsf{base}(\mathcal{N})$ *s.t.* $\kappa_i, I_{|i} \models B$ **do**
7 $\quad\quad$ $\lfloor$ $I \leftarrow I \cup \{i : B\}$
8 **until** $I = I'$
9 **if** $\vec{\kappa}$ *has some* κ_i *such that* $\kappa_i, I_{|i} \models \bot$ **then**
10 $\quad$ $\lfloor$ **return** "not exists"
11 **return** I

Theorem 4. *Assume a set $\mathcal{L} \subseteq \mathcal{K}$ in $\mathcal{C}$, and consider KBNs $\mathcal{N} = (\vec{\kappa}, \mathcal{R}, \sigma)$ over $\mathcal{L}$ and where the arity of relations occurring in $\mathcal{R}$ is bounded by a constant. Then deciding $\mathcal{N} \not\models j : B$ has the following complexity:*

(a) For $\mathcal{C} = \mathrm{NExp}$, it is in $\mathrm{NP}^{\mathrm{NExp}}$; if in addition $\mathcal{R}$ is positive, it is in NExp.
(b) For $\mathcal{C} = \mathrm{Exp}$, it is in Exp.
(c) For $\mathcal{C} = \Sigma_i^p$, it is in Σ_{i+1}^p if $i \geq 1$ and in Σ_2^p for $i = 0$.
(d) For $\mathcal{C} = \Sigma_i^p$ and positive $\mathcal{R}$, it is in Σ_i^p if $i \geq 2$, (ii) in P^{NP} if $i=1$ (i.e. $\mathcal{C}=\mathrm{NP}$), and (iii) in co-NP if $i=0$ (i.e. $\mathcal{C}=\mathrm{P}$).

Intuitively, the results are explained as follows. The guess for I in Algorithm 1 is polynomial due to bounded arities, and if $\mathrm{Exp} \subseteq \mathcal{C}$, we can run Algorithm 2 on $\mathcal{N}^I$ in polynomial time (without explicit grounding), where a $\mathcal{C}$ oracle is used to decide rule applicability in line 4 and conditional entailment in lines 6 and 9. This shows $\mathrm{NP}^{\mathrm{NExp}}$ and Exp upper bounds in (a) and (b). If $\mathcal{C}$ is Σ_i^p, we cannot ground $\mathcal{R}$ in polynomial time, but we can test rule applicability in $\mathcal{R}$ relative to I using an NP oracle. In case of a positive $\mathcal{R}$, it is sufficient to guess some $I' \supseteq I$ that is closed under $\mathcal{N}$ and $\mathcal{R}$ such that $\kappa_i, I'|_i \not\models B$; such a guess I' can be made and verified in NExp (resp., Σ_i^p), if $\mathcal{C}$ is NExp (resp., Σ_i^p, if $i > 1$); as checking whether I is closed under $\mathcal{R}$ is co-NP-complete, this guessing approach does not work for $\mathcal{C} = \Sigma_1^p = \mathrm{NP}$, but I can be computed there using an NP oracle in Algorithm 2. For $\mathcal{C} = \Sigma_0^p$, non-entailment amounts in essence to datalog non-entailment, which is co-NP-complete.

These upper bounds are tight, i.e., for all entries in Theorem 4 one can find $\mathcal{L}$ with complexity in $\mathcal{C}$ such that nonentailment in KBNs over $\mathcal{L}$ is hard for the respective class.

3.2 Data Complexity

The above results are on the *combined complexity* of non-entailment. We next consider the *data complexity* of this task, which we measure in the size of facts of an input KBN $\mathcal{N} = (\vec{\kappa}, \mathcal{R}, \sigma)$, i.e. by assuming that the size of $\vec{\kappa}$, the combined size of all rules ρ in $\mathcal{R}$ with non-empty bodies and the size of the input query is fixed by a constant.

Theorem 5. *Assume a set $\mathcal{L} \subseteq \mathcal{K}$ with data complexity in $\mathcal{C}$, and consider only KBNs $\mathcal{N}$ over $\mathcal{L}$. Then deciding $\mathcal{N} \not\models j : B$ has a data complexity as follows:*

(a) For $\mathcal{C} = \Sigma_i^p$ with $i \geq 0$, the data complexity is in Σ_{i+1}^p.
(b) For $\mathcal{C} = \Sigma_i^p$ with $i \geq 0$ and positive $\mathcal{N}$, the data complexity is in $\mathcal{C}$.

Different from above, grounding $\mathcal{R}$ is always feasible in polynomial time, which then leads to the upper bounds; in particular, for positive rules, the guess for I' can always be verified in Σ_i^p. Again, hardness holds in all cases (which follows easily from the complexity of model checking for second-order logic over finite structures).

3.3 Ontology Networks

As a concrete showcase, we discuss the complexity of KBNs whose KBs are ontologies expressed in DLs. We refer to [1] for an introduction to DLs. A DL ontology $\mathcal{O}$ can be simply seen as a theory in first-order logic (FOL) built using constants (a.k.a. *individuals*), and unary and binary relation symbols (a.k.a. *concept names* and *role names*, respectively). Various syntax restrictions on $\mathcal{O}$ give rise to a variety of DLs, such as $\mathcal{SHOIQ}$, $\mathcal{SHIQ}$, $\mathcal{EL}$, DL-Lite, etc. Like for full FOL, the semantics for a DL ontology $\mathcal{O}$ can be given in terms of interpretations as usual.

We consider here *ontology networks* (ONs), which are KBNs $\mathcal{N} = (\vec{\mathcal{O}}, \mathcal{R}, \sigma)$ such that $\vec{\mathcal{O}}$ is a tuple of DL ontologies, and σ is the set of constants in $\vec{\mathcal{O}}$ and $\mathcal{R}$. Since σ is determined by $\vec{\mathcal{O}}$ and $\mathcal{R}$, we will simply write $\mathcal{N} = (\vec{\mathcal{O}}, \mathcal{R})$ for ONs. Note that relations in $\mathcal{R}$ are not restricted to concept and role names, and thus can have any arity. Table 1 gives some complexity results for reasoning in ONs with various DLs. The upper bounds follow from Theorems 4 and 5. Hardness in (†) and (‡) follow from [10] and [20], respectively, and the rest from the complexity of DLs and ASP (see [8] for bounded predicate arities in ASP). We note that conditional entailment in DLs reduces to *instance checking*.

Table 1. Complexity (completeness) of atom non-entailment in ONs with various DLs (bounded arities / data complexity).

Description Logics	Positive KBNs	Normal KBNs
DL-Lite, $\mathcal{EL}$	co-NP / P	Σ_2^P / NP
$\mathcal{SHIQ}$	Exp / NP	Exp / Σ_2^p (‡)
$\mathcal{SHOIQ}$	NExp / NP	NP^{NExp} (†) / Σ_2^p (‡)

4   Encoding into DL-programs

The finite representation result given in Theorem 2 opens a way to implementing a procedure for reasoning in KBNs by a translation into (extensions of) answer set programming. We show here how entailment of ground atoms from a given ON can be reduced to entailment of atoms in a *dl-program* [10].

For presentation purposes, we define a variant of dl-programs that supports only the "addition" operator $\uplus$ but allows queries over multiple ontologies; this generalization is trivial and available in the dlvhex suite.[1]

We first recall dl-programs. A *dl-atom* α is an expression of the form

$$DL[\lambda; k : Q](\vec{t}), \qquad (2)$$

where $k \geq 1$, $\lambda = S_1 \uplus R_1, \ldots, S_n \uplus R_n$, $n \geq 0$, and (i) Q and every S_i is either a concept or a role name, (ii) if S_i is a concept (resp., role) name, then R_i is a unary (resp., binary) relation symbol, and (iii) $|\vec{t}|$ matches the arity of Q.

A *positive dl-literal* L is a dl-atom α or an atom B, and a *negative dl-literal* has the form *not* L. A *dl-rule* ρ is an expression $L_0 \leftarrow L_1, \ldots, L_n$, where all L_i are dl-literals. A *dl-program* is then a pair $P = (\vec{\mathcal{O}}, \mathcal{R})$ where $\vec{\mathcal{O}} = \mathcal{O}_1, \ldots, \mathcal{O}_n$ are ontologies and $\mathcal{R}$ is a set of dl-rules; it is *positive*, if "*not*" does not occur in $\mathcal{R}$. Ground dl-literals and grounding w.r.t. a set of constants are as for bridge rules in Sect. 2.

The semantics of a dl-program is given by *strong answer sets*, which are sets I of ground literals on the rule predicates that satisfy all rules (i.e., are models), and fulfill a stability condition. We say I *satisfies* a ground atom B, if $B \in I$. In addition, I *satisfies* a dl-atom α of form (2), if $\mathcal{O}_k \cup \mathcal{A} \models Q(\vec{t})$, where $\mathcal{A}$ is the DL *ABox* $\mathcal{A} = \{S_i(\vec{c}) \mid R_i(\vec{c}) \in I, 1 \leq i \leq n)\}$. We write $I \models^{\vec{\mathcal{O}}} L$ if a literal L (α or B) is satisfied. Satisfaction $I \models^{\vec{\mathcal{O}}} \beta$ then naturally extends to ground positive dl-rules and dl-programs β.

Each ground positive dl-program $P = (\vec{\mathcal{O}}, \mathcal{R})$ has a unique least model, which is the set I of ground atoms such that $I \models^{\vec{\mathcal{O}}} \mathcal{R}$ and $J \not\models^{\vec{\mathcal{O}}} \mathcal{R}$ for every $J \subset I$. We can now define the notion of *strong answer set* of dl-programs.

Definition 15. *Assume a dl-program* $P = (\vec{\mathcal{O}}, \mathcal{R})$ *and a set* I *of ground atoms. Let* Const_P *denote the set of constants occurring in* P. *We let* $\mathcal{R}^I$ *be the rule set obtained from* $\mathsf{ground}(\mathcal{R}, \mathsf{Const}_P)$ *by deleting*

(i) every rule with a body literal not L s.t. $I \models^{\vec{\mathcal{O}}} L$,
(ii) all negative body literals in the remaining rules.

Then I *is a strong answer set of* P *if* I *is the least model of the dl-program* $P^I = (\vec{\mathcal{O}}, \mathcal{R}^I)$.

Note that strong answer sets involve like stable models of KBNs a Gelfond-Lifschitz style reduct. In fact, we give a surprisingly simple translation from ONs into dl-programs.

---

[1] www.kr.tuwien.ac.at/research/systems/dlvhex.

In what follows, we view $k : R$ for $R \in \mathsf{Rel}$ and $k \geq 1$ as an "indexed" version of predicate symbol R, and identify $not\ k : B$ with $k : not\ B$.

Definition 16. *For an ON $\mathcal{N} = (\vec{\mathcal{O}}, \mathcal{R})$ let $P_{\mathcal{N}} = (\vec{\mathcal{O}}, \mathcal{R} \cup \mathcal{R}')$, where $\mathcal{R}'$ contains for each $1 \leq k \leq |\vec{\mathcal{O}}|$ the following rules:*

(i) $k : R_j(\vec{x}) \leftarrow DL[R_1 \uplus k : R_1, \ldots, R_m \uplus k : R_m; k : R_j](\vec{x})$, *and*
(ii) $f \leftarrow not\ f, DL[R_1 \uplus k : R_1, \ldots, R_m \uplus k : R_m; k : \bot](x)$,

where $1 \leq j \leq m$, $\vec{x}$ is a variable tuple, f is a fresh predicate symbol, and $R_1, \ldots, R_m$ are the roles and concept names of $\mathcal{O}_k$ occurring in $\mathcal{R}$.

Intuitively, the rules in $\mathcal{R}'$ require that a candidate strong answer set I is closed w.r.t. $\mathcal{N}$, i.e. they capture item (A) of Definition 14. The strong answer set semantics ensures then items (B) and (C). Thus the encoding generates stable knowledge states; combined with Theorem 2, we obtain:

Theorem 6. *Let $\mathcal{N} = (\vec{\mathcal{O}}, \mathcal{R})$ be an ON. Then a set I of indexed atoms is a strong answer set of $P_{\mathcal{N}}$ iff $\mathcal{N}$ has some stable model $\vec{\mathbb{I}}$ such that $I = K_{\mathcal{N}}(\vec{\mathbb{I}})$.*

5 Related Work and Conclusion

Information Integration. The first group of related works are formalisms where the semantics of a "composite" KB is given by a collection of models of the constituent KBs. In this way, they remain in the open-world setting. Examples of this kind are works on *Distributed DLs* [3,4,14], *Contextualized Knowledge Repositories* [24], and $\mathcal{E}$-connections [17]. The latter allows to integrate theories in different logics, but restricted to modal logics. To ensure decidability, the connections between theories in the above approaches are quite limited compared to the flexible rules in KBNs.

The second group consists of epistemic approaches. KBNs are closest in spirit to the multi-context systems of [21], which defined semantics in terms of maximization of sets of models. This prior work deals only with the propositional setting and does not explore the complexity of reasoning. The first-order setting of our work is more challenging in terms of computability and complexity. Another work that employs epistemic principles similar to ours is the work on peer-to-peer information integration of [6]. In their model, peers exchange knowledge by posing epistemic queries, possibly involving negation as failure. The semantics is defined in terms of a specially tailored epistemic multimodal logic. However, a global knowledge state and stability are not considered.

Donini et al. have introduced DLs with epistemic operators [7]. Such DLs have been advocated as a suitable tool for closed-world reasoning about OWL ontologies [13,15].

Relations to MCSs. Knorr et al. have recently shown how ground hybrid MKNF KBs of [20] can be supported in MCSs of [5] without the need for dedicated hybrid MKNF contexts [16]. In addition, the authors consider MCSs with contexts containing general FOL theories and DL ontologies, where acceptable belief sets are defined as possibly infinite deductively closed sets of formulas. The infinity of acceptable belief sets makes the decidability and complexity of reasoning in such MCSs non-obvious.

An encoding of KBNs into MCSs is also not straightforward. E.g., positive KBNs cannot be readily encoded into MCSs using *monotonic* contexts and negation-free rules. Indeed, the latter MCSs always have grounded equilibria, while a positive KBN may not have a stable model; even if all individual KBs of a KBN are consistent, negation-free bridge rules may still lead to the non-existence of a stable model.

Relations to Hybrid MKNF. The connection between KBNs and hybrid MKNF is strong due to the common roots in ASP. Assume a disjunction-free, DL-safe or ground hybrid MKNF KB $(P, \mathcal{O})$. We assume that only standard atoms of the form $R(\vec{t})$ occur in P.[2] It is not hard to see that $(P, \mathcal{O})$ can be viewed as a singleton KBN $\mathcal{N} = (\mathcal{O}, P, \sigma)$, where σ are the constants of $\mathcal{O}$ and P. In particular, if $\mathcal{O}$ does not use the congruence relation $\approx$, then $(P, \mathcal{O})$ and $\mathcal{N}$ entail the same ground atoms.[3] Together with Theorem 6 this provides an embedding of DL-safe and ground hybrid MKNF KBs into dl-programs, and complements the result in [20] that dl-programs without the "⋒" operator can be embedded into hybrid MKNF. This relationship is somewhat surprising, given the quite different setup of the formalisms (cf. [20]).

The encoding into a KBN fails in case $(P, \mathcal{O})$ is neither DL-safe nor ground, witnessed by the undecidability of general hybrid MKNF KBs. The reason for undecidability can be traced to the requirement that hybrid MKNF interpretations rigidly interpret an infinite set of constants (note, in addition, that this enforces infinite interpretations). In contrast, each KBN explicitly states a finite set σ of constants that have to be interpreted. We believe this is a sensible approach; we expect σ will often be inferred from the KBs and rules of a KBN. In addition, requiring infinite interpretations as in hybrid MKNF means that one cannot accommodate expressive KBs whose semantics allows to enforce finite models (e.g. this applies to $\mathcal{SHOIQ}$, see [19] for a similar observation).

Outlook. We believe that Knowledge Base Networks (KBNs) is a powerful formalism for interlinking open-world first-order KBs using nonmonotonic rules, while still having good decidability properties. We note that KBNs and Theorem 2 can be easily generalized to support disjunctive bridge rules. Our translation from hybrid MKNF to dl-programs can then be lifted to inputs with disjunctive rules (disjunctive dl-programs of [9] can be used as a target language). The main challenge for future work is to find ways to include into KBNs also KBs operating under the closed-world assumption, in addition to the open-world KBs considered here. Finally, it is important to understand how notions of stratification known from ASP can be transferred to KBNs.

---

[2] This was the setting of the initial version of [20]. A so-called *generalized atom*, i.e. a complex formula, can be replaced by a standard atom whenever it is expressible in the considered DL.

[3] If $\approx$ is present in $\mathcal{O}$, then its axiomatization must be added to P to preserve the correspondence. This is a general way to simulate equality under the SNA.

Acknowledgments. This work was supported by the Austrian Science Fund projects P25207 and P24090, and the Vienna Science and Technology Fund project ICT12-15.

References

1. Baader, F.: The Description Logic Handbook: Theory, Implementation, and Applications. Cambridge University Press, New York (2003)
2. Bikakis, A., Antoniou, G.: Defeasible contextual reasoning with arguments in ambient intelligence. IEEE Trans. Knowl. Data Eng. **22**(11), 1492–1506 (2010)
3. Borgida, A., Serafini, L.: Distributed description logics: assimilating information from peer sources. In: Spaccapietra, S., March, S., Aberer, K. (eds.) Journal on Data Semantics I. LNCS, vol. 2800, pp. 153–184. Springer, Heidelberg (2003)
4. Bouquet, P., Giunchiglia, F., van Harmelen, F., Serafini, L., Stuckenschmidt, H.: Contextualizing ontologies. J. Web Sem. **1**(4), 325–343 (2004)
5. Brewka, G., Eiter, T.: Equilibria in heterogeneous nonmonotonic multi-context systems. In: Proceedings of AAAI 2007 (2007)
6. Calvanese, D., De Giacomo, G., Lembo, D., Lenzerini, M., Rosati, R.: Inconsistency tolerance in P2P data integration: an epistemic logic approach. Inf. Syst. **33**(4–5), 360–384 (2008)
7. Donini, F.M., Nardi, D., Rosati, R.: Description logics of minimal knowledge and negation as failure. ACM Trans. Comput. Log. **3**(2), 177–225 (2002)
8. Eiter, T., Faber, W., Fink, M., Woltran, S.: Complexity results for answer set programming with bounded predicate arities and implications. Ann. Math. Artif. Intell. **51**(2–4), 123–165 (2007)
9. Eiter, T., Ianni, G., Krennwallner, T., Schindlauer, R.: Exploiting conjunctive queries in description logic programs. Ann. Math. Artif. Intell. **53**(1–4), 115–152 (2008)
10. Eiter, T., Ianni, G., Lukasiewicz, T., Schindlauer, R., Tompits, H.: Combining answer set programming with description logics for the semantic web. Artif. Intell. **172**(12–13), 1495–1539 (2008)
11. Gelfond, M., Lifschitz, V.: The stable model semantics for logic programming. In: Proceedings of ICLP/SLP 1988, pp. 1070–1080. MIT Press (1988)
12. Giunchiglia, F., Serafini, L.: Multilanguage hierarchical logics or: how we can do without modal logics. Artif. Intell. **65**(1), 29–70 (1994)
13. Grimm, S., Motik, B.: Closed world reasoning in the semantic web through epistemic operators. In: Proceedings of OWLED 2005. CEUR-WS.org (2005)
14. Homola, M.: Semantic Investigations in Distributed Ontologies. Ph.D. thesis, Comenius University, Bratislava, Slovakia, April 2010
15. Knorr, M., Hitzler, P., Maier, F.: Reconciling OWL and non-monotonic rules for the semantic web. In: Proceedings of ECAI 2012. IOS Press (2012)
16. Knorr, M., Slota, M., Leite, J., Homola, M.: What if no hybrid reasoner is available? Hybrid MKNF in multi-context systems. J. Logic Comput. (2014)
17. Kutz, O., Lutz, C., Wolter, F., Zakharyaschev, M.: E-connections of abstract description systems. Artif. Intell. **156**(1), 1–73 (2004)
18. Lifschitz, V.: Minimal belief and negation as failure. Artif. Intell. **70**(1–2), 53–72 (1994)
19. Mehdi, A., Rudolph, S.: Revisiting semantics for epistemic extensions of description logics. In: Proceedings of AAAI 2011 (2011)
20. Motik, B., Rosati, R.: Reconciling description logics and rules. J. ACM **57**(5), 1–62 (2010)

21. Roelofsen, F., Serafini, L.: Minimal and absent information in contexts. In: Proceedings of IJCAI 2005 (2005)
22. Rosati, R.: Reasoning about minimal belief and negation as failure. J. Artif. Intell. Res. (JAIR) **11**, 277–300 (1999)
23. Rosati, R.: Minimal belief and negation as failure in multi-agent systems. Ann. Math. Artif. Intell. **37**(1–2), 5–32 (2003)
24. Serafini, L., Homola, M.: Contextualized knowledge repositories for the semantic web. Web Semant. **12–13**, 64–87 (2012)

ASP, Amalgamation, and the Conceptual Blending Workflow

Manfred Eppe[1,4(✉)], Ewen Maclean[2], Roberto Confalonieri[1],
Oliver Kutz[3], Marco Schorlemmer[1], and Enric Plaza[1]

[1] IIIA-CSIC, Barcelona, Spain
{meppe,confalonieri,marco,enric}@iiia.csic.es
[2] University of Edinburgh, Edinburgh, UK
emaclea2@inf.ed.ac.uk
[3] Free University of Bozen-Bolzano, Bolzano, Italy
oliver.kutz@unibz.it
[4] International Computer Science Institute, Berkeley, USA

Abstract. We present a framework for *conceptual blending* – a concept invention method that is advocated in cognitive science as a fundamental, and uniquely human engine for creative thinking. Herein, we employ the search capabilities of ASP to find commonalities among input concepts as part of the blending process, and we show how our approach fits within a generalised conceptual blending workflow. Specifically, we orchestrate ASP with imperative Python programming, to query external tools for theorem proving and colimit computation. We exemplify our approach with an example of creativity in mathematics.

1 Introduction, Preliminaries and Motivation

Creativity is an inherent human capability, that is crucial for the development and invention of new ideas and concepts [2]. This paper addresses a kind of creativity which [2] calls *combinational*, and which has been studied by Fauconnier and Turner [4] in their framework of *conceptual blending*. In brief, conceptual blending is a process where one combines two input concepts to invent a new one, called the *blend*.

As a classical example of blending, consider the concepts *house* and *boat* (e.g. [4,7]): A possible result is the invention of a *house-boat* concept, where the medium on which a house is situated (land) becomes the medium on which boat is situated (water), and the inhabitant of the house becomes the passenger of the boat. A sub-task of conceptual blending is to find a common ground, called *generic space*, between the input concepts [4]. For example, the *house-boat* blend has the generic space of a person using an object which is not situated on any medium. Once the generic space has been identified, one can develop possible blends by specialising the generic space with elements from the input concepts in a meaningful way. This is not trivial because the naive 'union' of input spaces can lead to inconsistencies. For example, the medium on which an object is situated can not be land and water at the same time. Hence, before combining

© Springer International Publishing Switzerland 2015
F. Calimeri et al. (Eds.): LPNMR 2015, LNAI 9345, pp. 309–316, 2015.
DOI: 10.1007/978-3-319-23264-5_26

the input concepts, it is necessary to generalise, and to remove at least one medium assignment.

Finding the generic space of two concepts is a non-monotonic search problem, and it is well-known that *Answer Set Programming* (ASP) (see e.g. [5]) is a successful tool to cope with such problems. In this paper, we present a computational framework for blending, that addresses the following question: *"How can we use ASP as a non-monotonic search engine to find a generic space of input concepts, and how can we orchestrate this search process with external tools to produce meaningful blends within a computationally feasible system?"* Towards this, we use a mixed declarative-imperative *amalgams* process known from case-based reasoning [14], which coordinates the generalisation and combination of input concepts.

Concept Blending as Colimit of Algebraic Specifications. Goguen [7] proposes to model the input concepts of blending as *algebraic specifications* enriched by priority information about their elements, which he calls *semiotic systems*. This algebraic view on blending suggests to compute the blend of input specifications as their categorical *colimit* – a general unification operation for categories, similar to the *union* operation for sets. In our case the colimit unifies algebraic signatures (see [12,17] for category theoretical details). We represent semiotic systems by using the Common Algebraic Specification Language (CASL) [13]. CASL allows us to state first-order logical specifications, which consists of four kinds of elements, namely *sorts, operators, predicates* and first order logical *axioms*. Operators are functions that map a list of arguments of a certain sort to a range sort, and predicates are functions that map arguments to boolean values. Such a representation language lets us define more than just concepts, namely full first order theories. As an example, consider the following specifications that represent the mathematical theories of natural numbers and lists.

spec NAT =		
sort	Nat	p:3
ops	$zero : Nat;$	p:2
	$s : Nat \rightarrow Nat$	p:3
	$sum : Nat \rightarrow Nat$	p:2
	$qsum : Nat \times Nat \rightarrow Nat$	p:2
	$plus : Nat \times Nat \rightarrow Nat$	p:1
$\forall\ x,\ y : Nat$		
(0)	$.\ sum(zero) = zero$	p:2
(1)	$.\ sum(s(x)) = plus(s(x),\ sum(x))$	p:2
(2)	$.\ qsum(s(x),\ y) =$	p:2
	$\qquad qsum(x,\ plus(s(x),\ y))$	
(3)	$.\ qsum(zero,\ x) = x$	p:2
(4)	$.\ plus(zero,\ x) = x$	p:1
(5)	$.\ plus(s(x),\ y) = s(plus(x,\ y))$	p:1
(NT)	$.\ sum(x) = qsum(x,\ zero)$	p:3
(NL)	$.\ plus(sum(x),\ y) = qsum(x,\ y)$	p:3
end		

spec LIST =		
sorts	El	p:3
	L	p:3
ops	$nil : L;$	p:2
	$cons : El \times L \rightarrow L;$	p:3
	$app : L \times L \rightarrow L;$	p:2
	$rev : L \rightarrow L;$	p:2
	$qrev : L \times L \rightarrow L$	p:2
$\forall\ x,\ y : L;\ h : El$		
(6)	$.\ rev(nil) = nil$	p:2
(7)	$.\ rev(cons(h,\ x)) =$	p:2
	$\qquad app(rev(x),\ cons(h,\ nil))$	
(8)	$.\ qrev(nil,\ x) = x$	p:2
(9)	$.\ qrev(cons(h,\ x),\ y) =$	p:2
	$\qquad qrev(x,\ cons(h,\ y))$	
(10)	$.\ app(nil,\ x) = x$	p:1
(11)	$.\ app(cons(h,\ x),\ y) =$	p:1
	$\qquad cons(h,\ app(x,\ y))$	
(LT)	$.\ rev(x) = qrev(x,\ nil)$	p:3
end		

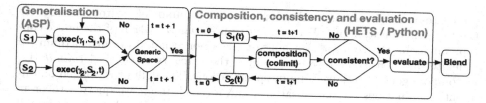

Fig. 1. Amalgamation workflow

For example, in LIST, the operator *cons* maps an object of the sort *El* (element) and an object of the sort *L* (list) to an object of the sort *L*. That is, *cons* constructs lists by appending one element to a list. The *rev* operator is a recursive reverse function on lists, and the *qrev* is a tail-recursive version of the reverse function. Similarly, in NAT, *s* is a successor function, *sum* denotes a recursive function to obtain the cumulative sum of a number (e.g. $sum(3) = 1+2+3 = 6$), and *qsum* is a tail-recursive version of *sum*. We enrich CASL specifications by considering priority information for the individual elements. We denote such specifications as *prioritised CASL specifications* (PCS). For example, the 'p:3' behind the *cons* operator declaration denotes that *cons* has a relatively high priority of 3, and analogously for the other operators, sorts and axioms.

Motivating Example – Discovering Eureka Lemmas by Blending. Of particular interest in the above theories are **(NT)** and **(LT)**. These theorems state that the recursive functions *sum* and *rev* are equivalent to the computationally less expensive tail-recursive quick-functions *qsum* and *qrev*. Proving such theorems by induction is very hard due to the absence of a universally quantified variable in the second argument of the tail-recursive version [9]. An expert's solution here is to use a lemma that generalises the theorem. An example of such a generalisation is the eureka lemma **(NL)** in the naturals, which we assume to be known in this scenario. Discovering such lemmas is a challenging well-known problem (see e.g. [10,11]), and we demonstrate how blending is used to obtain an analogous eureka lemma for lists as an example application.

2   ASP-Driven Blending by Amalgamation

We employ an interleaved declarative-imperative amalgamation process to search for generalisations of input spaces that produce and evaluate logically consistent blends.

System Description. The workflow of our system is depicted in Fig. 1. First, the input PCS $\mathfrak{s}_1$, $\mathfrak{s}_2$ are translated into ASP facts. Then, $\mathfrak{s}_1$, $\mathfrak{s}_2$ are iteratively generalised by an iterative ASP solver until a generic space is found. Each generalisation is represented by a fact $exec(\gamma, \mathfrak{s}, t)$, where t is an iterator and γ is a generalisation operator that, e.g., removes an axiom or renames a sort, as described below. The execution of generalisation operators is repeated until the generalised versions of the input specifications have the same sorts, operators, predicates and axioms, i.e., until a generic space is found. We write $\mathfrak{s}(t)$ to denote the t-th generalisation of $\mathfrak{s}$. For example, a first generalisation of the house concept might be the concept of a house that is not situated on any medium. In order to find consistent blends, we apply the category-theoretical *colimit* [12] to compose generalisations of input specifications. The colimit is applied on different combinations of generalisations, and for each result we query a theorem prover for logical consistency. To eliminate uninteresting blends from our search process, we consider that the more promising blends require less generalisations. Consequently, we go from less general generalisations to more general generalisations and stop when a consistent colimit is achieved. Thereafter, the result is evaluated using certain metrics that are inspired by Fauconnier and Turner [4]'s so-called *optimality principles* of blending to assess the quality of the blend (due to lack of space, we refer to the literature for details on those principles). Note that different stable models, and therefore different generalisations, can be found by the ASP solver, which lead to different blends.

Modelling Algebraic Specifications in ASP. First, we translate PCS to ASP facts, with atoms like $sort(\mathfrak{s}, s, t)$ that denote that s is a sort of the specification $\mathfrak{s}$ at a step t. Operators and predicates are declared similarly. Arguments of operators are defined by atoms $opHasSort(\mathfrak{s}, o, s_i, i, t)$ that denote that an operator o in a specification $\mathfrak{s}$ has the sort s_i as i-th argument. For each element e in a PCS specification $\mathfrak{s}$, we represent its priority v_p as a fact $priority(\mathfrak{s}, e, v_p)$.

Formalising Generalisation Operators in ASP. For the generalisation of PCS, we consider two kinds of generalisation operators. The first kind involves the removal of an element in a specification, denoted by rm predicates, and the second kind involves the renaming of an element, denoted by $rename$ predicates. We represent the execution of a generalisation operator with atoms $exec(\gamma, \mathfrak{s}, t)$, to denote that a generalisation operator γ was applied to $\mathfrak{s}$ at a step t. Each generalisation operator is defined via a precondition rule, and, in case of renaming operations, an effect rule. Preconditions are modelled with a predicate $poss/3$ that states when it is possible to execute a generalisation operation, and effect rules model how a generalisation operator changes an input specification. For example, the preconditions for removing and renaming operators are specified by the following rules:

$$poss(rm(e), \mathfrak{s}, t) \leftarrow op(\mathfrak{s}, e, t), exOtherSpecWithoutElem(\mathfrak{s}, e, t),$$

$$0\{ax(\mathfrak{s}, A, t) : axInvolvesElem(\mathfrak{s}, A, e, t)\}0 \tag{1a}$$

$$poss(rename(e, e', \mathfrak{s}'), \mathfrak{s}, t) \leftarrow op(\mathfrak{s}, e, t), op(\mathfrak{s}', e', t), not\ op(\mathfrak{s}, e', t), not\ op(\mathfrak{s}', e, t), \tag{1b}$$

$$not\ opSortsNotEquivalent(\mathfrak{s}, e, \mathfrak{s}', e', t), \mathfrak{s} \neq \mathfrak{s}'$$

For the removal of elements we have a condition $exOtherSpecWithout$ $Elem(\mathfrak{s}, e, t)$, which denotes that an element can only be removed if it is not involved in another specification. Such preconditions are required to allow only generic spaces that are *least general* for all input specifications, in the sense that elements can not be removed if they are contained in all specifications. We also require operators, predicates and sorts not to be involved in any axiom before they can be removed (denoted by $0\{ax(\mathfrak{s}, A, t) : axInvolvesElem(\mathfrak{s}, A, e, t)\}0$).

We also need rules to state when elements remain in a specification. This is expressed via *noninertial*/3 atoms as follows, where (2c) is an examplary case of operator elements of a specification.

$$noninertial(\mathfrak{s}, e, t) \leftarrow exec(rm(e), \mathfrak{s}, t) \tag{2a}$$

$$noninertial(\mathfrak{s}, e, t) \leftarrow exec(rename(e, e', \mathfrak{s}'), \mathfrak{s}, t) \tag{2b}$$

$$op(\mathfrak{s}, e, t + 1) \leftarrow not\ noninertial(\mathfrak{s}, e, t), op(\mathfrak{s}, e, t) \tag{2c}$$

For renaming, we also have effect rules that assign the new name for the respective element. For example, for renaming operators we have:

$$op(\mathfrak{s}, e', t + 1) \leftarrow exec(rename(e, e', \mathfrak{s}'), \mathfrak{s}, t), op(\mathfrak{s}, e, t) \tag{3}$$

Generalisation Search Process. ASP is employed to find a generic space, and generalised versions of the input specifications which lead to a consistent blend. This is done by successively generating generalisations of the input specifications. A sequence of generalisation operators defines a *generalisation path*, which is generated with the following choice rule:

$$0\{exec(a, \mathfrak{s}, t) : poss(a, \mathfrak{s}, t)\}1 \leftarrow not\ genericReached(t), spec(\mathfrak{s}). \tag{4}$$

Generalisation paths lead from the input specifications to a generic space, which is a generalised specification that describes the commonalities of the input specifications. *genericReached(t)* atoms determine if a generic space has been reached. This is the case if for two specifications $\mathfrak{s}_1$ and $\mathfrak{s}_2$, at step t, (i) sorts are equal, (ii) operator and predicate names are equal, and (iii) argument and range sorts of operators and predicates are equal, and (iv) axioms are equivalent.

Composition and Evaluation. The next step in the amalgamation process depicted in Fig. 1 is to compose generalised versions of input specifications to generate a candidate blend. The key component of this composition process is the categorical colimit [12] of the generalised specifications and the generic space. The colimit is then enriched with the priority information, which we compute as the sum of the priorities of the input elements. The composition is then evaluated according to several factors that reflect the rather informal optimality principles proposed by Fauconnier and Turner [4]. Our formal interpretation of these principles considers logical consistency and the following three evaluation metrics which are based on Fauconnier and Turner [4]'s informal descriptions of certain optimality principles for blending:

(a) We support blends that keep as much as possible from their input concepts by using the priority information of elements in the input concepts. This corresponds to *unpacking*, *web* and *integration* principles. Towards this, we compute the *amount of information* in a blend as the sum of the priorities of all of its elements.

(b) We support blends that maximise common relations among input concepts as a means to compress the structure of the input spaces. Relations are made common by appropriate renamings of elements in the input specification. This corresponds to the *vital relations* principle. Maximising common relations raises the *compression of structure* in a composition, which is computed as the sum of priorities of elements in the composition that have counterparts in both input specifications. For example, consider the predicate *liveIn* : *Person* × *House* of the *House* specification and the predicate *ride* : *Person* × *Boat* of the *Boat* specification. Both are mapped to the same element in the composition, i.e., the predicate *liveIn* : *Person* × *House*. The *liveIn* in the composition uses the same symbol as the one in *House*, but it carries more information because due to the renaming it now also represents the *ride* predicate. We account for this form of compression of information by adding the priority of *liveIn* to the compression value.

(c) We support blends where the amount of information from the input specifications is balanced. This corresponds to the *double-scope* property of blends, which is described by Fauconnier and Turner [4] as '... what we typically find in scientific, artistic, and literary discoveries and inventions.' Towards this, we consider a *balance penalty* of a blend, which we define as the difference between the amount of information from the input specifications as described in a).

Proof of Concept – Lemma Invention for Theorem Proving. To perform the blend of the theories of naturals and lists discussed in Sect. 1, our system first generates a generic space, which is achieved with the following generalisation paths:[1]

$P_{\mathrm{NAT}} = \{ exec(rename(Nat, L, \mathrm{LIST}), \mathrm{NAT}, 0), exec(rename(zero, nil, \mathrm{LIST}), \mathrm{NAT}, 1),$

$exec(rename(C, El, \mathrm{LIST}), \mathrm{NAT}, 2), exec(rename(s, cons, \mathrm{LIST}), \mathrm{NAT}, 3), \cdots$

$exec(rm(1), \mathrm{NAT}, 9), exec(rm(2), \mathrm{NAT}, 10), exec(rm(c), \mathrm{NAT}, 11), exec(rm(\mathbf{NL}), \mathrm{NAT}, 12)\}$

$P_{\mathrm{LIST}} = \{ exec(rm(10), \mathrm{LIST}, 0), \cdots, exec(rm(7), \mathrm{LIST}, 3)\}$

With this generalisation path, the sort L is mapped to the sort Nat, the terminal elements *nil* and *zero* are mapped to each other, the construction operator s is mapped to *cons*, *rev* is mapped to *sum*, *qrev* is mapped to *qsum*, and *app* is mapped to *plus*. Note, that the meaning of the *List*-symbols is now much more general because they map to both, the *List* and the *Nat* theory, and represent now analogies between both theories. After finding the generic space, our framework iterates over different combinations of generalised input specifications and computes the colimit. It then checks the colimits consistency and computes the blend value. In this example, the highest composition value for a consistent colimit is 90, where the 4th generalisation of LIST and the 8th generalisation of NAT is used as input. The result is a theory of lists with the newly invented lemma $app(rev(x), y) = qrev(x, y)$ which can be used successfully as a generalisation lemma to prove (**LT**).

---

[1] Note that for this example, we extend the unary constructor $s(n)$ in the naturals by an additional canonical argument c, so that the constructor becomes binary, i.e., $s(c, n)$. This is valid when considering a classical set theoretic construction of the naturals as the cardinality of a set (see [1] for example), where the theory of the naturals corresponds to a theory of lists of the same element.

3   Conclusion

We present a computational approach for conceptual blending where ASP plays a crucial role in finding the generic space and generalised input specifications. We implement the generalisation of algebraic specifications using a transition system semantics of preconditions and postconditions within ASP, which allows us to access generalised versions of the input specifications. These generalised versions of the input specifications let us find blends which are logically consistent. To the best of our knowledge, there exists currently no other blending framework that can resolve inconsistencies and automatically find a generic space, while using a representation language that is similarly expressive as ours. On top of the ASP-based implementation, we propose metrics to evaluate the quality of blends, based on the cognitive optimality principles by Fauconnier and Turner [4]. A number of researchers in the field of computational creativity have recognised the value of conceptual blending for building creative systems, and particular implementations of this cognitive theory have been proposed [3,6,8,15,16,18]. They are, however, mostly limited in the expressiveness of their representation language, and it is in most cases unclear how they deal with inconsistencies and how the generic space is computed. Furthermore, existing approaches lack a sophisticated evaluation to determine formally how 'good' a blend is. An exception is the very sophisticated framework in [15,16], which also has optimality criteria based on [4]'s theory. However, the authors do not say how to find the generic space automatically and how to deal with inconsistencies.

A prototypical implementation of our system can be accessed at https:// github.com/meppe/Amalgamation. It will be a core part of the bigger computational concept invention framework that is currently being built within the COINVENT project http://www.coinvent-project.eu.

Acknowledgements. This work is supported by the 7th Framework Programme for Research of the European Commission funded COINVENT project (FET-Open grant number: 611553). M. Eppe is supported by the German Academic Exchange Service.

References

1. Anderson, D., Zalta, E.: Frege, boolos and logical objects. J. Philos. Logic **33**, 1–26 (2004)
2. Boden, M.A.: Creativity. In: Boden, M.A. (ed.) Artificial Intelligence (Handbook of Perception and Cognition), pp. 267–291. Academic Press, London (1996)
3. Eppe, M., Confalonieri, R., Maclean, E., Kaliakatsos, M., Cambouropoulos, E., Schorlemmer, M., Kühnberger, K.-U.: Computational invention of cadences and chord progressions by conceptual chord-blending. In: IJCAI (2015, to appear)
4. Fauconnier, G., Turner, M.: The Way We Think: Conceptual Blending and the Mind's Hidden Complexities. Basic Books, New York (2002). ISBN 978-0-465-08785-3
5. Gebser, M., Kaminski, R., Kaufmann, B., Schaub, T.: Answer Set Solving in Practice. Morgan and Claypool, San Rafael (2012)

6. Goguen, J., Harrell, D.F.: Style: a computational and conceptual blending-based approach. In: Argamon, S., Burns, K., Dubnov, S. (eds.) The Structure of Style: Algorithmic Approaches to Understanding Manner and Meaning, pp. 291–316. Springer, Heidelberg (2010). doi:10.1007/978-3-642-12337-5_12. ISBN 978-3-642-12336-8

7. Goguen, J.: An introduction to algebraic semiotics, with application to user interface design. In: Nehaniv, C.L. (ed.) CMAA 1998. LNCS (LNAI), vol. 1562, pp. 242–291. Springer, Heidelberg (1999)

8. Guhe, M., Pease, A., Smaill, A., Martínez, M., Schmidt, M., Gust, H., Kühnberger, K.-U., Krumnack, U.: A computational account of conceptual blending in basic mathematics. Cogn. Syst. Res. 12(3–4), 249–265 (2011). doi:10.1016/j.cogsys.2011.01.004

9. Ireland, A., Bundy, A.: Productive use of failure in inductive proof. J. Autom. Reason. 16(1–2), 79–111 (1996)

10. Johansson, M., Dixon, L., Bundy, A.: Conjecture synthesis for inductive theories. J. Autom. Reason. 47, 251–289 (2011)

11. Montano-Rivas, O., McCasland, R., Dixon, L., Bundy, A.: Scheme-based synthesis of inductive theories. In: Sidorov, G., Hernández Aguirre, A., Reyes García, C.A. (eds.) MICAI 2010, Part I. LNCS, vol. 6437, pp. 348–361. Springer, Heidelberg (2010)

12. Mossakowski, T.: Colimits of order-sorted specifications. In: Parisi-Presicce, F. (ed.) WADT 1997. LNCS, vol. 1376, pp. 316–332. Springer, Heidelberg (1998)

13. Mosses, P.D. (ed.): CASL Reference Manual: The Complete Documentation of the Common Algebraic Specification Language. LNCS, vol. 2960. Springer, Heidelberg (2004)

14. Ontañón, S., Plaza, E.: Amalgams: a formal approach for combining multiple case solutions. In: Bichindaritz, I., Montani, S. (eds.) ICCBR 2010. LNCS, vol. 6176, pp. 257–271. Springer, Heidelberg (2010)

15. Pereira, F.C.: A computational model of creativity. PhD thesis, Universidade de Coimbra (2005)

16. Pereira, F.C.: Creativity and Artificial Intelligence: A Conceptual Blending Approach. Mouton de Gruyter, Berlin (2007)

17. Pierce, B.: Basic Category Theory for Computer Scientists. MIT Press, Cambridge (1991). ISBN 0262660717

18. Veale, T., Donoghue, D.O.: Computation and blending. Cogn. Linguist. 11(3–4), 253–282 (2000). doi:10.1515/cogl.2001.016

Diagnostic Reasoning for Robotics Using Action Languages

Esra Erdem[1]([✉]), Volkan Patoglu[1], and Zeynep Gozen Saribatur[2]

[1] Sabanci University, Istanbul, Turkey
esraerdem@sabanciuniv.edu
[2] Vienna University of Technology, Vienna, Austria

Abstract. We introduce a novel diagnostic reasoning method for robotic systems with multiple robots, to find the causes of observed discrepancies relevant for plan execution. Our method proposes (i) a systematic modification of the robotic action domain description by utilizing defaults, and (ii) algorithms to compute a smallest set of diagnoses (e.g., broken robots) by means of hypothetical reasoning over the modified formalism. The proposed method is applied over various robotic scenarios in cognitive factories.

Keywords: Diagnostic reasoning · Action languages · Answer set programming

1 Introduction

For the fault-awareness and reliability of robotic systems (e.g., in cognitive factories or service robotics), it is essential that the robots have cognitive skills, such as planning their own actions, identifying discrepancies between the expected states and the observed states during plan execution, checking whether these discrepancies would lead to a plan failure, diagnosing possible causes of relevant discrepancies, learning from earlier diagnoses, and finding new plans to reach their goals. In this paper, we focus on diagnostic reasoning for robotics.

Consider, for instance, a cognitive factory [11,32]. In a typical cognitive factory, each workspace is medium sized with 3–12 heterogeneous robots; there may be many workspaces, each focusing on a different task. On the other hand, each of these robots has many components, failure of which may cause abnormalities in the manufacturing process. Factory shut-downs for diagnosis and repairs are very costly; hence, it is required that diagnosis is performed accurately and fast. Furthermore, it is essential that the monitoring agent(s) have the capability of identifying further details on diagnosis (broken robot components, actions that could not be executed due to broken robots/components), and learning from earlier diagnoses and failures.

Z.G. Saribatur's work was carried out during her graduate studies at Sabancı University.

With these motivations, we introduce a diagnostic reasoning method to find a smallest set of broken robots (and their components) that cause the relevant discrepancies observed during plan execution. Our method starts with the robotic action domain description used to compute a plan, the part of the plan executed from the initial state until the current state, and a set of observations about the current state. The robotic domain description is represented in an action description language [16], like $\mathcal{C}+$ [18]. Then, our method (i) applies a systematic modification of the robotic action domain description by utilizing defaults, and (ii) computes a smallest set of diagnoses by means of hypothetical reasoning over the modified formalism.

The modification of the domain description is essential to be able to generate possible causes of the observed discrepancies that would lead to plan failures. The use of defaults help generation of possible causes for discrepancies (such as broken robots), as well as generation of further information (such as which actions are prevented from execution due to these causes). We have proven that our proposed modification conservatively extends the planning domain description, which is important for generating such further information.

The computation of smallest sets of diagnoses is done by means of "diagnostic queries" expressed in an action query language, like $\mathcal{Q}$ [16], and by using efficient automated reasoners, such as SAT solvers and ASP solvers. Two algorithms are proposed for this computation: one of them relies more on hypothetical reasoning and can be used in conjunction with both sorts of reasoners; the other algorithm relies on aggregates and optimization statements of answer set programming (ASP) [3, 22–24, 27].

Both algorithms synergistically integrate diagnostic reasoning with learning from earlier experiences and probabilistic geometric reasoning. As demonstrated by our experiments, learning improves the computational efficiency and quality of diagnoses, since it allows robots to utilize their previous experiences (e.g., which robots or robot components are more probable to be broken). As shown by examples, geometric reasoning improves accuracy of diagnoses by considering feasibility of robotic actions.

2   Preliminaries

We model dynamic domains like cognitive factories as transition diagrams – directed graphs where the nodes characterize world states and the edges represent transitions between states caused by (non)occurrences of actions. We represent transition diagrams formally in the nonmonotonic logic-based action description language $\mathcal{C}+$ [18].

Various sorts of reasoning tasks can be performed over transition diagrams, such as planning and prediction. We describe reasoning tasks by means of formulas in an action query language $\mathcal{Q}$ [16], and utilize SAT/ASP solvers to compute a solution.

This logic-based reasoning framework allows integration of low-level feasibility checks (e.g., collision checks for robots) performed externally by the state-of-the-art geometric reasoners. In this section, we will briefly describe these preliminaries over a cognitive toy factory scenario.

2.1 A Cognitive Toy Factory

As a running example, we consider the cognitive toy factory workspace of [12], where a team of multiple robots collectively works toward completion of an assigned manufacturing task. In particular, the team manufactures nutcracker toy soldiers through the sequential stages of cutting, carving and assembling. The workspace contains an assembly line to move the toys and a pit stop area where the worker robots can change their end-effectors. It also includes static obstacles.

The team is heterogeneous, composed of two types of robots with different capabilities. Worker robots operate on toys, they can configure themselves for different stages of processes; charger robots maintain the batteries of workers and monitor team's plan. All robots can move from any grid cell to another one following straight paths.

2.2 Representation of a Robotic Domain

The signature of an action domain description in $\mathcal{C}+$ consists of two sorts of multi-valued propositional constants, fluent constants $\mathcal{F}$ and action constants $\mathcal{A}$, along with a nonempty set $Dom(c)$ of possible values for each constant c. Atoms are of the form $c = v$ where c is a constant and v is a value in $Dom(c)$. If c is a boolean constant then we adopt the notations c and $\neg c$. We assume that action constants are boolean.

Transition diagrams modeling dynamic domains can be described by "causal laws" over such a signature. For instance, the following causal law describes a direct effect of a robot r moving in the $Right$ direction by u units from a location x on the x-axis: the location of the robot becomes $x + u$ at the next state after the execution of this action.

$$move(r, Right, u) \textbf{ causes } xpos(r) = x + u \textbf{ if } xpos(r) = x. \tag{1}$$

The causal law below describes a precondition of this action: a worker robot w cannot move u units in any direction d since its battery lasts for only $bl < u$ units of movement:

$$\textbf{nonexecutable } move(w, d, u) \textbf{ if } battery(w) = bl \quad (bl < u). \tag{2}$$

Similarly, causal laws can represent [18] ramifications of actions, noconcurrency constraints, state/transition constraints, the commonsense law of inertia.

We understand an action domain description as a finite set of definite causal laws.

2.3 Reasoning About a Robotic Domain

A *query* in $\mathcal{Q}$ [16] is a propositional combination of atomic queries of the two forms, F **holds at** t or A **occurs at** t, where F is a fluent formula, A is an action formula, and t is a time step.

The meaning of a query is defined in terms of histories. A *history* of an action domain description $\mathcal{D}$ is an alternating sequence $s_0, A_0, s_1, \ldots, s_{n-1}, A_{n-1}, s_n$ ($n \geq 0$) of states and actions, denoting a path in the transition diagram described by $\mathcal{D}$. States (resp. actions) can be considered as functions from fluent constants (resp. action constants) to their relevant domains of values. Then each state (resp. action) can be denoted by a set of fluent (resp. action) atoms.

A query Q of the form F **holds at** t (resp. A **occurs at** t) is *satisfied* by such a history if s_t satisfies F (resp. if A_t satisfies A). For non-atomic queries, satisfaction is defined by truth tables of propositional logic. A query Q is *satisfied* by an action description $\mathcal{D}$, if there is a history of $\mathcal{D}$ that satisfies Q.

Let F and G be fluent formulas representing an initial state and goal conditions. We can describe the problem of finding a plan of length k, with a query of the form F **holds at** $0 \wedge G$ **holds at** k. Similarly, we can describe the problem of predicting the resulting state after an execution of an action sequence $A_0, \ldots, A_{n-1}$ at a state described by a fluent formula F, with a query of the form

$$F \text{ holds at } 0 \wedge \wedge_i A_i \text{ occurs at i.} \tag{3}$$

It is shown [18] that an action description in $\mathcal{C}+$ can be transformed into a propositional theory and into an ASP program. Based on these sound and complete transformations, the software systems CCALC [18,25] and CPLUS2ASP [5] turn an action domain and a given query into the input languages of a SAT solver and an ASP solver.

2.4 Hybrid Robotic Planning

Geometric reasoning, such as motion planning and collision checks, can be integrated into an action description in $\mathcal{C}+$ by means of "external atoms" [9] (in the spirit of semantic attachments in theorem proving [31]). The idea is to compute the truth values of external atoms externally, and utilize these results, as needed, while computing plans.

For instance, consider an external atom $\& collision[r, x1, y1, x2, y2]$ that returns true if a robot r collides with some static obstacles while moving from $(x1, y1)$ to $(x2, y2)$. This atom can be evaluated by a program implemented in C++ utilizing a motion planner. With this atom, we can describe a precondition of a diagonal move action ("the robot can move diagonally if it does not collide with any static obstacles"):

nonexecutable $move(w, Right, u1) \wedge move(w, Up, u2)$
   if $xpos(w) = x1 \wedge ypos(w) = y1$ **where** $\& collision[w, x1, y1, x1+u1, y1+u2]$.

Further explanations and examples of the use of external atoms in robotic action domain descriptions, and a systematic analysis of various forms of integration of feasibility checks with planning can be found in [4,10,13].

3   Diagnostic Reasoning

To predict failures as early as possible and to recover from failures, values of some fluents can be monitored by an agent. If during the execution of a plan, a discrepancy is detected between the observed values of monitored fluents and the expected values of monitored fluents, then the monitoring agent can check whether the detected discrepancy might lead to any failures during the execution of the rest of the plan. If the discrepancy is relevant for the rest of the plan, then the monitoring agent has to make some decisions (e.g., replanning) to reach the goals. We propose the use of diagnostic reasoning to identify the cause of a discrepancy (e.g., robots may be broken), and then find a relevant recovery with the possibility of repairs.

3.1   Discrepancies

Let $\mathcal{D}$ be a domain description. Let $P = \langle A_0, A_1, \ldots, A_{n-1} \rangle$ be a plan that is being executed from an initial state s_0. We denote by P_t the sequence $A_0, A_1, \ldots, A_{t-1}$ of actions in the plan P executed until step t. Let o_t be an observed state of the world at step t, where the observed values of fluents can be obtained by sensors. The expected state e_t of the world at step t can be computed by a prediction query of the form (3) with the initial state s_0 and the sequence P_t of actions.

We say that there is a *discrepancy* at step t if the observed state and the expected state are different, $o_t \neq e_t$. A discrepancy is *relevant* to the rest of the plan if the rest of the plan can not be executed from the observed state or does not reach a goal state. This definition coincides with weakly k-irrelevant discrepancies when $k = 1$ [8] and with the definition of relevancy as in [17].

3.2   Diagnosis: Identifying Broken Robots

In a cognitive factory setting, a diagnosis for a discrepancy can be identified by a set of broken robots. To be able to find such a diagnosis, we first modify the action domain description to be able to perform diagnostic reasoning. Then, we define diagnosis and diagnostic queries to compute diagnoses over the modified domain description.

Modifying the Domain Description for Diagnosis. Let R denote the set of robots in a cognitive factory, that may get broken. Let *disables* : $2^R \times \mathcal{A} \times \mathcal{F}$ be a relation to describe which actions are affected and how, if a set of robots were broken: *disables*(X, a, F) expresses that if the set X of robots is broken, then the effect of action $a \in \mathcal{A}$ performed by some robots in X on a fluent F is disabled. Note that *disables* can be obtained automatically from the causal laws that describe effects of actions.

To find a diagnosis for the observed relevant discrepancies, we modify the domain description $\mathcal{D}$ in three stages as follows, and obtain a new domain description $\mathcal{D}_b$.

Step 1: To indicate whether a robot $r \in R$ is broken or not, we introduce a simple fluent constant of the form $broken(r)$.

By default, the robots in R are assumed to be not broken. We express this for every robot $r \in R$, with a causal law:

$$\textbf{default } \neg broken(r). \tag{4}$$

Meanwhile, every robot r may get broken at any time:

$$\textbf{caused } broken(r) \textbf{ if } broken(r) \textbf{ after } \neg broken(r) \tag{5}$$

and if a robot r becomes broken it remains broken:

$$\textbf{caused } broken(r) \textbf{ after } broken(r). \tag{6}$$

For every $r \in R$, we add the causal laws (4)–(6) to the domain description $\mathcal{D}$.

It is important to emphasize the usefulness of the nonmonotonicity of $\mathcal{C}+$, which allows us to represent defaults (4) as well as nondeterminism (5), and thus does not necessitate introduction of non-robotic actions like $break$.

Step 2: If a robot is broken, then it may affect preconditions and effects of the actions in the executed plan. It is good to know which actions could not be executed due to which sort of broken robots, so that we can learn/infer some new knowledge that might be used later for more accurate diagnoses and more effective repairs.

For that, for each (concurrent) action $A = \{a_0, ..., a_n\}$, we introduce a simple boolean fluent constant $pre(\{a_0, ..., a_n\})$ to express whether or not its preconditions hold; brackets are dropped when $|A| = 1$. The values of these fluents are considered as true by default; so we add to $\mathcal{D}$ the causal laws:

$$\textbf{default } pre(\{a_0, ..., a_n\}). \tag{7}$$

We then describe under which conditions an action's preconditions are violated, by replacing every causal law

$$\textbf{nonexecutable } \bigwedge_{a_i \in A} a_i \textbf{ if } G$$

in $\mathcal{D}$, where G is a fluent formula, with the causal law

$$\textbf{caused } \neg pre(\{a_0, ..., a_n\}) \textbf{ if } G. \tag{8}$$

For instance, the causal laws (2) are replaced by the causal laws:

$$\textbf{caused } \neg pre(move(w, d, u)) \textbf{ if } battery(w) = bl \quad (bl < u).$$

We then modify the causal laws describing the effects of actions, to express that they can be observed under the additional condition that the preconditions of these actions are satisfied. For that, we replace every causal law of the form

$$a \textbf{ causes } F \textbf{ if } G \tag{9}$$

in $\mathcal{D}$, where F and G are fluent formulas, with the dynamic causal law

$$a \textbf{ causes } F \textbf{ if } G \wedge pre(a). \tag{10}$$

For instance, the causal laws (1) describing some effects of the move action are replaced by the following:

$$move(r, Right, u) \textbf{ causes } xpos(r) = x + u \textbf{ if } xpos(r) = x \wedge pre(move(r, Right, u)).$$

<u>Step 3:</u> For every action a, let a_R denote the robots in R that take part in the execution of a. We reflect the influence of broken robots on direct effects of actions by replacing every causal law (9) in $\mathcal{D}$ such that $disables(a_R, a, F)$ holds, with the causal law

$$a \textbf{ causes } F \textbf{ if } G \wedge \bigwedge_{r \in a_R} \neg broken(r), \tag{11}$$

which expresses that the direct effects of actions are observed as expected only if the relevant robots are not broken.

In our cognitive factory scenario, a charger robot cannot dock to a worker robot w if the worker robot is broken, i.e., $disables(\{w\}, attach(w), attached(w))$ (the direct effect of $attach(w)$ action on the fluent $attached(w)$ is not observed if the worker robot w is broken). Then, the causal law

$$attach(w) \textbf{ causes } attached(w) \textbf{ if } pre(attach(w))$$

obtained after Step 2 is replaced by the causal law

$$attach(w) \textbf{ causes } attached(w) \textbf{ if } pre(attach(w)) \wedge \neg broken(w).$$

The following proposition shows that if a query is satisfied by $\mathcal{D}$ then it is also satisfied by $\mathcal{D}_b$; but not vice-versa. In that sense, $\mathcal{D}_b$ "conservatively extends" $\mathcal{D}$. This is particularly important to be able to reason about executions of a plan (earlier computed with $\mathcal{D}$) with respect to $\mathcal{D}_b$ to find diagnoses.

Proposition 1. *Every query satisfied by $\mathcal{D}$ is satisfied by $\mathcal{D}_b$.*

The proof of the proposition requires the following lemmas. Suppose that $\mathcal{D}$ is defined over a set $\mathcal{F}$ of fluent constants and a set $\mathcal{A}$ of boolean action constants. Let $\mathcal{M}$ be the set of all fluent constants of the forms $broken(r)$ and $pre(A)$ that are introduced to the signature of $\mathcal{D}$, to transform $\mathcal{D}$ into $\mathcal{D}_b$. Then, every state s' of $\mathcal{D}_b$ is a function mapping every constant c in $\mathcal{F} \cup \mathcal{M}$ to a value in $Dom(c)$.

Lemma 1. *For every state s of $\mathcal{D}$, there are states s_b of $\mathcal{D}_b$ such that (i) $s_b|_{\mathcal{F}} = s$, (ii) s_b maps every fluent constant $broken(r)$ to false, and (iii) for every causal law (8), s_b satisfies $G \supset \neg pre(\{a_0, \ldots, a_n\})$.*

Lemma 2. *For every transition $\langle s, A, s' \rangle$ of $\mathcal{D}$, there are transitions $\langle s_b, A, s'_b \rangle$ of $\mathcal{D}_b$ such that (i) $s_b|_{\mathcal{F}} = s$, (ii) $s'_b|_{\mathcal{F}} = s'$, (iii) both s_b and s'_b map every fluent constant $broken(r)$ to false, and (iv) s_b maps every fluent constant $pre(Z)$ ($Z \subseteq A$) to true.*

Diagnosis. Let us characterize a state s by the conjunction F_s of atoms in s; and an action A by the conjunction E_A of atoms in A.

Intuitively, a diagnosis for a discrepancy detected at time step t of a plan P is a set of robots that, when broken, provides a possible execution of the plan from the initial state to the observed state (i.e., they do not lead to an "inconsistency"). Formally:

Definition 1. *A* diagnosis problem, *DP, is characterized by a tuple* $\langle \mathcal{D}_b, R, s_0,$ $P_t, o_t \rangle$ *where* t *is the time step when a discrepancy is detected,* $P_t = \langle A_0, \ldots,$ $A_{t-1} \rangle$ *is the sequence of actions assumed to be executed at an initial state* s_0 *until time step* t, *and* o_t *is the observed state at time step* t. *A solution of DP is a set* $X \subseteq R$ *of broken robots such that* $\mathcal{D}_b$ *satisfies the query*

$$F_{s_0} \textbf{ holds at } 0 \wedge \bigwedge_{A_i \in P_t} E_{A_i} \textbf{occurs at } i \wedge F_{o_t} \textbf{holds at } t \wedge$$

$$\bigwedge_{r \in X} broken(r) \textbf{ holds at } t \wedge \bigwedge_{r \in R-X} \neg broken(r) \textbf{ holds at } t. \tag{12}$$

We also say that X *is a* diagnosis *of the discrepancy detected at time* t.

Note that further information about which actions' preconditions are violated can be obtained from the history of $\mathcal{D}_b$ that satisfies the query, since atoms of the form $\neg pre(A)$ are part of the state information.

The query (12) checks if the observed state o_t can be reached from the initial state s_0 by only executing the plan P_t, if the robots in X were broken. Note that, by the definition of a discrepancy, we know that (12) will not be satisfied by $\mathcal{D}_b$ if $X = \emptyset$, since the observed state is different from the state that is expected to be reached from s_0 by P_t.

Generally, there will be more than one diagnosis for a discrepancy according to the definition above. On the other hand, in practice, a discrepancy is caused by a small set of broken robots. Therefore, it is reasonable to find as few broken instances as possible to explain the discrepancy.

Definition 2. *A* diagnosis *X is a* minimum-cardinality *diagnosis if there does not exist any other diagnosis X' such that* $|X'| < |X|$.

3.3   Finding a Minimum-Cardinality Diagnosis

We introduce two algorithms to find minimum-cardinality diagnoses by means of diagnostic queries as described above, using SAT solvers or ASP solvers.

In the first algorithm, to find a minimum-cardinality diagnosis for a diagnosis problem $DP = \langle \mathcal{D}_b, R, s_0, P_t, o_t^M \rangle$, for each $X \subseteq R$ with increasing cardinalities, we iteratively check whether X is a solution for DP, i.e., whether $\mathcal{D}_b$ satisfies (12). We can utilize this algorithm with both SAT solvers and ASP solvers, thanks to the sound and complete transformations and software systems mentioned in the preliminaries.

The second algorithm is applicable with ASP solvers only, thanks to the possibility of representing aggregates and optimization statements. For instance, the following optimization statement minimizes the total number of broken robots:

$$\#minimize \; [broken(r) : robot(r)]. \tag{13}$$

4   Geometric Reasoning for Diagnosis

Embedding geometric reasoning in the domain description $\mathcal{D}$ (and thus $\mathcal{D}_b$) by means of external atoms, allows us to integrate diagnostic reasoning with geometric reasoning. Let us show the importance of such an integration over an example, as in [12]. Consider the execution of some part of a plan from a given state as in Fig. 2, and the observed state of the world at step 3 as in Fig. 1. There is a discrepancy at step 3. To find a diagnosis for this discrepancy, we ask diagnostic queries (12). To check whether the charger is broken, we ask "is it possible to reach from the initial state to the observed state by executing the plan, if the charger were broken?"

Without consideration of geometric reasoning, the answer to this question is "no" since the goal is not reachable when the plan is executed as shown in Fig. 3. However, such an execution is not feasible in the real world because the worker robot cannot move diagonally over the charger robot at Step 1 (due to collisions). On the other hand, with consideration of geometric reasoning, the answer to the diagnostic query is "yes" since the goal is reachable when the plan is executed as shown in Fig. 4. Note that, due to collisions, the preconditions of the action of the worker robot moving diagonally is not possible, i.e., $\neg pre(\{move(w, Up, 2), move(w, Right, 2)\})$ holds. Therefore, the detected discrepancy is correctly diagnosed with hybrid diagnostic reasoning and false negatives are avoided.

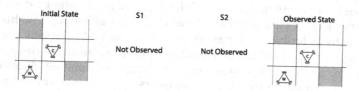

Fig. 1. Observed state of the world at step 3.

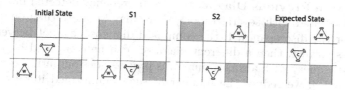

Fig. 2. Expected execution of a plan until step 3.

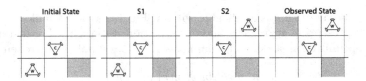

Fig. 3. Diagnostic reasoning without geometric reasoning

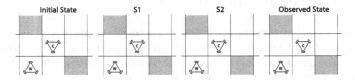

Fig. 4. Diagnostic reasoning with geometric reasoning

5   Extensions of Our Method

Further Diagnoses: Broken Robotic Parts. Usually robots are broken if some of their components are broken. Moreover, each broken component may play a role on the preconditions/effects of different actions. To extend diagnostic reasoning to robotic components, we modify the first and third steps of our procedure to obtain $\mathcal{D}_b$ from $\mathcal{D}$, as well as the algorithms to compute smallest diagnoses. The definition of a diagnosis remains the same.

For these modifications, we define R_p as the set of pairs (r, i) where $r \in R$ and i is a component of r that may be broken. For instance, for a cognitive toy factory scenario with one worker robot w and with one charger robot c, $R_p = \{(w, Arm), (w, Inlet), (c, Plug), (w, Base), (c, Base)\}$. We extend the *disables* relation: $disables_p : 2^{R_p} \times \mathcal{A} \times \mathcal{F}$. For instance, $disables_p(\{(w, Inlet)\}, charge(c), battery(w))$ expresses that, if the inlet of the worker robot is broken then the effect of charging will not be observed as expected on its battery level. In the representation $\mathcal{D}_b$, (i) we introduce new fluent constants of the form $broken_i(r)$ for every $(r, i) \in R_p$ and add relevant causal laws for them, and (ii) we modify the causal laws (11) according to $disables_p$. The algorithms are modified to ensure that the computed smallest diagnosis contains minimum number of broken components. Details are omitted for space restrictions.

Learning from Previous Diagnoses. If a discrepancy detected in a plan execution is associated with several robots, there may be more than one potential diagnosis for the discrepancy. If the computed diagnosis cannot be verified (by an agent) as correct, then a different diagnosis can be computed by modifying the diagnostic query with a constraint. This process goes on until a correct diagnosis. As in [12], we reduce the number of such uninformed iterations due to incorrect diagnoses, by utilizing learning from earlier correct diagnoses of the discrepancies. The idea is to maintain and update information about the likelihood of robots'/components' failures according to computed correct diagnoses;

and if there is a robot/component which has a history of being broken more often, then to consider it as part of a potential diagnosis before the others.

To describe how often a robot r (resp. a component i of a robot r) is diagnosed correctly as broken, we introduce atoms of the form $weight(r, w)$ (resp. $weight(r, i, w)$) where w is a number, which can store the number of times that r (resp. i of r) was correctly diagnosed to be broken, or the probability of r (resp. i of r) being diagnosed correctly so far. To find the most probable diagnosis, in the first algorithm proposed for finding a smallest diagnosis, we consider subsets of robots/components with larger weights before the others. In the second algorithm, we add a new optimization statement to the ASP program, which tries to maximize the total weight of broken components. This statement is added after the minimization statement (13), to maximize the total weight of broken components among the minimum-cardinality ones.

Extending the experimental evaluations of [12], we performed several experiments over various cognitive toy factory scenarios to show the usefulness of learning in the first algorithm for finding a minimum-cardinality diagnosis by using the SAT solver MiniSAT (Version 2.0) [7], and in the second algorithm by using the ASP solver CLASP (Version 2.1.3) [14]. Experiments are based on dynamic simulations, where kinematic and geometric constraints of robots are considered. For geometric reasoning, we use probabilistic motion planners and collision checkers available in open-source frameworks, like OPENRAVE [6]. In each scenario, we first performed the experiments without learning. Based on the computed diagnoses, we assigned the weights for robots/components, and then performed the experiments with learning. For each instance, the CPU time in seconds, and the number of iterations to compute the correct diagnosis with correct set of broken components are illustrated in Table 1 (which extends the results in Table 1 of [12]). The results are obtained on a Linux server with 16 Intel E5-2665 CPU cores (2.4 GHz) and 64 GB memory.

As in [12], we observe (1) in both algorithms, the number of iterations to find a correct diagnosis significantly decreases as learning is utilized; (2) as the size of the team and the cardinality of the diagnosis increase, the computation time to find a correct diagnosis increases. In addition to the observations of [12]: (3) The number of iterations in ASP is less than the number of iterations with SAT solver, since the optimization statement of ASP is effective in decreasing number of iterations to find a correct diagnosis.

Behavioral Modes. A further extension of our approach can be by behavioral modes [20]. Since our approach is based on representing actions and change in the environment (unlike the related work), behaviors of the robotic system are already modeled. We suppose each robot has three modes of behavior: normal mode (functions as expected), broken mode (does not function at all), or unknown mode. The broken modes are depicted by *disables* relation; the behaviors are described by causal laws. In the normal mode, the behavior (i.e., expected effects of relevant actions) is described by laws like (1). In the broken mode, the behavior (i.e., by default nothing changes) is described by the commonsense law of inertia. In the unknown mode, we assume by default nothing changes.

Table 1. Experimental evaluation of learning

Scenario	Number of iterations and CPU time [s]			
	ASP		SAT	
	w/o learning	w/ learning	w/o learning	w/ learning
1 charger, 1 wet, 2 dry	2	1	9	4
Discrepancy at Step 12, Diagnosis cardinality = 1	6.51 s	3.21 s	52.24 s	26.49 s
1 charger, 1 wet, 2 dry	2	1	21	11
Discrepancy at Step 8, Diagnosis cardinality = 2	4.14 s	2.06 s	67.43 s	39.85 s
1 charger, 1 wet, 2 dry	2	1	25	23
Discrepancy at Step 10, Diagnosis cardinality = 3	4.28 s	2.14 s	85.88 s	84.16 s
1 charger, 2 wet, 2 dry	2	1	7	4
Discrepancy at Step 8, Diagnosis cardinality = 1	7.66 s	3.85 s	83.70 s	42.97 s
1 charger, 2 wet, 2 dry	2	1	24	12
Discrepancy at Step 12, Diagnosis cardinality = 2	9.10 s	4.56 s	142.18 s	80.35 s
1 charger, 2 wet, 2 dry	2	1	39	25
Discrepancy at Step 13, Diagnosis cardinality = 3	9.86 s	4.93 s	197.01 s	127.25 s
2 charger, 2 wet, 2 dry	2	1	8	4
Discrepancy at Step 6, Diagnosis cardinality = 1	6.27 s	3.13 s	97.00 s	49.30 s
2 charger, 2 wet, 2 dry	2	1	31	13
Discrepancy at Step 14, Diagnosis cardinality = 2	13.29 s	6.63 s	195.17 s	102.48 s
2 charger, 2 wet, 2 dry	2	1	49	30
Discrepancy at Step 12, Diagnosis cardinality = 3	11.57 s	5.78 s	240.90 s	154.85 s

Considering failures of robotic components leads to more number of behavioral modes. The broken modes are specified by $disables_p$ relation, and behaviors are described by causal laws. Therefore, associating diagnoses explicitly with such behavioral modes may be possible; it is left as a future work.

6   Related Work

Our work on diagnosis is similar to conflict-based model-based diagnosis [21,28]: the diagnostic query checks whether broken robots would lead to an inconsistency with respect to the observations or not. It is also different in several ways. First, we consider dynamic domains with actions and change, rather than a static system like circuits. Second, our logical framework is nonmonotonic and thus the definition of a diagnosis is different from the existing definitions. Third, motivated by robotics applications, our approach to diagnostic reasoning is integrated with geometric reasoning and learning.

Later, the model-based diagnosis approach of [21,28] is extended to dynamic domains, using the logic-based action languages, such as situation calculus [19,26] and fluent calculus [30], and utilizing planning [29]. Our work is different from these approaches not only because of differences between the underlying formalisms but also due to diagnosis definition (faulty actions vs. components), assumptions (faulty components/actions may indirectly prevent execution of the rest of the plan, or not; our method is applicable in either case), methods (introducing abnormality predicates and utilizing nonmonotonicity by minimizing its extension, vs. expressing defaults; introducing further transformations to domain description beyond abnormality predicates).

A more closely related line of research is diagnostic reasoning in answer set programming and action languages [1,2,8,11,15,33] due to the common underlying formalisms. Our approach is different from these works in several ways. Eiter et al. [8] define a diagnosis as a "point of failure" which describes at which state and when the plan diverges from its expected evolution; we consider a diagnosis as a set of broken robots/components. Thanks to the modified domain description $\mathcal{D}$, we can identify which action has failed and when, as part of an answer to the diagnostic query. Balduccini and Gelfond [1], Baral et al. [2], and Gelfond and Kahl [15] (like [26]) define a diagnosis as a set of broken components; they introduce a "break" action to define what might be "broken". In $\mathcal{C}+$, no such non-robotic action is necessary; the causal laws (5) are sufficient. Gelfond and Kahl also suggest using consistency-restoring rules [1] to find minimal diagnoses. Erdem et al. [11] define a diagnosis as a set of broken robots, but their method does not generalize to scenarios where broken robots prevent the execution of an action in the rest of the plan and does not provide further information about broken components or failed actions. Zhang et al. [33] generate explanations for discrepancies in terms of exceptions to the defaults that hold at the initial state.

Our approach to diagnostic reasoning can be extended with repair planning by introducing "repair" actions into the robotic domain description $\mathcal{D}$ as in [1,2]. Such repairs can reduce the number of replanning needed to recover from plan failures. Extending our method with repairs is not discussed due to page limit.

7 Conclusion

We have introduced a diagnostic reasoning method which utilizes expressive formalisms of action languages and answer set programming, and efficient automated reasoners (SAT solvers and ASP solvers). This method integrates geometric reasoning (for feasibility checks of robotic actions), to eliminate false negatives and improve accuracy of diagnosis. It utilizes learning from earlier diagnoses and failures to improve the computation time required to find a correct diagnosis. Furthermore, causality-based hybrid planning/prediction is utilized for finding minimum-cardinality diagnoses.

In an accompanying work [12], we integrate our approach to diagnostic reasoning into a plan execution and monitoring framework, and perform further analysis to show the usefulness of diagnostic reasoning and repairs for replanning. We use the ASP formulation of the robotic domain, an ASP-based modification of the planning domain description for diagnosis, and the ASP solvers for hybrid planning and diagnosis. In that sense, the presented work complements the results in [12] by describing the formulations and transformations in action description languages, by performing reasoning tasks in action query languages with the possibility of using SAT solvers, by extending the experimental evaluations to SAT solvers, and by providing some theoretical guarantees over the proposed transformation of the robotic domain description.

Acknowledgements. Thanks to anonymous reviewers for useful comments. This work is partially supported by TUBITAK Grants 111E116, 113M422 and 114E491 (ChistEra COACHES).

References

1. Balduccini, M., Gelfond, M.: Diagnostic reasoning with A-Prolog. Theory Pract. Logic Program. **3**(4–5), 425–461 (2003)
2. Baral, C., McIlraith, S., Son, T.C.: Formulating diagnostic problem solving using an action language with narratives and sensing. In: Proceedings of KR (2000)
3. Brewka, G., Eiter, T., Truszczynski, M.: Answer set programming at a glance. Commun. ACM **54**(12), 92–103 (2011)
4. Caldiran, O., Haspalamutgil, K., Ok, A., Palaz, C., Erdem, E., Patoglu, V.: Bridging the gap between high-level reasoning and low-level control. In: Erdem, E., Lin, F., Schaub, T. (eds.) LPNMR 2009. LNCS, vol. 5753, pp. 342–354. Springer, Heidelberg (2009)
5. Casolary, M., Lee, J.: Representing the language of the causal calculator in answer set programming. In: Proceedings of ICLP (Technical Communications) (2011)
6. Diankov, R.: Automated construction of robotic manipulation programs. Ph.D. thesis, Carnegie Mellon University, Robotics Institute, August 2010
7. Eén, N., Sörensson, N.: An extensible SAT-solver. In: Giunchiglia, E., Tacchella, A. (eds.) SAT 2003. LNCS, vol. 2919, pp. 502–518. Springer, Heidelberg (2004)
8. Eiter, T., Erdem, E., Faber, W., Senko, J.: A logic-based approach to finding explanations for discrepancies in optimistic plan execution. Fundamenta Informaticae **79**, 25–69 (2007)
9. Eiter, T., Ianni, G., Schindlauer, R., Tompits, H.: A uniform integration of higher-orderreasoning and external evaluations in answer-set programming. In: Proceedings of IJCAI (2005)
10. Erdem, E., Haspalamutgil, K., Palaz, C., Patoglu, V., Uras, T.: Combining high-level causal reasoning with low-level geometric reasoning and motion planning for robotic manipulation. In: Proceedings of ICRA (2011)
11. Erdem, E., Haspalamutgil, K., Patoglu, V., Uras, T.: Causality-based planning and diagnostic reasoning for cognitive factories. In: Proceedings of ETFA (2012)
12. Erdem, E., Patoglu, V., Saribatur, Z.G.: Integrating hybrid diagnostic reasoning in plan execution monitoring for cognitive factories with multiple robots. In: Proceedings of ICRA (2015)
13. Erdem, E., Patoglu, V., Schüller, P.: A systematic analysis of levels of integration between high-level task planning and low-level feasibility checks. In: Proceedings of RCRA (2014)
14. Gebser, M., Kaufmann, B., Neumann, A., Schaub, T.: *clasp*: A conflict-driven answer set solver. In: Baral, C., Brewka, G., Schlipf, J. (eds.) LPNMR 2007. LNCS (LNAI), vol. 4483, pp. 260–265. Springer, Heidelberg (2007)
15. Gelfond, M., Kahl, Y.: Knowledge Representation, Reasoning, and the Design of Intelligent Agents: The Answer-Set Programming Approach. Cambridge University Press, New York (2014)
16. Gelfond, M., Lifschitz, V.: Action languages. ETAI **2**, 193–210 (1998)
17. Giacomo, G.D., Reiter, R., Soutchanski, M.: Execution monitoring of high-level robot programs. In: Proceedings of KR (1998)
18. Giunchiglia, E., Lee, J., Lifschitz, V., McCain, N., Turner, H.: Nonmonotonic causal theories. Artif. Intell. **153**, 49–104 (2004)

19. Iwan, G.: Explaining what went wrong in dynamic domains. In: Proceedings of CogRob (2000)
20. de Kleer, J., Williams, B.C.: Diagnosis with behavioral modes. In: Proceedings of IJCAI (1989)
21. Kleer, J.D., Mackworth, A.K., Reiter, R.: Characterizing diagnoses and systems. Artif. Intell. 56(2), 197–222 (1992)
22. Lifschitz, V.: Answer set programming and plan generation. Artif. Intell. 138, 39–54 (2002)
23. Lifschitz, V.: What is answer set programming? In: Proceedings of AAAI (2008)
24. Marek, V., Truszczyński, M.: Stable models and an alternative logic programming paradigm. In: Apt, K.R., et al. (eds.) The Logic Programming Paradigm: A 25-Year Perspective, pp. 375–398. Springer, Heidelberg (1999)
25. McCain, N.C.: Causality in commonsense reasoning about actions. Ph.D. thesis (1997)
26. McIlraith, S.A.: Explanatory diagnosis: conjecturing actions to explain observations. In: Proceedings of KR (1998)
27. Niemelä, I.: Logic programs with stable model semantics as a constraint programming paradigm. Ann. Math. Artif. Intell. 25, 241–273 (1999)
28. Reiter, R.: A theory of diagnosis from first principles (1987)
29. Sohrabi, S., Baier, J.A., McIlraith, S.A.: Diagnosis as planning revisited. In: Proceedings of KR (2010)
30. Thielscher, M.: A theory of dynamic diagnosis. ETAI 2(11) (1997)
31. Weyhrauch, R.W.: Prolegomena to a theory of formal reasoning. Technical report. Stanford University (1978)
32. Zaeh, M., Beetz, M., Shea, K., Reinhart, G., Bender, K., Lau, C., Ostgathe, M., Vogl, W., Wiesbeck, M., Engelhard, M., Ertelt, C., Rhr, T., Friedrich, M., Herle, S.: The cognitive factory. In: ElMaraghy, H.A. (ed.) Changeable and Reconfigurable Manufacturing Systems, pp. 355–371. Springer, London (2009)
33. Zhang, S., Sridharan, M., Gelfond, M., Wyatt, J.: Towards an architecture for knowledge representation and reasoning in robotics. In: Beetz, M., Johnston, B., Williams, M.-A. (eds.) ICSR 2014. LNCS, vol. 8755, pp. 400–410. Springer, Heidelberg (2014)

OOASP: Connecting Object-Oriented and Logic Programming

Andreas Falkner[1], Anna Ryabokon[2(✉)], Gottfried Schenner[1], and Kostyantyn Shchekotykhin[2]

[1] Siemens AG Österreich, Vienna, Austria
{andreas.a.falkner,gottfried.schenner}@siemens.com
[2] Alpen-Adria-Universität Klagenfurt, Klagenfurt, Austria
{anna.ryabokon,kostyantyn.shchekotykhin}@aau.at

Abstract. Most of contemporary software systems are implemented using an object-oriented approach. Modeling phases – during which software engineers analyze requirements to the future system using some modeling language – are an important part of the development process, since modeling errors are often hard to recognize and correct.

In this paper we present a framework which allows the integration of Answer Set Programming into the object-oriented software development process. OOASP supports reasoning about object-oriented software models and their instantiations. Preliminary results of the OOASP application in CSL Studio, which is a Siemens internal modeling environment for product configurators, show that it can be used as a lightweight approach to verify, create and transform instantiations of object models at runtime and to support the software development process during design and testing.

Keywords: Object-oriented modeling · Answer set programming · Product configuration · Software systems

1 Introduction

Object-oriented programming languages is de facto a standard approach to software development. Many systems are modeled and implemented using it. In practice of Siemens the object-oriented approach is also used in many domains among which development of product configurators is one of the prominent examples. A configurator is a software system that enables design of complex technical systems or services based on a predefined set of components. In modern configuration systems domain knowledge - comprising configuration requirements (product variability) and customer requirements - is expressed in terms of component types and relations between them. Each type is characterized by a set of attributes which specify functional and technical properties of real-world and

This research was funded by the Austrian Research Promotion Agency (grant number 840242) and Carinthian Science Fund (grant number KWF-3520/26767/38701).

F. Calimeri et al. (Eds.): LPNMR 2015, LNAI 9345, pp. 332–345, 2015.
DOI: 10.1007/978-3-319-23264-5_28

abstract components of a configurable product. An attribute takes values from a predefined domain. Furthermore, components are related/connected to each other in various ways.

Development of object-oriented configurators is a challenging task due to several important issues such as acquisition of configuration knowledge from domain experts, modeling of this knowledge, model verification and maintenance. Different types of errors might occur, for example, due to the complexity of configuration models or procedural approach of object-oriented languages. Moreover, a variety of problems arises when configurable products or services have a long life-span and requirements are not stable, but change over time - for instance, if some components of a product are not produced any more or if a new functionality has to be added to a system. Some typical challenges occurring when a configuration is changed are discussed in [6]. Configuration technologies which address these tasks enable efficient production processes and thus can help reduce the overall production costs.

Logic programming frameworks, such as Answer Set Programming (ASP), can improve the speed and quality of object-oriented development. These frameworks provide expressive and easily understandable knowledge representation language allowing declarative encodings of complex problems. Equipped with powerful solving algorithms the logic programming frameworks showed their applicability in both product configuration as well as software development domains. For instance, important practical and theoretical aspects of formalizing real-world (re)configuration scenarios using a logic-based formalism are discussed in [8]. The authors of [5] show how to support testing object-oriented and constraint-based configurators by automatically generating positive and negative test cases using ASP. A commercial ASP-based software for verification which makes the development of software easier and faster is suggested in [14].

In this paper we present an OOASP framework that uses a generic object-oriented configurator to encode its knowledge base and ASP for the computation of configurations. OOASP was implemented as an evaluation prototype for an extension to CSL Studio, an authoring environment for Configuration Specification Language (CSL) [3]. It aims at the improvement of the software development process during design and testing. We illustrate the mapping from an object-oriented formalism (UML) to logical descriptions using a simplified real-world example from Siemens. Additionally, the paper provides different insights on (re)configuration tasks such as validation, completion and reconciliation of a configuration which can be accomplished by our system.

The remainder of this paper is organized as follows. After a short ASP overview in Sect. 2, we describe in Sect. 3 how object-oriented knowledge bases can be specified using ASP within OOASP framework. In Sect. 4 we introduce CSL Studio and discuss various product (re)configuration scenarios. Finally, in Sect. 5 we conclude and discuss the future work.

2   Preliminaries

Answer set programming (ASP) is an approach to declarative problem solving which has its roots in logic programming and deductive databases. It is a decidable fragment of first-order logic interpreted under stable model semantics [11] and extended with default negation, aggregation, and optimization [15]. ASP allows modeling of a variety of (combinatorial) search and optimization problems in a declarative way using model-based problem specification methodology (see e.g. [1,4] for details).

An ASP program Π is a finite set of *normal rules* of the form:

$$h :\text{-} b_1, \ldots, b_m, \text{not } b_{m+1}, \ldots, \text{not } b_n. \tag{1}$$

where 'not' denotes *default negation*, b_i $(0 \leq i)$ and h are atoms. An *atom* is an expression of the form $p(t)$, where p is a predicate and t is a vector of terms, i.e. constants, variables or uninterpreted function symbols [9]. Extensions of ASP [15] allow specific forms of atoms. Thus, a *cardinality constraint* is an atom of the form $l\{h_1, \ldots, h_k\}u$, where $h_1, \ldots, h_k$ are atoms and l, u are non-negative integers. A *literal* is either an atom a or its negation not a. In rule (1) the set of atoms $H(r) = \{h\}$ is called *head*, whereas the sets $B(r)^+ = \{b_1, \ldots, b_m\}$ and $B(r)^- = \{b_{m+1}, \ldots, b_n\}$ are *positive* body and *negative* body, respectively. A *fact* is a rule r with $B(r)^+ \cup B(r)^- = \emptyset$; an *integrity constraint* is a rule r with $H(r) = \emptyset$; and a *choice* rule has a cardinality constraint as the head h. A literal, rule or program is *ground*, if it is variable-free. A non-ground program Π can be grounded by substituting variables with constants appearing in Π.

Semantics of a ground normal program Π is defined in terms of Gelfond-Lifschitz reduct. Let $A(\Pi)$ be a set of atoms appearing in Π, then $I \subseteq A(\Pi)$ is an *interpretation*. A Gelfond-Lifschitz reduct [11] of a program Π wrt. an interpretation I is defined as $\Pi^I = \{H(r) \leftarrow B^+(r) \mid r \in \Pi, I \cap B^-(r) = \emptyset\}$. An interpretation I is an *answer set* of Π, if I is a minimal model of Π^I. The semantics of a ground program Π with cardinality constraint atoms is defined similarly, since each rule with such atoms can be translated into a set of normal rules [15]. Informally, semantics of a cardinality constraint requires at least l and at most u atoms h_i to be in an answer set.

Moreover, ASP allows finding of preferred answer sets. The preferences are defined by *weak constraints* – a specific type of integrity constraints that can be violated. Each violation is penalized by a weight associated with a constraint. Given a program with weak constraints an ASP solver returns an answer set minimizing the sum of penalties.

3 OOASP Framework

The development of a object-oriented software is a complex and error-prone activity that requires careful modeling of an underlying problem. Siemens experience in the development of industrial applications shows that quite often incorrect models are responsible for faults in software artifacts that are hard to identify and debug. In this section we present the OOASP approach which allows to analyze object-oriented software models and their instances by means of ASP. In particular, we consider those models that can be described by a modeling language corresponding to a UML class diagram [13]. The latter is a language allowing a software developer to specify an object model and additional constraints that each valid instantiation of an object model must satisfy.

In order to reason about a software model, OOASP framework uses a *meta-programming* approach [17] which was successfully applied in a similar way, for instance, to debugging of ASP programs [10,12]. In our meta-programming approach an ASP program over a meta-language manipulates an ASP program describing a software model in terms of the Domain Description Language (DDL). In case of OOASP, all concepts of one or multiple software models as well as their instantiations are represented in OOASP-DDL as a set of rules of the form (1). Then, a meta-program, designed to accomplish a specific reasoning task, is applied to a program in OOASP-DDL. In a standard implementation of OOASP we provide meta-programs accomplishing the following tasks[1]:

Validation. Given an OOASP-DDL program describing an object-oriented model and its instantiation, a validation meta-program verifies whether all integrity and domain-specific constraints hold. The integrity constraints encode model requirements to relations between objects of an instantiation and are derived from the given model automatically. The domain-specific constraints ensure that some specific requirements to an instantiation of a model are satisfied. They can either be directly specified in the meta-program or imported from other languages. For instance, one could import domain-specific constraints defined in Object Constraint Language[2] (OCL), for which transformations to SAT [16] and constraints programming [2] exist.

Completion. Given an OOASP-DDL program describing an object-oriented model and its (partial) instantiation, the completion task is to find an extension of the instantiation that satisfies all constraints or to show that such extension does not exist. The latter may occur due to two main reasons: (i) the object-oriented model or the given (partial) instantiation are inconsistent and do not have a completion – an empty instantiation can be seen as a special case for the completion; and (ii) the extension of the given instantiation requires the creation of a number of objects that exceeds the given upper bounds for object instances.

Reconciliation. Given an OOASP-DDL program describing a legacy instantiation of an outdated object-oriented model, a new up-to-date model and a set of transformation rules, the goal of the reconciliation is to find a possibly preferred set of changes required to transform the legacy instantiation to a valid instantiation of the new model. The preferences in OOASP can be defined with domain-specific costs that assess the costs of required changes such as creation, reuse or disposal (deletion) of object instances.

If advanced features such as multiple inheritance, symmetry breaking, etc., are required, the default ASP encodings of reasoning tasks, outlined in this paper, must be replaced with alternative encodings, whereas the OOASP-DDL program remains the same.

---

[1] OOASP code and encodings are available upon request from the first author.
[2] OCL specification is available from http://www.omg.org/spec/OCL/2.4/PDF/.

Table 1. OOASP-DDL definitions for the encoding of models

$ooasp\_class(Id_M, Id_C)$	a class C is defined in a model M
$ooasp\_subclass(Id_M, Id_C, Id_{SC})$	defines a subclass relation between a class C and a super class CS in a model M
$ooasp\_assoc(Id_M, Id_A, Id_{C_1}, Min_{C_1},$ $Max_{C_1}, Id_{C_2}, Min_{C_2}, Max_{C_2})$	defines an association relation A between classes C_1 and C_2 with the given cardinalities, e.g. for every instance of the class C_1 at least Min_{C_2} and at most Max_{C_2} instances of the class C_2 must be associated
$ooasp\_attribute(Id_M, Id_C, Id_{AT},$ $\{$ "string", "integer", "boolean"$\})$	an attribute AT of a class C is defined to have one of the three possible types
$ooasp\_attribute\_minInclusive(Id_M,$ $Id_C, Id_{AT}, MinV)$	provides an optional minimum value $MinV$ for an integer attribute AT
$ooasp\_attribute\_maxInclusive(Id_M,$ $Id_C, Id_{AT}, MaxV)$	provides an optional maximum $MaxV$ for an integer attribute AT
$ooasp\_attribute\_enum(Id_M,$ $Id_C, Id_{AT}, Val)$	defines a possible value Val for a string attribute AT

Table 2. OOASP-DDL definitions for the encoding of instantiations

$ooasp\_instantiation(Id_M, Id_I)$	defines an instantiation I of a model M
$ooasp\_isa(Id_I, Id_C, Id_O)$	declares that an object O is an instance of the class C
$ooasp\_associated(Id_I, Id_A,$ $Id_{O_1}, Id_{O_2})$	objects O_1 and O_2 are associated by the association relation A
$ooasp\_attribute\_value(Id_I, Id_{AT},$ $Id_O, Val)$	assigns a value Val to an attribute AT of an object O

3.1 OOASP Domain Description Language

OOASP-DDL allows a software developer to define all standard concepts of object-oriented models such as classes, attributes and associations. Each concept of the model is translated to a corresponding OOASP-DDL atom, where each term Id_* is an identifier of a model, class, attribute, etc. In OOASP identifiers of models are globally unique, whereas all other identifiers are unique within a model. In the current version OOASP-DDL supports the definitions presented in Table 1. These definitions are sufficient to describe a subset of the object-oriented model of programming languages such as C++, Java, etc. Many features that can additionally be found in object-oriented models, e.g. initial values, constants, multi-valued attributes, ordered associations, etc., are currently not supported by the framework. This is because our main purpose was to provide a lightweight approach that, however, is able to capture most of the features commonly used in practice. The definition of an instantiation of an object-oriented model is done using OOASP-DDL in a similar way as the definition of the model. In particular, our language allows the definitions shown in Table 2.

Note that, OOASP-DDL is designed in a way to allow the definition of multiple models and their instantiation in one ASP program. This provides the necessary support for reconciliation and similar reasoning tasks that are applied to many models and/or their instantiations at once.

3.2   Definition of Constraints

Constraints allow a software developer to ensure that models and their instantiations are valid. In OOASP we support two types of constraints: integrity constraints and domain-specific constraints. The latter are used to verify some specific properties of a model and/or its instantiations. The definition of domain-specific constraints can be done by a developer directly in OOASP-DDL or by importing them from the input model, e.g. OCL constraints from a UML model. The integrity constraints, however, are included in the default OOASP implementation and capture the requirements of the input object-oriented model such as cardinality restrictions, typing, etc. For instance, in order to ensure that a minimal cardinality requirement of an association relation holds in a given instantiation, OOASP framework comprises the following rule[3]:

```
1  ooasp_cv(I,mincardviolated(O1,A)) :-
2     {ooasp_associated(I,A,O1,O2): ooasp_isa(I,C2,O2)} C2MIN-1, C2MIN>0,
3     ooasp_assoc(M,A,C1,C1MIN,C1MAX,C2,C2MIN,C2MAX),
4     ooasp_instantiation(M,I),
5     ooasp_isa(I,C1,O1).
```

The presence of an atom over *ooasp_cv* predicate in an answer set of an OOASP program indicates that a corresponding integrity constraint is violated by the given instantiation. In the sample rule above, the error atom is derived whenever less objects of type C_2 are associated with object O_1 than required by the cardinality restriction of the association.

4   System Description

OOASP was implemented as a potential extension to any object-oriented modeling environment and its practicability was evaluated together with CSL Studio [3]. The latter is a Siemens internal tool for the design of product configurators as Generative Constraints Satisfaction Problems (GCSPs) [7,18]. CSL (Configuration Specification Language) is a formal modeling language based on a standard object-oriented meta-model similar to Ecore[4] or MOF[5]. It provides all state-of-the-art features such as packages, interfaces, enumerations, classes

---

[3] In our examples we use the gringo [9] dialect of ASP that also allows usage of uninterpreted function symbols such as *mincardviolated*.

[4] Eclipse Modeling Framework https://www.eclipse.org/modeling/emf/.

[5] MetaObject Facility http://www.omg.org/mof/.

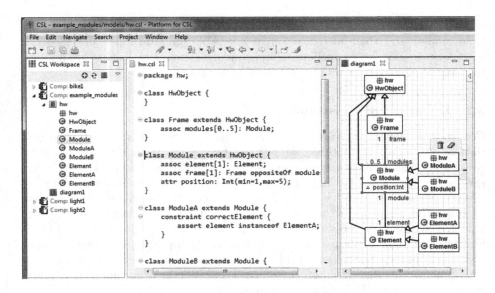

Fig. 1. CSL screenshot for the modules example

with attributes of various types, associations between classes, inheritance and aggregation relations. In addition, it offers reasoning methods such as rules and constraints which are not covered in this work. The reason is that they are not (yet) translated into OOASP domain-specific constraints.

A screenshot of CSL Studio, presented in Fig. 1, shows an example of a simple hardware configuration problem. A configuration problem corresponds to a composition activity in which a desired configurable product is assembled by relating individual components of predefined types. The components and relations between them are usually subject to constraints expressing their possible combinations allowed by the system's design. The types of the components, relations between them as well as additional constraints on sets of related components constitute *configuration requirements*. Many of those constraints can be expressed in an object-oriented model as cardinalities of association and aggregation relations.

The sample model shown in Fig. 1 describes a product configuration problem as a UML class diagram. In this problem the hardware product consists of a number of *Frames*. Each frame contains up to five *Modules* of types *ModuleA* or *ModuleB*, where each module occupies exactly one of the 5 positions in a frame. Moreover, each module has exactly one *Element* assigned to it. All elements are of one of two types *ElementA* or *ElementB*. The corresponding OOASP-DDL encoding for this example is automatically generated by CSL Studio. A part of the encoding excluding integrity constraints is shown in Listing 1.

Additionally to the integrity constraints, implied by the cardinalities of associations shown on the UML diagram, there are the following domain-specific constraints:

- Elements of type *ElementA* require a module of type *ModuleA*
- Elements of type *ElementB* require a module of type *ModuleB*
- Modules must occupy different positions in a frame

These constraints can easily be implemented in OOASP. For instance, the first
and the third can be formulated as shown in Listing 2.

```
 1  % modules example kb "v1"
 2  % classes
 3  ooasp_class("v1","HwObject").
 4  ooasp_class("v1","Frame").
 5  ooasp_class("v1","Module").
 6  ooasp_class("v1","ModuleA"). ooasp_class("v1","ModuleB").
 7  ooasp_class("v1","Element").
 8  ooasp_class("v1","ElementA"). ooasp_class("v1","ElementB").

 9  % class inheritance
10  ooasp_subclass("v1","Frame","HwObject").
11  ooasp_subclass("v1","Module","HwObject").
12  ooasp_subclass("v1","Element","HwObject").
13  ooasp_subclass("v1","ElementA","Element").
14  ooasp_subclass("v1","ElementB","Element").
15  ooasp_subclass("v1","ModuleA","Module").
16  ooasp_subclass("v1","ModuleB","Module").

17  % attributes and associations
18  % class Frame
19  ooasp_assoc("v1","Frame_modules","Frame",1,1,"Module",0,5).

20  % class Module
21  ooasp_attribute("v1","Module","position","integer").
22  ooasp_attribute_minInclusive("v1","Module","position",1).
23  ooasp_attribute_maxInclusive("v1","Module","position",5).

24  % class Element
25  ooasp_assoc("v1","Element_module","Element",1,1,"Module",1,1).
```

Listing 1. OOASP-DDL encoding of the Modules example shown in Fig. 1

A typical workflow of the product configurator development process in CSL
Studio and OOASP is depicted in Fig. 2. The development starts with a creation
of an initial configuration model in CSL. Then, the model can be exported to
OOASP and extended by the definition of domain-specific constraints. Finally,
the consistency of the developed model can be verified by execution of different reasoning tasks. For instance, the existence of model instantiations can be
checked by running a completion task with an empty instantiation. The validation task can be used to test whether some of the known valid product configurations are instantiations of the model. Moreover, OOASP can be used during the

implementation phase. Thus, CSL Studio allows a software developer to export a created model to a preferred object-oriented language as a set of classes. These generated classes must then be extended with the implementation of domain-specific constraints as well as additional methods and fields required for correct functionality of the software. In order to ensure that the software is implemented correctly, the software developer can export a (partial) instantiation generated by an object-oriented program to OOASP. In this case the completion reasoning task allows to test whether the obtained partial solution can be extended to a complete one, e.g. by creating missing modules for the elements as well as by adding missing frames and assigning the modules to them. In addition, if the software developer (tester) manipulates a completed configuration, for instance, by adding or removing elements, the configurator can restore consistency through reconciliation. The latter finds a set of changes that keep as much of the existing structure of the configured system as possible. In the following subsections we describe some use cases exemplifying OOASP applications during the development of configurators.

```
1  ooasp_cv(I,module_element_violated(M1,E1)) :-
2      ooasp_instantiation(M,I),
3      ooasp_associated(I,"Element_module",M1,E1),
4      ooasp_isa(I,"ElementA",E1),
5      not ooasp_isa(I,"ModuleA",M1).

6  ooasp_cv(I,alldiffviolated(M1,M2,F)) :-
7      ooasp_instantiation(M,I),
8      ooasp_isa(I,"Module",M1),
9      ooasp_isa(I,"Module",M2),
10     ooasp_attribute_value(I,"position",M1,P),
11     ooasp_attribute_value(I,"position",M2,P),
12     ooasp_associated(I,"Frame_modules",F,M1),
13     ooasp_associated(I,"Frame_modules",F,M2),
14     M1 != M2.
```

Listing 2. Sample domain-specific constraints in OOASP

4.1 Validation of a Configuration

The implementation of an object-oriented software requires continuous testing in order to identify and resolve faults early. The validation reasoning task provided by OOASP allows a software developer to verify whether an instantiation generated by the object-oriented code is consistent. Especially, the validation is important in the context of CSL Studio or similar systems while testing domain-specific constraints. Thus, in CSL Studio an instantiation of the object model provided by the software developer is automatically exported to OOASP and the validation meta-program is executed. The obtained answer set is then used to highlight the parts of the instantiation that violate requirements to a valid

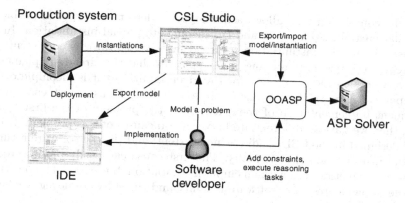

Fig. 2. Integration of OOASP in development of product configurators

configuration. Using this information, the developer can identify the faults in the software in a shorter period of time.

For instance, assume a software developer implemented a model designed in CSL Studio and the resulting program outputs an instantiation c2 comprising only one element of type ElementA. CSL Studio forwards this instantiation to OOASP which translates it to the OOASP-DDL program:

ooasp_isa("c2","ElementA",10).

For this input, execution of the validation task returns an answer set comprising:

ooasp_cv("c2",mincardviolated(10,"Element_module"))

This atom indicates that cardinality restrictions of the association between Element and Module classes are violated. The reason is that for the object with identifier 10 there is no corresponding object of the Module type.

Note that in the current OOASP prototype domain-specific constraints must be coded by a software developer manually and are not generated from the CSL (constraint language). However, this behavior was found to be advantageous in practice, since it provides a mechanism for the diverse redundancy [5]. The latter refers to the engineering principle that suggests application of two or more systems. These systems are built using different algorithms, design methodology, etc., to perform the same task. The main benefit of the diverse redundancy is that it allows software developers to find hidden faults caused by design flaws which are usually hard to detect. Generally, we found that software developers are able to formulate domain-specific constraints in OOASP after a short training. However, existence of ASP development environments supporting debugging and testing of ASP programs would greatly simplify this process.

4.2   Completion of an Instantiation

The completion task is often applied in situations when a software developer needs to generate a test case for a software that outputs an invalid instantiation.

Thus, the completion task allows a developer to detect two types of problems: (i) invalid partial instantiation and (ii) incomplete partial instantiation. In the last case, the partial instantiation returned by a configurator is consistent, but some required objects are missing. This indicates that the already implemented software works correctly, at least for the given input, but it is incomplete. The developer can export the obtained solution and use it as a test case during subsequent implementation of the software. If the problem of the first type is found, then we have to differentiate between two causes of this problem: (a) the model designed in the CSL Studio is inconsistent; and (b) the software returned a partial instantiation that is faulty. The first cause can easily be detected by running a completion task with an empty instantiation. If the model is consistent, then the implemented part software is faulty and the software developer has to correct it.

In order to execute the completion task the CSL Studio exports an instantiation obtained by an object-oriented system to OOASP-DDL. Then, this instantiation together with a corresponding meta-program is provided to an ASP solver. The returned answer sets are visualized by the system to the software developer. If needed, the developer can export the found complete instantiation to an instantiation of the object-oriented system. This translation is straight-forward due to the one-to-one correspondence between instances on the OOASP-level and the object-oriented system.

Consider an example in which a partially implemented configuration system returns an instantiation containing three instances of **ElementA** and two instances of **ElementB**.

```
1  % Partial configuration
2  ooasp_instantiation("v1","c1").
3  ooasp_isa("c3","ElementA",10). ooasp_isa("c3","ElementA",11).
4  ooasp_isa("c3","ElementA",12).
5  ooasp_isa("c3","ElementB",13). ooasp_isa("c3","ElementB",14).
```

In this case the completion task returns a solution visualized in Fig. 3. This solution comprises the existing objects with identifiers 10–14 as well as the new objects corresponding to a frame with object identifier 30 and five modules 20–24.

4.3   Reconciliation of an Inconsistent Instantiation

The reconciliation task deals with restoring consistency of an inconsistent (partial) instantiation given as an input. The problem arises in three scenarios: (1) the validation task finds an instantiation inconsistent; (2) the completion task detects that a model is consistent, but the given partial instantiation cannot be extended; and (3) the model is changed due to new requirements to a configurable product. In order to restore the consistency of an instantiation the reconciliation task comprises two meta-programs. One meta-program converts the input OOASP-DDL program into a reified form. This program comprises rules of the form:

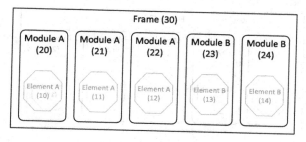

Fig. 3. Complete instantiation for the Modules example. The objects existing in the input instantiation are shown in gray.

$$fact(ooasp(t)) :\text{-} ooasp(t).$$

where $ooasp(t)$ stands for one of the OOASP-DDL atoms listed in Table 2. The second meta-program takes the output of the first one as an input and outputs a consistent instantiation as well as a set of changes applied to obtain it. The set of changes is obtained by the application of deletion/reuse rules of the form:

$$1\{reuse(ooasp(t)), delete(ooasp(t))\}1 :\text{-} fact(ooasp(t)).$$

$$ooasp(t) :\text{-} reuse(ooasp(t)).$$

A preferred solution can be found if a developer provides costs for reuse/delete actions performed by the reconciliation task.

For example, suppose that the developer created a configuration system that does not implement a domain-specific constraint preventing overheating of the system. Namely, this constraint avoids overheating by disallowing putting two modules of type ModuleA next to each other.

```
1  % do not put 2 modules of type ModuleA next to each other
2  ooasp_cv(IID,moduleANextToOther(M1,M2,P1,P2)):-
3    ooasp_instantiation("v2",IID),
4    ooasp_associated(IID,"Frame_modules",F,M1),
5    ooasp_associated(IID,"Frame_modules",F,M2),
6    ooasp_attribute_value(IID,"position",M1,P1),
7    ooasp_attribute_value(IID,"position",M2,P2),
8    M1!=M2,
9    ooasp_isa(IID,"ModuleA",M1),
10   ooasp_isa(IID,"ModuleA",M2),
11   P2=P1+1.
```

Due to the added constraint, the instantiation in Fig. 3 is no longer valid. The reconciliation task finds a required change by modifying the positions of modules with identifiers 21 and 24. The result of the reconciliation can be presented to a developer by OOASP framework as shown in Fig. 4.

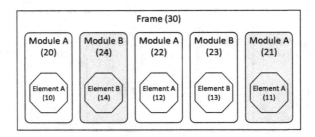

Fig. 4. Reconciled configuration for the Modules example

5   Conclusions

This paper demonstrates OOASP which integrates ASP into the object-oriented software development process using an industrial product configurator as an evaluation example. Our preliminary results are very encouraging and open a number of new directions for a tighter integration of object-oriented programming and ASP. Thus, our experiments with OOASP showed that checking constraints with respect to a given object-oriented model can be done efficiently by modern ASP solvers. However, execution of the reconciliation task still remains a challenge for large-scale instantiations [8]. It appears that the main obstacle for the approach based on ASP meta-programming is the explosion of grounding. In addition, the completion of large-scale instantiations indicated that a computation time for a solution can be improved by the application of domain-specific heuristics. The latter are often hard to implement for software developers, since they do not have enough experience in ASP. In our future work we are going to investigate these questions in more details.

References

1. Brewka, G., Eiter, T., Truszczynski, M.: Answer set programming at a glance. Commun. ACM **54**(12), 92–103 (2011)
2. Cabot, J., Clariso, R., Riera, D.: Verification of UML/OCL class diagrams using constraint programming. In: ICST Workshop, pp. 73–80 (2008)
3. Dhungana, D., Falkner, A.A., Haselböck, A.: Generation of conjoint domain models for system-of-systems. In: GPCE, pp. 159–168 (2013)
4. Eiter, T., Ianni, G., Krennwallner, T.: Answer set programming: a primer. In: Tessaris, S., Franconi, E., Eiter, T., Gutierrez, C., Handschuh, S., Rousset, M.-C., Schmidt, R.A. (eds.) Reasoning Web. LNCS, vol. 5689, pp. 40–110. Springer, Heidelberg (2009)
5. Falkner, A., Schenner, G., Friedrich, G., Ryabokon, A.: Testing object-oriented configurators with ASP. In: ECAI Workshop on Configuration, pp. 21–26 (2012)
6. Falkner, A., Haselböck, A.: Challenges of knowledge evolution in practice. AI Commun. **26**, 3–14 (2013)
7. Fleischanderl, G., Friedrich, G., Haselböck, A., Schreiner, H., Stumptner, M.: Configuring large systems using generative constraint satisfaction. IEEE Intell. Syst. **13**(4), 59–68 (1998)

8. Friedrich, G., Ryabokon, A., Falkner, A.A., Haselböck, A., Schenner, G., Schreiner, H.: (Re) configuration based on model generation. In: LoCoCo Workshop, pp. 26–35 (2011)
9. Gebser, M., Kaminski, R., König, A., Schaub, T.: Advances in *gringo* series 3. In: Delgrande, J.P., Faber, W. (eds.) LPNMR 2011. LNCS, vol. 6645, pp. 345–351. Springer, Heidelberg (2011)
10. Gebser, M., Pührer, J., Schaub, T., Tompits, H.: A meta-programming technique for debugging answer-set programs. In: AAAI, pp. 448–453 (2008)
11. Gelfond, M., Lifschitz, V.: The stable model semantics for logic programming. In: Logic Programming, pp. 1070–1080 (1988)
12. Oetsch, J., Pührer, J., Tompits, H.: Catching the Ouroboros: on debugging non-ground answer-set programs. TPLP 10(4–6), 513–529 (2010)
13. Rumbaugh, J., Jacobson, I., Booch, G.: The Unified Modeling Language Reference Manual, 2nd edn. Addison-Wesley, Boston (2005)
14. Schanda, F., Brain, M.: Using answer set programming in the development of verified software. In: ICLP, pp. 72–85 (2012)
15. Simons, P., Niemelä, I., Soininen, T.: Extending and implementing the stable model semantics. Artif. Intell. 138, 181–234 (2002)
16. Soeken, M., Wille, R., Kuhlmann, M., Gogolla, M., Drechsler, R.: Verifying UML/OCL models using Boolean satisfiability. In: DATE, pp. 1341–1344 (2010)
17. Sterling, L.S., Shapiro, E.Y.: The Art of Prolog: Advanced Programming Techniques. MIT Press, Cambridge (1994)
18. Stumptner, M., Friedrich, G., Haselböck, A.: Generative constraint-based configuration of large technical systems. AI EDAM 12, 307–320 (1998)

Reasoning with Forest Logic Programs Using Fully Enriched Automata

Cristina Feier[1]([✉]) and Thomas Eiter[2]

[1] Department of Computer Science, University of Oxford, Oxford, UK
cristina.feier@cs.ox.ac.uk
[2] Institute of Information Systems, Vienna University of Technology,
Vienna, Austria

Abstract. Forest Logic Programs (FoLP) are a decidable fragment of Open Answer Set Programming (OASP) which have the forest model property. OASP extends Answer Set Programming (ASP) with open domains—a feature which makes it possible for FoLPs to simulate reasoning with the description logic $\mathcal{SHOQ}$. In the past, several tableau algorithms have been devised to reason with FoLPs, the most recent of which established a NExpTime upper bound for reasoning with the fragment. While known to be ExpTime-hard, the exact complexity characterization of reasoning with FoLPs was still unknown. In this paper we settle this open question by a reduction of reasoning with FoLPs to emptiness checking of fully enriched automata which are known to be ExpTime-complete.

1 Introduction

Open Answer Set Programming (OASP) [8] extends (function-free) Answer Set Programming (ASP) [5] with an open domain semantics: programs are interpreted with respect to arbitrary domains that might contain individuals which do not occur explicitly in the program. This enables to state generic knowledge using OASP; at the same time, OASP inherits from ASP the negation under the stable model semantics.

While OASP is undecidable in general, several decidable fragments have been found by restricting the shape of the rules. One such fragment are Forest Logic Programs (FoLP), which enjoy the forest model property: a unary predicate is satisfiable iff it is satisfied by a model representable as a labeled forest. FoLPs are quite expressive; e.g., one can simulate satisfiability testing of an ontology in the Description Logic (DL) $\mathcal{SHOQ}$ by them [4]. This led to f-hybrid KBs, which combine rules and ontologies distinctly from other approaches like dl-safe rules [9], r-hybrid knowledge bases [10], or MKNF$^+$ knowledge bases, as the interaction between the signatures of the two components is not restricted.

The simulation of $\mathcal{SHOQ}$ implies that reasoning with FoLPs is EXPTIME-hard; however, the exact complexity was open. A tableau-based algorithm in [4]

Work supported by the EPSRC grants Score! and DBOnto and the FWF grant P24090.

F. Calimeri et al. (Eds.): LPNMR 2015, LNAI 9345, pp. 346–353, 2015.
DOI: 10.1007/978-3-319-23264-5_29

gave an 2NEXPTIME upper bound, which an improved algorithm in [2] lowered to NEXPTIME. In this paper, we close this gap and show that deciding satisfiability of unary predicates w.r.t. FoLPs is EXPTIME-complete, by reducing emptiness checking of Fully Enriched Automata (FEAs) to this problem; hence, adding FoLP rules to SHOQ ontologies does not make reasoning harder. An extended version of the paper can be found at http://www.kr.tuwien.ac.at/research/reports/rr1502.pdf.

2 Preliminaries

We assume countably infinite disjoint sets of constants, variables, and predicate symbols of positive arity. Terms and atoms are as usual. Atoms $p(\vec{t})$ are unary (resp. binary) if p is unary (resp. binary). A *literal* is an atom a or a negated atom $not\ a$. *Inequality literals* are of form $s \neq t$, where s and t are terms; all other literals are *regular*. For a set S of literals or (possibly negated) predicates, $S^+ = \{a \mid a \in S\}$ and $S^- = \{a \mid not\ a \in S\}$. If S is a set of (possibly negated) predicates of arity n and $\vec{t}$ are terms, then $S(\vec{t}) = \{l(\vec{t}) \mid l \in S\}$. For a set S of atoms, $not\ S = \{not\ a \mid a \in S\}$. A *program* is a countable set P of rules $r : \alpha \leftarrow \beta$, where α is a finite set of regular literals and β is a finite set of literals. We denote as $head(r)$ the set α, where α stands for a disjunction, and as $body(r)$ the set β, where β stands for a conjunction.

For R a rule, program, etc., let $vars(R)$, $preds(R)$, and $cts(R)$ be the sets of variables, predicates, and constants that occur in R, resp. A *universe* U for a program P is a non-empty countable set $U \supseteq cts(P)$. We let P_U be the grounding of P with U and let $\mathcal{B}_P$ be the set of regular atoms that can be formed from a ground program P.

An *interpretation* of a ground, i.e. variable free, program P is a subset I of $\mathcal{B}_P$. We write $I \models p(\vec{t})$ if $p(\vec{t}) \in I$ and $I \models not\ p(\vec{t})$ if $I \not\models p(\vec{t})$. For ground terms s, t, we write $I \models s \neq t$ if $s \neq t$. For a set of ground literals L, $I \models L$ if $I \models l$ for every $l \in L$. A ground rule $r : \alpha \leftarrow \beta$ is *satisfied* w.r.t. I, denoted $I \models r$, if $I \models l$ for some $l \in \alpha$ whenever $I \models \beta$. An interpretation I of a positive (i.e. not -free) ground program P is a *model* of P if I satisfies every rule in P; it is an *answer set* of P if it is a $\subseteq$- minimal model of P. For ground programs P with not, I is an answer set of P iff I is an answer set of $P^I = \{\alpha^+ \leftarrow \beta^+ \mid \alpha \leftarrow \beta \in P, I \models not\ \beta^-, I \models \alpha^-\}$.

An *open interpretation* of a program P is a pair (U, M) where U is a universe for P and M is an interpretation of P_U. An *open answer set* of P is an open interpretation (U, M) of P, with M an answer set of P_U.

Trees and Forests. Let $\mathbb{N}^+$ be the set of positive integers, and let $\langle \mathbb{N}^+ \rangle$ be the set of all sequences over $\mathbb{N}^+$, where ε is the empty sequence: for a sequence of constants and/or natural numbers s, $s \cdot \varepsilon = c$, where $\cdot$ is concatenation; also, by convention, $s \cdot c \cdot -1 = s \cdot c$, where c is a natural number, and $\varepsilon \cdot -1$ is undefined. A *tree* T with root c, also denoted as T_c, is a set of nodes, where each node is a sequence $c \cdot s$, where $s \in \langle \mathbb{N}^+ \rangle$, and for every $x \cdot d \in T_c$, $d \in \mathbb{N}^+$, $x \in T_c$. If c is irrelevant, we refer to T_c as T. Given a tree T, its arc set is $A_T = \{(x, y) \mid x, y \in$

$T, \exists n \in \mathbb{N}^+ . y = x \cdot n\}$. We denote with $succ_T(x) = \{y \in T \mid y = x \cdot i, i \in \mathbb{N}^+\}$ the successors of a node x in T and with $prec_T(x) = y$, where $x = y \cdot i \in T$, its predecessor.

A *forest* F is a set of trees $\{T_c \mid c \in C\}$, where C is a finite set of arbitrary constants. Its node set is $N_F = \cup_{T \in F} T$ and its arc set is $A_F = \cup_{T \in F} A_T$. For a node $x \in N_F$, $succ_F(x) = succ_T(x)$, and $prec_F(x) = prec_T(x)$, where $x \in T$ and $T \in F$. For a node $y = x \cdot i \in T$ and $T \in F$, $prec_F(y) = prec_T(y) = x$. An *interconnected forest* EF is a tuple (F, ES), where $F = \{T_c \mid c \in C\}$ is a forest and $ES \subseteq N_F \times C$. Its set of nodes is $N_{EF} = N_F$, and its set of arcs is $A_{EF} = A_F \cup ES$. A Σ-labelled forest is a tuple (F, f) where F is an interconnected forest/tree and $f : N_F \to \Sigma$ is a labelling function, where Σ is any set of symbols.

3   Forest Logic Programs

Forest Logic Programs (FoLPs) are a fragment of OASP which have the forest model property. They allow only for unary and binary predicates and tree-shaped rules.

Definition 1. *A forest logic program (FoLP) is an OASP with only unary and binary predicates, s.t. a rule is either:*

- *a free rule:* $\quad a(s) \vee not\ a(s) \leftarrow \quad (1) \quad or \quad f(s,t) \vee not\ f(s,t) \leftarrow \quad (2)$
- *a unary rule:* $a(s) \leftarrow \beta(s), \gamma_1(s,t_1), \dots, \gamma_m(s,t_m), \delta_1(t_1), \dots, \delta_m(t_m), \psi$ (3),
  with $\psi \subseteq \{t_i \neq t_j | 1 \leq i < j \leq m\}$ *and* $m \in \mathbb{N}$,
- *or a binary rule:* $\quad\quad\quad f(s,t) \leftarrow \beta(s), \gamma(s,t), \delta(t)$ (4),

where a is a unary predicate, and f is a binary predicate; s, t, and t_i-s are distinct terms; β, δ, and δ_i-s are sets of (possibly negated) unary predicates; γ, and γ_i-s are sets of (possibly negated) binary predicates; inequality does not appear in γ and γ_i; $\gamma_i^+ \neq \emptyset$, if t_i is a variable, for every $1 \leq i \leq m$, and $\gamma^+ \neq \emptyset$, if t is a variable.

A predicate q in a FoLP P is *free* if it occurs in a free rule in P. We denote with $upr(P)$, and $bpr(P)$ (resp. $urul(P)$, and $brul(P)$), the sets of unary and binary predicates (resp. unary and binary rules) which occur in P. The degree of a unary rule r of type (3), denoted $degree(r)$, is the number k of successor variables appearing in r. The degree of a free rule is 0. The degree of a FoLP P is $degree(P) = \sum_{p \in upr(P)} degree(p)$, where $degree(p) = max\{degree(r) \mid p \in preds(head(r))\}$.

A forest model of an OASP P that satisfies a unary predicate p is a forest which contains for each constant in P a tree having the constant as root, and possibly one more tree with an anonymous root; the predicate p is in the label of some root node.

Definition 2. *Let P be a program. A predicate $p \in upr(P)$ is forest satisfiable w.r.t. P if there exist an open answer set (U, M) of P; an interconnected forest $EF = (\{T_\rho\} \cup \{T_a \mid a \in cts(P)\}, ES)$, where ρ is a constant, possibly from*

$cts(P)$; and a labelling function $ef : \{T_\rho\} \cup \{T_a \mid a \in cts(P)\} \cup A_{EF} \to 2^{preds(P)}$ s. t. $p \in ef(\rho)$; $U = N_{EF}$; $ef(x) \in 2^{upr(P)}$, when $x \in T_\rho \cup \{T_a \mid a \in cts(P)\}$; $ef(x) \in 2^{bpr(P)}$, when $x \in A_{T_\rho}$; $M = \{p(x) \mid x \in N_{EF}, p \in ef(x)\} \cup \{f(x,y) \mid (x,y) \in A_{EF}, f \in ef(x,y)\}$; and for every $(z, z \cdot i) \in A_{EF}$: $ef(z, z \cdot i) \neq \emptyset$. We call such a pair (U, M) a forest model.

P has the forest model property if every unary predicate p that is satisfiable w.r.t. P, is forest satisfiable w.r.t. P; FoLPs enjoy this property [7].

4 Fully Enriched Automata

Fully enriched automata (FEAs) were introduced in [1] as a tool to reason in hybrid graded μ-calculus. They accept forests as input. We describe them following [1].

For a set Y, we denote with $B^+(Y)$ the set of positive Boolean formulas over Y, where $true$ and $false$ are also allowed and where $\wedge$ has precedence over $\vee$. For a set $X \subseteq Y$ and a formula $\theta \in B^+(Y)$, we say that X satisfies θ iff assigning true to elements in X and assigning false to elements in $Y \setminus X$ makes θ true. For $b > 0$, let $D_b = \{\langle 0 \rangle, \langle 1 \rangle, \ldots, \langle b \rangle\} \cup \{[0], [1], \ldots, [b]\} \cup \{-1, \varepsilon, \langle root \rangle, [root]\}$.

A fully enriched automaton (FEA) is a tuple $A = \langle \Sigma, b, Q, \delta, q_0, \mathcal{F} \rangle$, where Σ is a finite input alphabet, $b > 0$ is a counting bound, Q is a finite set of states, $\delta : Q \times \Sigma \to B^+(D_b \times Q)$ is a transition function, $q_0 \in Q$ is an initial state, and $\mathcal{F} = \{\mathcal{F}_1, \mathcal{F}_2, \ldots, \mathcal{F}_k\}$, where $\mathcal{F}_1 \subseteq \mathcal{F}_2 \subseteq \ldots \subseteq \mathcal{F}_k = Q$ is a parity acceptance condition. The number k of sets in $\mathcal{F}$ is the index of the automaton.

A run of a FEA on a labeled forest (F, V) is an $N_F \times Q$-labeled tree (T_c, r) s.t. $r(c) = (d, q_0)$, for some root d in F, and for all $y \in T_c$ with $r(y) = (x, q)$ and $\delta(q, V(x)) = \theta$, there is a (possibly empty) set $S \subseteq D_b \times Q$ such that S satisfies θ and for all $(d, s) \in S$, the following hold: (i) if $d \in \{-1, \varepsilon\}$, then $x \cdot d$ is defined and there is $j \in \mathbb{N}^+$ such that $y \cdot j \in T_c$ and $r(y \cdot j) = (x \cdot d, s)$; (ii) if $d = \langle n \rangle$, then there is a set $M \subseteq succ_F(x)$ of cardinality $n + 1$ s.t. for all $z \in M$, there is $j \in \mathbb{N}^+$ s.t. $y \cdot j \in T_c$ and $r(y \cdot j) = (z, s)$; (iii) if $d = [n]$, then there is a set $M \subseteq succ_F(x)$ of cardinality n s.t. for all $z \in succ_F(x) \setminus M$, there is $j \in \mathbb{N}^+$ s.t. $y \cdot j \in T_c$ and $r(y \cdot j) = (z, s)$; (iv) if $d = \langle root \rangle$ ($d = [root]$), then for some (all) root(s) $c \in F$ there exists $j \in \mathbb{N}^+$ s.t. $y \cdot j \in T_c$ and $r(y \cdot j) = (c, s)$;

If θ above is true, then y does not need to have successors. Moreover, since no set S satisfies $\theta = false$, there cannot be any run that takes a transition with $\theta = false$. A run is accepting if each of its infinite paths π is accepting, that is if the minimum i for which $Inf(\pi) \cap \mathcal{F}_i \neq \emptyset$, where $Inf(\pi)$ is the set of states occurring infinitely often in π, is even. The automaton accepts a forest iff there exists an accepting run of the automaton on the forest. The language of A, denoted $\mathcal{L}(A)$, is the set of forests accepted by A. We say that A is non-empty if $\mathcal{L}(A) \neq \emptyset$.

Theorem 1 (Corollary 4.3 [1]). *Given a FEA $A = \langle \Sigma, b, Q, \delta, q_0, \mathcal{F} \rangle$ with n states and index k, deciding whether $\mathcal{L}(A) = \emptyset$ is possible in time $(b + 2)^{\mathcal{O}(n^3 \cdot k^2 \cdot \log k \cdot \log b^2)}$.*

5   From Forest Logic Programs to Fully Enriched Automata

In this section we reduce satisfiability checking of unary predicates w.r.t. FoLPs to emptiness checking for FEAs. For a FoLP P and a unary predicate p, we introduce a class of FEAs $A_{\rho,\theta}^{p,P}$, where ρ is one of $cts(P)$ or a new anonymous individual and $\theta : cts(P) \cup \{\rho\} \rightarrow 2^{upr(P) \cup cts(P) \cup \{\rho\}}$ is s.t. $o_i \in \theta(o_i)$, and $o_j \notin \theta(o_i)$, for every $o_i, o_j \in cts(P) \cup \{\rho\}$, s.t. $o_i \neq o_j$. Furthermore, $p \in \theta(c)$, where c is one of $cts(P) \cup \{\rho\}$ and c is ρ if $\rho \notin cts(P)$. Intuitively, $A_{\rho,\theta}^{p,P}$ accepts forest models of p w.r.t. P encoded in a certain fashion: for every root in the forest model, the root node will appear in its own label; function θ fixes a content for the label of each root of accepted forest models.

Let $d = degree(P)$ and let $\text{PAT}_P = \{*\} \cup cts(P)$ be the set of *term patterns*, where $*$ stands for a generic anonymous individual: a term t matches a term pattern pt, written $t \mapsto pt$, iff $t = pt$, when t is a constant; if t is not a constant, the match trivially holds. We use term patterns as a unification mechanism: a variable matches with a constant or an anonymous individual, but a constant matches only with itself. $A_{\rho,\theta}^{p,P}$ will run on forests labelled using the following alphabet: $\Sigma = 2^S$, where $S = upr(P) \cup \{1,\ldots,d\} \cup cts(P) \cup \{\rho\} \cup \{\uparrow_f^o \mid f \in bpr(P)\} \cup \{\downarrow_f^t \mid f \in bpr(P), t \in \text{PAT}_P\}$.

Unlike forest models, arcs of forests accepted by FEAs are not labelled: as such, binary predicates occur in the label of nodes in an adorned form. These adorned predicates are of form $\downarrow_f^t$, in which case they represent an f-link between the predecessor of the labelled node, which has term pattern t and the node itself, or of form $\uparrow_f^o$, in which case the current node is linked to a constant o from P via the binary predicate f. Besides unary predicates, labels might contain natural numbers and constants, which will be used as an addressing mechanism for successors of a given node and nodes which stand for constants in accepted forests, resp. The set of states of the automaton are as follows: $Q = Q_i \cup Q_+ \cup Q_-$, with:

- $Q_i = \{q_0, q_1\} \cup \{q_o \mid o \in cts(P) \cup \{\rho\}\} \cup \{q_{\neg k} \mid 1 \leq k \leq d\}$,
- $Q_+ = \{q_{t,a}, q_{t,r_a}, q_{t_1,t_2,u}, q_{t_1,t_2,r_f}, q_{k,t,*,u} \mid t, t_1, t_2 \in \text{PAT}_P, a \in upr(P), f \in bpr(P), u$ is of form $a, f, not\ a$ or $not\ f, 1 \leq k \leq d, r_a \in urul(P), r_f \in brul(P)\}$,
- $Q_- = \{q_{\overline{t,a}}, q_{\overline{t,r_a}}, q_{\overline{t_1,t_2,u}}, q_{\overline{t_1,t_2,r_f}}, q_{\overline{k,t,*,u}} \mid t_1, t_2, t, a, f, u, k, r_a, r_f$ as above$\}$.

Positive states in Q_+ (resp. negative states in Q_-) are used to motivate the presence (resp. absence) of atoms in an open answer set. Q_i contains q_0, the initial state, q_1, a state which will be visited recursively in every node of the forest, q_o, a state corresponding to the initial visit of each constant node, and $q_{\neg k}$, a state which asserts that for every node in an accepted forest there must be at most one successor which has k in the label.

We next describe the transition function of $A_{\rho,\theta}^{p,P}$. The initial transition prescribes that the automaton visits a root of the forest in state q_o, for every $o \in cts(P) \cup \{\rho\}$:

$$\delta(q_0, \sigma) = \bigwedge_{o \in cts(P) \cup \{\rho\}} (\langle root \rangle, q_o) \tag{5}$$

In every such state q_o, it should hold that o and only o is part of the label. Furthermore, the automaton justifies the presence and absence of each unary predicate a and adorned upward binary predicate in the label[1] by entering states $q_{o,a}$, $q_{o,o',f}$, $q_{\overline{o,a}}$, and $q_{\overline{o,o',f}}$ resp. At the same time every successor of the constant node is visited in state q_1:

$$\delta(q_o, \sigma) = o \in \sigma \wedge \bigwedge_{o' \in cts(P) \cup \{\rho\} \setminus \{o\}} o' \notin \sigma \wedge \bigwedge_{a \in \theta(o)} (\varepsilon, q_{o,a}) \wedge \bigwedge_{a \notin \theta(o)} (\varepsilon, q_{\overline{o,a}})$$
$$\wedge \bigwedge_{\uparrow_f^{o'} \in \theta(o)} (\varepsilon, q_{o,o',f}) \wedge \bigwedge_{\uparrow_f^{o'} \notin \theta(o)} (\varepsilon, q_{\overline{o,o',f}}) \wedge ([0], q_1) \tag{6}$$

Whenever the automaton finds itself in state q_1 it tries to motivate the presence and absence of each unary and each adorned binary predicate in its label and then it recursively enters the same state into each successor of the current node:

$$\delta(q_1, \sigma) = \bigwedge_{a \in \sigma} (\varepsilon, q_{*,a}) \wedge \bigwedge_{a \notin \sigma} (\varepsilon, q_{\overline{*,a}}) \wedge \bigwedge_{\downarrow_f^t \in \sigma} (\varepsilon, q_{t,*,f}) \wedge \bigwedge_{\downarrow_f^t \notin \sigma} (\varepsilon, q_{\overline{t,*,f}}) \wedge$$
$$\bigwedge_{\uparrow_f^{o'} \in \sigma} (\varepsilon, q_{*,o',f}) \wedge \bigwedge_{\uparrow_f^{o'} \notin \sigma} (\varepsilon, q_{\overline{*,o',f}}) \wedge ([0], q_1) \wedge \bigwedge_{1 \le k \le d} ([1], q_{\neg k}) \tag{7}$$

It also ensures that for each integer $1 \le k \le d$, the labels of each but one successor do not contain k:

$$\delta(q_{\neg k}, \sigma) = k \notin \sigma \tag{8}$$

To motivate a predicate in a node label, the automaton checks whether it is free (using a test $free(.)$) or finds a supporting rule. We distinguish between unary and binary predicates and the term pattern for the node where the predicate has to hold. For unary predicates holding at a constant node, a first check is that we are at the right node - this is needed as later the automaton will visit all roots in this state. For binary predicates, depending on the term pattern, the label is checked for different types of adorned binary atoms.

$$\delta(q_{*,a}, \sigma) = a \in \sigma \wedge \left(free(a) \vee \bigvee_{r_a: a(s) \leftarrow \beta \in P} (\varepsilon, q_{*,r_a}) \right) \tag{9}$$

$$\delta(q_{o,a}, \sigma) = o \notin \sigma \vee a \in \theta(o) \wedge \left(free(a) \vee \bigvee_{r_a: a(s) \leftarrow \beta \in P, s \mapsto o} (\varepsilon, q_{o,r_a}) \right) \tag{10}$$

$$\delta(q_{t,*,f}, \sigma) = \downarrow_f^t \in \sigma \wedge \left(free(f) \vee \bigvee_{r_f: f(s,v) \leftarrow \beta \in P, s \mapsto t, v \mapsto *} (\varepsilon, q_{v,*,r_f}) \right) \tag{11}$$

$$\delta(q_{t,o,f}, \sigma) = \uparrow_f^o \in \sigma \wedge \left(free(f) \vee \bigvee_{r_f: f(s,v) \leftarrow \beta \in P, s \mapsto t, v \mapsto o} (\varepsilon, q_{t,o,r_f}) \right) \tag{12}$$

---

[1] Constants have no predecessors, hence there are no adorned downward predicates in the label.

Let $r_a : a(s) \leftarrow \beta(s), (\gamma_i(s, v_i), \delta_i(v_i))_{1 \leq i \leq m}, \psi$ be a unary rule. Then, we denote with J_{r_a} a multiset $\{j_i \mid 1 \leq i \leq m, j_i \in \{1, \ldots, d\} \cup cts(P)\}$ such that for every $j_i \in J_{r_a}$, $v_i \in cts(P)$ implies $j_i = v_i$, and for every $j_i, j_l \in J_{r_a}$, $v_i \neq v_l \in \psi$ implies $j_i \neq j_l$. A multiset provides a way to satisfy the successor part of a unary rule in a forest model by identifying the successor terms of the rule with successors of the current element in the model or constants in the program. Let $\mathcal{MJ}$ be the set of all such multisets. The following transition describes how the body of such a rule is checked to be satisfiable:

$$\delta(q_{t,r_a}, \sigma) = \bigwedge_{u \in \beta} (\varepsilon, q_{*,t,u}) \wedge \bigvee_{J_{r_a} \in \mathcal{MJ}} \left(\bigwedge_{k=1}^{d} \bigwedge_{j_i = k, j_i \in J_{r_a}} \bigwedge_{u \in \gamma_i \cup \delta_i} (\langle 0 \rangle, q_{k,t,*,u}) \wedge \right. \tag{13}$$
$$\left. \bigwedge_{o \in cts(P)} \bigwedge_{j_i = o, j_i \in J_{r_a}} \bigwedge_{u \in \gamma_i \cup \delta_i} (\varepsilon, q_{t,o,u}) \right)$$

State $q_{k,t,*,u}$ checks that the (possibly negated) unary or adorned binary predicate u is (is not) part of the label of the k-th successor of a given node:

$$\delta(q_{k,t_1,t_2,u}, \sigma) = k \in \sigma \wedge \bigwedge_{j \neq k} j \notin \sigma \wedge (\varepsilon, q_{t_1,t_2,u}) \tag{14}$$

State $q_{t_1,t_2,u}$ can be seen as a multi-state with different transitions depending on its arguments (two transitions have already been introduced as rules (11) and (12) above): if $t_2 = *$, one has the justify the presence/absence of u in the label of the current node; otherwise, when $t_2 = o$, one has to justify it from the label of the root node corresponding to constant o: note that, as it is not possible to jump directly to a given root node in the forest, nor to enforce that there will be a single root node corresponding to each constant, in transition (17) we visit each root node in state $q_{o,a}$:

$$\delta(q_{t_1,t_2,u}, \sigma) =$$

$(\varepsilon, q_{*,a})$,	if $t_2 = * \wedge u = a$ (15)	$a \notin \theta(o)$,	if $t_2 = o \wedge u = not\ a$ (16)
$([root], q_{o,a})$,	if $t_2 = o \wedge u = a$ (17)	$\downarrow_f^t \notin \sigma$,	if $t_2 = * \wedge u = not\ f$ (18)
$a \notin \sigma$,	if $t_2 = * \wedge u = not\ a$ (19)	$\uparrow_f^o \notin \theta(o)$,	if $t_2 = o \wedge u = not\ f$ (20)

For binary rules: $r_f : f(s, v) \leftarrow \beta(s), \gamma(s, v), \delta(v)$, where v is grounded using an anonymous individual, we also look at the predecessor node to see if the local part of the rule is satisfied. When v is grounded using a constant, the local part of the rule is checked at the current node and the successor part at the respective constant.

$$\delta(q_{t,*,r_f}, \sigma) = \bigwedge_{u \in \beta} (-1, q_{*,t,u}) \wedge \bigwedge_{u \in \gamma \cup \delta} (\varepsilon, q_{t,*,u}) \tag{21}$$

$$\delta(q_{t,o,r_f}, \sigma) = \bigwedge_{u \in \beta} (\varepsilon, q_{*,t,u}) \wedge \bigwedge_{u \in \gamma \cup \delta} (\varepsilon, q_{t,o,u}) \tag{22}$$

The transitions of the automaton in the negative states can be seen as dual versions of the ones for the counterpart positive states. They are presented in the technical report.

Finally we specify the parity acceptance condition. The index of the automaton is 2, with $\mathcal{F}_1 = \{q_{t,a}, q_{t_1,t_2,f} \mid a \in upr(P), f \in bpr(P), t, t_1, t_2 \in \text{PAT}_P\}$ and

$\mathcal{F}_2 = Q$. Intuitively, paths in a run of the automaton correspond to dependencies of literals in the accepted model and by disallowing the infinite occurrence on a path of states associated to atoms in the model we ensure that only well-supported models are accepted.

Theorem 2. *Let P be a FoLP and p be a unary predicate symbol. Then, p is satisfiable w.r.t. P iff there exists an automaton $A_{\rho,\theta}^{p,P}$ such that $\mathcal{L}(A_{\rho,\theta}^{p,P}) \neq \emptyset$.*

Theorem 3. *Satisfiability checking of unary predicates w.r.t. FoLPs is* EXPTIME-*complete.*

6 Discussion and Conclusion

We have described a reduction of the satisfiability checking task of unary predicates w.r.t. FoLPs to emptiness checking of FEAs. This enabled us to establish a tight complexity bound on this reasoning task for FoLPs. Other reasoning tasks like consistency checking of FoLPs and skeptical and brave entailment of ground atoms can be polynomially reduced to satisfiability checking of unary predicates [6]; thus, the complexity result applies to those tasks as well. Also, by virtue of the translation from fKBs to FoLPs, the result applies to fKBs as well: satisfiability checking of unary predicates w.r.t. fKBs is EXPTIME-complete. Thus, reasoning with FoLP rules and $\mathcal{SHOQ}$ ontologies is not harder than reasoning with $\mathcal{SHOQ}$ ontologies themselves.

Finally, as our result shows that FoLPs have the same complexity as CoLPs, we plan to further investigate the extension of the deterministic AND/OR tableau algorithm for CoLPs [3] to FoLPs. As explained in [3], such an extension is far from trivial.

References

1. Bonatti, P.A., Lutz, C., Murano, A., Vardi, M.Y.: The complexity of enriched μ-calculi. Log. Methods Comput. Sci. **4**(3), 1–27 (2008)
2. Feier, C.: Worst-case optimal reasoning with forest logic programs. In: Proceedings of the KR, 2012, 208–212 (2012)
3. Feier, C.: Reasoning with Forest Logic Programs, Ph.D thesis, TU Wien (2014)
4. Feier, C., Heymans, S.: Reasoning with forest logic programs and f-hybrid knowledge bases. TPLP **3**(13), 395–463 (2013)
5. Gelfond, M., Lifschitz, V.: The stable model semantics for logic programming. In: Proceedings of ICLP 1988, pp. 1070–1080 (1988)
6. Heymans, S.: Decidable Open Answer Set Programming. Ph.D thesis, Theoretical Computer Science Lab (TINF), Department of Computer Science, Vrije Universiteit Brussel (2006)
7. Heymans, S., Van Nieuwenborgh, D., Vermeir, D.: Open answer set programming for the semantic web. J. Appl. Logic **5**(1), 144–169 (2007)
8. Heymans, S., Van Nieuwenborgh, D., Vermeir, D.: Open answer set programming with guarded programs. Trans. Comput. Logic **9**(4), 1–53 (2008)
9. Motik, B., Sattler, U., Studer, R.: Query answering for OWL-DL with rules. J. Web Semant. **3**(1), 41–60 (2005)
10. Rosati, R.: On combining description logic ontologies and nonrecursive datalog rules. In: Proceedings of the RR, pp. 13–27 (2008)

ASP Solving for Expanding Universes

Martin Gebser[1,3], Tomi Janhunen[1], Holger Jost[3], Roland Kaminski[3],
and Torsten Schaub[2,3]([✉])

[1] Aalto University, HIIT, Espoo, Finland
tomi.janhunen@aalto.fi
[2] INRIA Rennes, Rennes, France
[3] University of Potsdam, Potsdam, Germany
{gebser,jost,kaminski,torsten}@cs.uni-potsdam.de

Abstract. Over the last years, Answer Set Programming has significantly extended its range of applicability, and moved beyond solving
static problems to dynamic ones, even in online environments. However,
its nonmonotonic nature as well as its upstream instantiation process
impede a seamless integration of new objects into its reasoning process,
which is crucial in dynamic domains such as logistics or robotics. We
address this problem and introduce a simple approach to successively
incorporating new information into ASP systems. Our approach rests
upon a translation of logic programs and thus refrains from any dedicated algorithms. We prove its modularity as regards the addition of
new information and show its soundness and completeness. We apply our
methodology to two domains of the Fifth ASP Competition and evaluate
traditional one-shot and incremental multi-shot solving approaches.

1 Introduction

Answer Set Programming (ASP; [1]) is deeply rooted in the paradigm of non-
monotonic reasoning. That is, conclusions can be invalidated upon the arrival of
new information. Unfortunately, this carries over to computationally relevant characterizations, involving completion and loop formulas, and thus extends to the data
structures capturing "nonmonotonicity" in modern ASP solvers. Hence, when solving in dynamic domains like logistics or robotics, the emergence of new properties or
even new objects cannot be accounted for in a modular way, since the existing structures become invalidated. This is different from monotonic (instantiation-based)
approaches, like the original DPLL procedure [2], where new objects can be modularly incorporated by adding the instantiations involving them. In fact, incremental satisfiability solving has been successfully applied in domains like finite model

This work was funded by DFG (SCHA 550/9), the Finnish Centre of Excel-
lence in Computational Inference Research (COIN) supported by the Academy
of Finland (AoF) under grant 251170, as well as DAAD and AoF under joint
project 57071677/279121. A draft version with proofs is available at http://www.cs.
uni-potsdam.de/wv/publications/.
T. Schaub—Affiliated with Simon Fraser University, Canada, and IIIS Griffith
University, Australia.

© Springer International Publishing Switzerland 2015
F. Calimeri et al. (Eds.): LPNMR 2015, LNAI 9345, pp. 354–367, 2015.
DOI: 10.1007/978-3-319-23264-5_30

finding [3], model checking [4], and planning [5], yet relying on application-specific instantiators rather than general-purpose grounding.

Incremental instantiation was so far neglected in ASP since traditional systems were designed for one-shot solving and thus needed to be relaunched whenever the problem specification changed. This is clearly disadvantageous in highly dynamic domains like logistics or robotics. Although new generation ASP systems, like *clingo* 4 [6], allow for multi-shot solving and thus abolish the need for relaunching, there is yet no principled way of modularly extending a problem specification upon the arrival of new objects.

In what follows, we address a rather general variant of this problem by allowing new information to successively expand the (Herbrand) universe. Our approach rests upon a simple translation of logic programs and thus refrains from dedicated algorithms (though it is only meaningful in the context of multi-shot ASP solving). We prove the modularity of our approach as regards the addition of new information and show its soundness and completeness. Finally, we illustrate our methodology, evaluate the resulting performance of solving approaches, and discuss related work.

2 Background

A signature $(\mathcal{P}, \mathcal{C}, \mathcal{V})$ consists of a set $\mathcal{P}$ of *predicate symbols*, a set $\mathcal{C}$ of *constant symbols*, also called Herbrand universe, and a set $\mathcal{V}$ of *variable symbols*; we usually omit the designation "symbol" for simplicity. The members of $\mathcal{C} \cup \mathcal{V}$ are *terms*. Given a predicate $p \in \mathcal{P}$ of arity n, also denoted as p/n, along with terms $t_1, \ldots, t_n$, $p(t_1, \ldots, t_n)$ is an *atom* over p/n. An atom a and $\sim a$ are (positive or negative, respectively) *literals*, where '$\sim$' stands for default negation; we sometimes (ab)use the same terminology for classical literals a and $\neg a$. Given a set $B = \{a_1, \ldots, a_m, \sim a_{m+1}, \ldots, \sim a_n\}$ of literals, $B^+ = \{a_1, \ldots, a_m\}$ and $B^- = \{a_{m+1}, \ldots, a_n\}$ denote the atoms occurring positively or negatively in B. A *logic program* R over $(\mathcal{P}, \mathcal{C}, \mathcal{V})$ is a set of *rules* $r = a \leftarrow B$, where a is an atom and B is a set of literals; if $B = \emptyset$, r is also called a *fact*. By $head(r) = a$ and $body(r) = B$, we refer to the *head* or *body* of r, respectively. We extend this notation to R by letting $head(R) = \{head(r) \mid r \in R\}$.

We denote the set of variables occurring in an atom a by $var(a)$. The variables in a rule r are $var(r) = var(head(r)) \cup \bigcup_{a \in body(r)^+ \cup body(r)^-} var(a)$. An atom, rule, or program is *non-ground* if it includes some variable, and *ground* otherwise. By $atom(\mathcal{P}, \mathcal{C}) = \{p(c_1, \ldots, c_n) \mid p/n \in \mathcal{P}, c_1 \in \mathcal{C}, \ldots, c_n \in \mathcal{C}\}$, we refer to the collection of ground atoms over predicates in $\mathcal{P}$. A *ground substitution* for a set V of variables is a mapping $\sigma : V \to \mathcal{C}$, and $\Sigma(V, \mathcal{C})$ denotes the set of all ground substitutions for V. The *instance* $a\sigma$ (or $r\sigma$) of an atom a (or a rule r) is obtained by substituting occurrences of variables in V according to σ. The *ground instantiation* of a program R is the set $grd(R, \mathcal{C}) = \{r\sigma \mid r \in R, \sigma \in \Sigma(var(r), \mathcal{C})\}$ of ground rules.

An *interpretation* $I \subseteq atom(\mathcal{P}, \mathcal{C})$ is a *supported model* [7] of a program R if $I = \{head(r) \mid r \in grd(R, \mathcal{C}), body(r)^+ \subseteq I, body(r)^- \cap I = \emptyset\}$. Moreover, I is a *stable model* [8] of R if I is a $\subseteq$-minimal (supported) model of the reduct $\{head(r) \leftarrow body(r)^+ \mid r \in grd(R, \mathcal{C}), body(r)^- \cap I = \emptyset\}$. Any stable model of R is a supported model of R as well, while the converse does not hold in general [9].

Supported and stable models can also be characterized in terms of classical models. To this end, given a set B of literals, let $bf(B) = (\bigwedge_{a \in B^+} a) \wedge (\bigwedge_{a \in B^-} \neg a)$ denote the *body formula* for B. Moreover, let $rf(r) = bf(body(r)) \rightarrow head(r)$ be the *rule formula* for a rule r. Then, we associate a ground logic program R with the set $RF(R) = \{rf(r) \mid r \in R\}$ of rule formulas. Given some ground atom a, the *completion formula* for a relative to R is $cf(R, a) = a \rightarrow \bigvee_{r \in R, head(r) = a} bf(body(r))$. For a set $\mathcal{A}$ of ground atoms, $CF(R, \mathcal{A}) = \{cf(R, a) \mid a \in \mathcal{A}\}$ denotes the corresponding set of completion formulas. The theory $RF(R) \cup CF(R, atom(\mathcal{P}, \mathcal{C}))$ is also known as *Clark's completion* [10], and its classical models coincide with the supported models of R. In order to extend the correspondence to stable models, for a set L of ground atoms, let $supp(R, L) = \{body(r) \mid r \in R, head(r) \in L, body(r)^+ \cap L = \emptyset\}$ denote the *external supports* for L [11]. Then, $lf(R, a, L) = a \rightarrow \bigvee_{B \in supp(R,L)} bf(B)$ is the *loop formula* for $a \in L$ relative to R [12]. Further distinguishing two sets $\mathcal{A}$ and $\mathcal{B}$ of ground atoms, we let $LF(R, \mathcal{A}, \mathcal{B}) = \{lf(R, a, L) \mid L \subseteq \mathcal{A} \cup \mathcal{B}, a \in L \cap \mathcal{A}\}$ be the corresponding set of loop formulas. Note that $LF(R, atom(\mathcal{P}, \mathcal{C}), \emptyset) \models CF(R, atom(\mathcal{P}, \mathcal{C}))$, and as shown in [11,12], the classical models of $RF(R) \cup LF(R, atom(\mathcal{P}, \mathcal{C}), \emptyset)$ match the stable models of R. Thus, when $\mathcal{A} = atom(\mathcal{P}, \mathcal{C})$ and $\mathcal{B} = \emptyset$, $LF(R, \mathcal{A}, \mathcal{B})$ yields the same as corresponding concepts from the literature, but we use $\mathcal{A}$ and $\mathcal{B}$ below to control the set $\mathcal{A}$ of atoms whose derivability is expressed by particular loop formulas.

3 Expanding Logic Programs

As in Datalog [13], we consider signatures $(\mathcal{P}_E \cup \mathcal{P}_I, \mathcal{C}, \mathcal{V})$ such that $\mathcal{P}_E \cap \mathcal{P}_I = \emptyset$. The part $\mathcal{P}_E$ includes *extensional* predicates provided by facts, while the *intensional* predicates in $\mathcal{P}_I$ are defined by rules. We thus deal with programs $F \cup R$ composed of (ground) facts F over $(\mathcal{P}_E, \mathcal{C}, \emptyset)$ and (non-ground) rules R over $(\mathcal{P}_E \cup \mathcal{P}_I, \emptyset, \mathcal{V})$ such that $\{p/n \mid p(X_1, \ldots, X_n) \in head(R)\} \subseteq \mathcal{P}_I$.

Example 1. For $\mathcal{P}_E = \{cs/1, st/1, in/2\}$ (for courses and students) and $\mathcal{P}_I = \{ok/1, ko/1\}$, the following non-ground rules R define the intensional predicates in $\mathcal{P}_I$:

$$ok(C) \leftarrow cs(C), st(S), in(S, C) \tag{1}$$
$$ko(C) \leftarrow cs(C), \sim ok(C) \tag{2}$$

Moreover, consider facts F over the extensional predicates in $\mathcal{P}_E$ and $\mathcal{C} = \{c_1, c_2, s_1, s_2\}$ as follows:

$$
\begin{array}{lll}
cs(c_1), & st(s_1), & in(s_1, c_1), \\
cs(c_2), & st(s_2), & in(s_2, c_1).
\end{array}
$$

The atoms over intensional predicates in $\mathcal{P}_I$ in the (unique) stable model of $F \cup R$ are $ok(c_1)$ and $ko(c_2)$. □

To characterize supported and stable models in terms of classical models relative to facts F over $(\mathcal{P}_E, \mathcal{C}, \emptyset)$, let $E(F, \mathcal{P}_E, \mathcal{C}) = F \cup \{\neg a \mid a \in atom(\mathcal{P}_E, \mathcal{C}) \setminus F\}$ denote the set of literals fixing atoms over extensional predicates. Then, supported models of $F \cup R$ match classical models of $RF(R') \cup CF(R', atom(\mathcal{P}_I, \mathcal{C})) \cup E(F, \mathcal{P}_E, \mathcal{C})$, where $R' = grd(R, \mathcal{C})$. Similarly, the latter theory augmented with $LF(R', atom(\mathcal{P}_I, \mathcal{C}), \emptyset)$ captures stable models of $F \cup R$.

For expressing the gradual expansion of an (infinite) Herbrand universe $\mathcal{C}$, we consider sequences over constants in $\mathcal{C}$.

Definition 1. *A constant stream over $\mathcal{C}$ is a sequence $(c_i)_{i \geq 1}$ such that $c_{i+1} \in \mathcal{C} \setminus \mathcal{C}_i$ for $i \geq 0$ and $\mathcal{C}_i = \{c_j \mid 1 \leq j \leq i\}$.*

Note that $\mathcal{C}_i \setminus \mathcal{C}_{i-1} = \{c_i\}$ for $i \geq 1$ and a constant $c_i \in \mathcal{C}$. Furthermore, given a set R of (non-ground) rules, each $\mathcal{C}_i$ yields a finite ground instantiation $grd(R, \mathcal{C}_i)$. While ground rules can simply be accumulated when the set of constants grows, completion (and loop) formulas cannot.

Example 2. Reconsider the rules R and facts F from Example 1. Relative to the constant stream $(c_1, s_1, s_2, c_2, \dots)$, the ground instantiations of R for $\mathcal{C}_1 = \{c_1\}$ and $\mathcal{C}_2 = \{c_1, s_1\}$, $R_1 = grd(R, \mathcal{C}_1)$ and $R_2 = grd(R, \mathcal{C}_2)$, are:

$$R_1 = \{ ok(c_1) \leftarrow cs(c_1), st(c_1), in(c_1, c_1) \qquad ko(c_1) \leftarrow cs(c_1), {\sim}ok(c_1) \}$$

$$R_2 = \begin{cases} ok(c_1) \leftarrow cs(c_1), st(c_1), in(c_1, c_1) & ko(c_1) \leftarrow cs(c_1), {\sim}ok(c_1) \\ ok(c_1) \leftarrow cs(c_1), st(s_1), in(s_1, c_1) & ko(s_1) \leftarrow cs(s_1), {\sim}ok(s_1) \\ ok(s_1) \leftarrow cs(s_1), st(c_1), in(c_1, s_1) & \\ ok(s_1) \leftarrow cs(s_1), st(s_1), in(s_1, s_1) & \end{cases}$$

Relative to R_1, we obtain

$$cf(R_1, ok(c_1)) = ok(c_1) \rightarrow (cs(c_1) \wedge st(c_1) \wedge in(c_1, c_1)).$$

Along with

$$E_1 = E(\{cs(c_1)\}, \mathcal{P}_E, \mathcal{C}_1) = \{cs(c_1), \neg st(c_1), \neg in(c_1, c_1)\}$$

and in view of $((cs(c_1) \wedge \neg ok(c_1)) \rightarrow ko(c_1)) \in RF(R_1)$, $RF(R_1) \cup CF(R_1, \{ok(c_1), ko(c_1)\}) \cup E_1$ entails $\neg ok(c_1)$ and $ko(c_1)$. Turning to $R_2 \supseteq R_1$, we have that $RF(R_1) \subseteq RF(R_2)$. However, the rules defining $ok(c_1)$ yield

$$cf(R_2, ok(c_1)) = ok(c_1) \rightarrow ((cs(c_1) \wedge st(c_1) \wedge in(c_1, c_1))$$
$$\vee (cs(c_1) \wedge st(s_1) \wedge in(s_1, c_1))),$$

so that $cf(R_2, ok(c_1)) \neq cf(R_1, ok(c_1))$. Moreover,

$$E_2 = E(\{cs(c_1), st(s_1), in(s_1, c_1)\}, \mathcal{P}_E, \mathcal{C}_2)$$
$$= E_1 \cup \{st(s_1), in(s_1, c_1), \neg cs(s_1), \neg in(c_1, s_1), \neg in(s_1, s_1)\}$$

and $((cs(c_1) \wedge st(s_1) \wedge in(s_1, c_1)) \rightarrow ok(c_1)) \in RF(R_2)$ entail $ok(c_1)$. Thus, $RF(R_2) \cup CF(R_1, \{ok(c_1), ko(c_1)\}) \cup E_2$ is unsatisfiable, and $cf(R_1, ok(c_1))$

must be replaced by $cf(R_2, ok(c_1))$ to obtain a (unique) model of $RF(R_2) \cup CF(R_2, \{ok(c_1), ko(c_1), ok(s_1), ko(s_1)\}) \cup E_2$, providing $ok(c_1)$, $\neg ko(c_1)$, $\neg ok(s_1)$, and $\neg ko(s_1)$ as conclusions.   □

In incremental CDCL-based Boolean constraint solvers (cf. [14]), a replacement as above amounts to the withdrawal of all conflict information relying on invalidated completion (and loop) formulas, essentially restricting the "memory" of an incremental solver to direct consequences of rules or rule formulas, respectively. In order to resolve this problem, we in the following provide a translation approach on the first-order level, leading to ground instantiations such that corresponding completion and loop formulas can be accumulated, even when expanding the underlying Herbrand universe.

Our translation approach successively extends the signature of (non-ground) rules. To this end, given a set $\mathcal{P}_I$ of intensional predicates, we let $\mathcal{P}_I^k = \{p^k/n \mid p/n \in \mathcal{P}_I\}$ be a corresponding set of new predicates labeled with k. For an atom $p(X_1, \ldots, X_n)$, we denote its labeled counterpart by $p(X_1, \ldots, X_n)^k = p^k(X_1, \ldots, X_n)$. Modifying the head of a rule r in this way yields $r^k = head(r)^k \leftarrow body(r)$.

The label k (or $k+1$) of a predicate serves as place holder for integers. Given $i \geq 0$, let $p^k[i] = p^i$ (or $p^{k+1}[i] = p^{i+1}$) if $p^k \in \mathcal{P}_I^k$ (or $p^{k+1} \in \mathcal{P}_I^{k+1}$) is labeled, while $p[i] = p$ for unlabeled predicates $p \in \mathcal{P}_E \cup \mathcal{P}_I$. We extend this notation to sets $\mathcal{P}$ of predicates and to atoms $p(X_1, \ldots, X_n)$ by letting $\mathcal{P}[i] = \{p[i] \mid p \in \mathcal{P}\}$ and $p(X_1, \ldots, X_n)[i] = p[i](X_1, \ldots, X_n)$. For a set R of rules, $R[i] = \{r[i] \mid r \in R\}$, where $r[i] = head(r)[i] \leftarrow \{a[i] \mid a \in body(r)^+\} \cup \{\sim a \mid a \in body(r)^-\}$.

Definition 2. *For a set R of rules over $(\mathcal{P}_E \cup \mathcal{P}_I, \emptyset, \mathcal{V})$, we define the sets $\Phi(R)$, $\Pi(\mathcal{P}_I)$, and $\Delta(\mathcal{P}_I)$ of rules as follows:*

$$\Phi(R) = \{r^k \mid r \in R\},$$
$$\Pi(\mathcal{P}_I) = \{p(X_1, \ldots, X_n) \leftarrow p^k(X_1, \ldots, X_n) \mid p/n \in \mathcal{P}_I\},$$
$$\Delta(\mathcal{P}_I) = \{p^k(X_1, \ldots, X_n) \leftarrow p^{k+1}(X_1, \ldots, X_n) \mid p/n \in \mathcal{P}_I\}.$$

Example 3. Labeling the heads of the rules in (1) and (2) leads to the following rules (without negative literals over labeled predicates) in $\Phi(R)$ for R from Example 1:

$$ok^k(C) \leftarrow cs(C), st(S), in(S, C)$$
$$ko^k(C) \leftarrow cs(C), \sim ok(C)$$

In view of $\mathcal{P}_I = \{ok/1, ko/1\}$, $\Pi(\mathcal{P}_I)$ consists of the rules:

$$ok(C) \leftarrow ok^k(C) \qquad\qquad ko(C) \leftarrow ko^k(C)$$

Moreover, the rules in $\Delta(\mathcal{P}_I)$ prepare definition expansions:

$$ok^k(C) \leftarrow ok^{k+1}(C) \qquad\qquad ko^k(C) \leftarrow ko^{k+1}(C) \qquad\qquad □$$

Given a constant stream $(c_i)_{i\geq 1}$, we aim at successive ground instantiations of $\Phi(R)$, $\Pi(\mathcal{P}_I)$, and $\Delta(\mathcal{P}_I)$ capturing the supported as well as the stable models of $F \cup R$ relative to each universe $\mathcal{C}_i$ and arbitrary facts F over $(\mathcal{P}_E, \mathcal{C}_i, \emptyset)$. To this end, we denote the ground substitutions and atoms that are *particular* to some $i \geq 0$ by $\Sigma(V, \mathcal{C}_i, i) = \{\sigma \in \Sigma(V, \mathcal{C}_i) \mid \max\{j \mid (X \mapsto c_j) \in \sigma\} = i\}$ and $atom(\mathcal{P}, \mathcal{C}_i, i) = \{p(X_1, \ldots, X_n)\sigma \mid p/n \in \mathcal{P}, \sigma \in \Sigma(\{X_1, \ldots, X_n\}, \mathcal{C}_i, i)\}$.[1] The resulting partition of substitutions (and atoms) forms the base for a selective instantiation of $\Phi(R)$, $\Pi(\mathcal{P}_I)$, and $\Delta(\mathcal{P}_I)$ relative to $(c_i)_{i\geq 1}$.

Definition 3. *For a set R of rules over $(\mathcal{P}_E \cup \mathcal{P}_I, \emptyset, V)$ and a constant stream $(c_i)_{i\geq 1}$ over $\mathcal{C}$, we define the expansible instantiation of R for $j \geq 0$ as $exp(R, j) = \bigcup_{i=0}^{j} R^i$, where*

$$R^i = \{(r[i])\sigma \mid r \in \Phi(R) \cup \Pi(\mathcal{P}_I), \sigma \in \Sigma(var(r), \mathcal{C}_i, i)\}$$
$$\cup \{(r[i])\sigma \mid r \in \Delta(\mathcal{P}_I), \sigma \in \Sigma(var(r), \mathcal{C}_i)\}.$$

Example 4. Starting with $\mathcal{C}_0 = \emptyset$, the rules $\Phi(R)$, $\Pi(\mathcal{P}_I)$, and $\Delta(\mathcal{P}_I)$ from Example 3 yield $exp(R, 0) = R^0 = \emptyset$ because each of them contains some variable. For $\mathcal{C}_1 = \{c_1\}$, however, we obtain the following set R^1 of ground rules:

$$R^1 = \left\{ \begin{array}{ll} ok^1(c_1) \leftarrow cs(c_1), st(c_1), in(c_1, c_1) \\ ko^1(c_1) \leftarrow cs(c_1), \sim ok(c_1) \\ ok(c_1) \leftarrow ok^1(c_1) \qquad ko(c_1) \leftarrow ko^1(c_1) \\ ok^1(c_1) \leftarrow ok^2(c_1) \qquad ko^1(c_1) \leftarrow ko^2(c_1) \end{array} \right\}$$

Observe that, beyond substituting variables with c_1, the label k (or $k+1$) is replaced by 1 (or 2). Also note that the atom $ok(c_1)$ over the original predicate $ok/1$ is derivable from $ok^1(c_1)$, an atom over the new predicate $ok^1/1$. Unlike the completion formula $cf(R_1, ok(c_1))$ from Example 2,

$$cf(R^1, ok^1(c_1)) = ok^1(c_1) \rightarrow ((cs(c_1) \wedge st(c_1) \wedge in(c_1, c_1)) \vee ok^2(c_1))$$

includes the yet undefined atom $ok^2(c_1)$ to represent derivations becoming available when another constant is added. In fact, such an additional derivation is contained in R^2, consisting of new ground rules relative to $\mathcal{C}_2 = \{c_1, s_1\}$:

$$R^2 = \left\{ \begin{array}{ll} ok^2(c_1) \leftarrow cs(c_1), st(s_1), in(s_1, c_1) \\ ok^2(s_1) \leftarrow cs(s_1), st(c_1), in(c_1, s_1) \\ ok^2(s_1) \leftarrow cs(s_1), st(s_1), in(s_1, s_1) \\ ko^2(s_1) \leftarrow cs(s_1), \sim ok(s_1) \\ ok(s_1) \leftarrow ok^2(s_1) \qquad ko(s_1) \leftarrow ko^2(s_1) \\ ok^2(c_1) \leftarrow ok^3(c_1) \qquad ko^2(c_1) \leftarrow ko^3(c_1) \\ ok^2(s_1) \leftarrow ok^3(s_1) \qquad ko^2(s_1) \leftarrow ko^3(s_1) \end{array} \right\}$$

While the first six ground rules in R^2, stemming from $\Phi(R)$ and $\Pi(\mathcal{P}_I)$, include the second constant, viz. s_1, two of the four instances of rules in $\Delta(\mathcal{P}_I)$ mention c_1 only.   □

---

[1] Letting $\max \emptyset = 0$, since $\mathcal{C}_0 = \emptyset$, we get $\Sigma(\emptyset, \mathcal{C}_0, 0) = \Sigma(\emptyset, \emptyset) = \{\emptyset\}$, and $atom(\mathcal{P}, \emptyset, 0) = \{p \mid p/0 \in \mathcal{P}\}$ consists of atomic propositions.

Intuitively, the substitutions applied to $\Phi(R)$ (and $\Pi(\mathcal{P}_I)$) aim at new rule instances mentioning the constant c_i at stream position i, and the replacement of labels k by i makes sure that the defined predicates are new. Via instances of rules in $\Pi(\mathcal{P}_I)$, the new predicates are mapped back to the original ones in $\mathcal{P}_I$, at position i concentrating on ground atoms including c_i. The purpose of $\Delta(\mathcal{P}_I)$, on the other hand, is to keep the definitions of atoms in $atom(\mathcal{P}_I, \mathcal{C}_i)$ expansible, and the rules with yet undefined body atoms provide an interface for connecting additional derivations.

For each $i \geq 0$ and R^i as in Definition 3, we have that $head(R^i) = atom(\mathcal{P}_I, \mathcal{C}_i, i) \cup atom(\mathcal{P}_I^k[i], \mathcal{C}_i)$. In view of distinct constants in arguments or different predicate names, respectively, $head(R^i) \cap head(R^j) = \emptyset$ for $i > j \geq 0$. Hence, letting $head(exp(R, -1)) = \emptyset$, it also holds that

$$RF(R^i) \cap RF(R^j) = \emptyset,$$
$$CF(R^i, head(R^i)) \cap CF(R^j, head(R^j)) = \emptyset,$$
$$LF(exp(R, i), head(R^i), head(exp(R, i-1)))$$
$$\cap LF(exp(R, j), head(R^j), head(exp(R, j-1))) = \emptyset.$$

As a consequence, the theories

$$RF^i(R) = \bigcup_{j=0}^i RF(R^j),$$
$$CF^i(R) = \bigcup_{j=0}^i CF(R^j, head(R^j)),$$
$$LF^i(R) = \bigcup_{j=0}^i LF(exp(R, j), head(R^j), head(exp(R, j-1)))$$

constitute disjoint unions. As shown next, they reproduce corresponding concepts for $exp(R, i)$ in a modular fashion.

Proposition 1. *Given a set R of rules over $(\mathcal{P}_E \cup \mathcal{P}_I, \emptyset, \mathcal{V})$ and a constant stream $(c_i)_{i \geq 1}$ over $\mathcal{C}$, we have for $j \geq 0$:*

$$RF^j(R) = RF(exp(R, j)),$$
$$CF^j(R) = CF(exp(R, j), head(exp(R, j))),$$
$$LF^j(R) \equiv LF(exp(R, j), head(exp(R, j)), \emptyset).$$

We now turn to the correspondence between supported as well as stable models of $F \cup R$ and $F \cup exp(R, i)$ for $i \geq 0$ and arbitrary facts F over $(\mathcal{P}_E, \mathcal{C}_i, \emptyset)$. To this end, we denote the *expansion atoms* over new predicates in $exp(R, i)$ by $expatom(\mathcal{P}_I, i) = \bigcup_{j=0}^i atom(\mathcal{P}_I^k[j], \mathcal{C}_j)$. For some $a \in atom(\mathcal{P}, \mathcal{C})$, by $\|a\| = \min\{j \geq 0 \mid a \in atom(\mathcal{P}, \mathcal{C}_j)\}$, we refer to the (unique) least j such that $a \in atom(\mathcal{P}, \mathcal{C}_j, j)$. Similarly, $\|r\| = \max\{\|a\| \mid a \in \{head(r)\} \cup body(r)^+ \cup body(r)^-\}$ denotes the smallest j such that all atoms in a ground rule r are contained in $atom(\mathcal{P}, \mathcal{C}_j)$. Moreover, we map any interpretation $I \subseteq atom(\mathcal{P}, \mathcal{C}_i)$ to an extended interpretation $I^* \subseteq atom(\mathcal{P}, \mathcal{C}_i) \cup expatom(\mathcal{P}_I, i)$ as follows:

$$I^* = I \cup \{head(r)^k[j] \mid r \in grd(R, \mathcal{C}_i), I \models bf(body(r)), \|head(r)\| \leq j \leq \|r\|\}.$$

That is, I^* augments a given I with expansion atoms for the heads of rules r whose bodies hold wrt I, where the label k is replaced by the integers from $\|head(r)\|$ to $\|r\|$. Note that the expansion atoms in $atom(\mathcal{P}_I^{k+1}[i], \mathcal{C}_i)$, which have no derivations in $exp(R, i)$, are fixed to false in I^* and any other interpretation $I' \subseteq atom(\mathcal{P}, \mathcal{C}_i) \cup expatom(\mathcal{P}_I, i)$.

The following result shows that our translation approach yields the intended semantics, viz. supported or stable models of a program $F \cup R$, relative to each universe $\mathcal{C}_i$ for $i \geq 0$.

Theorem 1. *Given a set R of rules over $(\mathcal{P}_E \cup \mathcal{P}_I, \emptyset, \mathcal{V})$ and a constant stream $(c_i)_{i \geq 1}$ over $\mathcal{C}$, we have for $j \geq 0$:*

1. *If $I \subseteq atom(\mathcal{P}, \mathcal{C}_j)$ is a supported (or stable) model of $(I \cap atom(\mathcal{P}_E, \mathcal{C}_j)) \cup R$, then I^* is a supported (or stable) model of $(I \cap atom(\mathcal{P}_E, \mathcal{C}_j)) \cup exp(R, j)$.*
2. *If $I' \subseteq atom(\mathcal{P}, \mathcal{C}_j) \cup expatom(\mathcal{P}_I, j)$ is a supported (or stable) model of $(I' \cap atom(\mathcal{P}_E, \mathcal{C}_j)) \cup exp(R, j)$, then $I' = I^*$ for the supported (or stable) model $I = I' \cap atom(\mathcal{P}, \mathcal{C}_j)$ of $(I' \cap atom(\mathcal{P}_E, \mathcal{C}_j)) \cup R$.*

Example 5. The ground rules R^1 and R^2 from Example 4 yield completion formulas $C_1 = CF(R^1, head(R^1))$ and $C_2 = CF(R^2, head(R^2))$ as follows:

$$C_1 = \begin{cases} ok^1(c_1) \to ((cs(c_1) \wedge st(c_1) \wedge in(c_1, c_1)) \vee ok^2(c_1)) \\ ko^1(c_1) \to ((cs(c_1) \wedge \neg ok(c_1)) \vee ko^2(c_1)) \\ ok(c_1) \to ok^1(c_1) \qquad\qquad ko(c_1) \to ko^1(c_1) \end{cases}$$

$$C_2 = \begin{cases} ok^2(c_1) \to ((cs(c_1) \wedge st(s_1) \wedge in(s_1, c_1)) \vee ok^3(c_1)) \\ ok^2(s_1) \to ((cs(s_1) \wedge st(c_1) \wedge in(c_1, s_1)) \\ \qquad\qquad \vee (cs(s_1) \wedge st(s_1) \wedge in(s_1, s_1)) \vee ok^3(s_1)) \\ ko^2(c_1) \to ko^3(c_1) \\ ko^2(s_1) \to ((cs(s_1) \wedge \neg ok(s_1)) \vee ko^3(s_1)) \\ ok(s_1) \to ok^2(s_1) \qquad\qquad ko(s_1) \to ko^2(s_1) \end{cases}$$

Along with literals E_1 and E_2 as in Example 2, fixing atoms over extensional predicates, we obtain (supported) models $I_1^* = \{cs(c_1), ko(c_1), ko^1(c_1)\}$ and $I_2^* = \{cs(c_1), st(s_1), in(s_1, c_1), ok(c_1), ok^1(c_1), ok^2(c_1)\}$ of $RF(R^1) \cup C_1 \cup E_1$ or $RF(R^1) \cup RF(R^2) \cup C_1 \cup C_2 \cup E_2$, respectively. In the transition from I_1^* to I_2^*, $ko(c_1)$ is withdrawn and exchanged with $ok(c_1)$, as with R_1 and R_2 from Example 2. In contrast to the latter, however, the completion formulas C_2 do not invalidate C_1, but rather their (disjoint) union can be used. $\square$

The benefit of expansible instantiation, $exp(R, i)$, in comparison to plain instantiation, $grd(R, \mathcal{C}_i)$, is that completion (and loop) formulas remain intact and can, like ground rules, be accumulated during the successive evolvement of a Herbrand universe. On the other hand, the downside is that, beyond the $\mathcal{O}(|grd(R, \mathcal{C}_i)|)$ ground rules stemming from $\Phi(R)$ and $\Pi(\mathcal{P}_I)$, additional $\mathcal{O}(i \times |atom(\mathcal{P}_I, \mathcal{C}_i)|)$ instances of rules in $\Delta(\mathcal{P}_I)$ are introduced for propagating derivations via expansion atoms. However, for an intensional predicate $p/n \in \mathcal{P}_I$ such that $var(r) = var(head(r))$ for all $r \in R$ with $head(r) = p(X_1, \ldots, X_n)$, definitions of atoms over p/n stay local because rule instances relying on a new constant c_i only provide derivations for atoms including c_i. In view of this, the introduction of a respective labeled predicate and corresponding rules in $\Phi(R)$, $\Pi(\mathcal{P}_I)$, and $\Delta(\mathcal{P}_I)$ is unnecessary, and the original rule(s), such as (2) for $ko/1$ in Example 1, can be instantiated (like members of $\Phi(R) \cup \Pi(\mathcal{P}_I)$) instead.

Example 6. Consider the following non-ground rules R over $\mathcal{P}_E = \{arc/2, vtx/1, init/1\}$ and $\mathcal{P}_I = \{cycle/2, other/2, reach/1\}$, aiming at directed Hamiltonian cycles:

$$R = \left\{ \begin{array}{c} cycle(X,Y) \leftarrow arc(X,Y), \sim other(X,Y) \\ other(X,Y) \leftarrow arc(X,Y), cycle(X,Z), Y \neq Z \\ reach(Y) \leftarrow cycle(X,Y), init(X) \\ reach(Y) \leftarrow cycle(X,Y), reach(X) \\ reach(Y) \leftarrow vtx(Y), \sim reach(Y) \end{array} \right\}$$

Since the variables X and Y occur in the head $cycle(X,Y)$ of the first rule, expansion atoms for $cycle/2$ and respective rules can be omitted. Keeping the original rule, a simplified set R^1 of ground rules is obtained relative to $\mathcal{C}_1 = \{v_1\}$:[2]

$$R^1 \cong \left\{ \begin{array}{c} cycle(v_1, v_1) \leftarrow arc(v_1, v_1), \sim other(v_1, v_1) \\ reach^1(v_1) \leftarrow cycle(v_1, v_1), init(v_1) \\ reach^1(v_1) \leftarrow cycle(v_1, v_1), reach(v_1) \\ reach^1(v_1) \leftarrow vtx(v_1), \sim reach(v_1) \\ other(v_1, v_1) \leftarrow other^1(v_1, v_1) \\ reach(v_1) \leftarrow reach^1(v_1) \\ other^1(v_1, v_1) \leftarrow other^2(v_1, v_1) \\ reach^1(v_1) \leftarrow reach^2(v_1) \end{array} \right\}$$

Given $F_1 = \{arc(v_1, v_1), vtx(v_1)\}$, $F_1 \cup R^1$ has $I_1^* = F_1 \cup \{cycle(v_1, v_1), reach(v_1), reach^1(v_1)\}$ as supported model that is not stable because, for $L_1 = \{reach(v_1), reach^1(v_1)\}$,

$$lf(R^1, reach(v_1), L_1) = reach(v_1) \rightarrow (reach^2(v_1) \vee$$
$$(cycle(v_1, v_1) \wedge init(v_1)) \vee (vtx(v_1) \wedge \neg reach(v_1)))$$

belongs to $LF(R^1, head(R^1), \emptyset)$. While $I_1^* \models RF(R^1) \cup CF(R^1, head(R^1))$, there is no model I' of $RF(R^1) \cup LF(R^1, head(R^1), \emptyset)$ such that $F_1 \subseteq I' \subseteq F_1 \cup head(R^1)$.

   Letting $R_1^2 = R^1 \cup R^2$, where R^2 is the set of new ground rules for $\mathcal{C}_2 = \{v_1, v_2\}$, along with $F_2 = F_1 \cup \{arc(v_1, v_2), arc(v_2, v_1), vtx(v_2)\}$, $F_2 \cup R_1^2$ yields the supported model

$$I_2^* = F_2 \cup \{cycle(v_1, v_2), cycle(v_2, v_1), other(v_1, v_1), other^1(v_1, v_1), other^2(v_1, v_1),$$
$$reach(v_1), reach(v_2), reach^1(v_1), reach^2(v_1), reach^2(v_2)\},$$

---

[2] The condition $Y \neq Z$ filters admissible ground substitutions.

which is not stable either. For $L_2 = \{reach(v_1), reach(v_2), reach^1(v_1), reach^2(v_1), reach^2(v_2)\}$, since the loop formula

$$lf(R_1^2, reach(v_2), L_2) = reach(v_2) \rightarrow ((\bigvee_{i=1}^{2} reach^3(v_i)) \vee$$
$$(cycle(v_1, v_1) \wedge init(v_1)) \vee (cycle(v_2, v_1) \wedge init(v_2)) \vee$$
$$(cycle(v_1, v_2) \wedge init(v_1)) \vee (cycle(v_2, v_2) \wedge init(v_2)) \vee$$
$$(vtx(v_1) \wedge \neg reach(v_1)) \vee (vtx(v_2) \wedge \neg reach(v_2)))$$

is contained in $LF(R_1^2, head(R^2), head(R^1))$, $I_2^* \not\models LF(R_1^2, head(R^2), head(R^1))$. However, when considering loop formulas for atoms defined by R^1 and R^2 in isolation, one can check that $I_2^* \models LF(R^1, head(R^1), \emptyset) \cup LF(R^2, head(R^2), \emptyset)$. In fact, positive dependencies between the atoms in L_2 involve rules from both R^1 and R^2. That is, R^1 and R^2 are not mutually independent in the sense of [15, 16].   □

4   Solving Expansible Programs

For an empirical evaluation of our translation approach, we modeled two benchmark domains, *Graph Coloring* and *Partner Units*, of the Fifth ASP Competition [17] by expansible programs in the language of *clingo* 4 [6]. Starting from an empty graph and no colors, the expansible program for *Graph Coloring* allows for a successive incorporation of vertices, arcs, and colors. While the addition of vertices and arcs constrains admissible colorings, colors serve as resources that must be increased whenever a coloring task turns out to be unsatisfiable. In *Partner Units*, pairwisely connected zones and sensors need to be mapped to units, where two units are partners if the zone of a connected pair is mapped to one of them and the corresponding sensor to the other. Moreover, at most two zones and two sensors may share a unit, and the number of partners per unit must not exceed a given threshold, varying between two and four in problem instances of the Fifth ASP Competition. That is, the demand for units increases whenever the capacities are exceeded upon the successive addition of zones and sensors.

For both benchmark domains, the idea is to gradually expand and solve instances over arbitrarily many objects by introducing the objects, along with respective data, one after the other. Similarly, resources such as colors or units are increased on demand, rather than fixing and thus limiting them a priori. For instance, the following sequence of facts induces four successive *Graph Coloring* tasks: $F_1 = \{vtx(v_1, 1)\}$, $F_2 = \{col(n_1, 2)\}$, $F_3 = \{vtx(v_2, 3), arc(v_1, v_2, 3)\}$, and $F_4 = \{col(n_2, 4)\}$. While introducing the first vertex v_1 in F_1 yields an unsatisfiable task, a coloring is obtained after supplying color n_1 in F_2. However, one color is no longer sufficient when adding the second vertex v_2 and an arc to v_1 in F_3. Thus, F_4 provides another color n_2, leading to colorings mapping v_1 to n_1 (or n_2) and v_2 to n_2 (or n_1). Note that each of the above facts includes as last argument the maximum position of mentioned vertices or colors in the constant stream $(v_1, n_1, v_2, n_2, \dots)$. For one, this enables a reuse of constants for referring to vertices and colors, and w.l.o.g. we may assume that colors are denoted by consecutive integers starting with 1, viz. $n_1 = 1$, $n_2 = 2$, and so on. For another, the arguments indicating stream positions can be explored to distinguish corresponding rules in a parametrized *clingo* 4 program as follows:

$$col(C) \leftarrow col(C, k) \tag{3}$$

$$vtx(X) \leftarrow vtx(X, k) \tag{4}$$

$$arc(X, Y) \leftarrow arc(X, Y, k) \tag{5}$$

$$new(X, C, k) \leftarrow vtx(X, k), col(C) \tag{6}$$

$$new(X, C, k) \leftarrow vtx(X), col(C, k) \tag{7}$$

$$\{map(X, C)\} \leftarrow new(X, C, k) \tag{8}$$

$$\leftarrow map(X, C), map(Y, C), arc(X, Y, k), col(C) \tag{9}$$

$$\leftarrow map(X, C), map(Y, C), arc(X, Y), col(C, k) \tag{10}$$

$$has(X, C) \leftarrow new(X, C, k), map(X, C) \tag{11}$$

$$has(X, C) \leftarrow new(X, C, k), has(X, C{+}1) \tag{12}$$

$$\leftarrow new(X, C, k), map(X, C{-}1), has(X, C) \tag{13}$$

$$\leftarrow vtx(X, k), {\sim}has(X, 1) \tag{14}$$

Assuming that the program parameter k is successively replaced by the stream positions of objects in supplied facts, the rules in (3)–(5) provide projections dropping the positions from respective colors, vertices, or arcs. The auxiliary predicate $new/3$, defined in (6) and (7), indicates pairs of vertices and colors such that either of them is introduced at a stream position substituted for parameter k. Given this, applicable instances of the choice rule (cf. [18]) in (8) have distinct heads at different positions, so that expansion atoms can be omitted for $map/2$. The integrity constraints (i.e., rules with false heads) in (9) and (10) deny choices of the same color for adjacent vertices, where the body atoms $arc(X, Y, k)$ or $col(C, k)$ confine applicable instances to new arcs or colors, respectively. The purpose of atoms of the form $has(v, n)$ is to indicate that a vertex v is mapped to some color $m \geq n$. When either v or n is introduced at a stream position, the rule in (11) captures the case that v is mapped to n, while colors added later on are addressed by the rule in (12). Making use of the convention that colors are denoted by consecutive integers, the latter includes expansion atoms of the form $has(v, n{+}1)$, rather than $has^{k+1}(v, n)$ or a corresponding *clingo* 4 representation $has(v, n, k{+}1)$, respectively. Note that, by saving an argument for the predicate label, the number of introduced expansion atoms remains linear, thus avoiding a quadratic blow-up as discussed below Example 5. The integrity constraints in (13) and (14) further investigate atoms over $has/2$ to make sure that each vertex is mapped to exactly one color. Finally, to keep *clingo* 4 off discarding body atoms that are not necessarily defined when instances of (12) and (14) are introduced, the program in (3)–(14) has to be accompanied by *#external* $has(X, C{+}1) : new(X, C, k)$ and *#external* $has(X, 1) : vtx(X, k)$. Without going into details, we note that *Partner Units* can be modeled in a similar way, where zones and sensors amount to vertices and units to colors.

Table 1(a) and (b) provide experimental results of running *clingo* 4 (version 4.4.0) on the *Graph Coloring* and *Partner Units* instances of the Fifth ASP Competition. In particular, we used the Python interface of *clingo* 4 to successively

Table 1. One- vs. multi-shot: (a) *Graph Coloring* and (b) *Partner Units*

Instance	#S	ØS	#U	ØU	#S	ØS	#U	ØU
04	125	0.28	5	0.10	125	0.15	5	0.04
05	125	0.13	5	0.12	125	0.03	5	0.08
07	125	0.13	5	0.16	125	0.02	5	0.13
08	125	0.14	5	0.20	125	0.04	5	0.16
13	130	0.13	5	0.11	130	0.02	5	0.04
21	121	0.26	5	0.10	121	0.27	5	0.03
22	121	0.29	5	0.09	121	0.80	5	0.02
23	135	1.17	5	0.11	135	2.07	5	0.05
25	125	0.21	5	0.11	125	0.03	5	0.04
32	140	0.22	6	68.59	140	0.07	6	95.33
36	128	0.52	5	0.18	128	0.55	5	0.14
39	124	0.15	5	0.15	124	0.07	5	0.11
40	121	0.59	5	0.13	121	2.14	5	0.07
46	124	0.69	5	0.15	124	0.50	5	0.10
47	132	0.68	5	0.12	132	0.63	5	0.04
48	128	1.81	5	0.10	128	1.76	5	0.02
50	130	0.88	5	0.12	130	0.49	5	0.04
56	139	2.17	5	0.17	139	2.43	5	0.11
59	128	0.35	5	0.21	128	0.09	5	0.15
60	118	0.16	5	0.11	118	0.04	5	0.02

Instance	#S	ØS	#U	ØU	#S	ØS	#U	ØU
026	40	0.10	10	34.69	40	0.04	10	3.00
052	40	0.10	10	15.29	40	0.03	10	5.58
058	40	0.10	10	22.97	40	0.04	10	5.67
069	40	0.11	10	12.87	40	0.04	10	5.98
091	40	0.10	10	3.71	40	0.04	10	8.42
099	40	0.10	10	10.49	40	0.04	10	5.41
100	40	0.09	10	57.05	40	0.04	10	2.13
102	40	0.10	10	13.29	40	0.04	10	8.45
114	40	0.10	10	16.95	40	0.04	10	7.13
115	40	0.10	10	19.02	40	0.04	10	3.98
119	40	0.10	10	26.01	40	0.04	10	5.65
127	40	0.10	10	4.99	40	0.04	10	9.38
153	40	0.11	10	32.08	40	0.04	10	6.04
154	40	0.10	10	22.75	40	0.04	10	10.23
156	40	0.10	10	11.63	40	0.04	10	9.41
161	40	0.11	10	24.56	40	0.04	10	5.10
175	40	0.12	10	48.44	40	0.04	10	4.86
180	36	0.08	9	2.02	36	0.03	9	1.84
188	40	0.11	10	54.67	40	0.03	10	2.69
196	40	0.07	10	26.80	40	0.03	10	13.97

(a) One- vs. multi-shot: *Graph Coloring*   (b) One- vs. multi-shot: *Partner Units*

introduce objects, viz. vertices or zones and sensors, and instantiate respective rules in (parametrized) expansible programs.[3] Whenever this leads to an unsatisfiable task, another color or unit is added in turn, thus obtaining sequences of satisfiable as well as unsatisfiable tasks of gradually increasing size. With sequential runtimes restricted to 10 min per instance on a Linux PC equipped with 2.4 GHz processors, columns headed by #S and ØS report the number of solved satisfiable tasks along with the corresponding average runtime in seconds, and columns headed by #U and ØU provide the same information for unsatisfiable tasks. We compare the performance of *clingo* 4 in two operation modes: one-shot solving, in which each task is processed independently from scratch, and multi-shot solving, where rule instances are successively accumulated and conflict information can be carried over between tasks. Experimental results with one-shot solving are given in the middle parts of Table 1(a) and (b), followed by multi-shot solving on the right. Notably, as instantiation times are negligible, the comparison primarily contrasts the search performance in solving successive tasks either independently or incrementally.

Given the combinatorial nature of the benchmark domains, one- and multi-shot solving scale up to tasks of similar size, leading to the same number of solved tasks. The gap between both solving approaches turns out to be small on the instances of *Graph Coloring*, where instance 32 yields an exceptionally hard unsatisfiable task that can still be solved in time. On the *Partner Units* instances in Table 1(b), however, we observe that multi-shot solving consistently reduces the runtime for rather easy satisfiable tasks. Except for two instances (091 and 127), it also yields shorter runtimes for unsatisfiable tasks, even by an order of magnitude in several cases (026, 100, 175, and 188). Comparing the numbers of conflicts revealed that, in these cases, multi-shot solving indeed encounters an

---

[3] http://svn.code.sf.net/p/potassco/code/trunk/gringo/examples/.

order of magnitude fewer conflicts, viz. about 200,000 vs. 2,000,000 on average with one-shot solving. The other way round, the difference amounts to a maximum factor of 2 (roughly 250,000 vs. 500,000 conflicts) between single- and multi-shot solving on instance 091. This indicates an increased robustness due to the reuse of conflict information in multi-shot solving.

5   Discussion

We introduced a simple approach to successively incorporating new objects into (multi-shot) ASP solving. Our approach rests upon a translation and refrains from dedicated algorithms. Also, it is modular and thus allows for adding new information without altering the existing ground program or underlying constraints, respectively. Hence, our approach enables incremental finite model finding [19], even under nonmonotonicity faced with supported or stable instead of classical models. Technically, it employs a less restrictive notion of modularity than [15,16]: Proposition 1 applies to successive ground programs of an expansible instantiation (according to Definition 3) for the rules in Example 6, although the ground programs are not mutually independent. In view of a lacking theoretical elaboration, incremental ASP systems like *clingo* 4 do not yet provide automatic support for expanding the definitions of atoms or handling mutual positive dependencies between successive ground programs. Our work thus outlines potential future system refinements in these regards.

A related approach is *dlvhex* [20] using external sources for value invention. This is accomplished by dedicated grounding algorithms incorporating external objects. Once grounding is completed, no further objects are taken into account. Unlike this, ASP systems relying on lazy grounding, like *asperix* [21], *gasp* [22], and *omiga* [23], aim at instantiating variables on demand. However, all of them rely on a fixed Herbrand universe and use dedicated grounding and solving algorithms, whose performance does not match that of modern ASP systems. Although the lazy grounding approach in [24] is tailored for query answering, interestingly, it introduces so-called Tseitin variables resembling ground expansion atoms in partial instantiations.

References

1. Baral, C.: Knowledge Representation, Reasoning and Declarative Problem Solving. Cambridge University Press, New York (2003)
2. Davis, M., Putnam, H.: A computing procedure for quantification theory. J. ACM **7**, 201–215 (1960)
3. Claessen, K., Sörensson, N.: New techniques that improve MACE-style finite model finding. In: Proceedings of the Workshop on Model Computation (MODEL 2003) (2003)
4. Eén, N., Sörensson, N.: Temporal induction by incremental SAT solving. Electron. Notes Theoret. Comput. Sci. **89**(4), 543–560 (2003)
5. Rintanen, J., Heljanko, K., Niemelä, I.: Planning as satisfiability: parallel plans and algorithms for plan search. Artif. Intell. **170**(12–13), 1031–1080 (2006)

6. Gebser, M., Kaminski, R., Kaufmann, B., Schaub, T.: Clingo = ASP + control: Preliminary report [25]. http://arxiv.org/abs/1405.3694v1

7. Apt, K., Blair, H., Walker, A.: Towards a theory of declarative knowledge. In: Minker, J. (ed.) Foundations of Deductive Databases and Logic Programming, pp. 89–148. Morgan Kaufmann, San Mateo (1987)

8. Gelfond, M., Lifschitz, V.: The stable model semantics for logic programming. In: Proceedings of ICLP 1988, pp. 1070–1080. MIT Press (1988)

9. Fages, F.: Consistency of Clark's completion and the existence of stable models. J. Methods Logic Comput. Sci. 1, 51–60 (1994)

10. Clark, K.: Negation as failure. In: Gallaire, H., Minker, J. (eds.) Logic and Data Bases, pp. 293–322. Plenum, New York (1978)

11. Lee, J.: A model-theoretic counterpart of loop formulas. In: Proceedings of the Nineteenth International Joint Conference on Artificial Intelligence (IJCAI 2005), pp. 503–508 (2005)

12. Lin, F., Zhao, Y.: ASSAT: computing answer sets of a logic program by SAT solvers. Artif. Intell. 157(1–2), 115–137 (2004)

13. Ullman, J.: Principles of Database and Knowledge-Base Systems. Computer Science Press, Rockville (1988)

14. Marques-Silva, J., Lynce, I., Malik, S.: Conflict-driven clause learning SAT solvers. In: Handbook of Satisfiability, pp. 131–153. IOS (2009)

15. Oikarinen, E., Janhunen, T.: Modular equivalence for normal logic programs. In: Proceedings of the European Conference on Artificial Intelligence (ECAI 2006), pp. 412–416. IOS (2006)

16. Janhunen, T., Oikarinen, E., Tompits, H., Woltran, S.: Modularity aspects of disjunctive stable models. J. Artif. Intell. Res. 35, 813–857 (2009)

17. Calimeri, F., Gebser, M., Maratea, M., Ricca, F.: The design of the fifth answer set programming competition [25]. http://arxiv.org/abs/1405.3710v4

18. Simons, P., Niemelä, I., Soininen, T.: Extending and implementing the stable model semantics. Artif. Intell. 138(1–2), 181–234 (2002)

19. Gebser, M., Sabuncu, O., Schaub, T.: An incremental answer set programming based system for finite model computation. AI Commun. 24(2), 195–212 (2011)

20. Eiter, T., Fink, M., Krennwallner, T., Redl, C.: Grounding HEX-programs with expanding domains. In: Proceedings of GTTV 2013, pp. 3–15 (2013)

21. Lefèvre, C., Nicolas, P.: The first version of a new ASP solver: ASPeRiX. In: Erdem, E., Lin, F., Schaub, T. (eds.) LPNMR 2009. LNCS, vol. 5753, pp. 522–527. Springer, Heidelberg (2009)

22. Dal Palù, A., Dovier, A., Pontelli, E., Rossi, G.: Answer set programming with constraints using lazy grounding. In: Hill, P.M., Warren, D.S. (eds.) ICLP 2009. LNCS, vol. 5649, pp. 115–129. Springer, Heidelberg (2009)

23. Dao-Tran, M., Eiter, T., Fink, M., Weidinger, G., Weinzierl, A.: OMiGA: an open minded grounding on-the-fly answer set solver. In: del Cerro, L.F., Herzig, A., Mengin, J. (eds.) JELIA 2012. LNCS, vol. 7519, pp. 480–483. Springer, Heidelberg (2012)

24. De Cat, B., Denecker, M., Stuckey, P.: Lazy model expansion by incremental grounding. In: Technical Communications of ICLP 2012, pp. 201–211. LIPIcs (2012)

25. Leuschel, M., Schrijvers, T. (eds.): Technical Communications of the Thirtieth International Conference on Logic Programming (ICLP 2014). Theory and Practice of Logic Programming, Online Supplement (2014)

Progress in *clasp* Series 3

Martin Gebser[1,3], Roland Kaminski[3], Benjamin Kaufmann[3], Javier Romero[3],
and Torsten Schaub[2,3]([✉])

[1] Aalto University, HIIT, Espoo, Finland
[2] INRIA Rennes, Rennes, France
[3] University of Potsdam, Potsdam, Germany
{gebser,kaminski,kaufmann,javier,torsten}@cs.uni-potsdam.de

Abstract. We describe the novel functionalities comprised in *clasp*'s series 3. This includes parallel solving of disjunctive logic programs, parallel optimization with orthogonal strategies, declarative support for specifying domain heuristics, a portfolio of prefabricated expert configurations, and an application programming interface for library integration. This is complemented by experiments evaluating *clasp* 3's optimization capacities as well as the impact of domain heuristics.

1 Introduction

The success of Answer Set Programming (ASP; [1]) is largely due to the availability of effective solvers. Early ASP solvers *smodels* [2] and *dlv* [3] were followed by SAT-based ones, such as *assat* [4] and *cmodels* [5], before genuine conflict-driven ASP solvers like *clasp* [6] and *wasp* [7] emerged. In addition, there is a continued interest in mapping ASP onto solving technology in neighboring fields [8,9].

In what follows, we provide a comprehensive description of *clasp*'s series 3 (along with some yet unpublished features already in *clasp* 2). Historically, *clasp* series 1 [6] constitutes the first genuine conflict-driven ASP solver, featuring conflict-driven learning and back-jumping. *clasp* series 2 [10] supports parallel search via shared memory multi-threading. *clasp* series 3 further extends its predecessors by integrating various advanced reasoning techniques in a uniform parallel setting. The salient features of *clasp* 3 include parallel solving of disjunctive logic programs, parallel optimization with orthogonal strategies, declarative support for specifying domain heuristics, a portfolio of prefabricated expert configurations, and an application programming interface for library integration. We detail these functionalities in Sects. 2–6 from a system- and user-oriented viewpoint. Section 7 is dedicated to an empirical study comparing the various optimization strategies of *clasp* 3. Also, we demonstrate the impact of domain heuristics and contrast their performance to using disjunctive logic programs when enumerating inclusion minimal answer sets.

We refer the interested reader to [11] for the formal foundations and basic algorithms underlying conflict-driven ASP solving as used in *clasp*. This includes

---

T. Schaub—Affiliated with Simon Fraser University, Canada, and IIIS Griffith University, Australia.

F. Calimeri et al. (Eds.): LPNMR 2015, LNAI 9345, pp. 368–383, 2015.
DOI: 10.1007/978-3-319-23264-5_31

basic concepts like completion and loop nogoods as well as algorithmic ones related to conflict-driven constraint learning (CDCL).

2 Disjunctive Solving

Solving disjunctive logic programs leads to an elevated level of complexity [12] because unfounded-set checking becomes a co-NP-complete problem [13]. As a consequence, corresponding systems combine a solver generating solution candidates with another testing them for unfounded-freeness. For instance, *dlv* [3] carries out the latter using a SAT solver [14], *claspD* [15] uses *clasp* for both purposes. Common to both is that search is driven by the generating solver and that the testing solver is merely re-invoked on demand. Such repeated invocations bear a great amount of redundancy, in particular, when using conflict-learning solvers because learned information is lost.

Unlike this, *clasp* 3 enables an equitable interplay between generating and testing solvers. Solver units are launched initially with their respective Boolean constraint problems, and they subsequently communicate in a bidirectional way. Constraints relevant for the unfounded set check are enabled using *clasp*'s interface for solving under assumptions (cf. Sect. 6). This allows both units to benefit from conflict-driven learning over whole runs. The theoretical foundations for this approach are laid in [16].[1]

clasp 3 decomposes the unfounded set problem based on the strongly connected components (SCCs) of the program's positive dependency graph. Head cycle components (SCCs sharing two head atoms of a rule) are associated with testing solvers responsible for complex unfounded set checks. Head cycle free components are checked using *clasp*'s tractable unfounded set checking procedure for normal programs. The generator combines propagation via completion nogoods with these unfounded set checks. Tractable unfounded set checks are performed once at each decision level. Because complex unfounded set checks are expensive, their frequency is limited by default (this is configurable using option `--partial-check`); only checks for total assignments are mandatory before a model is accepted. Finally, it is worth mentioning that *clasp* 3 propagates top-level assignments from generators to testers. That is, whenever a variable's truth value is determined by the generator, it is also fixed in the tester.

Building on *clasp*'s multi-threaded architecture, the assembly of generating and testing solvers is reproduced to obtain n threads running in parallel. This results in n generating and $k \times n$ testing solvers (given k head cycle components), all of which can be separately configured, for instance, by specifying portfolios of search strategies for model generation and/or unfounded set checking (cf. Sect. 5). Notably, different generators as well as testers solving the same unfounded set sub-problem share common data, rather than copying it n times. The testing solver can be configured via option `--tester`, accepting a string of *clasp* options. The individual performance of the $k \times n$ solvers and their respective problem statistics can be inspected by option `--stats=2`.

---

[1] [16] also contains experiments done with an early and restricted prototype of *clasp* 3.

Another advance in *clasp* 3 is its extension of preprocessing to disjunctive ASP. Preprocessing starts with the identification of equivalences and resulting simplifications on the original program (controlled by option `--eq`). This extends the techniques from [17] to disjunctive logic programs. A subsequent dependency analysis of the program results in the aforementioned decomposition in head cycle components. This is followed by a translation of recursive weight constraints. In contrast to previous approaches, *clasp* 3 restricts this translation to weight constraints belonging to some head cycle component. Finally, representations for completion nogoods and unfounded set nogoods for each head cycle component are created (according to [16]). Notably, *clasp* 3's preprocessor can be decoupled with option `--pre`, providing a mapping between two disjunctive logic programs in *smodels* format. In this way, it can be used as a preprocessor for other ASP solvers for (disjunctive) logic programs relying on *smodels* format. For example, the program

```
a ; b.                c :- a.    a :- c.    d :- not c.
```

is translated by calling 'clasp --pre' (and conversion to human-readable form) into

```
a :- not b. b :- not a. c :- a.              d :- not a.
```

Here, *clasp* 3's preprocessor turns the disjunction 'a ; b' into two normal rules due to a missing head cycle and identifies the equivalence between a and c.

3 Optimization

Lexicographic optimization of linear objective functions is an established component of ASP solvers, manifested by #*minimize* statements [2] and weak constraints [3]. Traditionally, optimization is implemented in ASP solvers via branch-and-bound search. As argued in [18], this constitutes a *model-guided* approach that aims at successively producing models of descending costs until an optimal model is found (by establishing the unsatisfiability of the problem with a lower cost). Since series 2, *clasp* features several corresponding strategies and heuristics [19], including strategies that allow for non-uniform descents during optimization. For instance, in multi-criteria optimization, this enables *clasp* to optimize criteria in the order of significance, rather than pursuing a rigid lexicographical descent. *clasp* 3 complements this with so-called *core-guided* optimization techniques originating in the area of MaxSAT [14]. Core-guided approaches rely on successively identifying and relaxing unsatisfiable cores until a model is obtained. The implementation in *clasp* 3 seamlessly integrates the core-guided optimization algorithms oll^2 [20] and *pmres* [21]. Both algorithms can be (optionally) combined with disjoint core preprocessing [22], which calculates an initial set of unsatisfiable cores to initialize the algorithms, and as a side effect provides an approximation of the optimal solution. Furthermore, whenever an algorithm relaxes an unsatisfiable core, constraints have to be added to the solver. These constraints can be represented using either equivalences or

---

[2] [20] contains experiments with an early prototype called *unclasp*.

implications. The former offers a slightly stronger propagation at the expense of adding more constraints.

The specific optimization strategy is configured in *clasp* 3 via option --opt-strategy. While its first argument distinguishes between model- and core-guided optimization, the second one handles the aforementioned refinements.

Building on *clasp*'s multi-threaded architecture, model- and core-guided optimization techniques can be combined. As detailed in Sect. 5, *clasp* 3 supports optimization portfolios for running several threads in parallel with different approaches, strategies, and heuristics, exchanging lower and upper bounds of objective functions (in addition to conflict nogoods). This combination of model- and core-guided optimization makes the overall optimization process more robust, as we empirically show in Sect. 7.

Moreover, *clasp* 3 adds a new reasoning mode for enumerating optimal models via option --opt-mode=optN. As usual, the number of optimal answer sets can be restricted by adding an integer to the command line. Interestingly, this option can also be combined with the intersection and union of answer sets (cf. option --enum-mode), respectively. This is of great practical relevance whenever it comes to identifying atoms being true or false in all optimal answer sets.

Finally, it is worth mentioning that *clasp* 3's optimization capacities can also be used for solving PB and (weighted/partial) MaxSAT problems. In fact, *clasp* 3 won the second place in the *Unweighted Max-SAT - Industrial* category by using its core-guided optimization in the Max-SAT Evaluation in 2014.

4   Heuristics

In many domains, general-purpose solving capacities can be boosted by domain-specific heuristics. To this end, *clasp* 3 provides a general declarative framework for incorporating such heuristics into ASP solving. The heuristic information is exploited by a dedicated **domain** heuristic in *clasp* when it comes to non-deterministically assigning a truth value to an atom. In fact, *clasp*'s decision heuristic is modifiable from within a logic program as well as from the command line. This allows for specifying context-dependent activation of heuristic biases interacting with a problem's encoding. This approach was formally introduced in [23][3] and extended in *clasp* 3 as described below. On the other hand, *clasp* 3's command line options allow us to directly refer to structural components of the program (optimization statements, strongly connected components, etc.) and do not require any additional grounding. The domain heuristic is enabled by setting option --heuristic to **domain**, which extends *clasp*'s activity-based vsids heuristic.

Heuristic information is represented by means of the predicate _heuristic. The ternary version of this predicate takes a reified atom, a heuristic modifier, and an integer to quantify a heuristic modification. There are four primitive

---

[3] [23] also contains experiments done with an early and restricted prototype, called *hclasp*.

heuristic modifiers, viz. `sign`, `level`, `init`, and `factor`. The modifier `sign` allows for controlling the truth value assigned to variables subject to a choice. For example, the program

```
{a}.   _heuristic(a,sign,1).
```

produces the answer set containing a first. The modifier `level` allows for using integers to rank atoms. Atoms at higher levels are decided before atoms with lower level. The default level for each atom is 0. Atoms sharing the same level are decided by their `vsids` score. Modifiers `true` and `false` are defined in terms of `level` and `sign`.

```
_heuristic(X,level,Y)  :- _heuristic(X,true,Y).
_heuristic(X,sign,1)   :- _heuristic(X,true,Y).
_heuristic(X,level,Y)  :- _heuristic(X,false,Y).
_heuristic(X,sign,-1)  :- _heuristic(X,false,Y).
```

The modifiers `init` and `factor` allow us to modify the scores of the underlying `vsids` heuristic. Unlike `level`, they only bias the search without establishing a strict ranking among atoms. The modifier `init` allows us to add a value to the initial heuristic score of an atom that decays as any `vsids` score, while `factor` allows us to multiply the `vsids` scores of atoms by a given value. For example, the following rule biases the solver to choosing $p(T-1)$ whenever $p(T)$ is true.

```
_heuristic(p(T-1),factor,2)  :- p(T).
```

clasp 3's structure-oriented heuristics are supplied via the command line. Apart from supplying `--heuristic=domain`, the heuristic modifications are specified by option `--dom-mod=`m, p, where m ranges from 0 to 5 and specifies the modifier:

m	Modifier	m	Modifier	m	Modifier
0	None	1	level	2	sign (positive)
3	true	4	sign (negative)	5	false

and p specifies bit-wisely the atoms to which the modification is applied:

0	Atoms only
1	Atoms that belong to strongly connected components
2	Atoms that belong to head cycle components
4	Atoms that appear in disjunctions
8	Atoms that appear in optimization statements
16	Atoms that are shown

Whenever m equals 1, 3, or 5, the level of the selected atoms depends on p. For example, with option `--dom-mod=2,8`, we apply a positive `sign` to atoms appearing in optimization statements, and with option `--dom-mod=1,20`, we apply modifier `level` to both atoms appearing in disjunctions as well as shown atoms. In this case, atoms satisfying both conditions are assigned a higher level than those that are only shown, and these get a higher level than those only appearing in disjunctions.

Compared to programmed heuristics, the command line heuristics do not allow for applying modifiers `init` or `factor` and cannot represent dynamic heuristics. But they allow us to directly refer to structural components of the program and do not require any additional grounding. When both methods are combined, the choices modified by the _heuristic predicate are not affected by the command line heuristics. When launched with option --stat, *clasp* 3 prints the number of modified choices.

Apart from boosting solver performance, domain specific heuristics can be used for computing inclusion minimal answer sets [24,25]. This can be achieved by ranking choices over shown atoms highest and setting their sign modifier to false. As an example, consider the following program

```
1 {a(1..3)}.    a(2) :- a(3).    a(3) :- a(2). {b(1)}.    #show a/1.
```

Both the command line option '--dom-mod=5,16' as well as the addition of the heuristic fact '_heuristic(a(1..3),false,1).' guarantee that the first answer set produced is inclusion minimal wrt. the atoms of predicate a/1. Moreover, both allow for enumerating all inclusion minimal solutions in conjunction with option --enum-mod=domRec. In our example, we obtain the answer sets {a(1)} and {a(2), a(3)}. Note that this enumeration mode relies on solution recording and is thus prone to an exponential blow-up in space. However, this often turns out to be superior to enumerating inclusion minimal model via disjunctive logic programs, which is guaranteed to run in polynomial space. We underpin this empirically in Sect. 7.

Independent of the above domain-specific apparatus, *clasp* provides means for configuring the sign heuristics, fixing which truth values get assigned to which type of variables. In general, *clasp* selects signs based on a given sign heuristics. For instance, this can be progress saving (--save-progress; [26]) or an optimization-oriented heuristic (--opt-heuristic). Also, each decision heuristic in *clasp* implements a sign heuristic. For example, *clasp*'s vsids heuristic prefers the sign of a variable according to the frequency of the corresponding literal in learned nogoods. Whenever no sign heuristic applies, e.g. in case of ties, the setting of option --sign-def determines the sign; by default, it assigns atoms to false and bodies to true. Other options are 1 (assign true), 2 (assign false), 3 (assign randomly), and 4 (assign bodies and atoms in disjunctions true). Finally, the option --sign-fix permits to disable all sign heuristics and enforce the setting of --sign-def.

5   Configuration

Just as any modern conflict-driven ASP, PB, or SAT solver, *clasp* is sensitive to search configurations. In order to relieve users from extensive parameter tuning, *clasp* offers a variety of prefabricated configurations that have shown to be effective in different settings. A specific configuration is selected by means of option --configuration, taking one of the following arguments:

frumpy Use conservative defaults similar to those used in earlier *clasp* versions
  jumpy Use more aggressive defaults (than frumpy)
tweety Use defaults geared towards typical ASP problems
trendy Use defaults geared towards industrial problems
crafty Use defaults geared towards crafted problems
  handy Use defaults geared towards large problems
<file> Use configuration file to configure solver(s)

The terms 'industrial' and 'crafted' refer to the respective categories at SAT competitions; 'aggressive defaults' restart search and erase learned nogoods quite frequently. Unlike previous *clasp* series relying on the default configuration frumpy aiming at highest robustness, the one of *clasp* 3, viz. tweety, was automatically identified by *piclasp*[4] (a configurator for *clasp* based on *smac* [27]) and manually smoothened afterwards. Such an automatic approach is unavoidable in view of *clasp* 3's huge space of 10^{60} configurations composed of more than 90 parameters.

Note that using a configuration file enables freely customizable solver portfolios in parallel solving. We rely on this for tackling optimization problems in Sect. 7 by running complementary optimization strategies in parallel. For an example of such a portfolio, call *clasp* with option --print-portfolio. The result constitutes a portfolio of complementary default configurations for parallel ASP solving. This also extends to the disjunctive case, where the configuration of testing solvers can be configured by option '--tester=--configuration=<file>' to apply the portfolio in <file> to the tester. Options given on the command-line are added to all configurations in a configuration file. If an option is given both on the command-line and in a configuration file, the one from the command-line takes precedence.

When solving in parallel, the configurations in the portfolio are assigned to threads in a round-robin fashion. That is, *clasp* runs with the configuration from the first line in thread 0, with the one from the second line in thread 1, etc., until all threads are (circularly) assigned configurations from the portfolio. The mapping of portfolios to threads is used for providing thread-specific solver statistics. That is, launching *clasp* 3 with --stats=2 does not only provide statistics aggregated over all threads but also for each individual one. Moreover, the winning thread[5] is identified by this mapping (and printed after 'Winner:').

Furthermore, *clasp*'s multi-threaded architecture was extended to handle more complex forms of nogood exchange. In addition to the global distribution scheme described in [10], *clasp* 3 implements a new thread-local scheme. The scheme can be configured by option --dist-mode. In the new scheme, each thread has a (lock-free) multi-producer/single-consumer queue. For distributing nogoods, threads push "interesting" nogoods onto the queues of their peers. For integrating nogoods, threads pop nogoods from their local queues. On the

---

[4] http://www.cs.uni-potsdam.de/piclasp.
[5] The winning thread either exhausts the search space or produces the last model if no complete search space traversal is necessary.

other hand, *clasp* 3 now supports topology-based nogood exchange. To this end, option `--integrate` allows for exchanging nogoods among `all` peers or those connected in the form of a `ring`, hyper `cube`, or extended hyper cube (`cubex`). With the global scheme, nogoods are distributed among all threads but only integrated by threads from their peers (via a peer check upon receive). With the thread-local scheme, threads distribute nogoods only to peers (via a peer check upon send). Hence, the thread-local scheme is more suited for a topology-based exchange.

6   Library

While *clasp* is a versatile stand-alone system, it can also be integrated as a library. To this end, *clasp* 3 provides various interfaces for starting and controlling operative solving processes. This includes interfaces for incremental solving, updating a logic program, managing solver configurations, and for (asynchronous as well as iterative) solving under assumptions. Furthermore, recorded nogoods, heuristic values, and other dynamic information can either be kept or removed after each solving step.

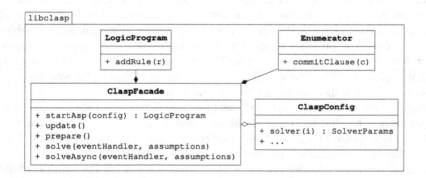

Fig. 1. Class diagram for excerpt of *clasp*'s C++ library

Figure 1 illustrates *clasp*'s C++ library. At its center is the `ClaspFacade` class, which provides a simplified interface to the various classes used for solving. The typical workflow for using the *clasp* library is as follows.

1. Construct a `ClaspFacade` and a `ClaspConfig` object.
2. Configure search, preprocessing, etc. options in the configuration object.
3. Obtain a `LogicProgram` object by calling `startAsp` with the respective configuration object.
4. Add (ground) rules to the logic program by calling `addRule`.
5. Call `prepare` for performing preprocessing and necessary initialization tasks, like creating and configuring solver objects.
6. Finally, call `solve` to start searching for models.

This workflow covers a single-shot solving process. For multi-shot solving, method `update` has to be called, which allows for continuing the above process at step 4. This is especially interesting when combined with solving under assumptions (second parameter of `solve`). For example, planning problems typically require an a priori unknown horizon to find a solution. With the above workflow, the horizon can be extended at each step and assumptions can be used to check the goal situation at the current horizon. Also note that the configuration object can be updated at each step; the changes are propagated when calling the `update` method. For example, *clasp* 3 allows us to control the information kept between successive calls via attribute `solver(i).forgetSet` of `ClaspConfig`, which can be configured for each solver thread i individually. This includes heuristic scores, nogood activities, signs, and learnt nogoods.[6] For instance, re-assigning previous truth values by keeping `heuristic scores` and `signs` usually makes the solver stay in similar areas of the search space.

Another interesting feature is asynchronous solving, using method `solveAsync`. This allows for starting a search in the background, which can be interrupted at any time. Use cases for this are applications that require to react to external events, as in assisted living or robotics scenarios.

Furthermore, the `solve` and `solveAsync` methods take an event handler as argument. This handler receives events at specific parts of the search, like the beginning and end of the search, as well as when a model is found. A model event is reported along with a reference to the underlying `Enumerator` object. At this point, it is possible to use the enumerator to add clauses over internal solver literals[7] to the current search. This is rather effective because it avoids program updates and preprocessing. And it is often sufficient for synthesizing a constraint, for instance, from the last obtained model.

Paired with corresponding interfaces of *gringo* 4, the extended low level interface of *clasp* 3 has led to *clingo* 4's higher level application programming interfaces (API) in *lua* and *python* [28].[8] Further applications using the *clasp* libray include the hybrid solvers *clingcon* [29] and *dlvhex* [30].

A final detail worth mentioning is that *clasp* 3 supports changing optimization statements between successive solving steps. This includes the extension and contraction of objective functions by adding or deleting weighted atoms from them. This is, for instance, relevant in planning domains whenever the horizon is extended.

7   Experiments

For studying the interplay of the various techniques discussed above, we conduct an empirical study on optimization problems. Despite their great practical relevance, only few such studies exist in ASP [7,19]. Moreover, optimization problems are not only more complex than their underlying decision problems,

---

[6] In fact, these parameters are also controllable in *clingo* by option `--forget-on-step`.

[7] The `LogicProgram` class provides methods to map atoms to solver literals.

[8] The API reference can be found at http://potassco.sourceforge.net/gringo.html.

but they also present quite an algorithmic challenge since solving them requires solving a multitude of SAT and UNSAT problems. More specifically, we carry out two series of experiments, one on sum-based optimization problems and another on inclusion minimality-based problems. In the first series, we investigate different optimization strategies, including core- and model-guided strategies as well as the impact of domain heuristics and multi-threading. The second series compares the use of domain heuristics with that of disjunctive logic programs for computing inclusion minimal stable models.

All experiments were run with *clasp* 3.1.2 on a Linux machine with two Intel Quad-Core Xeon E5520 2.27 GHz processors, imposing a limit of 600 s wall-clock time and 6 GB of memory per run. A timeout is counted as 600 s. For capturing not only the successful solution of an optimization process but also its convergence, we regard the quality of solutions too. To be more precise, we extend the scoring used in the 2014 ASP competition by considering runtime whenever two solvers yield the same solution quality (see (iii) below). Let m be the number of participant systems, then the score s of a solver for an instance i in a domain p featuring n instances is computed as $s(p,i) = \frac{m_s(i)\cdot 100}{m\cdot n}$ where $m_s(i)$ is (i) 0, if s does neither provide a solution, nor report unsatisfiability, or (ii) the number of solvers that do not provide a strictly better result than s, where a confirmed optimum solution is considered strictly better than an unconfirmed one. Furthermore, (iii) for two equally good solutions, one is considered strictly better, if it is computed at least 30 s faster than the other one.

Accordingly, each entry in Table 1 gives average time, number of time-outs, and score wrt the considered set of instances (except for column *multi*). The benchmark classes are given in the first column, which also includes the number of instances and their source. Also, we indicate via superscripts mn and w, whether a class comprises a *multi*-objective optimization problem with n objectives and whether its functions are *w*eight-based. The body of Table 1 gives the results obtained by evaluating *clasp*'s optimization strategies on 636 benchmark instances from various sources.[9] The first three data columns give the results obtained for model-, core-, and heuristic-guided strategies relying on *clasp*'s default configuration `tweety`, viz. plain model-guided optimization (`--opt-strategy=bb`), core-guided optimization using the *oll* algorithm (`--opt-strategy=usc`), and model-guided optimization using heuristics preferring minimized atoms and assigning them to false (`--opt-strategy=bb` `--dom-mod=5,8`).[10] The starred columns reflect the best configurations obtained for each optimization strategy, viz. model-guided optimization with exponentially increasing steps (`--opt-strategy=bb,2`) using configuration `trendy`, core-guided optimization algorithm *oll*, disjoint core pre-processing, and problem relaxation (cf. Sect. 3; `--opt-strategy=usc,3`) using `crafty`, and hierarchic model-guided optimization with heuristics preferring to assign false to minimized atoms (cf. Sect. 4; `--opt-strategy=bb,1` `--dom-mod=4,8`) using `trendy`.

---

[9] The benchmark set is available at http://www.cs.uni-potsdam.de/clasp/?page= experiments.

[10] Combining core-guided optimization with domain heuristics deteriorates results.

Table 1. Results for sum-based optimization

Benchmark		model			core			heuristic			model*			core*			heuristic*			multi	
15-puzzle	(16)	260	5	90	45	0	100	425	9	62	266	5	83	21	0	100	249	5	88	9	0
Fastfood^w	(29)	9	0	100	290	13	55	30	0	100	22	0	100	290	14	67	10	0	100	7	0
Labyrinth	(29)	445	18	75	299	11	62	365	14	84	395	15	79	250	10	66	442	19	58	229	9
Sokoban	(28)	1	0	100	1	0	100	1	0	100	1	0	100	1	0	100	1	0	100	0	0
Tsp^w	(29)	600	29	57	600	29	0	600	29	100	600	29	70	600	29	32	600	29	73	600	29
Wbds	(29)	600	29	70	421	19	34	600	29	82	600	29	31	394	17	67	600	29	72	397	17
Abstract^{m2}	(30)	19	0	100	99	0	100	311	13	57	20	0	100	73	2	94	21	0	100	6	0
Connected	(26)	513	22	75	476	20	23	513	22	89	531	23	52	474	20	51	514	22	93	479	20
Crossing	(30)	372	16	78	177	5	83	451	20	66	381	17	61	174	6	88	367	16	86	162	5
MaxClique	(30)	593	29	20	50	0	100	528	23	61	370	13	75	23	0	100	313	8	91	21	0
Valves^w	(30)	508	24	79	543	27	10	561	28	7	515	25	87	561	28	55	513	25	92	518	25
Asped^{m2,w}	(30)	57	0	100	540	27	38	490	21	42	89	1	99	470	23	54	64	0	100	65	0
Expansion	(30)	103	3	92	1	0	100	40	0	100	63	2	96	1	0	100	30	0	100	0	0
Repair	(30)	113	1	97	0	0	100	10	0	100	32	0	100	1	0	100	44	0	100	1	0
Iscas85	(30)	129	4	96	0	0	100	158	7	88	134	4	92	0	0	100	306	13	71	0	0
Paranoid^{m2}	(30)	377	8	79	1	0	100	103	4	92	80	3	94	1	0	100	59	2	98	1	0
Trendy^{m4,w}	(30)	485	19	47	4	0	100	241	11	80	254	11	82	6	0	100	219	10	87	6	0
Metro^w	(30)	42	0	100	237	7	77	325	14	59	45	0	100	162	4	93	29	0	100	21	0
PartnerUnits	(30)	234	5	94	111	2	93	150	4	87	225	8	82	103	1	97	251	9	83	97	0
Ricochet	(30)	86	0	100	85	0	100	97	0	100	167	2	95	88	0	100	136	1	97	21	0
ShiftDesign^{m3}	(30)	600	30	19	23	0	100	105	5	86	436	16	67	44	1	99	351	13	80	29	0
Timetabling^w	(30)	407	17	63	8	0	100	205	10	84	208	10	84	31	1	97	280	11	73	4	0
SUM	(636)	6553	259	1731	4011	160	1676	6307	263	1724	5435	213	1829	3768	**156**	1859	5397	212	**1942**	2674	105
AVG		298	12	79	182	7	76	287	12	78	247	10	83	171	7	85	245	10	88	122	5

And the last column shows results obtained with multi-threading. Scores are only computed among single-threaded configurations.

First of all, we observe that core-guided optimization solves the highest number of optimization problems. The fact that *core** solves more problems than *core* is due to `crafty`'s slow restart strategy that is advantageous when solving UNSAT problems, which are numerous in core-guided optimization.[11] While the latter seems to have an edge over the model-guided strategy whenever the optimum can be established, it is vacuous when the optimum is out of reach since it lacks the anytime behavior of model-guided search. This is nicely reflected by the score of 0 obtained by *core* on *TSP*, where no model at all is outputted. The best anytime behavior is obtained by boosting model-guided search with *heuristics*. Although no variant proves any optimum for *TSP*, the two *heuristic* strategies give the highest scores for *TSP*, reflecting the best solution quality. Interestingly, the less manipulative strategy of *heuristic** yields a better score. In fact, heuristic-guided search procedures show the best convergence to the optimum but often fall short in establishing its optimality. Otherwise, model-guided optimization appears to benefit from faster restart strategies, as comprised in `trendy`, since it involves solving several SAT problems. All in all, we observe that core-guided strategies dominate in terms of solved problems, while heuristic-guided ones yield the best solution quality. To have the cake and eat it, too, we can take advantage of *clasp*'s multi-threading capacities. To this end, we combined the three starred configurations with one running core-guided optimization with disjoint core pre-processing using `jumpy`.[12] The results are given in column *multi* and reflect a significant edge over each individual configuration. Interestingly, the number of timeouts is less than that of taking the ones of the respective best solver, viz. 106, and also surpasses this virtually best solver (taking 2785 seconds) as regards runtime.[13] In fact, the *multi* configuration performs at least as good as the other configurations on all but two benchmark classes. We trace this superior behavior back to exchanging bounds and constraints among the threads.

Our second series of experiments contrasts the usage of domain heuristics with that of disjunctive logic programs for computing inclusion-minimal stable models. All results are based upon *clasp*'s default configuration `tweety` and given in Table 2. The standard technique for encoding inclusion minimality in ASP is to use saturation-based, disjunctive encodings (cf. [12]). We generate the resulting programs automatically from the respective benchmarks with the *metasp* system [31] and solve them with the disjunctive solving techniques described in Sect. 2. The results are given in the columns headed by *meta*. The first three data columns give average time and number of timeouts for computing one inclusion-minimal answer set. The column headed *heuristic* accomplishes this via the heuristic approach described in Sect. 4, viz. option `--dom-mod=5,16`. We

---

[11] In fact, using `trendy` with `crafty`'s restart strategy performs even slightly better.
[12] This provides us with an approximate solution and complements *core** due to fast restarts.
[13] Running the four best configurations from Table 1 yields 2668/108.

Table 2. Results for inclusion minimality-based optimization

Benchmark	meta		heuristic		meta-heur.		meta			heuristic			meta-heuristic			meta-heur.-rec		
15-puzzle (16)	25	0	14	0	23	0	321	7	91	408	9	75	354	7	69	444	9	38
Fastfood (29)	1	0	0	0	0	0	356	14	59	210	9	100	348	14	65	268	10	71
Labyrinth (29)	356	16	84	3	347	15	600	29	72	600	29	91	600	29	73	600	29	61
Sokoban (28)	22	0	1	0	12	0	22	0	95	1	0	100	23	0	96	12	0	98
Tsp (29)	7	0	0	0	7	0	600	29	48	600	29	100	600	29	58	600	29	44
Wbds (29)	219	7	23	1	38	1	600	29	53	600	29	82	600	29	72	600	29	49
Connected (26)	109	3	0	0	61	2	532	23	35	532	23	100	532	23	60	532	23	70
Crossing (30)	98	1	14	0	14	0	600	30	32	600	30	99	600	30	42	600	30	76
MaxClique (30)	189	3	0	0	3	0	600	30	25	600	30	100	600	30	50	600	30	75
Valves (30)	600	30	560	28	600	30	600	30	98	560	28	100	600	30	98	600	30	98
Aspeed (30)	600	30	4	0	581	29	600	30	73	600	30	100	600	30	74	600	30	75
Expansion (30)	600	30	0	0	600	30	600	30	75	298	14	100	600	30	75	600	30	75
Repair (30)	552	26	0	0	5	0	595	29	25	438	20	100	589	29	50	481	21	77
Iscas85 (30)	60	3	0	0	0	0	600	30	25	600	30	100	600	30	50	600	30	75
Paranoid (30)	191	6	1	0	16	0	600	30	25	600	30	100	600	30	50	600	30	75
Trendy (30)	411	18	3	0	133	0	581	29	27	580	29	100	581	29	51	581	29	75
Metro (30)	126	5	54	1	33	1	571	27	42	576	28	70	581	28	65	573	27	78
PartnerUnits (30)	600	30	168	4	507	9	600	30	42	168	4	98	596	29	61	501	9	78
Ricochet (30)	405	16	57	0	266	10	388	14	46	56	0	100	285	11	77	264	10	83
Timetabling (30)	600	30	16	0	85	1	600	30	27	283	14	98	600	30	51	336	15	82
SUM (576)	5773	254	999	**37**	3332	128	10568	500	1013	8908	**415**	**1913**	10490	497	1285	9991	450	1453
AVG	289	13	50	2	167	6	528	25	51	445	21	96	525	25	64	500	22	73

see that this is an order of magnitude better than the disjunctive approach in terms of both runtime and timeouts. Clearly, this is because the former deals with normal programs only, while the latter involves intractable unfounded set tests. Although the frequency of such tests can be reduced by guiding the generating solver by the same heuristics, it fails to catch up with a purely heuristic approach (see column *meta-heur.*). The picture changes slightly when it comes to enumerating inclusion-minimal answer sets. Here, *heuristic* faces an exponential space complexity, while *meta* runs in polynomial space.[14] The remaining columns summarize enumeration results, and add the score as an indicative measure by taking as objective value the number of enumerated models. Although the differences are smaller, *heuristic* still outperforms all variants of *meta*. Again, adding heuristic support improves the performance of *meta*. More surprisingly, the best *meta* configuration is obtained by abolishing polynomial space guarantees and using *clasp*'s solution recording for enumeration, viz. *meta-heur.-rec*. In fact, the added (negated) solutions further focus the search of the generator and thus lead to fewer unfounded set tests.

8   Discussion

We presented distinguishing features of the *clasp* 3 series. And we evaluated their interplay by an empirical analysis on optimization problems. Comparative studies contrasting *clasp* with other systems can be found in the ASP competition series. Many of *clasp*'s features can be found in one form or another in other ASP, SAT, or PB solvers. For instance, *dlv* features several dedicated interfaces, *wasp* [18] also implements core- and model-guided optimization, Rintanen uses heuristics in [32] to improve SAT planning, etc. However, the truly unique aspect of *clasp* 3 is its wide variety of features combined in a single framework. We demonstrated the resulting added value by the combinations in our experiments.

Acknowledgments. This work was funded by AoF (251170) and DFG (SCHA 550/8 and 550/9).

References

1. Baral, C.: Knowledge Representation, Reasoning and Declarative Problem Solving. Cambridge University Press, Cambridge (2003)
2. Simons, P., Niemelä, I., Soininen, T.: Extending and implementing the stable model semantics. Artif. Intell. **138**(1–2), 181–234 (2002)
3. Leone, N., Pfeifer, G., Faber, W., Eiter, T., Gottlob, G., Perri, S., Scarcello, F.: The DLV system for knowledge representation and reasoning. ACM TOCL **7**(3), 499–562 (2006)

---

[14] Also, *meta* allows for query-answering, while *heuristic* requires a generate-and-test approach.

4. Lin, F., Zhao, Y.: ASSAT: computing answer sets of a logic program by SAT solvers. Artif. Intell. **157**(1–2), 115–137 (2004)
5. Giunchiglia, E., Lierler, Y., Maratea, M.: Answer set programming based on propositional satisfiability. J. Autom. Reason. **36**(4), 345–377 (2006)
6. Gebser, M., Kaufmann, B., Schaub, T.: Conflict-driven answer set solving: from theory to practice. Artif. Intell. **187–188**, 52–89 (2012)
7. Alviano, M., Dodaro, C., Ricca, F.: Preliminary report on WASP 2.0. In: Proceedings of NMR (2014)
8. Janhunen, T., Niemelä, I., Sevalnev, M.: Computing stable models via reductions to difference logic, pp. 142–154 [33]
9. Liu, G., Janhunen, T., Niemelä, I.: Answer set programming via mixed integer programming. In: Proceedings of KR, pp. 32–42. AAAI Press (2012)
10. Gebser, M., Kaufmann, B., Schaub, T.: Multi-threaded ASP solving with clasp. Theory Pract. Logic Program. **12**(4–5), 525–545 (2012)
11. Gebser, M., Kaminski, R., Kaufmann, B., Schaub, T.: Answer Set Solving in Practice. Morgan and Claypool Publishers, San Rafael (2012)
12. Eiter, T., Gottlob, G.: On the computational cost of disjunctive logic programming: propositional case. Ann. Math. Artif. Intell. **15**(3–4), 289–323 (1995)
13. Leone, N., Rullo, P., Scarcello, F.: Disjunctive stable models: unfounded sets, fixpoint semantics, and computation. Inf. Comput. **135**(2), 69–112 (1997)
14. Biere, A., Heule, M., van Maaren, H., Walsh, T. (eds.): Handbook of Satisfiability. IOS Press, Amsterdam (2009)
15. Drescher, C., Gebser, M., Grote, T., Kaufmann, B., König, A., Ostrowski, M., Schaub, T.: Conflict-driven disjunctive answer set solving. In: Proceedings of KR, pp. 422–432. AAAI Press (2008)
16. Gebser, M., Kaufmann, B., Schaub, T.: Advanced conflict-driven disjunctive answer set solving. In: Proceedings of IJCAI. IJCAI/AAAI, pp. 912–918 (2013)
17. Gebser, M., Kaufmann, B., Neumann, A., Schaub, T.: Advanced preprocessing for answer set solving. In: Proceedings of ECAI, pp. 15–19. IOS (2008)
18. Alviano, M., Dodaro, C., Marques-Silva, J., Ricca, F.: On the implementation of weak constraints in wasp. In: Proceedings of ASPOCP (2014)
19. Gebser, M., Kaminski, R., Kaufmann, B., Schaub, T.: Multi-criteria optimization in answer set programming. In: Technical Communications of ICLP. LIPIcs, pp. 1–10 (2011)
20. Andres, B., Kaufmann, B., Matheis, O., Schaub, T.: Unsatisfiability-based optimization in clasp. In: Technical Communications of ICLP. LIPIcs, pp. 212–221 (2012)
21. Narodytska, N., Bacchus, F.: Maximum satisfiability using core-guided maxsat resolution. In: Proceedings AAAI, pp. 2717–2723. AAAI Press (2014)
22. Marques-Silva, J., Planes, J.: On using unsatisfiability for solving maximum satisfiability (2007). CoRR abs/0712.1097
23. Gebser, M., Kaufmann, B., Otero, R., Romero, J., Schaub, T., Wanko, P.: Domain-specific heuristics in answer set programming. In: Proceedings AAAI, pp. 350–356. AAAI Press (2013)
24. Castell, T., Cayrol, C., Cayrol, M., Le Berre, D.: Using the Davis and Putnam procedure for an efficient computation of preferred models. In: Proceedings ECAI, pp. 350–354. Wiley (1996)
25. Di Rosa, E., Giunchiglia, E., Maratea, M.: Solving satisfiability problems with preferences. Constraints **15**(4), 485–515 (2010)

26. Pipatsrisawat, K., Darwiche, A.: A lightweight component caching scheme for satisfiability solvers. In: Marques-Silva, J., Sakallah, K.A. (eds.) SAT 2007. LNCS, vol. 4501, pp. 294–299. Springer, Heidelberg (2007)

27. Hutter, F., Hoos, H.H., Leyton-Brown, K.: Sequential model-based optimization for general algorithm configuration. In: Coello, C.A.C. (ed.) LION 2011. LNCS, vol. 6683, pp. 507–523. Springer, Heidelberg (2011)

28. Gebser, M., Kaminski, R., Kaufmann, B., Schaub, T.: Clingo = ASP + control: preliminary report. In: Technical Communications of ICLP (2014)

29. Ostrowski, M., Schaub, T.: ASP modulo CSP: the clingcon system. Theory Pract. Logic Program. 12(4–5), 485–503 (2012)

30. Eiter, T., Ianni, G., Schindlauer, R., Tompits, H.: DLVHEX: a prover for semantic-web reasoning under the answer-set semantics. In: Proceedings WI, pp. 1073–1074. IEEE (2006)

31. Gebser, M., Kaminski, R., Schaub, T.: Complex optimization in answer set programming. Theory Pract. Logic Program. 11(4–5), 821–839 (2011)

32. Rintanen, J.: Planning as satisfiability: heuristics. Artif. Intell. 193, 45–86 (2012)

33. Erdem, E., Lin, F., Schaub, T. (eds.): LPNMR 2009. LNCS, vol. 5753. Springer, Heidelberg (2009)

Combining Heuristics for Configuration Problems Using Answer Set Programming

Martin Gebser[1,2], Anna Ryabokon[3](✉), and Gottfried Schenner[4]

[1] HIIT, Aalto University, Espoo, Finland
martin.gebser@aalto.fi
[2] University of Potsdam, Potsdam, Germany
[3] Alpen-Adria-Universität Klagenfurt, Klagenfurt, Austria
anna.ryabokon@aau.at
[4] Siemens AG Österreich, Vienna, Austria
gottfried.schenner@siemens.com

Abstract. This paper describes an abstract problem derived from a combination of Siemens product configuration problems encountered in practice. Often isolated parts of configuration problems can be solved by mapping them to well-studied problems for which efficient heuristics exist (graph coloring, bin-packing, etc.). Unfortunately, these heuristics may fail to work when applied to a problem that combines two or more subproblems. In the paper we show how to formulate a combined configuration problem in Answer Set Programming (ASP) and to solve it using heuristics à la hclasp. In addition, we present a novel method for heuristic generation based on a combination of greedy search with ASP that allows to improve the performance of an ASP solver.

Keywords: Configuration problem · Heuristics · Answer Set Programming

1 Introduction

Configuration is a design activity aiming at creation of an artifact from given components such that a set of requirements reflecting individual needs of a customer and compatibility of the system's structures are satisfied. Configuration is a fully or partly automated approach supported by a knowledge-based information system called *configurator*. Originally configurators appeared because configurable products were expensive and very complex. They were developed by a significant number of highly qualified workers by order and single-copy. Emerging research on expert systems in the 1980s resulted in a number of approaches to knowledge-based configuration, such as McDermott's R1/XCON 'configurer' [13]. Since then many companies such as ConfigIt, Oracle, SAP, Siemens or Tacton have developed configurators for large complex systems reducing the production costs significantly. With the lapse of time the focus has been shifted more in the direction of mass customization. Currently configurators

© Springer International Publishing Switzerland 2015
F. Calimeri et al. (Eds.): LPNMR 2015, LNAI 9345, pp. 384–397, 2015.
DOI: 10.1007/978-3-319-23264-5_32

cover a wide range of customers and can be found in practically every price segment. One can configure a car, a computer, skis and even a forage for a dog.

Researchers in academia and industry have tried different approaches to configuration knowledge representation and reasoning, including production rules, constraints languages, heuristic search, description logics, etc.; see [10,17,19] for surveys. Although constraint-based methods remain de facto standard, ASP has gained much attention over the last years because of its expressive high-level representation abilities. Normal rules as well as rules including weight and cardinality atoms were used in the first application of ASP to configuration problems [18]. Regarding knowledge representation, [21] suggests a high-level object-oriented modeling language and a web-based graphical user interface to simplify the modeling of requirements.

In [5] important aspects for formalizing and tackling real-world configuration scenarios with ASP are discussed. Recently a framework for describing object-oriented knowledge bases was presented in [15]. The authors suggested a general mapping from an object-oriented formalism to ASP for S'UPREME based configurators. S'UPREME is a configuration engine of Siemens AG, which is applied to configure complex large-scale technical systems such as railway safety systems within Siemens. In fact, more than 30 applications are based on this system [9].

As evaluation shows ASP is a compact and expressive method to capture configuration problems [10], i.e. it can represent configuration knowledge consisting of component types, associations, attributes, and additional constraints. The declarative semantics of ASP programs allows a knowledge engineer to freely choose the order in which rules are written in a program, i.e. the knowledge about types, attributes, etc. can be easily grouped in one place and modularized. Sound and complete solving algorithms allow to check a configuration model and support evolution tasks such as reconfiguration. However, empirical assessments indicate that ASP has limitations when applied to large-scale product configuration instances [1,5]. The best results in terms of runtime and solution quality were achieved when domain-specific heuristics were used [14,20].

In this paper we introduce a combined configuration problem that reflects typical requirements frequently occurring in practice of Siemens. The parts of this problem correspond (to some extent) to classical computer science problems for which there already exist some well-known heuristics and algorithms that can be applied to speed up computations and/or improve the quality of solutions.

As the main contribution, we present a novel approach on how problem-specific heuristics generated by a greedy solver can be incorporated in an ASP program to improve computation time (and obtain better solutions). The application of domain-specific knowledge formulated succinctly in an ASP heuristic language [8] allows for better solutions within a shorter solving time, but it strongly deteriorates the search when additional requirements (conflicting with the formulated heuristics) are included. On the other hand, the formulation of complex heuristics might be cumbersome using greedy methods. Therefore, we exploit a combination of greedy methods with ASP for the generation of heuristics and integrate them to accelerate an ASP solver. We evaluate the method on a set of instances derived from configuration scenarios encountered by us in practice and in general. Our evaluation shows that solutions for three sets of instances can be found an order of magnitude faster than compared to a plain ASP encoding.

In the following, Sect. 2 introduces a combined configuration problem (CCP) which is exemplified in Sect. 3. Its ASP encoding is shown in Sect. 4. Section 5 discusses heuristics for solving the CCP problem and we present our evaluation results in Sect. 6. Finally, in Sect. 7 we conclude and discuss future work.

2 Combined Configuration Problem

The Combined Configuration Problem (CCP) is an abstract problem derived from a combination of several problems encountered in Siemens practice (railway interlocking systems, automation systems, etc.). A CCP instance is defined by a directed acyclic graph (DAG). Each vertex of the DAG has a type and each type of the vertices has a particular size. In addition, each instance comprises two sets of vertices specifying two vertex-disjoint paths in the DAG. Furthermore, an instance contains a set of areas, sets of vertices defining possible border elements of each area and a maximal number of border elements per area. Finally, a number of available colors as well as a number of available bins and their capacity are given.

Given a CCP instance, the goal is to find a solution that satisfies a set of requirements. All system requirements are separated into the corresponding subproblems which must be solved together or in particular combinations:

- **P1 Coloring.** *Every vertex must have exactly one color.*
- **P2 Bin-Packing.** *For every color a Bin-Packing problem must be solved, where the same number of bins are available for each color. Every vertex must be assigned to exactly one bin of its color and for every bin, the sum of sizes must be smaller or equal to the bin capacity.*
- **P3 Disjoint Paths.** *Vertices of different paths cannot be colored in the same color.*
- **P4 Matching.** *Each border element must be assigned to exactly one area such that the number of selected border elements of an area does not exceed the maximal number of border elements and all selected border elements of an area have the same color.*
- **P5 Connectedness.** *Two vertices with the same color must be connected via a path that contains only vertices of that color.*

Origin of the Problem. The considered CCP originates in the *railway domain.* The given DAG represents a track layout of a railway line. A coloring **P1** can then be thought as an assignment of resources (e.g. computers) to the elements of the railway line. In real-world scenarios different infrastructure elements may require different amounts of a resource that is summarized in **P2**. This may be hardware requirements (e.g. a signal requiring a certain number of hardware

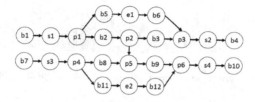

Fig. 1. Input CCP graph and a trivial solution of Coloring (**P1**)

parts) or software requirements (e.g. an infrastructural element requiring a specific processing time). The requirements of **P1** and **P2** are frequently used in configuration problems during an assignment of entities of one type to entities of another type [5,12]. The constraint of **P3** increases availability, i.e. in case one resource fails it should still be possible to get from a source vertex (no incoming edges) of the DAG to a target vertex (no outgoing edges) of the DAG. In the general version of this problem one has to find n paths that maximize availability. The CCP uses the simplified problem where 2 vertex-disjoint paths are given. **P4** stems from detecting which elements of the graph are occupied. The border elements function as detectors for an object leaving or entering an area. The PUP problem [1,2] is a more elaborate version of this problem. **P5** arises in different scenarios. For example, if communication between elements controlled by different resources is more costly, then neighboring elements should be assigned to the same resource whenever possible.

3 Example

Figure 1 shows a sample input CCP graph. In this section we illustrate how particular requirements can influence a solution. Namely, we add the constraints of each subproblem one by one. If only **P1** is active, any graph corresponds to a trivial solution of **P1** where all vertices are colored white.

Let us consider the input graph as a Bin-Packing problem instance with four colors and three bins per color of a capacity equal to five. The vertices of type b, e, s and p have the sizes 1, 2, 3 and 4 respectively. A sample solution of Coloring and Bin-Packing (**P1-P2**) is presented in Figs. 2 and 3.

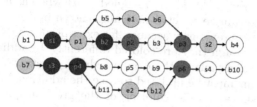

Fig. 2. Used colors in a solution of the Coloring and Bin-Packing problems (**P1-P2**)

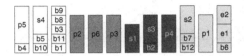

Fig. 3. Used bins in a solution of the Coloring and Bin-Packing problems (**P1-P2**)

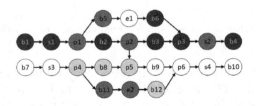

Fig. 4. Solution of the Coloring, Bin-Packing and Disjoint Paths problems (**P1-P3**)

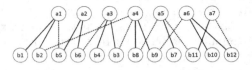

Fig. 5. A sample input and solution graphs for **P4**. The selected edges of the input graph are highlighted with solid lines.

For instance, when activating the Disjoint Paths constraint (**P3**), two vertex-disjoint paths $path1 = \{b1, s1, p1, b2, p2, b3, p3, s2, b4\}$ as well as $path2 = \{b7, s3, p4, b8, p5, b9, p6, s4, b10\}$ may be declared. Consequently, in this case the solution shown in Fig. 2 violates the constraint and must be modified as displayed in Fig. 4, where the vertices of different paths are colored with different colors (*path1* with dark grey and grey whereas white and light grey are used for *path2*).

Figure 5 shows a Matching example (**P4**). There are seven areas in the matching input graph, each corresponding to a subgraph surrounded with border elements (Fig. 1). For example, area $a1$ represents the subgraph $\{b1, s1, p1, b2, b5\}$ and area $a2$ the subgraph $\{b5, e1, b6\}$. The corresponding border elements are $\{b1, b2, b5\}$ and $\{b5, b6\}$ (Fig. 5).

Assume that an area can have at most 2 border elements assigned to it. In the resulting matching (Fig. 5) $b1$, $b2$ are assigned to $a1$ whereas $b5$, $b6$ are assigned to $a2$. Note that the sample selected matching shown in Fig. 5 is not valid with the coloring presented previously, because, for example, $b5$ and $b6$ are assigned to the same area $a2$ although they are colored differently. In addition, the coloring solution shown in Fig. 4 violates the Connectedness constraint (**P5**). Therefore, the previous solutions must be updated to take the additional requirements into account. Figure 6 shows a valid coloring of the given graph that satisfies all problem conditions (**P1-P5**).

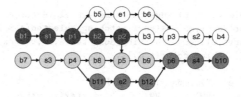

Fig. 6. A valid solution for **P1-P5**

4   ASP Encoding of the Combined Configuration Problem

A CCP instance is defined using the following atoms. An edge between two vertices in the DAG is defined by *edge(Vertex1, Vertex2)*. For each vertex, *type(Vertex, Type)* and *size(Vertex, Size)* are declared. *pathN(Vertex)* expresses that a vertex belongs to a particular path. In addition, each border element must be connected to one of the possible areas given by *edge_matching(Area, Vertex)* whereas each area can control at most *maxborder(C)* border elements. The number of colors and bins are defined using *nrofcolors(Color)* and *nrofbins(Bin)*. Finally, the capacity of a bin is fixed by *maxbinsize(Capacity)*.

Our ASP encoding for the CCP is shown in Listing 1. Line 1-5 implements Coloring (**P1**), assigning colors to vertices. The atoms *vertex_color(Vertex, Color)* and *usedcolor(Color)* express that a *Vertex* is connected to a *Color*, i.e. used in a solution via *usedcolor(Color)*. An assignment of a *Vertex* to a *Bin*, i.e. Bin-Packing problem (**P2**), is accomplished using Line 6-10, where the atoms *vertex_bin(Vertex, Bin)* and *usedbin(Bin)* represent a solution. Further, the atoms *bin(Color, Bin, Vertex)* represent a combined solution for **P1** and **P2**. The Disjoint paths constraint (**P3**) is stated in Line 11. In accordance with Matching (**P4**), i.e. Line 12-17, one has to find a matching between areas and border elements using *edge_matching_selected(Area, Vertex)* atoms. Finally, the Connectedness requirement (**P5**) is ensured in line 18-24.

```
 1 vertex(V):-type(V,_). vertex(V):-size(V,_). % P1
 2 vertex(V):-edge(V,_). vertex(V):-edge(_,V).
 3 color(1..MaxC):-nrofcolors(MaxC).
 4 1{vertex_color(V,C):color(C)}1:-vertex(V).
 5 usedcolor(C):-vertex_color(V,C).

 6 1{vertex_bin(V,B):B=1..K}1 :- vertex(V), nrofbins(K). % P2
 7 bin(C,B,V):-vertex_color(V,C),vertex_bin(V,B).
 8 :-color(C),nrofbins(K),maxbinsize(MaxS), B=1..K,
 9   MaxS+1 #sum{S,V:bin(C,B,V),size(V,S)}.
10 usedbin(B):-bin(C,B,V).

11 :-path1(V1),path2(V2),vertex_color(V1,C),vertex_color(V2,C).%P3

12 area(A):-edge_matching(A,B). % P4
13 borderelement(B):-edge_matching(A,B).
```

```
14 1{edge_matching_selected(A,B):edge_matching(A,B)}1 :- borderelement(B).
15 :-area(A),maxborder(MaxB),MaxB+1{edge_matching_selected(A,B)}.
16 edge_matching_color(A,C):-edge_matching_selected(A,B),vertex_color(B,C).
17 :-area(A), 2{edge_matching_color(A,C)}.

18 e(X,Y):-edge(X,Y). e(X,Y):-edge(Y,X). % P5
19 pred(V1,V2):-vertex(V1;V2),V1 < V2,V <= V1:vertex(V),V<V2.
20 first(C,V2):-color(C), pred(V1,V2),
21          not vertex_color(V1,C),first(C,V1):pred(V,V1).
22 reach_col(C,V1):-color(C),vertex(V1),first(C,V1):pred(V,V1).
23 reach_col(C,V2):-reach_col(C,V1),e(V1,V2),vertex_color(V1,C).
24 :-vertex_color(V,C),not reach_col(C,V).
```

Listing 1. ASP encoding for the Combined Configuration Problem

5   Combining Heuristics for Configuration Problems

To formulate a heuristic within ASP we use the declarative heuristic framework developed by Gebser et al. [8]. In this formalism the heuristics are expressed using atoms $\_heuristic(a, m, v, p)$, where a denotes an atom for which a heuristic value is defined, m is one of four modifiers (init, factor, level and sign), and v, p are integers denoting a value and a priority, respectively, of the definition. A number of shortcuts are available, e.g. $\_heuristic(a, v, l)$, where a is an atom, v is its truth value and l is a level. The heuristic atoms modify the behavior of the VSIDS heuristic [11]. Thus, if a $\_heuristic$ atom is true in some interpretation, then the corresponding atom a might be preferred by the ASP solver at the next decision point. For instance, given the choice rule $1\{vertex\_color(V, C) : color(C)\}1 :\!- vertex(V)$. and adding only the atom $\_heuristic(vertex\_color(`b1', 1), true, 1))$ to a program, the solver prefers the atom $vertex\_color(`b1', 1)$ over all other atoms $vertex\_color(`b1', X)$ for $X \neq 1$. If several atoms $vertex\_color/2$ are provided, the atom with the higher level l is preferred.

There are different ways to incorporate heuristics in a program. The standard approach [8] requires an implementation of a heuristic at hand using a pure ASP encoding, whereas the idea of our method is to delegate the (expensive) generation of a heuristic to an external tool and then to extend the program with generated heuristic atoms to accelerate the ASP search. Below we exemplify how both approaches can be applied.

5.1   Standard Generation of Heuristics in ASP

Several heuristics can be used for the problems that compose the CCP, e.g. for the coloring of vertices (**P1**) we seek to use as few colors as possible by the following rule:

```
1 _heuristic(vertex_color(V,C),true,MC-C) :- vertex(V), color(C),
    nrofcolors(MC).
```

Listing 2. Heuristic for an assignment of colors to vertices

Additionally, we can apply well-known Bin-Packing heuristics for the placement of colored vertices into the bins of specified capacity (**P2**). The Bin-Packing problem is known to be an NP-hard combinatorial problem. However, there is a number of approximation algorithms (construction heuristics) that allow efficient computation of good approximations of a solution [6], e.g. Best/First/Next-Fit heuristics. They can, of course, be used as heuristics for the CCP. As shown in Listing 3, given a (decreasing) order of vertices using $order(V, O)$ atoms, we can force the solver to place vertex V_i into the lowest-indexed bin for which the size of already placed vertices does not exceed the capacity, i.e. in a first-fit bin:

```
1 binDomain(1..NB) :- nrofbins(NB). offset(NB+1) :- nrofbins(NB).
2 _heuristic(vertex_bin(V,B),true,M+O*NB-B) :- binDomain(B), nrofbins(NB),
    order(V,O), offset(M).
```

Listing 3. First-Fit heuristic for an assignment of vertices to bins

The heuristic never uses a new bin until all the non-empty bins are full and it can be expressed by rules that generate always a higher level for the bins with smaller number. It is also possible (with an intense effort) to express other heuristics for **P1-P5** that guide the search appropriately and allow to speed up the computation of solutions if we solve these problems separately. However, as our experiments show, the inclusion of heuristics for different problem at the same time might drastically deteriorate the performance for real-world CCP instances.

5.2 Greedy Search

From our observations in the context of product configuration, it is relatively easy to devise a greedy algorithm to solve a part of a configuration problem. This is often the case in practice, because products are typically designed to be easily configurable. The hard configuration instances usually occur when new constraints arise due to the combination of existing products and technologies.

The same can be said for the CCP problem. Whereas it is easy to develop greedy search algorithms for the individual subproblems, it becomes increasingly difficult to come up with an algorithm that solves the combined problem. Algorithm 1 shows a greedy method that solves the Matching problem of the CCP (**P4**). For every vertex v it finds a related area a with the fewest assigned vertices so far and matches v with a. The algorithm assumes that all border elements are colored with one color, as it trivially satisfies the coloring

---

Algorithm 1. GreedyMatching

Input: A bipartite graph $G_A = (BE, A, E)$, where BE is a set of border
elements, A is a set of areas and $E \subseteq BE \times A$ is a set of edges
Output: A matching set M

1 $M \leftarrow \emptyset$;
2 **foreach** $v \in BE$ **do**
      // *Select areas with the minimum number of matched elements*
3       $A' \leftarrow \arg\min_{a \in A} |\{v' \mid v' \in BE, (v', a) \in M, (v, a) \in E\}|$;
4       $a \leftarrow \mathrm{pop}(A')$;
5       $M \leftarrow M \cup \{(v, a)\}$;
6 **return** M;

Algorithm 2. GreedyColoringBinPackingConnectedness

Input: A graph $G = (V, E)$, a maximum number of bins K for each color and a
bin capacity C
Output: A set B that comprises all bins of a solution

1 $B \leftarrow \emptyset$; $color \leftarrow 1$; $Q \leftarrow \emptyset$;
2 **while** $V \neq \emptyset$ **do**
3      $q \leftarrow \mathrm{pop}(V)$; $Q \leftarrow \{q\}$;
4   **while** $Q \neq \emptyset$ **do**
5          $v \leftarrow \mathrm{pop}(Q)$;
6   `labelVertexWithColor`$(v, color)$;
7          $B \leftarrow$ `assignVertexToBin`(B, v, C, K);   // *v is ignored, if it does not fit*
8   **if** $\exists b \in B$ $(v \in b)$ **then**
9             $V \leftarrow V \setminus \{v\}$;
10           $Q \leftarrow Q \cup$ `popNeigbours`(v, G);
11     $color \leftarrow color + 1$;
12 **return** B;

---

requirement of the matching problem. Algorithm 2 shows a greedy approach
to solving the CCP wrt. Coloring, Bin-Packing and Connectedness (**P1**, **P2**
and **P5**). Every call to pop returns and removes the first element v of the set
V and all corresponding edges. Then, the vertex v is assigned a color and is
put into a bin according to some heuristic Bin-Packing algorithm. For instance,
one can use classic heuristics as First-Fit or Best-Fit [6]. Our implementation of
assignVertexToBin puts vertices of only one color into a bin. If the number of
bins K is not enough to pack a vertex, then the set of bins B is not modified
and the vertex is ignored. In case the vertex was placed into a bin, Algorithm 2
retrieves and removes from G all vertices adjacent to v. The loop continues until
all vertices that can be reached from v are colored and assigned to some bin.
Finally, the number of colors is increased and the algorithm colors and removes
another subgraph of G until no vertices in G are left.

Suppose one wants to combine these two algorithms. One strategy would be to run greedy Matching and then solve the Bin-Packing problem taking matchings into account. Thus, a combined algorithm first calls Algorithm 1 and gets a set of matchings $M = \{(v_1, a_1), \ldots, (v_n, a_m)\}$. Then, for each vertex v_i of the input graph G the algorithm (i) assigns a new color to v_i, if v_i has no assigned color, and (ii) puts v_i into a bin, as in Algorithm 2. In case v_i is a border element, the combined algorithm retrieves an area a_j that matches v_i in M and colors all vertices of this area in the same color as v_i.

The combined algorithm might violate the Connectedness property, because it colors all border vertices assigned to an area with the same color. However, these vertices are not necessarily connected. That is, there might be a solution with a different matching, but the greedy algorithm tests only one of all possible matchings. Moreover, there is no obvious way how to create an algorithm solving all 3 problems efficiently. This is a clear disadvantage of using ad-hoc algorithms in contrast to the usage a logic-based formalism like ASP, where the addition of constraints is just a matter of adding some rules to an encoding. On the other hand, domain-specific algorithms are typically faster and scale better than ASP-based or SAT-based approaches that cannot be used for very large instances. For instance, the memory demand of the greedy Algorithm 2 is almost independent of graph size.

5.3   Combining Greedy Search and ASP

One way to let a complete ASP solver and a greedy search algorithm benefit from each other is to use the greedy algorithm to compute upper bounds for the problem to solve. The tighter upper bound usually means smaller grounding size and shorter solving time because the greedy solver being domain-specific usually outperforms ASP for the relaxed version of the problem. For instance, running the greedy algorithm for the Bin-Packing problem and Matching problem gives upper bounds for the maximal number of colors, i.e. number of different Bin-Packing problems to solve. The same applies to the Matching problem. This kind of application of greedy algorithms has a long tradition in branch and bound search algorithms, where greedy algorithms are used to compute the upper bound of a problem. For an example see [22], where a greedy coloring algorithm is used to find an upper bound for the clique size in a graph in order to compute maximum cliques. In this paper we investigate a novel way to combine greedy algorithms and ASP (Algorithm 3). Given all required inputs, first, a greedy algorithm is used to solve the Matching and Bin-Packing problems. The greedy algorithm typically solves a relaxed version of the problem, therefore, the solution found by the greedy algorithm may not be a consistent solution for ASP. This solution is converted into a heuristic for an ASP solver by giving the atoms of the solution a higher heuristic value.

As an example for solving the complete CCP problem, we can, first, find an unconnected solution for the combination of Coloring, Bin-Packing, Disjoint paths and Matching problems (**P1-P4**), and then, use the ASP solver to fix

---

Algorithm 3. Greedy & ASP

Input: A problem P, an ASP program Π solving the problem P
Output: A solution S
1 $GreedySolution \leftarrow$ solveGreedy(P);
2 $H \leftarrow$ generateHeuristic($GreedySolution$);
3 return solveWithASP(Π, H);

---

the Connectedness property (**P5**). The idea of combining local search with a complete solver is also found in large neighborhood search [4].

6 Experimental Results

Experiment1. In our evaluation we compared a plain ASP encoding of the CCP with an ASP encoding extended with domain-specific knowledge. The Bin-Packing problem (**P2**) of the CCP corresponds to the classic Bin-Packing problem and the same heuristics can be applied. We implemented several Bin-Packing heuristics such as First/Best/Next-Fit (Decreasing) heuristics using ASP as shown in Sect. 5.1. For the evaluation we took 37 publicly available Bin-Packing problem instances[1], for which the optimal number of bins *optnrofbins* is known, and translated them to CCP instances. The biggest instance of the set includes 500 vertices and 736 bins of the capacity 100. In the experiment, the maximal number of colors was set to 1 and the maximal number of bins was set to $2 \cdot optnrofbins$. All instances were solved by both approaches[2]. For a plain ASP encoding the solver required at most 27 s to find a solution whereas for the heuristic ASP program solving took at most 6 s, which is 4.5 times faster. The best results for the heuristic approach were obtained using the First-Fit heuristic with the decreasing order of vertices. Corresponding solutions utilized less bins then the ones obtained with the plain ASP program. Moreover, using First-Fit heuristic, for 23 from 37 instances a solution with optimal number of bins was found and for 13 other instances at most 4 bins more were required. The plain ASP encoding resulted in solutions that used on average 4 bins more than corresponding solutions of the heuristic approach. Only for 1 instance the heuristic program generated a worse solution than the plain ASP encoding.

Experiment2. In the next experiment we tested the same Bin-Packing heuristics implemented in ASP for the combined CCP, i.e. when all subproblems **P1-P5** are active, on 100 real-world test instances of moderate size (maximally 500 vertices in an input). The instances in this experiment were derived from a number of

---

[1] The instances were taken from: http://www.wiwi.uni-jena.de/Entscheidung/binpp/ index.htm.

[2] The evaluation was performed using clingo version 4.3.0 from the Potassco ASP collection [7] on a system with Intel i7-3030K CPU (3.20 GHz) and 64 GB of RAM, running Ubuntu 11.10.

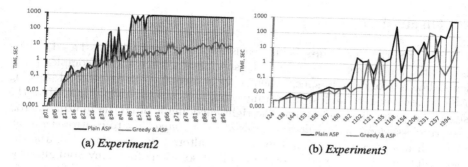

(a) *Experiment2*   (b) *Experiment3*

Fig. 7. Evaluation results using Plain ASP and Greedy & ASP

real-world configurations. Neither the plain program nor the heuristic programs were able to improve runtime/quality of solutions. Moreover, our greedy method described in Sect. 5.2 also failed to find a connected solution, i.e. when **P5** is active. For this reason, we investigated the combined approach (Greedy & ASP) described in Sect. 5.3. This approach uses the greedy method to generate a partial solution ignoring the Connectedness constraint and provides this solution as _heuristic atoms to the ASP solver. Our experiments show (see Fig. 7a) that the combined approach can solve all 100 benchmarks from the mentioned set, whereas the plain encoding presented in Sect. 4 solves only 54 instances (the time frame was set to 900 s in this and the next experiment). Moreover, for those instances which were solved using both approaches, the quality of solutions measured in terms of used bins and colors was the same. However, the runtime of the combined approach was 18 times faster on average and required at most 24 s instead of 848 s needed for the plain ASP encoding.

Experiment3. In addition, we tested more complex real-world instances (max-imally 1004 vertices in an input)[3] which we have also submitted to the ASP competition 2015. Similarly to *Experiment2* we compared the plain ASP encod-ing from Sect. 4 to the combined approach in Sect. 5.3. Again, regarding the quality of solutions, both approaches are comparable, i.e. they use on aver-age the same number of colors and bins, with the combined approach having a slight edge. Generally, from 48 instances considered in this experiment, 36/38 instances were solved using the plain/combined encoding, respectively. On aver-age/maximally the plain encoding needed 69/887 s to find a solution whereas the combined method took 14/196 s, respectively, which is about 5 times faster. Figure 7b shows the influence of heuristics on the performance for the instances from *Experiment3* that were solved by both approaches within 900 s. Although the grounding time is not presented for both experiments, we note that it requires about 10 s using both approaches for the biggest instance when all subproblems **P1-P5** are active.

---

[3] The instances are available at: http://isbi.aau.at/hint/problems.

7 Discussion

Choosing the right domain-specific heuristics for simple backtrack-based solvers is essential for finding a solution at all, especially for large and/or complex problems. The role of domain-specific heuristics in a conflict-driven nogood learning ASP solver seems to be less important when it comes to solving time. Here the size of the grounding and finding the right encoding is often the limiting factor. Nevertheless, *domain-specific heuristics are very important* to control the order in which answer sets are found and are an alternative to optimization statements. The latter hinder the computation of solutions for many configuration problem instances in a time which is reasonable for the application domain [1,5]. As we have shown, domain-specific heuristics also provide a mechanism to combine greedy algorithms with ASP solvers, which opens up the possibility to use ASP in a meta-heuristic setting. However, the possible applications go beyond this. The same approach could be used to repair an infeasible assignment using an ASP solver. This is currently a field of active research for us and has applications in the context of product reconfiguration. Reconfiguration occurs when a configuration problem is not solved from scratch, but some parts of an existing configuration have to be taken into account.

An open question is how to combine heuristics for different subproblems in a modular manner without the adaptation of every domain-specific heuristic. Here approaches like search combinators [16] from the constraint programming community might be useful. Another interesting topic for future research would be how to learn heuristics from an ASP solver, i.e. to investigate the variable-/value order chosen by an ASP solver for medium size problem instances and use them as heuristics in a backtrack solver for larger instances that are out of scope of an ASP solver due to the grounding size. Some aspects of this topic were discussed in [3]. Moreover, it is worthwhile to investigate how our method can be generalized to other application domains and whether we will be generally able to gain better performance if more heuristics are combined.

Acknowledgments. This work was funded by COIN and AoF under grant 251170 as well as by FFG under grant 840242. The authors would like to thank all anonymous reviewers for their comments and Konstantin Schekotihin for helpful discussions on the subject of this paper.

References

1. Aschinger, M., Drescher, C., Friedrich, G., Gottlob, G., Jeavons, P., Ryabokon, A., Thorstensen, E.: Optimization methods for the partner units problem. In: Achterberg, T., Beck, J.C. (eds.) CPAIOR 2011. LNCS, vol. 6697, pp. 4–19. Springer, Heidelberg (2011)
2. Aschinger, M., Drescher, C., Gottlob, G., Jeavons, P., Thorstensen, E.: Tackling the partner units configuration problem. In: Proceedings of IJCAI, pp. 497–503 (2011)

3. Balduccini, M.: Learning and using domain-specific heuristics in ASP solvers. AI Commun. **24**(2), 147–164 (2011)
4. Cipriano, R., Di Gaspero, L., Dovier, A.: A hybrid solver for large neighborhood search: mixing gecode and easylocal++. In: Hybrid Metaheuristics, pp. 141–155 (2009)
5. Friedrich, G., Ryabokon, A., Falkner, A.A., Haselböck, A., Schenner, G., Schreiner, H.: (Re) configuration based on model generation. In: LoCoCo Workshop, pp. 26–35 (2011)
6. Garey, M.R., Johnson, D.S.: Computers and Intractability: A Guide to the Theory of NP-Completeness. W. H. Freeman, New York (1979)
7. Gebser, M., Kaminski, R., Kaufmann, B., Schaub, T.: Answer Set Solving in Practice. Morgan & Claypool Publishers, San Rafael (2012)
8. Gebser, M., Kaufmann, B., Romero, J., Otero, R., Schaub, T., Wanko, P.: Domain-specific heuristics in answer set programming. In: Proceedings of AAAI (2013)
9. Haselböck, A., Schenner, G.: S'UPREME. In: Knowledge-Based Configuration: From Research to Business Cases, pp. 263–269 (2014)
10. Hotz, L., Felfernig, A., Stumptner, M., Ryabokon, A., Bagley, C., Wolter, K.: Configuration knowledge representation and reasoning. In: Knowledge-Based Configuration: From Research to Business Cases, pp. 41–72 (2014)
11. Madigan, C., Malik, S., Moskewicz, M., Zhang, L., Zhao, Y.: Chaff: engineering an efficient SAT solver. In: Proceedings of DAC (2001)
12. Mayer, W., Bettex, M., Stumptner, M., Falkner, A.: On solving complex rack configuration problems using CSP methods. In: Proceedings of the IJCAI Workshop on Configuration (2009)
13. McDermott, J.: R1: a rule-based configurer of computer systems. Artif. Intell. **19**(1), 39–88 (1982)
14. Ryabokon, A., Friedrich, G., Falkner, A.A.: Conflict-based program rewriting for solving configuration problems. In: Cabalar, P., Son, T.C. (eds.) LPNMR 2013. LNCS, vol. 8148, pp. 465–478. Springer, Heidelberg (2013)
15. Schenner, G., Falkner, A., Ryabokon, A., Friedrich, G.: Solving object-oriented configuration scenarios with ASP. In: Proceedings of the Configuration Workshop, pp. 55–62 (2013)
16. Schrijvers, T., Tack, G., Wuille, P., Samulowitz, H., Stuckey, P.J.: Search combinators. Constraints **18**(2), 269–305 (2013)
17. Sinz, C., Haag, A.: Configuration. IEEE Intell. Syst. **22**(1), 78–90 (2007)
18. Soininen, T., Niemelä, I., Tiihonen, J., Sulonen, R.: Representing configuration knowledge with weight constraint rules. In: Proceedings of the Workshop on ASP, pp. 195–201 (2001)
19. Stumptner, M.: An overview of knowledge-based configuration. AI Commun. **10**(2), 111–125 (1997)
20. Teppan, E.C., Friedrich, G., Falkner, A.A.: QuickPup: a heuristic backtracking algorithm for the partner units configuration problem. In: Proceedings of IAAI, pp. 2329–2334 (2012)
21. Tiihonen, J., Heiskala, M., Anderson, A., Soininen, T.: WeCoTin - a practical logic-based sales configurator. AI Commun. **26**(1), 99–131 (2013)
22. Tomita, E., Kameda, T.: An efficient branch-and-bound algorithm for finding a maximum clique with computational experiments. J. Global Optim. **37**(1), 95–111 (2007)

Infinitary Equilibrium Logic and Strong Equivalence

Amelia Harrison[1]([⊠]), Vladimir Lifschitz[1], David Pearce[2],
and Agustín Valverde[3]

[1] University of Texas, Austin, TX, USA
{ameliaj,vl}@cs.utexas.edu
[2] Universidad Politécnica de Madrid, Madrid, Spain
david.pearce@upm.es
[3] University of Málaga, Málaga, Spain
a_valverde@ctima.uma.es

Abstract. Strong equivalence of logic programs is an important concept in the theory of answer set programming. Equilibrium logic was used to show that propositional formulas are strongly equivalent if and only if they are equivalent in the logic of here-and-there. We extend equilibrium logic to formulas with infinitely long conjunctions and disjunctions, define and axiomatize an infinitary counterpart to the logic of here-and-there, and show that the theorem on strong equivalence holds in the infinitary case as well.

1 Introduction

The original definition of a stable model [3] is restricted to Prolog-style rules—implications with a conjunction of literals in the antecedent and an atom in the consequent. Extending it to arbitrary propositional formulas has been accomplished by two equivalent constructions: using equilibrium logic [13] and using modified reducts [2]. Equilibrium logic served as the basis for the characterization of strong equivalence of logic programs [10] in terms of the logic of Kripke models with two worlds, "the logic of here-and-there." The first axiomatization of that logic was given without proof by Łukasiewicz [11]: add the axiom schema

$$(\neg F \to G) \to (((G \to F) \to G) \to G)) \tag{1}$$

to propositional intuitionistic logic. This axiomatization was rediscovered and proved complete by Thomas [15]. (In the notation of that paper, axiom schema (1) is $3''_2$.) A few years earlier, Umezawa [17] had proposed a simpler axiom schema

$$F \vee (F \to G) \vee \neg G \tag{2}$$

that can be used to axiomatize the logic of here-and-there instead of (1). The completeness of this axiomatization was proved by Hosoi [6].

The definition of a stable model for propositional formulas [2] was extended to formulas with infinitely long conjunctions and disjunctions by Truszczynski [16].

F. Calimeri et al. (Eds.): LPNMR 2015, LNAI 9345, pp. 398–410, 2015.
DOI: 10.1007/978-3-319-23264-5_33

Harrison et al. [4] introduced a deductive system that includes an infinitary counterpart of axiom schema (2) and proved that if two infinitary formulas are equivalent in that system then they are strongly equivalent. Whether the converse holds is posed in that paper as an open question.

In this note, our goals are

(i) to define the infinitary version of the logic of here-and-there,
(ii) to define its nonmonotonic counterpart—the infinitary version of equilibrium logic,
(iii) to verify that stable models of infinitary formulas in the sense of Truszczynski can be characterized in terms of infinitary equilibrium logic,
(iv) to verify that infinitary propositional formulas are strongly equivalent to each other iff they are equivalent in the infinitary logic of here-and-there,
(v) to find an axiomatization of that logic.

The results of this note give a positive answer to the open question mentioned above. Moreover, they show that some axiom schemas introduced by Harrison et al. are redundant.

We will see in Sects. 2–5 that achieving goals (i)–(iv) is straightforward, given the work done earlier for finite formulas. Goal (v) is more challenging; see Sects. 6, 7. Early work on deductive systems of infinitary propositional formulas [8,14] was restricted to classical logic. Infinitary intuitionistic logic was studied by Nadel [12]. We are not aware of published work on extending intermediate systems, such as the logic of here-and-there, to infinitary formulas. Additional difficulties arise in connection with the fact that we allow uncountable conjunctions and disjunctions, not covered by Nadel's work.

The main reason why we are interested in stable models of infinitary propositional formulas is that they can be used to define the semantics of the input language of the ASP grounder GRINGO. Consider, for instance, the aggregate expression

```
#count{X:p(X)}==1 .
```

Intuitively, it says that the cardinality of the set $\{X \mid p(X)\}$ is 1. If there are infinitely many possible values for X (for instance, if the program uses integers or terms containing function symbols) then this meaning cannot be expressed using a propositional formula. Aggregate expressions like this can be represented by first-order formulas [9], but that method has significant limitations. For example, it is not clear how to apply it to the expression

```
#count{X:p(X)}==Y .
```

Such expressions are included, however, in the subset of the input language of GRINGO studied by Harrison et al. [5], who approached the problem of defining the semantics of that language using infinitary propositional formulas. That direction of research shows that the study of strong equivalence of infinitary propositional formulas may be essential for answer set programming.

A preliminary version of this paper was presented at the 2014 Workshop on Answer Set Programming and Other Computing Paradigms.

2   Review: Infinitary Formulas and Their Stable Models

Let σ be a propositional signature, that is, a set of propositional atoms. For every nonnegative integer r, (*infinitary propositional*) formulas (over σ) of rank r are defined recursively, as follows:

- every atom from σ is a formula of rank 0,
- if $\mathcal{H}$ is a set of formulas, and r is the smallest nonnegative integer that is greater than the ranks of all elements of $\mathcal{H}$, then $\mathcal{H}^\wedge$ and $\mathcal{H}^\vee$ are formulas of rank r,
- if F and G are formulas, and r is the smallest nonnegative integer that is greater than the ranks of F and G, then $F \to G$ is a formula of rank r.

We will write $\{F, G\}^\wedge$ as $F \wedge G$, and $\{F, G\}^\vee$ as $F \vee G$. The symbols $\top$ and $\bot$ will be understood as abbreviations for $\emptyset^\wedge$ and $\emptyset^\vee$ respectively; $\neg F$ stands for $F \to \bot$, and $F \leftrightarrow G$ stands for $(F \to G) \wedge (G \to F)$. These conventions allow us to view finite propositional formulas over σ as a special case of infinitary formulas.

A set or family of formulas is *bounded* if the ranks of its members are bounded from above. For any bounded family $(F_\alpha)_{\alpha \in A}$ of formulas, we denote the formula $\{F_\alpha : \alpha \in A\}^\wedge$ by $\bigwedge_{\alpha \in A} F_\alpha$, and similarly for disjunctions.

Subsets of a signature σ will be also called *interpretations* of σ. The satisfaction relation between an interpretation and a formula is defined recursively, as follows:

- For every atom p from σ, $I \models p$ if $p \in I$.
- $I \models \mathcal{H}^\wedge$ if for every formula F in $\mathcal{H}$, $I \models F$.
- $I \models \mathcal{H}^\vee$ if there is a formula F in $\mathcal{H}$ such that $I \models F$.
- $I \models F \to G$ if $I \not\models F$ or $I \models G$.

The *reduct* F^I of a formula F w.r.t. an interpretation I is defined recursively, as follows:

- For every atom p from σ, p^I is p if $p \in I$, and $\bot$ otherwise.
- $(\mathcal{H}^\wedge)^I = \{G^I \mid G \in \mathcal{H}\}^\wedge$.
- $(\mathcal{H}^\vee)^I = \{G^I \mid G \in \mathcal{H}\}^\vee$.
- $(G \to H)^I$ is $G^I \to H^I$ if $I \models G \to H$, and $\bot$ otherwise.

An interpretation I is a *stable model* of a set $\mathcal{H}$ of formulas if it is minimal w.r.t. set inclusion among the interpretations satisfying the reducts F^I of all formulas F from $\mathcal{H}$.

3   Infinitary Logic of Here-and-There

An *HT-interpretation* of σ is an ordered pair $\langle I, J \rangle$ of interpretations of σ such that $I \subseteq J$. Intuitively, such a pair describes "two worlds": the atoms in I are true "here" ("in the world H"), and the atoms in J are true "there" ("in the world T").

The satisfaction relation between an HT-interpretation and a formula is defined recursively, as follows:

- For every atom p from σ, $\langle I, J \rangle \models p$ if $p \in I$.
- $\langle I, J \rangle \models \mathcal{H}^\wedge$ if for every formula F in $\mathcal{H}$, $\langle I, J \rangle \models F$.
- $\langle I, J \rangle \models \mathcal{H}^\vee$ if there is a formula F in $\mathcal{H}$ such that $\langle I, J \rangle \models F$.
- $\langle I, J \rangle \models F \to G$ if
  (i) $\langle I, J \rangle \not\models F$ or $\langle I, J \rangle \models G$, and
  (ii) $J \models F \to G$.

An *HT-model* of a set $\mathcal{H}$ of infinitary formulas is an HT-interpretation that satisfies all formulas in $\mathcal{H}$.

About a formula F we say that it is *forced in the world* H of an HT-interpretation $\langle I, J \rangle$ if it is satisfied by $\langle I, J \rangle$; we will say that it is *forced in the world* T if it is satisfied by J. The set of worlds in which F is forced will be called the *truth value* of F with respect to $\langle I, J \rangle$. It is easy to check by induction on the rank that every formula that is forced in H is forced in T as well. Consequently, the only possible truth values of a formula are $\emptyset$, $\{T\}$, and $\{H, T\}$.

4   Equilibrium Models

An HT-interpretation $\langle I, J \rangle$ is *total* if $I = J$. It is clear that a total HT-interpretation $\langle J, J \rangle$ satisfies F iff J satisfies F.

An *equilibrium model* of a set $\mathcal{H}$ of infinitary formulas is a total HT-model $\langle J, J \rangle$ of $\mathcal{H}$ such that for every proper subset I of J, $\langle I, J \rangle$ is not an HT-model of $\mathcal{H}$.

The following proposition is similar to Theorem 1 from [2].

Theorem 1. *An interpretation J is a stable model of a set $\mathcal{H}$ of infinitary formulas iff $\langle J, J \rangle$ is an equilibrium model of $\mathcal{H}$.*

Lemma 1. *For any infinitary formula F and any HT-interpretation $\langle I, J \rangle$,*

$$I \models F^J \text{ iff } \langle I, J \rangle \models F.$$

The lemma can be proved by strong induction on the rank of F.

Proof of Theorem 1. It follows from the lemma that a total HT-interpretation $\langle J, J \rangle$ is an equilibrium model of $\mathcal{H}$ iff

- J satisfies all formulas from $\mathcal{H}$, and
- there is no proper subset I of J such that I satisfies the reducts F^J of all formulas F from $\mathcal{H}$.

This condition expresses that J is a stable model of $\mathcal{H}$.

5   Strong Equivalence

About sets $\mathcal{H}_1$, $\mathcal{H}_2$ of infinitary formulas we say that they are *strongly equivalent* to each other if, for every set $\mathcal{H}$ of infinitary formulas, the sets $\mathcal{H}_1 \cup \mathcal{H}$ and $\mathcal{H}_2 \cup \mathcal{H}$ have the same stable models. About formulas F and G we say that they are *strongly equivalent* if the singleton sets $\{F\}$ and $\{G\}$ are strongly equivalent.

A *unary formula* is an atom or a formula of the form $p \to q$, where p and q are atoms. The following theorem is similar to the main theorem from [10].

Theorem 2. *For any sets $\mathcal{H}_1$, $\mathcal{H}_2$ of infinitary formulas, the following conditions are equivalent:*

(i) $\mathcal{H}_1$ is strongly equivalent to $\mathcal{H}_2$,
(ii) for every set $\mathcal{H}$ of unary formulas, sets $\mathcal{H}_1 \cup \mathcal{H}$ and $\mathcal{H}_2 \cup \mathcal{H}$ have the same stable models;
(iii) sets $\mathcal{H}_1$ and $\mathcal{H}_2$ have the same HT-models.

Proof. Clearly, (i) implies (ii). To see that (iii) implies (i), observe that if sets $\mathcal{H}_1$ and $\mathcal{H}_2$ have the same HT-models then $\mathcal{H}_1 \cup \mathcal{H}$ and $\mathcal{H}_2 \cup \mathcal{H}$ have the same HT-models, and consequently have the same equilibrium models. It follows by Theorem 1 that $\mathcal{H}_1 \cup \mathcal{H}$ and $\mathcal{H}_2 \cup \mathcal{H}$ have the same stable models.

It remains to check that (ii) implies (iii). Suppose $\langle I, J \rangle$ is an HT-model of $\mathcal{H}_1$ but not an HT-model of $\mathcal{H}_2$. We will show how to find a set $\mathcal{H}$ of unary formulas such that $\langle J, J \rangle$ is an equilibrium model of one of the sets $\mathcal{H}_1 \cup \mathcal{H}, \mathcal{H}_2 \cup \mathcal{H}$ but not the other. It will follow that the interpretation J is a stable model of one but not the other.

Case 1: $\langle J, J \rangle$ is not an HT-model of $\mathcal{H}_2$. Since $\langle I, J \rangle$ is an HT-model of $\mathcal{H}_1$, it is easy to see that $\langle J, J \rangle$ must be an HT-model of $\mathcal{H}_1$ as well. Then we can take $\mathcal{H} = J$. Indeed, it is clear that $\langle J, J \rangle$ is an HT-model of $\mathcal{H}_1 \cup J$. Furthermore, for any I that is a proper subset of J, $\langle I, J \rangle$ cannot be an HT-model of $\mathcal{H}_1 \cup J$, so that $\langle J, J \rangle$ is an equilibrium model of $\mathcal{H}_1 \cup J$. On the other hand, since $\langle J, J \rangle$ is not a HT-model of $\mathcal{H}_2$, it cannot be an HT-model of $\mathcal{H}_2 \cup J$.

Case 2: $\langle J, J \rangle$ is an HT-model of $\mathcal{H}_2$. Let $\mathcal{H}$ be the set

$$I \cup \{p \to q \mid p, q \in J \setminus I\}.$$

Since $\langle J, J \rangle$ satisfies every formula in $\mathcal{H}$, it is an HT-model of $\mathcal{H}_2 \cup \mathcal{H}$. To see that it is an equilibrium model, consider any HT-model $\langle K, J \rangle$ of $\mathcal{H}_2 \cup \mathcal{H}$. Clearly, K must contain I. But it cannot be equal to I, since $\langle I, J \rangle$ is not an HT-model of $\mathcal{H}_2$. Thus $I \subset K \subset J$. Consider an atom p in $K \setminus I$ and an atom q in $J \setminus K$. For these atoms, $p \to q$ belongs to $\mathcal{H}$. But $\langle K, J \rangle$ does not satisfy this implication, contrary to the assumption that it is an HT-model of $\mathcal{H}_2 \cup \mathcal{H}$. We may conclude that $\langle J, J \rangle$ is an equilibrium model of $\mathcal{H}_2 \cup \mathcal{H}$. Finally, we will check that $\langle J, J \rangle$ is not an equilibrium model of $\mathcal{H}_1 \cup \mathcal{H}$. Consider the HT-model $\langle I, J \rangle$ of $\mathcal{H}_1$. Clearly, it is an HT-model of I. Moreover, it satisfies each implication $p \to q$ in $\mathcal{H}$: $\langle I, J \rangle$ does not satisfy p because $p \notin I$, and J satisfies q because $q \in J$.

We see that $\langle I, J \rangle$ satisfies all formulas in $\mathcal{H}$, so that it is an HT-model of $\mathcal{H}_1 \cup \mathcal{H}$. Furthermore, I is different from J since $\langle J, J \rangle$ is an HT-model of $\mathcal{H}_2$ and $\langle I, J \rangle$ is not. Consequently, I is a proper subset of J, and we may conclude that $\langle J, J \rangle$ is not an equilibrium model of $\mathcal{H}_1 \cup \mathcal{H}$.

A part of any formula can be replaced with a strongly equivalent formula without changing the set of stable models. For instance, it is easy to check that the formulas $p \wedge \neg p$ and $\bot$ are strongly equivalent to each other; it follows that the formulas

$$F \wedge (q \rightarrow (p \wedge \neg p)) \qquad \text{and} \qquad F \wedge \neg q \qquad (3)$$

have the same stable models. Corollary 1 expresses a more general fact: several parts (even infinitely many) can be simultaneously replaced by strongly equivalent formulas. Its statement uses the following definitions [4]. Let σ and σ' be disjoint signatures. A *substitution* is a bounded family of formulas over σ with index set σ'. For any substitution ϕ and any formula F over the signature $\sigma \cup \sigma'$, ϕF stands for the formula over σ formed as follows:

- If $F \in \sigma$ then $\phi F = F$.
- If $F \in \sigma'$ then $\phi F = \phi_F$.
- If F is $\mathcal{H}^{\wedge}$ then $\phi F = \{\phi G \mid G \in \mathcal{H}\}^{\wedge}$.
- If F is $\mathcal{H}^{\vee}$ then $\phi F = \{\phi G \mid G \in \mathcal{H}\}^{\vee}$.
- If F is $G \rightarrow H$ then $\phi F = \phi G \rightarrow \phi H$.

For instance, if $\sigma' = \{r\}$, $\phi_r = p \wedge \neg p$, and $\psi_r = \bot$, then $\phi(F \wedge (q \rightarrow r))$ and $\psi(F \wedge (q \rightarrow r))$ are the formulas (3).

Corollary 1. *Let ϕ and ψ be substitutions such that for all $p \in \sigma'$, ϕ_p is strongly equivalent to ψ_p. Then for any formula F, ϕF is strongly equivalent to ψF, so that ϕF and ψF have the same stable models.*

Proof. By Theorem 2, the assertion of the corollary can be stated as follows: if for all $p \in \sigma'$, ϕ_p and ψ_p are satisfied by the same HT-interpretations, then for any formula F, ϕF and ψF are satisfied by the same HT-interpretations. This is easy to check by induction on the rank of F.

6   An Axiomatization of the Infinitary Logic of Here-and-There

We present an axiomatization HT^{∞} of the infinitary logic of here-and-there. The derivable objects in HT^{∞} are *(infinitary) sequents*—expressions of the form $\Gamma \Rightarrow F$, where F is an infinitary formula, and Γ is a finite set of infinitary formulas ("F under assumptions Γ"). To simplify notation, we will write Γ as a list. We will identify a sequent of the form $\Rightarrow F$ with the formula F.

The inference rules are the introduction and elimination rules for the propositional connectives

$$(\wedge I)\ \frac{\Gamma \Rightarrow H \quad \text{for all } H \in \mathcal{H}}{\Gamma \Rightarrow \mathcal{H}^\wedge} \qquad (\wedge E)\ \frac{\Gamma \Rightarrow \mathcal{H}^\wedge}{\Gamma \Rightarrow H}\ (H \in \mathcal{H})$$

$$(\vee I)\ \frac{\Gamma \Rightarrow H}{\Gamma \Rightarrow \mathcal{H}^\vee}\ (H \in \mathcal{H}) \qquad (\vee E)\ \frac{\Gamma \Rightarrow \mathcal{H}^\vee \quad \Delta, H \Rightarrow F \quad \text{for all } H \in \mathcal{H}}{\Gamma, \Delta \Rightarrow F}$$

$$(\to I)\ \frac{\Gamma, F \Rightarrow G}{\Gamma \Rightarrow F \to G} \qquad (\to E)\ \frac{\Gamma \Rightarrow F \quad \Delta \Rightarrow F \to G}{\Gamma, \Delta \Rightarrow G},$$

where $\mathcal{H}$ is a bounded set of formulas, and the weakening rule

$$(W)\ \frac{\Gamma \Rightarrow F}{\Gamma, \Delta \Rightarrow F}.$$

The set of axioms in HT^∞ is a subset of the set of axioms introduced in the extended system of natural deduction from [4]. HT^∞ includes three axiom schemas:

$$F \Rightarrow F,$$

$$F \vee (F \to G) \vee \neg G, \tag{4}$$

and

$$\bigwedge_{\alpha \in A} \bigvee_{F \in \mathcal{H}_\alpha} F \to \bigvee_{(F_\alpha)_{\alpha \in A}} \bigwedge_{\alpha \in A} F_\alpha \tag{5}$$

for every non-empty family $(\mathcal{H}_\alpha)_{\alpha \in A}$ of sets of formulas such that its union is bounded; the disjunction in the consequent of (5) extends over all elements $(F_\alpha)_{\alpha \in A}$ of the Cartesian product of the family $(\mathcal{H}_\alpha)_{\alpha \in A}$. Axiom schema (4) was mentioned in the introduction in connection with the problem of axiomatizing the logic of here-and-there in the finite case, but now F and G can be infinitary formulas. Axiom schema (5) generalizes (one direction of) the distributivity of conjunction over disjunction to infinitary formulas: if $A = \{1, 2\}$, $\mathcal{H}_1 = \{F_1, G_1\}$, and $\mathcal{H}_2 = \{F_2, G_2\}$, then (5) turns into

$$(F_1 \vee G_1) \wedge (F_2 \vee G_2) \to (F_1 \wedge F_2) \vee (F_1 \wedge G_2) \vee (G_1 \wedge F_2) \vee (G_1 \wedge G_2).$$

The set of *theorems of* HT^∞ is the smallest set of sequents that includes the axioms of the system and is closed under the application of its inference rules. We say that formulas F and G are *equivalent in* HT^∞ if $F \leftrightarrow G$ is a theorem of HT^∞.

The following theorem expresses the soundness and completeness of HT^∞.

Theorem 3. *An infinitary formula F is a theorem of HT^∞ iff it is satisfied by all HT-interpretations.*

The proof of soundness is straightforward. The proof of completeness given in the next section is analogous to the proof of completeness for classical propositional logic from [7].

From Theorems 2 and 3 we conclude:

Corollary 2. *Bounded sets $\mathcal{H}_1$, $\mathcal{H}_2$ of infinitary formulas are strongly equivalent iff $\mathcal{H}_1^\wedge$ is equivalent to $\mathcal{H}_2^\wedge$ in HT^∞.*

7   Proof of Completeness

In the proof of completeness, we use the following construction, due to Cabalar and Ferraris [1, Section 5]. Let $\langle I, J \rangle$ be an HT-interpretation. We define the set M_{IJ} to be

$$I \cup \{\neg\neg p \mid p \in J\} \cup \{\neg p \mid p \in \sigma \setminus J\} \cup \{p \to q \mid p, q \in J \setminus I\}$$

(recall that σ is the set of all atoms). By $v_{IJ}(F)$ we denote the truth value of F with respect to $\langle I, J \rangle$ (see Sect. 3). We will omit the subscripts I, J in M_{IJ} and $v_{IJ}(F)$ when it is clear which HT-interpretation we refer to.

Lemma 2. *For any infinitary formula F and HT-interpretation $\langle I, J \rangle$,*

  (i) if $v(F) = \emptyset$ then $M^{\wedge} \Rightarrow \neg F$ is a theorem of HT^{∞};
  (ii) if $v(F) = \{T\}$ then for every atom q in $J \setminus I$, $M^{\wedge} \Rightarrow F \leftrightarrow q$ is a theorem of HT^{∞};
  (iii) if $v(F) = \{H, T\}$ then $M^{\wedge} \Rightarrow F$ is a theorem of HT^{∞}.

Proof. We will prove the claim by strong induction on the rank of F. We assume the claim holds for all formulas with rank less than n and show that it holds for a formula F of rank n. We consider cases corresponding to the different possible forms of F and truth values $v(F)$. Note that if $v(F)$ is $\{T\}$ then the set $J \setminus I$ is non-empty. Indeed, if $I = J$ then the truth value of any formula is either $\emptyset$ or $\{H, T\}$.

Case 1: F is an atom.
Case 1.1: $v(F) = \emptyset$. Then $F \in \sigma \setminus J$, and $\neg F \in M$.
Case 1.2: $v(F) = \{T\}$. Then $F \in J \setminus I$, and for every atom q in $J \setminus I$, the implications $F \to q$ and $q \to F$ are in M.
Case 1.3: $v(F) = \{H, T\}$. Then $F \in M$.

Case 2: F is of the form $\mathcal{H}^{\wedge}$. The induction hypothesis is then applicable to all formulas in $\mathcal{H}$.
Case 2.1: $v(F) = \emptyset$. Then there exists a formula G in $\mathcal{H}$ such that $v(G)$ is $\emptyset$. By the induction hypothesis, $M^{\wedge} \Rightarrow \neg G$ is a theorem of HT^{∞}. From this we can derive $M^{\wedge} \Rightarrow \neg(\mathcal{H}^{\wedge})$.
Case 2.2: $v(F) = \{T\}$. Let $\mathcal{H}_1$ be the set of all formulas in $\mathcal{H}$ with truth value $\{T\}$, and $\mathcal{H}_2$ be the set of all formulas in $\mathcal{H}$ with truth value $\{H, T\}$. It is clear that $\mathcal{H}_1 \cup \mathcal{H}_2 = \mathcal{H}$ and that $\mathcal{H}_1$ is non-empty. Consider an arbitrary element q of $J \setminus I$. By the induction hypothesis $M^{\wedge} \Rightarrow G \leftrightarrow q$ is a theorem for every G in $\mathcal{H}_1$, and $M^{\wedge} \Rightarrow G$ is a theorem for every G in $\mathcal{H}_2$. From these we can derive $M^{\wedge} \Rightarrow \mathcal{H}_1^{\wedge} \leftrightarrow q$ and $M^{\wedge} \Rightarrow \mathcal{H}_2^{\wedge}$. Then we can derive $M^{\wedge} \Rightarrow \mathcal{H}^{\wedge} \leftrightarrow q$.
Case 2.3: $v(F) = \{H, T\}$. Then for each element G in $\mathcal{H}$, $v(G) = \{H, T\}$, and by the induction hypothesis $M^{\wedge} \Rightarrow G$ is a theorem. From these sequents we can derive $M^{\wedge} \Rightarrow \mathcal{H}^{\wedge}$.

Case 3: F is of the form $\mathcal{H}^{\vee}$. The induction hypothesis is then applicable to all formulas in $\mathcal{H}$.

Case 3.1: $v(F) = \emptyset$. Then for each element G in $\mathcal{H}$, $v(G) = \emptyset$, and by the induction hypothesis $M^\wedge \Rightarrow \neg G$ is a theorem. From these sequents we can derive $M^\wedge \Rightarrow \neg(\mathcal{H}^\vee)$.

Case 3.2: $v(F) = \{T\}$. Let $\mathcal{H}_1$ be the set of all formulas in $\mathcal{H}$ with truth value $\{T\}$, and $\mathcal{H}_2$ be the set of all formulas in $\mathcal{H}$ with truth value $\emptyset$. It is clear that $\mathcal{H}_1 \cup \mathcal{H}_2 = \mathcal{H}$ and that $\mathcal{H}_1$ is non-empty. Consider an arbitrary element q of $J \setminus I$. By the induction hypothesis $M^\wedge \Rightarrow G \leftrightarrow q$ is a theorem for every G in $\mathcal{H}_1$, and $M^\wedge \Rightarrow \neg G$ is a theorem for every G in $\mathcal{H}_2$. From these we can derive $M^\wedge \Rightarrow \mathcal{H}_1^\vee \leftrightarrow q$ and $M^\wedge \Rightarrow \neg(\mathcal{H}_2^\vee)$. Then we can derive $M^\wedge \Rightarrow \mathcal{H}^\vee \leftrightarrow q$.

Case 3.3: $v(F) = \{H, T\}$. Then there exists a formula G in $\mathcal{H}$ such that $v(G)$ is $\{H, T\}$. By the induction hypothesis, $M^\wedge \Rightarrow G$ is a theorem. From this we can derive $M^\wedge \Rightarrow \mathcal{H}^\vee$.

Case 4: F is of the form $F_1 \to F_2$. The induction hypothesis is then applicable to F_1 and F_2.

Case 4.1: $v(F) = \emptyset$. Then $v(F_1)$ is non-empty and $v(F_2)$ is empty.

Case 4.1.1: $v(F_1) = \{T\}$. By the induction hypothesis $M^\wedge \Rightarrow \neg F_2$ is a theorem, as is $M^\wedge \Rightarrow F_1 \leftrightarrow q$ for any q in $J \setminus I$. Consider an atom q in $J \setminus I$. By the construction of M, we know that $\neg\neg q$ is an element of M. From the sequents $M^\wedge \Rightarrow F_1 \leftrightarrow q$, $M^\wedge \Rightarrow \neg F_2$, and $M^\wedge \Rightarrow \neg\neg q$, we can derive $M^\wedge \Rightarrow \neg(F_1 \to F_2)$.

Case 4.1.2: $v(F_1) = \{H, T\}$. By the induction hypothesis, both $M^\wedge \Rightarrow F_1$ and $M^\wedge \Rightarrow \neg F_2$ are theorems. From these sequents we can derive $M^\wedge \Rightarrow \neg(F_1 \to F_2)$.

Case 4.2: $v(F) = \{T\}$. Then $v(F_1) = \{H, T\}$ and $v(F_2) = \{T\}$. By the induction hypothesis $M^\wedge \Rightarrow F_2 \leftrightarrow q$ is a theorem for any $q \in J \setminus I$, and $M^\wedge \Rightarrow F_1$ is a theorem as well. From these two sequents we can derive $M^\wedge \Rightarrow (F_1 \to F_2) \leftrightarrow q$.

Case 4.3: $v(F) = \{H, T\}$.

Case 4.3.1: $v(F_1) = \emptyset$. Then by the induction hypothesis $M^\wedge \Rightarrow \neg F_1$ is a theorem. From this we can derive $M^\wedge \Rightarrow F_1 \to F_2$.

Case 4.3.2: $v(F_2) = \{H, T\}$. Then by the induction hypothesis $M^\wedge \Rightarrow F_2$ is a theorem. From this we can derive $M^\wedge \Rightarrow F_1 \to F_2$.

Case 4.3.3: $v(F_1) \neq \emptyset$ and $v(F_2) \neq \{H, T\}$. Since $v(F)$ is $\{H, T\}$, $v(F_1)$ is different from $\{H, T\}$ and therefore must be equal to $\{T\}$. It follows that $v(F_2)$ is different from $\emptyset$, and therefore must be $\{T\}$ also. Consider an element q in $J \setminus I$. By the induction hypothesis both $M^\wedge \Rightarrow F_1 \leftrightarrow q$ and $M^\wedge \Rightarrow F_2 \leftrightarrow q$ are theorems. From these two sequents we can derive $M^\wedge \Rightarrow F_1 \to F_2$.

Note that in the proof of the lemma we did not refer to axiom schemas (4) and (5); the assertion of the lemma would hold even if those axioms were removed from HT^∞.

Lemma 3. *The disjunction of the formulas $M_{IJ}^\wedge$ over all HT-interpretations $\langle I, J \rangle$ is a theorem of HT^∞.*

Proof. Let Q stand for the set of disjunctions

$$p \vee (p \to q) \vee \neg q, \tag{6}$$

$$\neg p \vee \neg\neg p \tag{7}$$

for all p, q from σ. Let $(\mathcal{H}_D)_{D \in Q}$ be the following family of sets:

$$\mathcal{H}_D = \{p, p \to q, \neg q\} \quad \text{if } D = p \vee (p \to q) \vee \neg q;$$
$$\mathcal{H}_D = \{\neg p, \neg\neg p\} \quad \text{if } D = \neg p \vee \neg\neg p.$$

Then the formula

$$\bigwedge_{D \in Q} \bigvee_{S \in \mathcal{H}_D} S \to \bigvee_{(S_D)_{D \in Q}} \bigwedge_{D \in Q} S_D,$$

(where the disjunction in the consequent extends over all elements $(S_D)_{D \in Q}$ of the Cartesian product of the family $(\mathcal{H}_D)_{D \in Q}$) is an instance of axiom schema (5). Since the antecedent of this implication is the conjunction of all formulas in Q, it is a theorem of HT^∞. It follows that the consequent is a theorem as well. To complete the proof it is sufficient to show that for every disjunctive term

$$\bigwedge_{D \in Q} S_D \tag{8}$$

of the consequent there exists an HT-interpretation $\langle I, J \rangle$ such that the sequent

$$\bigwedge_{D \in Q} S_D \Rightarrow M_{IJ}^\wedge \tag{9}$$

is a theorem.

Consider one of the conjunctions (8), and let C be set of its conjunctive terms. The elements of C are formulas of the forms

$$p, \ \neg p, \ \neg\neg p, \ p \to q.$$

If C contains both a formula and its negation then (9) is a theorem for every $\langle I, J \rangle$. Otherwise, let I denote the set of all atoms in C, and J denote the set of all atoms p such that $\neg\neg p$ is in C. Let us check that $I \subseteq J$. Assume $p \in I$ so that $p \in C$. Since C is consistent, it does not contain $\neg p$, and since it contains a term from each disjunction (7), it contains $\neg\neg p$. So $\langle I, J \rangle$ is an HT-interpretation.

We will show that every formula from M_{IJ} belongs to C. By the choice of I, $I \subseteq C$. By the choice of J, $\{\neg\neg p \mid p \in J\} \subseteq C$. Consequently $\{\neg p \mid p \in \sigma \backslash J\} \subseteq C$, because C contains one term from each disjunction (7). Finally, we need to check that $\{p \to q \mid p, q \in J \backslash I\} \subseteq C$. Consider a pair of atoms p, q that occur in J but not in I. By the choice of I, p is not in C, and by the choice of J, $\neg q$ is not in C. Since C contains one term from each of the disjunctions (6) and contains neither p nor $\neg q$, C must contain $p \to q$.

Proof of Completeness. Let F be an infinitary formula over signature σ that is satisfied by all HT-interpretations of σ. By Lemma 2(iii), $M_{IJ} \Rightarrow F$ is a theorem of HT^∞ for all HT-interpretations $\langle I, J \rangle$. By Lemma 3, it follows that F is a theorem also.

It is clear from the proof that HT^∞ will remain complete if we require that formulas F and G in axiom schema (4) must be literals, and that the sets $\mathcal{H}_i$ in axiom schema (5) must be finite.

8   Example: Infinitary De Morgan's Law

As observed in Sect. 6, the set of axioms in HT^∞ is a subset of the set of axioms introduced in the extended system of natural deduction from [4]. From the results presented in this note it is clear that the other axioms in the extended system are redundant. The infinitary De Morgan's law,

$$\neg \bigwedge_{F \in \mathcal{H}} F \to \bigvee_{F \in \mathcal{H}} \neg F, \tag{10}$$

is one of these redundant axioms. In this section, we show directly, without a reference to the general completeness theorem, how to prove (10) in HT^∞.

Let Q stand for the set of disjunctions

$$F \vee (F \to G) \vee \neg G, \tag{11}$$

for all formulas F, G from $\mathcal{H}$. Let $(\mathcal{H}_D)_{D \in Q}$ be the following family of sets:

$$\mathcal{H}_D = \{F, F \to G, \neg G\}.$$

Then the formula

$$\bigwedge_{D \in Q} \bigvee_{S \in \mathcal{H}_D} S \to \bigvee_{(S_D)_{D \in Q}} \bigwedge_{D \in Q} S_D, \tag{12}$$

(where the disjunction in the consequent extends over all elements $(S_D)_{D \in Q}$ of the Cartesian product of the family $(\mathcal{H}_D)_{D \in Q}$) is an instance of axiom schema (5). Since the antecedent of this implication is the conjunction of all formulas in Q, it is a theorem of HT^∞. It follows that the consequent is a theorem as well. To complete the proof it is sufficient to show that from the antecedent of (10) and any disjunctive term

$$\bigwedge_{D \in Q} S_D \tag{13}$$

of the consequent of (12), we can derive the consequent of (10). Consider one of the conjunctions (13), and let C be set of its conjunctive terms. The elements of C are formulas of the forms

$$F, \ F \to G, \ \neg G.$$

If C contains $\neg F$ for some formula F then the consequent of (10) follows immediately. Otherwise, we will show that assuming $C^\wedge$ and any element F of $\mathcal{H}$ we can derive

$$\bigwedge_{F \in \mathcal{H}} F, \tag{14}$$

contradicting the antecedent of (10), and allowing us to derive $\neg F$ from $C^\wedge$ and the antecedent of (10). If C contains every formula F in $\mathcal{H}$ then (14) is immediate. Otherwise, there is some G from $\mathcal{H}$ which is not in C. Assume G. Since G is not in C and C does not contain the negation of any formula, we may conclude that C contains $G \to F$ for all formulas F from $\mathcal{H}$. It follows that from G and $C^\wedge$ we can derive (14).

9   Conclusion

Under the stable model semantics, two sets of propositional formulas are strongly equivalent if and only if they are equivalent in the logic of here-and-there. This theorem was originally proved using equilibrium logic in [10]. In this paper, we extended equilibrium logic to infinitary formulas; we defined an infinitary counterpart to the logic of here-and-there and introduced an axiomatization, HT^∞, of that system; finally, we showed that bounded sets of infinitary propositional formulas are strongly equivalent if and only if they are equivalent in HT^∞.

Acknowledgements. Thanks to Pedro Cabalar and Yuliya Lierler, and to the anonymous referees for helpful comments. The first two authors were partially supported by the National Science Foundation under Grant IIS-1422455.

References

1. Cabalar, P., Ferraris, P.: Propositional theories are strongly equivalent to logic programs. Theor. Pract. Logic Program. **7**, 745–759 (2007)
2. Ferraris, P.: Answer sets for propositional theories. In: Baral, C., Greco, G., Leone, N., Terracina, G. (eds.) LPNMR 2005. LNCS (LNAI), vol. 3662, pp. 119–131. Springer, Heidelberg (2005)
3. Gelfond, M., Lifschitz, V.: The stable model semantics for logic programming. In: Kowalski, R., Bowen, K. (eds.) Proceedings of International Logic Programming Conference and Symposium, pp. 1070–1080. MIT Press, Cambridge (1988)
4. Harrison, A., Lifschitz, V., Truszczynski, M.: On equivalence of infinitary formulas under the stable model semantics. Theor. Pract. Logic Program. **15**(1), 18–34 (2015)
5. Harrison, A., Lifschitz, V., Yang, F.: The semantics of Gringo and infinitary propositional formulas. In: Proceedings of International Conference on Principles of Knowledge Representation and Reasoning (KR) (2014)
6. Hosoi, T.: The axiomatization of the intermediate propositional systems S_n of Gödel. J. Fac. Sci. Univ. Tokyo **13**, 183–187 (1966)
7. Kalmár, L.: Zurückführung des Entscheidungsproblems auf den Fall von Formeln mit einer einzigen, bindren, Funktionsvariablen. Compositio Mathematica **4**, 137–144 (1936)
8. Karp, C.R.: Languages with Expressions of Infinite Length. North-Holland, Amsterdam (1964)
9. Lee, J., Lifschitz, V., Palla, R.: A reductive semantics for counting and choice in answer set programming. In: Proceedings of the AAAI Conference on Artificial Intelligence (AAAI), pp. 472–479 (2008)
10. Lifschitz, V., Pearce, D., Valverde, A.: Strongly equivalent logic programs. ACM Trans. Comput. Logic **2**, 526–541 (2001)
11. Łukasiewicz, J.: Die Logik und das Grundlagenproblem. In: Les Entretiens de Zürich sue les Fondements et la méthode des sciences mathématiques 1938, pp. 82–100. Leemann, Zürich (1941)
12. Nadel, M.: Infinitary intuitionistic logic from a classical point of view. Ann. Math. Logic **14**, 159–191 (1978)

13. Pearce, D.: A new logical characterisation of stable models and answer sets. In: Dix, J., Przymusinski, T.C., Moniz Pereira, L. (eds.) NMELP 1996. LNCS, vol. 1216, pp. 57–70. Springer, Heidelberg (1997)
14. Scott, D., Tarski, A.: The sentential calculus with infinitely long expressions. Colloquium Math. **6**, 165–170 (1958)
15. Thomas, I.: Finite limitations on dummett's LC. Notre Dame J. Formal Logic **III**(3), 170–174 (1962)
16. Truszczynski, M.: Connecting first-order ASP and the logic FO(ID) through reducts. In: Erdem, E., Lee, J., Lierler, Y., Pearce, D. (eds.) Correct Reasoning. LNCS, vol. 7265, pp. 543–559. Springer, Heidelberg (2012)
17. Umezawa, T.: On intermediate many-valued logics. J. Math. Soc. Jpn. **11**(2), 116–128 (1959)

On the Relationship Between Two Modular Action Languages: A Translation from MAD into $\mathcal{ALM}$

Daniela Inclezan[✉]

Miami University, Oxford, OH 45056, USA
inclezd@MiamiOH.edu

Abstract. Modular action languages MAD and $\mathcal{ALM}$ share the goal of providing means for the reuse of knowledge in order to facilitate the creation of libraries of knowledge. They differ substantially in their underlying assumptions (*Causality Principle* for MAD, *Inertia Axiom* for $\mathcal{ALM}$) and in the constructs that enable the reuse of knowledge, especially the mechanisms used to declare actions in terms of already described actions. In this paper, we investigate the relationship between the two action languages by providing a translation from MAD into $\mathcal{ALM}$. We specify a condition that ensures that, for a specific class of MAD action descriptions, our translation produces a transition diagram isomorphic to the original one, modulo the common vocabulary.

1 Introduction

This paper investigates the relationship between two action languages, MAD and $\mathcal{ALM}$, by providing a translation from MAD into $\mathcal{ALM}$. Action languages [6] are high-level declarative languages dedicated to the concise and elegant representation of dynamic systems. By a dynamic system we mean a system that can be represented by a transition diagram whose nodes correspond to possible states of the system and arcs are labeled by actions.

Currently, several action languages exist (e.g., [6,8,9,12,14]), which address significant problems from the field of reasoning about actions and change. A next challenge seems to be the creation of *libraries of commonsense knowledge* and *large knowledge bases* about dynamic domains, an effort that can contribute greatly to automating reasoning tasks such as natural language understanding. Traditional action languages do not provide the means for addressing this issue, given that they do not address the *structuring* and *reuse* of knowledge. Action languages MAD [2,13] and $\mathcal{ALM}$ [4,10,11] were designed to target this problem.

Both of these languages use *modularity* to facilitate organization and reuse, where a module is a coherent and reusable piece of knowledge on a specific theme. However, the actual reuse of knowledge is achieved using different means. In $\mathcal{ALM}$ ($\mathcal{A}$ction $\mathcal{L}$anguage with $\mathcal{M}$odules), objects of the domain, including actions, are grouped into sorts organized in a sort hierarchy. This is defined using the *specialization* construct, whose semantics specifies that sorts inherit

© Springer International Publishing Switzerland 2015
F. Calimeri et al. (Eds.): LPNMR 2015, LNAI 9345, pp. 411–424, 2015.
DOI: 10.1007/978-3-319-23264-5_34

attributes (i.e., intrinsic properties) and behavior (i.e., axioms) from supersorts. This allows describing, for example, action *carry* in terms of *move*, a common practice in natural language where *carry* is defined as *to move while holding*.

In MAD (Modular Action Description), the concept of a *sort* also exists, but it does not apply to actions – there is a built-in sort *action*, but it has no further subsorts. The reuse of knowledge is achieved via *import statements* possibly containing *renaming clauses*, a construct similar to that of *bridge rules* [2] in action language $\mathcal{C}+$ [8]. Sorts, properties of the domain, and actions can be renamed. For instance, an action *walk* can be defined by importing the module containing the declaration of action *move* and renaming *move* as *walk*.

Given the similar goals of $\mathcal{ALM}$ and MAD, it is important to study the relationship between the two languages and especially between *their mechanisms for the reuse of knowledge*. This will allow knowledge engineers to incorporate knowledge modules written in one language when creating system descriptions in the other. Thus, our work has intentions similar to, and relies on, the work of Gelfond and Lifschitz on the common core of $\mathcal{B}$ and $\mathcal{C}$ [7].

In this paper, we approach this topic by proposing a translation from MAD into $\mathcal{ALM}$. We review the syntax and semantics of the two languages. We illustrate our translation via examples first and then introduce it formally. Finally, we define a class of MAD action descriptions for which our translation produces transition diagrams isomorphic to the original, modulo the common signature.

2   Language $\mathcal{ALM}$

A dynamic system is represented in $\mathcal{ALM}$ by a *system description* that consists of two parts: a *general theory* (i.e., a collection of modules with a common theme organized in a hierarchy) and a *structure* (i.e., an interpretation of some of the symbols in the theory). A *module* is a collection of declarations of sorts and functions together with a set of axioms. The purpose of a module is to allow the organization of knowledge into smaller reusable pieces of code. Modules serve a similar role to that of procedures in procedural languages and can be organized in a hierarchy (a DAG) such that, if module M_1 depends on module M, then the declarations and axioms of M are implicitly part of M_1. We briefly illustrate the syntax of $\mathcal{ALM}$ via some examples. Boldface symbols denote keywords of the language; identifiers starting with a lowercase letter denote constant symbols; and identifiers starting with an uppercase letter denote variables.

Sorts (i.e., types, classes) are organized in a hierarchy with root *universe*. The hierarchy contains pre-defined sorts *actions* and *booleans*. The sort hierarchy is specified in $\mathcal{ALM}$ via the specialization construct "::". For instance, we say that *points* and *things* are subsorts of *universe* and *agents* is a subsort of *things* by:

> *points*, *things* :: *universe*
> *agents* :: *things*

We use the same construct to define action classes as special cases of other action classes. For instance, the statements:

> move :: *actions*
> > **attributes**
> > > *actor* : *agents*
> > > *origin, dest* : *points*
> > *carry* :: *move*
> > **attributes**
> > > *carried_thing* : *carriables*

define action *move* as having three attributes (i.e., three intrinsic properties) – *actor*, *origin*, and *dest* – that are (possibly partial) functions mapping elements of *move* into elements of sorts *agents*, *points*, and *points* respectively; *carry* is a special case of *move*, meaning that it inherits the attributes of *move* and has an additional attribute, *carried_thing*, mapping elements of *carry* into a new sort *carriables* assumed to be declared in the same module. The axioms written for *move* will apply to actions of sort *carry* as well, as all instances of *carry* are also instances of *move*. For example, in the axiom (dynamic causal law):

$$occurs(X) \textbf{ causes } loc\_in(A) = D \textbf{ if } instance(X, move),$$
$$actor(X) = A, \; dest(X) = D.$$

the variable X will be replaced by objects defined in the structure that belong to the interpretation of sort *move*, which includes instances of *carry*.

Properties of objects of a dynamic system are represented using functions. Functions are partitioned in $\mathcal{ALM}$ into *fluents* (those that can be changed by actions) and *statics* (those that cannot); each of these two sets are further divided into *basic* and *defined*, where defined functions can be viewed just as a means to facilitate knowledge encoding. Basic fluents are subject to the law of inertia.

The second part of a system description is its *structure*, which represents information about a specific domain: instances of sorts (including actions) and values of statics. For example, a domain that is about John and Bob, and their movements between two points, London and Paris, may be described as follows:

> *john, bob* **in** *agents*
> *london, paris* **in** *points*
> *go*(A, P) **in** *move*
> > *actor* = A
> > *dest* = P

Action $go(A, P)$ is an instance schema that stands for all actions of this form obtained by replacing A and P with instances of *agents* and *points*, respectively.

The semantics of $\mathcal{ALM}$ is given by defining the states and transitions of the transition diagram defined by a system description. For that purpose, we encode statements of the system description into a logic program of ASP{f} [1], an extension of Answer Set Prolog [5] by non-Herbrand functions. The states and transitions of the corresponding transition diagram are determined by parts of the answer sets of this logic program. As an example, the dynamic causal law about actions of the type *move* shown above is encoded as the ASP{f} rule:

$$loc\_in(A, I + 1) = D \leftarrow instance(X, move),\ occurs(X, I),$$
$$actor(X) = A,\ dest(X) = D.$$

The structure is encoded using statements like:

$is\_a(john, agents).$
$is\_a(go(john, london), move).\ actor(go(john, london)) = john.\ \ldots$

where the function $instance$ is pre-defined as the transitive closure of is_a.

3   Language MAD

Dynamic systems are described in MAD by *action descriptions*, which consist of declarations of sorts and their subsort relations, followed by one or more modules. A module contains the declarations of objects, actions, fluents (which correspond to functions of $\mathcal{ALM}$), and variables; import statements; and axioms. Actions are represented using terms and, while there is a built-in sort $action$, special case actions are not sorts. Import statements allow the renaming of sorts, fluents, and actions. Conventionally, identifiers used for variables start with a lowercase letter while those for constants start with an uppercase letter, the opposite of $\mathcal{ALM}$.

We illustrate the syntax of MAD on an action description, MBP, extracted from the encoding of the Monkey and Bananas Problem in [3]. We only represent the monkey's action of walking to a desired location. MBP includes two library modules, $ASSIGN$ and $MOVE$, and starts with the section declaring sorts.

```
1   sorts Domain; Range; Thing; Place;
```

Module $ASSIGN$ shown next defines an action $Assign$ with two parameters ranging over sorts $Domain$ and $Range$. The action is exogenous (i.e., it does not need a cause to occur) and may change the fluent $Value$ that maps elements of sort $Domain$ into elements of $Range$. Line 7 says that the law of inertia, not part of the semantics of MAD by default, applies to this fluent. Note that in MAD the sort of a variable is explicitly given, whereas in $\mathcal{ALM}$ the sort of a variable is inferred from the literals in which it appears, for each axiom.

```
2   module ASSIGN;
3       actions     Assign(Domain, Range);
4       fluents     Value(Domain) : simple(Range);
5       variables   x : Domain;  y : Range;
6       axioms
7           inertial Value(x);
8           exogenous Assign(x, y);
9           Assign(x, y) causes Value(x) = y;
```

Next is the module $MOVE$, containing a new action, fluent, and axiom, and an import statement for module $ASSIGN$ with several renaming clauses. These clauses should be seen as directives indicating that, when the module is imported, occurrences of sort $Domain$ are to be replaced by $Thing$, and $Range$ by $Place$; fluent $Value$ is equivalent to $Location$ and action $Assign$ to $Move$.

```
10    module MOVE;
11        actions      Move(Thing, Place);
12        fluents      Location(Thing) : simple(Place);
13        variables    x : Thing;  y : Place;
14        import ASSIGN;
15            Domain is Thing;
16            Range is Place;
17            Value(x) is Location(x);
18            Assign(x, p) is Move(x, p);
19        axioms
20            nonexecutable Move(x, p) if Location(x) = p;
```

A final module, MB, defines objects $Monkey$ of sort $Thing$ and P_1, P_2 of sort $Place$, and declares a new action, $Walk$. Module $MOVE$ is imported here to say that $Walk(p)$ is equivalent to $Monkey$ moving to place p.

```
21    module MB;
22        objects      Monkey : Thing;  P_1, P_2 : Place;
23        actions      Walk(Place);
24        variables    p : Place;
25        import MOVE;
26            Move(Monkey, p) is Walk(p);
```

The semantics of a MAD action description is given by first flattening it, which means eliminating import statements and producing a uni-module action description. Afterward, the semantics is given via a translation into $\mathcal{C}+$. In the flattening process, names of renamed sorts are replaced by their new names; variables, renamed fluents and actions receive a prefix of the type "In." (where n is the smallest positive natural number not yet used); and axioms are added to capture the renaming of fluents and actions. For instance, the flat version of module MB will contain actions I2.I1.$Assign(Thing, Place)$, I2.$Move(Thing, Place)$, and $Walk(Place)$; inertial fluents I1.$Value$ and $Location$, mapping $Things$ into $Places$; and the following axioms:

I2.I1.$Assign$(I2.I1.x, I2.I1.y) **causes** I1.$Value$(I2.I1.x) = I2.I1.y;
I1.$Value$(I2.x) $\equiv Location$(I2.x);
I2.I1.$Assign$(I2.x, I2.p) $\equiv$ I2.$Move$(I2.x, I2.p);
I2.$Move(Monkey, p) \equiv Walk(p)$;

4   Informal Translation

We illustrate our translation from MAD into $\mathcal{ALM}$ on the action description MBP from Sect. 3. An action description of MAD corresponds to a system description of $\mathcal{ALM}$. We call mbp^1 the system description resulting from MBP and use the same name for its theory:

---

[1] Recall the difference in the capitalization of identifiers between $\mathcal{ALM}$ and MAD.

system description *mbp*
theory *mbp*

In $\mathcal{ALM}$, sorts are always declared within the module in which they are used. Thus, we ignore for now the sort declarations in lines 1–2 of *MBP* and skip to the translation of the first module, *ASSIGN*. Our $\mathcal{ALM}$ theory, *mbp*, will now contain a new module, *assign*, that starts with the declarations of those sorts that are referenced in the MAD module. The MAD action description contains no information about subsort relationships, but in $\mathcal{ALM}$ all user-defined sorts are expected to be subsorts of the pre-defined sort *universe*:

module *assign*
sort declarations
  domain, range :: *universe*

We continue with action *Assign(Domain, Range)*. In $\mathcal{ALM}$, actions are not described using terms like in MAD, but by using action classes (i.e., subsorts of the pre-defined sort *actions*) and instances of action classes; attributes are used to describe the intrinsic properties of action classes. For that reason, we add two attributes to the $\mathcal{ALM}$ action class *assign*, one for each parameter of the term.

  assign : *actions*
  attributes
    attr1_assign : *domain*
    attr2_assign : *range*

Next, we focus on the inertial fluent *Value(Domain)* : *simple(Range)*. Simple inertial fluents of MAD correspond to basic fluents of $\mathcal{ALM}$; rigid fluents to basic statics, and statically determined fluents to defined fluents. Therefore, we declare *Value* as a basic fluent in $\mathcal{ALM}$:

function declarations
  fluents basic total *value* : *domain* $\rightarrow$ *range*

Finally, we translate the axioms of module *ASSIGN*. The axiom on line 7 is covered by the semantics of basic fluents in $\mathcal{ALM}$. The axiom on line 8 does not need to be translated, as all actions are exogenous in $\mathcal{ALM}$ (they do not need a cause in order to be executed). We translate the axiom on line 9 as follows:

axioms
  $occurs(A)$ **causes** $value(X) = Y$ **if** $instance(A, assign)$,
                                $attr1\_assign(A) = X$,
                                $attr2\_assign(A) = Y$.

Note that, in $\mathcal{ALM}$, we distinguish between the occurrence of an action and the action's name by using the expression $occurs(a)$ in the former case; MAD does not make this distinction – action terms are used in both cases. As we translated parameters of a MAD action via attributes, we needed to match variables X and Y in $value(X) = Y$ with the correct attributes of the action, $attr1\_assign$ and $attr2\_assign$ respectively. This concludes the translation of module *ASSIGN*.

To translate module $MOVE$, we begin by addressing its import statement. In $\mathcal{ALM}$, we say that module *move* depends on module *assign*, i.e., all the declarations and axioms from the latter are implicitly part of the former.

> **module** *move*
>> **depends on** *assign*

The MAD import statement says that sorts *Domain* and *Thing* are synonyms; also *Range* and *Place*. There is no direct translation for this in $\mathcal{ALM}$. The $\mathcal{ALM}$ feature that could mimic synonymy is the specialization construct for describing subsort relations. We cannot declare *Domain* as a special case of *Thing* and vice-versa, as the sort hierarchy would no longer be a DAG. Thus, we only declare *Thing* as a subsort of *Domain* (similarly for *Place* and *Range*), which will result in all axioms defined on instances of *Domain* to be applicable to *Thing*s as well, including the axiom about the direct effect of action *assign*.

> **sort declarations**
>> *thing* :: *domain*
>> *place* :: *range*

Action *Move* is defined as equivalent to action *Assign* in the MAD import statement, with the goal to reuse the axioms of *Assign* for *Move*. The same goal is achieved in $\mathcal{ALM}$ by declaring *move* as a special case of action *assign*:

> *move* :: *assign*

Actions of sort *move* will inherit the attributes of the supersort *assign*. However, the sorts of these attributes do not match exactly the sorts of the MAD action *Move(Thing, Place)*; in our translation, they are supersorts of the corresponding sorts *thing* and *place*. Rather than restricting the range of attributes to the appropriate sorts via axioms, which hides details from the declaration of *move*, we add two new attributes to *move*, as follows:

> **attributes**
>> $attr1\_move$: *thing*
>> $attr2\_move$: *place*

and add two axioms to the appropriate section of the *move* module to connect the new attributes to the attributes inherited from *assign*:

> $attr1\_assign(A) = X$ **if** $instance(A, move), attr1\_move(A) = X.$
> $attr2\_assign(A) = X$ **if** $instance(A, move), attr2\_move(A) = X.$

This allows axioms for action *assign* to apply as expected to instances of *move*.

Fluent *Location* is translated next, similarly to *Value*:

> **function declarations**
>> **fluents basic total** *location* : *thing* $\rightarrow$ *place*

In the import statement, *Location* is declared as equivalent to *Value* from module *ASSIGN*. MAD fluent renaming clauses cannot be translated using the specialization construct of $\mathcal{ALM}$, which only applies to sorts. Instead, we add axioms to say that fluents *location* and *value* must have the same values in every state of the transition diagram, and then translate the axiom on line 20:

axioms

$location(X_1) = X_2$ **if** $value(X_1) = X_2$.
$value(X_1) = X_2$ **if** $location(X_1) = X_2$.

impossible $occurs(A)$ **if** $instance(A, move)$,
$attr1\_move(A) = X$, $attr2\_move(A) = P$,
$location(X) = P$.

Finally, we translate module MB. As it contains object constants that do not appear in any axiom, we translate it as the *structure* of our system description, called mbp, in which the objects are defined as instances of the appropriate sorts:

structure mbp
instances
$monkey$ **in** $thing$
p_1, p_2 **in** $place$

Action $Walk$ and its definition as equivalent to $Move(Monkey, p)$ in the action renaming clause of the import statement for module $MOVE$ are translated by defining an instance $walk(P)$ of action class $move$ with attribute $attr1\_move$ mapped into $monkey$ and $attr2\_move$ into variable P ranging over $places$.

$walk(P)$ **in** $move$
$attr1\_move = monkey$
$attr2\_move = P$

This concludes the translation of action description MBP.

The translation does not cover a couple of situations that we describe next. We propose changes to the translation to accommodate these issues.

Issue 1: Imagine that module MB also contained an object Box of sort $Thing$ and an action $PushBox(Place)$ defined in terms of action $Move$ [3] as:

import $MOVE$;
$Move(Monkey, p)$ **is** $PushBox(p)$;

import $MOVE$;
$Move(Box, p)$ **is** $PushBox(p)$;

According to our solution, $pushbox$ (the translation of $PushBox$) belongs to the structure; both the $monkey$ and the box should be values of its $attr1\_move$, which is impossible as attributes are functions. To remedy this problem, we expand the declaration of attribute $attr1\_move$ by turning it into a boolean function $attr1\_move : place \rightarrow booleans$ and we expand literals in axioms accordingly. The attribute definitions for $pushbox$ will now contain $attr1\_move(monkey) = true$ and $attr1\_move(box) = true$. Similarly for other attributes.

Issue 2: Assume that module MB contained an additional sort $Supporter$ that is a supersort of $Thing$; a new fluent $Support(Thing) : simple(Supporter)$; and a new action $Mount(Thing, Supporter)$. Moreover, imagine that the module MB

imported module $ASSIGN$ and defined fluent $Support$ as equivalent to $Value$ and action $Mount$ as equivalent to $Assign$ (see [3]) as follows:

> **import** $ASSIGN$;
> $Domain$ **is** $Thing$; $Range$ **is** $Supporter$;
> $Value(t)$ **is** $Support(t)$; $Assign(t, s)$ **is** $Mount(t, s)$;

where variable t ranges over $Things$ and s over $Supporters$. Given our translation, the $\mathcal{ALM}$ system description would contain axioms

> $location(X_1) = X_2$ **if** $value(X_1) = X_2$.
> $value(X_1) = X_2$ **if** $location(X_1) = X_2$.
> $support(X_1) = X_2$ **if** $value(X_1) = X_2$.
> $value(X_1) = X_2$ **if** $support(X_1) = X_2$.

where variable X_1 ranges over $things$; X_2 ranges over $places$ in the first two axioms and over $supporters$ in the last two. This leads to inconsistency as $value$ is a function but it may have two conflicting values for the same $thing$, one representing the thing's location, and the other one its support.

To solve the problem, in our $\mathcal{ALM}$ translation we replace every $renamed$ function $f : s_1 \times \cdots \times s_n \rightarrow s_{n+1}$ by the expanded boolean function $f : s_1 \times \cdots \times s_n \times s_{n+1} \rightarrow booleans$ and change literals in axioms accordingly. In our example, function $value$ would be expanded, but not $location$ nor $support$ as they do not appear on the left-hand side of renaming clauses.

If the interpretations of sorts $place$ and $supporter$ are $disjoint$, this transformation can be seen as partitioning the domain of function $value$ into two: for pairs of $things$ and $places$ it has the same meaning as $location$ and its value may be affected by occurrences of action $move$; for pairs of $things$ and $supporters$ it has the same meaning as $support$ and its value may be affected by occurrences of action $mount$. This is equivalent to what is achieved in MAD by adding prefixes of the form "In." when flattening modules. If sorts $place$ and $supporter$ do not have disjoint interpretations, the occurrence of a $move$ action may unintentionally change the value of fluent $support$.

5   Formal Translation

We limit ourselves to action descriptions of MAD in which:

- Each action is exogenous and declared explicitly in a single module.
- Each simple fluent is inertial and is defined in a single module, in terms of at most one fluent. Each statically determined fluent is boolean.
- Axioms do not contain object constants and are of the type "a **causes** l **if** $cond$;" "l **if** $cond$;" or "**nonexecutable** a **if** $cond$" (where a is an action term, l is a fluent atom, and $cond$ is a collection of fluent literals).
- Renaming clauses do not contain "**case**" clauses.
- Terms on the two sides of "**is**" in a renaming clause contain the same variables; fluent terms have the same arity and do not contain object constants.
- Fluents are either renamed with every import of their module, or not at all.

- An action may appear in axioms only if none of the renaming clauses in which it appears on the right-hand side of "**is**" contains object constants.
- Variables appearing on the right-hand side of "**is**" in a fluent or action renaming clause range over the whole sort of the corresponding parameter.

Most of these limitations are not serious and are meant to simplify presentation.

We introduce some notation for syntactic transformations: $\alpha(x)$ is the string obtained by converting x to uppercase if x is a variable, and to lowercase otherwise (e.g., $\alpha(p) =_{def} P$; $\alpha(Value) =_{def} value$; $\alpha(MB) =_{def} mb$). If x is a string and n is a number, $\gamma(x, n)$ is the string obtained by concatenating "$attr$", "n", "$\_$", and $\alpha(x)$ (e.g., $\gamma(Move, 1) =_{def} attr1\_move$).

Let AD be an action description of MAD consisting of modules $M_1, \ldots, M_n$. The translation of AD into an $\mathcal{ALM}$ system description $\alpha(AD)$ is defined as:

Step 1: Create the system description $\alpha(AD)$ with a theory and a structure with the same name. Create section **instances** for the structure. For each module M_k of AD, where $1 \leq k \leq n$, add to $\alpha(AD)$ a module called $\alpha(M_k)$ with sections **sort declarations**, **function declarations**, and **axioms**.

Step 2: Translate the **sorts** and **inclusions** sections of AD: For every sort s, every inclusion statement of the type $s << s_1$ and every module M_k in which s is used, add to the sort declaration section of module $\alpha(M_k)$ the statement

$$\alpha(s) :: \alpha(s_1).$$

If there are no inclusion statements of this type, add the statement

$$\alpha(s) :: universe.$$

Step 3: For every module M_k, translate its various parts:

(a) Objects. For every object declaration $o : s$, add to the instances part of the structure a statement

$$\alpha(o) \text{ in } \alpha(s)$$

(b) Actions. The translation of an action declaration $a(s_1, \ldots, s_n)$ belongs to the theory of $\alpha(AD)$ if there are axioms about this action or the action appears on the right-hand side of a renaming clause that does not contain object constants (e.g., $Assign$, $Move$ in MBP). Otherwise, (given the restrictions we placed on actions) the translation belongs to the structure (e.g., $PushBox$ in MBP).

Case 1: The translation belongs to the theory.

If a is not defined as a special case of some other action, then add to sort declarations of module $\alpha(M_k)$:

$$\alpha(a) :: actions$$
$$\textbf{attributes}$$
$$\gamma(a, 1) : \alpha(s_1) \rightarrow booleans$$
$$\ldots$$
$$\gamma(a, n) : \alpha(s_n) \rightarrow booleans$$

If a is defined in terms of action $b(z_1, \ldots, z_m)$ via a renaming clause of the form $b(y_1, \ldots, y_m)$ **is** $a(x_1, \ldots, x_n)$, then replace the supersort *actions* by $\alpha(b)$ in the action declaration above. Also, for every $x_i = y_j$ in the action renaming clause, where $1 \leq i \leq n$ and $1 \leq j \leq m$, add the following axioms to module $\alpha(M_k)$:

$$\gamma(b, j)(A, X) \text{ if } instance(A, \alpha(a)), \; \gamma(a, i)(A, X).$$

Case 2: The translation belongs to the structure.

If a does not appear on the right-hand side of any action renaming clause, add to the structure of $\alpha(AD)$ the instance definition (where $\beta(s_k)$ is the string obtained by appending the index k to the string s_k, for $1 \leq k \leq n$):

$$\alpha(a)(\; \beta(s_1), \ldots, \beta(s_n) \;) \text{ in } actions$$
$$\gamma(a, 1)\; (\; \beta(s_1) \;) = true$$
$$\ldots$$
$$\gamma(a, n)\; (\; \beta(s_n) \;) = true$$

If a appears on the right-hand side of an action renaming clause of the form $b(y_1, \ldots, y_m)$ **is** $a(x_1, \ldots, x_n)$, then add the instance definition:

$$\alpha(a)(\alpha(x_1), \ldots, \alpha(x_n)) \text{ in } \alpha(b)$$
$$\gamma(b, 1)\; (\alpha(y_1)) = true$$
$$\ldots$$
$$\gamma(b, m)\; (\alpha(y_m)) = true$$

(Given our restrictions, for $1 \leq j \leq m$, either y_j is an object constant or $\exists i, 1 \leq i \leq n$ such that $y_j = x_i$.) If there are multiple such renaming clauses, add to $\alpha(b)$ the names of the other actions that appear on the left-hand side of **is** in the renaming clauses, and add more definitions of attributes, if needed.

(c) Fluents. For every fluent declaration $f(s_1, \ldots, s_n) : \langle type \rangle \; (s_{n+1})$, if the fluent does not appear on the left-hand side of any fluent renaming clause in any of the modules of AD, then translate it as:

$$\textbf{total } \alpha(f) : \alpha(s_1) \times \cdots \times \alpha(s_n) \rightarrow s_{n+1}$$

Otherwise, translate it as

$$\alpha(f) : \alpha(s_1) \times \cdots \times \alpha(s_n) \times \alpha(s_{n+1}) \rightarrow booleans$$

We call *expanded* the functions in this second category and the literals built with them. Add the declaration to the appropriate part of the function declarations of module $\alpha(M_k)$, where $\alpha(f)$ is a basic fluent if $\langle type \rangle = simple$, a defined fluent if $\langle type \rangle = staticallyDetermined$, and a basic static if $\langle type \rangle = rigid$.

(d) Import Statements. For every module M_i imported in M_k, add to $\alpha(M_k)$

$$\textbf{depends on } \alpha(M_i)$$

Sort and action renaming clauses were handled when translating the sort and inclusions sections and the action section of AD, respectively. For every fluent renaming clause $g(y_1, \ldots, y_n)$ **is** $f(x_1, \ldots, x_n)$ in M_k, if f is not an expanded function ranging over sort s, then add to the axioms of $\alpha(M_k)$:

$$\alpha(g)(\alpha(y_1), \ldots, \alpha(y_n), V) \quad \textbf{if } \alpha(f)(\alpha(x_1), \ldots, \alpha(x_n)) = V.$$
$$\alpha(f)(\alpha(x_1), \ldots, \alpha(x_n)) = V \ \textbf{if } \alpha(g)(\alpha(y_1), \ldots, \alpha(y_n), V).$$

$$\neg\alpha(g)(X_1, \ldots, X_n, V) \ \textbf{if } \alpha(f)(X_1, \ldots, X_n) \neq V.$$
$$\neg\alpha(g)(X_1, \ldots, X_n, V_1) \ \textbf{if } \alpha(g)(X_1, \ldots, X_n, V_2), \ V_1 \neq V_2, \qquad (1)$$
$$instance(V_1, \alpha(s)), \ instance(V_2, \alpha(s)).$$

If f is expanded, expand the corresponding literals and omit the last axiom.

(e) Axioms. If r is an axiom of AD, then by $vars(r)$ we denote the collection of atoms of the type $instance(\alpha(x), \alpha(s))$ for every variable x in r defined as $x : s$. If l is a fluent atom of AD of the form $f(x_1, \ldots, x_n) = x_{n+1}$, then by $\alpha(l)$ we denote $\alpha(f)(\alpha(x_1), \ldots, \alpha(x_n)) = \alpha(x_{n+1})$ if f is not an expanded fluent, and $\alpha(f)(\alpha(x_1), \ldots, \alpha(x_n), \alpha(x_{n+1}))$ otherwise. We extend the notation α to arithmetic comparisons, literals, and collections of literals in a natural way. For every axiom r in M_k, add to the axioms of $\alpha(M_k)$:

(i) If r is a dynamic causal law of the form $a(x_1, \ldots, x_n)$ **causes** l **if** $cond$:

$$occurs(A) \textbf{ causes } \alpha(l) \ \textbf{if } instance(A, \alpha(a)), \ \alpha(cond), \ vars(r),$$
$$\gamma(a, 1)(A, \alpha(x_1)), \ \ldots, \ \gamma(a, n)(A, \alpha(x_n)).$$

(ii) If r is a state constraint of the form l **if** $cond$:

$$\alpha(l) \ \textbf{if } \alpha(cond), \ vars(r).$$

(iii) If r is an executability constraint **nonexecutable** $a(x_1, \ldots, x_n)$ **if** $cond$:

$$\textbf{impossible } occurs(A) \ \textbf{if } instance(A, \alpha(a)), \ \alpha(cond), \ vars(r),$$
$$\gamma(a, 1)(A, \alpha(x_1)), \ \ldots, \ \gamma(a, n)(A, \alpha(x_n)).$$

where A is a variable such that $\alpha(A)$ does not appear in r.

Step 4: Add to $\alpha(AD)$ a module *main* that depends on all other modules. For every expanded function $f : s_1 \times \cdots \times s_{n+1} \to \textit{booleans}$ of $\alpha(AD)$ and every combination $(z_1, \ldots, z_{n+1})$ of leaves of the sort hierarchy of $\alpha(AD)$ such that no function that is a renaming of f is defined on it, add to the axioms of *main*:

$$\neg dom_f(X_1, \ldots, X_{n+1}) \ \textbf{if } instance(X_1, z_1), \ \ldots, \ instance(X_{n+1}, z_{n+1}). \quad (2)$$

6 Properties of the Translation

Given the translation described in Sect. 5, we are interested in determining a class of MAD action descriptions that have transition diagrams isomorphic to those of their $\mathcal{ALM}$ counterparts, modulo the common vocabulary. Specifically, problems may result from our solution for translating sort and fluent renaming clauses, which allows actions to cause unintended changes to unrelated fluents, as in the example at the end of Sect. 4. In MAD, this is prevented by adding a prefix of the type "In." in front of the name of renamed fluents. To mimic this behavior, we restrict ourselves to action descriptions in which the domains of the special cases of every fluent are disjoint. We start with some definitions.

Let AD be a MAD action description that meets the restrictions described at the beginning of Sect. 5, and let $\alpha(AD)$ be its translation into $\mathcal{ALM}$. Let τ_m and τ_a denote the transition diagrams described by AD and $\alpha(AD)$, respectively.

Definition 1 (Well-defined Function). A function $f(s_1, \ldots, s_n)$: $\langle type \rangle$ (s_{n+1}) in AD is well-defined if for every interpretation $\mathcal{I}$ of $\alpha(AD)$ and every pair of functions $g(z_1, \ldots, z_n)$: $\langle type \rangle (z_{n+1})$ and $h(c_1, \ldots, c_n)$: $\langle type \rangle (c_{n+1})$ of AD, such that both g and h are special cases of f, $\exists k, 1 \leq k \leq n + 1$, such that $\mathcal{I}(z_k) \cap \mathcal{I}(c_k) = \emptyset$.

Definition 2 (Basic Action Description). AD is a basic action description if **(i)** every function of $\alpha(AD)$ is well-defined; **(ii)** the values of all defined fluents in $\alpha(AD)$ are fully determined by the values of basic statics and basic fluents (i.e., $\alpha(AD)$ is well-founded - see Definition 16 in [11]); and **(iii)** all instances belong to leaves of the sort hierarchy of $\alpha(AD)$.

We formally define what we mean here by *isomorphic* transition diagrams:

Definition 3 (MA-Isomorphic Transition Diagrams). Transition diagrams τ_m and τ_a are ma-isomorphic if their states and transitions only differ as follows:

1. If we removed prefixes of the type "In." from MAD-fluent literals in τ_m and considered that some fluents were expanded during the translation, states would only differ by the additional static atoms derived from $\mathcal{ALM}$'s predefined functions (is_a, instance, subsort, etc.).
2. For each action a labeling an arc in τ_a and for each action b in AD that is renamed as a (i.e., a is a special case of b), the corresponding arc in τ_m is labeled by b as well, where b is preceded by prefix(es) of the type "In.".

Now we can formulate our proposition:

Proposition 1. If AD is a basic action description, τ_m and τ_a are ma-isomorphic.

Sketch: The requirement on well-defined functions prevents actions from having unintended effects on unrelated fluents. Axioms (1) and (2) ensure that there are no additional atoms constructed from renamed functions in states of τ_a. We also rely on results on the common core of languages $\mathcal{B}$ and $\mathcal{C}$ from [7].

7   Conclusions

In this paper we have investigated the relationship between modular action languages MAD and $\mathcal{ALM}$ by introducing a translation from MAD into $\mathcal{ALM}$. We have proposed that this translation produces isomorphic transition diagrams, modulo the common vocabulary, for a particular class of MAD action descriptions. *This is an important result, as it allows for libraries of knowledge developed in MAD (e.g., [3]) to be seamlessly combined with knowledge modules written in $\mathcal{ALM}$*, thus facilitating the knowledge representation task. The current translation focused on understanding the *correspondence between the* MAD *and $\mathcal{ALM}$ constructs for the reuse of knowledge and the description of actions as special cases of other actions* (import statements and renaming clauses in MAD vs. specialization construct and instances of sorts in $\mathcal{ALM}$). Future work will address other constructs of MAD that our translation did not cover.

References

1. Balduccini, M.: ASP with non-Herbrand partial functions: a language and system for practical use. TPLP **13**(4–5), 547–561 (2013)
2. Erdoğan, S., Lifschitz, V.: Actions as special cases. In: Proceedings of the International Conference on Principles of Knowledge Representation and Reasoning, pp. 377–387. AAAI Press (2006)
3. Erdoğan, S.T.: A library of general-purpose action descriptions. Ph.D. thesis, University of Texas at Austin, Austin, TX, USA (2008)
4. Gelfond, M., Inclezan, D.: Yet Another Modular Action Language. In: Proceedings of SEA 2009, pp. 64–78. University of Bath Opus: Online Publications Store (2009)
5. Gelfond, M., Lifschitz, V.: Classical negation in logic programs and disjunctive databases. New Gener. Comput. **9**(3/4), 365–386 (1991)
6. Gelfond, M., Lifschitz, V.: Action languages. Electron. Trans. AI **3**(16), 193–210 (1998)
7. Gelfond, M., Lifschitz, V.: The common core of action languages B and C. In: Proceedings of the 14th International Workshop on Non-Monotonic Reasoning (NMR 2012) (2012)
8. Giunchiglia, E., Lee, J., Lifschitz, V., McCain, N., Turner, H.: Nonmonotonic causal theories. Artif. Intell. **153**(1–2), 105–140 (2004)
9. Giunchiglia, E., Lifschitz, V.: An Action language based on causal explanation: preliminary report. In: Proceedings of National Conference on Artificial Intelligence (AAAI), pp. 623–630. AAAI Press (1998)
10. Inclezan, D., Gelfond, M.: Representing biological processes in modular action language ALM. In: Proceedings of the 2011 AAAI Spring Symposium on Formalizing Commonsense, pp. 49–55. AAAI Press (2011)
11. Inclezan, D., Gelfond, M.: Modular action language ALM - accepted for publication in TPLP (2015) http://arxiv.org/abs/1505.05022
12. Lee, J., Lifschitz, V., Yang, F.: Action language BC: preliminary report. In: The 23rd International Joint Conference on Artificial Intelligence (IJCAI 2013) (2013)
13. Lifschitz, V., Ren, W.: A Modular action description language. In: Proceedings of the 21st National Conference on Artificial Intelligence (AAAI), pp. 853–859 (2006)
14. Turner, H.: Representing actions in logic programs and default theories: a situation calculus approach. J. Logic Program. **31**(1–3), 245–298 (1997)

Compacting Boolean Formulae for Inference in Probabilistic Logic Programming

Theofrastos Mantadelis[1]([✉]), Dimitar Shterionov[2], and Gerda Janssens[2]

[1] CRACS & INESC TEC, Faculty of Sciences, University of Porto,
Rua do Campo Alegre 1021/1055, 4169-007 Porto, Portugal
`theo.mantadelis@dcc.fc.up.pt`
[2] Department of Computer Science, KU Leuven, Celestijnenlaan 200A,
2402 3001 Heverlee, Belgium
`{Dimitar.Shterionov,Gerda.Janssens}@cs.kuleuven.be`

Abstract. Knowledge compilation converts Boolean formulae for which some inference tasks are computationally expensive into a representation where the same tasks are tractable. ProbLog is a state-of-the-art Probabilistic Logic Programming system that uses knowledge compilation to reduce the expensive probabilistic inference to an efficient weighted model counting. Motivated to improve ProbLog's performance we present an approach that optimizes Boolean formulae in order to speed-up knowledge compilation. We identify 7 Boolean subformulae patterns that can be used to re-write Boolean formulae. We implemented an algorithm with polynomial complexity which detects and compacts 6 of these patterns. We employ our method in the inference pipeline of ProbLog and conduct extensive experiments. We show that our compaction method improves knowledge compilation and consecutively the overall inference performance. Furthermore, using compaction reduces the number of time-outs, allowing us to solve previously unsolvable problems.

1 Introduction

Knowledge compilation [6] encompasses a set of methods to compile a Boolean formula for which some inference tasks are computationally expensive into a Negation Normal Form (NNF) with special properties that allow to solve the same tasks efficiently. Knowledge compilation finds application in planning [21], computer-aided design [20], probabilistic reasoning [6,8]. Even if solving those problems on the compiled Boolean formulae is efficient, knowledge compilation itself is an #P-complete problem [27].

State-of-the-art Probabilistic Logic Programming systems like ProbLog [7,11] use knowledge compilation approaches to reduce the expensive inference task to a weighted model counting (WMC) problem. Motivated to solve larger problems in ProbLog, in this paper we present an optimization method that compacts Boolean formulae in order to speed-up knowledge compilation. While we implemented our approach in the scope of ProbLog and used common ProbLog problems to evaluate its effectiveness our approach is more general and any application using Boolean formulae to represent knowledge could benefit from it.

© Springer International Publishing Switzerland 2015
F. Calimeri et al. (Eds.): LPNMR 2015, LNAI 9345, pp. 425–438, 2015.
DOI: 10.1007/978-3-319-23264-5_35

Our first contribution is the identification of seven Boolean subformulae patterns that can be detected and used to re-write Boolean formulae in order to improve knowledge compilation. Our detected patterns fall into two types: one type that retains equivalence with the input Boolean formulae and a second type that reduces the number of Boolean variables contained in the formulae. The latter type patterns correspond to AND/OR clusters [16]. While they do not preserve the equivalence directly, an application specific equivalence can be defined and computed. In the context of ProbLog we preserve the weighted model count of the Boolean formulae.

Our second contribution is the implementation of an efficient algorithm that detects and compacts the presented patterns. Our implementation is independent from any ProbLog system. We incorporated it in two different implementations of ProbLog: MetaProbLog [17] and ProbLog2 [8] and evaluated it extensively with 7 different benchmarks. Further than the empirical evaluation of our approach we also provide a complexity analysis that shows that our algorithm is polynomial.

This paper builds on and extends the work presented in [25]. We introduced two new patterns, namely the minimal proof and OR-Cluster II; we improved the performance of the implementation; allowed it to work with multiple queries and evidence; and performed extensive experiments within the scope of ProbLog.

The paper is structured as follows: in Sect. 2 we present background and discuss related work; Sect. 3 describes the patterns while Sect. 4 gives an overview of the algorithm we implemented to detect and compact them; in Sect. 5 we analyze the effects of our compaction on inference with ProbLog; finally, we present our conclusions in Sect. 7.

2 Background

2.1 The Probabilistic Logic Programming Language ProbLog

ProbLog [7,11] is a general purpose Probabilistic Logic Programming (PLP) language. It extends Prolog with probabilistic facts which encode uncertain knowledge. Probabilistic facts have the form $p_i :: f_i$, where p_i is the probability label of the fact f_i. Prolog rules define the logic consequences of the probabilistic facts. No probabilistic fact can unify with a head of a rule in a ProbLog program. A simple ProbLog program is shown in Example 1.

Example 1. The following ProbLog program encodes a probabilistic graph. The predicate e/2 encodes a probabilistic edge between two nodes; the predicate p/2 defines a path between nodes.

```
0.6::e(a, b).  0.3::e(a, d).  0.8::e(b, c).  0.2::e(e, f).    p(X, Y):- e(X, Y).
0.7::e(c, d).  0.4::e(d, f).  0.4::e(d, e).                   p(X, Y):- e(X, X1), p(X1, Y).
```

For a ProbLog program L each ground probabilistic fact[1] f_i can be *true* with probability p_i or *false* with probability $(1 - p_i)$. A particular decision d on

---

[1] Probabilistic facts can be ground or non-ground. [11] proves that finitely many groundings of non-ground probabilistic facts are sufficient to compute probabilities. That is why we restrict our discussion to programs with ground probabilistic facts.

the truth values of all probabilistic facts determines a unique logic program L^d. For N probabilistic facts there exist 2^N such logic programs. Each probabilistic fact can be seen as an independent random variable. ProbLog then defines a distribution over the logic programs L^d as:

$$P(L^d) = \prod_{f_i \in L^d} p_i \prod_{f_i \in L \setminus L^d} (1 - p_i) \qquad (1)$$

ProbLog systems[2] provide a wide choice of inference and learning algorithms, which are used in applications like system prognostics and diagnostics [28], link and node prediction in biological data [10], robotics [19]. ProbLog focuses on two main inference tasks: (a) computing the probability that a query is true for a given ProbLog program, namely the marginal (MARG) probability of a query; and (b) computing the conditional probability (COND) of a query for a ProbLog program given some evidence, i.e. a set of facts for which the truth values have been decided.

Computing the MARG task boils down to determining all logic programs L^d which entail the query and summing their probabilities as computed by Eq. 1. Similar, for the COND task ProbLog needs to determine the logic programs L^d which entail the query but also the evidence. An exhaustive enumeration of these programs is infeasible but for the tiniest problems. That is why ProbLog's inference mechanism employs a step-wise procedure called an *inference pipeline* [26] that reduces the inference task into a WMC problem. First, given a ProbLog program L, a set of queries and evidence, ProbLog uses SLD [12] or SLG [4] resolution on the logical part of L, that is, ignoring the label of probabilistic facts, in order to determine the ground logic program relevant to the queries and the evidence (Ground LP) [8]. Then, the Ground LP is converted to an equivalent with respect to the WMC Boolean formula. During this process any cycles that occur in the Ground LP are handled. We use the *Proof-Based* approach [15] for this. Next, using knowledge compilation the Boolean formula is compiled into a negation normal form (NNF) with special properties which allows efficient WMC. Two target compilation languages have been considered so far: *ROBDDs* [3] and *sd-DNNFs* [6]. The NNF is then associated with the probabilities of L and used for WMC.

2.2 AND-OR Graphs

We represent Boolean formulae as AND-OR graphs. An AND-OR graph is a directed graph composed by AND and OR nodes. An AND node indicates that all child nodes must be true, while an OR node indicates that at least one of the child nodes must be true. An AND-OR graph is a suitable representation for a ground logic program relative to a query q. The different clauses $(q_{i \in 1..m} :- r_{i,1}, ..., r_{i,n}.)$ of the predicate q are processed as follows: for each clause q_i all literals $r_{i,j}$ in the body are grouped as children of an AND node. The different AND nodes then are grouped as children of an OR node labeled with q. Next,

---

[2] When it is clear from the context we use the term ProbLog to refer to either the language or the system. Otherwise we state it explicitly.

each literal $r_{i,j}$ is treated as a new query. An AND-OR graph of a query has the following characteristics: cycles that appear in the logic program also appear in the AND-OR graph; for each subgoal g there is only one OR node; an OR-node has multiple parents if the subgoal is repeated and goals proven as facts are represented by special OR nodes without children, called terminal nodes. The edge from a child node to a parent node states that the parent depends on the child node.

Definition 1. *An **AND-OR graph** for a query q is a directed graph $G = (V_{and}, V_{or}, V_{term}, E)$ with V_{and} a set of AND nodes, V_{or} a set of labeled OR nodes, $V_{term} \subset V_{or}$ a set of terminal nodes, $V_{nonterm} = V_{or} \setminus V_{term}$ and $E \subseteq R$ a set of directed edges, where $R = (V_{and} \times V_{or}) \cup (V_{nonterm} \times V_{and}) \cup (V_{nonterm} \times V_{or})$. The root of the graph is an OR node labeled with q.*

Example 2. For the ProbLog program in Example 1 and the query p(a, f). the corresponding AND-OR graph is:

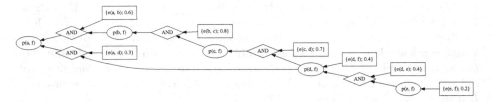

Ellipses depict OR nodes, diamonds depict AND nodes and rectangles terminal nodes. OR nodes are labeled with the goal they prove. Note that in the context of ProbLog terminal nodes have attached probabilities.

3   Compactable Patterns

We identify 7 patterns that appear in AND-OR graphs and present how we use them in order to compact the graph. The patterns we identify and their compacted form are illustrated in Table 1. Patterns 1 to 4 maintain the Boolean formulae equivalence. The compaction of patterns 5 to 7 removes Boolean variables and introduces a new Boolean variable to represent them. These compactions do not directly maintain the equivalence of the Boolean formulae. Application specific problems require a special calculation for the introduced representative Boolean variable. For correct ProbLog inference we need to maintain the WMC. That requires to calculate the probability of the representative Boolean variable. Proof of correctness for these compactions appears in [16].

1. **Single Variable:** There are an OR node A and a terminal node B, such that A depends only on B. **Compaction:** Node A and the edge from B to A are deleted. The edges starting from A now start from B.
2. **Single Branch I:** There are a node A, an OR node B and an AND node C, such that B depends only on C and A depends on B. **Compaction:** If A is an OR node then node B and the edge from C to B are deleted. A new edge

from C to A is created. If A is an AND node then nodes B and C are deleted together with the edge from C to B. All children of C are connected to A.

3. **Single Branch II**: There are two OR nodes A and B, such that A depends on B and no other node depends on B. **Compaction**: Node B and the edge from B to A are deleted. All children of B are connected to A.

4. **Minimal Proof**: There are an OR node A, two AND nodes B_1 with a set of children Ch_{B_1} and B_2 with a set of children Ch_{B_2} such that $Ch_{B_1} \subseteq Ch_{B_2}$. **Compaction**: Node B_2 and all edges from the child nodes in Ch_{B_2} to B_2 are deleted. The edge from B_2 to A is deleted as well.

5. **AND-Cluster**: There are an AND node A, a set of nodes $Ch'_A \subseteq Ch_A$, where Ch_A are all terminal nodes A depends on, such that $Ch'_A = Ch_A \setminus \{C | \exists B, B \neq A, B \text{ depends on } C\}$. **Compaction**: All terminal nodes $C_i \in Ch'_A$ are deleted, together with the edges from C_i to A. A new terminal node C_t is created together with an edge from C_t to A. A joint probability $p_t = \prod_{C_i \in Ch'_A} p_i$, where C_i is a terminal node with probabilistic label p_i is calculated. The probabilistic label p_t is attached to node C_t.

6. **OR-Cluster I**: There are an OR node A, a set of nodes $Ch'_A \subseteq Ch_A$, where Ch_A are all terminal nodes A depends on, such that $Ch'_A = Ch_A \setminus \{C | \exists B, B \neq A, B \text{ depends on } C\}$. **Compaction**: All terminal nodes $C_i \in Ch'_A$ are deleted, together with the edges from C_i to A. A new terminal node C_t is created together with an edge from C_t to A. A joint probability p_t is calculated as $p_t = (..((p_1 * (1 - p_2) + p_2) * (1 - p_3) + p_3).. + p_n)$, where p_i is the probability labeled in $C_i \in Ch'_A$, $i = 1..|Ch'_A|$. The probability p_t is attached to node C_t.

7. **OR-Cluster II**: There are an OR node A, that depends on n AND nodes $B_1...B_n$ that each has exactly one different terminal child node $Ch_1...Ch_n$ and all the rest child nodes (denoted as node C) are common. **Compaction**: All AND nodes $B_1...B_n$ and all terminal nodes $Ch_1...Ch_n$ are deleted. A new terminal node Ch is created. A joint probability p_t is calculated as $p_t = (..((p_1 * (1 - p_2) + p_2) * (1 - p_3) + p_3).. + p_n)$, where p_i is probabilistic part of the label of $Ch_i, i = 1..n$. The probabilistic label p_t is attached to node Ch. A new AND node B that contains Ch, C is created, finally, an edge from B to A is created.

4 Algorithm

Our algorithm iterates over patterns 1 to 6 in the order presented in Table 1. As soon as a pattern is detected the corresponding compaction is applied. According to the order we choose the detection and compaction of one pattern allows the detection and compaction of the next one in the same iteration. This ensures the minimum number of iterations required to compact a graph. Our algorithm terminates once no patterns can be detected. Algorithm 1 presents the pseudo-code for detecting patterns 1 to 6.

Table 1. AND-OR graph patterns and the compacting transformations. We denote with "..." multiple possible nodes to/from which exists an edge. With octagons we represent nodes that can be of any type (terminal, AND or OR).

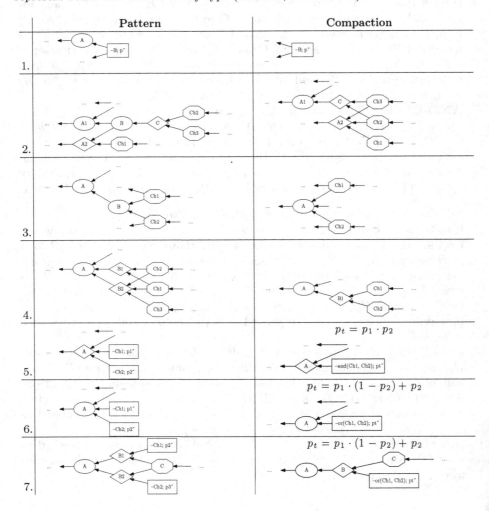

Completeness: Our algorithm does neither detect nor compact pattern 7. We are also confident that there exist more patterns which we do not consider. Thus, AND-OR graphs which include at least one of these patterns will not be fully compacted. Therefore, our algorithm is not complete.

Data: An AND-OR graph
Result: Detected Node, Nodes to be compacted

detect_single_variable(NodeA, Terminal) ←
  or_edge(NodeA, Terminal),
  is_terminal(Terminal),
  ∄ and_edge(NodeA, _),
  ∄ (or_edge(NodeA, Any), Terminal ≠ Any).

detect_single_branch1(NodeB, NodeC) ←
  or_edge(NodeB, NodeC),
  and_edge(NodeC, _),
  ∄ and_edge(NodeB, _),
  ∄ (or_edge(NodeB, Any), NodeC ≠ Any).

detect_single_branch2(NodeA, NodeB) ←
  or_edge(NodeA, NodeB),
  or_edge(NodeB, _),
  ∄ and_edge(_, NodeB),
  ∄ (or_edge(Any, NodeB), NodeA ≠ Any).

detect_minimal_proof(NodeB1, NodeB2) ←
  or_edge(NodeA, NodeB1),
  and_edge(NodeA, NodeB1, _),
  or_edge(NodeA, NodeB2),
  and_edge(NodeB2, _),
  NodeB1 ≠ NodeB2,
  all(Child, and_edge(NodeB1, Child),
                ChildsB1),
  all(Child, and_edge(NodeB2, Child),
                ChildsB2),
$ChildsB1 \subseteq ChildsB2$.

detect_and_cluster(RefChilds) ←
  and_edge(NodeA, _),
  all(Terminal, (
    and_edge(NodeA, Terminal),
    is_terminal(Terminal),
    ∄ or_edge(_, Terminal)
  ), Childs),
  get_all_and_edge_sets(ChildSets),
  refine_cluster(ChildSets, Childs, RefChilds),
  $RefChilds \neq \emptyset$.

detect_or_cluster1(RefChilds) ←
  or_edge(NodeA, _),
  all(Terminal, (
    or_edge(NodeA, Terminal),
    is_terminal(Terminal),
    ∄ and_edge(_, Terminal)
  ), Childs),
  get_all_or_edge_sets(ChildSets),
  refine_cluster(ChildSets, Childs, RefChilds),
  $RefChilds \neq \emptyset$.

refine_cluster([], RefChilds, RefChilds)
refine_cluster([Set|ChildSets], Childs,
                   RefChilds) ←
  $NewChilds = Set \wedge Childs$,
  refine_cluster(ChildSets, NewChilds,
                   RefChilds).

Algorithm 1: The 6 pattern detection algorithms.

Complexity:[3] The compaction operations are very efficient ($O(N)$ with N the number of edges affected). Detecting and verifying a pattern is computationally expensive and deserves a thorough analysis.

For an arbitrary AND-OR graph G we denote with N_{or} the number of *OR* edges, with N_{and} the number of *AND* edges and with N_{term} the number of terminal nodes. We assume that a node always contains N_{term} children; this is a high upper bound assumption but does not affect the complexity class.

The complexity for detecting and verifying all patterns 1 to 3 in an AND-OR graph is $O(N_{or} \cdot (log(N_{or}) + log(N_{and})))$; for all patterns 4 the complexity is $O(N_{or} \cdot (log(N_{or}) + log(N_{and}) + N_{term}))$; for all patterns 5 the complexity is $O(N_{and}^2 \cdot N_{term})$; finally, for all patterns 6 the complexity is $O(N_{or}^2 \cdot N_{term})$.

We illustrate the steps taken when applying our compaction algorithm to an AND-OR graph derived from the ProbLog program in Example 3.

Example 3. We apply our compaction algorithm on the graph in Example 2. In the 1st iteration it detects 1 **Single Variable** of $p(e, f)$ and 2 **Single Branch I** of $p(b, f)$ and $p(c, f)$ resulting in Graph 1 in the following table; and 2 AND-Clusters resulting in Graph 2. In the 2nd iteration 1 **OR-Cluster I** and 1 **Single Variable** of $p(d, f)$ are detected resulting in Graph 3 and Graph 4 accordingly.

---

[3] More details for the algorithm and the full complexity analysis can be found at: https://lirias.kuleuven.be/bitstream/123456789/500398/5/appendix.pdf.

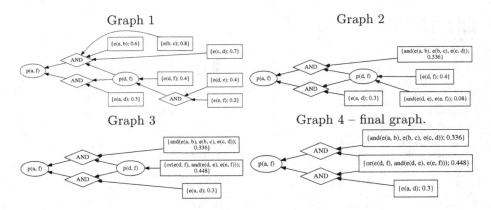

The final AND-OR graph forms 1 **OR-Cluster II** pattern. If we detected and compacted OR-Cluster II patterns, it would enable a final AND-Cluster compaction to fully compact the AND-OR graph into a single terminal node containing the probability of the query.

The implementation neither detects nor compacts pattern 7; pattern 7 may correspond to complex subgraphs with unreasonably high detection cost. By using indexing we decreased the complexity of our previous implementation [25] from $O(N^2)$ to $O(N \cdot log(N))$ for several tasks. We also added support for multiple queries and evidence. The implementation of the detection/compaction algorithm is a stand-alone Prolog program.

5   Compacting ProbLog Programs

Section 2.1 presents the general scheme of a ProbLog inference pipeline. We focus on 4 particular ProbLog pipelines, based on two mainstream ProbLog implementations – MetaProbLog [17] and ProbLog2 [8]. These inference pipelines differ with respect to (a) representation of the Ground LP and the Boolean formulae: ProbLog2 uses AND-OR graphs and CNF DIMACS, while MetaProbLog uses Nested Tries [15] and BDD scripts [14]; (b) ways of preprocessing the Boolean formulae: ProbLog2 uses Boolean subformulae repetition detection and MetaProbLog uses the recursive node merging method presented in [18]; and (c) in the knowledge compilation method: ProbLog2 uses the sd-DNNF compiler c2d [5] and MetaProbLog uses the SimpleCUDD [14] compiler for ROBDDs. The 4 pipelines we use for our experiments are listed in Table 2. The pipeline implementations of ProbLog allow us to employ our detection/compaction algorithm (a) before and (b) after the cycle handling processing of the Boolean formula in any ProbLog pipeline. In (a), called the *prior* compaction, the Ground LP is represented as an AND-OR graph and then processed by our algorithm. In ProbLog2 the loop-breaking mechanism applies directly on the AND-OR graph and generates a loop-free AND-OR graph. In MetaProbLog the loop-breaking operates on the nested trie structure and produces a BDD Script which is easily rewritten as an AND-OR graph. This allows (b), that is, to invoke the compaction algorithm again and attempt a further optimization of the AND-OR graph before the knowledge compilation step. We call this the *post* compaction. Furthermore, we can invoke the *prior* and *post* compactions in the same pipeline; we refer to this compaction setting as *both*.

Table 2. ProbLog pipelines.

Pipeline	Ground LP representation	Cycle handling	Boolean formulae representation	Compilation language
ProbLog2/sd-DNNF[a]	AND-OR	Proof-Based	AND-OR→CNF DIMACS	sd-DNNF
ProbLog2/ROBDD	AND-OR	Proof-Based	AND-OR→BDD script	ROBDD
MetaProbLog/sd-DNNF	Nested Tries	Proof-Based + [18]	BDD script→CNF DIMACS	sd-DNNF
MetaProbLog/ROBDD[b]	Nested Tries	Proof-Based + [18]	BDD script	ROBDD

[a] ProbLog2 and [b] MetaProbLog default pipelines.

5.1   Experimental Set-Up

We experiment with 7 benchmark sets with a total of 738 programs. These benchmark sets have been previously used for testing the performance of different ProbLog implementations. The variety of these benchmarks and the different inference tasks ensure a realistic estimate of the gain or the loss in the performance of ProbLog pipelines due to our compaction algorithm.

In order to present our data in a more comprehensive way, we divide our benchmarks in three groups: 387 *easy* programs which consume less than 10 s; 99 *medium* programs which consume between 10 and 60 s and 150 *hard* programs which consume more than 60 s. To classify a program we use the total run time for the MetaProbLog/ROBDD pipeline without compaction – the default MetaProbLog pipeline.

Each program is executed with the 4 ProbLog pipelines and the 4 compaction settings – *none*, *prior*, *post* and *both*. Their combination results in 16 different ProbLog pipelines to run each benchmark program with. We chose a time-out of 540 s for each test run. We managed to solve 636 out of the 738 programs within the 540 s time-out with at least one of the 16 pipelines.

The c2d compiler is non-deterministic [5], meaning that for the same CNF the compiled sd-DNNFs may differ. That is why we run each test invoking c2d 5 times (8 pipelines use c2d). Then we report the average time consumed by the test. From previous experiments we have determined this number to give a reliable estimate of the performance of c2d.

We run our experiments on 17 computers with Intel® quad-core 64-bit CPU at 2.83GHz, 8GBs of RAM running Ubuntu 12.04 LTS (under normal load). The chosen time-out ensured our experiments to terminate within at most 14 days.

5.2   Experimental Results

In our experiments we collect the *total* run time for executing a benchmark program (including the compaction time). We use the time results to determine

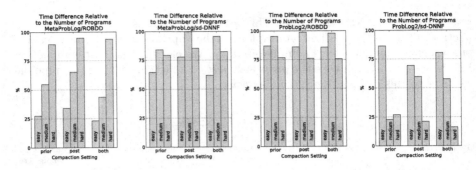

Fig. 1. Relative time gain due to a specific compaction.

the compaction setting which leads to (a) the lowest run times; and (b) to the lowest number of time-outs for each of the pipelines and each benchmark set.

For each ProbLog pipeline and each compaction $c \in \{prior, post, both\}$ we sum (a) the gain in the total run time for each benchmark compared with the run time of no compaction ($c = none$) when the compaction performs better: T_C^g (this is the total gain for compaction); (b) the gain in the total run time when no compaction performs better: T_N^g (this is the total loss for compaction). We normalize each gain by dividing by the total number of programs within a benchmark set to compensate for the fact that some of them contain more programs. For example, consider a particular benchmark with programs $\{b_1, b_2, b_3\}$ each with run times: with compaction 20, 30 and 40 s and with no compaction 10, 25, 70 s. Then $T_C^g = 70 - 40 = 30$ and $T_N^g = (20 - 10) + (30 - 25) = 15$ showing that in total the gain with compaction exceeds the loss due to compaction. We exclude programs for which both inference with compaction and with no compaction times out. We chose this measurement because it shows the overall gain in run time due to compaction. We present the gain due to compaction relative to the total gain $\frac{T_C^g}{T_C^g + T_N^g}$ in Fig. 1 as percentage. Detailed results are given in our online appendix. We base our conclusions on all the results.

5.3   Experimental Conclusions

First, our algorithm improves the knowledge compilation time for the majority of the benchmarks, between 75 % to 85 % for ROBDDs and between 55 % to 65 % for sd-DNNFs. Our intuition is that ROBDDs benefit more than sd-DNNFs because ROBDDs use a general Boolean formula for input while sd-DNNFs require a conversion to a CNF Boolean formula. Compaction is beneficial for most of the medium and hard problems but not that much for the easy problems. It was expected that the time spend to perform our algorithm would not be compensated from the gain in knowledge compilation for small problems.

Second, regarding the used pipelines from Fig. 1 we conclude that the highest gain from our algorithm was for the ProbLog2/ROBDD pipeline having an almost 100 % gain for all compaction settings (on medium problems). Using compaction is preferable to no compaction for the MetaProbLog/ROBDD, MetaProbLog/ sd-DNNF and ProbLog2/ROBDD pipelines and complex problems. We also note that the ProbLog2/sd-DNNF pipeline in overall does not benefit from our approach.

Third, none of the compaction settings (i.e. the *prior*, *post* or *both*) outperforms the other compaction settings. For MetaProbLog/ROBDD the *both* compaction is preferable; for MetaProbLog/sd-DNNF and ProbLog2/ROBDD pipelines preferable is the *post* compaction; for ProbLog2/sd-DNNF the *prior* compaction. The *post* and *both* compactions often yield the same Boolean formulae. In such cases *both* spends unnecessary extra time for *prior* compaction. We also note that the actual cost to perform detection and compaction is generally small. In particular, compacting AND-OR graphs generated from a Ground LP consumed at most 18.25 s for a program with total run time of 264.27 s; compacting AND-OR graphs generated from Nested Tries consumed at most 64.95 s for a program with total run time of 297.79 s.

Fourth, compaction allowed us to solve significantly more problems that would otherwise timeout. Particularly, in the best case, MetaProbLog/ROBDD with *both* compaction, we can solve 38 % more programs; ProbLog2/ROBDD with *post* compaction can solve 37 % more programs. The two pipelines which use compilation to sd-DNNF benefit less from compaction, MetaProbLog/sd-DNNF with *post* compaction can solve 6 % more programs while for ProbLog2/sd-DNNF compaction introduces up to 12 % more timeouts. The extra time-outs occur for benchmarks that contain multiple queries and evidence. Often the query and evidence atoms appear also as subgoals. Queries and evidence are required for the final step of WMC thus they should not be removed from the Boolean formula. Therefore there are less patterns that can be compacted in the case of COND with respect to MARG tasks. For the other benchmarks compaction reduces the overall amount of time-outs.

Finally, following from all our results, we must indicate that there is not one best performing pipeline over all benchmarks. On average, the pipeline with the least timeouts was ProbLog2/ROBDD with *post* compaction. The gain due to compaction (*prior*, *post* or *both*) on the hard problems and the decrease of timeouts indicate that our approach improves the performance of the system at problems that it was poor before.

6   Related Work

Rewriting a Boolean formula to improve the performance of knowledge compilation in the scope of ProbLog had first been investigated in [18]. [18] shows that feeding a rewritten Boolean formulae instead of a non optimized one reduces the operations needed by the knowledge compilation step and consequently the knowledge compilation run time. The work we present in this paper, focuses on optimizing even further the Boolean formula and works in parallel with these Boolean formulae rewrites. Boolean formulae rewriting, in the scope of assessing the Probability of a Sum-of-Products, has been investigated also in [24].

Detecting regularities such as AND/OR-Clusters on a Boolean formula in normal form (e.g., DNF), has been investigated in [16]. Our approach performs similar transformations on an AND-OR graph instead of a Boolean formula in normal form. [16] proves completeness of detecting AND/OR-Clusters in Boolean formulae but faces some practical limitations. The most important of which is that ProbLog using *tabling* and *cycle handling* as presented in [15] would generate a Boolean formula that is not in normal form.

Hintsanen [9] argues that structural properties are important for finding the most reliable subgraph. He calculates the probability of subgraphs connecting two nodes and searching for the subgraph with the maximum probability. The paper identifies as a special case the series-parallel subgraphs for which they can compute the probability polynomially. These series-parallel subgraphs have similarities with the AND/OR-Clusters.

Our work is also similar to [13] which presents a preprocessing of propositional formulae to optimize model counting. Their approach optimizes CNF Boolean formulae by using seven preprocessing methods. Similar to our work, some of their preprocessing methods maintain equivalence and others not. In contrast to our approach some of their methods increase the size of the Boolean formulae which is an interesting point for us to look upon. There exist several other related works from other fields such as variable ordering approaches for BDDs [22,23] or preprocessing methods used in SAT solving [1,2].

7   Conclusion and Future Work

This paper presented a pattern-based approach for compacting Boolean formulae. It detects and compacts 6 (out of 7 identified) patterns – 4 that preserve logic equivalence and 2 that preserve equivalence with respect to the weighted model count. Our approach aims to improve probabilistic inference that uses knowledge compilation and weighted model counting. It targets but is not limited to the probabilistic logic programming system ProbLog and its underlying implementations.

We performed experiments with 4 different ProbLog pipelines and 3 compaction settings on 7 benchmark sets with 738 benchmarks in total. Our results show that compaction improves knowledge compilation to ROBDDs as well as to sd-DNNFs. The gain in the total run time due to compaction is most salient for harder problems. The decreased amount of time-outs proves that our approach enables inference on problems unsolved before (i.e. without compaction).

In the future we want to investigate also non-compacting transformations that could aid (thus improve) the knowledge compilation. In addition, we plan to extend our algorithm to handle problems outside the domain of ProbLog. We aim to test it on benchmarks from [13] in order to determine its general effects.

Acknowledgments. We want to thank the anonymous reviewers for their comments and help to improve our paper. Theofrastos Mantadelis is funded by the Portuguese

Foundation for Science and Technology (FCT) within the projects SIBILA NORTE-07-0124-FEDER-000059 and UID/EEA/50014/2013. Dimitar Shterionov is funded by KU Leuven within the project GOA 13/010.

References

1. Aloul, F.A., Markov, I.L., Sakallah, K.A.: Faster SAT and smaller BDDs via common function structure. In: ICCAD, pp. 443–448 (2001)
2. Bacchus, F., Winter, J.: Effective preprocessing with hyper-resolution and equality reduction. In: Giunchiglia, E., Tacchella, A. (eds.) SAT 2003. LNCS, vol. 2919, pp. 341–355. Springer, Heidelberg (2004)
3. Bryant, R.E.: Graph-based algorithms for boolean function manipulation. IEEE Trans. Comput. **35**(8), 677–691 (1986)
4. Chen, W., Swift, T., Warren, D.S.: Efficient top-down computation of queries under the well-founded semantics. JLP **24**(3), 161–199 (1995)
5. Darwiche, A.: A compiler for deterministic, decomposable negation normal form. In: AAAI/IAAI, pp. 627–634. AAAI Press/MIT Press (2002)
6. Darwiche, A., Marquis, P.: A knowledge compilation map. JAIR **17**, 229–264 (2002)
7. De Raedt, L., Kimmig, A., Toivonen, H.: ProbLog: a probabilistic Prolog and its application in link discovery. In: IJCAI, pp. 2468–2473. AAAI Press (2007)
8. Fierens, D., Van den Broeck, G., Renkens, J., Shterionov, D.S., Gutmann, B., Thon, I., Janssens, G., De Raedt, L.: Inference and learning in probabilistic logic programs using weighted boolean formulas. TPLP **15**(3), 358–401 (2015)
9. Hintsanen, P.: The most reliable subgraph problem. In: Kok, J.N., Koronacki, J., Lopez de Mantaras, R., Matwin, S., Mladenič, D., Skowron, A. (eds.) PKDD 2007. LNCS (LNAI), vol. 4702, pp. 471–478. Springer, Heidelberg (2007)
10. Kimmig, A., Costa, F.: Link and node prediction in metabolic networks with probabilistic logic. In: Berthold, M.R. (ed.) Bisociative Knowledge Discovery. LNCS, vol. 7250, pp. 407–426. Springer, Heidelberg (2012)
11. Kimmig, A., Demoen, B., De Raedt, L., Santos Costa, V., Rocha, R.: On the implementation of the probabilistic logic programming language ProbLog. TPLP **11**, 235–262 (2011)
12. Kowalski, R.A.: Predicate logic as programming language. In: IFIP Congress, pp. 569–574 (1974)
13. Lagniez, J., Marquis, P.: Preprocessing for propositional model counting. In: AAAI, pp. 2688–2694 (2014)
14. Mantadelis, T., Demoen, B., Janssens, G.: A simplified fast interface for the use of CUDD for binary decision diagrams (2008). http://people.cs.kuleuven.be/theofrastos.mantadelis/tools/simplecudd.html
15. Mantadelis, T., Janssens, G.: Dedicated tabling for a probabilistic setting. In: ICLP (Technical Communications), LIPIcs, vol. 7, pp. 124–133 (2010)
16. Mantadelis, T., Janssens, G.: Variable compression in ProbLog. In: Fermüller, C.G., Voronkov, A. (eds.) LPAR-17. LNCS, vol. 6397, pp. 504–518. Springer, Heidelberg (2010)
17. Mantadelis, T., Janssens, G.: Nesting probabilistic inference (2011). CoRR, abs/1112.3785
18. Mantadelis, T., Rocha, R., Kimmig, A., Janssens, G.: Preprocessing boolean formulae for BDDs in a probabilistic context. In: Janhunen, T., Niemelä, I. (eds.) JELIA 2010. LNCS, vol. 6341, pp. 260–272. Springer, Heidelberg (2010)

19. Moldovan, B., van Otterlo, M., Moreno, P., Santos-Victor, J., De Raedt, L.: Statistical relational learning of object affordances for robotic manipulation. In: ILP, p. 6 (2012)

20. Mostow, J.: Towards automated development of specialized algorithms for design synthesis: knowledge compilation as an approach to computer-aided design. Res. Eng. Des. 1(3–4), 167–186 (1990)

21. Namioka, Y., Tanaka, T.: Knowledge compilation for interactive design of sequence control programs. In: IEA/AIE, pp. 363–368 (1996)

22. Narodytska, N., Walsh, T.: Constraint and variable ordering heuristics for compiling configuration problems. In: IJCAI, pp. 149–154 (2007)

23. Panda, S., Somenzi, F.: Who are the variables in your neighborhood. In: ICCAD, pp. 74–77 (1995)

24. Rauzy, A., Châtelet, E., Dutuit, Y., Bérenguer, C.: A practical comparison of methods to assess sum-of-products. Reliab. Eng. Syst. Saf. **79**(1), 33–42 (2003)

25. Shterionov, D., Mantadelis, T., Janssens, G.: Pattern-based compaction for ProbLog inference. In: ICLP (Technical Communications), pp. 1–4 (2013)

26. Shterionov, D., Janssens, G.: Implementation and performance of probabilistic inference pipelines. In: Pontelli, E., Son, T.C. (eds.) PADL 2015. LNCS, vol. 9131, pp. 90–104. Springer, Heidelberg (2015)

27. Valiant, L.G.: Why is Boolean complexity theory difficult? In: London Mathematical Society Symposium on Boolean Function Complexity, pp. 84–94 (1992)

28. Vlasselaer, J., Meert, W.: Statistical relational learning for prognostics. In: Belgian-Dutch Conference on Machine Learning, pp. 45–50 (2012)

Multi-level Algorithm Selection for ASP

Marco Maratea[1]([✉]), Luca Pulina[2], and Francesco Ricca[3]

[1] DIBRIS, Univ. degli Studi di Genova, Viale F. Causa 15, 16145 Genova, Italy
marco@dist.unige.it
[2] POLCOMING, Univ. degli Studi di Sassari, Viale Mancini 5, 07100 Sassari, Italy
lpulina@uniss.it
[3] Dip. di Matematica ed Informatica, Univ. della Calabria, Via P. Bucci,
87030 Rende, Italy
ricca@mat.unical.it

Abstract. Automated algorithm selection techniques have been applied successfully to Answer Set Programming (ASP) solvers. ASP computation includes two levels of computation: variable substitution, called grounding, and propositional answer set search, called solving. In this paper we present ME-ASP^{ML}, an extended ASP system applying algorithm selection techniques to both levels of computation in order to choose the most promising solving strategy. Experiments conducted on benchmarks and solvers of the Fifth ASP Competition shows that ME-ASP^{ML} is able to solve more instances than state-of-the-art systems.

1 Introduction

Answer Set Programming (ASP) [8,13] is a declarative programming paradigm developed in the area of logic programming and non-monotonic reasoning. The evaluation of ASP programs usually includes two levels of computation, called grounding and solving. At the first level, a propositional program is obtained from a non-ground specification by applying intelligent techniques that eliminate variables; then, at the second level, the propositional program is fed to an ASP solver to produce answer sets.

Automated algorithm selection techniques have been applied in ASP to obtain evaluation of programs in reasonable time. The idea is to automatically select the "best" computation strategy on the basis of features computed on a training set of instances. In the literature there are several different proposals, and among them we mention the portfolio solver CLASPFOLIO ver. 1 [11], which then evolved into a framework combining different approaches in CLASPFOLIO ver. 2 [15], the multi-engine approach implemented in ME-ASP [19], the techniques for learning heuristic orders presented in [3], and the work in [14,21] that employs parameters tuning and/or design a solvers schedule. However, to the best of our knowledge, in ASP the application of automated selection techniques is typically limited to the evaluation of propositional programs, thus the choice of algorithms has been limited to the solving level. A preliminary contribution that exploits the features of non-ground programs was presented in [18], but also in this case the application was limited to only one level, namely the choice of the most promising grounding tool.

© Springer International Publishing Switzerland 2015
F. Calimeri et al. (Eds.): LNMR 2015, LNAI 9345, pp. 439–445, 2015.
DOI: 10.1007/978-3-319-23264-5_36

In this paper we present ME-ASP^{ML}, an extension of the multi-engine ASP system ME-ASP, that applies algorithm selection techniques to before each level of computation of answer sets, with the goal of selecting the most promising computation strategy. ME-ASP^{ML} supports the new standard ASPCore 2.0 [5] and selects among the systems that participated to the Fifth ASP Competition [6].

The new architecture of ME-ASP^{ML} takes advantage from the extraction of syntactic features of non-ground programs at the first level, so to identify a number of classes of non-ground programs. Then, ME-ASP^{ML} (possibly) applies to each class identified at the first level an additional phase of algorithm selection, which exploits the features of ground programs measured after running the grounder GRINGO [12]. A key-enabler in achieving good performance in ME-ASP^{ML} is the extraction of cheap-to-compute features. These can be obtained at the price of a minimum overhead also in case of large input programs. The features employed at the first level are able to characterize a program w.r.t. relevant subclasses of programs (featuring different complexity of the evaluation), and can even identify a class of programs where a specific grounder is the most promising (as done in [18]). The algorithm selection approach of ME-ASP [19] is then applied in the second level to the classes of programs identified in the first level, allowing for a more accurate selection of the solver to be employed, given that different classes of programs are usually characterized by different sets of meaningful features. Notably, the set of features used in the second level of ME-ASP^{ML} is a strict superset of the ones employed in ME-ASP [19], extended to deal with ASPCore 2.0 [5] programs.

An experimental analysis conducted on benchmarks and solvers of the Fifth ASP Competition shows that ME-ASP^{ML} is able to solve more instances than: (i) any solver that participated to the competition, (ii) the mere update of the (single-level) ME-ASP [19] system, and (iii) the state-of-the-art system CLASP-FOLIO ver. 2.2. The results hold also considering separately each track of the competition. Such analysis, thus, suggests that the application of a multi-level algorithm selection strategy, also exploiting the features of non-ground programs, can lead to a performance that cannot be matched by any existing system applying algorithm selection only to propositional programs.

2   Architecture and Implementation

Architecture. Figure 1 presents the architecture of ME-ASP^{ML} (available for download at www.mat.unical.it/ricca/downloads/measpml.tar.gz). Looking at the figure, we can see that ME-ASP^{ML} is composed of six main modules. NGFE (Non-Ground Feature Extractor) aims at computing features from the input (non-ground) program that are "pragmatically" cheap-to-compute, such as the number of Head-Cycle Free components, presence of queries, and stratification property. Such features are passed to NON-GROUND MANAGER, that is devoted to identify the class of the input ASP program. GROUNDER takes as input the non-ground ASP program and returns the related grounded instance. The next module, namely GFE (Ground Feature Extractor), aims at computing the syntactic

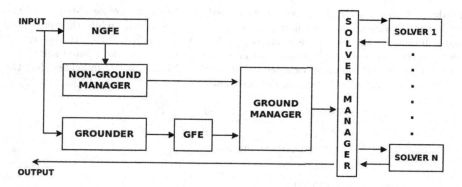

Fig. 1. The architecture of ME-ASP^{ML}. Solid boxes represent the modules, while arrows denote functional connections between them.

features of the input ground program. We used the features detailed in [19], with the addition of some ASPCore 2.0 specific features such as the number of choice rules, number of aggregates, and number of weak constraints. GROUND MANAGER is devoted to the prediction of the solver to run. It contains the inductive models related to the considered classes. Its working process can be divided in two steps, i.e., (i) given the input received by NON-GROUND MANAGER, it selects the proper inductive model; and (ii) given the features computed in GFE, it outputs to SOLVER MANAGER the name of the predicted solver. Finally, SOLVER MANAGER manages the interaction with the engines. At the end of the engine computation, SOLVER MANAGER returns as output the result given by the solver.

Implementation. In ME-ASP^{ML}, algorithm selection is implemented by means of multinomial classification. In a few words, given a set of patterns, i.e., input vectors $X = \{\underline{x}_1, \ldots \underline{x}_k\}$ with $\underline{x}_i \in \mathbb{R}^n$, and a corresponding set of labels, i.e., output values $Y \in \{1, \ldots, m\}$, where Y is composed of values representing the m classes of the multinomial classification problem, in our modeling, the m classes are m ASP solvers. Given a set of patterns X and a corresponding set of labels Y, the task of a multinomial classifier c is to construct c from X and Y so that when we are given some $\underline{x}^\star \in X$ we should ensure that $c(\underline{x}^\star)$ is equal to $f(\underline{x}^\star)$, which is the unknown function we are extrapolating. This task is called *training*, and the pair (X, Y) is called the *training set*. Concerning the training set, we selected instances and encodings involved in the Fifth ASP Competition [6]. The considered pool of benchmarks is composed of 26 domains organized into tracks, which are based on both complexity issues and language constructs of ASPCore 2.0. Starting from a total amount of 8572 instances, we pragmatically randomly split the amount of instances in each domain, using 50 % of the total amount for training purpose, and the remaining ones for testing – the full list is available at www.mat.unical.it/ricca/downloads/measpmlts.tar.gz. Concerning the NON-GROUND MANAGER (see Fig. 1), we used a list of if-then-else rules obtained running the PART decision list generator [9] on the training instances. About

Table 1. Considered ASP solvers that entered the Single Processor category of the Fifth ASP Competition. The first column contains the solvers, while the remaining four columns are related to the inductive models. A "✓" indicates that the solver has been selected as ME-ASP^{ML} engine. An empty cell means that the related solver has been evaluated but it is not included in ME-ASP^{ML}. Finally, a "–" indicates that the related solver can not compete on the program class.

Solver	T_1	T_2	T_3	T_4	Q
CLASP [7]	✓	✓	✓	✓	–
LP2BV2+BOOLECTOR [20]		✓	–	–	–
LP2GRAPH [10]	✓		–	–	–
LP2MAXSAT+CLASP [4]		✓		–	–
LP2NORMAL2+CLASP [4]	✓	✓	✓	✓	–
LP2SAT3+GLUCOSE [16]			–	–	–
LP2SAT3+LINGELING [16]			–	–	–
WASP1 [1]			✓		
WASP1.5 [1]					✓
WASP2 [1]	✓	✓		–	–

the labels, we considered five program classes, namely the queries (Q) and ASP competition tracks names (in the following, for short, T_i, $i \in \{1, \ldots, 4\}$). Notice that this module goes beyond selection of the grounder as done in [18], but in principle the approach of [18] can be implemented in our architecture.

Considering GROUNDER module, it is actually implemented using GRINGO ver. 4. Regarding GROUND MANAGER, in the current version of ME-ASP^{ML} is composed of four different inductive models, i.e., models obtained training a classifier. Models are related to the program classes $T_1, \ldots, T_4$ and are computed using the training sets mentioned above. The pattern comprised in the training set is composed of the values related to the same features as computed in GFE, while the label corresponds to the best solver – in terms of CPU time – on the given instance. In Table 1 we show the solvers that could be used as engine of ME-ASP^{ML}. (We have not considered LP2MIP2 given that we did not receive the license of CPLEX on time.) Considering that using all the solvers altogether increases the probability of getting a bad prediction because of aliasing, for each T_i we chosen different subsets of them as follows. First, we computed the total amount of training instances solved by the state-of-the-art solver (SOTA), i.e., given an instance, the oracle that always fares the best among all the solvers. Second, we calculated the minimum number of solvers such that the total amount of instances solved by the pool is at least 90 % of the SOTA solver on the training instances.

Looking at Table 1, we can see the results of this process, as well as the involved solvers considering each T_i. Notice that in the case of Q (i.e., the class of query problems) we had only one "label", namely WASP1.5, which internally calls

Table 2. Results of the experiments grouped according to 5th competition Tracks. The first column contains the various solvers considered, plus ME-ASP^{ML}. The remaining four columns contain the results for Track 1 to Track 4. Each of these columns is then divided into two subcolumns, containing the number of solved instances within the time limit, and the sum of their CPU times in seconds, respectively. If, for a track, both subcolumns contain "–", this means that the related solver can not compete on the track. Note that query problems are included in their original track.

Solver	Track 1		Track 2		Track 3		Track 4	
	#	Time	#	Time	#	Time	#	Time
CLASP	362	12318	1241	41049	154	4578	503	8078
LP2BV2+BOOLECTOR	205	19396	822	43124	–	–	–	–
LP2GRAPH	324	23592	1030	50341	–	–	–	–
LP2MAXSAT+CLASP	264	18537	1066	60185	74	5548	–	–
LP2NORMAL2+CLASP	342	18263	1252	60031	119	9379	496	12921
LP2SAT3+GLUCOSE	278	25149	1027	50170	–	–	–	–
LP2SAT3+LINGELING	256	23682	1108	80465	–	–	–	–
WASP1	268	15155	719	52260	88	3558	238	18951
WASP1.5	242	3782	1042	32159	23	754	238	19285
WASP2	317	13622	1146	41140	24	750	–	–
ME-ASP^{ML}	376	15391	1341	46143	231	5632	532	8960

DLV with magic sets. Finally, the multinomial classification algorithm employed was k-Nearest Neighbors.

3   Experiments and Conclusion

We assessed the performance of ME-ASP^{ML} on the Fifth ASP Competition benchmarks. All the experiments run on a cluster of Intel Xeon E31245 PCs at 3.30 GHz equipped with 64 bit Ubuntu 12.04, granting 600 seconds of CPU time and 2GB of memory to each solver.

The results of the analysis are presented in Table 2. We first note that ME-ASP^{ML} can solve more instances than all its engines in all tracks, followed by CLASP in Tracks 1, 3 and 4, and by LP2NORMAL+CLASP in Track 2. In sum, ME-ASP^{ML} solves 2480 instances, while the second overall best, which is CLASP, solves a total of 2260 instances.

An aggregate picture of the performance of competing systems is presented in the cactus plot of Fig. 2. This plot also includes ME-ASP and CLASPFOLIO ver. 2.2[1] for a direct comparison with approaches of algorithm selection that only

---

[1] CLASPFOLIO has been run with its default setting, and with CLASP ver. 3 as a backend solver. This improved version has been provided by Marius Lindauer, who is thanked.

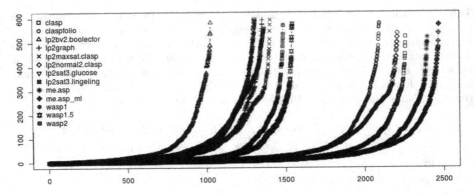

Fig. 2. Results of the solvers in Table 2, plus ME-ASP and CLASPFOLIO ver. 2.2, showed with a cactus plot. In the x-axis it is shown the total amount of solved instances, while the y-axis reports the CPU time in seconds.

exploit ground features. From the figure we can see that ME-ASP^{ML} solves more instances also in comparison with its previous version ME-ASP and the state-of-the-art system CLASPFOLIO ver. 2.2, other than all its engines. In particular, CLASPFOLIO solves 358, 1148, 118 e 471 instances on the four tracks, respectively. We can note that, consistently with the information provided in the CLASPFOLIO web page, CLASPFOLIO performance are not optimized on Track 3: indeed, this is the track where it shows the biggest performance gap (as percentage of solved instances) w.r.t. ME-ASP^{ML}.

To sum up, the extended approach implemented in ME-ASP^{ML}, which applies algorithm selection to both levels of computation, performs very well, being able to solve more instances than (i) its engines, (ii) its previous version ME-ASP, and (iii) CLASPFOLIO ver. 2.2, in all tracks of the Fifth ASP Competition.

References

1. Alviano, M., Dodaro, C., Faber, W., Leone, N., Ricca, F.: WASP: a native ASP solver based on constraint learning. In: Cabalar, P., Son, T.C. (eds.) LPNMR 2013. LNCS, vol. 8148, pp. 54–66. Springer, Heidelberg (2013)
2. Alviano, M., Faber, W., Greco, G., Leone, N.: Magic sets for disjunctive datalog programs. Artif. Intell. **187**, 156–192 (2012)
3. Balduccini, M.: Learning and using domain-specific heuristics in ASP solvers. AI Commun. **24**(2), 147–164 (2011)
4. Bomanson, J., Janhunen, T.: Normalizing cardinality rules using merging and sorting constructions. In: Cabalar, P., Son, T.C. (eds.) LPNMR 2013. LNCS, vol. 8148, pp. 187–199. Springer, Heidelberg (2013)
5. Calimeri, F., Faber, W., Gebser, M., Ianni, G., Kaminski, R., Krennwallner, T., Leone, N., Ricca, F., Schaub, T.: Asp-core-2 input language format (since 2013). https://www.mat.unical.it/aspcomp2013/ASPStandardization
6. Calimeri, F., Gebser, M., Maratea, M., Ricca, F.: The design of the fifth answer set programming competition. ICLP 2014 TC abs/1405.3710 (2014). http://arxiv.org/abs/1405.3710

7. Drescher, C., Gebser, M., Grote, T., Kaufmann, B., König, A., Ostrowski, M., Schaub, T.: Conflict-driven disjunctive answer set solving. In: Proceedings of KR 2008, pp. 422–432. AAAI Press (2008)
8. Eiter, T., Gottlob, G., Mannila, H.: Disjunctive datalog. ACM TODS **22**(3), 364–418 (1997)
9. Frank, E., Witten, I.H.: Generating accurate rule sets without global optimization. In: ICML 1998, p. 144. Morgan Kaufmann Publisher (1998)
10. Gebser, M., Janhunen, T., Rintanen, J.: Answer set programming as sat modulo acyclicity. In: Proceedings of ECAI 2014, FAIA, vol. 263, pp. 351–356. IOS Press (2014)
11. Gebser, M., Kaminski, R., Kaufmann, B., Schaub, T., Schneider, M.T., Ziller, S.: A portfolio solver for answer set programming: preliminary report. In: Delgrande, J.P., Faber, W. (eds.) LPNMR 2011. LNCS, vol. 6645, pp. 352–357. Springer, Heidelberg (2011)
12. Gebser, M., Schaub, T., Thiele, S.: GrinGo: a new grounder for answer set programming. In: Baral, C., Brewka, G., Schlipf, J. (eds.) LPNMR 2007. LNCS (LNAI), vol. 4483, pp. 266–271. Springer, Heidelberg (2007)
13. Gelfond, M., Lifschitz, V.: Classical negation in logic programs and disjunctive databases. NGC **9**, 365–385 (1991)
14. Hoos, H., Kaminski, R., Lindauer, M.T., Schaub, T.: ASPeed: solver scheduling via answer set programming. TPLP **15**(1), 117–142 (2015)
15. Hoos, H., Lindauer, M.T., Schaub, T.: claspfolio 2: Advances in algorithm selection for answer set programming. TPLP **14**(4–5), 569–585 (2014)
16. Janhunen, T.: Some (in)translatability results for normal logic programs and propositional theories. J. Appl. Non Class. Logics **16**, 35–86 (2006)
17. Leone, N., Pfeifer, G., Faber, W., Eiter, T., Gottlob, G., Perri, S., Scarcello, F.: The DLV system for knowledge representation and reasoning. ACM TOCL **7**(3), 499–562 (2006)
18. Maratea, M., Pulina, L., Ricca, F.: Automated selection of grounding algorithm in answer set programming. In: Baldoni, M., Baroglio, C., Boella, G., Micalizio, R. (eds.) AI*IA 2013. LNCS, vol. 8249, pp. 73–84. Springer, Heidelberg (2013)
19. Maratea, M., Pulina, L., Ricca, F.: A multi-engine approach to answer-set programming. TPLP **14**(6), 841–868 (2014)
20. Nguyen, M., Janhunen, T., Niemelä, I.: Translating answer-set programs into bit-vector logic. In: Tompits, H., Abreu, S., Oetsch, J., Pührer, J., Seipel, D., Umeda, M., Wolf, A. (eds.) INAP/WLP 2011. LNCS, vol. 7773, pp. 91–109. Springer, Heidelberg (2013)
21. Silverthorn, B., Lierler, Y., Schneider, M.: Surviving solver sensitivity: an ASP practitioner's guide. In: ICLP 2012, LIPIcs, vol. 17, pp. 164–175 (2012)

Clause-Learning for Modular Systems

David Mitchell[(✉)] and Eugenia Ternovska

Computational Logic Laboratory, Simon Fraser University, Burnaby, Canada
{mitchell,ter}@cs.sfu.ca

Abstract. We present an algorithm, CDCL-AMS, for solving Modular Systems consisting of a set of modules where, for each module, we have a simple "black-box" solver. The algorithm is based on the Conflict-Directed Clause Learning algorithm for SAT, and communicates asynchronously with the black-box solvers to accommodate high variability in response latencies.

1 Introduction

In many modern contexts, finding a solution to a problem amounts to solving a combinatorial search problem where the constraints are implicit in a collection of more-or-less independent modules, each of which is a knowledge base or problem-solving system in its own right, and typically presented via network connections. The conflict-directed clause learning (CDCL) algorithm [5] is the basis of SAT solvers with impressive performance on many constraint problems. However, in the multi-module context we consider here, it is often undesirable or even impossible to transform each module into a set of explicit constraints. In this context, each module is a black box which answers queries, for example of the form "do you have a solution consistent with partial solution X?" In many settings, the response latencies for modules will be substantial, highly variable, and largely un-correlated. The main purpose of this paper is to present a CDCL-based algorithm suited to this context.

Modular Systems. A general logic-based formalization of problem solving in a multi-module context is provided by the notion of Modular Systems, as defined in [8]. This generalizes the formalization of a decision problem as a class of structures, or of a search problem as model expansion [6]. Formally, a *module* is a class of structures for a fixed vocabulary. Modular systems are defined by combining primitive modules with an algebra of modular systems. The algebra is similar to Codd's relational algebra, but defined on *classes of structures* rather than on relational tables. The operations are Sequential Composition, Union, Complementation, Projection and Feedback.

In this short paper, we restrict our attention to systems which are conjunctions of "primitive" modules. In the context of solving a particular problem instance. The algorithm is based on the idea of querying modules with a partial structures, which for a particular problem instance are on a fixed universe.

© Springer International Publishing Switzerland 2015
F. Calimeri et al. (Eds.): LPNMR 2015, LNAI 9345, pp. 446–452, 2015.
DOI: 10.1007/978-3-319-23264-5_37

In this setting, queries are essentially propositional, so for simplicity we present our algorithms in purely propositional form. Our setting is as follows.

1. Each module M_i is a set of truth assignments for propositional vocabulary σ_i. We say assignment α satisfies M_i if $\alpha \in M_i$, and we say module M_i implies clause C iff every assignment in M_i satisfies C.
2. Modular System $\mathcal{M}$ is a conjunction of modules. Its vocabulary is the union of those of its modules. $\mathcal{M}$ is satisfiable if there is a truth assignment for σ which satisfies every module. $\mathcal{M}$ implies clause C if every satisfying assignment for $\mathcal{M}$ satisfies C.
3. For each module M_i, we have a solver S_i that answers queries in the form of partial truth assignments for σ_i. We consider only solvers which are propagators, that is, when queried with partial assignment α for σ_i, in finite time return one of:
   (a) $\langle \text{Reject}, \text{Reason} \rangle$, where Reason is a set of clauses which are implied by M_i and false in α.
   (b) $\langle \text{Accept}, \text{Advice} \rangle$, where Advice is a set of clauses implied by M_i, with exactly one literal not defined by α and all other literals false in α.

Remark 1. It is possible for Advice or Reason to be empty. In particular, a solver S_i for which Advice and Reason are *always* empty is simply verifier for M_i. It follows that any verifier can be wrapped as a (not very helpful) propagator.

The CDCL Algorithm. We assume the reader is familiar with CDCL, but review some aspects here, and also fix some notation. Given CNF formula Γ, the algorithm incrementally constructs a sequence of literals α (the "assignment stack") which defines a partial truth assignment for Γ. (We henceforth gloss over the distinction between assignments and sequences of literals.) The algorithm alternately adds an unassigned literal l to α (called a "decision"), and then extends α with any literals newly determined by unit propagation. The sub-sequence of α consisting of only decision literals is the "decision sequence" corresponding to α, and will be denoted δ. Each literal l of α that is not a decision is labelled with the clause that the unit propagation engine used to set it true, called the "reason for l". This guess-propagate process continues until either $\alpha \models \Gamma$, in which case the algorithm halts, or the unit propagation engine determines that a literal l must be true, but $\neg l$ is already in α, called a "conflict".

For clause set Γ and set δ of literals, we write $l \in \text{UP}(\Gamma, \delta)$ if setting the literals of δ true and running unit propagation on Γ results in l being set true, and $C \in \text{UP}(\Gamma, \delta)$ if C is obtained from a clause of Γ when, for every literal $l \in \text{UP}(\Gamma, \delta)$ we delete from δ any clause containing l, and delete from each clause of δ any occurrence of $\neg l$. Notice that, if α is an assignment stack, and δ the corresponding decision sequence then $\alpha = \text{UP}(\Gamma, \delta)$. Also, we denote the empty clause by $\square$, so $\square \in \text{UP}(\Gamma, \delta)$ that executing unit propagation from Γ and δ produces a conflict.

Upon a conflict, CDCL derives a clause C by resolution, using the reasons labelling α. The clause C is used to determine a proper prefix α' of α, which replaces α as the current assignment stack (called "back-jumping"), and C is added to Γ (it is called a "learned" clause). The clause C and assignment stack α bear a particular relationship: C is an "asserting clause" α' and Γ, as defined next.

A clause C is a conflict clause for decision sequence δ and clause set Γ iff: (1) $\square \in \mathrm{UP}(\Gamma, \overline{C})$, and (2) For each literal $l \in C$, $\overline{l} \in \mathrm{UP}(\Gamma, \delta)$. C is an asserting clause for Γ and δ if it is a conflict clause and also satisfies: (3) For exactly one literal $l \in C$, $\overline{l} \notin UP(\Gamma, \delta^-)$, where δ^- is δ with its last element removed.

Observe that, when C is a conflict clause for α and Γ, $\mathrm{UP}(\Gamma \cup \{C\}, \delta)$ contains at least one literal not in α, so immediately after back-jumping, the new assignment stack is extended by unit propagation, which could in turn lead to a new conflict. If a state is reached where this does not happen, the algorithm returns to extending α with decisions.

Synchronous CDCL for Sets of Modules. We can make a version of CDCL for sets of modules very simply, provided we query them synchronously and the response times are sufficiently fast. To do this, we simply query each solver every time we extend the current assignment stack, adding all reasons and advice returned by the solvers to Γ. However, we are interested in the case when these properties do not hold, in which case it is unreasonable to make so many queries, and unreasonable to have the algorithm wait for all solvers to respond to a given query before proceeding.

2 Asynchronous CDCL for Modular Systems

Our asynchronous clause-learning algorithm, which we call CDCL-AMS, has three processes (not including the solvers): The CDCL Engine, the Query Handler and the Response Handler. These communicate via four data objects: clause set Γ; set QUERIES of available solver queries; set HOLD decision sequences corresponding to pending queries; and set CONTINUE of query responses waiting to be handled by the Engine. The Engine is a CDCL solver that tries to decide satisfiability of $\mathcal{M}$. It generates models of Γ, which become queries to solvers. Each query is extended until either it is rejected by a solver, at which time at least one clause is added to Γ, or is accepted by all solvers. The other processes handle the communication between the Engine and the solvers.

Γ is initially empty, and is extended by clauses obtained from solvers in response to queries, and by the standard clause learning mechanism. The algorithm proceeds as in standard CDCL, extending its assignment stack until either a conflict or a satisfying assignment for Γ is obtained. Upon conflict, learned clause derivation and back-jumping are carried out. This process is identical to standard CDCL, except for the role of HOLD, which is discussed below. If $\alpha \models \Gamma$, α becomes a query to send to a solver, and the Engine:

1. Adds α to QUERIES;
2. Adds δ, the decision sequence corresponding to α, to HOLD;

3. Replaces δ with some proper subsequence, and modifies α accordingly. (This is back-jumping in the absence of a true conflict. Correctness requires only that the back-jump is proper, i.e., removes at least one decision from α.)

When a solver S becomes available, the Query Handler selects and removes an appropriate query from QUERIES and submits it to S. (A query is appropriate for S if S could reject it. Slightly more precisely: the vocabulary of α, less any prefix of α that has previously been accepted by S, has non-empty intersection with the vocabulary of S.) When S responds to query α, the Response Handler:

1. Adds all clauses in the returned Reasons or Advice to Γ. If the response is Reject, but Reasons is empty, it adds to Γ the clause consisting of the disjunction of the negations of the literals in δ;
2. Marks α to indicate that S has accepted it;
3. Removes the δ corresponding to α from HOLD.

A solution to $\mathcal{M}$ is an assignment that has been accepted by every solver. In the algorithm, a query is sent only to one solver before being returned to the main engine (via CONTINUE). This keeps all reasoning in one algorithm, and also ensures that information obtained by the main engine is exploited as soon as possible, avoiding, for example, submitting a query to a solver when clauses returned by other solvers already imply its rejection. To keep track of which solvers have accepted a query, each assignment stack has a mark for each solver indicating its largest prefix which has been accepted by that solver. When a query is returned to the main engine, via CONTINUE, it may be extended, both by unit propagation involving new clauses and by new decisions, and these marks are maintained.

The purpose of the HOLD set is to ensure that, while a response to a query based on decision sequence δ is pending (in QUERIES, or being handled by a solver), the engine does not generate any query which is "no better than" δ (i.e., is a superset of δ). To this end, each decision sequence δ in HOLD is treated by the unit propagation engine as the clause $\overline{\delta}$, the disjunction of complements of literals in δ.

The clauses in HOLD cannot be used as "reasons" for the purpose of asserting clause derivation because they are guesses and might not be implied by $\mathcal{M}$. In the derivation, any literal set by unit propagation by using a clause of HOLD must be treated as a decision literal. The standard methods for asserting clause derivation ensure every learned clauses is unique. This now fails, and it is possible that an execution of the body of the main loop of the Engine fails to generate a new query or add a new clause. Loop iterations which are essentially "wheel-spinning" may result.While undesirable, it seems that this cannot entirely be avoided: If solvers take sufficiently long to respond to queries, the main engine will generate all possible resolvents and all possible queries based on Γ, and can make no further progress until some further query response arrives.

The CDCL-AMS Engine is given by Algorithm 1.

Algorithm 1. CDCL-AMS Engine

Input: Vocabulary σ
Output: SAT or UNSAT

1   $\delta \leftarrow$ any non-empty decision sequence
2   **repeat**
3   **if** CONTINUE $\neq \emptyset$ **then**
4   Remove one query γ from CONTINUE
5   **if** γ *satisfied all modules* **then**
6   **return** *SAT*
7   **end**
8   Remove γ from HOLD
9       $\delta \leftarrow \gamma$
10   **end**
11    $\delta \leftarrow$ **ExtendAndLearn-AMS**(δ, Γ)
12   **if** $\delta = \langle \rangle$ **then**
13   **return** *UNSAT*
14   **end**
15   **if** $UP(\Gamma, \delta) \models \Gamma$ **then**
16   Add δ to HOLD
17   Add δ to QUERIES
18      $\delta \leftarrow$ a proper sub-sequence of δ
19   **end**
20 **end**

Algorithm 1 is described in terms of decision sequences (δ and γ), but the corresponding assignment stack, labelled with both reasons and marks from accepting solvers, is implicitly maintained, and is in fact what is being operated on. (For example, in adding δ to QUERIES, it is clear that the labelled assignment stack must be intended. The exception is line 16, where it is indeed just the decision sequence δ that is added to HOLD.)

In line 11, **ExtendAndLearn-AMS**(δ, Γ) carries out the process of extending decision sequence δ until either it satisfies Γ, or generates a conflict. In the latter case, it may generate multiple conflicts (and learned clauses), eventually returning a decision sequence δ such that either $UP(\Gamma, \delta) \models \Gamma$ or computing $UP(\Gamma, \delta)$ does *not* generate a conflict. (This may correspond to many iterations of the main loop of CDCL as it is usually presented.) At line 12, if $\delta = \langle \rangle$, the last learned clause was $\square$.

Remark 2. If at any point Γ becomes unsatisfiable, the algorithm essentially becomes a standard CDCL solver. In particular, it generates no new queries, and once existing contents of QUERIES have been exhausted, and their responses all handled, the algorithm becomes CDCL proving unsatisfiability of Γ.

In Line 5, "γ satisfied all modules" means that every solver S accepted (some previous version of) γ at a time when it was (already) total for the vocabulary of S.

Correctness. If the algorithm returns SAT, then some assignment is total for, and accepted by, every solver, so $\mathcal{M}$ is satisfiable. If the algorithm returns UNSAT, it is because the empty clause has been derived from Γ. Since $\mathcal{M} \models \Gamma$, $\mathcal{M}$ is unsatisfiable. It remains to establish termination.

Observe that the main engine is simply a CDCL engine which generates all satisfying assignments for Γ. (Γ grows monotonically, so some generated satisfying assignments later are not satisfying, but this does not affect the argument.) To see that the algorithm makes progress, we observe that every assignment α generated by the engine is eventually extended until it either becomes a satisfying assignment for $\mathcal{M}$, or is "killed" by generation of a clause which is implied by $\mathcal{M}$, but is false in α. We need to see that, in each iteration of the Engine main loop, progress is made either by δ being killed by a new clause added to Γ, or by "getting closer" to satisfying every module. For the moment, set aside the question of the "wheel-spinning" iterations mentioned above. Consider line 11, where δ is re-assigned by the call to **ExtendAndLearn-AMS**(δ, Γ). To avoid ambiguity, let's write β for the new value of δ. If at line 3 CONTINUE was empty, then the standard CDCL process will ensure that either β is a proper extension of δ, or that a clause that "kills" δ was added to Γ, and β is new. So we have progress. If CONTINUE was not empty, δ could be killed either because the response for δ was Reject (in which case a killing clause was added to Γ), or because unit propagation from δ produced a conflict based on other clauses that have been added to Γ since δ was generated as a query. If δ does not get killed, and is not total, then β is set to a proper extension of δ. If δ is not killed and is total, then there is some solver S which has accepted a larger prefix of α than the previous time this query was in QUERIES, and α is marked with this information. In each case, we have made progress. It remains to verify that every query is eventually responded to, and that "wheel-spinning" iterations do not prevent eventual progress. There are finitely many possible queries, and each solver responds to each query in finite time, so each query is eventually responded to by a solver. Wheel-spinning iterations are only possible if HOLD is not empty, and since every query is eventually responded to, every element of HOLD is eventually deleted, at which point progress is ensured.

3   Discussion

Related Work. Many related algorithms have been presented in the literature. These include propagation via lazy clause generation [7], algorithms used in used in "lazy" SMT solvers, methods for supporting external constraints in SMT and ASP solvers [1–3], and distributed and parallel CSP and SAT algorithms [4]. An abstract algorithmic scheme for Modular System solvers was given in [8]. We will give detailed comparisons in a longer paper.

Future Work. A number of details of this algorithm warrant more careful discussion, and there are many refinements and heuristics to consider when contemplating implementation, even without taking into account the practical complexity of interacting with real on-line solvers. We will discuss some of these in a longer paper. In future work, we intend to examine the extension of these algorithms to the full algebra of modular systems; develop versions for use with solvers which are more than just propagators; study the relationship between problem structure and algorithm complexity; develop versions which make use of FO vocabulary of modules as classes of structures; attend more closely to issues that need to be addressed for implementability; and develop versions for use in distributed and many-core computational environments.

References

1. de Moura, L., Bjørner, N.S.: Z3: an efficient SMT solver. In: Ramakrishnan, C.R., Rehof, J. (eds.) TACAS 2008. LNCS, vol. 4963, pp. 337–340. Springer, Heidelberg (2008)
2. Eiter, T., Fink, M., Krennwallner, T., Redl, C.: Conflict-driven ASP solving with external sources. Theory Pract. Logic Program. $12(4–5)$, 659–679 (2012)
3. Gebser, M., Ostrowski, M., Schaub, T.: Constraint answer set solving. In: Hill, P.M., Warren, D.S. (eds.) ICLP 2009. LNCS, vol. 5649, pp. 235–249. Springer, Heidelberg (2009)
4. Manthey, N.: Towards Next Generation Sequential and Parallel SAT Solvers. Ph.D. thesis, TU Dresden (2014)
5. Marques-Silva, J.P., Sakallah, K.A.: Grasp: a search algorithm for propositional satisfiability. IEEE Trans. Comput. $48(5)$, 506–521 (1999)
6. Mitchell, D.G., Ternovska, E.: A framework for representing and solving NP search problems. In: Veloso, M.M., Kambhampati, S. (eds.) Proceedings The Twentieth National Conference on Artificial Intelligence and the Seventeenth Innovative Applications of Artificial Intelligence Conference, 9–13 July, 2005, Pittsburgh, Pennsylvania, USA, pp. 430–435. AAAI Press/The MIT Press (2005)
7. Ohrimenko, O., Stuckey, P.J., Codish, M.: Propagation = lazy clause generation. In: Bessière, C. (ed.) CP 2007. LNCS, vol. 4741, pp. 544–558. Springer, Heidelberg (2007)
8. Tasharrofi, S., Ternovska, E.: A semantic account for modularity in multi-language modelling of search problems. In: Tinelli, C., Sofronie-Stokkermans, V. (eds.) FroCoS 2011. LNCS, vol. 6989, pp. 259–274. Springer, Heidelberg (2011)

Solving Disjunctive Fuzzy Answer Set Programs

Mushthofa Mushthofa[1,3]([✉]), Steven Schockaert[2], and Martine De Cock[1,4]

[1] Department of Applied Mathematics, Computer Science and Statistics,
Ghent University, Ghent, Belgium
{Mushthofa.Mushthofa,Martine.DeCock}@UGent.be
[2] School of Computer Science and Informatics, Cardiff University, Cardiff, UK
SchockaertS1@cardiff.ac.uk
[3] Department of Computer Science, Bogor Agricultural University, Bogor, Indonesia
mush@ipb.ac.id
[4] Center for Data Science, University of Washington Tacoma, Tacoma, USA
mdecock@uw.edu

Abstract. Fuzzy Answer Set Programming (FASP) is an extension of
the popular Answer Set Programming (ASP) paradigm which is tailored
for continuous domains. Despite the existence of several prototype implementations, none of the existing solvers can handle disjunctive rules in
a sound and efficient manner. We first show that a large class of disjunctive FASP programs called the *self-reinforcing cycle-free* (SRCF)
programs can be polynomially reduced to normal FASP programs. We
then introduce a general method for solving disjunctive FASP programs,
which combines the proposed reduction with the use of mixed integer
programming for minimality checking. We also report the result of the
experimental benchmark of this method.

1 Introduction

Answer Set Programming (ASP) is one of the most popular and well-studied
declarative programming paradigms [2, 11]. Based on the *stable model semantics*
[15], ASP allows an intuitive encoding of combinatorial search and optimization
problems [20]. Due to the availability of fast and efficient solvers, such as clasp [14]
and DLV [19], ASP found practical applications in many fields [11, 13]. However,
because of the fact that it relies on Boolean logic, ASP is not directly suitable
for encoding problems in continuous domains.

Fuzzy Answer Set Programming (FASP) [30] is a form of declarative programming that extends classical ASP by allowing graded truth values in atomic
propositions and extending classical Boolean operators to fuzzy logic connectives.
Although work has done on the theoretical aspects of FASP, e.g., [5, 6, 18, 22–24, 28],
progress on the development of FASP solvers has not yet reached the maturity level
of ASP solvers. Several notable results about FASP are: (1) the development of a
reduction method for FASP into bilevel linear programming [5], (2) a FASP solver
based on answer set approximation operators [1] and (3) a solver for FASP based on
a reduction to classical ASP [25]. The prototype solver developed in [1] only deals
with normal rules and does not allow disjunctions at all. The method proposed

© Springer International Publishing Switzerland 2015
F. Calimeri et al. (Eds.): LNMR 2015, LNAI 9345, pp. 453–466, 2015.
DOI: 10.1007/978-3-319-23264-5_38

by [5] only allows disjunctions in the head, while the evaluation method described in [25] only allows disjunctions in the body. No evaluation method/solver for FASP currently allows disjunctions in the body and the head of the rules.

Allowing disjunctions both in the head and the body of FASP rules enables us to represent and solve a broader class of problems. As an example, consider the following fuzzy graph colorability problem: given a graph $G = \langle V, E \rangle$ with weighted edges, can we color each node with a fuzzy color (intuitively, a grey level value between fully black and fully white) such that connected nodes are colored with *sufficiently* differing colors? Formally, let V be the set of the nodes and $E : V \times V \rightarrow [0, 1]$ be the set of fuzzy edges. We want to find functions $f_b : V \rightarrow [0, 1]$ and $f_w : V \rightarrow [0, 1]$ such that: (i) $f_b(x) \oplus f_w(x) = 1, \forall x \in V$ and (ii) $f(x) \otimes f(y) \otimes e(x, y) = 0, \forall x, y \in V, f \in \{f_b, f_w\}$, where $\oplus$ and $\otimes$ represent disjunction and conjunction in Łukasiewicz logic, respectively (see Sect. 2). Given the input graph as facts of the form $node(x) \leftarrow \overline{1}$ and $edge(x, y) \leftarrow \overline{c}$ for $c \in [0, 1]$[1], the following FASP program solves the problem (see Sect. 2 for formal definitions):

$$b(X) \oplus w(X) \leftarrow node(X)$$
$$edge(X, Y) \leftarrow edge(Y, X)$$
$$\overline{0} \leftarrow b(X) \otimes b(Y) \otimes edge(X, Y)$$
$$\overline{0} \leftarrow w(X) \otimes w(Y) \otimes edge(X, Y)$$

Suppose we further require that for some subset $S \subseteq V$ of the nodes, the color assigned must be either fully black or fully white (i.e., either $f_w(v) = 1$ or $f_b(v) = 1$ for $v \in S$). We can encode this requirement by adding the so-called *saturation* rules $\{b(v) \leftarrow b(v) \oplus b(v), w(v) \leftarrow w(v) \oplus w(v) \mid v \in S\}$, which force the atoms $b(v)$ and $b(w)$ to be Boolean. In this example, we see how disjunctions in the head and in the body of the rules appear naturally in the problem's representation.

It is well known that the presence of disjunctions in (F)ASP programs can increase the computational complexity of various reasoning tasks. In [5], the complexity of answer set existence, set-membership and set-entailment in various classes of FASP under the Łukasiewicz semantics was studied. Interestingly, it was shown that for the class of *strict* FASP programs, where only the conjunction, maximum and negation operators are allowed in the body, disjunctions in the head do not increase the complexity of the reasoning tasks, which are still within the first level of the Polynomial Hierarchy (PH), even for disjunctive strict FASP. However, when disjunctions are allowed in the head as well as in the body, the complexity increases to the second level of PH.

In classical ASP, allowing disjunctions in the head has been shown to increase the complexity of the reasoning tasks, from NP-complete and coNP-complete for *brave* and *cautious* reasoning, respectively, to Σ_2^P and Π_2^P (see e.g., [7]). However, [3] has shown that the semantics of a large class of disjunctive ASP programs, called the *head-cycle free* (HCF) programs, can be efficiently expressed in non-disjunctive propositional logic, effectively reducing their complexity to the first

---

[1] The symbol $\overline{c}$ for a numeric value c represents a truth-value constant in a program.

level of PH. Furthermore, by a sequence of applications of the so-called *shift operators* described in [8], one can reduce any HCF disjunctive ASP program into an equivalent normal program.

In this paper, we show that a large class of disjunctive FASP programs can be similarly reduced to normal programs. Unlike in classical ASP, however, the reduction is not based on head-cycle freeness, but rather on the concept we define as *self-reinforcing cycle freeness*, which covers a much larger subclass of disjunctive FASP programs. More specifically, we will show that the shifting method for HCF disjunctive programs as being used in classical ASP also applies for FASP programs. However, the class of disjunctive FASP programs that can be shifted is strictly larger than the class of HCF programs. In fact, we can show that every disjunctive strict FASP program can be reduced to a normal FASP program, even if head cycles occur. Subsequently, we introduce a general method for finding answer sets of disjunctive FASP programs (allowing disjunctions in the body, as well), which combines the proposed reduction with an additional minimality check based on mixed integer programming.

2 Preliminaries

2.1 Fuzzy Answer Set Programming

We follow the definition of FASP syntax as described in [5] and consider only the Łukasiewicz operators and semantics. As is the case in classical ASP, the syntax of a FASP program $\mathcal{P}$ is based on atoms drawn from either a propositional or a first-order Herbrand base $\mathcal{B}_\mathcal{P}$. For simplicity, in this paper we consider only propositional atoms. A (classical) literal is either an atom a or a classical negation literal $\neg a$. An extended literal is a classical literal a or a NAF literal **not** a. A *head/body expression* is a formula defined recursively as follows:

- a constant $\bar{c}, c \in [0, 1] \cap \mathbb{Q}$, and a classical literal a are head expressions.
- a constant $\bar{c}, c \in [0, 1] \cap \mathbb{Q}$, and an extended literal a are body expressions.
- if α and β are head/body expressions, then $\alpha \otimes \beta$, $\alpha \oplus \beta$, $\alpha \veebar \beta$ and $\alpha \barwedge \beta$ are also head/body expressions, respectively.

We denote by $Lit(E)$ the set of classical literals appearing in an expression E. A FASP program is a finite set of rules, where a rule r is of the form $\alpha \leftarrow \beta$. Here, α is a head expression (called the *head* of r) and β is a body expression (called the *body* of r). We also write $Head(r)$ and $Body(r)$ to denote the head and body of a rule r, respectively. A FASP rule of the form $a \leftarrow \bar{c}$ for an atom a and a constant c is called a fact. A FASP rule of the form $\bar{c} \leftarrow \beta$ is called a constraint. A rule which does not contain any application of the operator **not** is called a *positive* rule. A rule which only has one literal in the head is called a *normal* rule. A rule which contains the application of the operator $\oplus$ in the head is called a *disjunctive* rule. A FASP program is called [*positive, normal*] iff it contains only [*positive, normal*] rules, respectively. A FASP program which contains disjunctive rules is called a disjunctive program. A FASP program whose only connectives in the body are **not**, $\otimes$ and $\overline{\wedge}$, and has only disjunctions in the head is called a *strict* FASP program. Intuitively, the class of strict FASP programs contains those programs whose syntax corresponds to that of classical ASP programs. In this paper, we restrict the discussion to FASP syntax that allows only disjunction in the head (but no restrictions for the connectives in the body). Furthermore, following the rule rewriting technique described in [25], we may assume w.l.o.g. that the rules in a FASP program only contain at most one application of the Łukasiewicz connectives, either in the body or in the head of the rules.

The semantics of FASP is traditionally defined on a complete truth lattice $\mathcal{L} = \langle L, \leq_L \rangle$ [4]. In this paper, we consider two types of truth-lattice: the infinite-valued lattice $\mathcal{L}_\infty = \langle [0, 1], \leq \rangle$ and the finite-valued lattices $\mathcal{L}_k = \langle \mathbb{Q}_k, \leq \rangle$, where $\mathbb{Q}_k = \{\frac{i}{k} \mid 0 \leq i \leq k\}$, and $k \geq 1$ is a positive integer such that $c \in \mathbb{Q}_k$ for every constant c in the program. An interpretation of a FASP program $\mathcal{P}$ is a function $I : \mathcal{B}_\mathcal{P} \rightarrow \mathcal{L}^2$, which can be extended to expressions and rules as follows:

- $I(\bar{c}) = c$, for any constant c in the program.
- $I(\alpha \otimes \beta) = \max(I(\alpha) + I(\beta) - 1, 0)$.
- $I(\alpha \oplus \beta) = \min(I(\alpha) + I(\beta), 1)$.
- $I(\alpha \veebar \beta) = \max(I(\alpha), I(\beta))$.
- $I(\alpha \overline{\wedge} \beta) = \min(I(\alpha), I(\beta))$.
- $I(\mathbf{not}\ \alpha) = 1 - I(\alpha)$.
- $I(\alpha \leftarrow \beta) = \min(1 - I(\beta) + I(\alpha), 1)$.

for appropriate expressions α and β. Here, the operators **not**, $\otimes, \oplus, \veebar, \overline{\wedge}$ and $\leftarrow$ denote the Łukasiewicz negation, t-norm (conjunction), t-conorm (disjunction), maximum, minimum and implication, respectively.

An interpretation I is consistent iff $I(l) + I(\neg l) \leq 1$ for each $l \in \mathcal{B}_\mathcal{P}$. We say that a consistent interpretation I of $\mathcal{P}$ satisfies a FASP rule r iff $I(r) = 1$. This condition is equivalent to $I(Head(r)) \geq I(Body(r))$. An interpretation is a model of a program $\mathcal{P}$ iff it satisfies every rule in $\mathcal{P}$. For interpretations I_1, I_2, we write $I_1 \leq I_2$ iff $I_1(l) \leq I_2(l)$ for each $l \in \mathcal{B}_\mathcal{P}$, and $I_1 < I_2$ iff $I_1 \leq I_2$ and $I_1 \neq I_2$. We call a model I of $\mathcal{P}$ a *minimal* model if there is no other model J of $\mathcal{P}$ such that $J < I$.

For a positive FASP program $\mathcal{P}$, a model I of $\mathcal{P}$ is called an *answer set* of $\mathcal{P}$ iff it is a minimal model of $\mathcal{P}$. For a non-positive FASP program $\mathcal{P}$, a generalization of the so-called Gelfond-Lifschitz reduct is defined in [23] as follows: the reduct of a rule r w.r.t. an interpretation I is the positive rule r^I obtained by replacing each occurrence of **not** a by the constant $\overline{I(\mathbf{not}\ a)}$. The reduct of a FASP program $\mathcal{P}$ w.r.t. an interpretation I is then defined as the positive program $\mathcal{P}^I = \{r^I \mid r \in \mathcal{P}\}$. A model I of $\mathcal{P}$ is called an answer set of $\mathcal{P}$ iff I is an answer set of $\mathcal{P}^I$. We say that an answer set I is k-valued if $I(a) \in \mathcal{L}_k$ for every literal a; we say that an answer set is finite valued if it is k-valued for some $k \in \mathbb{N}$.

Example 1. *Consider the disjunctive FASP program $\mathcal{P}_1$ which has the following rules:*

$$\{a \leftarrow \mathbf{not}\ c, b \leftarrow \mathbf{not}\ c, c \leftarrow a \oplus b, d \oplus e \leftarrow c\}$$

---

2 Note that by $\mathcal{L}$ here, we mean the "set" part of the lattice $\mathcal{L}$.

One can check that under both the truth-lattices $\mathcal{L}_3$ and $\mathcal{L}_\infty$, the interpretation $I_x = \{(a, \frac{1}{3}), (b, \frac{1}{3}), (c, \frac{2}{3}), (d, \frac{2}{3} - x), (e, x)\}$ is a minimal model of $\mathcal{P}_1^{I_x}$ for any $0 \leq x \leq \frac{2}{3}$, and hence it is an answer set of $\mathcal{P}_1$. However, $\mathcal{P}_1$ has no answer sets under any $\mathcal{L}_k$ where k is not a multiple of 3.

2.2   Finite-Valued Evaluation of FASP Programs

We briefly recall the method from [25] to evaluate FASP programs based on a reduction to classical ASP. The method relies on a procedure $Tr(\cdot, k)$ that translates a FASP program into a classical ASP program whose answer sets correspond to the k-valued answer sets of the original FASP program. For normal programs, every k-valued answer set of the program can be found by using the translation. Furthermore, any answer set obtained from the translation is guaranteed to be a k-valued answer set of the original FASP program.

For disjunctive FASP programs, however, the result does not necessarily hold, as illustrated in the next example.

Example 2. *Program $\mathcal{P}_2$ has the following rules: $\{a \oplus b \leftarrow \bar{1}, a \leftarrow b, b \leftarrow a\}$. The finite-valued answer set obtained by applying the translation method to $\mathcal{P}_2$ using $k = 1$ is $A_1 = \{(a, 1), (b, 1)\}$. However, A_1 is not an answer set of $\mathcal{P}_2$ in $\mathcal{L}_\infty$. In fact, the only answer set of $\mathcal{P}_2$ in $\mathcal{L}_\infty$ is $A_2 = \{(a, 0.5), (b, 0.5)\}$, which is obtained using the translation method when $k = 2$.*

To ensure that each k-valued answer set obtained is indeed an answer set of the FASP program, [25] suggested conducting an extra minimality check, which can however be costly. In this paper, we show how to reduce a large class of disjunctive FASP programs into normal programs, allowing us to avoid the minimality check.

3   Evaluating Disjunctive Rules

In this section we will identify a large fragment of the class of disjunctive FASP programs which can be reduced in polynomial time to a normal FASP program. Subsequently, we will show how this reduction can be used to develop a sound method for finding answer sets of general disjunctive FASP programs.

Following [3], the head-cycle free (HCF) ASP programs are programs whose positive dependency graphs (see Sect. 3.3) do not contain cycles that go through two literals occurring in the head of a rule. In [8], it was shown that any HCF program can be reduced to an equivalent program using the *shift* operator. Briefly, the shift operator replaces any rule $a_1 \vee \ldots \vee a_n \leftarrow B$ with the set of rules $R = \{a_i \leftarrow B \wedge NB_i \mid 1 \leq i \leq n\}$, where $NB_i = \bigwedge_{1 \leq j \leq n, j \neq i} \text{not } a_j$. For example, the program $\{a \vee b \leftarrow\}$ can be reduced to the equivalent program $\{a \leftarrow \text{not } b, b \leftarrow \text{not } a\}$. However, when we introduce head cycles, such as in the program $\mathcal{P}_3 = \{a \vee b \leftarrow, a \leftarrow b, b \leftarrow a\}$, shifting is no longer guaranteed to produce an equivalent normal program. Interestingly, in the case of FASP programs, the syntactically similar program $\mathcal{P}_4 = \{a \oplus b \leftarrow \bar{1}, a \leftarrow b, b \leftarrow a\}$ is equivalent to the shifted version: $\mathcal{P}'_4 = \{a \leftarrow \textbf{not } b, b \leftarrow \textbf{not } a, a \leftarrow b, b \leftarrow a\}$. In fact, we will show that any strict disjunctive FASP program can be reduced to an equivalent normal FASP program in this way. This explains the observation in [5] that allowing disjunction in the head does not affect the computational complexity of strict FASP programs. For programs with disjunction in the body, shifting does not always yield an equivalent FASP program, for e.g., $\mathcal{P}_4 \cup \{a \leftarrow a \oplus a\}$ is not equivalent to $\mathcal{P}'_4 \cup \{a \leftarrow a \oplus a\}$. Intuitively, we can safely shift disjunctive rules if there is no interaction between disjunctions in the body and a head cycle. We will now formalize this idea based on the notion of a self-reinforcing cycle.

3.1   SRCF Programs

We first extend the notion of *proof* for classical disjunctive programs as defined in [3]. Let $\mathbf{0}$ denotes the interpretation that assigns zeros to all atoms. Let I be an interpretation of a program $\mathcal{P}$, and let a be any atom such that $I(a) > 0$. Then, a support of a in $\mathcal{P}$ w.r.t. I is defined as a sequence of rules $r_1, r_2, \ldots, r_n \in \mathcal{P}^I$ such that:

1. $\mathbf{0}(Body(r_1)) > 0$
2. $a \in Head(r_n)$
3. $\sum_{a \in Lit(Head(r_i))} I(a) = I(Body(r_i))$ for all $1 \leq i \leq n$
4. For every $x \in Lit(Body(r_i))$ there exists a $j < i$ such that $x \in Lit(Head(r_j))$

We characterize the non-existence of self-reinforcing cyclic rules in a program using the following definition: for a FASP program $\mathcal{P}$, we say that $\mathcal{P}$ is self-reinforcing cycle free (SRCF) w.r.t. an atom a, iff we can find a stratification function $f : \mathcal{B}_{\mathcal{P}} \rightarrow \mathbb{N}$, such that for every rule $r \in \mathcal{P}$ which contains a, it holds that:

1. $f(x) \geq f(y)$ for every $x \in Lit(Head(r))$ and $y \in Lit(Body(r))$
2. If $r \equiv x \leftarrow y \oplus z$, then $f(x) > f(y)$ and $f(x) > f(z)$.

Intuitively, we can see that a program $\mathcal{P}$ is SRCF w.r.t. atom a iff the dependency graph does not contain any cycle which goes through a and involves at least one rule with disjunction in the body. We say that a program $\mathcal{P}$ is SRCF iff it is SRCF w.r.t every atom $a \in \mathcal{B}_{\mathcal{P}}$. The following theorem characterizes the notion of support for SRCF programs.

Theorem 1 (Support). *Let $\mathcal{P}$ be an SRCF program and let I be a consistent interpretation. Then I is an answer set of $\mathcal{P}$ iff:*

1. I is a model of $\mathcal{P}$.
2. Every $a \in \{x \mid I(x) > 0\}$ has a support in $\mathcal{P}$ w.r.t. I.

The proof runs parallel to the proof of Theorem 2.3 in [3] by noting that support plays a similar role for the answer sets of SRCF FASP programs as proof does for HCF ASP programs. Due to space constraints, we omit the details.

Example 3. *Consider program* $\mathcal{P}_5 = \{a \oplus b \leftarrow \bar{1}, c \leftarrow b \otimes \textbf{not } a, c \leftarrow a\}$. *It is clear that* $I = \{(a, 0.3), (b, 0.7), (c, 0.4)\}$ *is an answer set of* $\mathcal{P}_5$. *In accordance with Theorem 1, for each of* a, b *and* c, *we can take* $r_1 = a \oplus b \leftarrow \bar{1}, r_2 = c \leftarrow b \otimes \overline{0.7}$ *as the support of these atoms in* $\mathcal{P}_5$ *w.r.t.* I. *Furthermore, any* $J > I$ *obtained by increasing the truth value of* a, b *or* c *will not have a support for that atom. On the other hand, the non-SRCF program* $\mathcal{P}_6 = \{a \oplus b \leftarrow \bar{1}, a \leftarrow b, b \leftarrow a, a \leftarrow a \oplus a\}$ *has only one answer set, namely* $I = \{(a, 1), (b, 1)\}$. *One can check that there is no support for each of* a *and* b *in* $\mathcal{P}_6$ *w.r.t.* I, *since in this case,* $I(a) + I(b) > 1$.

The following lemma holds in both classical and fuzzy ASP.

Lemma 1 (Locality). *Let* $\mathcal{P}'$ *be any subset of a program* $\mathcal{P}$. *If* I *is an answer set of* $\mathcal{P}'$ *and it satisfies all the rules in* $\mathcal{P} - \mathcal{P}'$, *then* I *is also an answer set of* $\mathcal{P}$.

Proof. Since I is an answer set of $\mathcal{P}'$ and I satisfies every rule of $\mathcal{P} - \mathcal{P}'$, I also satisfies every rule in $\mathcal{P}$. Then clearly, I satisfies $\mathcal{P}^I$ as well. Suppose that I is not the minimal model of $\mathcal{P}^I$, i.e., that there is another model $J < I$ of $\mathcal{P}^I$. Since $\mathcal{P}' \subseteq \mathcal{P}$ (and hence $\mathcal{P}'^I \subseteq \mathcal{P}^I$), it must also be the case that J satisfies $\mathcal{P}'^I$. But this means that I is not the minimal model of $\mathcal{P}'^I$, contradicting the assumption that I is an answer set of $\mathcal{P}'$.

We now present the main result for this section.

Theorem 2. *Let* $\mathcal{P}_1 = \mathcal{P} \cup \{a \oplus b \leftarrow c\}$ *be any SRCF program w.r.t.* a, b *and* c. *Then, an interpretation* I *is an answer set of* $\mathcal{P}_1$ *iff it is also an answer set of* $\mathcal{P}_2 = \mathcal{P} \cup \{a \leftarrow c \otimes \textbf{not } b, b \leftarrow c \otimes \textbf{not } a\}$.

Proof. (a) "If"-part: Let I be an answer set of $\mathcal{P}_2$. Then I is a minimal model of $\mathcal{P}^I \cup \{a \leftarrow c \otimes (\overline{1 - I(b)}), b \leftarrow c \otimes (\overline{1 - I(a)})\}$. Clearly, we have $I(a) + I(b) \geq I(c)$. We consider two cases:

(i) $I(a) + I(b) = I(c)$. Let $p \in \{x \mid I(x) > 0\}$. By Theorem 1, there is a support R_p of p in $\mathcal{P}_2$ w.r.t I. If $\{a \leftarrow c \otimes (\overline{1 - I(b)}), b \leftarrow c \otimes (\overline{1 - I(a)})\} \cap R_p = \emptyset$, then we must have $R_p \subseteq \mathcal{P}^I$. This means that R_p is a support for p in $\mathcal{P}$ w.r.t I. On the other hand, if any (or both) of $\{a \leftarrow c \otimes (\overline{1 - I(b)}), b \leftarrow c \otimes (\overline{1 - I(a)})\}$ occurs in R_p, we can replace it (them) with the rule $a \oplus b \leftarrow c$, to obtain the set R'_p which can serve as a support for p in $\mathcal{P}_1$ w.r.t. I. In any case, each support R_p in $\mathcal{P}_2$ can be replaced with a support for p in $\mathcal{P}_1$. By Theorem 1, this means that every answer set of $\mathcal{P}_2$ is also an answer set of $\mathcal{P}_1$.

(ii) $I(a) + I(b) > I(c)$. In this case, we have that $\{a \leftarrow c \otimes (\overline{1 - I(b)}), b \leftarrow c \otimes (\overline{1 - I(a)})\} \not\subseteq R_p$ for any support R_p of any $p \in \{x \mid I(x) > 0\}$, since it does not satisfy the first condition in the definition of support. Therefore, we have that $R_p \subseteq \mathcal{P}$ for any p, which, by Theorem 1 means that I is also an answer set of $\mathcal{P}$. Since I definitely satisfies $a \oplus b \leftarrow c$, by Lemma 1, I is also an answer set of $\mathcal{P}_1$.

(b) "Only if"-part: Similar to the previous part, let I be an answer set of $\mathcal{P}_1$. Then I is a minimal model of $\mathcal{P}^I \cup \{a \oplus b \leftarrow c\}$, and also $I(a) + I(b) \geq I(c)$. As before, we consider two cases:

(i) $I(a) + I(b) = I(c)$. Let $p \in \{x \mid I(x) > 0\}$. By Theorem 1, there is a support R_p of p in $\mathcal{P}_2$ w.r.t I. If $a \oplus b \leftarrow c \notin R_p$, then we must have $R_p \subseteq \mathcal{P}^I$, which means that R_p is a support for p w.r.t. $\mathcal{P}$. On the other hand, if $a \oplus b \leftarrow c \in R_p$, we can replace it with the two rules $\{a \leftarrow c \otimes \overline{1 - I(b)}, b \leftarrow c \otimes \overline{1 - I(a)}\}$ to obtain the set R'_p which can serve as a support for p in $\mathcal{P}_2$ w.r.t I. In any case, each support R_p in $\mathcal{P}_1$ w.r.t. I can be replaced with a support for p in $\mathcal{P}_2$ w.r.t I. By Theorem 1, this means that every answer set of $\mathcal{P}_1$ is also an answer set of $\mathcal{P}_2$.

(ii) $I(a) + I(b) > I(c)$. Similar to case (a)(ii), here we have that $a \oplus b \leftarrow c \notin R_p$ for any support R_p of any $p \in \{x \mid I(x) > 0\}$. Again, using Theorem 1, we get that every answer set of $\mathcal{P}_1$ is also an answer set of $\mathcal{P}_2$.

This result allows us to reduce an SRCF disjunctive FASP program to an equivalent normal program by performing the shifting operations, thus allowing the use of evaluation methods geared towards normal programs.

Example 4. *The program $\mathcal{P}_5$ with the following rules:*

$$\{a \oplus b \leftarrow \overline{1}, c \leftarrow b, c \leftarrow d \oplus e\}$$

is SRCF, since we can assign the stratification function $f(a) = f(b) = f(d) = f(e) = 1$ and $f(c) = 2$. Hence, by Theorem 2, it is equivalent to the normal program:

$$\{a \leftarrow \textbf{not}\, b, b \leftarrow \textbf{not}\, a, c \leftarrow b, c \leftarrow d \oplus e\}$$

However, program $\mathcal{P}_5 \cup \{d \leftarrow c\}$ is not SRCF, and the shifting method does not work.

As a corollary of Theorem 2, any strict disjunctive FASP program can be reduced to a normal FASP program by shifting.

Example 5. *Consider program $\mathcal{P}_2$ from Example 2. It is a strict disjunctive FASP program, and hence it can be reduced to the equivalent normal program $\{a \leftarrow \textbf{not}\, b, b \leftarrow \textbf{not}\, a, a \leftarrow b, b \leftarrow a\}$.*

3.2 Non-SRCF Programs

For non-SRCF programs, finding an answer set in $\mathcal{L}_\infty$ requires finding an answer set I in $\mathcal{L}_k$ for some $k \geq 1$, and checking whether I is also an answer set for $\mathcal{L}_\infty$. We show in this section how the last step can be implemented using Mixed Integer Programming (MIP). For some background on MIP, one can consult, e.g., [17,29].

In [16], a representation of infinitely-valued Łukasiewicz logic using MIP was proposed by defining a translation of each of the Łukasiewicz expressions $x \oplus y$, $x \otimes y$ and $\neg x$ into a set of MIP inequality constraints characterizing the value of each of the expressions. Given a FASP program $\mathcal{P}$ and an interpretation I, we can use the MIP representation of $\mathcal{P}$ (denoted as $MIP(\mathcal{P})$) based on the representations proposed by [16] to check whether I is the minimal model of $\mathcal{P}^I$, as follows:

1. For each atom a in $\mathcal{P}^I$, we use a MIP variable $v_a \in [0,1]$ in $MIP(\mathcal{P})$ to express the truth value that a can take.
2. For any expression $e \in \{a \oplus b, a \otimes b, a \veebar b, a \barwedge b\}$ in any rule of $\mathcal{P}^I$, we create the appropriate set of constraints in MIP to represent the value of the expression, as suggested in [16]. For example, for $a \oplus b$, we have the following MIP constraints:

$$v_a + v_b + z_{a \oplus b} \geq v_{a \oplus b}$$
$$v_a + v_b - z_{a \oplus b} \leq v_{a \oplus b}$$
$$v_a + v_b - z_{a \oplus b} \geq 0$$
$$v_a + v_b - z_{a \oplus b} \leq 1$$
$$v_{a \oplus b} \geq z_{a \oplus b}$$

In each case, z_e is a 0–1 variable and v_e is a variable representing the value of the expression e.
3. For each rule $\alpha \leftarrow \beta \in \mathcal{P}^I$, we add the constraint $v_\alpha \geq v_\beta$, where v_α and v_β are the variables corresponding to the values of the atoms/expressions α and β, respectively.
4. For each atom a, we add the constraint $v_a \leq I(a)$.
5. We set the objective function of the MIP program to minimise the value $\sum_{a \in \mathcal{B}_\mathcal{P}} v_a$.

Theorem 3. *The interpretation I is the minimal model of $\mathcal{P}^I$ iff the solution returned in $MIP(\mathcal{P})$ is equal to I.*

3.3   Incorporating Program Decomposition

While in practical applications FASP programs will not always be SRCF, often it will be possible to decompose programs such that many of the resulting components are SRCF. In this section, we show how we can apply the reduction from Sect. 3.1 to these individual components, and thus efficiently solve the overall program.

Program modularity and decomposition using dependency analysis have been extensively studied and implemented in classical ASP. In [21], the concept of splitting sets for decomposing an ASP program was introduced. Dependency analysis and program decomposition using strongly connected components (SCC) was described in [10,27], and has been used as a framework for efficient evaluation of logic programs, such as in [9,12,26]. In this section, we build on this idea to develop a more efficient evaluation framework for FASP programs by exploiting the program's modularity/decomposability.

For a (ground) FASP program $\mathcal{P}$, consider a directed graph $G_\mathcal{P} = \langle V, E \rangle$, called the *dependency graph* of $\mathcal{P}$, defined as follows: (i) $V = \mathcal{B}_\mathcal{P}$ and (ii) $(a, b) \in E$ iff there exists a rule $r \in \mathcal{P}$ s.t. $a \in Lit(Body(r))$ and $b \in Lit(Head(r))$. By SCC analysis, we can decompose $G_\mathcal{P}$ into SCCs $C_1, \ldots, C_n$. With each SCC C_i, we associate a *program component* $PC_i \subseteq \mathcal{P}$, defined as the maximal set of rules such that for every $r \in PC_i$, the literals in PC_i are contained in C_i. The set of all program components of $\mathcal{P}$ is denoted as $PC(\mathcal{P})$. We define dependency between program components PC_i and PC_j as follows: PC_i depends on PC_j iff there is an atom a in PC_i and atom b in PC_j such that a depends on b in $G_{\mathcal{P}}$. The program component graph $C = \langle PC(\mathcal{P}), E_C \rangle$ is defined according to the dependency relation between the program components.

Similar to the case in classical ASP, the program component graph of a FASP program allows us to decompose the program into "modular components" that can be separately evaluated. For non-disjunctive components, the evaluation method described in [25] can be directly used. For SRCF disjunctive components, we can perform the shifting method as described in Sect. 3.1 to reduce the component into a normal program, and again use the evaluation method for normal programs. For non-SRCF disjunctive components, an extra minimality check as defined in Sect. 3.2 is needed after finding a k-answer set. Evaluation proceeds along the program components according to the topological sorting of the components in the program component graph, feeding the "partial answer sets" obtained from one component into the next. If a "complete answer set" is found, we stop. Otherwise, we backtrack to the previous component(s), obtaining k-answer sets for the next values of k until the stopping criterion is met.

Proposition 1. *Label an edge in $G_{\mathcal{P}}$ with the symbol $\oplus$ if the edge corresponds to a rule containing a disjunction in the body. A component is non-SRCF w.r.t. the atoms in that component iff there is a cycle in the component containing a labelled edge.*

Example 6. *Consider the program $\mathcal{P}_6$ containing the rules:*

$$\{a \leftarrow b \oplus c, b \leftarrow a \otimes \overline{0.5}, c \leftarrow \overline{0.7}, d \oplus e \leftarrow a\}$$

Program $\mathcal{P}_6$ is not SRCF, hence we cannot directly use Theorem 2 to perform shifting. However, using SCC program decomposition, we obtain three components $PC_1 = \{a \leftarrow b \oplus c, b \leftarrow a \otimes \overline{0.5}\}$, $PC_2 = \{c \leftarrow \overline{0.7}\}$ and $PC_3 = \{d \oplus e \leftarrow a\}$. PC_1 only depends on PC_2, PC_2 has no dependencies, while PC_3 depends only on PC_1. Proceeding according to the topological order of the program components, we start by evaluating PC_2 and obtain the partial answer set $\{(c, 0.7)\}$. We feed this partial answer set into the next component PC_1. This program is normal and hence requires no minimality check associated to disjunctive programs. We obtain the partial answer set $\{(a, 1), (b, 0.5), (c, 0.7)\}$. The last component PC_3 is disjunctive, but it is also SRCF w.r.t. its atoms, and hence we can perform shifting to obtain the normal program $\{d \leftarrow a \otimes \textbf{not } e, e \leftarrow a \otimes \textbf{not } d\}$, which again can be evaluated without using minimality checks.

4   Experimental Benchmark

In this section, we experimentally evaluate the effectiveness of the proposed method. We implemented our method on top of the solver developed in [25][3].

---

[3] https://github.com/mushthofa/ffasp.

We used clingo from the Potassco project [14] as the underlying ASP solver for finding k-answer sets, and the Coin-OR Cbc[4] solver as the MIP program solver for minimality checking.

For this benchmark, we used two problems: (1) the fuzzy graph colorability as given in the introduction, and (2) the fuzzy set covering problem, which is a generalization of the classical set covering problem, defined as follows: A fuzzy set F is defined as a function $F : U \to [0, 1]$, where U is the universe of discourse, and $F(u)$ for $u \in U$ is the degree of membership of u in F. A fuzzy subset S of F is a fuzzy set such that $S(u) \leq F(u), \forall u \in U$. Given a fuzzy set F and a collection of subsets $C = \{S_1, \ldots, S_n\}$ of F, the problem asks whether we can find a fuzzy sub-collection of C, such that every member of F is covered *sufficiently* by the subsets selected from C, and that the degree to which a subset S_i is selected is below a given threshold. We encode the problem in FASP as follows: the fuzzy set F is given by a set of facts of the form $f(x) \leftarrow \overline{a}$, the subsets S_i given by facts of the form $subset(s_i)$ and their membership degrees by $member(s_i, x) \leftarrow \overline{b}$. The maximum degree to which a subset S_i can be selected is denoted by a constraint $\overline{c} \leftarrow in(s_i)$. The following FASP program encodes the problem goal and constraints:

$$in(S) \oplus out(S) \leftarrow subset(S)$$
$$covered(X) \leftarrow (in(s_1) \otimes member(s_1, X)) \oplus \ldots \oplus (in(s_n) \otimes member(s_n, X))$$
$$\overline{0} \leftarrow f(X) \otimes \textbf{not } covered(X)$$

For both benchmark problems, instances are generated randomly with no attempt to produce "hard" instances. Constant truth values for fuzzy facts (e.g., for edge weights) are drawn randomly from the set $\mathbb{Q}_{10}$. Two types of instances are considered: (1) where no saturation rules are present, which means that the program is an SRCF program, and (2) where the saturation rules are added randomly with a 0.1 probability for each $b(x)$ and $w(x)$ atoms (in the graph coloring problem instances) and each $in(x)$ atoms (in the set covering problem instances). Since fuzzy answer set evaluation using finite-valued translation such as the one used in [25] cannot, in principle, be used to prove inconsistency, we opted to generate only instances that are known to be satisfiable.

To be able to see the advantage of applying our approach, we run the solver on all instances both with and without employing SRCF detection and shifting to reduce to normal programs. When SRCF detection is not employed, a minimality check has to be performed to verify that the answer sets obtained in any disjunctive component of the program are indeed minimal. Thus, our experiment will be useful to see the effectiveness of the proposed reduction over the baseline method of computing answer sets and checking for minimality.

The experiment was conducted on a Macbook with OS X version 10.8.5 running on Intel Core i5 2.4 GHz with 4 GB of memory. Execution time for each instance is limited to 2 min, while memory usage is limited to 1 GB. Table 1 presents the results of the experiment. Each value is an average over ten repeats.

---

[4] https://projects.coin-or.org/Cbc.

Table 1. Values in the cells indicate average execution times (over ten instances) in seconds for the non-timed-out executions. Cells labeled with '(TO)' indicates that all executions of corresponding instances exceeded time/memory limit.

Problem	Fuzzy graph coloring					Fuzzy set cover				
Saturation		no		yes			no		yes	
Method		δ	σ	δ	σ		δ	σ	δ	σ
1	$n = 20$	2.9	1.7	2.7	1.8	$n = 10$	7.9	5.2	8.2	7.9
2	$n = 30$	6.6	3.8	6.1	3.8	$n = 15$	13.4	6.5	17.3	17.1
3	$n = 40$	10.6	5.7	10.7	6.1	$n = 20$	18.2	9.6	17.4	17.3
4	$n = 50$	19.8	11.5	23.0	11.0	$n = 25$	29.8	13.4	30.1	29.9
5	$n = 60$	34.8	17.7	36.0	20.4	$n = 30$	71.4	17.6	71.4	70.6
6	$n = 70$	53.4	25.4	55.3	28.2	$n = 35$	(TO)	22.3	(TO)	(TO)
7	$n = 80$	74.8	33.9	76.1	41.1	$n = 40$	(TO)	27.8	(TO)	(TO)

δ = no shifting, σ = with shifting applied

From the result, we can see that when SRCF detection and shifting are used, execution times are generally lower than when only minimality checks are used, even when saturation rules are present. This is especially true for the instances of the fuzzy graph coloring problem. The use of program decomposition/modularity analysis to separate program components that are SRCF from those that are non-SRCF can be beneficial since this means we can isolate the need for minimality checks to only those non-SRCF components, while the rest can be evaluated as normal programs after performing the shifting operation. For the set covering problem, we see no significant improvement in the running time when using the shifting operator for instances with saturation, one of the reasons being that the instances are such that minimality checks are needed regardless of what method is used (due to the fact that most components are non-SRCF). However, for the non-saturated instances, again we still see a clear advantage of using SRCF detection and shifting.

5   Conclusion

In this paper, we have identified a large class of disjunctive FASP programs, called SRCF programs, which can be efficiently evaluated by reducing them to equivalent normal FASP programs. We also proposed a method to perform a minimality check to determine the answer sets of non-SRCF programs based on a MIP formulation, and we showed how we can decompose FASP programs to further increase the applicability of our approach. We have implemented our approach on top of a current FASP solver, and integrated a MIP solver into the system to allow efficient minimality checking. To the best of our knowledge, our implementation represents the first FASP solver to allow evaluation rules with disjunctions in the body and/or in the head. We have also performed a benchmark testing of the proposed methods to measure their computational efficiency. Our result indicates that identifying SRCF components of a FASP program allows us to evaluate the program more efficiently.

References

1. Alviano, M., Peñaloza, R.: Fuzzy answer sets approximations. Theor. Pract. Logic Program. **13**(4–5), 753–767 (2013)
2. Baral, C.: Knowledge Representation, Reasoning and Declarative Problem Solving. Cambridge University Press, New York (2003)
3. Ben-Eliyahu, R., Dechter, R.: Propositional semantics for disjunctive logic programs. Ann. Math. Artif. Intell. **12**(1–2), 53–87 (1994)
4. Blondeel, M., Schockaert, S., Vermeir, D., De Cock, M.: Fuzzy answer set programming: an introduction. In: Yager, R.R., Abbasov, A.M., Reformat, M., Shahbazova, S.N. (eds.) Soft Computing: State of the Art Theory. STUDFUZZ, vol. 291, pp. 209–222. Springer, Heidelberg (2013)
5. Blondeel, M., Schockaert, S., Vermeir, D., De Cock, M.: Complexity of fuzzy answer set programming under Łukasiewicz semantics. Int. J. Approximate Reasoning **55**(9), 1971–2003 (2014)
6. Viegas Damásio, C., Moniz Pereira, L.: Antitonic logic programs. In: Eiter, T., Faber, W., Truszczyński, M. (eds.) LPNMR 2001. LNCS (LNAI), vol. 2173, pp. 379–393. Springer, Heidelberg (2001)
7. Dantsin, E., Eiter, T., Gottlob, G., Voronkov, A.: Complexity and expressive power of logic programming. ACM Comput. Surv. **33**(3), 374–425 (2001)
8. Dix, J., Gottlob, G., Marek, W.: Reducing disjunctive to non-disjunctive semantics by shift-operations. Fundamenta Informaticae **28**(1), 87–100 (1996)
9. Eiter, T., Faber, W., Mushthofa, M.: Space efficient evaluation of ASP programs with bounded predicate arities. In: 24th AAAI Conference on Artificial Intelligence, pp. 303–308 (2010)
10. Eiter, T., Gottlob, G., Mannila, H.: Disjunctive datalog. ACM Trans. Database Syst. **22**(3), 364–418 (1997)
11. Eiter, T., Ianni, G., Krennwallner, T.: Answer set programming: a primer. In: Tessaris, S., Franconi, E., Eiter, T., Gutierrez, C., Handschuh, S., Rousset, M.-C., Schmidt, R.A. (eds.) Reasoning Web. LNCS, vol. 5689, pp. 40–110. Springer, Heidelberg (2009)
12. Eiter, T., Ianni, G., Schindlauer, R., Tompits, H.: DLVHEX: a prover for semantic-web reasoning under the answer-set semantics. In: Proceedings of the IEEE/WIC/ACM International Conference on Web Intelligence, pp. 1073–1074 (2006)
13. Erdem, E.: Theory and Applications of Answer Set Programming. Ph.D. thesis. the University of Texas at Austin (2002)
14. Gebser, M., Kaufmann, B., Kaminski, R., Ostrowski, M., Schaub, T., Schneider, M.: Potassco: the Potsdam answer set solving collection. AI Commun. **24**(2), 107–124 (2011)
15. Gelfond, M., Lifschitz, V.: The stable model semantics for logic programming. In: Proceedings of the Fifth International Conference and Symposium on Logic Programming, vol. 88, pp. 1070–1080 (1988)
16. Hähnle, R.: Proof theory of many-valued logic-linear optimization-logic design: connections and interactions. Soft Comput. **1**(3), 107–119 (1997)

17. Jeroslow, R.G.: Logic-Based Decision Support: Mixed Integer Model Formulation. Elsevier, Amsterdam (1989)
18. Lee, J., Wang, Y.: Stable models of fuzzy propositional formulas. In: Fermé, E., Leite, J. (eds.) JELIA 2014. LNCS, vol. 8761, pp. 326–339. Springer, Heidelberg (2014)
19. Leone, N., Pfeifer, G., Faber, W., Eiter, T., Gottlob, G., Perri, S., Scarcello, F.: The DLV system for knowledge representation and reasoning. ACM Trans. Comput. Logic **7**(3), 499–562 (2006)
20. Lifschitz, V.: What is answer set programming? In: Proceedings of the 23rd AAAI Conference in Artificial Intelligence, vol. 8, pp. 1594–1597 (2008)
21. Lifschitz, V., Turner, H.: Splitting a logic program. In: Proceedings of the 11th International Conference on Logic Programming, pp. 23–37 (1994)
22. Loyer, Y., Straccia, U.: Epistemic foundation of stable model semantics. Theor. Pract. Logic Program. **6**(4), 355–393 (2006)
23. Lukasiewicz, T., Straccia, U.: Tightly integrated fuzzy description logic programs under the answer set semantics for the semantic web. In: Marchiori, M., Pan, J.Z., Marie, C.S. (eds.) RR 2007. LNCS, vol. 4524, pp. 289–298. Springer, Heidelberg (2007)
24. Madrid, N., Ojeda-Aciego, M.: Measuring inconsistency in fuzzy answer set semantics. IEEE Trans. Fuzzy Syst. **19**(4), 605–622 (2011)
25. Mushthofa, M., Schockaert, S., De Cock, M.: A finite-valued solver for disjunctive fuzzy answer set programs. In: Proceedings of European Conference in Artificial Intelligence 2014, pp. 645–650 (2014)
26. Oikarinen, E., Janhunen, T.: Achieving compositionality of the stable model semantics for smodels programs. Theor. Pract. Logic Program. **8**(5–6), 717–761 (2008)
27. Ross, K.A.: Modular stratification and magic sets for datalog programs with negation. In: Proceedings of the 9th ACM SIGACT-SIGMOD-SIGART Symposium on Principles of Database Systems, pp. 161–171 (1990)
28. Schockaert, S., Janssen, J., Vermeir, D.: Fuzzy equilibrium logic: declarative problem solving in continuous domains. ACM Trans. Comput. Logic **13**(4), 33:1–33:39 (2012)
29. Schrijver, A.: Theory of Linear and Integer Programming. Wiley, Chichester (1998)
30. Van Nieuwenborgh, D., De Cock, M., Hadavandi, E.: Fuzzy answer set programming. In: Fisher, M., van der Hoek, W., Konev, B., Lisitsa, A. (eds.) JELIA 2006. LNCS (LNAI), vol. 4160, pp. 359–372. Springer, Heidelberg (2006)

Characterising and Explaining Inconsistency in Logic Programs

Claudia Schulz[1]([⊠]), Ken Satoh[2], and Francesca Toni[1]

[1] Department of Computing, Imperial College London, London SW7 2AZ, UK
{claudia.schulz,f.toni}@imperial.ac.uk
[2] National Institute of Informatics, Tokyo 101-8430, Japan
ksatoh@nii.ac.jp

Abstract. A logic program under the answer set semantics can be inconsistent because its only answer set is the set of all literals, or because it does not have any answer sets. In both cases, the reason for the inconsistency may be (1) only explicit negation, (2) only negation as failure, or (3) the interplay between these two kinds of negation. Overall, we identify four different inconsistency cases, and show how the respective reason can be further characterised by a set of culprits using semantics which are weaker than the answer set semantics. We also provide a technique for explaining the set of culprits in terms of trees whose nodes are derivations. This can be seen as an important first step towards debugging inconsistent logic programs.

Keywords: Logic programming · Inconsistency · Explanation

1 Introduction

A *logic program* represents knowledge in the form of rules made of statements which can be negated in two ways: using *explicit negation*, expressing that the statement does not hold, or *negation as failure* (NAF), expressing that the statement cannot be proven to hold. If no negation of either kind is present, a logic program will always be consistent under the *answer set* semantics [8]. However, if negation is used in a logic program, inconsistency may arise in one of two different ways: either the only answer set of the logic program is the set of all literals, or the logic program has no answer sets at all.

Efficient solvers have been developed for computing the answer sets of a given logic program [6,10,11]. However, in the case of an inconsistent logic program these solvers do not provide any classification of the inconsistency, or explanation thereof. Especially when dealing with a large inconsistent logic program or if the inconsistency is unexpected, understanding why the inconsistency arises and which part of the logic program is responsible for it is an important first step towards debugging the logic program in order to restore consistency. Various approaches have been developed for finding the source of inconsistency and even for suggesting ways of debugging the logic program. In particular, [7,12,17]

© Springer International Publishing Switzerland 2015
F. Calimeri et al. (Eds.): LPNMR 2015, LNAI 9345, pp. 467–479, 2015.
DOI: 10.1007/978-3-319-23264-5_39

feed the inconsistent logic program into a meta logic program whose answer set describes the structure of and potential mistakes in the original logic program. In another recent approach [13], the user assigns truth values to literals in the inconsistent logic program step-by-step until encountering a conflict.

These debugging approaches assume explicitly or implicitly the existence of an intended answer set. We propose a new method for identifying the reason of inconsistency in a logic program without the need of an intended answer set, based on the well-founded [19] and M-stable [3] model semantics. These semantics are "weaker" than answer sets in that they are 3-valued rather than 2-valued. In contrast to some previous approaches [1,7,17], we consider logic programs that may comprise both explicit negation and NAF. We prove that the two ways in which a logic program may be inconsistent are further divided into four inconsistency cases which have different reasons for the inconsistency: one where only explicit negation is responsible and the only answer set is the set of all literals, one where only NAF is responsible and the logic program has no answer sets, and two where an interplay of explicit negation and NAF is responsible and the logic program has no answer sets. We show how in each of these inconsistency cases the reason of the inconsistency can be refined to a characteristic set of "culprits". These "culprits" can then be used to construct trees whose nodes hold derivations. These trees explain why the inconsistency arises and which part of the logic program is responsible. Furthermore, we show how the inconsistency case and the respective set of culprits can be determined using the aforementioned "weaker" semantics.

2   Background

A *logic program* $\mathcal{P}$ is a (finite) set of ground clauses[1] of the form $l_0 \leftarrow l_1, \ldots, l_m,$ $not\ l_{m+1}, \ldots, not\ l_{m+n}$ with $m, n \geq 0$. All l_i $(1 \leq i \leq m)$ and all l_j $(m+1 \leq j \leq m+n)$ are classical literals, i.e. atoms a or *explicitly negated atoms* $\neg a$, and $not\ l_j$ are *negation as failure* (NAF) literals. We will use the following notion of dependency inspired by [20]: l_0 is *positively dependent* on l_i and *negatively dependent* on l_j. A *dependency path* is a chain of positively or negatively dependent literals. A *negative dependency path* is obtained from a dependency path by deleting all literals l in the path such that some k in the path is positively dependent on l, e.g. if p, q, r is a dependency path where p is positively dependent on q, and q is negatively dependent on r then p, r is a negative dependency path.

$\mathcal{HB}_\mathcal{P}$ is the *Herbrand Base* of $\mathcal{P}$, i.e. the set of all ground atoms of $\mathcal{P}$, and $Lit_\mathcal{P} = \mathcal{HB}_\mathcal{P} \cup \{\neg a \mid a \in \mathcal{HB}_\mathcal{P}\}$ consists of all classical literals of $\mathcal{P}$. $NAF_{\mathcal{HB}_\mathcal{P}} = \{not\ a \mid a \in \mathcal{HB}_\mathcal{P}\}$ consists of all NAF literals of atoms of $\mathcal{P}$ and $NAF_{Lit_\mathcal{P}} = \{not\ l \mid l \in Lit_\mathcal{P}\}$ of all NAF literals of classical literals of $\mathcal{P}$. An atom a and the explicitly negated atom $\neg a$ are called *complementary literals*.

$\vdash_{MP}$ denotes derivability using *modus ponens* on $\leftarrow$ as the only inference rule, treating $l \leftarrow$ as $l \leftarrow true$, where $\mathcal{P} \vdash_{MP} true$ for any $\mathcal{P}$. For a logic program

---

[1] Clauses containing variables are used as shorthand for all their ground instances over the Herbrand Universe of the logic program.

$\mathcal{P}$ and $\Delta \subseteq NAF_{Lit_{\mathcal{P}}}$, $\mathcal{P} \cup \Delta$ denotes $\mathcal{P} \cup \{not\ l \leftarrow |not\ l \in \Delta\}$. When used on such $\mathcal{P} \cup \Delta$, $\vdash_{MP}$ treats NAF literals syntactically as in [4] and thus $\mathcal{P} \cup \Delta$ can be seen as a logic program. $l \in Lit_{\mathcal{P}}$ is *strictly derivable* from $\mathcal{P}$ iff $\mathcal{P} \vdash_{MP} l$, and *defeasibly derivable* from $\mathcal{P}$ iff $\mathcal{P} \nvdash_{MP} l$ and $\exists \Delta \subseteq NAF_{Lit_{\mathcal{P}}}$ such that $\mathcal{P} \cup \Delta \vdash_{MP} l$. l is *derivable* from $\mathcal{P}$ iff l is strictly or defeasibly derivable from $\mathcal{P}$.

Answer Sets [8]. Let $\mathcal{P}$ be a logic program without NAF literals. The *answer set* of $\mathcal{P}$, denoted $\mathcal{AS}(\mathcal{P})$, is the smallest set $S \subseteq Lit_{\mathcal{P}}$ such that:

(1) for any clause $l_0 \leftarrow l_1, \ldots, l_m$ in $\mathcal{P}$: if $l_1, \ldots, l_m \in S$ then $l_0 \in S$; and
(2) $S = Lit_{\mathcal{P}}$ if S contains complementary literals.

For a logic program $\mathcal{P}$, possibly with NAF literals, and any $S \subseteq Lit_{\mathcal{P}}$, the *reduct* $\mathcal{P}^S$ is obtained from $\mathcal{P}$ by deleting:

– all clauses containing *not l* where $l \in S$, and
– all NAF literals in the remaining clauses.

Then S is an answer set of $\mathcal{P}$ if it is the answer set of the reduct $\mathcal{P}^S$, i.e. if $S = \mathcal{AS}(\mathcal{P}^S)$. $\mathcal{P}$ is *inconsistent* if it has no answer sets or if its only answer set is $Lit_{\mathcal{P}}$, else it is *consistent*.

3-Valued Models [15]. Let $\mathcal{P}$ be a logic program with no explicitly negated atoms. A *3-valued interpretation* of $\mathcal{P}$ is a pair $\langle \mathcal{T}, \mathcal{F} \rangle$, where $\mathcal{T}, \mathcal{F} \subseteq \mathcal{HB}_{\mathcal{P}}$, $\mathcal{T} \cap \mathcal{F} = \emptyset$, and $\mathcal{U} = \mathcal{HB}_{\mathcal{P}} \backslash (\mathcal{T} \cup \mathcal{F})$. The *truth value* of $a \in \mathcal{HB}_{\mathcal{P}}$ and *not a* $\in NAF_{\mathcal{HB}_{\mathcal{P}}}$ with respect to $\langle \mathcal{T}, \mathcal{F} \rangle$ is:
$val(a) = T$, if $a \in \mathcal{T}$; $val(not\ a) = T$, if $a \in \mathcal{F}$;
$val(a) = F$, if $a \in \mathcal{F}$; $val(not\ a) = F$, if $a \in \mathcal{T}$;
$val(a) = U$, if $a \in \mathcal{U}$; $val(not\ a) = U$, if $a \in \mathcal{U}$;
The truth values are ordered by $T > U > F$ and naturally $val(T) = T$, $val(F) = F$, and $val(U) = U$. A 3-valued interpretation $\langle \mathcal{T}, \mathcal{F} \rangle$ *satisfies* a clause $a_0 \leftarrow a_1, \ldots, a_m, not\ a_{m+1}, \ldots, not\ a_{m+n}$ if $val(a_0) \geq min\{val(a_1), \ldots, val(a_{m+n})\}$. $\langle \mathcal{T}, \mathcal{F} \rangle$ satisfies $a_0 \leftarrow$ if $val(a_0) = T$. The *partial reduct* $\frac{\mathcal{P}}{\langle \mathcal{T}, \mathcal{F} \rangle}$ of $\mathcal{P}$ with respect to a 3-valued interpretation $\langle \mathcal{T}, \mathcal{F} \rangle$ is obtained by replacing each NAF literal in every clause of $\mathcal{P}$ by its respective truth value.

– A 3-valued interpretation $\langle \mathcal{T}, \mathcal{F} \rangle$ of $\mathcal{P}$ is a *3-valued model* of $\mathcal{P}$ iff $\langle \mathcal{T}, \mathcal{F} \rangle$ satisfies every clause in $\mathcal{P}$.
– A 3-valued model $\langle \mathcal{T}, \mathcal{F} \rangle$ of $\mathcal{P}$ is a *3-valued stable model* of $\mathcal{P}$ iff it is a minimal 3-valued model of $\frac{\mathcal{P}}{\langle \mathcal{T}, \mathcal{F} \rangle}$, i.e. if $\nexists \langle \mathcal{T}_1, \mathcal{F}_1 \rangle$ which is a 3-valued model of $\frac{\mathcal{P}}{\langle \mathcal{T}, \mathcal{F} \rangle}$ such that $\mathcal{T}_1 \subseteq \mathcal{T}$ and $\mathcal{F}_1 \supseteq \mathcal{F}$ and $\mathcal{T} \neq \mathcal{T}_1$ or $\mathcal{F} \neq \mathcal{F}_1$.
– A 3-valued stable model $\langle \mathcal{T}, \mathcal{F} \rangle$ of $\mathcal{P}$ is the *well-founded model* of $\mathcal{P}$ if $\mathcal{U}$ is maximal (w.r.t. $\subseteq$) among all 3-valued stable models of $\mathcal{P}$. The well-founded model always exists for logic programs without explicitly negated atoms.
– A 3-valued stable model $\langle \mathcal{T}, \mathcal{F} \rangle$ of $\mathcal{P}$ is a *3-valued M-stable model* (Maximal-stable) of $\mathcal{P}$ if $\nexists \langle \mathcal{T}_1, \mathcal{F}_1 \rangle$ which is a 3-valued stable model of $\mathcal{P}$ such that $\mathcal{T} \subseteq \mathcal{T}_1$ and $\mathcal{F} \subseteq \mathcal{F}_1$ and $\mathcal{T} \neq \mathcal{T}_1$ or $\mathcal{F} \neq \mathcal{F}_1$ [3].

The *translated logic program* $\mathcal{P}'$ of a logic program $\mathcal{P}$ is obtained by substituting every explicitly negated atom $\neg a$ in $\mathcal{P}$ with a new atom $a' \notin \mathcal{HB}_\mathcal{P}$ [8,15]. Then a' (resp. a) is the *translated literal* of the *original literal* $\neg a$ (resp. a). The 3-valued stable models of a logic program $\mathcal{P}$, possibly containing explicitly negated atoms, are defined in terms of the 3-valued stable models of $\mathcal{P}'$ [15]. For every 3-valued stable model $\langle \mathcal{T}', \mathcal{F}' \rangle$ of $\mathcal{P}'$ the *corresponding 3-valued stable model* $\langle \mathcal{T}, \mathcal{F} \rangle$ of $\mathcal{P}$ is obtained from $\langle \mathcal{T}', \mathcal{F}' \rangle$ by replacing every translated literal by its original literal. The *3-valued stable models* of $\mathcal{P}$ are those corresponding 3-valued stable models where $\mathcal{T}$ does not contain complementary literals. Note that $\mathcal{P}'$ always has a well-founded model but that $\mathcal{P}$ might not have a well-founded model. Furthermore, note that a 3-valued stable model of $\mathcal{P}$ is an answer set of $\mathcal{P}$ iff $\mathcal{U} = \emptyset$.

From here onwards, and if not stated otherwise, we assume as given an inconsistent logic program $\mathcal{P}$ and the translated logic program $\mathcal{P}'$, where a', a'_i, a are the translated literals of $\neg a$, $\neg a_i$, and a, respectively.

3   Characterising the Type of Inconsistency

We first show how to identify in which way a logic program is inconsistent, i.e. if its only answer set is the set of all literals or if it has no answer sets at all, assuming that we only know what an answer set solver gives us, i.e. that the logic program is inconsistent. This identification is based on whether or not the logic program has a well-founded model, which can be computed in polynomial time [19]. Our results show that even though a logic program can only be inconsistent in two ways, in fact there are three different inconsistency cases which arise due to different reasons (see Sect. 4). The three inconsistency cases are:

- $\mathcal{P}$ has no well-founded model and:

  (1) the only answer set of $\mathcal{P}$ is $Lit_\mathcal{P}$;
  (2) $\mathcal{P}$ has no answer sets.

- $\mathcal{P}$ has a well-founded model and

  (3) $\mathcal{P}$ has no answer sets.

In the following, we prove that these three cases are the only ones, and characterise them in more detail.

Example 1. Let $\mathcal{P}_1$ be the following logic program:

$$p \leftarrow q \qquad\qquad q \leftarrow r,s \qquad\qquad r \leftarrow \qquad\qquad s \leftarrow$$
$$u \leftarrow not\ t \qquad\qquad t \leftarrow not\ u \qquad\qquad \neg p \leftarrow$$

$\mathcal{P}_1$ has no well-founded model and its only answer set is $Lit_{\mathcal{P}_1}$, so $\mathcal{P}_1$ falls into inconsistency case 1. The reason that the only answer set is $Lit_{\mathcal{P}_1}$ is that for any $S \subseteq Lit_{\mathcal{P}_1}$ satisfying the conditions of an answer set, $s, r, \neg p \in S$, then $q, p \in S$, and thus S contains the complementary literals p and $\neg p$. Note that NAF literals do not play any role in the inconsistency of $\mathcal{P}_1$; an atom and its explicitly negated atom, both strictly derivable, are responsible for the inconsistency.

The observations in Example 1 agree with a well-known result about logic programs whose only answer set it the set of all literals (Proposition 6.7 in [9]).

Lemma 1. *The only answer set of* $\mathcal{P}$ *is* $Lit_\mathcal{P}$ *iff* $\exists a \in \mathcal{HB}_\mathcal{P}$ *such that* $\mathcal{P} \vdash_{MP} a$ *and* $\mathcal{P} \vdash_{MP} \neg a$.

Example 2. Let $\mathcal{P}_2$ be the following logic program:

$$q \leftarrow not\ r \qquad \neg q \leftarrow \neg s, not\ p \qquad r \leftarrow not\ \neg t \qquad \neg s \leftarrow \qquad \neg t \leftarrow$$

$\mathcal{P}_2$ has no well-founded model and no answer sets, so $\mathcal{P}_2$ falls into inconsistency case 2. The reason that $\mathcal{P}_2$ has no answer sets is an interplay of explicit negation and NAF: for any $S \subseteq Lit_{\mathcal{P}_2}$ satisfying the conditions of an answer set, $\neg t, \neg s \in S$, and thus $r \leftarrow not\ \neg t$ is always deleted in $\mathcal{P}_2{}^S$ and both $q \leftarrow$ and $\neg q \leftarrow \neg s$ are always part of $\mathcal{P}_2{}^S$. Consequently, for any such S it holds that $q, \neg q \in \mathcal{AS}(\mathcal{P}_2{}^S)$, meaning that the only possible answer set is $Lit_{\mathcal{P}_2}$. However, since $r, p, \neg t \in Lit_{\mathcal{P}_2}$ the reduct will only consist of $\neg t \leftarrow$ and $\neg s \leftarrow$, so that $\mathcal{AS}(\mathcal{P}_2{}^{Lit_{\mathcal{P}_2}}) = \{\neg t, \neg s\}$ which does not contain complementary literals. Consequently, $\mathcal{P}_2$ has no answer sets at all. Even though both here and in $\mathcal{P}_1$ the inconsistency arises due to complementary literals, the difference lies in their derivations: here the complementary literals are defeasibly derivable, i.e. not only explicit negation but also NAF involved in the derivation is responsible for the inconsistency.

The following Theorem characterises inconsistency cases 1 and 2 illustrated in Examples 1 and 2.

Theorem 1. *If* $\mathcal{P}$ *has no well-founded model then*

1. *the only answer set of* $\mathcal{P}$ *is* $Lit_\mathcal{P}$ *iff* $\exists a \in \mathcal{HB}_\mathcal{P}$ *such that* $\mathcal{P} \vdash_{MP} a$ *and* $\mathcal{P} \vdash_{MP} \neg a$;
2. $\mathcal{P}$ *has no answer sets iff* $\nexists a \in \mathcal{HB}_\mathcal{P}$ *such that* $\mathcal{P} \vdash_{MP} a$ *and* $\mathcal{P} \vdash_{MP} \neg a$.

Proof. From Lemma 1.

Example 3. Let $\mathcal{P}_3$ be the following logic program:

$$r \leftarrow not\ s \qquad q \leftarrow not\ s \qquad p \leftarrow not\ r$$
$$s \leftarrow not\ r \qquad \neg q \leftarrow not\ s \qquad \neg p \leftarrow not\ r$$

The well-founded model of $\mathcal{P}_3$ is $\langle \emptyset, \emptyset \rangle$ but $\mathcal{P}_3$ has no answer sets, so it falls into inconsistency case 3. The reason that $\mathcal{P}_3$ has no answer sets is an interplay of explicit negation and NAF similar to Example 2. From the first two clauses, it follows that any potential answer set $S \subseteq Lit_{\mathcal{P}_3}$ cannot contain both s and r. If $r \notin S$ then $p, \neg p \in S$; if $s \notin S$ then $q, \neg q \in S$, and thus the only possible answer set is $Lit_{\mathcal{P}_3}$. However, $\mathcal{P}_3{}^{Lit_{\mathcal{P}_3}}$ is empty, so $\mathcal{AS}(\mathcal{P}_3{}^{Lit_{\mathcal{P}_3}}) = \emptyset$, which does not contain complementary literals. Thus, $\mathcal{P}_3$ has no answer sets. As in Example 2, the inconsistency is due to an atom and its explicitly negated atom being defeasibly derivable, but in contrast to $\mathcal{P}_2$ here the derivations of the complementary literals involve NAF literals which form an even-length negative dependency loop, namely s and r.

Theorem 2 characterises inconsistency case 3, illustrated in Example 3.

Theorem 2. *If $\mathcal{P}$ has a well-founded model then $\mathcal{P}$ has no answer sets.*

Proof. Assume that $\exists a \in \mathcal{HB}_{\mathcal{P}}$ s.t. $\mathcal{P} \vdash_{MP} a$ and $\mathcal{P} \vdash_{MP} \neg a$. Then a and a' are in the well-founded model of $\mathcal{P}'$ (by the alternating fixpoint definition of well-founded models [18]) and thus a and $\neg a$ are contained in the corresponding well-founded model of $\mathcal{P}$, so $\mathcal{P}$ has no well-founded model (contradiction). Thus, $\not\exists a \in \mathcal{HB}_{\mathcal{P}}$ s.t. $\mathcal{P} \vdash_{MP} a$ and $\mathcal{P} \vdash_{MP} \neg a$, so by Lemma 1 it is not the case that the only answer set of $\mathcal{P}$ is $Lit_{\mathcal{P}}$. Consequently, $\mathcal{P}$ has no answer sets.

In summary, if $\mathcal{P}$ has no well-founded model then its only answer set is $Lit_{\mathcal{P}}$ – caused by explicit negation – or it has no answer sets – caused by the interplay of explicit negation and NAF. If $\mathcal{P}$ has a well-founded model then it definitely has no answer sets – caused by the interplay of explicit negation and NAF.

4   Characterising Culprits

In the examples in Sect. 3, we already briefly discussed that the reasons for the inconsistency are different in the three inconsistency cases: either only explicit negation or the interplay of explicit negation and NAF. In this section, we show that inconsistency case 3 can in fact be further split into two sub-cases: one where the interplay of explicit negation and NAF is responsible as seen in Example 3 (case 3a), and one where only NAF is responsible for the inconsistency (case 3b). Furthermore, we characterise the different reasons of inconsistency in more detail in terms of "culprit" sets, which are sets of literals included in the well-founded (cases 1,2) or 3-valued M-stable (case 3b) model of $\mathcal{P}$, or in the answer sets of $\mathcal{P}'$ (case 3a). In other words, culprits can be found in "weaker" models.

Definition 1 (culprit sets). *Let $\mathcal{P}$, $\langle \mathcal{T}'_w, \mathcal{F}'_w \rangle$ be the well-founded model of $\mathcal{P}'$, $S'_1, \ldots, S'_n$ $(n \geq 0)$ its answer sets, and $\langle \mathcal{T}'_M, \mathcal{F}'_M \rangle$ one of its 3-valued M-stable models with $\mathcal{U}'_M$ the set of undefined atoms.*

– *If $\mathcal{P}$ has no well-founded model then*
  - *$\{a, \neg a\}$ is a culprit set of $\mathcal{P}$ iff $a, a' \in \mathcal{T}'_w$ and a and a' are strictly derivable from $\mathcal{P}'$ (**case 1**).*
  - *$\{a, \neg a\}$ is a culprit set of $\mathcal{P}$ iff $a, a' \in \mathcal{T}'_w$ and one of them is defeasibly derivable from $\mathcal{P}'$ and the other one is derivable from $\mathcal{P}'$ (**case 2**).*
– *If $\mathcal{P}$ has a well-founded model and*
  - *$\mathcal{P}'$ has n answer sets $(n \geq 1)$, then $\{a_1, \neg a_1, \ldots, a_n, \neg a_n\}$ is a culprit set of $\mathcal{P}$ iff $\forall a_i, \neg a_i$ $(1 \leq i \leq n)$: $a_i, a'_i \in S'_i$ and one of them is defeasibly derivable from $\mathcal{P}'$ and the other one is derivable from $\mathcal{P}'$ (**case 3a**).*
  - *$\mathcal{P}'$ has no answer sets, then C is a culprit set of $\mathcal{P}$ iff for some $a_1 \in \mathcal{U}'_M$ there exists a negative dependency path $a_1, \ldots, a_m, b_1, \ldots, b_o$ $(m, o \geq 1)$, in $\mathcal{P}'$ such that all a_h $(1 \leq h \leq m)$ and b_j $(1 \leq j \leq o)$ are in $\mathcal{U}'_M$, o is odd, $a_m = b_o$, and C consists of the original literals of the translated literals $b_1, \ldots, b_o$ (**case 3b**).*

We now show that for every inconsistency case at least one culprit set exists.

Example 4. The well-founded model of the translated logic program $\mathcal{P}_1{}'$ (see $\mathcal{P}_1$ in Example 1) is $\langle\{p, p', q, r, s\}, \emptyset\rangle$. $p, p' \in T'_w$ and both of them are strictly derivable from $\mathcal{P}'$. Thus, $\{p, \neg p\}$ is a culprit set of $\mathcal{P}_1$, which confirms our observation that $Lit_{\mathcal{P}_1}$ is the only answer set of $\mathcal{P}_1$ because every potential answer set contains both p and $\neg p$ (see Example 1). Note that it is not only the literals in the culprit set which characterise this inconsistency case, it is the derivation of the literals, i.e. that both are strictly derivable.

Theorem 3 states the existence of a culprit set in inconsistency case 1.

Theorem 3. *Let $\mathcal{P}$ have no well-founded model and let its only answer set be $Lit_\mathcal{P}$. Then $\mathcal{P}$ has a case 1 culprit set $\{a, \neg a\}$.*

Proof. By Lemma 1, $\exists a, a' \in \mathcal{HB}_{\mathcal{P}'}$ s.t. $\mathcal{P}' \vdash_{MP} a$ and $\mathcal{P}' \vdash_{MP} a'$. By definition of well-founded model (as an alternating fixpoint [18]), $a, a' \in T'_w$ where $\langle T'_w, \mathcal{F}'_w \rangle$ is the well-founded model of $\mathcal{P}'$. By Definition 1, $\{a, \neg a\}$ is a case 1 culprit set.

Example 5. The well-founded model of $\mathcal{P}'_2$ (see $\mathcal{P}_2$ in Example 2) is $\langle\{q, q', s', t'\}, \{p, r\}\rangle$. $q, q' \in T'_w$ and here even both of them are defeasibly derivable. Thus, $\{q, \neg q\}$ is a culprit set of $\mathcal{P}_2$ which confirms our observation that the reason for the inconsistency of $\mathcal{P}_2$ is that every potential answer set contains both q and $\neg q$, but $Lit_{\mathcal{P}_2}$ is not an answer set due to the NAF literals involved in the derivations of q and $\neg q$. Note that even though the culprit sets of $\mathcal{P}_1$ and $\mathcal{P}_2$ are very similar – both consist of complementary literals – the difference lies in the derivations of the literals in the culprit set: here the literals are not both strictly derivable, so the reason of the inconsistency is both that complementary literals are derivable (explicit negation) and that their derivations involve NAF literals.

Theorem 4 proves the existence of a culprit set in inconsistency case 2.

Theorem 4. *Let $\mathcal{P}$ have no well-founded model and no answer sets. Then $\mathcal{P}$ has a case 2 culprit set $\{a, \neg a\}$.*

Proof. Let $\langle T'_w, \mathcal{F}'_w \rangle$ be the well-founded model of $\mathcal{P}'$. Since $\mathcal{P}$ has no well-founded model, T'_w must contain some a, a'. Since every answer set is a superset of the well-founded model (Corollary 5.7 in [19]), every potential answer set of $\mathcal{P}$ contains a and $\neg a$, meaning that the only possible answer set is $Lit_\mathcal{P}$. From the assumption that $\mathcal{P}$ has no answer sets, we can conclude that $\mathcal{AS}(\mathcal{P}^{Lit_\mathcal{P}})$ does not contain a and $\neg a$. Thus, all of the rules needed for the derivation of either a or $\neg a$ are deleted in $\mathcal{P}^{Lit_\mathcal{P}}$, meaning that a or $\neg a$ is defeasibly derivable. Trivially, the other literal is also derivable as $a, a' \in T'_w$. Then by Definition 1, $\{a, \neg a\}$ is a case 2 culprit set of $\mathcal{P}$.

Example 6. $\mathcal{P}'_3$ (see $\mathcal{P}_3$ in Example 3) has two answer sets $S'_1 = \{q, q', r\}$ and $S'_2 = \{p, p', s\}$, so $\mathcal{P}_3$ falls into inconsistency case 3a. q, q', p, p' are all defeasibly derivable from $\mathcal{P}_3{}'$ and thus $\{q, \neg q, p, \neg p\}$ is a culprit set of $\mathcal{P}_3$. This confirms our observation that the reason for the inconsistency of $\mathcal{P}_3$ is that the two potential answer sets both contain complementary literals but that $Lit_{\mathcal{P}_3}$ is not an answer set due to the NAF literals involved in the derivations of the complementary literals. Thus, as in Example 5 the inconsistency is due to the interplay of explicit negation and NAF with the difference of the even-length loop described in Example 2. Due to this difference in the derivations, here the well-founded model of the translated logic program does not provide any information about culprits, but the answer sets do.

Theorem 5 states the existence of a culprit set in inconsistency case 3a.

Theorem 5. *Let $\mathcal{P}$ have a well-founded model and let $\mathcal{P}'$ have $n \geq 1$ answer sets. Then, $\mathcal{P}$ has a case 3a culprit set $\{a_1, \neg a_1, \ldots, a_n, \neg a_n\}$.*

Proof. By Theorem 2, $\mathcal{P}$ has no answer sets, so all $S_i \subseteq Lit_{\mathcal{P}}$ with $S_i = \mathcal{AS}(\mathcal{P}^{S_i})$ contain complementary literals a_i and $\neg a_i$, but $\mathcal{AS}(\mathcal{P}^{Lit_{\mathcal{P}}})$ does not contain complementary literals. Thus, all S_i' with $S_i' = \mathcal{AS}(\mathcal{P}'^{S_i'})$ contain a_i and a_i', so a_i and a_i' must be derivable from $\mathcal{P}'$. Assume that $\mathcal{P}' \vdash_{MP} a_i$ and $\mathcal{P}' \vdash_{MP} a_i'$. Then by Lemma 1 the only answer set of $\mathcal{P}$ is $Lit_{\mathcal{P}}$ (contradiction). Thus, at least one of a_i and a_i' is defeasibly derivable from $\mathcal{P}'$. Then by Definition 1, $\{a_1, \neg a_1, \ldots, a_n, \neg a_n\}$ is a case 3a culprit set of $\mathcal{P}$. $\qquad\square$

Example 7. Let $\mathcal{P}_4$ be the following logic program:

$s \leftarrow w$	$w \leftarrow not\ t$	$t \leftarrow \neg x$	$\neg x \leftarrow not\ \neg u$
$\neg u \leftarrow not\ v$	$v \leftarrow not\ t, not\ x$	$x \leftarrow$	$y \leftarrow not\ x$

$\mathcal{P}_4$ has a well-founded model and $\mathcal{P}_4'$ has no answer sets, so $\mathcal{P}_4$ falls into inconsistency case 3b. The only 3-valued M-stable model of $\mathcal{P}_4'$ is $\langle\{x\}, \{y\}\rangle$, where $\mathcal{U}_M' = \{s, t, u', v, w, x'\}$. For $s \in \mathcal{U}_M'$ there exists a negative dependency path s, t, u', v, t of atoms in $\mathcal{U}_M'$, where u', v, t is an odd-length loop. Thus, $C = \{\neg u, v, t\}$ is a culprit set of $\mathcal{P}_4$. Note that this culprit set is found no matter with which atom in $\mathcal{U}_M'$ the negative dependency path is started. This example shows that in inconsistency case 3b the inconsistency is due to NAF on its own; explicit negation plays no role.

Theorem 6 states not only the existence of a culprit set in inconsistency case 3b, but also characterises how to find a culprit set. This extends the results of [20] about odd-length loops.

Theorem 6. *Let $\mathcal{P}$ have a well-founded model and let $\mathcal{P}'$ have no answer sets. Let $\langle T_M', \mathcal{F}_M' \rangle$ be a 3-valued M-stable model of $\mathcal{P}'$ with $\mathcal{U}_M'$ the set of undefined atoms. Then, for any $a_1 \in \mathcal{U}_M'$ there exists a negative dependency path $a_1, \ldots, a_n, b_1, \ldots, b_m$ such that the set C consisting of the original literals of the translated literals $b_1, \ldots, b_m$ is a case 3b culprit set of $\mathcal{P}$.*

Proof (Sketch). By definition of 3-valued stable models any undefined atom is negatively dependent on an undefined atom. Thus, there is a negative dependency path of undefined atoms $a_1, \ldots, a_n$, which must lead to a negative dependency loop $a_n, b_1, \ldots, b_m$ ($a_n = b_m$) where no b_j is part of another negative-dependency path containing literals other than the ones in the loop. Assuming that the negative-dependency loop and all sub-loops are of even length, the loop on its own has a 2-valued stable model. We can show (omitted for lack of space) that it is possible to change the truth values of atoms not in the loop and in $\mathcal{U}'_M$ to T or F in such a way that if combined with the new truth values of the loop, a 3-valued stable model $\langle T', \mathcal{F}' \rangle$ of $\mathcal{P}'$ is created. Clearly $T'_M \subset T'$ and $\mathcal{F}'_M \subset \mathcal{F}'$, so $\langle T'_M, \mathcal{F}'_M \rangle$ is not a 3-valued M-stable model. Contradiction.

Note that in each of the three inconsistency cases discussed in Sect. 3, the translated logic program $\mathcal{P}'$ might or might not have answer sets. However, regarding culprit sets this distinction only makes a difference in inconsistency case 3.

It follows directly from the previous theorems that the culprit sets we identified are indeed responsible for the inconsistency, i.e. if no culprit sets exist then the logic program is not inconsistent, which is an essential first step for a user to understand what causes the inconsistency in a logic program.

Corollary 1. *Let $\mathcal{P}$ be a (possibly consistent) logic program. If there exists no culprit set of inconsistency cases 1, 2, 3a, or 3b of $\mathcal{P}$, then $\mathcal{P}$ is consistent.*

5 Explaining Culprits

As pointed out in the previous sections, even though we identify culprits as sets of literals, the reason of the inconsistency is mostly the way in which these literals are derivable from the logic program. In order to make the reason of the inconsistency more understandable for the user, we now show how explanations of the inconsistency can be constructed in terms of trees whose nodes are derivations. For this purpose, we define derivations with respect to a 3-valued interpretation $\langle T, \mathcal{F} \rangle$. We call a derivation *true* with respect to $\langle T, \mathcal{F} \rangle$ if all NAF literals *not* k used in the derivation are true with respect to the interpretation, i.e. the literals k are false in the interpretation. We call a derivation *false* with respect to $\langle T, \mathcal{F} \rangle$ if there exists a NAF literal *not* k used in the derivation which is false with respect to the interpretation, i.e. k is true in the interpretation.

Definition 2 (true/false derivation). *Let $\langle T, \mathcal{F} \rangle$ be a 3-valued interpretation of $\mathcal{P}$, $l \in Lit_\mathcal{P}$, and $\Delta \subseteq NAF_{Lit_\mathcal{P}}$.*

1. *$\mathcal{P} \cup \Delta \vdash_{MP} l$ is a true derivation of l w.r.t. $\langle T, \mathcal{F} \rangle$ if $\forall not\ k \in \Delta : k \in \mathcal{F}$.*
2. *$\mathcal{P} \cup \Delta \vdash_{MP} l$ is a false derivation of l w.r.t. $\langle T, \mathcal{F} \rangle$ if $\exists not\ k \in \Delta : k \in T$.*

Example 8. Consider $\mathcal{P}_4$ from Example 7. $\mathcal{P}_4 \cup \{not\ t, not\ x\} \vdash_{MP} v$ is a true derivation w.r.t. $\langle \{s\}, \{t, x\} \rangle$, a false derivation w.r.t $\langle \{s, t\}, \{x\} \rangle$, and neither a true nor a false derivation w.r.t. $\langle \{s\}, \{x\} \rangle$.

An explanation of inconsistency cases 1–3a illustrates why the literals in a culprit set are contained in the respective 3-valued stable model $\langle T, \mathcal{F} \rangle$ used to identify this culprit set, which is due to the literals' derivations. Thus, an explanation starts with a true derivation of a literal in the culprit set with respect to $\langle T, \mathcal{F} \rangle$. The explanation then indicates why this derivation is true, i.e. why all NAF literals *not k* are true with respect to $\langle \mathcal{T}, \mathcal{F} \rangle$. The reason why *not k* is true is that some derivation of k is false, i.e. a NAF literal *not m* in a derivation of k is false with respect to $\langle \mathcal{T}, \mathcal{F} \rangle$. This in turn is explained in terms of why m is the true with respect to $\langle \mathcal{T}, \mathcal{F} \rangle$, and so on.

Definition 3 (explanation). *Let $\langle \mathcal{T}, \mathcal{F} \rangle$ be a 3-valued stable model of $\mathcal{P}$ and let $l \in Lit_{\mathcal{P}}$. An explanation of l w.r.t. $\langle \mathcal{T}, \mathcal{F} \rangle$ is a tree such that:*

1. *Every node holds either a true or a false derivation w.r.t. $\langle \mathcal{T}, \mathcal{F} \rangle$.*
2. *The root holds a true derivation of l w.r.t. $\langle \mathcal{T}, \mathcal{F} \rangle$.*
3. *For every node N holding a true derivation $\mathcal{P} \cup \Delta \vdash_{MP} k$ w.r.t. $\langle \mathcal{T}, \mathcal{F} \rangle$ and for every not $m \in \Delta$: every false derivation of m w.r.t. $\langle \mathcal{T}, \mathcal{F} \rangle$ is held by a child of N.*
4. *For every node N holding a false derivation $\mathcal{P} \cup \Delta \vdash_{MP} k$ w.r.t. $\langle \mathcal{T}, \mathcal{F} \rangle$: N has exactly one child holding a true derivation of some m w.r.t. $\langle \mathcal{T}, \mathcal{F} \rangle$ such that not $m \in \Delta$.*
5. *There are no other nodes except those given in 1-4.*

Since culprit sets are determined with respect to different 3-valued stable models in the different inconsistency cases, explanations are constructed with respect to these different models, too.

Definition 4 (inconsistency explanation - cases 1,2). *Let $\mathcal{P}$ have no well-founded model and let $\langle \mathcal{T}'_w, \mathcal{F}'_w \rangle$ be the well-founded model of $\mathcal{P}'$. Let $\{a, \neg a\}$ be a culprit set of $\mathcal{P}$. A* translated inconsistency explanation *of $\mathcal{P}$ consists of an explanation of a w.r.t. $\langle \mathcal{T}'_w, \mathcal{F}'_w \rangle$ and an explanation of a' w.r.t. $\langle \mathcal{T}'_w, \mathcal{F}'_w \rangle$. An* inconsistency explanation *of $\mathcal{P}$ is derived by replacing every translated literal in the translated inconsistency explanation by its respective original literal.*

Since explanations are trees, they can be easily visualised, as shown for $\mathcal{P}_2$ (see Examples 2 and 5) in Fig. 1.

Definition 5 (inconsistency explanation - case 3a). *Let $\mathcal{P}$ have a well-founded model and let $S'_1, \ldots, S'_n$ ($n \geq 1$) be the answer sets of $\mathcal{P}'$. Let $\{a_1, \neg a_1, \ldots, a_n, \neg a_n\}$ be a culprit set of $\mathcal{P}$. A* translated inconsistency explanation *of $\mathcal{P}$ consists of an explanation of all a_i and a'_i ($1 \leq i \leq n$) w.r.t. $\langle S'_i, (\mathcal{HB}_{\mathcal{P}'} \setminus S'_i) \rangle$. An* inconsistency explanation *of $\mathcal{P}$ is derived by replacing every translated literal in the translated inconsistency explanation by its respective original literal.*

$$\mathcal{P}_2 \cup \{not\ r\} \vdash_{MP} q \qquad \mathcal{P}_2 \cup \{not\ p\} \vdash_{MP} \neg q$$
$$\uparrow$$
$$\mathcal{P}_2 \cup \{not\ \neg t\} \vdash_{MP} r$$
$$\uparrow$$
$$\mathcal{P}_2 \cup \emptyset \vdash_{MP} \neg t$$

Fig. 1. The inconsistency explanation of $\mathcal{P}_2$ (Examples 2, 5).

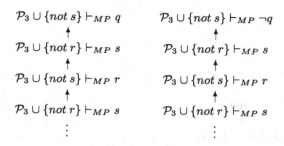

$$\mathcal{P}_3 \cup \{not\ s\} \vdash_{MP} q \qquad\qquad \mathcal{P}_3 \cup \{not\ s\} \vdash_{MP} \neg q$$
$$\uparrow \qquad\qquad\qquad\qquad\qquad \uparrow$$
$$\mathcal{P}_3 \cup \{not\ r\} \vdash_{MP} s \qquad\qquad \mathcal{P}_3 \cup \{not\ r\} \vdash_{MP} s$$
$$\uparrow \qquad\qquad\qquad\qquad\qquad \uparrow$$
$$\mathcal{P}_3 \cup \{not\ s\} \vdash_{MP} r \qquad\qquad \mathcal{P}_3 \cup \{not\ s\} \vdash_{MP} r$$
$$\uparrow \qquad\qquad\qquad\qquad\qquad \uparrow$$
$$\mathcal{P}_3 \cup \{not\ r\} \vdash_{MP} s \qquad\qquad \mathcal{P}_3 \cup \{not\ r\} \vdash_{MP} s$$
$$\vdots \qquad\qquad\qquad\qquad\qquad \vdots$$

Fig. 2. Part of the inconsistency explanation of $\mathcal{P}_3$ explaining q and $\neg q$. The full inconsistency explanation also comprises similar explanations for p and $\neg p$.

Figure 2 shows part of the inconsistency explanation of $\mathcal{P}_3$ (see Examples 3 and 6). It also illustrates the difference between the reasons of inconsistency in $\mathcal{P}_2$ and $\mathcal{P}_3$, namely the negative dependency loop of s and r in $\mathcal{P}_3$.

For inconsistency case 3b, where the literals in a culprit set form an odd-length negative dependency loop, the inconsistency explanation is a tree whose nodes hold derivations. However, since all literals in a culprit set are undefined with respect to a 3-valued M-stable model, an explanation is constructed with respect to the set of undefined atoms $\mathcal{U}$ rather than $\mathcal{T}$ and $\mathcal{F}$. In particular, the reason that a literal is undefined is that its derivation contains a NAF literal $not\ k$ which is undefined. Then $k \in \mathcal{U}$ which again is due to the derivation containing some undefined NAF literal, and so on. Thus, an explanation of inconsistency case 3b is a tree of negative derivations with respect to $\mathcal{U}$.

Definition 6 (inconsistency explanation - case 3b). *Let $\mathcal{P}$ have a well-founded model and let $\mathcal{P}'$ have no answer sets. Let $\langle \mathcal{T}'_M, \mathcal{F}'_M \rangle$ be a 3-valued M-stable model of $\mathcal{P}'$ with $\mathcal{U}'_M$ the set of undefined atoms. Let C be a culprit set of $\mathcal{P}$ and $a \in C$. A translated inconsistency explanation of $\mathcal{P}$ is a tree such that:*

1. *Every node holds a false derivation w.r.t. $\langle \mathcal{U}'_M, \emptyset \rangle$.*
2. *The root holds a false derivation of a w.r.t. $\langle \mathcal{U}'_M, \emptyset \rangle$.*
3. *For every node N holding a false derivation $\mathcal{P} \cup \Delta \vdash_{MP} b$ w.r.t. $\langle \mathcal{U}'_M, \emptyset \rangle$: N has exactly one child node holding a false derivation of some m w.r.t. $\langle \mathcal{U}'_M, \emptyset \rangle$ such that not $m \in \Delta$ and $m \in C$.*
4. *There are no other nodes except those given in 1–3.*

An inconsistency explanation of $\mathcal{P}$ is derived by replacing every translated literal in the translated inconsistency explanation by its respective original literal.

Figure 3 illustrates the inconsistency explanation of $\mathcal{P}_4$ (see Example 7), showing the responsible odd-length loop. It also illustrates how the derivations in an inconsistency explanation can be expanded to derivation trees, which can also be done for cases 1–3a.

Note that in all our examples, the culprit set is unique. However, in general a logic program may have various culprit sets (from the same inconsistency case) resulting in various inconsistency explanations. Moreover, there may be various inconsistency explanations for a given culprit set.

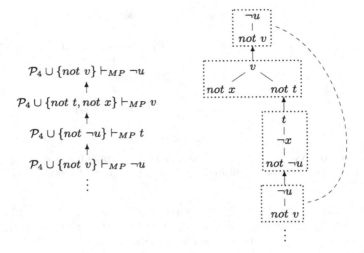

$$\mathcal{P}_4 \cup \{not\ v\} \vdash_{MP} \neg u$$
$$\uparrow$$
$$\mathcal{P}_4 \cup \{not\ t, not\ x\} \vdash_{MP} v$$
$$\uparrow$$
$$\mathcal{P}_4 \cup \{not\ \neg u\} \vdash_{MP} t$$
$$\uparrow$$
$$\mathcal{P}_4 \cup \{not\ v\} \vdash_{MP} \neg u$$
$$\vdots$$

Fig. 3. The inconsistency explanation of $\mathcal{P}_4$ (left) and the version where derivations are expanded to trees (right).

6   Conclusion

We showed that the two ways in which a logic program may be inconsistent – it has no answer sets or its only answer set is the set of all literals – can be determined using the well-founded model semantics and further divided into four inconsistency cases: one where only explicit negation is responsible, one where only NAF is responsible, and two where the interplay of explicit negation and NAF is responsible for the inconsistency. Each of these cases is characterised by a different type of culprit set, containing literals which are responsible for the inconsistency due to the way in which they are derivable. These culprit sets can be identified using "weaker" semantics than answer sets and can be used to explain the inconsistency in terms of trees whose nodes are derivations.

Our approach is related to early work on characterising logic programs with respect to the existence of answer sets [2,5,20]. However, none of these considers the properties of explicit negation in addition to NAF. It should also be pointed out that our explanations are related to the graphs used in [14,16] for explaining answer sets. In comparison to debugging approaches [1,7,12,13,17], our approach detects reasons for the inconsistency in terms of culprit sets, which is independent of an intended answer set. This naturally leads to the questions how to perform debugging based on the culprit sets, as well as how to deal with of multiple culprit sets for a logic program, which will be addressed in the future.

Since answer set programming for real-world applications often involves more complicated language constructs, e.g. constraints or aggregates, future work involves the extension of our approach to characterising inconsistency in logic programs using these constructs.

References

1. Brain, M., De Vos, M.: Debugging logic programs under the answer set semantics. In: Vos, M.D., Provetti, A. (eds.) ASP 2005. CEUR Workshop Proceedings, vol. 142, pp. 141–152. CEUR-WS.org (2005)
2. Dung, P.M.: On the relations between stable and well-founded semantics of logic programs. Theoret. Comput. Sci. **105**(1), 7–25 (1992)
3. Eiter, T., Leone, N., Saccà, D.: On the partial semantics for disjunctive deductive databases. Ann. Math. AI **19**(1–2), 59–96 (1997)
4. Eshghi, K., Kowalski, R.A.: Abduction compared with negation by failure. In: Levi, G., Martelli, M. (eds.) ICLP 1989, pp. 234–254. MIT Press (1989)
5. Fages, F.: Consistency of clark's completion and existence of stable models. Methods Logic CS **1**(1), 51–60 (1994)
6. Gebser, M., Kaufmann, B., Kaminski, R., Ostrowski, M., Schaub, T., Schneider, M.T.: Potassco: the Potsdam answer set solving collection. AI Commun. **24**(2), 107–124 (2011)
7. Gebser, M., Pührer, J., Schaub, T., Tompits, H.: A meta-programming technique for debugging answer-set programs. In: Fox, D., Gomes, C.P. (eds.) AAAI 2008, pp. 448–453. AAAI Press (2008)
8. Gelfond, M., Lifschitz, V.: Classical negation in logic programs and disjunctive databases. New Gener. Comput. **9**(3–4), 365–385 (1991)
9. Inoue, K.: Studies on Abductive and Nonmonotonic Reasoning. Ph.D. thesis, Kyoto University (1993)
10. Leone, N., Pfeifer, G., Faber, W., Eiter, T., Gottlob, G., Perri, S., Scarcello, F.: The dlv system for knowledge representation and reasoning. ACM Trans. Comput. Logic **7**(3), 499–562 (2006)
11. Niemelä, I., Simons, P., Syrjänen, T.: Smodels: A system for answer set programming. In: Baral, C., Truszczynski, M. (eds.) NMR 2000, vol. cs.AI/0003033. CoRR (2000)
12. Oetsch, J., Pührer, J., Tompits, H.: Catching the ouroboros: on debugging non-ground answer-set programs. TPLP **10**(4–6), 513–529 (2010)
13. Oetsch, J., Pührer, J., Tompits, H.: Stepping through an answer-set program. In: Delgrande, J.P., Faber, W. (eds.) LPNMR 2011. LNCS, vol. 6645, pp. 134–147. Springer, Heidelberg (2011)
14. Pontelli, E., Son, T.C., Elkhatib, O.: Justifications for logic programs under answer set semantics. TPLP **9**(1), 1–56 (2009)
15. Przymusinski, T.C.: Stable semantics for disjunctive programs. New Gener. Comput. **9**(3–4), 401–424 (1991)
16. Schulz, C., Toni, F.: Justifying answer sets using argumentation. TPLP FirstView, 1–52 (2015)
17. Syrjänen, T.: Debugging inconsistent answer set programs. In: Dix, J., Hunter, A. (eds.) NMR 2006. Technical report Series, vol. IfI-06-04, pp. 77–83. Clausthal University of Technology, Institute of Informatics (2006)
18. Van Gelder, A.: The alternating fixpoint of logic programs with negation. J. Comput. Syst. Sci. **47**(1), 185–221 (1993)
19. Van Gelder, A., Ross, K.A., Schlipf, J.S.: The well-founded semantics for general logic programs. J. ACM **38**(3), 619–649 (1991)
20. You, J.H., Yuan, L.Y.: A three-valued semantics for deductive databases and logic programs. J. Comput. Syst. Sci. **49**(2), 334–361 (1994)

An Implementation of Consistency-Based Multi-agent Belief Change Using ASP

Paul Vicol[1], James Delgrande[1], and Torsten Schaub[2]([⊠])

[1] Simon Fraser University, Burnaby, BC V5A 1S6, Canada
{pvicol,jim}@sfu.ca
[2] Universität Potsdam, August-Bebel-Strasse 89, 14482 Potsdam, Germany
torsten@cs.uni-potsdam.de

Abstract. This paper presents an implementation of a general framework for consistency-based belief change using Answer Set Programming (ASP). We describe Equibel, a software system for working with belief change operations on arbitrary graph topologies. The system has an ASP component that performs a core maximization procedure, and a Python component that performs additional processing on the output of the ASP solver. The Python component also provides an interactive interface that allows users to create a graph, set formulas at nodes, perform belief change operations, and query the resulting graph.

Keywords: Belief change · Belief merging · Answer set programming · Python

1 Introduction

We present an implementation of the consistency-based framework for multi-agent belief change discussed in [2]. In a network of connected agents, each with a set of beliefs, it is important to determine how the beliefs of the agents change as a result of incorporating information from other agents. We represent such a network by an undirected graph $G = \langle V, E \rangle$, where vertices represent agents and edges represent communication links by which agents share information. Associated with each agent is a *belief base* expressed as a propositional formula. Beliefs are shared among agents via a maximization procedure, wherein each agent incorporates as much information as consistently possible from other agents.

Before delving into the implementation, we introduce a motivating example:

Example 1. Consider a group of drones searching for missing people in a building. Each of the drones has some initial beliefs regarding where the missing people might be. Drone 1 believes that there is a person in the bookstore, as well as one in the atrium; drone 2 believes that there cannot be missing people in both the atrium *and* the bookstore; drone 3 just believes that there is a person in the cafeteria. The drones communicate with one another, and each is willing to incorporate new information that does not conflict with its initial beliefs. Our goal is to determine what each drone will believe following the communication.

© Springer International Publishing Switzerland 2015
F. Calimeri et al. (Eds.): LPNMR 2015, LNAI 9345, pp. 480–487, 2015.
DOI: 10.1007/978-3-319-23264-5_40

We have developed a software system called Equibel that can be used to simulate the above scenario, and determine where each drone would look for the missing people. More generally, Equibel allows for experimentation with belief sharing in arbitrary networks of agents. It uses Answer Set Programming (ASP) to perform the maximization step, and Python to manage the solving process and provide programmatic and interactive interfaces. The software is available online at www.github.com/asteroidhouse/equibel.

2   Related Work

Many methods have been proposed to deal with belief change involving multiple sources of information. Classical approaches to belief merging, such as [6] and [7], start with a set of belief bases and produce a single, merged belief base. Our approach differs in that we update multiple belief bases simultaneously.

The BReLS system [8] implements a framework for integrating information from multiple sources. In BReLS, pieces of information may have different degrees of reliability and may be believed at different discrete time points. Revision, update, and merging operations are each restrictions of the full semantics. The REV!GIS system [10] deals with belief revision in the context of geographic information systems, using information in a certain region to revise adjacent regions. There have been many approaches to iterative multi-agent belief sharing, including the iterated merging conciliation operators introduced in [4], and Belief Revision Games (BRGs) introduced in [9]. BRGs are ways to study the evolution of beliefs in a network of agents over time. While sharing a graph-based model, our framework differs from BRGs in two ways. First, we use a *consistency-based* approach, which is distinct from any of the revision policies in [9]. Second, we describe a "one-shot" method for belief sharing, rather than an iterated method.

The consistency-based framework we employ here has been developed in a series of papers, including [1–3].

3   The Consistency-Based Belief Change Framework

We work with a propositional language $\mathcal{L}_{\mathcal{P}}$, defined over an alphabet $\mathcal{P} = \{p, q, r, \dots\}$ of propositional atoms, using the connectives $\neg, \wedge, \vee, \rightarrow$ and $\equiv$ to construct formulas in the standard way. For $i \geq 0$, we define $\mathcal{P}^i = \{p^i \mid p \in \mathcal{P}\}$ containing superscripted versions of the atoms in $\mathcal{P}$, and define $\mathcal{L}^i$ to be the corresponding language. We denote the original, non-superscripted language by $\mathcal{L}^0$. We denote formulas by Greek letters α, β, etc. Given a formula $\alpha^i \in \mathcal{L}^i$, $\alpha^j \in \mathcal{L}^j$ is the formula obtained by replacing all occurrences of $p^i \in \mathcal{P}^i$ by $p^j \in \mathcal{P}^j$. For example, if $\alpha^1 = (p^1 \wedge \neg q^1) \rightarrow r^1$, then $\alpha^2 = (p^2 \wedge \neg q^2) \rightarrow r^2$.

Our implementation uses the maximization approach to belief sharing described in [2]; here we recall the terminology and notation for maximal equivalence sets, and we refer the reader to [2] for more details.

Definition 1 (G-scenario). *Let $G = \langle V, E \rangle$ be a graph with $|V| = \{1, 2, \ldots, n\}$. A G-scenario Σ_G is a vector of formulas $\langle \varphi_1, \ldots, \varphi_n \rangle$. The notation $\Sigma_G[i]$ denotes the ith component, φ_i.*

An agent starts with some initial beliefs that she does not want to give up, and then "includes" as much information as consistently possible from other agents. This is done as follows. Each agent expresses her beliefs in a distinct language, such that the languages used by any two agents are isomorphic. Specifically, agent i expresses her beliefs as formulas of the language $\mathcal{L}^i$. Because the agents' languages are disjoint, $\bigcup_{1 \leq i \leq n} \varphi_i^i$ is trivially consistent. For each agent, we want to find out what "pieces" of beliefs from other agents she can incorporate. To do this, we assert that the languages of adjacent agents agree on the truth values of corresponding atoms as much as consistently possible. The equivalences between the atoms of agents i and j tell us what those agents can and cannot agree on, and provide a means to "translate" formulas between those agents' languages. If agents i and j cannot agree on the truth value of p, then we can translate formulas from $\mathcal{L}^i$ to $\mathcal{L}^j$ by replacing p^i by $\neg p^j$. This process is formalized below.

Definition 2 (Equivalence sets, Fits, Maximal fits). *Let $G = \langle V, E \rangle$ be a graph and $\mathcal{P}$ be an alphabet.*

- *An equivalence set EQ is a subset of $\{p^i \equiv p^j \mid \langle \{i, j\}, p \rangle \in E \times \mathcal{P}\}$.*
- *Given a G-scenario $\Sigma_G = \langle \varphi_1, \ldots, \varphi_n \rangle$, a fit for Σ_G is an equivalence set EQ such that $EQ \cup \bigcup_{i=1}^{n} \varphi_i^i$ is consistent.*
- *A maximal fit for Σ_G is a fit EQ such that for all fits $EQ' \supset EQ$, we have that $EQ' \cup \bigcup_{i=1}^{n} \varphi_i^i$ is inconsistent.*

Let $\Sigma_G = \langle \varphi_1, \ldots, \varphi_n \rangle$ be a G-scenario and **F** be the set of maximal fits for Σ_G. Informally, the *completion* of Σ_G, denoted $\Theta(\Sigma_G)$, is a G-scenario $\Sigma'_G = \langle \varphi'_1, \ldots, \varphi'_n \rangle$ consisting of updated formulas for each agent following a belief sharing procedure. [2] gives both semantic and syntactic characterizations of the completion. Here we state the syntactic characterization, based on translation.

Definition 3 (Substitution function). *Let $G = \langle V, E \rangle$ be a graph, and EQ be an equivalence set. Let R^* denote the transitive closure of a binary relation R. Then, for $i, j \in V$, we define a substitution function $s_{i,j}^{EQ} : \mathcal{P}^i \to \{l(p^j) \mid p^j \in \mathcal{P}^j\}$, where $l(p^j)$ is either p^j or $\neg p^j$, as follows:*

$$s_{i,j}^{EQ}(p^i) = \begin{cases} p^j : (p^i \equiv p^j) \in EQ \\ \neg p^j : \{i, j\} \in E^*, (p^i \equiv p^j) \notin EQ \end{cases}$$

Given a formula α^i, $s_{i,j}^{EQ}(\alpha^i)$ is the formula that results from replacing each atom p^i in α^i by its unique counterpart $s_{i,j}^{EQ}(p^i)$. Thus, $s_{i,j}^{EQ}(\alpha^i)$ is a *translation* of α^i into the language of agent j, that is consistent with agent j's initial beliefs.

Proposition 1. *Let* $G = \langle V, E \rangle$ *be a graph and* $\Sigma_G = \langle \varphi_1, \ldots, \varphi_n \rangle$ *be a G-scenario. Let* $\Theta(\Sigma_G) = \langle \varphi'_1, \ldots, \varphi'_n \rangle$ *be the completion of* Σ_G *, and let* $\boldsymbol{F}$ *be the set of maximal fits of* Σ_G. *Then, we find* φ'_j, *for* $j \in \{1, \ldots, n\}$, *as follows:*

$$\varphi'_j \equiv \bigvee_{EQ \in F} \left(\bigwedge_{\{i,j\} \in E^*} (s^{EQ}_{i,j}(\varphi^i_i))^0 \right)$$

4   System Design

Equibel is split into two architectural layers: an ASP layer, which performs the core maximization procedure, and a Python layer, which performs post-processing of answer sets and provides programmatic and interactive user interfaces to experiment with belief sharing on custom graphs. Equibel provides a Python package (`equibel`) that allows users to perform belief change operations in programs, and a user-friendly command-line interface (CLI) that allows for real-time experimentation. The CLI allows users to enter commands to create agents, edges, and formulas, execute belief change operations and query the resulting graph. A query might ask what a particular agent believes, or what the common knowledge is (the disjunction of all agents' beliefs).

There are three major stages to computing the completion of a G-scenario: (1) finding maximal sets of equivalences between atoms of adjacent agents; (2) translating beliefs between the languages of adjacent agents; and (3) combining beliefs that result from different maximal equivalence sets. The first two steps are done in ASP, while the third is done in Python.

The ASP layer consists of a set of logic programs that can be combined in different ways to achieve different functionality. The core of Equibel is the `eq_sets.lp` logic program that finds maximal sets of equivalences of the form $p^i \equiv p^j$ between atoms at neighbouring agents i and j. We use the logic program `translate.lp` to translate formulas between the languages of connected agents based on the equivalence sets. Each optimal answer set gives the new information incorporated by each agent, based on a specific maximal EQ set. We use the ASP grounder/solver `clingo`, from the Potsdam Answer Set Solving Collection [5].

The Python component combines formulas that occur in different answer sets. The Python `clingo` interface also manages the solving state, by loading specific combinations of logic modules. This allows the system to find either cardinality- or containment-maximal EQ sets, and potentially perform iterated belief sharing. We also designed a file format for specifying belief change problems, called the Belief Change Format (BCF). This is an extension of the DIMACS graph format, and is a standard for communication within our system.

5   ASP Implementation

Encoding Graphs. Encoding a graph involves creating agents, assigning formulas to the agents, and setting up connections between the agents. We declare agents using the `node/1` predicate, and declare edges using `edge/2`. We assign formulas to agents using `formula/2`, where the first argument is a formula built using the function symbols `and/2`, `or/2`, `implies/2`, `iff/2`, and `neg/1`, and the second argument is an integer identifying an agent. For example, we can assign the formula $(p \land q) \lor \neg r$ to agent 1 with `formula(or(and(p,q),neg(r)),1)`.

Finding Maximal EQ Sets. Maximal equivalence sets are found by `eq_sets.lp`, which: (1) *generates* candidate equivalence sets; (2) *tests* the equivalence sets by attempting to find a truth assignment, constrained by the equivalences, that satisfies all agents' initial beliefs; and (3) *optimizes* the results to find containment- or cardinality-maximal sets.

In order to check whether a truth assignment is satisfying, we first break each formula down into its subformulas. After truth values have been assigned to the atoms, an agent's beliefs are built back up from its subformulas; this allows us to determine whether an assignment models the original beliefs of each agent. We classify each subformula as either a compound or an atomic proposition:

```
atom(P,X) :- subform(P,X), not compound_prop(P,X).
atom(P)   :- atom(P,_).
```

Candidate EQ sets are generated by:

```
{ eq(P,X,Y) : atom(P), edge(X,Y), X < Y }.
```

The predicate `eq(P,X,Y)` expresses that $P^X \equiv P^Y$. The condition $X < Y$ halves the search space over edges; this is justified because edges are undirected. After we generate a candidate EQ set, we check whether it is possible to assign truth values to all atoms, restricted by the equivalences, such that the agents' original formulas are satisfied. We assign a truth value to each atom, with the constraint that atoms linked by an equivalence must have the same truth value:

```
1 { truth_value(P,X,true), truth_value(P,X,false) } 1 :-
        atom(P), node(X).
:- eq(P,X,Y), truth_value(P,X,V), truth_value(P,Y,W), V != W.
```

Now we build up the original formulas from their subformulas, to see if the assignment is satisfying. A sample of the code used to build up the original formulas starting from the atoms is shown below:

```
sat(F,X) :- F = and(A,B), sat(A,X), sat(B,X),
            subform(F,X), subform(A,X), subform(B,X).
```

For an EQ set to be acceptable, it must be possible to find a truth assignment that satisfies all the original formulas. Thus, we introduce the constraint:

```
:- formula(F,X), not sat(F,X).
```

There are two types of maximality, each requiring a different program statement and solving configuration. The standard `#maximize` statement in `clingo` finds EQ sets that are maximal with respect to cardinality. To find *containment-maximal* EQ sets, however, we use a domain-specific heuristic:

```
_heuristic(eq(P,X,Y), true, 1) :- atom(P), edge(X,Y), X < Y.
```

The **true** heuristic modifier tells the solver to decide first on **eq** atoms, and to set them to **true**. The solver initially makes all **eq** atoms true, and then "whittles down" the set, producing containment-maximal sets.

For our one-shot approach to belief sharing, we take the transitive closure of the **eq/3** predicates. This allows for an agent to learn from other agents throughout the graph, not just from its immediate neighbours. The module **translate.lp** translates formulas between the languages of connected agents, and outputs **new_formula/2** predicates that indicate the new information obtained by an agent from its neighbours.

Consider an equivalence set EQ and an agent i. The new belief of i, based on EQ, is the conjunction of translated beliefs from all agents connected to i. But we may have multiple maximal equivalence sets, each of which represents an equally plausible way to share information. Thus, we combine beliefs that result from different equivalence sets by taking their disjunction. Both of these steps are performed in Python, using the output of **translate.lp**.

We now look at how the system works on our opening example. Let the atomic propositions a, b, and c denote the facts that there are missing people in the atrium, bookstore, and cafeteria, respectively. The network of drones is represented by a complete graph on three nodes, numbered 1 to 3, and the associated G-scenario is $\Sigma_G = \langle a \wedge b, \neg a \vee \neg b, c \rangle$. Solving with **eq_sets.lp**, we find four maximal EQ sets. Based on the first set, $\{a_1 \equiv a_2, a_1 \equiv a_3, a_2 \equiv a_3, b_1 \equiv b_3, c_1 \equiv c_2, c_1 \equiv c_3, c_2 \equiv c_3\}$, the new beliefs of the drones would be $\langle a \wedge b \wedge c, a \wedge \neg b \wedge c, a \wedge b \wedge c \rangle$. Taking the disjunction of formulas obtained from different EQ sets, the final beliefs of the drones are $\langle a \wedge b \wedge c, (a \equiv \neg b) \wedge c, (a \vee b) \wedge c \rangle$.

6 Expressing Revision and Merging

The **equibel** Python module allows users to perform belief change operations such as revision and merging, without explicitly creating graph topologies. For these operations, the user only needs to specify formulas representing belief bases to be operated on, and the system constructs an *implicit* graph topology, finds the completion, and returns either a formula of a specific agent in the completion, or the invariant knowledge. In this section, we show how belief revision and two types of merging are expressed in our framework.

For belief revision, we need to consistently introduce a new belief α into a belief base K, while retaining as much of K as possible. Belief revision can be modeled as a two-agent graph $G = \langle V, E \rangle$, with $V = \{1, 2\}$, $E = \{\{1, 2\}\}$, and $\Sigma_G = \langle K, \alpha \rangle$. Through belief sharing, agent 2 will incorporate as much information from agent 1 as possible, while maintaining consistency with α. The belief of agent 2 in the completion, $\Theta(\Sigma_G)[2]$, is the result of the revision, $K * \alpha$. This corresponds to consistency-based revision, as defined in [1].

Turning to belief merging, multiple, potentially mutually inconsistent, bodies of knowledge need to be combined into a coherent whole. Two approaches to merging are described in [3]. The first approach is a generalization of belief revision called *projection*. Given a multiset $\mathcal{K} = \langle K_1, \ldots, K_n \rangle$ and a constraint

μ, the contents of each belief base K_i are projected onto a distinguished belief base which initially contains just μ. This is expressed in our framework using a star graph, such that the central agent initially believes μ, and incorporates as much information as possible from each of its neighbours. Formally, $G = \langle V, E \rangle$, where $V = \{0, 1, \ldots, n\}$, $E = \{\{0, i\} \mid i \in V \setminus \{0\}\}$, and $\Sigma = \langle \mu, K_1, \ldots, K_n \rangle$. The merged belief base $\Delta_\mu(\mathcal{K})$ is the belief of the central agent in the completion, $\Theta(\Sigma)[0]$. The second approach to merging, called *consensus merging*, involves "pooling" together information from the belief bases. Let G be a complete graph and let Σ be a G-scenario. Information is pooled by taking the invariant of the completion $\Theta(\Sigma) = \langle \varphi'_1, \ldots, \varphi'_n \rangle$, so that $\Delta(\mathcal{K}) = \bigvee_{i=1}^{n} \varphi'_i$.

7 Conclusion

In this paper, we introduce Equibel, a software system for experimenting with consistency-based multi-agent belief sharing. We model networks of communicating agents using arbitrary undirected graphs, where each node is associated with a belief base represented by a propositional formula. Each agent shares information with its neighbours, and learns as much as possible from connected agents, while not giving up her initial beliefs. Belief sharing is carried out via a global procedure that maximizes similarities between belief bases of adjacent agents. We describe Equibel's architecture, examine how maximal equivalence sets are found using ASP, and look at how Equibel handles belief revision and merging by constructing implicit graph topologies. Other operations, such as belief extrapolation, can also be expressed within this framework. We are working on expanding the system to support iterated change and agent expertise.

Acknowledgements. Financial support was gratefully received from the Natural Sciences and Engineering Research Council of Canada.

References

1. Delgrande, J., Schaub, T.: A consistency-based approach for belief change. Artif. Intell. **151**(1–2), 1–41 (2003)
2. Delgrande, J.P., Lang, J., Schaub, T.: Belief change based on global minimisation. In: IJCAI, Hyderabad, India (2007)
3. Delgrande, J.P., Schaub, T.: A consistency-based framework for merging knowledge bases. J. Appl. Logic **5**(3), 459–477 (2006)
4. Gauwin, O., Konieczny, S., Marquis, P.: Iterated belief merging as conciliation operators. In: 7th International Symposium on Logical Formalizations of Commonsense Reasoning, Corfu, Greece, pp. 85–92 (2005)
5. Gebser, M., Kaminski, R., Kaufmann, B., Schaub, T.: Clingo = ASP + control: Preliminary report. volume arXiv:1405.3694v1
6. Konieczny, S., Pino Pérez, R.: Merging information under constraints: a logical framework. J. Logic Comput. **12**(5), 773–808 (2002)
7. Liberatore, P., Schaerf, M.: Arbitration: a commutative operator for belief revision. In: Proceedings of 2nd WOCFAI 1995, pp. 217–228 (1995)

8. Liberatore, P., Schaerf, M.: Brels: a system for the integration of knowledge bases. In: Proceedings of the 7th International Conference on Principles of KR&R, pp. 145–152 (2000)
9. Schwind, N., Inoue, K., Bourgne, G., Konieczny, S., Marquis, P.: Belief revision games. In: AAAI (2015)
10. Würbel, E., Jeansoulin, R., Papini, O.: Revision: an application in the framework of GIS. In: Proceedings of the 7th International Conference on Principles of KR&R, pp. 505–515 (2000)

ASPMT(QS): Non-Monotonic Spatial Reasoning with Answer Set Programming Modulo Theories

Przemysław Andrzej Wałęga[1]($\boxtimes$), Mehul Bhatt[2], and Carl Schultz[2]

[1] Institute of Philosophy, University of Warsaw, Warsaw, Poland
przemek.walega@wp.pl
[2] Department of Computer Science, University of Bremen, Bremen, Germany

Abstract. The systematic modelling of *dynamic spatial systems* [9] is a key requirement in a wide range of application areas such as comonsense cognitive robotics, computer-aided architecture design, dynamic geographic information systems. We present ASPMT(QS), a novel approach and fully-implemented prototype for non-monotonic spatial reasoning — a crucial requirement within dynamic spatial systems– based on Answer Set Programming Modulo Theories (ASPMT). ASPMT(QS) consists of a (qualitative) spatial representation module (QS) and a method for turning tight ASPMT instances into Sat Modulo Theories (SMT) instances in order to compute stable models by means of SMT solvers. We formalise and implement concepts of default spatial reasoning and spatial frame axioms using choice formulas. Spatial reasoning is performed by encoding spatial relations as systems of polynomial constraints, and solving via SMT with the theory of real nonlinear arithmetic. We empirically evaluate ASPMT(QS) in comparison with other prominent contemporary spatial reasoning systems. Our results show that ASPMT(QS) is the only existing system that is capable of reasoning about indirect spatial effects (i.e. addressing the ramification problem), and integrating geometric and qualitative spatial information within a non-monotonic spatial reasoning context.

Keywords: Non-monotonic spatial reasoning · Answer set programming modulo theories · Declarative spatial reasoning

1 Introduction

Non-monotonicity is characteristic of commonsense reasoning patterns concerned with, for instance, making default assumptions (e.g., about spatial inertia), counterfactual reasoning with hypotheticals (e.g., what-if scenarios), knowledge interpolation, explanation & diagnosis (e.g., filling the gaps, causal links), belief revision. Such reasoning patterns, and therefore non-monotonicity, acquires a special significance in the context of *spatio-temporal dynamics*, or computational commonsense *reasoning about space, actions, and change* as applicable within areas as disparate as geospatial dynamics, computer-aided design, cognitive vision, commonsense cognitive robotics [6]. Dynamic spatial systems are characterised

© Springer International Publishing Switzerland 2015
F. Calimeri et al. (Eds.): LPNMR 2015, LNAI 9345, pp. 488–501, 2015.
DOI: 10.1007/978-3-319-23264-5_41

by scenarios where spatial configurations of objects undergo a change as the result of interactions within a physical environment [9]; this requires explicitly identifying and formalising relevant actions and events at both an ontological and (qualitative and geometric) spatial level, e.g. formalising *desertification* and *population displacement* based on spatial theories about *appearance, disappearance, splitting, motion,* and *growth* of regions [10]. This calls for a deep integration of spatial reasoning within KR-based non-monotonic reasoning frameworks [7].

We select aspects of a theory of *dynamic spatial systems*—pertaining to *(spatial) inertia, ramifications, causal explanation*— that are inherent to a broad category of dynamic spatio-temporal phenomena, and require non-monotonic reasoning [5,9]. For these aspects, we provide an operational semantics and a computational framework for realising fundamental non-monotonic spatial reasoning capabilities based on Answer Set Programming Modulo Theories [3]; ASPMT is extended to the qualitative spatial (QS) domain resulting in the non-monotonic spatial reasoning system ASPMT(QS). Spatial reasoning is performed in an analytic manner (e.g. as with reasoners such as CLP(QS) [8]), where spatial relations are encoded as systems of polynomial constraints; the task of determining whether a spatial graph G is consistent is now equivalent to determining whether the system of polynomial constraints is satisfiable, i.e. Satisfiability Modulo Theories (SMT) with real nonlinear arithmetic, and can be accomplished in a sound and complete manner. Thus, ASPMT(QS) consists of a (qualitative) spatial representation module and a method for turning tight ASPMT instances into Sat Modulo Theories (SMT) instances in order to compute stable models by means of SMT solvers.

In the following sections we present the relevant foundations of stable model semantics and ASPMT, and then extend this to ASPMT(QS) by defining a (qualitative) spatial representations module, and formalising spatial default reasoning and spatial frame axioms using choice formulas. We empirically evaluate ASPMT(QS) in comparison with other existing spatial reasoning systems. We conclude that ASMPT(QS) is the only system, to the best of our knowledge, that operationalises dynamic spatial reasoning within a KR-based framework.

2   Preliminaries

2.1   Bartholomew – Lee Stable Models Semantics

We adopt a definition of stable models based on syntactic transformations [2] which is a generalization of the previous definitions from [13,14,19]. For predicate symbols (constants or variables) u and c, expression $u \leq c$ is defined as shorthand for $\forall \mathbf{x}(u(\mathbf{x}) \rightarrow c(\mathbf{x}))$. Expression $u = c$ is defined as $\forall \mathbf{x}(u(\mathbf{x}) \equiv c(\mathbf{x}))$ if u and c are predicate symbols, and $\forall \mathbf{x}(u(\mathbf{x}) = c(\mathbf{x}))$ if they are function symbols. For lists of symbols $\mathbf{u} = (u_1, \ldots, u_n)$ and $\mathbf{c} = (c_1, \ldots, c_n)$, expression $\mathbf{u} \leq \mathbf{c}$ is defined as $(u_1 \leq c_1) \wedge \cdots \wedge (u_n \leq c_n)$, and similarly, expression $\mathbf{u} = \mathbf{c}$ is defined as $(u_1 = c_1) \wedge \cdots \wedge (u_n = c_n)$. Let $\mathbf{c}$ be a list of distinct predicate and function constants, and let $\widehat{\mathbf{c}}$ be a list of distinct predicate and function variables corresponding to c. By $\mathbf{c}^{pred}$ ($\mathbf{c}^{func}$, respectively) we mean the list of all predicate constants (function constants, respectively) in $\mathbf{c}$, and by $\widehat{\mathbf{c}}^{pred}$ ($\widehat{\mathbf{c}}^{func}$, respectively) the list of the corresponding predicate variables (function variables, respectively) in $\widehat{\mathbf{c}}$. In what follows, we refer to function constants and predicate constants of arity 0 as object constants and propositional constants, respectively.

Definition 1 *(Stable model operator SM). For any formula F and any list of predicate and function constants $\mathbf{c}$ (called intensional constants), $SM[F;\mathbf{c}]$ is defined as*

$$F \wedge \neg\exists\widehat{\mathbf{c}}(\widehat{\mathbf{c}} < \mathbf{c} \wedge F^*(\widehat{\mathbf{c}})), \tag{1}$$

where $\widehat{\mathbf{c}} < \mathbf{c}$ is a shorthand for $(\widehat{\mathbf{c}}^{pred} \leq \mathbf{c}^{pred}) \wedge \neg(\widehat{\mathbf{c}} = \mathbf{c})$ and $F^*(\widehat{\mathbf{c}})$ is defined recursively as follows:

- *for atomic formula F, $F^* \equiv F' \wedge F$, where F' is obtained from F by replacing all intensional constants $\mathbf{c}$ with corresponding variables from $\widehat{\mathbf{c}}$,*
- $(G \wedge H)^* = G^* \wedge H^*, \quad (G \vee H)^* = G^* \vee H^*,$
- $(G \rightarrow H)^* = (G^* \rightarrow H^*) \wedge (G \rightarrow H),$
- $(\forall x G)^* = \forall x G^*, \quad (\exists x G)^* = \exists x G^*.$

$\neg F$ *is a shorthand for $F \rightarrow \bot$, $\top$ for $\neg\bot$ and $F \equiv G$ for $(F \rightarrow G) \wedge (G \rightarrow F)$.*

Definition 2 *(Stable model). For any sentence F, a stable model of F on $\mathbf{c}$ is an interpretation I of underlying signature such that $I \models SM[F;\mathbf{c}]$.*

2.2 Turning ASPMT into SMT

It is shown in [3] that a tight part of ASPMT instances can be turned into SMT instances and, as a result, off-the-shelf SMT solvers (e.g. Z3 for arithmetic over reals) may be used to compute stable models of ASP, based on the notions of Clark normal form, Clark completion.

Definition 3 *(Clark normal form). Formula F is in Clark normal form (relative to the list $\mathbf{c}$ of intensional constants) if it is a conjunction of sentences of the form (2) and (3).*

$$\forall\mathbf{x}(G \rightarrow p(\mathbf{x})) \qquad (2) \qquad\qquad \forall\mathbf{x}y(G \rightarrow f(\mathbf{x}) = y) \qquad (3)$$

one for each intensional predicate p and each intensional function f, where $\mathbf{x}$ is a list of distinct object variables, y is an object variable, and G is an arbitrary formula that has no free variables other than those in $\mathbf{x}$ and y.

Definition 4 *(Clark completion). The completion of a formula F in Clark normal form (relative to $\mathbf{c}$), denoted by $Comp_c[F]$ is obtained from F by replacing each conjunctive term of the form (2) and (3) with (4) and (5) respectively*

$$\forall\mathbf{x}(G \equiv p(\mathbf{x})) \qquad (4) \qquad\qquad \forall\mathbf{x}y(G \equiv f(\mathbf{x}) = y). \qquad (5)$$

Definition 5 *(Dependency graph). The dependency graph of a formula F (relative to $\mathbf{c}$) is a directed graph $DG_c[F] = (V, E)$ such that:*

1. V consists of members of c,
2. for each $c, d \in V$, $(c, d) \in E$ whenever there exists a strictly positive occurrence of $G \rightarrow H$ in F, such that c has a strictly positive occurrence in H and d has a strictly positive occurrence in G,

where an occurrence of a symbol or a subformula in F is called strictly positive in F if that occurrence is not in the antecedent of any implication in F.

Definition 6 (Tight formula).   *Formula F is* tight *(on c) if $DG_c[F]$ is acyclic.*

Theorem 1 (Bartholomew, Lee).   *For any sentence F in Clark normal form that is tight on c, an interpretation I that satisfies $\exists xy(x = y)$ is a model of $SM[F;c]$ iff I is a model of $Comp_c[F]$ relative to c.*

3   ASPMT with Qualitative Space – ASPMT(QS)

In this section we present our spatial extension of ASPMT, and formalise spatial default rules and spatial frame axioms.

3.1   The Qualitative Spatial Domain $\mathcal{QS}$

Qualitative spatial calculi can be classified into two groups: topological and positional calculi. With topological calculi such as the *Region Connection Calculus* (RCC) [25], the primitive entities are spatially extended regions of space, and could possibly even be 4D spatio-temporal histories, e.g., for *motion-pattern* analyses. Alternatively, within a dynamic domain involving translational motion, point-based abstractions with orientation calculi could suffice. Examples of orientation calculi include: the Oriented-Point Relation Algebra ($\mathcal{OPRA}_m$) [22], the Double-Cross Calculus [16]. The qualitative spatial domain ($\mathcal{QS}$) that we consider in the formal framework of this paper encompasses the following ontology:

QS1.   Domain Entities in $\mathcal{QS}$. Domain entities in $\mathcal{QS}$ include *circles, triangles, points* and *segments*. While our method is applicable to a wide range of 2D and 3D spatial objects and qualitative relations, for brevity and clarity we primarily focus on a 2D spatial domain. Our method is readily applicable to other 2D and 3D spatial domains and qualitative relations, for example, as defined in [8, 11, 12, 23, 24, 26, 27].

- a *point* is a pair of reals x, y
- a *line segment* is a pair of end points p_1, p_2 $(p_1 \neq p_2)$
- a *circle* is a centre point p and a real radius r $(0 < r)$
- a *triangle* is a triple of vertices (points) p_1, p_2, p_3 such that p_3 is *left of* segment p_1, p_2.

QS2.   Spatial Relations in $\mathcal{QS}$. We define a range of spatial relations with the corresponding polynomial encodings. Examples of spatial relations in $\mathcal{QS}$ include:

Relative Orientation.   Left, right, collinear orientation relations between *points* and *segments*, and *parallel, perpendicular* relations between *segments* [21].

Mereotopology.   Part-whole and contact relations between regions [25, 28].

3.2   Spatial Representations in ASPMT(QS)

Spatial representations in ASPMT(QS) are based on parametric functions and qualitative relations, defined as follows.

Definition 7 (Parametric function). *A parametric function is an n-ary function $f_n : D_1 \times D_2 \times \cdots \times D_n \to \mathbb{R}$ such that for any $i \in \{1 \ldots n\}$, D_i is a type of spatial object, e.g., Points, Circles, Polygons, etc.*

Example 1. Consider following parametric functions $x : Circles \to \mathbb{R}$, $y : Circles \to \mathbb{R}$, $r : Circles \to \mathbb{R}$ which return the position values x, y of a circle's centre and its radius r, respectively. Then, circle $c \in Cirlces$ may be described by means of parametric functions as follows: $x(c) = 1.23 \wedge y(c) = -0.13 \wedge r(c) = 2$.

Definition 8 (Qualitative spatial relation). *A qualitative spatial relation is an n-ary predicate $Q_n \subseteq D_1 \times D_2 \times \cdots \times D_n$ such that for any $i \in \{1 \ldots n\}$, D_i is a type of spatial object. For each Q_n there is a corresponding formula of the form*

$$\forall d_1 \in D_1 \ldots \forall d_n \in D_n \Big(Q_n(d_1, \ldots, d_n) \leftarrow p_1(d_1, \ldots, d_n) \wedge \cdots \wedge p_m(d_1, \ldots, d_n) \Big) \quad (6)$$

where $m \in \mathbb{N}$ and for any $i \in \{1 \ldots n\}$, p_i is a polynomial equation or inequality.

Proposition 1. *Each qualitative spatial relation according to Definition 8 may be represented as a tight formula in Clark normal form.*

Proof. Follows directly from Definitions 3 and 8.

Thus, qualitative spatial relations belong to a part of ASPMT that may be turned into SMT instances by transforming the implications in the corresponding formulas into equivalences (Clark completion). The obtained equivalence between polynomial expressions and predicates enables us to compute relations whenever parametric information is given, and vice versa, i.e. computing possible parametric values when only the qualitative spatial relations are known.

Many relations from existing qualitative calculi may be represented in ASPMT(QS) according to Definition 8; our system can express the polynomial encodings presented in e.g. [8,11,12,23,24]. Here we give some illustrative examples.

Proposition 2. *Each relation of Interval Algebra (IA) [1] and Rectangle Algebra (RA) [20] may be defined in ASPMT(QS).*

Proof. Each IA relation may be described as a set of equalities and inequalities between interval endpoints (see Fig. 1 in [1]), which is a conjunction of polynomial expressions. RA makes use of IA relations in 2 and 3 dimensions. Hence, each relation is a conjunction of polynomial expressions [27].

Proposition 3. *Each relation of RCC-5 in the domain of convex polygons with a finite maximum number of vertices may be defined in ASPMT(QS).*

Proof. Each RCC–5 relation may be described by means of relations $P(a, b)$ and $O(a, b)$. In the domain of convex polygons, $P(a, b)$ is true whenever all vertices of a are in the interior (inside) or on the boundary of b, and $O(a, b)$ is true if there exists a point p that is inside both a and b. Relations of a point being inside, outside or on the boundary of a polygon can be described by polynomial expressions e.g. [8]. Hence, all RCC–5 relations may be described with polynomials, given a finite upper limit on the number of vertices a convex polygon can have.

Proposition 4. *Each relation of Cardinal Direction Calculus (CDC) [15] may be defined in ASPMT(QS).*

Proof. CDC relations are obtained by dividing space with 4 lines into 9 regions. Since halfplanes and their intersections may be described with polynomial expressions, then each of the 9 regions may be encoded with polynomials. A polygon object is in one or more of the 9 cardinal regions by the topological *overlaps* relation between polygons, which can be encoded with polynomials (i.e. by the existence of a shared point) [8].

3.3   Choice Formulas in ASPMT(QS)

A choice formula [14] is defined for a predicate constant p as $\mathtt{Choice}(p) \equiv \forall \mathbf{x}(p(\mathbf{x}) \vee \neg p(\mathbf{x}))$ and for function constant f as $\mathtt{Choice}(f) \equiv \forall \mathbf{x}(f(\mathbf{x}) = y \vee \neg f(\mathbf{x}) = y)$, where $\mathbf{x}$ is a list of distinct object variables and y is an object variable distinct from $\mathbf{x}$. We use the following notation: $\{F\}$ for $F \vee \neg F$, $\forall \mathbf{x}y\{f(\mathbf{x}) = y\}$ for $\mathtt{Choice}(f)$ and $\forall \mathbf{x}\{p(\mathbf{x})\}$ for $\mathtt{Choice}(p)$. Then, $\{\mathbf{t} = \mathbf{t}'\}$, where $\mathbf{t}$ contains an intentional function constant and $\mathbf{t}'$ does not, represents the default rule stating that $\mathbf{t}$ has a value of $\mathbf{t}'$ if there is no other rule requiring $\mathbf{t}$ to take some other value.

Definition 9 *(Spatial choice formula).* *The* spatial choice formula *is a rule of the form (8) or (7):*

$$\{f_n(d_1, \ldots, d_n) = x\} \leftarrow \alpha_1 \wedge \alpha_2 \wedge \cdots \wedge \alpha_k, \tag{7}$$

$$\{Q_n(d_1, \ldots, d_n)\} \leftarrow \alpha_1 \wedge \alpha_2 \wedge \cdots \wedge \alpha_k. \tag{8}$$

where f_n is a parametric function, $x \in \mathbb{R}$, Q_n is a qualitative spatial relation, and for each $i \in \{1, \ldots, k\}$, α_i is a qualitative spatial relation or expression of a form $\{f_r(d_k, \ldots, d_m) = y\}$ or a polynomial equation or inequality, whereas $d_i \in D_i$ is an object of spatial type D_i.

Definition 10 *(Spatial frame axiom).* *The* spatial frame axiom *is a special case of a spatial choice formula which states that, by default, a spatial property remains the same in the next step of a simulation. It takes the form (9) or (10):*

$$\{f_n(d_1, \ldots, d_{n-1}, s+1) = x\} \leftarrow f_n(d_1, \ldots, d_{n-1}, s) = x, \tag{9}$$

$$\{Q_n(d_1, \ldots, d_{n-1}, s+1)\} \leftarrow Q_n(d_1, \ldots, d_{n-1}, s). \tag{10}$$

where f_n is a parametric function, $x \in \mathbb{R}$, Q_n is a qualitative spatial relation, and $s \in \mathbb{N}$ represents a step in the simulation.

Corollary 1. *One spatial frame axiom for each parametric function and qualitative spatial relation is enough to formalise the intuition that spatial properties, by default, do not change over time.*

The combination of spatial reasoning with stable model semantics and arithmetic over the reals enables the operationalisation of a range of novel features within the context of dynamic spatial reasoning. We present concrete examples of such features in Sect. 5.

4   System Implementation

We present our implementation of ASPMT(QS) that builds on ASPMT2SMT [4] – a compiler translating a tight fragment of ASPMT into SMT instances. Our system consists of an additional module for spatial reasoning and Z3 as the SMT solver. As our system operates on a tight fragment of ASPMT, input programs need to fulfil certain requirements, described in the following section. As output, our system either produces the stable models of the input programs, or states that no such model exists.

4.1   Syntax of Input Programs

The input program to our system needs to be *f-plain* to use Theorem 1 from [2].

Definition 11 *(f-plain formula).* *Let f be a function constant. A first–order formula is called f-plain if each atomic formula:*

- *does not contain f, or*
- *is of the form $f(\mathbf{t}) = u$, where $\mathbf{t}$ is a tuple of terms not containing f, and u is a term not containing f.*

Additionally, the input program needs to be *av-separated*, i.e. no variable occurring in an argument of an uninterpreted function is related to the value variable of another uninterpreted function via equality [4]. The input program is divided into declarations of:

- `sorts` (data types);
- `objects` (particular elements of given types);
- `constants` (functions);
- `variables` (variables associated with declared types).

The second part of the program consists of clauses. ASPMT(QS) supports:

- connectives: `&`, `|`, `not`, `->`, `<-`, and
- arithmetic operators: `<`, `<=`, `>=`, `>`, `=`, `!=`, `+`, `=`, `*`, with their usual meaning.

Additionally, ASPMT(QS) supports the following as native / first-class entities:

- **sorts** for geometric objects types, e.g., **point**, **segment**, **circle**, **triangle**;
- parametric functions describing objects parameters e.g., $x($**point**$)$, $r($**circle**$)$;
- qualitative relations, e.g., **rccEC**(**circle**, **circle**), **coincident**(**point**, **circle**).

▷ **Example 1: combining topology and size.** Consider a program describing three circles a, b, c such that a is discrete from b, b is discrete from c, and a is a proper part of c, declared as follows:

```
:- sorts
   circle.
:- objects
   a, b, c      :: circle.
:- constants
   .
:- variables
   C, C1, C2    :: circle.

{x(C)=X}. {y(C)=X}. {r(C)=X}.
rccDR(a,b)=true. rccDR(b,c)=true. rccPP(a,c)=true.
```

ASPMT(QS) checks if the spatial relations are satisfiable. In the case of a positive answer, a parametric model and computation time are presented. The output of the above mentioned program is:

```
r(a) = 0.5        r(b) = 1.0        r(c) = 0.25
x(a) = 1.0        x(b) = 1.0        x(c) = 1.0
y(a) = 3.0        y(b) = 1.0        y(c) = 3.0
```

This example demonstrates that ASPMT(QS) is capable of computing *composition tables*, in this case the RCC–5 table for circles [25]. Now, consider the addition of a further constraint to the program stating that circles a, b, c have the same radius:

```
<- r(a)=R1 & r(b)=R2 & r(c)=R3 & (R1!=R2 | R2!=R3 | R1!=R3).
```

This new program is an example of combining different *types* of qualitative information, namely topology and size, which is a non-trivial research topic within the relation algebraic spatial reasoning community; relation algebraic-based solvers such as GQR [17, 29] will not correctly determine inconsistencies in general for arbitrary combinations of different types of relations (orientation, shape, distance, etc.). In this case, ASPMT(QS) correctly determines that the spatial constraints are inconsistent:

```
UNSATISFIABLE; Z3 time in milliseconds: 10; Total time in milliseconds: 946
```

▷ **Example 2: combining topology and relative orientation.** Given three circles a, b, c let a be proper part of b, b discrete from c, and a in contact with c, declared as follows:

Fig. 1. Reasoning about consistent and refinement by combining topology and relative orientation.

```
:- sorts
    circle.
:- objects
    a, b, c        :: circle.
:- constants
    .
:- variables
    C, C1, C2      :: circle.

{x(C)=X}. {y(C)=X}. {r(C)=X}.
rccPP(a,b)=true. rccDR(b,c)=true. rccC(a,c)=true.
```

Given this basic qualitative information, ASPMT(QS) is able to refine the topological relations to infer that (Fig. 1a): (i) *a* must be a *tangential proper part* of *b* (ii) both *a* and *b* must be *externally connected* to *c*.

```
r(a) = 1.0        r(b) = 2.0        r(c) = 1.0
x(a) = 1.0        x(b) = 0.0        x(c) = 3.0
y(a) = 0.0        y(b) = 0.0        y(c) = 0.0
rccTPP(a,b) = true    rccEC(a,c) = true    rccEC(b,c) = true
```

We then add an additional constraint that the centre of *a* is *left of* the segment between the centres *b* to *c*.

```
...
left_of(center(a),center(b),center(c)).
```

ASPMT(QS) determines that this is inconsistent, i.e., the centres must be *collinear* (Fig. 1b).

```
UNSATISFIABLE;
```

5 Empirical Evaluation and Examples

In this section we present an empirical evaluation of ASPMT(QS) in comparison with other existing spatial reasoning systems. The range of problems demonstrate the unique, non-monotonic spatial reasoning features that ASPMT(QS) provides beyond what is possible using other currently available systems. Table 1 presents run times obtained by Clingo – an ASP grounder and solver [18],

Table 1. Cumulative results of performed tests. "—" indicates that the problem can not be formalised, "^{I}" indicates that indirect effects can not be formalised, "^{D}" indicates that default rules can not be formalised.

Problem	Clingo	GQR	CLP($\mathcal{QS}$)	ASPMT(QS)
Growth	$0.004s^I$	$0.014s^{I,D}$	$1.623s^D$	$0.396s$
Motion	$0.004s^I$	$0.013s^{I,D}$	$0.449s^D$	$15.386s$
Attach I	$0.008s^I$	—	$3.139s^D$	$0.395s$
Attach II	—	—	$2.789s^D$	$0.642s$

GQR – a binary constraint calculi reasoner [17], CLP($\mathcal{QS}$) – a declarative spatial reasoning system [8] and our ASPMT(QS) implementation. Tests were performed on an Intel Core 2 Duo 2.00 GHZ CPU with 4 GB RAM running Ubuntu 14.04. The polynomial encodings of the topological relations have not been included here for space considerations.

5.1 Ramification Problem

The following two problems, *Growth* and *Motion*, were introduced in [5]. Consider the initial situation S_0 presented in Fig. 2, consisting of three cells: a, b, c, such that a is a non-tangential proper part of b: $\mathtt{rccNTPP}(a, b, 0)$, and b is externally connected to c: $\mathtt{rccEC}(b, c, 0)$.

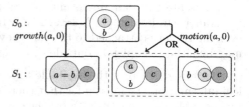

Fig. 2. Indirect effects of $growth(a, 0)$ and $motion(a, 0)$ events.

▷ *Growth.* Let a grow in step S_0; the event $\mathtt{growth}(a, 0)$ occurs and leads to a successor situation S_1. The direct effect of $growth(a, 0)$ is a change of a relation between a and b from $\mathtt{rccNTPP}(a, b, 0)$ to $\mathtt{rccEQ}(a, b, 1)$ (i.e. a is equal to b). No change of the relation between a and c is directly stated, and thus we must derive the relation $\mathtt{rccEC}(a, c, 1)$ as an indirect effect.

▷ *Motion.* Let a move in step S_0; the event $\mathtt{motion}(a, 0)$ leads to a successor situation S_1. The direct effect is a change of the relation $\mathtt{rccNTPP}(a, b, 0)$ to $\mathtt{rccTPP}(a, b, 1)$ (a is a tangential proper part of b). In the successor situation S_1 we must determine that the relation between a and c can only be either $\mathtt{rccDC}(a, c, 1)$ or $\mathtt{rccEC}(a, c, 1)$.

GQR provides no support for domain-specific reasoning, and thus we encoded the problem as two distinct qualitative constraint networks (one for each simulation step) and solved them independently i.e. with no definition of *growth* and *motion*. Thus, GQR is not able to produce any additional information about indirect effects. As Clingo lacks any mechanism for analytic geometry, we implemented the RCC8 composition table and thus it inherits the incompleteness of relation algebraic reasoning. While CLP(QS) facilitates the modelling of domain rules such as *growth*, there is no native support for default reasoning and thus we forced b and c to remain unchanged between simulation steps, otherwise all combinations of spatially consistent actions on b and c are produced without any preference (i.e. leading to the frame problem).

In contrast, ASPMT(QS) can express spatial inertia, and derives indirect effects directly from spatial reasoning: in the *Growth* problem ASPMT(QS) abduces that a has to be concentric with b in S_0 (otherwise a *move* event would also need to occur). Checking global consistency of scenarios that contain interdependent spatial relations is a crucial feature that is enabled by a support polynomial encodings and is provided only by CLP(QS) and ASPMT(QS).

5.2   Geometric Reasoning and the Frame Problem

In problems *Attachment I* and *Attachment II* the initial situation S_0 consists of three objects (circles), namely car, trailer and garage as presented in Fig. 3. Initially, the trailer is attached to the car: rccEC(car, trailer, 0), attached(car, trailer, 0). The successor situation S_1 is described by rccTPP(car, garage, 1). The task is to infer the possible relations between the trailer and the garage, and the necessary actions that would need to occur in each scenario.

There are two domain-specific actions: the car can move, move(car, X), and the trailer can be detached, detach(car, trailer, X) in simulation step X. Whenever the trailer is attached to the car, they remain rccEC. The car and the trailer may be either completely outside or completely inside the garage.

▷ *Attachment I.* Given the available topological information, we must infer that there are two possible solutions (Fig. 3); (a) the car was detached from the trailer and then moved into the garage: (b) the car, together with the *trailer* attached to it, moved into the garage:

▷ *Attachment II.* We are given additional geometric information about the objects' size: $r(car) = 2$, $r(trailer) = 2$ and $r(garage) = 3$. Case (b) is now inconsistent, and we must determine that the only possible solution is (a).

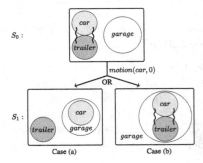

Fig. 3. Non-monotonic reasoning with additional geometric information.

These domain-specific rules require default reasoning: "*typically* the trailer remains in the same position" and "*typically* the trailer remains attached to the car". The later default rule is formalised in ASPMT(QS) by means of the spatial defaul.: The formalisation of such rules addresses the frame problem. GQR is not capable of expressing the domain-specific rules for detachment and attachment in *Attachment I* and *Attachment II*. Neither GQR nor Clingo are capable of reasoning with a combination of topological and numerical information, as required in *Attachment II*. As CLP(QS) cannot express default rules, we can not capture the notion that, for example, the trailer should typically remain in the same position unless we have some explicit reason for determining that it moved; once again this leads to an exhaustive enumeration of all possible scenarios without being able to specify preferences, i.e. the frame problem, and thus CLP(QS) will not scale in larger scenarios.

The results of the empirical evaluation show that ASPMT(QS) is the only system that is capable of (a) non-monotonic spatial reasoning, (b) expressing domain-specific rules that also have spatial aspects, and (c) integrating both qualitative and numerical information. Regarding the greater execution times in comparison to CLP(QS), we have not yet implemented any optimisations with respect to spatial reasoning; this is one of the directions of future work.

6   Conclusions

We have presented ASPMT(QS), a novel approach for reasoning about spatial change within a KR paradigm. By integrating dynamic spatial reasoning within a KR framework, namely answer set programming (modulo theories), our system can be used to model behaviour patterns that characterise high-level processes, events, and activities as identifiable with respect to a general characterisation of commonsense *reasoning about space, actions, and change* [6,9]. ASPMT(QS) is capable of sound and complete spatial reasoning, and combining qualitative and quantitative spatial information when reasoning non-monotonically; this is due to the approach of encoding spatial relations as polynomial constraints, and solving using SMT solvers with the theory of real nonlinear arithmetic. We have demonstrated that no other existing spatial reasoning system is capable of supporting the key non-monotonic spatial reasoning features (e.g., spatial inertia, ramification) provided by ASPMT(QS) in the context of a mainstream knowledge representation and reasoning method, namely, answer set programming.

Acknowledgments. This research is partially supported by: (a) the Polish National Science Centre grant 2011/02/A/HS1/0039; and (b). the DesignSpace Research Group www.design-space.org.

References

1. Allen, J.F.: Maintaining knowledge about temporal intervals. Commun. ACM **26**(11), 832–843 (1983)
2. Bartholomew, M., Lee, J.: Stable models of formulas with intensional functions. In: KR (2012)
3. Bartholomew, M., Lee, J.: Functional stable model semantics and answer set programming modulo theories. In: Proceedings of the Twenty-Third International Joint Conference on Artificial Intelligence, pp. 718–724. AAAI Press (2013)
4. Bartholomew, M., Lee, J.: System ASPMT2SMT: computing ASPMT theories by SMT solvers. In: Fermé, E., Leite, J. (eds.) JELIA 2014. LNCS, vol. 8761, pp. 529–542. Springer, Heidelberg (2014)

5. Bhatt, M.: (Some) Default and non-monotonic aspects of qualitative spatial reasoning. In: AAAI 2008 Technical reports, Workshop on Spatial and Temporal Reasoning, pp. 1–6 (2008)

6. Bhatt, M.: Reasoning about space, actions and change: a paradigm for applications of spatial reasoning. In: Qualitative Spatial Representation and Reasoning: Trends and Future Directions. IGI Global, USA (2012)

7. Bhatt, M., Guesgen, H., Wölfl, S., Hazarika, S.: Qualitative spatial and temporal reasoning: emerging applications, trends, and directions. Spat. Cogn. Comput. **11**(1), 1–14 (2011)

8. Bhatt, M., Lee, J.H., Schultz, C.: CLP(QS): a declarative spatial reasoning framework. In: Egenhofer, M., Giudice, N., Moratz, R., Worboys, M. (eds.) COSIT 2011. LNCS, vol. 6899, pp. 210–230. Springer, Heidelberg (2011)

9. Bhatt, M., Loke, S.: Modelling dynamic spatial systems in the situation calculus. Spat. Cogn. Comput. **8**(1), 86–130 (2008)

10. Bhatt, M., Wallgrün, J.O.: Geospatial narratives and their spatio-temporal dynamics: commonsense reasoning for high-level analyses in geographic information systems. ISPRS Int. J. Geo-Inf. **3**(1), 166–205 (2014)

11. Bouhineau, D.: Solving geometrical constraint systems using CLP based on linear constraint solver. In: Pfalzgraf, J., Calmet, J., Campbell, J. (eds.) AISMC 1996. LNCS, vol. 1138. Springer, Heidelberg (1996)

12. Bouhineau, D., Trilling, L., Cohen, J.: An application of CLP: checking the correctness of theorems in geometry. Constraints **4**(4), 383–405 (1999)

13. Ferraris, P.: Answer sets for propositional theories. In: Baral, C., Greco, G., Leone, N., Terracina, G. (eds.) LPNMR 2005. LNCS (LNAI), vol. 3662, pp. 119–131. Springer, Heidelberg (2005)

14. Ferraris, P., Lee, J., Lifschitz, V.: Stable models and circumscription. Artif. Intell. **175**(1), 236–263 (2011)

15. Frank, A.U.: Qualitative spatial reasoning with cardinal directions. In: Kaindl, H. (ed.) 7. Österreichische Artificial-Intelligence-Tagung/Seventh Austrian Conference on Artificial Intelligence. Informatik-Fachberichte, vol. 287, pp. 157–167. Springer, Heidelberg (1991)

16. Freksa, C.: Using orientation information for qualitative spatial reasoning. In: Frank, A.U., Formentini, U., Campari, I. (eds.) GIS 1992. LNCS, vol. 639, pp. 162–178. Springer, Heidelberg (1992)

17. Gantner, Z., Westphal, M., Wölfl, S.: GQR-A fast reasoner for binary qualitative constraint calculi. In: Proceedings of AAAI, vol. 8 (2008)

18. Gebser, M., Kaminski, R., Kaufmann, B., Schaub, T.: Clingo= ASP+ control: Preliminary report. arXiv preprint arXiv:1405.3694 (2014)

19. Gelfond, M., Lifschitz, V.: The stable model semantics for logic programming. ICLP/SLP, vol. 88, pp. 1070–1080 (1988)

20. Guesgen, H.W.: Spatial reasoning based on Allen's temporal logic. Technical report TR-89-049, International Computer Science Institute Berkeley (1989)

21. Lee, J.H.: The complexity of reasoning with relative directions. In: 21st European Conference on Artificial Intelligence (ECAI 2014) (2014)

22. Moratz, R.: Representing relative direction as a binary relation of oriented points. In: Brewka, G., Coradeschi, S., Perini, A., Traverso, P. (eds.) ECAI. Frontiers in Artificial Intelligence and Applications, vol. 141, pp. 407–411. IOS Press (2006)

23. Pesant, G., Boyer, M.: QUAD-CLP(R): adding the power of quadratic constraints. In: Borning, A. (ed.) PPCP 1994. LNCS, vol. 874, pp. 95–108. Springer, Heidelberg (1994)

24. Pesant, G., Boyer, M.: Reasoning about solids using constraint logic programming. J. Autom. Reasoning **22**(3), 241–262 (1999)

25. Randell, D.A., Cui, Z., Cohn, A.G.: A spatial logic based on regions and connection. In: KR, vol. 92, pp. 165–176 (1992)

26. Schultz, C., Bhatt, M.: Towards a declarative spatial reasoning system. In: 20th European Conference on Artificial Intelligence (ECAI 2012) (2012)

27. Schultz, C., Bhatt, M.: Declarative spatial reasoning with boolean combinations of axis-aligned rectangular polytopes. In: ECAI 2014–21st European Conference on Artificial Intelligence, pp. 795–800 (2014)

28. Varzi, A.C.: Parts, wholes, and part-whole relations: the prospects of mereotopology. Data Knowl. Eng. **20**(3), 259–286 (1996)

29. Wölfl, S., Westphal, M.: On combinations of binary qualitative constraint calculi. In: Proceedings of the 21st International Joint Conference on Artificial Intelligence. IJCAI 2009, Pasadena, California, USA, 11–17 July 2009, pp. 967–973 (2009)

Mobile Robot Planning Using Action Language $\mathcal{BC}$ with an Abstraction Hierarchy

Shiqi Zhang[1]([✉]), Fangkai Yang[2], Piyush Khandelwal[1], and Peter Stone[1]

[1] Department of Computer Science, The University of Texas at Austin,
2317 Speedway, Stop D9500, Austin, TX 78712, USA
{szhang,piyushk,pstone}@cs.utexas.edu
[2] Schlumberger Software Technology, Schlumberger Ltd,
5599 San Felipe Rd, Houston, TX 77056, USA
fkyang@cs.utexas.edu

Abstract. Planning in real-world environments can be challenging for intelligent robots due to incomplete domain knowledge that results from unpredictable domain dynamism, and due to lack of global observability. Action language $\mathcal{BC}$ can be used for planning by formalizing the preconditions and (direct and indirect) effects of actions, and is especially suited for planning in robotic domains by incorporating defaults with the incomplete domain knowledge. However, planning with $\mathcal{BC}$ is very computationally expensive, especially when action costs are considered. We introduce algorithm *PlanHG* for formalizing $\mathcal{BC}$ domains at different abstraction levels in order to trade optimality for significant efficiency improvement when aiming to minimize overall plan cost. We observe orders of magnitude improvement in efficiency compared to a standard "flat" planning approach.

1 Introduction

To operate in real-world environments, intelligent robots need to represent and reason with a large amount of domain knowledge about robot actions and environments. However, domain knowledge given to the robot is usually incomplete (due to unpredictable domain dynamism) and defeasible (i.e., usually true but not always). From STRIPS [4] to PDDL [17], many action languages (and their extensions) have been developed to support automated plan generation by formalizing action preconditions and effects. While some action languages support reasoning about the knowledge not directly related to actions, e.g., PDDL has semantics to reason with axioms [25], most action languages lack a strong capability of reasoning with incomplete knowledge in dynamic domains, making it difficult to embrace rich domain knowledge into planning scenarios. Action language $\mathcal{BC}$ can be used for planning with guaranteed soundness by formalizing the preconditions and (direct and indirect) effects of actions [14]. $\mathcal{BC}$ inherits the knowledge representation and reasoning (KRR) advantages from action languages $\mathcal{B}$ [9] and $\mathcal{C}+$ [10], and is especially suited for planning in robotic domains.

Unfortunately, in robotic domains where action costs need to be considered, planning with action language $\mathcal{BC}$ is very computationally expensive.

© Springer International Publishing Switzerland 2015
F. Calimeri et al. (Eds.): LPNMR 2015, LNAI 9345, pp. 502–516, 2015.
DOI: 10.1007/978-3-319-23264-5_42

For instance, in the office domain presented in [13], generating the optimal plan to visit three people in different rooms takes more than 5 min on a powerful desktop machine (details in Sect. 5), where the optimal plan has about 30 actions. Such long planning time prevents the robot from being deemed useful in real-world environments.

Hierarchical planning has been studied for years and people have developed many algorithms including Hierarchical Task Network (HTN) [3] and Hierarchical Planning in the Now (HPN) [12]. Different from existing work on hierarchical planning that aims to reduce the amount of search with guaranteed optimality (e.g., [16]), we trade optimality for significant improvements in efficiency (similar to HPN). We adapt the idea of describing task domains at different abstraction levels [6] and propose an algorithm to enable hierarchical planning with action language $\mathcal{BC}$ in real-world robotic domains.

This algorithm has been fully implemented in simulation and on a physical robot. Experiments on a mail collection problem show 2 orders of magnitude improvements of efficiency with a 11.25 % loss in optimality, compared to a baseline algorithm that plans with a non-hierarchical domain description in $\mathcal{BC}$ [13]. To the best of our knowledge, this is the first work that combines the KRR advantages of a modern action language and the efficiency of hierarchical planning to enable mobile robots to compute provably sound plans in real-world environments.

2 Related Work

This work is closely related to research areas including action languages and hierarchical planning. We select representative research on these topics.

Action Languages: The planning domain definition language (PDDL) has been widely applied to planning problems [17]. One of the most appealing advantages of (the official versions of) PDDL is its syntax, which despite being simple supports important features of STRIPS [4], ADL [20], and other features such as conditional action effects (PDDL1.2) and numeric fluents (PDDL2.1). Furthermore, advanced planning algorithms such as Fast-Foward [19] and Fast-Downward [11] have been implemented in existing planning systems supporting PDDL.

While PDDL is strong in efficient plan generation, the official versions of PDDL do not focus on reasoning with default knowledge, which is important for robots to plan with incomplete knowledge in dynamic environments. Action language $\mathcal{C}+$ supports the representation and reasoning with defaults [10], but does not allow recursively defined fluents that are frequently needed in robotic domains (action language $\mathcal{B}$ does), as will be shown in Sect. 3. $\mathcal{BC}$, an action language recently developed based on answer set semantics [8], can be used to compute provably sound plans while supporting representation of and reasoning with defaults with exceptions at different levels [14].

Recently, a two-level architecture has been developed for KRR in robotics [26], where the high level uses action language AL for symbolic planning and the low level uses probabilistic algorithms for modeling uncertainties. In that work, each default is associated with a consistency-restoring rule for restoring consistency in history. In contrast, we intentionally make our robots memoryless to avoid reasoning about history, i.e., whenever robot observations have conflicts with defaults, our robot starts over by replanning with defaults and the observed "facts".

Hierarchical Planning: In existing research on hierarchical planning, the hierarchy is frequently constructed through setting up connections either between actions or between states. For instance, macro-actions (also called *complex* or *composite* actions) are described as a sequence of primitive actions and possibly some imperative constructs, e.g., hierarchical task network [3], planning with composite actions [1], planning with complex actions [18], ordered task decomposition [2], and hierarchical planning in the now [12]. These macro-actions are either directly expanded after a plan is generated, or expanded in the reasoning process using a predefined structure. These macro-actions limit the flexibility of reducing plan costs at a finer abstraction level.

Another way of constructing the hierarchy is to describe the domain at different abstraction levels through setting up connections between states, where a state at a coarser (higher) level includes a set of states at a finer (lower) level [6,23,24]. Planning in such systems happens in a top-down manner and constraints extracted from coarser levels help improve the efficiency in computing plans at finer levels. This mechanism allows more flexibility in planning at finer levels, compared to macro-based hierarchical planning algorithms. In this paper, we introduce action costs to such abstraction-based hierarchical planning algorithms and implement the algorithm using action language $\mathcal{BC}$ on a real robot system.

3   Abstraction Hierarchy Formalization

A $\mathcal{BC}$ action description D denotes a transition system $T(D)$, which is a digraph whose vertices are *states*, which is a set of atoms, and whose edges are *actions*. A transition in $T(D)$ is of the form $\langle s, a, s' \rangle$, where a is an action constant, and s and s' are states before and after executing a. A path $P(n)$ of length n in the transition system is of the form:

$$\langle s_0, a_0, \ldots, s_{n-1}, a_{n-1}, s_n \rangle$$

where s_i $(0 \leq i \leq n)$ are states and a_i $(0 \leq i \leq n-1)$ are actions. We use $Len(P)$ to denote the length of a path. $P^s(i)$ denotes state s_i and $P^a(i)$ denotes action a_i. We use $f(D)$ to represent the set of fluents occurring in D, and $a(D)$ to represent the set of actions occurring in D. To define the notion of abstraction hierarchy, we first define the cost function C that maps a tuple (s, a) to an integer $C(s, a)$ that denotes the cost of executing action a at state s. Furthermore, $cost(P(n))$ is the cost of path $P(n)$:

$$cost(P(n)) = \Sigma_{0 \leq i < n} C(s_i, a_i) \tag{1}$$

Given an action description D and a cost function C, its *abstraction hierarchy* $\mathcal{H}$ is a tuple $(\mathcal{D}, \mathcal{L})$: $\mathcal{D}$ is a list of action descriptions $D_1, D_2, \ldots, D_d$ such that $f(D_i) \subseteq f(D_j)$ for $1 \le i < j \le d$, where $D_d = D$ and d is the depth of $\mathcal{H}$; and $\mathcal{L}$ is the *step bound estimation function*.

$$\mathcal{L}(a) = \max_{\langle s, a, s' \rangle \, \in \, T(D_i)} \Big(Len\big(\hat{P}(s, s') \big) \Big) \qquad (2)$$

Given an action constant $a \in a(D_i)$, $\mathcal{L}$ maps a to an integer $\mathcal{L}(a)$ representing the minimum number of steps needed to ensure that the effect of a can be optimally achieved using actions in $a(D_{i+1})$ as shown in Eq. 2, where $\mathcal{L}$ is independent of s and s', and is precomputed to reduce the planning time[1]. $\hat{P}$ represents the path of the plan that leads the transition from s to s' with minimum plan cost. If we use $A(s)$ to represent the set of literals that specify state s, $\hat{P}(s, s')$ can be computed by:

$$\hat{P}(s, s') = \underset{\substack{P(n) \in T(D_{i+1}), n \in \mathbb{N}, \\ A(P^s(0)) \subseteq A(s), A(P^s(n)) \subseteq A(s')}}{\arg\min} \Big(cost\big(P(n) \big) \Big) \qquad (3)$$

Note that states s and s' and action a are at level i while path P is at level $i + 1$. Intuitively, the abstraction hierarchy $\mathcal{H}$ contains a set of action descriptions where each description formalizes the same dynamic domain at a different granularity. The hierarchy is organized from the most coarse description D_1 to the most concrete description D_d. Different from existing work on hierarchical planning using macro actions, we use function $\mathcal{L}$ to provide step bounds in the search for plans at lower levels. This is an important criterion of our approach as it provides flexibility in reducing overall plan costs in lower levels. As an example, we next apply this hierarchy to a real-world robot planning problem in $\mathcal{BC}$.

Mail Collection Problem: A mobile robot drops by offices at 2 pm every day to collect outgoing mail from the residents. However, some people may not be in their offices at that time, so they can pass their outgoing mail to colleagues in other offices, and send this information to the robot. When the robot collects the mail, it should obtain it while only visiting people as necessary. An example floor plan is shown in Fig. 1. We will use meta-variables $E, E_1, E_2, \ldots$ to denote people (*alice*, *bob*, *carol*, *daniel* and *erin*), $R, R_1, R_2, \ldots$ to denote rooms, and $K, K_1, K_2, \ldots$ to denote doors. Specifically, o_1, o_2, o_3, o_4 are offices, lab_1 is a lab and *cor* is a room, where offices and labs are sub-sorts of room.

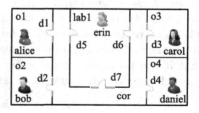

Fig. 1. Example floor plan.

---

[1] As a preprocessing step, computing $\mathcal{L}$ does not affect the runtime efficiency, so we leave the discussion of its complexity to future work.

This example domain has been formulated at three levels of abstraction. The fluents at the most abstract level primarily describe how mail is passed from one person to another and if mail has been collected from each person. At the middle level, we add fluents to describe the connections of rooms through doors, but still disregard the details about the robot's more refined position in a room and if the doors are open or not. Finally, all domain details are added into the bottom level. An action in the bottom level must be primitive (i.e., can be physically executed by the robot) and currently this hierarchy is manually constructed.

Action Description D_1: In D_1, we use $passto(E_1, E_2)$ to describe E_1's mail has been passed to E_2. The current locations of the robot and a person E are described by $loc = R$ and $inside(E, R)$ respectively. Whether the robot has collected mail from person E is represented by $mailcollected(E)$. For instance, the static law below states: if E_1's mail has been passed to E_2 and that the robot has collected mail from E_2, then E_1's mail has been collected as well using a recursive definition of fluent $mailcollected$:

$$mailcollected(E_1) \text{ if } mailcollected(E_2), \ passto(E_1, E_2).$$

The two laws below state that person E cannot be in two different rooms at the same time and that by default E's location does not change over time (a commonsense law of inertia), where $inside$ is an *inertial fluent*.

$$\sim inside(E, R_2) \text{ if } inside(E, R_1) \ (R_1 \neq R_2).$$
$$\textbf{inertial } inside(E, R).$$

Action $serve$ in D_1 states that serving person E in room R causes $mailcollected(E)$ to be true and the robot to be in R.

$$serve(E) \textbf{ causes } mailcollected(E).$$
$$serve(E) \textbf{ causes } loc = R \textbf{ if } inside(E, R).$$

Action Description D_2: D_2 inherits all fluents and corresponding non-action rules from D_1 (actions of D_1 are discarded) and further adds fluents to describe whether a room has a door using $hasdoor(R, K)$ and whether two adjacent rooms are connected through a door using $acc(R_1, K, R_2)$. The static laws below state that if two rooms share the same door, then they are accessible to each other through the door and that acc is symmetric.

$$acc(R_1, K, R_2) \textbf{ if } hasdoor(R_1, K), \ hasdoor(R_2, K).$$
$$acc(R_1, K, R_2) \textbf{ if } acc(R_2, K, R_1).$$

We add defaults to reason with incomplete knowledge. For instance, rooms R_1 and R_2 are not accessible through door K by default. This default value can be reverted if there is evidence supporting the opposite.

$$\textbf{default } \sim acc(R_1, K, R_2).$$

Using the fluents in D_2, we can formalize action $collectmail(E)$ that is similar to action $serve$ in D_1, and action $cross(K)$ that allows the robot to cross door K

to move from room R_1 to room R_2, if R_2 is accessible from R_1 through door K. There is a restriction on the executability of $cross(K)$: the robot cannot cross a door if that door is not accessible from the robot's current location.

$$cross(K) \textbf{ causes } loc = R_2 \textbf{ if } loc = R_1, acc(R_1, K, R_2).$$
$$\textbf{nonexecutable } cross(K) \textbf{ if } loc = R, \sim hasdoor(R, K).$$

Action Description D_3: D_3 inherits all fluents and corresponding non-action rules from D_2 (actions of D_2 are discarded) and further introduces fluents $beside(K)$ to describe whether the robot is beside door K, $facing(K)$ to describe whether the robot is beside and facing door K, and $open(K)$ to describe if door K is open. We use Monte Carlo Localization [5] to estimate the robot's exact position (including orientation) in physical environments. Using an occupancy-grid map with manually added semantic labels, this exact position specifies the values of loc, $beside$ and $facing$ and is also used for path planning. Using the fluents in D_3, we can formalize the primitive actions $approach(K)$, $opendoor(K)$, $gothrough(K)$, and $collectmail(E)$. D_3 corresponds to the "flat" action description presented in previous work [13].

Action descriptions D_1, D_2 and D_3 together determine $\mathcal{D}$, the first element of the abstraction hierarchy $\mathcal{H}$. The other element is the step bound estimation function $\mathcal{L}$, which is partially decided by the cost function C, as presented Eq. 1. The value of $C(s, a)$ is assigned empirically based on robot experiments using existing approach [13]. As an illustrative example, let us consider the calculation of $\mathcal{L}(serve)$ using Eqs. 2 and 3. Since $serve$ is an action in D_1, we first collect all possible $\langle s, serve, s' \rangle \in T(D_1)$. The longest path in D_2 that can be used to achieve the same effect as a $serve$ action in D_1 occurs when $loc = o_3 \in s$ and $mailcollected(alice) \in s'$. The corresponding path $P_2(5)$ includes the following actions in the order of execution:

$$cross(d_3), cross(d_6), cross(d_5), cross(d_1), collectmail(alice).$$

Consequently, $\mathcal{L}(serve) = 5$. Similarly, $\mathcal{L}(cross) = 3$.

4 Planning Using an Abstraction Hierarchy

In this section, we formally define two planning problems that aim to minimize the plan length (Type-I) and plan cost (Type-II) respectively, where the first is a special case of the second. Then we propose two algorithms to solve Type-II problems using an abstraction hierarchy.

Type-I Problem: A Type-I planning problem aims at minimizing the plan length (i.e., the number of actions), and is defined as a tuple (D, S, G). D is an action description; S is a *state constraint set* including state constraints of the form $i : A_i$, where i is an integer denoting the *timestamp* at which A_i (a set of fluent atoms) needs to be met; and G is a list of fluent atoms G_i, which are *goals*. The initial system state is specified as a part of the state constraint set as $0 : A_0 \in S$.

Given (D, S, G), a *satisfactory path* is a path $P(n)$ (defined in Sect. 3) of the transition system $T(D)$ such that $A_i \subset s_i$ for every $i : A_i \subset S$, and $G_i \in s_n$ for every $G_i \in G$. The *satisfactory plan* is the list $(a_0, \cdots, a_{n-1})$ obtained from $P(n)$. A satisfactory plan is a *shortest plan* if the length of the satisfactory path is minimal among all satisfactory paths. Algorithms that solve this problem do not consider the overall cost of plans. To find the shortest-length plan in a Type-I problem, we incrementally increase the plan length in solvers until a satisfactory plan is found.

Type-II Problem: A Type-II problem aims at minimizing overall plan cost, and can be defined as a tuple (D, S, G, C), where C is the cost function of actions. Given an optimizing planning problem, an *optimal path* $P(n)$ is a satisfactory path of the satisfactory planning problem (D, S, G) such that the overall cost of the path, $cost(P(n))$, is minimal among all satisfactory paths of (D, S, G). Incrementally lengthening the plan length will not necessarily lead to the optimal plan because a very long plan could have the lowest cost. Algorithms that solve this problem compute plans toward minimizing the overall cost of the plan.

Without concurrent actions, a Type-I problem can be reduced to a Type-II problem by using unit cost for any (s, a) in function C. Therefore, we will focus on applying the abstraction hierarchy to Type-II problems.

4.1   PlanHG: The Proposed Planning Algorithm

Given a Type-II problem (D, S, G, C) and hierarchy $\mathcal{H} = (\mathcal{D}, \mathcal{L})$, for a state $P^s(i)$ in a path $P(n)$, we define its *shifted timestamp* in Eq. 4. The shifted timestamp for state $P^s(i)$ is the timestamp when this state constraint needs to be achieved when $P(n)$ is further elaborated at the next level $i + 1$. State constraint $sh(i) : P^s(i)$ is functionally a "bottleneck" that guides the solution path in the next level of hierarchy by reducing the search space.

$$sh(i) = \sum_{a_j \in P(n),\ j < i} \mathcal{L}\big(P^a(j)\big) - 1 \tag{4}$$

Furthermore, we impose the restriction that the only constraint contained in S is the initial state that can be sensed by the robot. This restriction allows us to easily project S and G on to each level of the hierarchy as S_i and G_i, respectively. As a result, we obtain an optimizing planning problem at each abstraction level: (D_1, S_1, G_1, C), (D_2, S_2, G_2, C), ..., (D_d, S_d, G_d, C). For a Type-II problem (D_i, S_i, G_i, C) at the ith level, let the path obtained from level $i - 1$ be $P_{i-1}(n)$. We define the *extended state constraint set* at the ith level, S_i', in Eq. 5. Therefore, a *guided* Type-II problem (D_i, S_i', G_i, C) is formed at the ith level using (D_i, S_i, G_i, C) and state constraints extracted from level $i - 1$. We call it a "guided" problem because the state constraints reduce the search space in planning at the ith level.

$$S_i' = S_i \cup \bigcup_{1 \le j \le n-1} sh(j) : P_{i-1}^s(j). \tag{5}$$

Algorithm 1. PlanHG: Planning using $\mathcal{H}$ while applying $\mathcal{L}$ globally

---

Input: Type-II problem (D, S, G, C), and abstraction hierarchy $\mathcal{H} = (\mathcal{D}, \mathcal{L})$, where
$\quad\mathcal{D} = (D_1, \ldots, D_d)$, and $D_d = D$
1: create a list of problems $(D_i, S_i, G_i, C), 1 \leq i \leq d$ using (D, S, G, C) and $\mathcal{D}$
2: generate path P_1 for (D_1, S_1, G_1, C)
3: **for** level $i \in \{2, \ldots, d\}$ **do**
4: $\quad$ compute S_i' based on S_i and P_{i-1}, using Eq. (5)
5: $\quad$ generate path P_i for (D_i, S_i', G_i, C)
6: **end for**
7: **return** the plan obtained from P_d

Solving Type-II problems directly using the optimization function of answer set solvers may require prohibitively long time. Using an abstraction hierarchy, we can obtain a list of guided Type-II problems. In practice, each level has action $noop(I)$ of zero cost representing no operation at timestamp I. The optimal path generated at a higher level is passed down as "bottlenecks" such that the Type-II problem at a lower level becomes a guided Type-II problem, until the bottom level is reached. This approach guarantees the soundness of generated plans but may lead to sub-optimal results. We present Algorithm 1 that solves Type-II problems using an abstraction hierarchy $\mathcal{H} = (\mathcal{D}, \mathcal{L})$. We call this algorithm *PlanHG* to identify the use of the hierarchy and *global* minimization of plan costs at each level.

In the mail collection domain, the robot can obtain the initial state constraint from its internal knowledge base and sensor readings. For instance, initially the robot can perceive that it is located in *cor* and beside d_4. Such information is used to automatically create a state constraint set S:

$$\{0 : loc = cor, \ 0 : ibeside(d_4), \ 0 : \sim facing(D)\}.$$

Given a goal G of *mailcollected(erin)*, at level 1 the projection S_1 becomes $\{0 : loc = cor\}$ and the goal $G_1 = G$. The solver returns the optimal path:

$$\langle s_0 = \{loc = cor, \sim mailcollected(erin)\}, \ a_0 = \{serve(erin)\},$$
$$s_1 = \{loc = lab_1, mailcollected(erin)\}\rangle$$

Now, we can compute the shifted timestamps for s_0 and s_1 given $\mathcal{L}(serve) = 5$ (Sect. 3) and we obtain the guided state constraint set S_2':

$$0 : loc = cor, 0 : \sim mailcollected(erin),$$
$$5 : loc = lab_1, 5 : mailcollected(erin).$$

The guided Type-II problem (D_2, S_2', G_2, C) aims to find an optimal plan such that at time 0 the robot is in *cor*, at time 5 the robot is in lab_1 and Erin's mail is collected, and the goal of Erin's mail being collected is achieved. Indeed, the optimal plan generated at this level consists of two actions: $cross(d_7), collectmail(erin)$. Note that $cross(d_7)$ is selected because it has a lower cost than $cross(d_5)$ and $cross(d_6)$. Using this plan, we can generate the

---

Algorithm 2. PlanHL: Planning using $\mathcal{H}$ while applying $\mathcal{L}$ locally

---
Input: Type-II problem (D, S, G, C), and abstraction hierarchy $\mathcal{H} = (\mathcal{D}, \mathcal{L})$, where $\mathcal{D} = (D_1, \ldots, D_d)$, and $D_d = D$

1: generate path P' for (D_i, S, G, C), where, in a top-down manner, $i = 1$ at the first call.
2: **if** P' includes only primitive actions **then**
3:    **return** P'
4: **end if**
5: generate a list of optimizing planning problems using P' and $(\mathcal{D}, \mathcal{L})$: $(D_i, j_k : s_k, s_{k+1}, C)$, where $k \in \{1, \ldots, l-1\}$, and l is the length of P'
6: **for** $k \in \{1, \ldots, l-1\}$ **do**
7:    call Algorithm 2 to solve $(D_{i+1}, j_k : s_k, s_{k+1}, C)$, and compute P'_k
8: **end for**
9: **return** $P = (P'_1, \ldots, P'_{l-1})$

---

next level of state constraints that require the robot to be in lab_1. At level 3, the robot will execute $approach(d_6)$, $open(d_6)$, $gothrough(d_6)$ instead of going through d_7 because this plan meets the state constraint requirements, but is cheaper due to the robot's current position (beside d_4). This flexibility is attributed to the strategy that instead of expanding macro-actions, we generate plans for the same problem described at different abstraction levels and meet the requirement of state constraints.

We will use *PlanFG* to represent a special form of algorithm PlanHG that does not pass state constraints to lower levels but simply plans at the bottom level. PlanFG is a "flat" planning algorithm as presented in [13].

4.2   PlanHL: A Baseline Planning Algorithm

Alternatively, instead of satisfying all state constraints simultaneously, we can treat each pair of consecutive state constraints as a specification of a sub-problem. In this case, the step bound estimation function $\mathcal{L}$ is used for finding local optimal plans. Following this idea, a guided Type-II problem at the ith level, (D_i, S'_i, G_i, C), can be split into a sequence of Type-II subproblems $(D_i, j_k : s_k, s_{k+1}, C)$ for $0 \leq k \leq l - 1$, where S'_i is of the form: $\{j_0 : s_0, \ldots, j_l : s_l\}$, where $j_1 < j_2 < \ldots < j_l$.

The optimal paths of these problems are then joined to obtain the solution to the original Type-II problem. This algorithm is presented in Algorithm 2, where the implementation uses depth-first search to recursively call itself until reaching the bottom level. We name this algorithm *PlanHL* to identify the use of $\mathcal{H}$ and *local* minimization of plan costs using function $\mathcal{L}$ at each level. In comparison to PlanHL, algorithm PlanHG (proposed) does not decompose the original problem to subproblems at each level. Instead, it generates paths for the original problem to simultaneously satisfy all state constraints (by applying step bound estimation function $\mathcal{L}$ globally) at each level, toward minimizing the overall plan cost. Since both the algorithms are sacrificing plan quality for efficiency in solving Type-II problem, neither of the algorithms can guarantee the optimality (i.e. minimal cost) of generated plans, but the provably sound semantics of action language $\mathcal{BC}$ ensures the soundness of PlanHG (and PlanHL).

5   Experiments

The abstraction hierarchy $\mathcal{H}$ (Sect. 3) and the planning algorithms (Sect. 4) have been fully implemented in simulation and on a real robot using the mail collection problem domain. This section describes the results of experiments evaluating the efficiency, solution quality, and scalability of the proposed algorithm. Generally, a planning algorithm's quality can be measured by optimality and efficiency. Since we trade optimality for significant efficiency improvement in this work, our hypotheses are: 1) PlanHG can solve planning problems that existing "flat" algorithms cannot solve in reasonable time (PlanHG vs. PlanFG); and 2) PlanHG can generate better-quality plans than the ones generated by existing hierarchical algorithms (PlanHG vs. PlanHL).

5.1   Experiments in Simulation

The simulated domain used in experiments consists of 10 people, 20 rooms and 25 doors in an office environment, where mail needs to be collected from everyone inside the building. No two people are in the same room. We vary how mail is passed between people such that the number of people that need to be visited to collect all the mail varies from 1 to 10. Initially, the robot is placed in the corridor beside a randomly-selected door. Each data point is an average of 1000 trials. If the trials take more than 5 h, we terminate the trials and take the average over the available data. Action descriptions in $\mathcal{BC}$ are translated into logic programming, and the algorithms are implemented natively in CLINGO 4.3 [7]. Unless otherwise stated, experiments were conducted on a 32-bit laptop machine with 4 G memory and 2.0 GHz Dual Core processor.

PlanHG vs. PlanFG on Type-I Problems: We first compared PlanHG against PlanFG on the efficiency of solving Type-I problems that aim at minimizing the length of plans. The approach of applying PlanFG on Type-I problems was presented in previous research [15]. The planning time is plotted in Fig. 2a. Not surprisingly, PlanHG leads to significantly reduced planning time over PlanFG

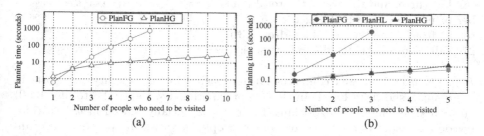

Fig. 2. (a) PlanHG vs. PlanFG in efficiency on the Type-I problems (i.e., minimizing plan length); and (b) PlanHG vs. PlanFG in efficiency on Type-II problems (i.e., minimizing plan cost).

by inserting state constraints (as "bottlenecks") at lower levels. For instance, creating a plan to visit six people (in six different rooms) takes PlanHG less than 20 seconds, but requires more than 13 min for PlanFG. Therefore, PlanHG can significantly reduce the planning time in solving Type-I planning problems. While PlanHG is not guaranteed to find the shortest length plan, both PlanHG and PlanFG find the shortest length plan in our testing domain.

PlanHG vs. PlanFG on Type-II Problems: As shown in Fig. 1, multi-entrance rooms make the shortest plan not necessarily the lowest-cost plan. We compare PlanHG against PlanFG on Type-II problems that aim at minimizing overall plan costs. Previous research has studied applying the PlanFG algorithm on Type-II problems [13], where the lowest-cost plan is found by searching among all plans of length less than a user specified upper-bound. Instead, we use Eqs. 2 and 3 of PlanHG to estimate this upper bound. The efficiency has been significantly improved, because instead of directly solving the Type-II problem, PlanHG solves a set of low-weight *guided* Type-II problems generated using the abstraction hierarchy.

Figure 2b shows the significant improvement in efficiency against PlanFG. To run larger numbers of trials, the experiments were conducted on a powerful desktop machine with 15 G memory and Intel Core i7 CPU at 3.40 GHz. For instance, to create a plan visiting three people, PlanFG needs 5.98 min, while PlanHG requires less than one second. When a small number of people need to be visited, PlanHL took more time than PlanHG, because PlanHL calls the ASP solver more frequently. Although PlanHG (the proposed approach) becomes slower than PlanHL while planning for visiting more than three people, both require significantly less time than PlanFG. PlanHL's significant loss in plan quality will be discussed.

Scalability of PlanHG on Type-II Problems: We next evaluate the scalability of PlanHG to learn how the planning time changes given different problem domains. We keep the number of people who need to be visited fixed at three, and then vary the total number of people in the building from 5 to 15, rooms from 10 to 25, and doors from 13 to 27. Table 1 presents the planning time as the size of the domain increases.

Table 1. Scalability of PlanHG on Type-II problems with three people need to be visited (in seconds).

# of ppl.	Number of rooms			
	10	15	20	25
5	1.41	2.86	5.65	10.28
10	1.83	4.18	7.69	11.56
15	-	6.20	9.93	14.58

Plan Quality: Figure 3a compares all approaches in plan quality (i.e. cost) in the domain shown in Fig. 1. In this set of experiments, the trials (totally 1000) are paired for different algorithms: the robot is initially placed in the corridor beside a randomly-selected door; and n people (n varies from 2 to 4) are randomly selected to need the robot's visit. Realistic action costs are learned and associated with the actions using algorithms presented in [13]. We observe that algorithms solving a Type-I problem do not perform as well as those solving the

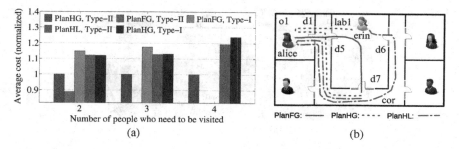

Fig. 3. (a) Evaluation of plan quality in minimizing plan cost using different planning algorithms (normalized, paired)—results of trials that would require longer then five hours to complete were not included; and (b) Visual illustration of computed plans (worst case) in a test case.

corresponding Type-II problem as they do not attempt to minimize overall plan cost, but in turn have much faster execution times.

Figure 3a shows that PlanFG, the "flat" planning approach that computes optimal plans, produces plans of the best quality in overall plan cost, but cannot solve Type-II problems with more than two people in reasonable time (as shown in Fig. 2b). Comparing with PlanFG on Type-II problem with two people, we find PlanHG has only a 11.25 % loss in optimality. Compared to PlanHL, the baseline hierarchical planning algorithm, PlanHG significantly improved the quality of generated plans—when compared over 1000 trials using a student's t-test with p-$value < 10^{-50}$.

Figure 3b presents a test case of planning to visit two people, to demonstrate why PlanHG can produce lower cost plans than PlanHL, where the robot is initially beside d_7 at the corridor, and the goal is to collect mail from *Alice* and *Erin*. We present the plans in the worst case. As expected, algorithm PlanFG generated the optimal plan with the minimum cost (195). While planning with PlanHG, the robot decided to visit room o_1 first (suboptimal) because the robot's finer position (e.g., $beside(d_7)$) could not be represented at level 1—D_1 only "knows" the robot is in the corridor. While planning with PlanHL, the robot decided to go through d_6 because the subproblem is to find the optimal plan going into lab_1. As a result, PlanHG and PlanHL produce plans with costs of 205 and 345 respectively. Without minimizing plan cost globally, the robot could not know going through d_5 could reduce the overall cost.

5.2 Illustrative Trials of PlanHG on a Robot

Algorithm PlanHG has been implemented on an autonomous Segway-based robot—see Fig. 4b. The robot uses a Hokuyo URG-04LX LIDAR and a Kinect RGB-D camera for sensing and navigation. The robot moves in indoor environments at a maximum speed of 0.7m/s. Figure 4a shows part of the real world map generated using a simultaneous localization and mapping (SLAM) algorithm. Since manipulation tasks are not the focus of this paper, similar to [22],

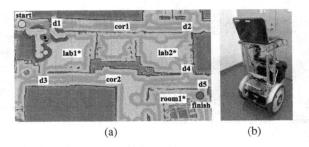

(a)   (b)

Fig. 4. (a) Part of the inflated occupancy-grid map with a path (green bubbles) planned for going through a door; and (b) The robot platform used in experiments.

the robot simply asks help from humans to open doors. The system architecture has been implemented using Robot Operating System (ROS) [21].

As a trial, we initially placed the robot at a position labeled by the yellow dot in Fig. 4a and asked the robot to collect mail from three people in lab_1, lab_2 and $room_1$ respectively. Using PlanHG, the robot found the plan within 2 seconds. In contrast, the robot needed more than 5 min to find the plan when PlanFG was used (see Fig. 2b). The plan suggests following this path: $start \xrightarrow{d_1} lab_1* \xrightarrow{d_1} cor_1 \xrightarrow{d_2} lab_2* \xrightarrow{d_4} cor_2 \xrightarrow{d_5} room_1*$, where mail was collected at the rooms labeled with the star sign. The red dot in Fig. 4a shows the position where the robot finished the task. It should be noted that there are multiple plans of similar lengths leading to the goal. For instance, the robot can cross d_3 after serving the first person in lab_1. This plan is not preferred, because d_3 is a narrow door and has a high cost of navigating through it. A video of the robot's performance can be viewed online.[2]

6 Conclusions

In this paper, we present algorithm PlanHG for robotic task planning using an abstraction hierarchy represented in action language $\mathcal{BC}$. The hierarchy is obtained by composing additional domain descriptions at coarser granularities and plans computed at coarser levels are used to generate "bottlenecks" in the form of search depth bounds at lower levels. This work combines the KRR advantages of $\mathcal{BC}$ and the efficiency of hierarchical planning to enable mobile robots to compute provably sound plans in real-world environments. The hierarchy and algorithm have been fully implemented in simulation and on real robots. We observed orders of magnitude improvements in efficiency with only a 11.25 % loss in optimality compared to a "flat" planning approach.

Acknowledgments. This work has taken place in the Learning Agents Research Group (LARG) at the Artificial Intelligence Laboratory, The University of Texas at

---

[2] https://youtu.be/-QpFj7BbiRU.

Austin. LARG research is supported in part by grants from the National Science Foundation (CNS-1330072, CNS-1305287), ONR (21C184-01), AFOSR (FA8750-14-1-0070, FA9550-14-1-0087), and Yujin Robot.

References

1. Chen, X., Jin, G., Yang, F.: Extending C+ with composite actions for robotic task planning. In: ICLP (Technical Communications), pp. 404–414 (2012)
2. Dix, J., Kuter, U., Nau, D.S.: Planning in answer set programming using ordered task decomposition. In: Günter, A., Kruse, R., Neumann, B. (eds.) KI 2003. LNCS (LNAI), vol. 2821, pp. 490–504. Springer, Heidelberg (2003)
3. Erol, K., Hendler, J.A., Nau, D.S.: HTN planning: complexity and expressivity. In: National Conference on Artificial Intelligence (AAAI) (1994)
4. Fikes, R.E., Nilsson, N.J.: Strips: a new approach to the application of theorem proving to problem solving. Artif. Intell. **2**(3), 189–208 (1972)
5. Fox, D., Burgard, W., Dellaert, F., Thrun, S.: Monte carlo localization: efficient position estimation for mobile robots. In: National Conference on Artificial Intelligence (AAAI) (1999)
6. Galindo, C., Fernandez-Madrigal, J.A., Gonzalez, J.: Improving efficiency in mobile robot task planning through world abstraction. IEEE Trans. Robot. **20**(4), 677–690 (2004)
7. Gebser, M., Kaminski, R., Kaufmann, B., Schaub, T.: Clingo= asp+ control: preliminary report. In: arXiv preprint (2014) arXiv:1405.3694
8. Cabalar, P.: Answer set; programming? In: Balduccini, M., Son, T.C. (eds.) Logic Programming, Knowledge Representation, and Nonmonotonic Reasoning. LNCS, vol. 6565, pp. 334–343. Springer, Heidelberg (2011)
9. Gelfond, M., Lifschitz, V.: Action languages. Electron. Trans. Artif. Intell. (ETAI) **3**, 193–210 (1998)
10. Giunchiglia, E., Lee, J., Lifschitz, V., McCain, N., Turner, H.: Nonmonotonic causal theories. Artif. Intell. (AIJ) **153**(1–2), 49–104 (2004)
11. Helmert, M.: The fast downward planning system. J. Artif. Intell. Res. **26**, 191–246 (2006)
12. Kaelbling, L.P., Lozano-Pérez, T.: Hierarchical task and motion planning in the now. In: IEEE International Conference on Robotics and Automation (ICRA). IEEE (2011)
13. Khandelwal, P., Yang, F., Leonetti, M., Lifschitz, V., Stone, P.: Planning in action language $\mathcal{BC}$ while learning action costs for mobile robots. In: International Conference on Automated Planning and Scheduling (ICAPS) (2014)
14. Lee, J., Lifschitz, V., Yang, F.: Action Language $\mathcal{BC}$: a preliminary report. In: International Joint Conference on Artificial Intelligence (IJCAI) (2013)
15. Lifschitz, V.: Answer set programming and plan generation. Artif. Intell. **138**, 39–54 (2002)
16. Marthi, B., Russell, S., Wolfe, J.: Angelic hierarchical planning: optimal and online algorithms (revised). Technical Report, UCB/EECS-2009-122, EECS Department, University of California, Berkeley (2009)
17. McDermott, D., Ghallab, M., Howe, A., Knoblock, C., Ram, A., Veloso, M., Weld, D., Wilkins, D.: PDDL-the planning domain definition language (1998)
18. McIlraith, S.A., Fadel, R.: Planning with complex actions. In: International Workshop on Non-Monotonic Reasoning (NMR) (2002)

19. Nebel, B.: The FF planning system: fast plan generation through heuristic search. J. Artif. Intell. Res. **14**, 253–302 (2001)
20. Pednault, E.: ADL: exploring the middle ground between STRIPS and the situation calculus. In: International Conference on Principles of Knowledge Representation and Reasoning (KR) (1989)
21. Quigley, M., Conley, K., Gerkey, B., Faust, J., Foote, T., Leibs, J., Wheeler, R., Ng, A.Y.: ROS: an open-source robot operating system. In: Open Source Software in Robotics Workshop (2009)
22. Rosenthal, S., Veloso, M.M.: Mobile robot planning to seek help with spatially-situated tasks. In: AAAI (2012)
23. Sacerdoti, E.D.: Planning in a hierarchy of abstraction spaces. Artif. Intell. **5**(2), 115–135 (1974)
24. Tenenberg, J.D.: Abstraction in planning. Ph.D. thesis, University of Rochester (1988)
25. Thiébaux, S., Hoffmann, J., Nebel, B.: In defense of PDDL axioms. In: International Joint Conference on Artificial Intelligence (IJCAI) (2003)
26. Zhang, S., Sridharan, M., Gelfond, M., Wyatt, J.: Towards an architecture for knowledge representation and reasoning in robotics. In: Beetz, M., Johnston, B., Williams, M.-A. (eds.) ICSR 2014. LNCS, vol. 8755, pp. 400–410. Springer, Heidelberg (2014)

Logic Programming with Graded Modality

Zhizheng Zhang$^{(\boxtimes)}$ and Shutao Zhang

School of Computer Science and Engineering, Southeast University,
Nanjing, China
{seu_zzz,stzhang}@seu.edu.cn

Abstract. Logic programs with graded modality (*LPGMs*) combine ideas underlying graded modal logic and answer set programming. Logic programming under answer set semantics is extended with a new graded modality $M_{[lb:ub]}$ where lb and ub are natural numbers satisfying $lb \leq ub$. The modality is used to precede a literal in rules bodies, and thus allows for the representation of graded introspections: $M_{[lb:ub]}F$ intuitively means: it is known that the number of belief sets where F is true is between lb and ub. We define the semantics of Logic programs with graded modality, give an algorithm for computing solutions of *LPGMs*, and show the effectiveness of the formalism for modeling two problems.

Keywords: Logic programming · Graded modality · Answer set

1 Introduction

Modalities and negation as failure have proved to be intuitive and powerful for the declarative representation of modal concepts and defaults respectively. Several epistemic nonmonotonic formalisms have been developed to support qualitative modalities and negation as failure. Examples include the logic of grounded knowledge [6], the logic of MKNF [7], and epistemic specifications [8] etc. Those formalisms have proved to be potentials in dealing with some important issues in the field of knowledge representation and reasoning, for instance the correct representation of incomplete information in the presence of multiple belief sets [9], epistemic queries [10], commonsense reasoning [9], formalization for conformant planning [11], meta-reasoning [12] etc. Recently, there is increasing research in this direction, e.g. [13–16]. In those formalisms, two qualitative modalities are usually considered: Kp:"p is known"(*p is true in all belief sets of the agent*), Mp:"p may be true"(*p is true in some belief sets of the agent*), and their negations: ¬Kp:"p is not known"(*p is not true in some belief sets of the agent*), and ¬Mp:"p is never true"(*p is not true in every belief sets of the agent*).

Graded modalities are quantitative modalities that are widely used to represent the modal concepts: "at least as many as..." and "at most as many as..." (based on counting). This is especially useful in knowledge representation because humans tend to describe objects by the number of other objects they are related to. Several formalisms are proposed to extend classical logics

© Springer International Publishing Switzerland 2015
F. Calimeri et al. (Eds.): LPNMR 2015, LNAI 9345, pp. 517–530, 2015.
DOI: 10.1007/978-3-319-23264-5_43

with graded modalities. Examples include the family of graded modal logics containing the logic GrK defined in [5], the logic Gr(S5) proposed in [17], and the logic $\overline{T}$ in [18] etc., and the description logic based $\mathcal{ALCQ}$ in [19]. In those logics, the graded modality $\Diamond_n F$ ($n = 1, 2, ...$) is usually interpreted in a possible world framework as "there are more than n accessible worlds where F is true.". However those formalisms do not support negation as failure.

In this paper, our purpose is to represent a formalism that is able to support not only the declarative representation of counting based gradation of an agents belief but also the declarative representation of defaults. We present a new logic programming language that combines ideas underlying graded modal logics and answer set programming that adopts negation as failure *not* to represent defaults declaratively. A new graded modality $M_{[lb:ub]}$ is proposed to precede a literal in rules bodies, and thus allows for the representation of graded introspections: $M_{[lb:ub]}F$ intuitively means: it is known that the number of belief sets where F is true is between lb and ub.

The rest of the paper is organized as follows. In the next section, we review the basic principles underlying the answer set semantics of logic programs and the basic idea underlying graded modal logics, we show how they can be merged resulting in $LPGMs$. In Sect. 3, we introduce syntax and semantics of $LPGMs$. In Sect. 4, we consider several properties of $LPGMs$. In Sect. 5, we consider the relationship between $LPGMs$ and epistemic specifications. In Sect. 6, we give an algorithm for computing solutions of $LPGMs$ and an analysis of its complexity. In Sect. 7, we show the effectiveness of the formalism for modeling two graph problems. We conclude in Sect. 8 with some further discussion.

2   Preliminaries

Throughout this paper, we assume a finite first-order signature σ that contains no function constants of positive arity. There are finitely many Herbrand interpretations of σ, each of which is finite as well.

2.1   Answer Set

A logic program over σ is a collection of rules of the form

$$l_1 \; or \; ... \; or \; l_k \leftarrow l_{k+1}, ..., l_m, not \; l_{m+1}, ..., not \; l_n.$$

where ls are literals of σ, *not* is called negation as failure and *not* l is often read as "it is not believed that l is true.", *or* is epistemic disjunction. $l_1 \; or \; l_1$ can be read as "l_1 is believed to be true or l_2 is believed to be true." The left-hand of a rule is called the *head* and the right-hand is called the *body*. A rule is called a fact if its body is empty and its head contains only one literal, and a rule is called a denial if its head is empty. A logic program is called ground if it contains no variables. [2] intuitively interprets that an answer set associated with a ground logic program is a set of beliefs (collection of ground literals) and is formed by a rational reasoner guided by three principles:

- Believe in the head of a rule if you believe in its body (*Rule's Satisfiability principle*).
- Do not believe in contradictions (*Consistency principle*).
- Believing nothing you are not forced to believe (*Rationality Principle*).

For example for the program

$$p \leftarrow not\ a. \quad q \leftarrow not\ b. \quad a\ or\ b.$$

The first rule means "If a does not belong to your set of beliefs, then p must.", the second rule means "If b does not belong to your set of beliefs, then q must.", the last rule means "Believe a or believe b." Clearly, by the three principles, the last says that a is possible to be believed i.e., it may belong to an answer set of the program. This means that the body of the first rule is not satisfied in the answer set, thus the first rule will not contribute to form such answer set. Since we cannot be forced to believe b, the body of the second rule is satisfied, and hence q must belong to the answer set. Thus, the belief set $\{a, q\}$ is an answer set of the program. In a similar way, we can construct another answer set $\{b, p\}$ of the program.

The definition of the answer set is extended to any non-ground program by identifying it with the ground program obtained by replacing every variable with every ground term of σ.

2.2 Graded Modal Logic

We introduce the basic ideas underlying graded modal logic by a concise review of the language and semantics of (GrK) defined in [5].

A GrK formula is constructed from a set of propositional variable $p_1, ..., p_n$, the usual logic connectives: the binary operator $\vee$ and the unary operator $\neg$, and a necessity modality $\Box$ and graded modalities $\Diamond_n$ ($n = 1, 2, ...$).

The semantics of GrK is defined on the possible world framework $< W, R, \pi >$, where W is a non-empty set and R is a binary relation on W, π is a map from variable set to 2^W. Then the satisfiability of GrK is defined as below: for $w \in W$

- for a propositional variable p, $w \models p$ iff $w \in \pi(p)$,
- for a formula F, $w \models \neg F$ iff not $w \models F$,
- for formulas F_1 and F_2, $w \models F_1 \vee F_2$ iff $w \models F_1$ or $w \models F_2$,
- for a formula F, $w \models \Box F$ iff not $v \models F$ for all v such that wRv,
- for a formula F, $w \models \Diamond_n F$ iff $|\{v | wRv, v \models F\}| \geq n$,

The basic idea of GrK is to allow propositional formulas preceded by graded modalities which are interpreted by counting possible worlds where the proposition is true. The principles of defining answer sets provide an rational approach to the construction of all possible worlds where the logic program possible having negation as failure can be satisfied.

3 Logic Programs with Graded Modality

3.1 Syntax

A *LPGM* program Π is a finite collection of rules of the form

$$l_1 \ or \ ... \ or \ l_k \leftarrow e_1, ..., e_m, s_1, ..., s_n.$$

where $k \geq 0$, $m \geq 0$, $n \geq 0$, ls are literals in first order logic language and are called objective literals here, es are extended literals which are objective literals possibly preceded by a negation as failure operator *not*, ss are subjective literals of the form $M_\omega e$ where e is an extended literal and M_ω is a modality where ω is of the form $[lb : ub]$ or $[lb :]$ where lb and ub are natural numbers satisfying $lb \leq ub$. As in usual logic programming, a rule is called a fact if its body is empty and its head contains only one literal, and a rule is called a denial if its head is empty. We use $head(r)$ to denote the set of objective literals in the head of a rule r and $body(r)$ to denote the set of extended literals and subjective literals in the body of r. Sometimes, we use $head(r) \leftarrow body(r)$ to denote a rule r. The positive body of a rule r is composed of the extended literals containing no *not* in its body. We use $body^+(r)$ to denote the positive body of r. r is said to be *safe* if each variable in it appears in the positive body of the rule.

It is clear that a *LPGM* containing no subjective literals is a disjunctive logic program that can be dealt with by ASP solvers like DLV [3], CLASP [4].

3.2 Semantics

We will restrict our definition of the semantics to ground programs. However, we admit rule schemata containing variables bearing in mind that these schemata are just convenient representations for the set of their ground instances. In the following definitions, l is used to denote a ground objective literal, e is used to denote a ground extended literal with or without one *not*.

Models of *LPGM*s. Let W be a non-empty collection of consistent sets of ground objective literals, $< W, w >$ is a pointed structure of W where $w \in W$. We call w as a belief set in W. W is a model of a program Π if for each rule r in Π, r is satisfied by every pointed structure of W. The notion of satisfiability is defined below.

- $< W, w > \models l$ if $l \in w$
- $< W, w > \models not\ l$ if $l \notin w$
- $< W, w > \models M_{[lb:ub]}e$ if $lb \leq |\{w \in W | < W, w > \models e\}| \leq ub$
- $< W, w > \models M_{[lb:]}e$ if $|\{w \in W | < W, w > \models e\}| \geq lb$

Then, for a rule r in Π, $< W, w > \models r$ if

- $\exists l \in head(r)$: $< W, w > \models l$
- $\exists t \in body(r)$: $< W, w > \not\models t$.

The satisfiability of a subjective literal does not depend on a specific belief set w in W, hence we can simply write $W \models K_\omega\, e$ if $< W, w > \models M_\omega\, e$ and say the subjective literal $M_\omega\, e$ is satisfied by W. For convenient description, if an extended literal e can be satisfied by $< W, w >$, we also say that e is believed with regard to w.

World Views of $LPGM$s. The definition of the world view consists of two parts. The first part is for disjunctive logic programs. The second part is for arbitrary $LPGM$s.

Definition 1. *Let Π be a disjunctive logic program, the world view of Π is the non-empty set of all its answer sets, written as $AS(\Pi)$.*

We immediately have a conclusion below.

Theorem 1. *Π has an unique world view if it is a consistent disjunctive logic program.*

Now, we give the second part of the definition of the world view to address graded modalities in an arbitrary $LPGM$.

Definition 2. *Let Π be an arbitrary $LPGM$, and W is a non-empty collection of consistent sets of ground objective literals in the language of Π, we use Π^W to denote the disjunctive logic program obtained by removing graded modalities using the following reduct laws*

1. *removing from Π all rules containing subjective literals not satisfied by W.*
2. *replacing all other occurrences of subjective literals of the form $M_{[lb:ub]}\, l$ or $M_{[lb:]}\, l$ where $lb = |W|$ with l.*
3. *removing all occurrences of subjective literals of the form $M_{[lb:ub]}$ not l or $M_{[lb:]}$ not l where $lb = |W|$.*
4. *replacing other occurrences of subjective literals of the form $M_{[0:0]}\, e$ with e^{not}.*
5. *replacing other occurrences of subjective literals of the form $M_\omega\, e$ with e and e^{not} respectively.*

where e^{not} is l if e is not l, and e^{not} is not l if e is l. Then, W is a world view of Π if W is a world view of Π^W.

Π^W is said to be the *introspective reduct* of Π with respect to W. Such a reduct process eliminates graded modalities so that the belief sets in the model are identified with the answer sets of the program obtained by the reduct process. The intuitive meanings of the reduct laws can be described as follows:

- The first reduct law directly comes from the *Rule's Satisfiability principle* and *Rationality Principle* which means if a rule's body cannot be satisfied (believed in), the rule will contribute nothing;
- The second reduct law states that, if it is known that there are at least lb number of belief sets where l is true and there are totally lb belief sets in W, then, by the meaning of the gradation based on counting, l must be believed

with regard to each belief set in W. Then, by the *Rationality Principle* in answer set semantics, you are forced to believe l with regard to each belief set. Hence, $M_{[lb:ub]} \, l$ (or $M_{[1b:]} \, l$) should be replaced by l (instead of being removed) to avoid self-support;

- The third law states that, if it is known that there are at least lb number of belief sets where l is not true and there are totally lb belief sets in W, then *not* l is satisfied by each belief set in W. Hence, Removing $M_{[lb:ub]} not \, l$ or $M_{[lb:]} not \, l$ in a rule will not effect the satisfiability of the rule[1].
- The fourth law states that if it is known that e is not believed with regard to each belief set in W, then e^{not} must be believed with regard to each belief set in W.
- The last law states that, if it is known that there are at least lb number of belief sets where e is believed, and the number of belief sets in W is strict greater than lb, then e may be believed or may not be believed with regard to a belief set in W. Hence, $M_\omega \, e$ where ω is $[lb : ub]$ or $[lb :]$ should be replaced by e and e^{not} respectively.

Example 1. Consider a *LPGM* Π containing rules:

$$a \; or \; b \leftarrow . \qquad c \leftarrow M_{[1:1]}a.$$

Let $W = \{\{a, c\}, \{b, c\}\}$. Because of $W \models M_{[1:1]}a$, Π^W is a disjunctive logic program contains rules:

$$a \; or \; b \leftarrow . \qquad c \leftarrow a. \qquad c \leftarrow not \; a.$$

It is easy to see that W is a world view of Π^W, hence, W is a world view of Π.

Combine the two parts of the definition of the world view, it is easy to get the following conclusion.

Theorem 2. *For an arbitrary LPGM Π, a non-empty set W is its world view if and only if $W = AS(\Pi^W)$.*

Definition 3. *A LPGM is said to be consistent if it has at least one world view. Otherwise, it is said to be inconsistent.*

Example 2. An inconsistent program:

$$\neg p \leftarrow M_{[1:1]}p. \qquad p \leftarrow not \; \neg p. \qquad \leftarrow p.$$

Some *LPGM*s have two or more world views. We use $WV(\Pi)$ to denote the set of all world views of a program Π.

---

[1] Remove $M_\omega not \, l$ in a denial of the form $\leftarrow M_\omega not \, l$, we will get a denial with empty body, and such a denial is considered as an unsatisfiable rule in the answer set program. For example, for a logic program with a denial with empty body, CLASP answers "UNSATISFIABLE".

Example 3. A program Π with Multiple world views:

$f \leftarrow p. \quad f \leftarrow q. \quad p \text{ or } q \leftarrow . \quad \leftarrow M_{[1:]} not\ p, p. \quad \leftarrow M_{[1:]} not\ q, q. \quad \leftarrow M_{[1:]} not\ f.$

Π has two world views: $\{\{p, f\}\}$ and $\{\{q, f\}\}$.

Now, we give a program that has a world view containing multiple belief sets.

Example 4. Consider Π: $\quad p \text{ or } \neg p \leftarrow M_{[2:3]} not\ r. \quad \{\{p\}, \{\neg p\}\}$ is a world view of Π.

4   Some Properties

We first consider the introspective ability of a *LPGM*s based agent.

Definition 4. *Let Π be a LPGM, for a pair (l, n) where l is a ground objective literal and $n \in \mathbb{N}$*

- *$(l, = n)$ is true in Π written by $\Pi \models_n l$ if $\forall W \in WV(\Pi)\ W \models M_{[n,n]}l$.*
- *$(l, \geq n)$ is true in Π written by $\Pi \models_{\geq n} l$ if $\forall W \in WV(\Pi)\ W \models M_{[n:]}l$.*
- *$(l, \leq n)$ is true in Π written by $\Pi \models_{\leq n} l$ if $\forall W \in WV(\Pi)\ W \models M_{[0:n]}l$.*

Theorem 3. *Let Π be a LPGM, l and l' be ground objective literals and $n \in \mathbb{N}$*

- *$\Pi \cup \{\leftarrow M_{[n,n]}l.\}$ is inconsistent if $\Pi \models_n l$.*
- *$\Pi \models_{\geq n-1} l$ if $\Pi \models_{\geq n} l$ for $n \geq 1$.*
- *$\Pi \models_{\leq n+1} l$ if $\Pi \models_{\leq n} l$ for $n \geq 0$.*
- *$\Pi \cup \{l' \leftarrow l\} \models_n l'$ if $\Pi \cup \{l' \leftarrow l\} \models_n l$*

Second, we find that the division of the counting intervals in the subjective literals will not change the world views of a *LPGM*. This is in line with our intuition. For convenient description, we consider $M_{[lb:]}e$ as an abbreviation of $M_{[lb:\infty]}e$.

Definition 5. *A pair (r_1, r_2) of rules is called a substitute of a rule r if there exists exactly one subjective literal $M_{[lb:ub]}e$ in r such that: (1) r_1 is obtained from r by replacing $M_{[lb:ub]}e$ with a subjective literal $M_{[lb_1:ub_1]}e$, and r_2 is obtained from r by replacing $M_{[lb:ub]}e$ with a subjective literal $M_{[lb_2:ub_2]}e$, and (2) $min\{lb_1, lb_2\} = lb$ and $max\{ub_1, ub_2\} = ub$.*

Theorem 4. *If a LPGM Π' is obtained from a LPGM Π by replacing a rule r of Π with rules in a substitute of r, then $WV(\Pi') = WV(\Pi)$.*

Example 5. Consider a *LPGM* Π containing one rule:

$$p \leftarrow M_{[1:]} not\ \neg p.$$

Let Π' consist of two rules in a substitute of the rule of Π:

$$p \leftarrow M_{[1:2]} not\ \neg p. \quad p \leftarrow M_{[3:]} not\ \neg p.$$

It is easy to see that $WV(\Pi') = WV(\Pi) = \{\{\{p\}\}\}$.

5   Relation to Epistemic Specifications

In this section we will explore the relationship between the language of *LPGM*s and the language of epistemic specifications. We show that the language of epistemic specifications recently defined in [11] can be viewed as *LPGM*s.

An epistemic specification is a set of rules of the form

$$h_1 \ or \ ... \ or \ h_k \leftarrow b_1, ..., b_m.$$

where $k \geq 0, m \geq 0$, hs are objective literals, and each b is an objective literal possible preceded by a negation as failure operator *not*, an modal operator K or M, or a combination operator *not* K or *not* M. The semantics of an epistemic specification is defined by a notion of **world view**[2].

Theorem 5. *From an epistemic specification Π, a LPGM Π' is obtained by*

- *Replacing all occurrences of literals of the form Kl in Π by "$M_{[0,0]}not \ l$".*
- *Replacing all occurrences of literals of the form Ml in Π by "$M_{[1:]}l$" and "not not l"[3] respectively.*
- *Replacing all occurrences of literals of the form not Kl in Π by "$M_{[1:]}not \ l$" and not l respectively.*
- *Replacing all occurrences of literals of the form not Ml in Π by $M_{[0:0]}l$.*

*then W is a world view of Π' iff W is a **world view** of Π.*

Example 6. Consider an epistemic specification Π:

$$p \leftarrow Mp.$$

Π has an unique **world view** $\{\{p\}\}$. By the above theorem, Π' contains two rules:

$$p \leftarrow M_{[1:]}p. \quad p \leftarrow \ not \ not \ p.$$

It is easy to see that Π' has an unique world view $\{\{p\}\}$.

6   An Algorithm for Computing World Views

For a given *LPGM* Π, we use $WV_i(\Pi)$ ($i \in \mathbb{N}$ and $i \geq 1$) to denote the set of world views of Π which contain exactly i belief sets, i.e., $WV_i(\Pi) = \{W \in WV(\Pi)||W| = i\}$. Obviously, we have $WV(\Pi) = \bigcup_{i \geq 1} WV_i(\Pi)$. In addition, we use $WV_{>i}(\Pi)$ to denote the set of world views of Π which contain strictly more than i belief sets. Then we have $WV(\Pi) = (\bigcup_{1 \leq i \leq k} WV_i(\Pi)) \cup WV_{>k}(\Pi)$

---

[2] To distinguish the world view semantics defined in [11] from the world view semantics in this paper, we use bold face **world view** to denote the former.

[3] Here, we view *not not l* as a representation of *not l'* where we have $l' \leftarrow not \ l$ and l' is a fresh literal. It is worthwhile to note that CLINGO [21] is able to deal with *not not*.

for any natural number k and $k \geq 1$. Based on this idea, we propose an algorithm for finding world views of $LPGM$s composed of safe rules. At a high level of abstraction, the algorithm for computing world views of a $LPGM$ Π is as showed in Algorithm 1. In Algorithm 1, **LPGMSolver** first computes a dividing line n. Then, **LPGMSolver** adopts a $FOR\text{-}LOOP$ to compute all world views of size less than or equal to n for Π by calling a function $WViSolver$ that computes all world views of size i for Π. After that, **LPGMSolver** calls a function $WVgiSolver$ that computes all world views of size strict greater than n for Π.

Algorithm 1. LPGMSolver.

Input:

  Π: A $LPGM$;

Output:

  All world views of Π;

1: $n = max\{lb | M_{[lb:ub]}e$ or $M_{[lb:]}e$ in $\Pi\}$ {computes the maximal lb of subjective literals in Π}
2: $WV = \emptyset$
3: **for** every natural number $1 \leq k \leq n$ **do**
4:   $WV_k = $ WViSolver(Π, k) {computes all world views of size k for Π}
5:    $WV = WV \cup WV_k$
6: **end for**
7: $WV_{>n} = $ WVgiSolver(Π, n) {computes all world views of size strict greater than n for Π}
8: $WV = WV \cup WV_{>n}$
9: output WV

For ease of description of $WViSolver$ and $WVgiSolver$, $m\_C\_lb\_U\_V\_p$ will be used to denote the fresh atom obtained from a subjective literals $M_\omega e$, where p is the atom in e, and in the prefixes, V is t if e is p, V is f if e is $\neg p$, V is nt if e is not p, V is nf if e is not $\neg p$, C is 0 if ω is of the form $[lb : ub]$, C is 1 if ω is of the form $[lb :]$, and U is ub if ω is of the form $[lb : ub]$, and U is o if ω is of the form $[lb :]$. Thus, $m\_C\_lb\_U\_V\_p$ is called a *denoter* of $M_\omega e$ and also recorded as m_ω_l for convenience. We assume prefixes used here do not occur in Π. Other fresh atoms may be used to avoid conflicts.

  $WViSolver(\Pi, k)$ includes the following steps.

1. Create a disjunctive logic program Π' from Π.
   Rules without subjective literals are left unchanged. For each rule r containing a subjective literal $M_\omega e$
   (a) Eliminate $M_\omega e$ by the following laws:
       i. if ω is $[0 : 0]$, replace $M_\omega e$ with m_ω_e, e^{not}.
       ii. if ω is $[k :]$ or $[k : ub]$, and e is of the form not l, replace $M_\omega e$ with m_ω_e.
       iii. if ω is $[k :]$ or $[k : ub]$, and e is a literal l, replace $M_\omega e$ with m_ω_e, l.
       iv. otherwise,
           A. Add a rule obtained from r by replacing $M_\omega e$ by m_ω_e, e, and
           B. Add a rule obtained from r by replacing $M_\omega e$ by m_ω_e, e^{not}

(b) Add rules
$$m\_\omega\_e \leftarrow body^+(r), not\ \neg m\_\omega\_e.$$

and
$$\neg m\_\omega\_e \leftarrow body^+(r), not\ m\_\omega\_e.$$

2. Compute the set $AS(\Pi')$ of answer sets of Π' using ASP grounder-solver like DLV, CLINGO etc.
3. Generate a set $CWV(\Pi)$ of candidate world views of k-size from $AS(\Pi')$. Group the answer sets in $AS(\Pi')$ by common $m\_$ and $\neg m\_$–literals. Each group is said to be a candidate world view.
4. Generate k-size world views of Π by checking each candidate world view in $CWV(\Pi)$.
   For each candidate world view W, check that the following condition are met
   - $|W| = k$
   - if m_ω_e is a common literal in W, then $W \models M_\omega e$ is true.
   - if $\neg m\_\omega\_e$ is a common literal in W, then $W \models M_\omega e$ is false.
   Let W_S denote the set of literals with a prefixes $m\_$ or $\neg m\_$ in W. $\{A|\exists B \in W, A = B - W_S\}$ is a world view of Π if the above two conditions are met.

$WVgiSolver(\Pi, n)$ includes the following steps.

1. Create a disjunctive logic program Π' from Π.
   Rules without subjective literals are left unchanged. For each rule r containing a subjective literal $M_\omega e$
   (a) Eliminate $M_\omega e$ by the following laws:
      i. if $\omega = [0:0]$, replace $M_\omega e$ with m_ω_e, e^{not}.
      ii. otherwise,
         A. Add a rule obtained from r by replacing $M_\omega e$ with m_ω_e, e, and
         B. Add a rule obtained from r by replacing $M_\omega e$ with m_ω_e, e^{not}
   (b) Add rules
   $$m\_\omega\_e \leftarrow body^+(r), not\ \neg m\_\omega\_e.$$

   and
   $$\neg m\_\omega\_e \leftarrow body^+(r), not\ m\_\omega\_e.$$

2. Compute the set $AS(\Pi')$ of answer sets of Π' using ASP grounder-solver like DLV, CLINGO etc.
3. Generate a set $CWV(\Pi)$ of candidate world views of size $> n$ from $AS(\Pi')$. Group the answer sets in $AS(\Pi')$ by common $m\_$ and $\neg m\_$–literals. Each group is said to be a candidate world view.
4. Generate $> n$ size world views of Π by checking each candidate world view in $CWV(\Pi)$.
   For each candidate world view W, check that the following condition are met
   - $|W| > n$
   - if m_ω_e is a common literal in W, then $W \models M_\omega e$ is true.
   - if $\neg m\_\omega\_e$ is a common literal in W, then $W \models M_\omega e$ is false.
   Let W_S denote the set of literals with a prefixes $m\_$ or $\neg m\_$ in W. $\{A|\exists B \in W, A = B - W_S\}$ is a world view of Π if the above two conditions are met.

Theorem 6. LPGMSolver *is sound and complete for computing world views.*

Now, we consider the complexity of **LPGMSolver** informally. Let $\mathcal{L}$ be the set of all ground literals in Π. It is easy to see that: step 1 of $WViSolver(\Pi, k)$ only takes linear time and needs a polynomial space; step 2 calls an ASP solver where deciding whether a given disjunctive logic program has some answer sets is $\sum_{2}^{P}$-complete and needs a polynomial space [20]; step 3 and step 4 generates and checks each collection of the subsets of $\mathcal{L}$ that costs $\mathcal{O}(2^{3|\mathcal{L}|})$, but uses a polynomial space. In summary, $WViSolver(\Pi, k)$ is in PSPACE and $\mathcal{O}(2^{3|\mathcal{L}|})$. It is easy to see that $WViSolver(\Pi, k)$ and $WVgiSolver(\Pi, n)$ have same time and space complexity. Therefore, **LPGMSolver** is in PSPACE while the time complexity of **LPGMSolver** depends on the number of calling $WViSolver$ by the FOR loop from step 3 to step 6 in Algorithm 1. The number is determined in step 1 of Algorithm 1 by computing the maximal lb of all subjective literals of the form $M_{[lb:ub]}$ or $M_{[lb:]}$ in the program. Theoretically, that loops number may be anyone of the natural numbers.

7 Modeling with *LPGM*s: A Case Study

We will now present two graph problems which illustrate the applications of the language of *LPGM*s we developed in the previous sections. We first consider a modification of the *critical edge* problem[4] related to the existence of Hamiltonian cycles.

- *n-critical edge* problem: Given a directed graph $G = \{V, E\}$ where V is the set of vertices of G and E is the set of edges of G, find the set of all edges that belong to n or more hamiltonian cycles in G.

Represent G using a set of facts:

$$D(G) = \{vertex(x).|x \in V\} \cup \{edge(x,y).|(x,y) \in V\}$$

Encode the definition of Hamiltonian cycle using a disjunctive logic program HC below.

$inhc(X, Y) \ or \ \neg inhc(X, Y) \leftarrow edge(X, Y).$
$\leftarrow inhc(X1, Y1), inhc(X2, Y1), X1 \neq X2.$
$\leftarrow inhc(X1, Y1), inhc(X1, Y2), Y1 \neq Y2.$
$reachable(X, X) \leftarrow vertex(X).$
$reachable(X, Y) \leftarrow inhc(X, Z), reachable(Z, Y).$
$\leftarrow vertex(X), vertex(Y), not \ reachable(X, Y).$

We represent the definition of $n - critical$ edge by a rule r:

$$ncritical(X, Y) \leftarrow M_{[n:]}inhc(X, Y), edge(X, Y).$$

Then, we have the following result.

---

[4] *critical edge* problem: given a directed graph G, find all critical edges, i.e., the edges that belong to every hamiltonian cycle in G. The encoding of *critical edge* problem is given in [12] by using the language of epistemic specifications.

Theorem 7. *Let $G = \{V, E\}$ be a directed graph, n a natural number. An edge $(x, y) \in E$ is a n-critical edge iff $HC \cup D(G) \cup \{r\}$ has a world view W such that $ncritical(x, y) \in w$ is true for any $w \in W$.*

The second problem is related to the paths. Let G be a directed graph, a and b be two different vertices in G, n a nonzero natural number. Two paths are exclusive if they have no common edges. We say that it is reliable with a degree n with regard to m-lane edges from a to b if there are n or more paths between a and b, and there are no edge belonging to m or more of the paths (m-exclusive paths). Then, we have

– *n-exclusive paths* problem: Given a directed graph $G = \{V, E\}$ where V is the set of vertices of G and E is the set of edges of G, decide whether there are n number of m-exclusive paths from a vertex a to another vertex b.

Represent G using $D(G)$ defined above.

Encode the definition of path between a and b using a disjunctive logic program $PATH(a, b)$ below.

$inpath(X, Y)$ or $\neg inpath(X, Y) \leftarrow edge(X, Y)$.
$\leftarrow inpath(X1, Y1), inpath(X2, Y1), X1 \neq X2$.
$\leftarrow inpath(X1, Y1), inpath(X1, Y2), Y1 \neq Y2$.
$reachable(X, X) \leftarrow vertex(X)$.
$reachable(X, Y) \leftarrow inpath(X, Z), reachable(Z, Y)$.
$path \leftarrow reachable(a, b)$.
$\leftarrow not\ path$.

We represent the constraint of "n or more paths" by $nPath(n)$ containing:

$npath \leftarrow \mathrm{M}_{[n:]}path$.
$\leftarrow not\ npath$.

We represent the "m-exlusive" condition by a denial r:

$$\leftarrow \mathrm{M}_{[m:]}inpath(X, Y), edge(X, Y).$$

Then, we have the following result.

Theorem 8. *Let $G = \{V, E\}$ be a directed graph, n a nonzero natural number, a and b vertices. From a to b is safe with n degree with regard to m-lane edges iff $D(G) \cup PATH(a, b) \cup nPath(n) \cup \{r\}$ is consistent.*

8   Conclusion and Future Work

We present a logic programming formalism capable of reasoning that combines nonmonotonic reasoning, graded epistemic reasoning via introspections. The restriction to logic programming gives us a computationally attractive framework built on the existing efficient answer set programs grounders and solvers. This makes it an elegant way to formalize some non-trivial problems with default knowledge and graded introspections.

Actually, besides graded modalities, there have been a lot of quantitative modalities proposed to express modal concepts like certainty, confidence, likelihood etc. We expect many results established in logic programming with quantitative modalities, which may provide more powerful paradigms for commonsense reasoning.

Acknowledgments. This work was supported by the National Science Foundation of China (Grant No. 60803061), the National Science Foundation of China (Grant No. 61272378), Natural Science Foundation of Jiangsu (Grant No. BK2008293), and the National High Technology Research and Development Program of China (Grant No. 2015AA015406).

References

1. Baral, C., Gelfond, M., Rushton, J.N.: Probabilistic reasoning with answer sets. TPLP **9**(1), 57–144 (2009)
2. Gelfond, M., Kahl, Y.: Knowledge Representation, Reasoning, and the Design of Intelligent Agents: The Answer-Set Programming Approach. Cambridge University Press, Cambridge (2014)
3. Faber, W., Pfeifer, G., Leone, N., Dellarmi, T., Ielpa, G.: Design and implementation of aggregate functions in the DLV system. Theory Pract. Log. Program. **8**(5–6), 545–580 (2008)
4. Gebser, M., Kaufmann, B., Schaub, T.: Conflict-driven answer set solving: from theory to practice. Artif. Intell. **187–188**, 52–89 (2012)
5. Fine, K.: In so many possible worlds. Notre Dame J. Formal Logic **13**, 516–520 (1972)
6. Lin, F., Shoham, Y.: Epistemic semantics for fixed-points non-monotonic logics. In: TARK-1990, pp. 111–120 (1990)
7. Lifschitz, V.: Nonmonotonic databases and epistemic queries. In: IJCAI-1991, pp. 381–386 (1991)
8. Gelfond, M.: Strong introspection. In: AAAI-1991, pp. 386–391 (1991)
9. Gelfond, M.: Logic programming and reasoning with incomplete information. Ann. Math. Artif. Intell. **12**(1–2), 89–116 (1994)
10. Gomes, A.S., Alferes, J.J., Swift, T.: Implementing query answering for hybrid MKNF knowledge bases. In: Carro, M., Peña, R. (eds.) PADL 2010. LNCS, vol. 5937, pp. 25–39. Springer, Heidelberg (2010)
11. Kahl, P., Watson, R., Gelfond, M., Zhang, Y.: A refinement of the language of epistemic specifications. In: Workshop ASPOCP-2014 (2014)
12. Truszczyński, M.: Revisiting epistemic specifications. In: Balduccini, M., Son, T.C. (eds.) Logic Programming, Knowledge Representation, and Nonmonotonic Reasoning. LNCS, vol. 6565, pp. 315–333. Springer, Heidelberg (2011)
13. Motik, B., Rosati, R.: Reconciling description logics and rules. J. ACM **57**, 5 (2010)
14. Zhang, Y.: Epistemic reasoning in logic programs. In: IJCAI, pp. 647–653 (2007)
15. Gelfond, M.: New semantics for epistemic specifications. In: Delgrande, J.P., Faber, W. (eds.) LPNMR 2011. LNCS, vol. 6645, pp. 260–265. Springer, Heidelberg (2011)
16. Zhang, Z., Zhao, K.: Esmodels: an epistemic specification inference. In: ICTAI-2013, pp. 769–774 (2013)
17. van der Hoek, W., Meyer, J.J.C.: Graded modalities in epistemic logic. Logique et Analyse **34**(133–134), 251–270 (1991)

18. Fattorosi-Barnaba, T., de Caro, F.: Graded modalities I. Stud. Logica. **44**, 197–221 (1985)
19. Baader, F., Nutt, W.: Basic description logics. In: Description Logic Handbook, pp. 43–95. Cambridge University Press (2003)
20. Eiter, T., Gottlob, G.: On the computational cost of disjunctive logic programming: propositional case. Ann. Math. Artif. Intell. **15**(3/4), 289–323 (1995)
21. http://sourceforge.net/projects/potassco/files/guide/2.0/guide-2.0.pdf

The Design of the Sixth Answer Set Programming Competition
– Report –

Martin Gebser[1], Marco Maratea[2], and Francesco Ricca[3]([✉])

[1] Helsinki Institute for Information Technology HIIT,
Aalto University, Espoo, Finland
[2] DIBRIS, Università di Genova, Genova, Italy
[3] Dipartimento di Matematica e Informatica,
Università della Calabria, Rende Cs, Italy
ricca@mat.unical.it

Abstract. Answer Set Programming (ASP) is a well-known paradigm of declarative programming with roots in logic programming
and non-monotonic reasoning. Similar to other closely-related problem-
solving technologies, such as SAT/SMT, QBF, Planning and Scheduling, advances in ASP solving are assessed in competition events. In
this paper, we report about the design of the Sixth ASP Competition,
which is jointly organized by the University of Calabria (Italy), Aalto
University (Finland), and the University of Genova (Italy), in affiliation with the 13th International Conference on Logic Programming and
Non-Monotonic Reasoning (LPNMR 2015). This edition maintains some
of the design decisions introduced in the last event, e.g., the design of
tracks, the scoring scheme, and the adherence to a fixed modeling language in order to push the adoption of the ASP-Core-2 standard. On the
other hand, it features also some novelties, like a benchmarks selection
stage to classify instances according to their expected hardness, and a
"marathon" track where the best performing systems are given more
time for solving hard benchmarks.

1 Introduction

Answer Set Programming [7, 13–15, 28, 29, 35, 41, 44] is a well-known declarative
programming approach to knowledge representation and reasoning, with roots in
the areas of logic programming and non-monotonic reasoning as well as close relationships to other formalisms such as SAT, SAT Modulo Theories, Constraint Programming, PDDL, and many others. With the exception of the fifth event,[1] which
was held in 2014 in order to join the FLoC Olympic Games at the Vienna Summer
of Logic,[2] ASP Competitions are biennial events organized in odd years. The goal

---

M. Gebser—Affiliated with the University of Potsdam, Germany.
[1] https://www.mat.unical.it/aspcomp2014/.
[2] http://vsl2014.at/.

© Springer International Publishing Switzerland 2015
F. Calimeri et al. (Eds.): LPNMR 2015, LNAI 9345, pp. 531–544, 2015.
DOI: 10.1007/978-3-319-23264-5_44

of the Answer Set Programming (ASP) Competition series is to access the state of the art in ASP solving (see, e.g., [1, 10, 12, 23, 25, 30, 31, 33, 36, 37, 42, 45] on challenging benchmarks.

In this paper, we report about the design of the Sixth ASP Competition,[3] jointly organized by the University of Calabria (Italy), Aalto University (Finland), and the University of Genova (Italy), in affiliation with the 13th International Conference on Logic Programming and Non-Monotonic Reasoning (LPNMR 2015).[4] This edition maintains some of the design decisions introduced in the last event, e.g., (*i*) the design of tracks, based on the "complexity" of the encoded problems (as in past events), but also considering the language features involved in encodings (e.g., choice rules, aggregates, presence of queries), (*ii*) the scoring scheme, which had been significantly simplified, and (*iii*) the adherence to a fixed modeling language in order to push the adoption of the ASP-Core-2 standard.[5] On the other hand, we also introduce novelties, some of them borrowed from past editions of the SAT and QBF Competitions, i.e., (*i*) a benchmarks selection stage to classify instances according to their expected hardness, in order to select instances from a broad range of difficulty, and (*ii*) a "marathon" track where the best performing systems are given more time for solving hard benchmarks, in order to check whether they are able to complete difficult instances in the long run.

The present report is structured as follows. First, Sect. 2 introduces the setting of the Sixth ASP Competition. Then, Sects. 3 and 4 present the problem domains and the instance selection process, respectively. Section 5 surveys the participants and systems registered for the competition. The report is concluded by final remarks in Sect. 6.

2   Format of the Sixth ASP Competition

In this section, we discuss the format of the competition event, describe categories and tracks, and recapitulate the scoring scheme along with general rules. Furthermore, we provide some information about the competition infrastructure.

As outlined in Sect. 1, the Sixth ASP competition maintains choices made in the last event, but also adds some novelties. First, the scoring scheme, which was significantly simplified in the last edition (cf. [11]), remains unchanged. In order to encourage new teams and research groups to join the event, we also maintain the division into tracks, primarily based on language features rather than inherent computational complexity, as in the last edition. Given this, preliminary or otherwise confined systems may take part in some tracks only, i.e., the ones featuring the subset of the language they support. Furthermore, the tracks draw a clearer and more detailed picture about what (combinations of) techniques work well for particular language features, which, in our opinion, is more interesting than merely reporting overall winners.

---

[3] https://aspcomp2015.dibris.unige.it/.

[4] http://lpnmr2015.mat.unical.it/.

[5] https://www.mat.unical.it/aspcomp2013/ASPStandardization/.

Competition Format. The competition is open to any general-purpose solving system, provided it is able to parse the **ASP-Core-2** input format. However, following the positive experience of 2014, we also plan to organize an on-site modeling event at LPNMR 2015, in the spirit of the Prolog contest. Regarding benchmarks, this year featured a call to submit new domains (see Sect. 3) that, together with the domains employed in the last event, are part of the benchmark collection of this edition. For the latter, the Fifth ASP Competition proposed and evaluated a new set of encodings: this year we fix the encodings to those that led to better performance in 2014. For new domains, we consider the encodings provided by benchmark contributors. The whole benchmark set undergoes a benchmark selection phase in order the classify instances based on their expected hardness, and then to pick instances of varying difficulty to be run in the competition (see Sect. 4 for details).

Competition Categories. The competition consists of *two categories*, depending on the computational resources made available to each running system:

- **SP**: One processor allowed;
- **MP**: Multiple processors allowed.

While the **SP** category aims at sequential solving systems, parallelism can be exploited in the **MP** category.

Competition Tracks. Both categories of the competition are structured into *four tracks*, which are described next:

- **Track #1**: *Basic Decision.* Encodings: normal logic programs, simple arithmetic and comparison operators.
- **Track #2**: *Advanced Decision.* Encodings: full language, with queries, excepting optimization statements and non-HCF disjunction.
- **Track #3**: *Optimization.* Encodings: full language with optimization statements, excepting non-HCF disjunction.
- **Track #4**: *Unrestricted.* Encodings: full language.

We also plan to introduce a **Marathon** track this year, thus analyzing participant systems along a different dimension. The idea, borrowed from past QBF Competitions, is to grant more time to the best solvers on a limited set of instances that proved to be difficult in regular tracks.

Scoring Scheme. The scoring scheme adopted is the same as in the Fifth ASP Competition. In particular, it considers the following factors:

- Problems are always weighted equally.
- If a system outputs an incorrect answer to some instance of a problem, this invalidates its score for the problem, even if other instances are correctly solved.
- In case of Optimization problems, scoring is mainly based on solution quality.

In general, 100 points can be earned for each benchmark problem. The final score of a solving system consists of the sum of scores over all problems.

Scoring Details. For *Decision and Query problems*, the score of a solver S on a problem P featuring N instances is computed as

$$S(P) = \frac{N_S * 100}{N}$$

where N_S is the number of instances solved within the allotted time and memory limits.

For *Optimization problems*, solvers are ranked by solution quality, in the spirit of the MANCOOSI International Solver Competition.[6] Given M participant systems, the score of a solver S for an instance I of a problem P featuring N instances is computed as

$$S(P, I) = \frac{M_S(I) * 100}{M * N}$$

where $M_S(I)$ is

- 0, if S did neither provide a solution, nor report unsatisfiability, or
- the number of participant systems that did not provide any strictly better solution than S, where a confirmed optimum solution is considered strictly better than an unconfirmed one, otherwise.

The score $S(P)$ of a solver S for problem P consists of the sum of scores $S(P, I)$ over all N instances I featured by P. Note that, as with Decision and Query problems, $S(P)$ can range from 0 to 100.

Global Ranking. The global ranking for each track, and the overall ranking, is obtained by awarding each participant system the sum of its scores over all problems; systems are ranked by their sums, in decreasing order. In case of a draw in terms of sums of scores, sums of runtimes are taken into account.

Competition Environment. The competition is run on a Debian Linux server (64bit kernel), featuring 2.30 GHz Intel Xeon E5-4610 v2 Processors with 16 MB of cache and 128 GB of RAM. Time and memory for each run are limited to 20 min and 12 GB, respectively. Participant systems can exploit up to 8 cores (i.e., up to 16 virtual CPUs since Intel Hyperthreading technology is enabled) in the **MP** category, whereas the execution is constrained to one core in the **SP** category. The execution environment is composed of a number of scripts, and performance is measured using the *pyrunlim* tool.[7]

3   Benchmark Suite

The benchmark domains considered in the Sixth ASP Competition include those from the previous edition, summarized first. Moreover, encodings and instances were provided for six new domains, introduced afterwards.

---

[6] http://www.mancoosi.org/misc/.

[7] https://github.com/alviano/python/.

Previous Domains. The Fifth ASP Competition featured 26 benchmark domains that had been submitted to earlier editions already, mainly in 2013 when the ASP-Core-2 standard input format was specified. In some domains, however, "unoptimized" encodings submitted by benchmark authors incurred grounding bottlenecks that made participant systems fail on the majority of instances. In view of this and in order to enrich the available benchmark collection, alternative encodings were devised and empirically compared last year for all but two domains dealing with Query answering, which were modeled by rather straightforward positive programs.

The first part of assembling the benchmark suite for the Sixth ASP Competition consisted in the choice of encodings for previously used domains. Table 1 gives an overview of these domains, outlining application-oriented problems, respective computational tasks, i.e., Decision, Optimization, or Query answering, and tracks. Most importantly, the fourth column indicates whether the encoding made available in 2013 or the alternative one provided last year has been picked for this edition of the ASP Competition. The selection was based on the results from 2014, favoring the encoding variant that exhibited better performance of participant systems in a benchmark domain.

For Decision problems in the *Hanoi Tower, Knight Tour with Holes, Stable Marriage, Incremental Scheduling, Partner Units, Solitaire, Weighted-Sequence Problem,* and *Minimal Diagnosis* domains, all systems benefited from the usage of alternative 2014 encodings. Although the results were not completely uniform, improvements of more systems or greater extent were obtained in *Graph Colouring, Visit-all, Nomystery, Permutation Pattern Matching,* and *Qualitative Spatial Reasoning.* On two remaining Decision problems, *Sokoban* and *Complex Optimization,* no significant performance gaps were observed, and 2014 encodings were picked as they simplify the original submissions, i.e., aggregates are omitted in *Sokoban* and redundant preconditions of rules dropped in *Complex Optimization.* In fact, due to similar simplifications, the Basic Decision track (#1) consists of six domains, while it previously included *Labyrinth* and *Stable Marriage* only. On the other hand, the encodings from 2013 were kept for domains where alternative variants did not lead to improvements or even deteriorated performance, as it was the case in *Graceful Graphs.*

In view of the relative scoring of systems on Optimization problems, the selection of encoding variants could not be based on (uniform) improvements in terms of score here. Rather than that, we investigated timeouts, runtimes, and solution quality of the top-performing systems from last year, thus concentrating on the feasibility of good but not necessarily optimal solutions. In this regard, the alternative 2014 encodings turned out to be advantageous in *Crossing Minimization* and *Maximal Clique,* while the original submissions led to better results in *Connected Still Life* and *Valves Location,* or essentially similar performance in *Abstract Dialectical Frameworks.*

Notably, this edition of the ASP Competition utilizes a revised formulation of *Connected Still Life* (thus marked by '*' in Table 1), where instances specify grid cells that must be "dead" or "alive" according to the Game of Life version considered in this domain. Respective conditions are addressed by side

Table 1. Encodings selected for benchmark domains from the Fifth ASP competition

Domain	App	Problem	Encoding	
Graph Colouring		Decision	2014	
Hanoi Tower		Decision	2014	Track #1
Knight Tour with Holes		Decision	2014	
Labyrinth		Decision	2013	
Stable Marriage		Decision	2014	
Visit-all		Decision	2014	
Bottle Filling		Decision	2013	
Graceful Graphs		Decision	2013	
Incremental Scheduling	√	Decision	2014	
Nomystery		Decision	2014	
Partner Units	√	Decision	2014	Track #2
Permutation Pattern Matching		Decision	2014	
Qualitative Spatial Reasoning		Decision	2014	
Reachability		Query	2013	
Ricochet Robots		Decision	2013	
Sokoban		Decision	2014	
Solitaire		Decision	2014	
Weighted-Sequence Problem		Decision	2014	
Connected Still Life*		Optimization	2013	
Crossing Minimization	√	Optimization	2014	Track #3
Maximal Clique		Optimization	2014	
Valves Location	√	Optimization	2013	
Abstract Dialectical Frameworks		Optimization	2013	
Complex Optimization	√	Decision	2014	Track #4
Minimal Diagnosis	√	Decision	2014	
Strategic Companies		Query	2013	

constraints added to the previously available encodings and enable a diversification of instances of same size, while size had been the only parameter for obtaining different instances before. In addition, benchmark authors provided new instance sets for the *Knight Tour with Holes*, *Stable Marriage*, *Ricochet Robots*, and *Maximal Clique* domains. For *Knight Tour with Holes*, the instances from last year were too hard for most participant systems, and too easy in the other three domains. Finally, recall that the 2013 encodings for Query problems in the *Reachability* and *Strategic Companies* domains are reused.

Six of the 26 benchmark domains stemming from earlier editions of the ASP competition are based on particular applications. In more detail, *Incremental Scheduling* [6] deals with assigning jobs to devices such that the makespan of a schedule stays within a given budget. The matching problem *Partner Units* [5] has applications in the configuration of surveillance, electrical engineering, computer network, and railway safety systems. The *Crossing Minimization* [17] domain aims at optimized layouts of hierarchical network diagrams in graph drawing. The hydroinformatics problem *Valves Location* [18] is concerned with designing water distribution systems such that the isolation in case of damages

Table 2. New benchmark domains of the sixth ASP competition

Domain	App	Problem	
Combined Configuration	√	Decision	Tr. #2
Consistent Query Answering	√	Query	
MaxSAT	√	Optimization	Track #3
Steiner Tree	√	Optimization	
System Synthesis	√	Optimization	
Video Streaming	√	Optimization	

is minimized. In contrast to objective functions considered in the Optimization track (#3), the *Complex Optimization* [22] domain addresses subset minimization in the contexts of biological network repair [19] and minimal unsatisfiable core membership [32]. Finally, *Minimal Diagnosis* [27] tackles the identification of minimal reasons for inconsistencies between biological networks and experimental data.

New Domains. Six new benchmark domains, all of which are application-oriented as indicated in Table 2, were submitted to the Sixth ASP Competition:

- *Combined Configuration* [26] is a Decision problem inspired by industrial product configuration tasks dealing with railway interlocking systems, automation systems, etc. In the considered scenario, orthogonal requirements as encountered in bin packing, graph coloring, matching, partitioning, and routing must be fulfilled by a common solution. Since the combined problem goes beyond its individual subtasks, specialized procedures for either of them are of limited applicability, and the challenge is to integrate all requirements into general solving methods.
- *Consistent Query Answering* [38] addresses phenomena arising in the integration of data from heterogeneous sources. The goal is to merge as much information as possible, even though local inconsistencies and incompleteness typically preclude a mere data fusion. In particular, the Query problem amounts to cautious reasoning, retrieving consequences that are valid under all candidate repairs of input data.
- *MaxSAT* [34] is the optimization variant of SAT, where so-called soft clauses may be violated to particular costs and the sum of costs ought to be minimal. Industrial instances, taken from the 2014 MaxSAT Evaluation,[8] are represented by facts and encoded as an Optimization problem.
- *Steiner Tree* [16] is concerned with connecting particular endpoints by a spanning tree. The domain deals with the rectilinear version of this problem, where points on a two-dimensional grid may be connected by horizontal or vertical line segments. This setting is of practical relevance as it corresponds to wire routing in circuit design. The accumulated line segments determine a wire length, which is subject to minimization in the considered Optimization problem.

---

[8] http://www.maxsat.udl.cat/14/index.html.

- *System Synthesis* [9] deals with the allocation of parallel tasks and message routing in integrated hardware architectures for target applications. On the one hand, the capacities of processing elements are limited, so that communicating tasks must be distributed. On the other hand, network communication shall avoid long routes to reduce delays. The Optimization problem combines three lexicographically ordered objectives: balancing the allocation of processing elements, minimizing network communication, and keeping routes short.
- *Video Streaming* [46] aims at an adaptive regulation of resolutions and bit rates in a content delivery network. While the bit rates of users and the number of different video formats that can be offered simultaneously are limited, service disruptions are admissible for a fraction of users only. The objective of the Optimization problem is to achieve high user satisfaction with respect to particular video contents.

4   Benchmark Selection

For an informed instance selection going beyond the random selection adopted in the 2014 edition of the ASP Competition or the solver-dependent criterion employed in 2013, we utilize an instance selection strategy inspired by the 2014 SAT Competition.[9] First, the empirical hardness of all available instances is evaluated by running the top-performing systems from last year, and then a balanced selection is made among instances of varying difficulty.

Top-performing Systems. We considered the best performing system per team that participated in the Fifth ASP Competition, corresponding to the systems taking the first three places last year, i.e., CLASP, LP2NORMAL2+CLASP, and WASP-1.5. This choice comes close to the ideal state-of-the-art solver that matches the best performing system on each instance.

Instance Classification. All instances available in the benchmark collection are classified according to the runtimes of the top-performing systems by picking the upmost applicable category as follows:

(non-groundable) Instances that could not be grounded by any of the top-performing systems within the timeout of 20 min.
(very easy) Instances solved by all top-performing systems in less than 20 s.
(easy) Instances solved by all top-performing systems in less than 2 min.
(medium) Instances solved by all top-performing systems within the timeout of 20 min.
(hard) Instances solved by at least one among the top-performing systems within 40 min, i.e., twice the timeout.
(too hard) Instances that could not be solved (no solution produced in case of Optimization problems) by any of the top-performing systems within 40 min.

---

[9] http://www.satcompetition.org/2014/index.shtml.

While non-groundable instances are basically out of reach for traditional ASP systems, very easy ones are highly unlikely to yield any relevant distinction between participant systems. Hence, instances falling into the first two categories are discarded and not run in the competition. Unlike that, easy, medium, and hard instances are expected to differentiate between unoptimized, average, and top-performing competition entries. Albeit they may not be solvable by any participant system within 20 min, too hard instances are included to impose challenges and are primary candidates for the Marathon track in which the timeout will be increased.

Instance Selection. Instances to be run in the competitions will be picked per benchmark domain, matching the following conditions as much as possible:

1. 20 instances are included in each domain.
2. Easy, medium, hard, too hard, and randomly picked (yet excluding non-groundable and very easy) instances shall evenly contribute 20 % (i.e., four) instances each.
3. Satisfiable and unsatisfiable instances should be balanced (if known/applicable).
4. The selection among candidate instances according to the previous conditions is done randomly, using the concatenation of winning numbers in the EuroMillions lottery of 23rd June 2015 as seed.

Further criteria will be taken into account to filter domains by need. That is, domains in which instances lack variety, i.e., all instances turn out as easy to medium or (too) hard, may be excluded in the competition. We do not impose strict conditions, however, as being new, application-oriented, based on ASP-specific language features (e.g., aggregates, recursion, or disjunction) or a particular computational task (Optimization or Query answering) may justify an interest beyond the scalability of available instances.

Preliminary Data. The instance classification process has been running at the time of writing this report. In the first stage, non-groundable instances were identified and discarded, thus dropping 88 of the available instances (86 from *Incremental Scheduling* and two from *Sokoban*). This leaves 4970 instances for running the three top-performing systems from last year, using a timeout of 40 min. We expect to obtain complete results of these runs, on which the instance selection will be based, by 22nd June 2015.

5   Participants

In this section, we briefly survey the participants and systems registered for the competition. In total, the competition features 13 systems coming from three teams:

- The Aalto team from Aalto University submitted nine systems, mainly working by means of translations [10,20,37,43]. Two systems, LP2SAT+LINGELING and LP2SAT+PLINGELING-MT, rely on translation to SAT, which includes the normalization of aggregates as well as the encoding of level mappings for non-tight problem instances. The latter are expressed in terms of acyclicity checking [20,21] on top of ASP, Pseudo-Boolean or SAT formulations, respectively, used in the systems LP2ACYCASP+CLASP, LP2ACYCPB+CLASP, LP2ACYCSAT+CLASP, and LP2ACYCSAT+GLUCOSE. While LP2SAT+LINGELING and LP2SAT+PLINGELING-MT do not support optimization and participate in the Basic and Advanced Decision tracks (#1 and #2) only, the latter systems compete also in the Optimization track (#3). The same applies to LP2MIP and LP2MIP-MT, which run CPLEX as a Mixed Integer Programming solver backend. Finally, LP2NORMAL+CLASP normalizes aggregates (of small to medium size) and uses CLASP as back-end ASP solver; LP2NORMAL+CLASP participates in all four tracks and thus also in the Unrestricted track (#4). All systems by the Aalto team utilize GRINGO-4 for grounding, and neither of them supports Query problems (*Consistent Query Answering*, *Reachability*, and *Strategic Companies*). The systems LP2SAT+PLINGELING-MT and LP2MIP-MT exploit multi-threading and run in the **MP** category, while the other, sequential systems participate in the **SP** category.
- The ME-ASP team from the University of Genova, the University of Sassari, and the University of Calabria submitted the multi-engine ASP solver ME-ASP [39,40]. ME-ASP applies a selection policy to decide what is the most promising ASP solver to run, given some characteristics of an input program. The pool of ASP solvers from which ME-ASP can choose is a selection of the solvers submitted to the Fifth ASP Competition, while input characteristics correspond to non-ground and ground features. The ME-ASP system utilizes GRINGO-4 for grounding and participates in all four tracks of the **SP** category.
- The Wasp team from the University of Calabria submitted two systems based on WASP [1,2,4], namely WASP and WASP+DLV, as well as the proof-of-concept prototype JWASP, written in Java. WASP is a native ASP solver based on conflict-driven learning, yet extended with techniques specifically designed for solving disjunctive logic programs. It utilizes GRINGO-4 for grounding and participates in all tracks, although with limited support for Query problems. On the other hand, WASP+DLV includes full functionalities for Query answering [3,33] and competes in all domains. The prototype system JWASP is based on the SAT4J [8] SAT solver and implements some of the algorithms employed in WASP for handling ASP-specific features, which enables its participation in the Basic and Advanced Decision tracks (#1 and #2). All systems by the Wasp team run in the **SP** category.

In sum, similar to past competitions, the vast majority of submitted systems is based on two main approaches to ASP solving: (*i*) "native" systems, which exploit techniques purposely conceived and/or adapted for dealing with logic programs under the stable models semantics, and (*ii*) "translation-based" systems, which (roughly) at some stage of the evaluation produce an intermediate specification in some different formalism that is then fed to a corresponding solver. The solvers submitted by the Wasp team as well as ME-ASP and LP2NORMAL+CLASP pursue a native approach, while the remaining systems by the Aalto team utilize translations.

The main novelty among competition entries this year is the "portfolio" solver ME-ASP. Its multi-engine approach differs from CLASPFOLIO [24], which participated last in the 2013 edition of the ASP Competition. Furthermore, it is worth mentioning that, in order to assess the improvements in ASP solving, we also consider the version of CLASP submitted in 2014 for reference, given that CLASP was the overall winner of the Fifth ASP Competition.

6 Conclusions

The Sixth ASP Competition is jointly organized by the University of Calabria (Italy), Aalto University (Finland), and the University of Genova (Italy), in affiliation with the 13th International Conference on Logic Programming and Non-Monotonic Reasoning (LPNMR 2015). The main goal is measuring advances of the state of the art in ASP solving, where native and translation-based systems constitute the two main approaches. This report presented the design of the event and gave an overview of benchmarks as well as participants. On the one hand, this edition of the ASP Competition maintains design decisions from 2014, e.g., tracks are conceived on the basis of language features. On the other hand, it also introduces some novelties, i.e., a benchmark selection phase and a Marathon track.

The competition results will be announced at LPNMR 2015, at which, following the positive experience of 2014, we also plan to organize another on-site modeling event in the spirit of the Prolog contest. This modeling competition to some extent replaces the Model&Solve track that was included last in the 2013 edition of the ASP Competition, yet the idea is to margin the effort of problem modeling.

References

1. Alviano, M., Dodaro, C., Faber, W., Leone, N., Ricca, F.: WASP: a native ASP solver based on constraint learning. In: Cabalar, P., Son, T.C. (eds.) LPNMR 2013. LNCS, vol. 8148, pp. 54–66. Springer, Heidelberg (2013)
2. Alviano, M., Dodaro, C., Leone, N., Ricca, F.: Advances in Wasp. In: Proceedings of LPNMR 2015. Springer (2015)
3. Alviano, M., Dodaro, C., Ricca, F.: Anytime computation of cautious consequences in answer set programming. Theory Pract. Logic Program. **14**(4–5), 755–770 (2014)
4. Alviano, M., Dodaro, C., Ricca, F.: Preliminary report on Wasp 2.0. In: Proceedings of NMR 2014, pp. 68–72. Vienna University of Technology (2014)
5. Aschinger, M., Drescher, C., Friedrich, G., Gottlob, G., Jeavons, P., Ryabokon, A., Thorstensen, E.: Optimization methods for the partner units problem. In: Achterberg, T., Beck, J.C. (eds.) CPAIOR 2011. LNCS, vol. 6697, pp. 4–19. Springer, Heidelberg (2011)

6. Balduccini, M.: Industrial-size scheduling with ASP+CP. In: Delgrande, J.P., Faber, W. (eds.) LPNMR 2011. LNCS, vol. 6645, pp. 284–296. Springer, Heidelberg (2011)
7. Baral, C.: Knowledge Representation Reasoning and Declarative Problem Solving. Cambridge University Press, Cambridge (2003)
8. Berre, D., Parrain, A.: The Sat4j library, release 2.2. J. Satisfiability Boolean Model. Comput. **7**, 59–64 (2010)
9. Biewer, A., Andres, B., Gladigau, J., Schaub, T., Haubelt, C.: A symbolic system synthesis approach for hard real-time systems based on coordinated SMT-solving. In: Proceedings of DATE 2015, pp. 357–362. ACM (2015)
10. Bomanson, J., Gebser, M., Janhunen, T.: Improving the normalization of weight rules in answer set programs. In: Fermé, E., Leite, J. (eds.) JELIA 2014. LNCS, vol. 8761, pp. 166–180. Springer, Heidelberg (2014)
11. Calimeri, F., Gebser, M., Maratea, M., Ricca, F.: The design of the fifth answer set programming competition. In: Technical Communications of ICLP 2014, http://arxiv.org/abs/1405.3710v4. CoRR (2014)
12. Palù, D., Dovier, A., Pontelli, E., Rossi, G.: GASP: Answer set programming with lazy grounding. Fundamenta Informaticae **96**(3), 297–322 (2009)
13. Eiter, T., Faber, W., Leone, N., Pfeifer, G.: Declarative problem-solving using the DLV system. In: Minker, J. (ed.) Logic-Based Artificial Intelligence, pp. 79–103. Kluwer Academic Publishers, Norwell (2000)
14. Eiter, T., Gottlob, G., Mannila, H.: Disjunctive datalog. ACM Trans. Database Syst. **22**(3), 364–418 (1997)
15. Eiter, T., Ianni, G., Krennwallner, T.: Answer set programming: a primer. In: Tessaris, S., Franconi, E., Eiter, T., Gutierrez, C., Handschuh, S., Rousset, M.-C., Schmidt, R.A. (eds.) Reasoning Web. LNCS, vol. 5689, pp. 40–110. Springer, Heidelberg (2009)
16. Erdem, E., Wong, M.D.F.: Rectilinear steiner tree construction using answer set programming. In: Demoen, B., Lifschitz, V. (eds.) ICLP 2004. LNCS, vol. 3132, pp. 386–399. Springer, Heidelberg (2004)
17. Fulek, R., Pach, J.: A computational approach to conway's thrackle conjecture. In: Brandes, U., Cornelsen, S. (eds.) GD 2010. LNCS, vol. 6502, pp. 226–237. Springer, Heidelberg (2011)
18. Gavanelli, M., Nonato, M., Peano, A., Alvisi, S., Franchini, M.: An ASP approach for the valves positioning optimization in a water distribution system. In: Proceedings of CILC 2012, pp. 134–148 (2012). http://www.CEUR-WS.org
19. Gebser, M., Guziolowski, C., Ivanchev, M., Schaub, T., Siegel, A., Thiele, S., Veber, P.: Repair and prediction (under inconsistency) in large biological networks with answer set programming. In: Proceedings of KR 2010, pp. 497–507. AAAI (2010)
20. Gebser, M., Janhunen, T., Rintanen, J.: Answer set programming as SAT modulo acyclicity. In: Proceedings of ECAI 2014, pp. 351–356. IOS (2014)
21. Gebser, M., Janhunen, T., Rintanen, J.: SAT modulo graphs: acyclicity. In: Fermé, E., Leite, J. (eds.) JELIA 2014. LNCS, vol. 8761, pp. 137–151. Springer, Heidelberg (2014)
22. Gebser, M., Kaminski, R., Schaub, T.: Complex optimization in answer set programming. Theory Pract. Logic Program. **11**(4–5), 821–839 (2011)
23. Gebser, M., Kaufmann, B., Schaub, T.: Advanced conflict-driven disjunctive answer set solving. In: Proceedings of IJCAI 2013, pp. 912–918. IJCAI/AAAI (2013)

24. Gebser, M., Kaminski, R., Kaufmann, B., Schaub, T., Schneider, M.T., Ziller, S.: A portfolio solver for answer set programming: preliminary report. In: Delgrande, J.P., Faber, W. (eds.) LPNMR 2011. LNCS, vol. 6645, pp. 352–357. Springer, Heidelberg (2011)

25. Gebser, M., Kaufmann, B., Schaub, T.: Conflict-driven answer set solving: from theory to practice. Artif. Intell. **187–188**, 52–89 (2012)

26. Gebser, M., Ryabokon, A., Schenner, G.: Combining heuristics for configuration problems using answer set programming. In: Proceedings of LPNMR 2015. Springer (2015)

27. Gebser, M., Schaub, T., Thiele, S., Veber, P.: Detecting inconsistencies in large biological networks with answer set programming. Theory Pract. Logic Program. **11**(2–3), 323–360 (2011)

28. Gelfond, M., Leone, N.: Logic programming and knowledge representation - the A-Prolog perspective. Artif. Intell. **138**(1–2), 3–38 (2002)

29. Gelfond, M., Lifschitz, V.: Classical negation in logic programs and disjunctive databases. New Gener. Comput. **9**, 365–385 (1991)

30. Giunchiglia, E., Lierler, Y., Maratea, M.: Answer set programming based on propositional satisfiability. J. Autom. Reason. **36**(4), 345–377 (2006)

31. Janhunen, T., Niemelä, I., Seipel, D., Simons, P., You, J.: Unfolding partiality and disjunctions in stable model semantics. ACM Trans. Comput. Logic **7**(1), 1–37 (2006)

32. Janota, M., Marques-Silva, J.: On deciding MUS membership with QBF. In: Lee, J. (ed.) CP 2011. LNCS, vol. 6876, pp. 414–428. Springer, Heidelberg (2011)

33. Leone, N., Pfeifer, G., Faber, W., Eiter, T., Gottlob, G., Perri, S., Scarcello, F.: The DLV system for knowledge representation and reasoning. ACM Trans. Comput. Logic **7**(3), 499–562 (2006)

34. Li, C., Manyà, F.: MaxSAT. In: Biere, A. (ed.) Handbook of Satisfiability, pp. 613–631. IOS Press, Amsterdam (2009)

35. Lifschitz, V.: Answer set programming and plan generation. Artif. Intell. **138**(1–2), 39–54 (2002)

36. Lin, F., Zhao, Y.: ASSAT: computing answer sets of a logic program by SAT solvers. Artif. Intell. **157**(1–2), 115–137 (2004)

37. Liu, G., Janhunen, T., Niemelä, I.: Answer set programming via mixed integer programming. In: Proceedings of KR 2012, pp. 32–42. AAAI (2012)

38. Manna, M., Ricca, F., Terracina, G.: Consistent query answering via ASP from different perspectives: theory and practice. Theory Pract. Logic Program. **13**(2), 227–252 (2013)

39. Maratea, M., Pulina, L., Ricca, F.: A multi-engine approach to answer-set programming. Theory Pract. Logic Program. **14**(6), 841–868 (2014)

40. Maratea, M., Pulina, L., Ricca, F.: Multi-level algorithm selection for ASP. In: Proceedings of LPNMR 2015. Springer (2015)

41. Marek, V., Truszczyński, M.: Stable models and an alternative logic programming paradigm. In: Apt, K.R., Marek, V.W., Truszczynski, M., Warren, D.S. (eds.) The Logic Programming Paradigm: A 25-Year Perspective, pp. 375–398. Springer, Heidelberg (1999)

42. Mariën, M., Wittocx, J., Denecker, M., Bruynooghe, M.: SAT(ID): satisfiability of propositional logic extended with inductive definitions. In: Kleine Büning, H., Zhao, X. (eds.) SAT 2008. LNCS, vol. 4996, pp. 211–224. Springer, Heidelberg (2008)

43. Nguyen, M., Janhunen, T., Niemelä, I.: Translating answer-set programs into bit-vector logic. In: Tompits, H., Abreu, S., Oetsch, J., Pührer, J., Seipel, D., Umeda, M., Wolf, A. (eds.) INAP/WLP 2011. LNCS, vol. 7773, pp. 91–109. Springer, Heidelberg (2013)
44. Niemelä, I.: Logic programs with stable model semantics as a constraint programming paradigm. Ann. Math. Artif. Intell. **25**(3–4), 241–273 (1999)
45. Simons, P., Niemelä, I., Soininen, T.: Extending and implementing the stable model semantics. Artif. Intell. **138**(1–2), 181–234 (2002)
46. Toni, L., Aparicio-Pardo, R., Simon, G., Blanc, A., Frossard, P.: Optimal set of video representations in adaptive streaming. In: Proceedings of MMSys 2014, pp. 271–282. ACM (2014)

Doctoral Consortium Extended Abstract: Planning with Concurrent Transaction Logic

Reza Basseda[✉]

Stony Brook University, Stony Brook, NY 11794, USA
rbasseda@cs.stonybrook.edu

Abstract. Automated planning has been the subject of intensive research and is at the core of several areas of AI, including intelligent agents and robotics. In this thesis proposal, we argue that Concurrent Transaction Logic ($\mathcal{CTR}$) is a natural specification language for planning algorithms, which enables one to see further afield and thus discover better and more general solutions than using one-of-a-kind formalisms. Specifically, we take the well-known $STRIPS$ planning strategy and show that $\mathcal{CTR}$ lets one specify the $STRIPS$ planning algorithm easily and concisely, and extend it in several respects. For instance, we show that extensions to allow indirect effects and to support action ramifications come almost for free. The original $STRIPS$ planning strategy is also shown to be incomplete. Using concurrency operators in $\mathcal{CTR}$, we propose a non-linear $STRIPS$ planning algorithm, which is proven to be complete. Moreover, this thesis proposal outlines several extensions of $STRIPS$ planning strategy. All of the extensions show that the use of $\mathcal{CTR}$ accrues significant benefits in the area of planning.

1 Introduction

The classical problem of automated planning has been used in a wide range of applications such as robotics, multi-agent systems, and more. Due to this wide range of applications, automated planning has become one of the most important research areas in Artificial Intelligence (AI). The history of using logical deduction to solve classical planning problems in AI dates back to the late 1960 s when situation calculus was applied in the planning domain [18]. There are several planners that encode planning problems into satisfiability problems [21], constraint satisfaction problems (CSP) [1,14,24], or answer set programming [16,17,22,23] and use logical deduction to solve the planning problems. Beside those planners, a number of deductive planning frameworks have been proposed over the years [5–7,13,19,20].

There are several reasons that make logical deduction suitable to be used by a classical planner: (1) Logic-based deduction used in planning can be cast as a formal framework that eases proving different planning properties such as completeness. (2) Logic-based systems naturally provide a declarative language that

Extended Abstract as part of the program of the 1st Joint ADT/LPNMR 2015 Doctoral Consortium co-chaired by Esra Erdem and Nicholas Mattei.

© Springer International Publishing Switzerland 2015
F. Calimeri et al. (Eds.): LPNMR 2015, LNAI 9345, pp. 545–551, 2015.
DOI: 10.1007/978-3-319-23264-5_45

simplifies the specification of planning problems. (3) Logical deduction is usually an essential component of intelligent and knowledge representation systems. Therefore, applying logical deduction in classical planning makes the integration of planners with such systems simpler. Despite these benefits of using logical deduction in planning, many of the above mentioned deductive planning techniques are not getting as much attention as algorithms specifically provided for planning problems. The most obvious reason for this shortcoming is as follows: These works generally show how they can represent and encode classical planning actions and rely on a theorem prover of some sort to find plans. Therefore, the planning techniques embedded in such planners are typically the simplest state space planning (e.g. forward state space search) that has a extremely large search space. Consequently, they cannot exploit planning heuristics and techniques.

In this thesis, we will show that *Concurrent Transaction Logic* (or $\mathcal{CTR}$) [10–12] addresses this issue and also provides multiple advantages for specifying, generalizing, and solving planning problems. To illustrate the point, we will take *STRIPS* planning technique and show that its associated planning algorithm easily and naturally lend itself to compact representation in $\mathcal{CTR}$.

The expressiveness of $\mathcal{CTR}$ let the *STRIPS* planning algorithm be naturally extended with intensional rules. Concurrency in $\mathcal{CTR}$ lets us introduce a non-linear *STRIPS* planning algorithm, which is also proven to be complete. After inspecting the logic rules that simulate the planning algorithm, we observe some heuristics that can be applied to reduce the search space. The elegant expression of *STRIPS* algorithm in $\mathcal{CTR}$ lets us first characterize the regression of literals through *STRIPS* actions, then enhance the proposed *STRIPS* algorithm with a regression analysis method. We also extend the *STRIPS* planning algorithm with a few extra rules to solve planning problems with negative derived literals.

This extended abstract paper is organized as follows. Section 2 briefly explains how we formally encode planning techniques in $\mathcal{CTR}$. Section 3 also provides the results of our simple experiments to illustrate the practical applications of this method. The last section concludes our paper.

2   Planning Using $\mathcal{CTR}$

We assume denumerable pairwise disjoint sets of variables $\mathcal{V}$, constants $\mathcal{C}$, extensional predicate symbols $\mathcal{P}_{ext}$, and intensional predicate symbols $\mathcal{P}_{int}$. Table 1 shows the syntax of our language. Our work is based on our formal definitions of the basics of *STRIPS* planning that can be found in [4].[1] Here, we just briefly remind basics of planning to the reader. A *STRIPS* **action** is a triple of the form $\alpha = \langle p_\alpha(X_1, ..., X_n), Pre_\alpha, E_\alpha \rangle$, where $p_\alpha(X_1, ..., X_n)$ denotes the action and, Pre_α and E_α are sets of literals representing the precondition and effects of α. A **planning problem** $\langle \mathbb{R}, \mathbb{A}, G, \mathbf{S} \rangle$ consists of a set of rules $\mathbb{R}$, a set of

---

[1] To understand this report, the reader is expected to be familiar with $\mathcal{CTR}$. We provide a brief introduction to the relevant *subset* of $\mathcal{CTR}$ in [4] that is needed for the understanding of this paper. More explanation about $\mathcal{CTR}$ can be found in [8–12].

Table 1. The syntax of the language for representing *STRIPS* planning problems.

Term	$t := V \mid c$	where $V \in \mathcal{V}, c \in \mathcal{C}$
Atom	$P_\tau := p(t_1, \ldots, t_k)$	where $p \in \mathcal{P}_\tau, \tau \in \{ext, int\}$
Literal	$L := P_{int} \mid P_{ext} \mid \neg P_{ext}$	
Rule	$R := P_{int} \leftarrow L_1 \wedge \cdots \wedge L_m$	where $m \geq 0$

STRIPS actions $\mathbb{A}$, a set of literals G, called the **goal** of the planning problem, and an **initial state** $\mathbf{S}$. A sequence of actions $\sigma = \alpha_1, \ldots, \alpha_n$ is a **planning solution** (or simply a **plan**) for the planning problem if there is a sequence of states $\mathbf{S}_0, \mathbf{S}_1, \ldots, \mathbf{S}_n$ such that

- $\mathbf{S} = \mathbf{S}_0$ and $G \subseteq \mathbf{S}_n$ (i.e., G is satisfied in the final state);
- for each $0 < i \leq n$, α_i is executable in state $\mathbf{S}_{i-1}$ and the result of that execution (for some substitution) is the state $\mathbf{S}_i$.[2]

As mentioned in Sect. 1, we define a set of $\mathcal{TR}$ clauses that simulate the well-known *STRIPS* planning algorithm and extend this algorithm to handle intentional predicates and rules. In essence, these rules are a natural (and much more concise and general) verbalization of the classical *STRIPS* algorithm [15]. However, unlike the original *STRIPS*, these rules constitute a *complete* planner when evaluated with the $\mathcal{CTR}$ proof theory. Using this encoding of *STRIPS* planning algorithm, we have extended *STRIPS* algorithm via extensions of our encoding with different respects. Our encoding is defined as follows.

Definition 1 ($\mathcal{TR}$ planning rules). *Let $\Pi = \langle \mathbb{R}, \mathbb{A}, G, \mathbf{S} \rangle$ be a STRIPS planning problem. We define a set of $\mathcal{TR}$ rules, $\mathbb{P}(\Pi)$, which provides a sound and complete solution to the STRIPS planning problem. $\mathbb{P}(\Pi)$ has three disjoint parts, $\mathbb{P}_\mathbb{R}$, $\mathbb{P}_\mathbb{A}$, and $\mathbb{P}_G$, that are briefly described below. The details of this definition can be found in [4].*

- *The $\mathbb{P}_\mathbb{R}$ part is an extension to the classical STRIPS planning algorithm and is intended to capture intentional predicates and ramification of actions.*
- *The part $\mathbb{P}_\mathbb{A} = \mathbb{P}_{actions} \cup \mathbb{P}_{atoms} \cup \mathbb{P}_{achieves}$ is constructed out of the actions in $\mathbb{A}$. $\mathbb{P}_{actions}$ are the $\mathcal{CTR}$ rules that maps actions in $\mathbb{A}$ to $\mathcal{CTR}$ transactions. $\mathbb{P}_{atoms} = \mathbb{P}_{achieved} \cup \mathbb{P}_{enforced}$ has two disjoint parts. $\mathbb{P}_{achieved}$ are $\mathcal{CTR}$ rules that say if an extensional literal is true in a state then that literal has already been achieved as a goal. $\mathcal{CTR}$ rules in $\mathbb{P}_{enforced}$ say that one way to achieve a goal that occurs in the effects of an action is to execute that action. $\mathbb{P}_{achieves}$ is a set of $\mathcal{CTR}$ rules saying that to execute an action, one must first achieve the precondition of the action and then perform the state changes prescribed by the action.*
- *The $\mathbb{P}_G$ part is showing how the goal will be achieved.* □

Given a set $\mathbb{R}$ of rules, a set $\mathbb{A}$ of *STRIPS* actions, an initial state $\mathbf{S}$, and a goal G, Definition 1 gives a set of $\mathcal{TR}$ rules that specify a planning strategy for that

---

[2] In this case we will also say that $\mathbf{S}_0, \mathbf{S}_1, \ldots, \mathbf{S}_n$ is an execution of σ.

problem. To find a solution for that planning problem, one simply needs to place the request

$$? - achieve_G . \tag{1}$$

at a desired initial state and use the $\mathcal{TR}$'s inference system to find a proof. As mentioned before, a solution plan for a *STRIPS* planning problem is a sequence of actions leading to a state that satisfies the planning goal. Such a sequence can be extracted by picking out the atoms of the form p_α from a successful derivation branch generated by the $\mathcal{TR}$ inference system. We provide a technique to extract that sequence of actions, called **pivoting sequence of actions**. Soundness of a planning strategy means that, for any *STRIPS* planning problem, if sequence of actions is extracted as a solution, then that sequence of actions is a *solution* to the planning problem. We also prove that the proposed technique is sound. That is, the pivoting sequence of actions is a solution to the planning problem. **Completeness** of a planning strategy means that, for any *STRIPS* planning problem, if there is a solution, the planner will find *at least one* plan. A stronger statement about completeness is called **comprehensive completeness**: if there is a *non-redundant* solution for a *STRIPS* planning problem, the planner will find *exactly that* plan. We also prove that the proposed $\mathcal{TR}$-based planner is comprehensively complete.

We also introduce *fSTRIPS* — a modification of the previously introduced *STRIPS* transform, which represents to a new planning strategy, which we call *fast STRIPS*. We show that although the new strategy explores a smaller search space, it is still sound and complete. Our experiments show that *fSTRIPS* can be orders of magnitude faster than *STRIPS*. The details of *fSTRIPS* can be found in [4].

The third part of our work shows how the simplicity embedded in $\mathcal{TR}$ encoding of planning strategies let one apply heuristics in planning mechanism. As an indirect result of our research, in this stage, we have shown that sophisticated planning heuristics, such as regression analysis, can be naturally represented in $\mathcal{TR}$ and that such representation can be used to express complex planning strategies such as *RSTRIPS* [3]. The simplicity of $\mathcal{TR}$ representation of *RSTRIPS* let us prove its completeness for the first time. We have also extended our above mentioned non-linear $\mathcal{TR}$-based *STRIPS* planner with regression analysis. The $\mathcal{TR}$ representation of those planning strategies rely on the concept of regression of a *STRIPS* action that is explained in [2]. The idea behind planning with regression is that the already achieved goals should be protected so that subsequent actions of the planner would not *unachieve* those goals. The details of these techniques can be found in [3].

In the last stage of our dissertation research, we have extended out above mentioned $\mathcal{TR}$-based *STRIPS* planner with respect to negative derived atoms. Our original domain specification language syntax reflected in Table 1 did not include negative derived atoms as none of the existing planners were able to solve planning problems with negative derived atoms. We also extend both the planning domain specification language and our above mentioned planner to solve planning problems with negative derived atoms. The idea behind this planner is that, in order to make a derived atom false, the planner disables every derivation can lead to make that atom true. The details of this extension along with the proofs of its soundness and completeness will be published in our under-preparation journal paper.

3   Experiments

In this section we first briefly report on our experiment. The first set of experiment compares *STRIPS* planning algorithm and forward state space search [2] that show that *STRIPS* can be faster than forward state space search up to three orders of magnitude. Our results in [4] also compare *STRIPS* and *fSTRIPS* and show that *fSTRIPS* can be two orders of magnitude faster than *STRIPS*. We compared *RSTRIPS* and *fSTRIPS* to demonstrate our proposed regression analysis mechanism can improve the performance of planning up to three orders of magnitude [3]. Due to space limitations we skip explanations of the test environment, test cases, and the complete result tables and the interested readers are referred to our previous papers [2–4]. We have also explained our PDDL2TR translator that maps planning problems to $\mathcal{TR}$ rules in [2].

4   Conclusion

This dissertation will demonstrate that the use of $\mathcal{CTR}$ and $\mathcal{TR}$ accrues significant benefits in the area of planning. As an illustration, we have shown that sophisticated planning strategies, such as *STRIPS*, not only can be naturally represented in $\mathcal{TR}$, but also such representation can be used to design new planning strategies such as non-linear *STRIPS*, *fSTRIPS*, non-linear *STRIPS* with regression analysis, *RSTRIPS*, and extensions with negative derived atoms. There are several promising directions to continue this work. One is to investigate other planning strategies and, hopefully, accrue similar benefits. We are also working to represent *GraphPlan* planning algorithm in $\mathcal{TR}$ to get same benefits.

Acknowledgments. This work was supported, in part, by the NSF grant 0964196. I also thank Prof. Michael Kifer for his great advises on my PhD research.

References

1. Barták, R., Toropila, D.: Solving sequential planning problems via constraint satisfaction. Fundam. Inform. **99**(2), 125–145 (2010). http://dx.doi.org/10.3233/FI-2010-242
2. Basseda, R., Kifer, M.: State space planning using transaction logic. In: Pontelli, E., Son, T.C. (eds.) PADL 2015. LNCS, vol. 9131, pp. 17–33. Springer, Heidelberg (2015)
3. Basseda, R., Kifer, M.: Planning with regression analysis in transaction logic. In: ten Cate, B., Mileo, A. (eds.) RR 2015. LNCS, vol. 9209, pp. 45–60. Springer, Heidelberg (2015)

4. Basseda, R., Kifer, M., Bonner, A.J.: Planning with transaction logic. In: Kontchakov, R., Mugnier, M.-L. (eds.) RR 2014. LNCS, vol. 8741, pp. 29–44. Springer, Heidelberg (2014)

5. Bibel, W.: A deductive solution for plan generation. New Gener. Comput. 4(2), 115–132 (1986)

6. Bibel, W.: A deductive solution for plan generation. In: Schmidt, J.W., Thanos, C. (eds.) Foundations of Knowledge Base Management. Topics in Information Systems, pp. 453–473. Springer, Heidelberg (1989)

7. Bibel, W., del Cerro, L.F., Fronhfer, B., Herzig, A.: Plan generation by linear proofs: on semantics. In: Metzing, D. (ed.) GWAI-89 13th German Workshop on Artificial Intelligence. Informatik-Fachberichte, vol. 216, pp. 49–62. Springer, Berlin Heidelberg (1989)

8. Bonner, A., Kifer, M.: Transaction logic programming. In: International Conference on Logic Programming, pp. 257–282. MIT Press, Budapest, Hungary, June 1993

9. Bonner, A., Kifer, M.: Transaction logic programming (or a logic of declarative and procedural knowledge). Technical report CSRI-323, University of Toronto, November 1995. http://www.cs.toronto.edu/~bonner/transaction-logic.html

10. Bonner, A., Kifer, M.: Concurrency and communication in transaction logic. In: Joint International Conference and Symposium on Logic Programming, pp. 142–156. MIT Press, Bonn, Germany, September 1996

11. Bonner, A., Kifer, M.: A logic for programming database transactions. In: Chomicki, J., Saake, G. (eds.) Logics for Databases and Information Systems, chap. 5, pp. 117–166. Kluwer Academic Publishers, March 1998

12. Bonner, A.J., Kifer, M.: An overview of transaction logic. Theor. Comput. Sci. 133(2), 205–265 (1994)

13. Cresswell, S., Smaill, A., Richardson, J.: Deductive synthesis of recursive plans in linear logic. In: Biundo, S., Fox, M. (eds.) Recent Advances in AI Planning. LNCS, vol. 1809, pp. 252–264. Springer, Berlin Heidelberg (2000)

14. Erol, K., Hendler, J.A., Nau, D.S.: UMCP: a sound and complete procedure for hierarchical task-network planning. In: Hammond, K.J. (ed.) Proceedings of the Second International Conference on Artificial Intelligence Planning Systems, University of Chicago, Chicago, Illinois, USA, 13–15 June 1994, pp. 249–254. AAAI (1994). http://www.aaai.org/Library/AIPS/1994/aips94-042.php

15. Fikes, R.E., Nilsson, N.J.: STRIPS: a new approach to the application of theorem proving to problem solving. Artif. Intell. 2(3–4), 189–208 (1971)

16. Gebser, M., Kaufmann, R., Schaub, T.: Gearing up for effective ASP planning. In: Erdem, E., Lee, J., Lierler, Y., Pearce, D. (eds.) Correct Reasoning. LNCS, vol. 7265, pp. 296–310. Springer, Heidelberg (2012). http://dl.acm.org/citation.cfm?id=2363344.2363364

17. Gelfond, M., Lifschitz, V.: Action languages. Electron. Trans. Artif. Intell. 2, 193–210 (1998). http://www.ep.liu.se/ej/etai/1998/007/

18. Green, C.: Application of theorem proving to problem solving. In: Proceedings of the 1st International Joint Conference on Artificial Intelligence, IJCAI 1969, pp. 219–239. Morgan Kaufmann Publishers Inc., San Francisco, CA, USA (1969). http://dl.acm.org/citation.cfm?id=1624562.1624585

19. Hölldobler, S., Schneeberger, J.: A new deductive approach to planning. New Gener. Comput. 8(3), 225–244 (1990)

20. Kahramanoğulları, O.: On linear logic planning and concurrency. In: Martín-Vide, C., Otto, F., Fernau, H. (eds.) LATA 2008. LNCS, vol. 5196, pp. 250–262. Springer, Heidelberg (2008)

21. Kautz, H., Selman, B.: Planning as satisfiability. In: Proceedings of the 10th European Conference on Artificial Intelligence, ECAI 1992, pp. 359–363. John Wiley & Sons Inc, New York, USA (1992). http://dl.acm.org/citation.cfm?id=145448.146725

22. Lifschitz, V.: Answer set programming and plan generation. Artif. Intell. 138(12), 39–54 (2002). http://www.sciencedirect.com/science/article/pii/S0004370202001868, knowledge Representation and Logic Programming

23. Son, T.C., Baral, C., Tran, N., Mcilraith, S.: Domain-dependent knowledge in answer set planning. ACM Trans. Comput. Logic 7(4), 613–657 (2006). http://doi.acm.org/10.1145/1183278.1183279

24. Stefik, M.J.: Planning with Constraints. Ph.D. thesis, Stanford, CA, USA (1980), aAI8016868

Doctoral Consortium Extended Abstract: Multi-context Systems with Preferences

Tiep Le[✉]

Department of Computer Science, New Mexico State University,
Las Cruces, NM, USA
tile@cs.nmsu.edu

Abstract. Multi-Context Systems (MCSs) have been introduced in [1] as a framework for integration of knowledge from different sources. This research formalizes *MCSs with preferences* (MCSPs) that allows to integrate preferences into an MCS at the context level and at the MCS level, and proposes novel distributed algorithms to compute their semantics.

1   Introduction and Problem Description

Multi-Context Systems (MCSs) has been introduced in [1] as a framework for integration of knowledge from different sources. Intuitively, an MCS consists of several theories, referred to as contexts. The contexts may be heterogeneous; each context could rely on a different logical language and a different inference system; e.g., propositional logic, first order logic, or logic programming. The information flow among contexts is modeled via *bridge rules*. Bridge rules allow for the modification of the knowledge of a context depending on the knowledge of other contexts. The semantics of MCSs in [1] is defined in terms of its equilibria.

By definition, an MCS is a general framework not tied to any specific language, inference system, or implementation. Whenever an MCS is used to model a concrete problem, it is necessary to specify which logics should be used in encoding its contexts. Moreover, in several applications, some contexts need to express their *preferences*. For instance, consider the following example.

Example 1 *[Dining Out Plan]. Two friends A and B will dine together. A can eat chicken (c) or lamb (l); B can eat fish (f) or steak (s). They share a bottle of wine, either red (r) or white (w) wine. A knows that white (resp. red) wine could go with chicken (resp. lamb); B knows that fish (resp. steak) should go with white (resp. red) wine. A prefers chicken to lamb, and B prefers steak to fish. The two cannot afford two bottles of wine, and neither of them can eat two dishes.*

The MCS in Fig. 1 represents the Example 1 where underlying logics in the contexts of A and B are *Answer Set Optimization* (ASO) [2] (a preference logic built over answer set programming [3]). In Fig. 1, the first five rules encode the choice of each person, the sixth rule captures the preferences, and the last two lines in each context encode its bridge rules.

Extended Abstract as part of the program of the 1st Joint ADT/LPNMR 2015 Doctoral Consortium co-chaired by Esra Erdem and Nicholas Mattei.

© Springer International Publishing Switzerland 2015
F. Calimeri et al. (Eds.): LPNMR 2015, LNAI 9345, pp. 552–558, 2015.
DOI: 10.1007/978-3-319-23264-5_46

Choices	$w \leftarrow c$ $r \leftarrow l$ $\leftarrow r, w$ $c \leftarrow not\ l$ $l \leftarrow not\ c$	$w \leftarrow f$ $r \leftarrow s$ $\leftarrow w, r$ $s \leftarrow not\ f$ $f \leftarrow not\ s$
Preferences	$c > l \leftarrow$	$s > f \leftarrow$
Bridges	$w \leftarrow (b : w)$ $r \leftarrow (b : r)$	$w \leftarrow (a : w)$ $r \leftarrow (a : r)$
	Context of A	**Context of B**

Fig. 1. Two contexts in ASO logic

Although expressing preferences is important and pervasive, integrating preferences into MCSs, in the literature, has not been investigated thoroughly. *One way* is to use *preference logics* that are able to express preferences (e.g., the MCS in Fig. 1) at the contexts of an MCS. However, since the semantics of a preference logic (e.g., ASO) usually tries to obtain its *most preferred solution* (*most preferred answer set* or *belief set* in ASO), such an MCS might be *inconsistent*; i.e., for such an MCS, it does not have any equilibrium. For example, the MCS in Fig. 1 has no equilibrium under the semantics of ASO. The context of A has a unique most preferred answer set (i.e., $\{c, w\}$), while that of B has a unique most preferred answer set (i.e., $\{s, r\}$). Neither of them can be used to construct any equilibrium. As a result, A and B cannot dine together.

We also observe that $\{l, r\}$ is not the most preferred answer set but the second most preferred one of A. Furthermore, if $\{l, r\}$ is considered as an alternative acceptable answer set of A, then the MCS will have $(\{l, r\}, \{s, r\})$ as an equilibrium. This means A and B can dine together if A (resp. B) has its second most (resp. most) preferred choice meal, respectively. Likewise, the MCS would have $(\{c, w\}, \{f, w\})$ as another equilibrium where A (resp. B) has its most (resp. second most) preferred meal, respectively. This discussion raises a question *"How would one express preferences locally in the contexts of an MCS, and redefine overall semantics in such a way that every context has its most preferred possible solution?"*

The above discussion allows to integrate preferences into MCSs in a *competitive* manner (i.e., each context expresses its own preferences, and expects to obtain its most preferred possible solution). On the other hand, an integration in a *cooperative* manner, where a preference order is defined over equilibria of an MCS aggregately, (e.g., to model voting problems) also needs to be considered.

Example 2. *From Example 1, we replace the preferences of A and B ($c > l$, and $s > f$) with the facts that "eating chickens costs A \$5", and "eating steak costs B \$7". Assume that the other types of food or wine are free, the goal now is to seek a solution where A and B dine together with the cheapest total dining cost.*

There are two possible cases for A and B to dine together, which correspond to two equilibria of the MCS in Fig. 1 without the sixth rules in both of the contexts; i.e., $(\{c, w\}, \{f, w\})$ and $(\{l, r\}, \{s, r\})$. Among these equilibria, the former one is more desirable since it costs, \$5, less than the later one, \$7. To solve this, one could find all of the equilibria of an MCS and then compute *centrally* the most preferred one. However, this approach is not desirable since it requires a central computation unit for the latter step, while original MCSs have been proposed for integration of distributed knowledge where it is not suitable for sharing all information among different contexts. This raises another question "*How would one define a preference order among the equilibria of an MCS, and compute distributedly the most preferred equilibrium?*"

My thesis aims to address the above two questions. Its main contributions are to

- define a general framework, *MCS with local preferences* (MCS-LP), that integrates preference logics at the context level, and
- define a general framework, *MCS with global preferences* (MCS-GP), that expresses preference order among equilibria of an MCS, and propose novel algorithms that solves MCS-GP in a distributed manner.

In addition, I also expect to define a generic framework that merges both MCS-LP and MCS-GP, and identify some applications to motivate MCSPs.

2   Background

In this section, I present some background on MCSs, and a typical preference logic, ASO, which is used in later sections.

2.1   Multi-context Systems[1]

Heterogeneous non-monotonic multi-context systems (MCS) were defined via a generic notion of a logic. A *logic* L is a tuple (KB_L, BS_L, ACC_L) where KB_L is the set of well-formed knowledge bases of L, each being a set of formulae. BS_L is the set of possible belief sets; each being a set of syntactic elements representing the beliefs L may adopt. $ACC_L : KB_L \to 2^{BS_L}$ describes the "*semantics*" of L by assigning to each element of KB_L a set of acceptable sets of beliefs.

An MCS $M = (C_1, \ldots, C_n)$ consists of contexts $C_i = (L_i, kb_i, br_i)$, $(1 \leq i \leq n)$, where $L_i = (KB_i, BS_i, ACC_i)$ is a logic, $kb_i \in KB_i$, and br_i is a set of L_i-bridge rules of the form: $s \leftarrow (c_1 : p_1), \ldots, (c_j : p_j), not\,(c_{j+1} : p_{j+1}), \ldots, not\,(c_m : p_m)$ where, for each $1 \leq k \leq m$, we have that: $1 \leq c_k \leq n$, p_k is an element of some belief set of L_{c_k}, and $kb_i \cup \{s\} \in KB_i$. Intuitively, a bridge rule r allows us to add s to a context, depending on the beliefs in the other contexts. Given a bridge rule r, we will denote by $head(r)$ the part s of r.

The semantics of an MCS is described by the notion of belief states. A *belief state* of an MCS $M = (C_1, \ldots, C_n)$ is a sequence $S = (S_1, \ldots, S_n)$ where each S_i is an element of BS_i.

---

[1] The following definitions are from [1].

Given a belief state $S = (S_1, \ldots, S_n)$ and a bridge rule r, r is *applicable* in S if $p_v \in S_{c_v}$ where $1 \le v \le j$ and $p_k \notin S_{c_k}$ where $j + 1 \le k \le m$. We denote by $app(B, S)$ the set of the bridge rules $r \in B$ that are applicable in S.

The semantics of an MCS M is defined via its equilibrium. A belief state $S = (S_1, \ldots, S_n)$ of M is an *equilibrium* if, for all $1 \le i \le n$, we have that $S_i \in ACC_i(kb_i \cup \{head(r) \mid r \in app(br_i, S)\})$.

2.2   Answer Set Optimization (ASO)[2]

An ASO program over a signature Σ [2] is a pair (P_{gen}, P_{pref}), where: P_{gen} is a logic program over Σ, called the *generating program*, and P_{pref} is a *preference program*. P_{gen} is used for generating answer sets which are required to be given in terms of sets of literals (or belief sets) associated with programs. Given a set of atoms A, a P_{pref} over A is a finite set of preference rules of the form $\Gamma_1 > \ldots > \Gamma_k \leftarrow a_1, \ldots, a_n, \text{ not } b_1, \ldots, \text{ not } b_m$ where a_i and b_j are literals over A, and Γ_l are boolean combinations over A. A boolean combination is a formula built from atoms in A by means of disjunction, conjunction, strong negation ($\neg$, which appears only in front of atoms), and default negation (*not*, which appears only in front of literals). Given a preference rule r, we denote $body(r) = a_1, \ldots, a_n, \text{ not } b_1, \ldots, \text{ not } b_m$. An answer set S satisfies the body of r, denoted by $S \models body(r)$ iff S contains $a_1, \ldots, a_n$ and does not contain any $b_1, \ldots, b_m$. The preference rule states that if an answer set S satisfies the body of r, then Γ_1 is preferred to Γ_2, Γ_2 to Γ_3, etc. The *satisfaction degree* of preference rule r in an answer set S (denoted $v_S(r)$) is:

- $v_S(r) = 1$, if **(i)** $S \not\models body(r)$, or **(ii)** $S \models body(r)$ and $S \not\models \Gamma_l$ for each $1 \le l \le k$;
- $v_S(r) = \min\{i : S \models \Gamma_i\}$, otherwise.

Given an ASO program $P = (P_{gen}, P_{pref})$, let S_1 and S_2 be answer sets of P_{gen}, and $P_{pref} = \{r_1, \ldots, r_n\}$. We write $S_1 < S_2$ if **(i)** $v_{S_1}(r_i) \ge v_{S_2}(r_i)$, for $1 \le i \le n$, and **(ii)** for some $1 \le j \le n$, $v_{S_1}(r_j) > v_{S_2}(r_j)$. We refer to this ordering as *preference order* between S_1 and S_2.

Example 3. *Consider the ASO program from the context B in Fig. 1. $P^B = (P_{gen}^B, P_{pref}^B)$, where P_{gen}^B is the program consisting of the first five rules under answer set semantics and $P_{pref}^B = \{s > f \leftarrow\}$. It is possible to show that P^B has two answer sets $S_1 = \{s, r\}$, $S_2 = \{f, w\}$, and $S_2 < S_1$.*

3   Research Accomplishments and Future

3.1   Research Accomplishments

In this subsection, I will briefly define the notion of an MCS-LP.

---

[2] The following definitions are from [2].

Definition 1 (Ranked Logic). *A ranked logic L is a tuple $(KB_L, BS_L, ACC_L, <_L)$ where*

- *(KB_L, BS_L, ACC_L) is an arbitrary logic; and*
- *$<_L \subseteq (KB_L, BS_L) \times (KB_L, BS_L)$ is a partial order over pairs of knowledge bases and belief sets satisfying the condition that if $((kb_1, b_1), (kb_2, b_2)) \in <_L$ then $b_i \in ACC_L(kb_i)$ for $i = 1, 2$.*

We often write $(kb_1, b_1) <_L (kb_2, b_2)$ instead of the $((kb_1, b_1), (kb_2, b_2)) \in <_L$. Given $b_i \in ACC_L(kb_i)$ for $i = 1, 2$, (kb_1, b_1) and (kb_2, b_2) are *incomparable*, denoted with $(kb_1, b_1) \sim (kb_2, b_2)$, if $(kb_1, b_1) \not<_L (kb_2, b_2)$ and $(kb_2, b_2) \not<_L (kb_1, b_1)$.

As an example, I will construct a ranked logic, denoted by L_{ASO}, from ASO. Given a signature Σ, $L_{ASO} = (\mathbf{KB_{ASO}}, \mathbf{BS_{ASO}}, \mathbf{ACC_{ASO}}, <_{ASO})$ is a ranked logic over Σ where

- $\mathbf{KB_{ASO}}$ is the set of ASO programs over Σ;
- $\mathbf{BS_{ASO}}$ is the set of answer sets of ASO programs over Σ;
- $\mathbf{ACC_{ASO}}$ maps each ASO program to its possible answer sets;
- $<_{ASO}$ is defined as follows: $(kb_1, S_1) <_{ASO} (kb_2, S_2)$ for $kb_i = (P_{gen}^i, P_{pref})$ and $S_i \in ACC_{ASO}(kb_i)$ where $i = 1, 2$ such that $S_2 > S_1$ with respect to the ASO preference order defined over the set of rules P_{pref}.

Example 4. *Consider the ASO program $P^B = (P_{gen}^B, P_{pref}^B)$ from the Example 3. P^B has two answer sets $S_1 = \{s, r\}$, $S_2 = \{f, w\}$, and $S_2 < S_1$. Thus, $(P^B, S_2) <_{ASO} (P^B, S_1)$. Furthermore, let $P^{B'} = (P_{gen}^B \cup \{w\}, P_{pref}^B)$ be another ASO program, and $P^{B'}$ has a unique answer set $S_3 = \{f, w\}$. It is possible to see that $(P^{B'}, S_3) <_{ASO} (P^B, S_1)$.*

Definition 2 (MCSs with Local Preferences (MCS-LP)). *A MCS-LP $M = (C_1, \ldots, C_n)$ consists of a collection of contexts $C_i = (L_i, kb_i, br_i)$ where $L_i = (KB_i, BS_i, ACC_i, <_i)$ is a ranked logic, kb_i is a knowledge base $kb_i \in KB_i$, and br_i is a set of L_i-bridge rules.*

Definition 3 (Preferred). *Let $M = (C_1, \ldots, C_n)$ be an MCS-LP, and let $S = (S_1, \ldots, S_n)$ and $E = (E_1, \ldots, E_n)$ be equilibria of M. We say that*

- *S is preferred to E, denoted with $E \prec S$, iff*
  - *$(kb_i^E, E_i) <_i (kb_i^S, S_i)$ or $(kb_i^S, S_i) \sim (kb_i^E, E_i)$, $\forall i\ 1 \le i \le n$; and*
  - *$(kb_j^E, E_j) <_j (kb_j^S, S_j)$ for some j, $1 \le j \le n$.*
- *S is incomparable to E, denoted with $S \sim E$, if $S \not\prec E$ and $E \not\prec S$.*

Definition 4 (Preferred Equilibrium[3]). *An equilibrium E of an MCS-LP M is a preferred equilibrium iff there is no equilibrium E' of M such that $E \prec E'$.* E.g., the MCS-LP in Fig. 1 has two preferred equilibria $(\{c, w\}, \{f, w\})$ and $(\{l, r\}, \{s, r\})$.

---

[3] The term "a preferred equilibrium" is used, rather than "the most preferred equilibrium", to show that there might not be a unique most preferred equilibrium.

3.2   Open Issues and Expected Achievements

The MCSs with global framework (MCS-GP) is described via the notion of a language $\mathcal{L}$ associated with an MCS M that defines a partial preference-order among the equilibria of M. $\mathcal{L}$ is made up of MCS atoms, weighted (non-weighted) formulas, preference elements, and preference statements.

MCS Atoms: Given an MCS $M = (C_1, \ldots, C_n)$, a *MCS atom* over M is of the form $(i : p_i)$ where p_i is an element of some belief sets in BS_i of L_{c_i}. We denote by $\mathcal{A}_M$ the set of all MCS atoms over M.

Weighted (Non-weighted) Formulas: Given an MCS M, a *weighted formula* is of the form $w_1, \ldots, w_l : \phi$, where each w_i is a term and ϕ is a Boolean expression over $\mathcal{A}_M$ with logical connectives $\top, \neg, \vee,$ and $\wedge$. Whenever $l = 0$, it derives a *non-weighted formula*, and, for short, it is denoted by ϕ.

Preference Elements: A *preference element* over an MCS M is of the form $\Phi_1 > \ldots > \Phi_m \| \phi$, where $m \geq 1$, ϕ is a non-weighted formula describing the context, and each Φ_r is a set of weighted formulas for $1 \leq r \leq m$. For short, if $\phi = \top$ which is tautological, we may drop "$\| \phi$", if Φ_r is a singleton set, we drop surrounding braces of such sets, and if $m = 1$, we drop "$>$". For example, a *preference element* $10 : (1 : a_1)$ stands for $\{10 : (1 : a_1)\} \| \top$.

Intuitively, r gives the rank of the respective set of weighted formulas, and each preference element provides a (possible) structure to a set of sets of weighted formulas by *(i.)* giving the context (in ϕ) by means of conditionalization, and *(ii.)* specifying the pre-orders among its members (sets of weighted formulas) using the symbolic way (in addition to the way of using weights). For example, rather than using the respective weights, we use $\{(1 : a_1), (2 : b_2)\} > \{(1 : c_1), (2 : d_2)\}$ to define the pre-order between $\{(1 : a_1), (2 : b_2)\}$ and $\{(1 : c_1), (2 : d_2)\}$.

Preference Statements: A *preference statement* over an MCS M is of the form $\#pref(t)\{e_1, \ldots, e_n\}$, where t are ground terms specifying the preference type, and each e_j is a preference element over M. Each preference type has its own semantics to declare the preference relation on equilibria of M. E.g., $\#pref(less(weight))\{5 : (A : c), 7 : (B : s)\}$ presents the preferences in Example 2

The semantics of an MCS-GP$=(M, \mathcal{L})$ is defined via preferred equilibria which are sorted based on the semantics of preference type in $\mathcal{L}$. For now, I have been implementing different preference types such as *more(weight)*, *less(weight)*, *more(cardinality)*, *less(cardinality)*, and *subset*. In future, I expect to have some more complicated preference types such as *pareto*.

Some preference types allow to compute preferred equilibria of an MCS-GP in a distributed way, and some others don't. For those that allow, I have been designing some algorithms to compute them distributedly. The idea lies in representing the structure of the MCS-GP as a *tree*. From that tree, nodes (contexts) exchange necessary information to compute preferred equilibria.

References

1. Brewka, G., Eiter, T.: Equilibria in heterogeneous nonmonotonic multi-context systems. In: Proceedings of the AAAI, pp. 385–390 (2007)
2. Brewka, G., Niemelä, I., Truszczynski, M.: Answer set optimization. In: Proceedings of the IJCAI 2003, pp. 867–872 (2003)
3. Gelfond, M., Lifschitz, V.: Logic programs with classical negation. In: Proceedings of the ICLP, pp. 579–597 (1990)

Doctoral Consortium Extended Abstract: Default Mappings in Ontology-Based Data Access

Daniel P. Lupp[✉]

Department of Informatics, University of Oslo, Oslo, Norway
danielup@ifi.uio.no

Abstract. *Ontology-based data access (OBDA)* is a technology where heterogeneous, distributed data is accessed through queries over an ontology which provides domain-specific knowledge. The ontology is connected to the stored data using mappings, which are commonly interpreted as first-order implications from the source language to the ontology language. I propose to generalize OBDA mappings using defaults from default logic. The resulting mapping language is much more expressive than standard mapping languages, while retaining all the desired properties related to query answering in OBDA systems. This would allow the statement of epistemic queries and greatly reduce maintenance cost and potential for error in lack of knowledge and exception handling.

1 Introduction

Ontology-based data access has in recent years established itself as a popular research topic in data transformation and retrieval. It enables the access of data through the use of an ontology and without any prior knowledge on the manner in which the data is stored. However, the languages used in order to have tractable query answering are very limited with regards to expressivity. It is not immediately possible in an OBDA setting to differentiate between knowledge derived from ontology reasoning and explicit information in the database. Current research regarding extending OBDA with nonmonotonic capabilities has focused on the ontology side, e.g., through modal description logics or inclusion of closed predicates [4, 7]. However the modal semantics are quite unintuitive and the modal ontology axioms do not behave well with nonmodal axioms, while [7] demonstrates that adding closed predicates usually results in intractability.

Rather than focusing on the ontology, I propose extending the mapping language to allow nonmonotonic behavior through the use of defaults. This has the potential for greatly reducing redundant querying as well as maintenance costs and potential for error.

Extended Abstract as part of the program of the 1st Joint ADT/LPNMR 2015 Doctoral Consortium co-chaired by Esra Erdem and Nicholas Mattei.

F. Calimeri et al. (Eds.): LPNMR 2015, LNAI 9345, pp. 559–564, 2015.
DOI: 10.1007/978-3-319-23264-5_47

2   Ontology-Based Data Access

Ontology-based data access (OBDA) [8] is a method for data integration, utilizing a semantic layer consisting of an ontology and a set of mappings on top of a database. An *ontology* is a machine-readable model designed to faithfully represent knowledge of a domain independently of the structure of the database; it is comprised of concepts and relationships between these concepts. These ontologies are often formulated using *description logics*, a class of decidable fragments of first-order logic, due to their desirable, application-dependent properties [2].

The process of translating a query over the ontology to a query over the database(s) can be seen in Fig. 1. Firstly, the ontology query is rewritten using the ontology in the rewriting step. Here, the query is extended by ontology reasoning. Secondly, in the unfolding step, the mappings are used to translate the extended query to a traditional query language, such as SQL.

To allow the translation of ontology queries to database queries, the language used to express the ontology must be first-order logic rewritable (FOL-rewritable). An ontology language is said to be *FOL-rewritable* if the resulting query of the rewrite step is always expressible as a conjunctive query. The necessity of this property stems from the fact that conjunctive queries are equivalent in expressiveness to the core of SQL. However, not all description logics have this property. A common class of ontology languages used in OBDA is the DL-lite family, a family of description logics tailored to be FOL-rewritable [2].

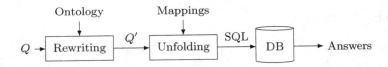

Fig. 1. OBDA architecture.

Following [6], an OBDA specification is a tuple $(\mathcal{K}, \mathcal{D}_\mathcal{S}, \mathcal{M})$ consisting of a knowledge base $\mathcal{K} = (\mathcal{O}, \mathcal{S})$ (the first-order ontology and source schema theories, respectively), a database instance $\mathcal{D}_\mathcal{S}$ over the schema $\mathcal{S}$, and a set $\mathcal{M}$ consisting of *mapping assertions* of the form $m : \varphi \rightsquigarrow \psi$, where φ and ψ are queries over the data source and ontology, respectively. Semantically, a mapping assertion is interpreted as the first-order formula

$$\pi(m) : \forall \mathbf{x}.(\varphi(\mathbf{x}) \rightarrow \exists \mathbf{z}.\psi(\mathbf{y}, \mathbf{z})) \tag{1}$$

where $\mathbf{y} \subseteq \mathbf{x}$. Then a model of an OBDA specification $(\mathcal{K}, \mathcal{D}_\mathcal{S}, \mathcal{M})$ is a first-order model of the theory $\mathcal{O} \cup \mathcal{D}_\mathcal{S} \cup \pi(\mathcal{M})$, where $\pi(\mathcal{M}) = \{\pi(m) \mid m \in \mathcal{M}\}$.

Example 1. Consider a database consisting of precisely one table **Persons**, with the columns Name, hasDriversLicense, Occupation, and CarRegistration. Furthermore, consider the following ontology:

$$Student \sqsubseteq Person$$
$$Consultant \sqsubseteq Person$$
$$TruckDriver \sqsubseteq \exists hasDriversLicense$$
$$\exists hasCar \sqsubseteq \exists hasDriversLicense$$
$$\exists hasDriversLicense \sqsubseteq Person$$

Suppose that we simply wish to query for all instances of $Person$ in the database. In the rewriting process, the query $Person(x)$ would be rewritten to

$$\exists y, z : Person(x) \sqcup Student(x) \sqcup Consultant(x) \sqcup TruckDriver(x)$$
$$\sqcup \exists hasDriversLicense(x, z) \sqcup hasCar(x, y)$$

while in the unfolding step, each of the above disjuncts would be expanded to a database query using the mapping assertions. For example, if there existed two mapping assertions $\varphi_1(x) \rightsquigarrow Student(x)$ and $\varphi_2(x) \rightsquigarrow Student(x)$, then the disjunct $Student(x)$ would be rewritten as $\varphi_1(x) \vee \varphi_2(x)$.

Example 1 demonstrates some of the current shortcomings of OBDA: due to its inherent, first-order nature, it is impossible to distinguish between reasoned and explicit knowledge in the database. In the above example, in the presence of a mapping assertion $m :$ SELECT x FROM Persons $\rightsquigarrow Person(x)$[1] the query $Person(x)$ would have sufficed without any ontology rewriting, since all desired information was contained in one table. Thus we have an entirely undesirable and redundant exponential blow-up in query size.

Another issue with the current approach is how exceptions and a lack of information are dealt with. Currently, one must keep track of exceptions manually by explicitly listing all exceptions to a rule. Furthermore, due to the closed world assumption (CWA) in the database, a lack of knowledge is interpreted as knowledge itself, e.g., if something is not contained in the Persons table, it is not a $Person$.

To address these issues, I propose to generalize the mappings in OBDA by considering them as defaults from default logic [9]. Since defaults support non-monotonic notions such as negation-as-failure, they could allow the distinction between explicit and implicit knowledge. This would help to avoid the redundant query blow-up from Example 1. Furthermore, in the presence of lack of knowledge, such as nulls in the database, defaults would allow a "loosening" of the closed world assumption: allowing for assumptions based on lack of information as opposed to hard negation. Finally, defaults would simplify the management of mappings: as opposed to classical first-order mappings, default mappings would not have to be updated every time a new type of exception is added to the database. This would greatly reduce maintenance cost and potential for error.

3 Three Approaches to Default Mappings

In the following, we consider mapping assertions of the form $m : \varphi(x) \rightsquigarrow \psi(x)$.

---

[1] The query SELECT x FROM Persons corresponds to $\exists y, z, w.$Persons(x, y, z, w) in first-order logic.

3.1   Rule-Based Approach

As before, let $\mathcal{K} = (\mathcal{O}, \mathcal{S})$ be a knowledge base consisting of an ontology and a source schema, where $\mathcal{O}$ contains $\top$. Furthermore, let $\mathcal{D}_\mathcal{S}$ be a database over $\mathcal{S}$.

Definition 1. A *default mapping assertion (DMA)* is a default

$$\frac{\varphi(x) : \rho(x)}{\psi(x)}$$

where φ is a formula over $\mathrm{sig}(\mathcal{S})$ and ρ, ψ are formulas over $\mathrm{sig}(\mathcal{O})$. If $\rho = \top$, we refer to it as a *crisp mapping assertion (CMA)*. A *default mapping* $\mathcal{M}$ is a finite set of DMA's over $\mathcal{K}$. It is *crisp* if it only contains CMA's.

Since our application is accessing data in a database, retrieving *correct* data is of utmost importance; we want to avoid false query results. To this end, we only consider cautious extensions of our database.

Definition 2. For a default OBDA specification $(\mathcal{K}, \mathcal{D}_\mathcal{S}, \mathcal{M})$ the *certain answer theory* $\mathcal{T}_{\langle \mathcal{K}, \mathcal{D}_\mathcal{S}, \mathcal{M} \rangle}$ is the intersection of all extensions of $(\mathcal{K}, \mathcal{D}_\mathcal{S})$ over $\mathcal{M}$.

Definition 3. A *model of* $(\mathcal{K}, \mathcal{D}_\mathcal{S}, \mathcal{M})$ is a first-order model of $\mathcal{T}_{\langle \mathcal{K}, \mathcal{D}_\mathcal{S}, \mathcal{M} \rangle}$.

By taking the intersection of all extensions in Definition 2, consequences of contradictory defaults in $\mathcal{M}$ are ignored. This guarantees that a model of $(\mathcal{K}, \mathcal{D}_\mathcal{S}, \mathcal{M})$ corresponds to the certain answers of $(\mathcal{K}, \mathcal{D}_\mathcal{S}, \mathcal{M})$, hence the choice in nomenclature.

Definition 4. A default mapping $\mathcal{M}$ is *consistent with* $(\mathcal{K}, \mathcal{D}_\mathcal{S})$ if $(\mathcal{K}, \mathcal{D}_\mathcal{S})$ has only one extension over $\mathcal{M}$ and a model of $(\mathcal{K}, \mathcal{D}_\mathcal{S}, \mathcal{M})$ exists. It is said to be *globally consistent* if it is consistent and every DMA in $\mathcal{M}$ is applied in the extension.

We can trivially translate a classical OBDA mapping $\mathcal{M}'$ into a corresponding default mapping

$$\mathrm{tr}(\mathcal{M}') = \left\{ \frac{\varphi(x) : \top}{\psi(x)} \mid m \in \mathcal{M}', m : \varphi(x) \rightsquigarrow \psi(x) \right\}.$$

The notion of global consistency of defaults in Definition 4 is a direct translation of the definition of global consistency for classical OBDA mappings introduced in [6].

Proposition 1. *A classical OBDA mapping $\mathcal{M}'$ is globally consistent with an OBDA setup $(\mathcal{K}, \mathcal{D}_\mathcal{S})$ if and only if its default translation $\mathrm{tr}(\mathcal{M}')$ is globally consistent with $(\mathcal{K}, \mathcal{D}_\mathcal{S})$.*

Similarly, a crisp default mapping $\mathcal{M}$ can be translated to a classical OBDA mapping

$$\mathrm{tr}^{-1}(\mathcal{M}) = \left\{ \varphi(x) \rightsquigarrow \psi(x) \mid \frac{\varphi(x) : \top}{\psi(x)} \in \mathcal{M} \right\}.$$

Clearly, tr and tr^{-1} are mutually inverse, justifying the notation.

Theorem 1. *Let M be a crisp default mapping consistent with $(\mathcal{K}, \mathcal{D_S})$. Then M is a model of the classical OBDA specification $(\mathcal{K}, \mathcal{D_S}, \mathrm{tr}^{-1}(\mathcal{M}))$ if and only if it is a model of $(\mathcal{K}, , \mathcal{D_S}, \mathcal{M})$. Similarly, if $\mathcal{M}'$ is a classical OBDA mapping such that $(\mathcal{K}, \mathcal{D_S}, \mathcal{M}')$ is consistent, then M is a model of $(K, \mathcal{D_S}, \mathcal{M}')$ if and only if it is a model of its default translation $(K, \mathcal{D_S}, \mathrm{tr}(\mathcal{M}'))$.*

Interpreting mappings as rules as opposed to formulas has the side-effect that inconsistency is not synonymous with unsatisfiability. Hence, if we have a inconsistent set of mappings together with a satisfiable theory $\mathcal{K}$, the OBDA specification will always have a model; one would simply not apply conflicting mappings.

Thus it might be wise to consider default mappings as formulas instead, retaining the classical sense of inconsistency.

3.2   Naive Model-Based Approach

A simple way of interpreting defaults as formulas is by taking the intuitive meaning of what a default is and translating that into logic. Since *consistency* of a theory is equivalent to *there exists a model* in first-order logic, we can apply this to our setting. Thus we can say that $\mathfrak{I}$ satisfies a default mapping assertion $\frac{\varphi:\rho}{\psi}$ if

$$\mathfrak{I} \vDash \varphi \text{ and } \exists \mathfrak{I}' : \mathfrak{I}' \vDash \rho \cup \mathcal{O} \text{ then } \mathfrak{I} \vDash \psi.$$

Then consistency of a set of DMA's can be defined in the usual way.

Definition 5. *A set $\mathcal{M}$ of DMA's is consistent if there exists a model $\mathfrak{I}$ that satisfies every $m \in M$.*

Definition 6. *A set $\mathcal{M}$ of DMA's is called strongly consistent if it is satisfiable by some $\mathfrak{I}$ such that*

1. $\mathfrak{I} \vDash \exists t \in \mathrm{eval}(\varphi, \mathcal{D_S}).\rho_m[t]$ for every $m \in \mathcal{M}$ (activation)
2. $\mathfrak{I} \vDash \rho_m[t] \Rightarrow \mathfrak{I} \vDash \psi_m[t]$ for every $m \in \mathcal{M}$ and $t \in \mathrm{eval}(\varphi, \mathcal{D_S})$ (consistency)

3.3   Modal and Epistemic Approach

The naive semantics for defaults from the previous section bear a certain resemblance to Kripke semantics; the truth of a default in a first order interpretation is dependent on other interpretations. It might be prudent to restrict the model $\mathfrak{I}'$ to have some relationship to $\mathfrak{I}$, i.e., an accessibility relation. I plan on further developing and extending the naive approach as needed, considering it is quite lightweight and very compatible with classical, crisp mappings. This could give rise to similarities to other, well-known logics that have a semantics for defaults [1,5]. Intuitively, this would correspond to considering the database as an autoepistemic agent and describing its knowledge (what is explicit in the database) and belief (what can be reasoned).

4   Conclusion

The goal for my thesis is to extend the OBDA setting to allow the expression of epistemic properties with respect to the database, with an emphasis on removing redundant query rewriting and unfolding. To that end, I propose extending the mapping language to include defaults. This has not only the potential to address the aforementioned problems but also to substantially simplify the mapping and OBDA specification maintenance with respect to incomplete knowledge and exceptions. Furthermore, I shall investigate a lightweight semantics for interpreting defaults as formulas that, for crisp mappings, reduces entirely to the classical setting. Finally, I will study the complexity of query answering with default mappings and investigate possibilities of reducing the complexity in practical applications through ontology and mapping approximation [3].

References

1. Ben-David, S., Ben-Eliyahu, R.: A modal logic for subjective default reasoning. In: Proceedings of Logic in Computer Science, LICS 1994, pp. 477–486. IEEE (1994)
2. Calvanese, D., De Giacomo, G., Lembo, D., Lenzerini, M., Rosati, R.: Tractable reasoning and efficient query answering in description logics: the DL-Lite family. J. Autom. Reasoning **39**(3), 385–429 (2007)
3. Console, M., Mora, J., Rosati, R., Santarelli, V., Savo, D.F.: Effective computation of maximal sound approximations of description logic ontologies. In: Mika, P., Tudorache, T., Bernstein, A., Welty, C., Knoblock, C., Vrandečić, D., Groth, P., Noy, N., Janowicz, K., Goble, C. (eds.) ISWC 2014, Part II. LNCS, vol. 8797, pp. 164–179. Springer, Heidelberg (2014)
4. Donini, F.M., Nardi, D., Rosati, R.: Description logics of minimal knowledge and negation as failure. ACM Trans. Comput. Logic **3**(2), 177–225 (2002)
5. Engan, I., Langholm, T., Lian, E.H., Waaler, A.: Default reasoning with preference within only knowing logic. In: Baral, C., Greco, G., Leone, N., Terracina, G. (eds.) LPNMR 2005. LNCS (LNAI), vol. 3662, pp. 304–316. Springer, Heidelberg (2005)
6. Lembo, D., Mora, J., Rosati, R., Savo, D.F., Thorstensen, E.: Towards mapping analysis in ontology-based data access. In: Kontchakov, R., Mugnier, M.-L. (eds.) RR 2014. LNCS, vol. 8741, pp. 108–123. Springer, Heidelberg (2014)
7. Lutz, C., Seylan, I., Wolter, F.: Ontology-based data access with closed predicates is inherently intractable (sometimes). In: Proceedings of the Twenty-Third International Joint Conference on Artificial Intelligence, IJCAI 2013, pp. 1024–1030. AAAI Press (2013)
8. Poggi, A., Lembo, D., Calvanese, D., De Giacomo, G., Lenzerini, M., Rosati, R.: Linking data to ontologies. In: Spaccapietra, S. (ed.) Journal on Data Semantics X. LNCS, vol. 4900, pp. 133–173. Springer, Heidelberg (2008)
9. Reiter, R.: A logic for default reasoning. Artif. Intell. **13**(1–2), 81–132 (1980)

Doctoral Consortium Extended Abstract: Nonmonotonic Qualitative Spatial Reasoning

Przemysław Andrzej Wałęga[(✉)]

University of Warsaw, Institute of Philosophy, Warsaw, Poland
przemek.walega@wp.pl

Abstract. My work on PhD thesis consists in nonmonotonic reasoning about spatial relations and how they change in time. Although there are several approaches concerning this topic, to the best of my knowledge, there is no general framework that provides nonmonotonic (qualitative) spatial reasoning. The work I have accomplished so far consists in introducing the so-called ASPMT(QS) system. It is based on a paradigm of Answer Set Programming Modulo Theories (ASPMT) and polynomial encodings of spatial relations. The system enables modelling of dynamically varying spatial information, as well as abductive reasoning, and its first version is already implemented. As a future work I consider extending ASPMT(QS) in order to perform more complex spatio-temporal reasoning and try to overcome limitations of the current implementation.

Keywords: Nonmonotonic spatial reasoning · Declarative spatial reasoning · Qualitative reasoning · Answer set programming modulo theories

1 Background

The notion of space has fascinated mathematicians, logicians and philosophers for ages. Of my main interest are formal methods that enable to model human-like spatial reasoning. Such formalisms are recently studied in Artificial Intelligence (AI), in particular in Knowledge Representation. As stated in [2]:

'Space, with its manifold layers of structure, has been an inexhaustible source of intellectual fascination since Antiquity. [...] In this long intellectual history, however, one relatively recent, yet crucial, event stands out: the rise of the logical stance in geometry.'

Although, there are numerous AI spatial reasoning systems (see [2]), human methods are still infeasible. Human-like reasoning about space, and how objects and spatial relations can change, is a key requirement in systems that aim to model a wide range of dynamic application domains. The role of space is ubiquitous, and thus myriad domain-specific conclusions drawn via, e.g., causal explanation [16] or default reasoning [15], must also be necessarily concerned with

Extended Abstract as part of the program of the 1st Joint ADT/LPNMR 2015 Doctoral Consortium co-chaired by Esra Erdem and Nicholas Mattei.

F. Calimeri et al. (Eds.): LPNMR 2015, LNAI 9345, pp. 565–571, 2015.
DOI: 10.1007/978-3-319-23264-5_48

spatial consistency – an explanatory hypothesis is not viable if the abduced spatial relations can not be physically realised.

A number of domain-independent approaches for modelling spatial change have been developed [9,11,12,19]. Shanahan ([17]) integrates a first-order theory of shape with the situation calculus for describing common-sense laws of motion. In [7] the authors present a framework for modelling dynamic spatial systems by integrating qualitative spatial theories into situation calculus. However no systems currently exist that are capable of efficient and general nonmonotonic spatial reasoning.

The aim of my thesis is twofolded. Firstly, to discuss how nonomotonic logics known from the literature can be extended to perform spatial reasoning in a qualitative manner. Secondly, to construct new approaches for nonmonotonic spatial reasoning and implement them.

2   Accomplished Work – ASPMT(QS) System

The main work accomplished so far consists in introduction and first implementation of the ASPMT(QS) system. ASPMT(QS) is a new approach for dynamic spatial reasoning (see [18]). It is based on Answer Set Programming Modulo Theories approach ([4]) extended to spatial domains. Spatial reasoning is performed in an analytic manner, i.e., as in reasoners such as CLP(QS) ([6]), where relations are encoded as polynomial constraints. The main reasoning task, i.e., determining whether a spatial configuration is consistent, is equivalent to determine whether the system of polynomial constraints is satisfiable. The reasoning method uses Satisfiability Modulo Theories (SMT) with real nonlinear arithmetic, and can be accomplished in a sound and complete manner. ASPMT(QS) implementation consists of a spatial representation module and a method for turning ASPMT instances into SMT that computes stable models by means of SMT solvers. ASPMT(QS) is the only spatial reasoning system that enables to perform nonmonotonic reasoning and uses the theory of arithmetic over real numbers. In what follows, I will briefly present the system's implementation and its applications.

2.1   ASPMT(QS) Implementation

ASPMT(QS) implementation is build on ASPMT2SMT ([5]) – a compiler that translates a tight fragment of ASPMT into SMT instances. Additionally, there is a module for spatial reasoning and Z3 as SMT solver. Input programs consist of declaration of sorts (data types), objects (particular elements of given types), constants (functions) and variables (variables associated with declared types). Afterwards, there are program's clauses.

The ASPMT(QS) supports standard connectives: &, |, not, ->, <- and arithmetic operators: <, <=, >=, >, =, !=, +, =, *, with their usual meaning. Additionally, spatial domain entities are available, namely spatial sorts for geometric objects types, e.g., point, segment, circle, triangle, functions describing objects parameters, e.g., x(point), r(circle), and spatial relations, e.g., coincident(point, circle) or qualitative Region Connection Calculus (RCC) relations [14], e.g., external connection (EC): rccEC(circle, circle). As an output, the system either produces an instance corresponding to a stable model of the input program, or states that there is no such model.

ASPMT(QS) uses polynomial encodings of spatial relations and as a result is able to express a number of relations from the well-known qualitative approaches, e.g., Interval Algebra [3], Rectangle Algebra [10], Region Connection Calculus, [14] and Cardinal Direction Calculus [8]. The following Proposition 1 has been proved in [18].

Proposition 1. *Each relation of Interval Algebra, Rectangle Algebra, Region Connection Calculus and Cardinal Direction Calculus may be defined in ASPMT(QS).*

As an example consider, RCC spatial relations in a domain of circles, as depicted in Fig. 1.

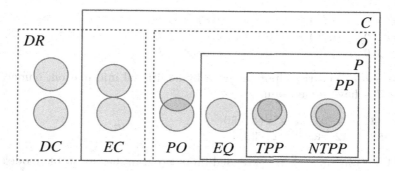

Fig. 1. Region connection calculus relations.

The abovementioned relations are encoded by means of polynomial systems. For instance, partially overlapping (PO) relation is encoded as follows.

```
rccPO(C1,C2)=true <- (x(C1)=X1 & y(C1)=Y1 & r(C1)=R1
& x(C2)=X2 & y(C2)=Y2 & r(C2)=R2)
& ( (X1-X2)*(X1-X2)+(Y1-Y2)*(Y1-Y2) > (R1-R2)*(R1-R2)
& (X1-X2)*(X1-X2)+(Y1-Y2)*(Y1-Y2) < (R1+R2)*(R1+R2)).

rccPO(C1,C2)=false <- (x(C1)=X1 & y(C1)=Y1 & r(C1)=R1
& x(C2)=X2 & y(C2)=Y2 & r(C2)=R2)
& not ( (X1-X2)*(X1-X2)+(Y1-Y2)*(Y1-Y2) > (R1-R2)*(R1-R2)
& (X1-X2)*(X1-X2)+(Y1-Y2)*(Y1-Y2) < (R1+R2)*(R1+R2)).
```

2.2 Program Example

In this section, I present a simple ASPMT(QS) program that models spatial configuration of three circles a, b, c, such that a is discrete from b (rccDR(a,b)),

b is discrete from c (rccDR(b,c)), and a is a proper part of c (rccPP(a,c)). The input program is as follows.

```
:- sorts
   circle.
:- objects
   a, b, c      ::  circle.
:- constants
   .
:- variables
   C, C1, C2    ::  circle.
```

```
{x(C)=X}. {y(C)=X}. {r(C)=X}.
rccDR(a,b)=true. rccDR(b,c)=true. rccPP(a,c)=true.
```

The ASPMT(QS) task is to determine an exact model for such a spatial configuration. The output of the abovementioned program is:

```
r(a) = 0.5    r(b) = 1.0    r(c) = 0.25
x(a) = 1.0    x(b) = 1.0    x(c) = 1.0
y(a) = 3.0    y(b) = 1.0    y(c) = 3.0
```

Now, let me extend the input program by additional information, namely that circles a, b, c have same radius:

```
<- r(a)=R1 & r(b)=R2 & r(c)=R3
 & (R1!=R2 | R2!=R3 | R1!=R3).
```

ASPMT(QS) infers that the extended input program has no stable models, i.e., the spatial configuration is impossible and the system of polynomials is unsatisfiable, therefore the output is:

```
UNSATISFIABLE;
```

This example shows that ASPM(QS) is able to check consistency of RCC relations. This task is also known as computing composition table of RCC relations (see [14]).

2.3   Abductive Reasoning

In this section, I show how abductive reasoning may be achieved in ASPMT(QS). Consider an application where spatial configuration of objects is recorded in discrete time points (e.g., geospatial information collected about cities or a dynamic environment observed by a mobile robot). Consider the situation presented in Fig. 2, where in time point t_1: c is a proper part of a and b is a proper part of a, in t_2: c is a proper part of a and b is externally connected with a, in t_3: c is a proper part of a and b is externally connected with a. Additionally, radius of b in t_2 is greater than its radius in t_1. The main part of input program is:

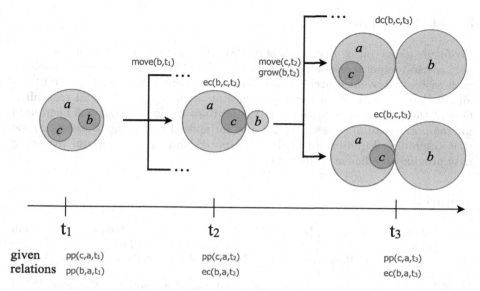

given relations

t_1	t_2	t_3
pp(c,a,t₁)	pp(c,a,t₂)	pp(c,a,t₃)
pp(b,a,t₁)	ec(b,a,t₂)	ec(b,a,t₃)

Fig. 2. Nonmonotonically inferring actions and qualitative spatial relations between spatial objects a, b, c.

```
pp(c,a,t1)=true.  pp(b,a,t1)=true.
pp(c,a,t2)=true.  ec(b,a,t2)=true.
pp(c,a,t3)=true.  pp(b,a,t3)=true.
<- r(b,t1)=R1 & r(b,t2)=R2 & R1>=R2.
```

Having a knowledge base about available actions that may be performed, namely move and grow, and a number of rules describing the spatial behaviour of objects, ASPMT(QS) infers what actions had to be performed between time points, namely:

```
move(b,t1)=true.    move(b,t2)=true.    grow(b,t2)=true.
```

Abductive reasoning requires default rules such as inertia which states that if no action is performed, then object locations and parameters remain the same by default. This rule enables the inference that the position of c with respect to a is the same in t_1 and t_2. Another important rule is action minimality, which is responsible for inferring only those actions that are necessary to be performed. Furthermore, the example illustrated in Fig. 2 shows that ASPMT(QS) checks the global consistency of various spatial relations. Therefore, it infers relations occurring as a result of performed actions (indirect effects) even if there is more than one possible solution, e.g., the relation between c and b in t_2 may be either $dc(b, c, t2)$ or $ec(b, c, t2)$.

3   Future Plans

My future plans consist in extending ASPMT(QS) in order to perform more complex spatio-temporal reasoning and applying ASPMT(QS) to practical problems such as computer-aided architecture design or mobile robots control. Additionally, I plan to introduce other approaches for nonmonotonic spatial reasoning. One of the promising approaches is Equilibrium Logic [13] that has been used for temporal reasoning (see [1]) but not for spatial reasoning. It seems to me that combining Equilibrium Logic with spatial and spatio-temporal reasoning may provide an efficient framework.

References

1. Aguado, F., Cabalar, P., Diéguez, M., Pérez, G., Vidal, C.: Temporal equilibrium logic: a survey. J. Appl. Non-Class. Logics **23**(1–2), 2–24 (2013)
2. Aiello, M., Pratt-Hartmann, I., van Benthem, J.F.: Handbook of Spatial Logics. Springer, Heidelberg (2007)
3. Allen, J.F.: Maintaining knowledge about temporal intervals. Commun. ACM **26**(11), 832–843 (1983)
4. Bartholomew, M., Lee, J.: Functional stable model semantics and answer set programming modulo theories. In: Proceedings of the Twenty-Third international Joint Conference on Artificial Intelligence, pp. 718–724. AAAI Press (2013)
5. Bartholomew, M., Lee, J.: System ASPMT2SMT: computing ASPMT theories by SMT solvers. In: Fermé, E., Leite, J. (eds.) JELIA 2014. LNCS, vol. 8761, pp. 529–542. Springer, Heidelberg (2014)
6. Bhatt, M., Lee, J.H., Schultz, C.: CLP(QS): a declarative spatial reasoning framework. In: Egenhofer, M., Giudice, N., Moratz, R., Worboys, M. (eds.) COSIT 2011. LNCS, vol. 6899, pp. 210–230. Springer, Heidelberg (2011)
7. Bhatt, M., Loke, S.: Modelling dynamic spatial systems in the situation calculus. Spat. Cogn. Comput. **8**(1–2), 86–130 (2008)
8. Frank, A.U.: Qualitative spatial reasoning with cardinal directions. In: Kaindl, H. (ed.) 7. Österreichische Artificial-Intelligence-Tagung/Seventh Austrian Conference on Artificial Intelligence. Informatik-Fachberichte, vol. 287, pp. 157–167. Springer, Heidelberg (1991)
9. Gooday, J., Cohn, A.G.: Conceptual neighbourhoods in temporal and spatial reasoning. Spat. Temporal Reasoning, ECAI **94** (1994)
10. Guesgen, H.W.: Spatial Reasoning Based on Allen's Temporal Logic. Technical report, International Computer Science Institute (1989)
11. Hazarika, S.M.: Qualitative spatial change: space-time histories and continuity. Ph.D. thesis, The University of Leeds (2005)
12. Muller, P.: A qualitative theory of motion based on spatio-temporal primitives. KR **98**, 131–141 (1998)
13. Pearce, D.: Equilibrium logic. Ann. Math. Artif. Intell. **47**(1–2), 3–41 (2006)
14. Randell, D.A., Cui, Z., Cohn, A.G.: A spatial logic based on regions and connection. KR **92**, 165–176 (1992)
15. Reiter, R.: A logic for default reasoning. Artif. Intell. **13**(1), 81–132 (1980)
16. Shanahan, M.: Prediction is deduction but explanation is abduction. IJCAI **89**, 1055–1060 (1989)

17. Shanahan, M.: Default reasoning about spatial occupancy. Artif. Intell. **74**(1), 147–163 (1995)
18. Cabalar, P.: Answer Set; Programming? In: Balduccini, M., Son, T.C. (eds.) Logic Programming, Knowledge Representation, and Nonmonotonic Reasoning. LNCS, vol. 6565, pp. 334–343. Springer, Heidelberg (2011)
19. Van de Weghe, N., Kuijpers, B., Bogaert, P., De Maeyer, P.: A qualitative trajectory calculus and the composition of its relations. In: Rodríguez, M.A., Cruz, I., Levashkin, S., Egenhofer, M. (eds.) GeoS 2005. LNCS, vol. 3799, pp. 60–76. Springer, Heidelberg (2005)

Author Index

Printed in the United States
By Bookmasters